Construction Planning, Equipment, and Methods

Eliezer Shapira, a civil engineer, general contractor, and father of Aviad Shapira. As a father and most loving teacher it was he who sparked Aviad's passion for construction. Over the years, Eliezer and Cliff have also shared adventures at equipment shows in Europe and enjoyed many an interesting construction story. This book is therefore dedicated to Eliezer Shapira—a constructor who has taught both of us an appreciation for meeting the challenges of construction.

Cliff Schexnayder
Aviad Shapira

McGraw-Hill Series in Civil Engineering

CONSULTING EDITORS
George Tchobanoglous, *University of California, Davis*
Raymond E. Levitt, *Stanford University*

Bailey and Ollis
Biochemical Engineering Fundamental

Banks
Introduction to Transportation Engineering

Barrie and Paulson
Professional Construction Management: Including CM, Design-Construct, and General Contracting

Benjamin
Water Chemistry

Bishop
Pollution Prevention: Fundamentals and Practice

Bockrath
Contracts and the Legal Environment for Engineers and Architects

Callahan, Quackenbush, and Rowlings
Construction Project Scheduling

Canter
Environmental Impact Assessment

Chanlett
Environmental Protection

Chapra
Applied Numerical Methods with MATLAB for Engineers and Scientists

Chapra
Surface Water-Quality Modeling

Chapra and Canale
Numerical Methods for Engineers

Chow, Maidment, and Mays
Applied Hydrology

Crites and Tchobanoglous
Small and Decentralized Wastewater Management Systems

Davis and Cornwell
Introduction to Environmental Engineering

Davis and Masten
Principles of Environmental Engineering and Science

de Nevers
Air Pollution Control Engineering

Eckenfelder
Industrial Water Pollution Control

Eweis, Ergas, Chang, and Schroeder
Bioremediation Principles

Finnemore and Franzini
Fluid Mechanics with Engineering Applications

Gaylord and Stallmeyer
Design of Steel Structures

Griffis and Farr
Construction Project Planning

Heerwagen
Passive and Active Environmental Controls

Hinze
Construction Contracts

LaGrega, Buckingham, and Evans
Hazardous Waste Management

Leet and Bernal
Reinforced Concrete Design

Leet and Uang
Fundamentals of Structural Analysis

Linsley, Franzini, Freyberg, and Tchobanoglous
Water Resources and Engineering

McGhee
Water Supply and Sewage

Metcalf & Eddy, Inc.
Wastewater Engineering: Collection and Pumping of Wastewater

Metcalf & Eddy, Inc.
Wastewater Engineering: Treatment, Disposal, Reuse

Meyer and Miller
Urban Transportation Planning

Nilson
Design of Concrete Structures

Nowak and Collins
Reliability of Structures

Oberlender
Project Management for Engineering and Construction

Peavy, Rowe, and Tchobanoglous
Environmental Engineering

Peurifoy and Oberlender
Estimating Construction Costs

Peurifoy, Schexnayder, and Shapira
Construction Planning, Equipment, and Methods

Rittmann and McCarty
Environmental Biotechnology: Principles and Applications

Rubin
Introduction to Engineering and the Environment

Sawyer, McCarty, and Parkin
Chemistry for Environmental Engineering

Schexnayder and Mayo
Construction Management Fundamentals

Streeter
Fluid Mechanics

Sturm
Open Channel Hydraulics

Tchobanoglous, Theisen, and Vigil
Integrated Solid Waste Management: Engineering Principles and Management Issues

Vinnakota
Steel Structures: Behavior and LRFD

Wentz
Safety, Health, and Environmental Protection

Wolf and Dewitt
Elements of Photogrammetry

Construction Planning, Equipment, and Methods

Seventh Edition

Robert. L. Peurifoy, P.E.
Late Consulting Engineer
Austin, Texas

Clifford J. Schexnayder, P.E., Ph.D.
Eminent Scholar Emeritus
Del E. Webb School of Construction
Arizona State University
Tempe, Arizona

Aviad Shapira, D.Sc.
Associate Professor
Faculty of Civil and Environmental Engineering
Technion–Israel Institute of Technology
Haifa, Israel

McGraw Hill Education (India) Private Limited
CHENNAI

McGraw Hill Education Offices
Chennai New York St Louis San Francisco Auckland Bogotá Caracas
Kuala Lumpur Lisbon London Madrid Mexico City Milan Montreal
San Juan Santiago Singapore Sydney Tokyo Toronto

McGraw Hill Education (India) Private Limited

CONSTRUCTION PLANNING, EQUIPMENT, AND METHODS, SEVENTH EDITION

Sixteenth reprint 2017
RCQYRRDMDRYAQ

McGraw Hill Education (India) Edition 2010

Reprinted in India by arrangement with McGraw-Hill Global Education Holdings, LLC, New York

Sales territories: India, Pakistan, Nepal, Bangladesh, Sri Lanka and Bhutan

ISBN-13: 978-0-07-070699-6
ISBN-10: 0-07-070699-9

Published by McGraw Hill Education (India) Private Limited, 444/1, Sri Ekambara Naicker Industrial Estate, Alapakkam, Porur, Chennai 600 116, Tamil Nadu, India, and printed at Rajkamal Electric Press, Plot No. 2, Phase-IV, Kundli, 131 028, Haryana

Visit us at: www.mheducation.co.in

ABOUT THE AUTHORS

R. L. Peurifoy (1902–1995), after serving as principal specialist in engineering education for the U.S. Office of Education during World War II, began teaching construction engineering at Texas A&M University in 1946. In the years that followed, Peurifoy led the transformation of the study of construction engineering into an academic discipline. In 1984 the Peurifoy Construction Research Award was instituted by the American Society of Civil Engineers upon recommendation of the Construction Research Council. This award was instituted to honor R. L. Peurifoy's exceptional leadership in construction education and research. The award recipients since the last edition of the book are:

2001 M. Dan Morris
2003 Jimmie W. Hinze, University of Florida
2004 David B. Ashley, University of California Merced
2005 Abraham Warszawski, Technion–Israel Institute of Technology

Clifford J. Schexnayder is an Eminent Scholar Emeritus at the Del E. Webb School of Construction, Arizona State University. He received his Ph.D. in Civil Engineering (Construction Engineering and Management) from Purdue University, and a Master's and Bachelor's in Civil Engineering from Georgia Institute of Technology. A construction engineer with over 35 years of practical experience, Dr. Schexnayder has worked with major heavy/highway construction contractors as field engineer, estimator, and corporate Chief Engineer.

As Chief Engineer he was the qualifying party for the company's Contractor's License and had direct line responsibility for the coordination and supervision of both the estimating and construction of projects. He provided management, administrative, and technical direction to the company's operations and represented the company in project meetings and negotiations.

Additionally, he served with the U.S. Army Corps of Engineers on active duty and in the reserves, retiring as a Colonel. His last assignment was as Executive Director, Directorate of Military Programs, Office of the Chief of Engineers, Washington, D.C.

He has taught construction equipment at Arizona State University, Louisiana Tech University, Purdue, Technion–Israel Institute of Technology, Universidad de Piura (Peru), the U.S. Air Force Academy, Universidad Tecnica Particuar de Loja (Equador), Virginia Polytechnic Institute and State University, and the U.S. Army Engineer School.

Dr. Schexnayder is a registered professional engineer in six states, as well as a member of the American Society of Civil Engineers. He served as chairman of the ASCE's Construction Division and on the task committee, which formed the ASCE Construction Institute. From 1997 to 2003 he served as chairman of the Transportation Research Board's Construction Section.

Aviad Shapira is an Associate Professor of Construction Engineering and Management in the Faculty of Civil and Environmental Engineering at the Technion–Israel Institute of Technology. He received his B.Sc., M.Sc., and D.Sc. degrees in Civil Engineering from the Technion. After completing his degrees, he spent one year as a post-doctoral fellow at the University of Illinois at Urbana–Champaign under a grant from the U.S. Air Force Civil Engineering Support Agency. In the 1990s he spent a year at the University of New Mexico in Albuquerque as the AGC Visiting Professor.

Dr. Shapira accrued his practical experience as a project engineer, project manager, and Chief Engineer in a general contracting firm prior to pursuing an academic career. During that period, he was in charge of the construction engineering for industrial, commercial, and public projects in Israel. His teaching, research, and consulting interests have taken him to construction projects around the world.

He has taught construction equipment and formwork design in Israel and the United States since 1985, and authored or co-authored the only texts addressing these subjects in Israel. His research has focused on formwork design and construction equipment for building construction. That work has covered equipment selection, operation, management, productivity, economics, and safety. He co-developed an innovative crane-mounted video camera that serves as an operator aid. This camera system has been used on most of the high-rise building projects built in Israel since 1998 and on several projects in Europe.

Dr. Shapira is a member of the American Society of Civil Engineers and the American Concrete Institute. He has been an active member of ACI Committee 347 Formwork for Concrete since 1997, and has also served on several ASCE and TRB construction equipment committees. Additionally, he is the Vice-Chair of Technical Committee 120 of the Standard Institution of Israel, which wrote the new Israeli formwork standard, first published in 1995 and revised in 1998.

CONTENTS

Preface xii

CHAPTER **1**

Machines Make It Possible 1

The History of Construction Equipment 1
Being Competitive 8
The Construction Industry 9
Safety 10
The Contracting Environment 11
Planning Equipment Utilization 12
Summary 14
Problems 14
References 15
Website Resources 15

CHAPTER **2**

Fundamental Concepts of Equipment Economics 17

Important Questions 17
Equipment Records 18
The Rent Paid for the Use of Money 19
Cost of Capital 25
Evaluating Investment Alternatives 26
Elements of Ownership Cost 28
Elements of Operating Cost 34
Cost for Bidding 39
Replacement Decisions 47
Rent and Lease Considerations 48
Summary 52
Problems 53
References 58
Website Resources 58

CHAPTER **3**

Planning for Earthwork Construction 60

Planning 60
Graphical Presentation of Earthwork 64
Earthwork Quantities 67
Mass Diagram 75
Using the Mass Diagram 77
Pricing Earthwork Operations 84
Summary 86
Problems 86
References 89
Website Resources 89

CHAPTER **4**

Soil and Rock 90

Introduction 90
Glossary of Terms 91
Soil and Rock Properties 91
COMPACTION SPECIFICATION AND CONTROL 101
Introduction 101
Compaction Tests 102
Soil Processing 106
Summary 111
Problems 112
References 113
Website Resources 114

CHAPTER **5**

Compaction and Stabilization Equipment 115

Compaction of Soil and Rock 115
Glossary of Terms 116
Types of Compacting Equipment 116
Roller Production Estimating 128

Dynamic Compaction 129
SOIL STABILIZATION 131
General Information 131
Stabilizing Soils with Lime 133
Cement-Soil Stabilization 134
  Summary 138
  Problems 138
  References 139
  Website Resources 139

CHAPTER 6
Machine Equipment Power Requirements 140
General Information 140
Required Power 141
Available Power 148
Usable Power 155
Performance Charts 158
  Summary 165
  Problems 165
  References 169
  Website Resources 169

CHAPTER 7
Dozers 171
Introduction 171
Performance Characteristics of Dozers 172
PUSHING MATERIAL 178
General Information 178
Blades 178
Project Employment 182
Dozer Production Estimating 185
Dozer Production Estimating Format 191
Dozer Safety 195
LAND CLEARING 196
Land-Clearing Operations 196
Types of Equipment Used 196
Land-Clearing Production Estimating 199
RIPPING ROCK 204
Rippers 204
Determining the Rippability of Rock 205
Determining the Thickness and Strength of Rock Layers 207
Ripper Attachments 209
Ripper Production Estimates 211
  Summary 213
  Problems 214
  References 220
  Website Resources 220

CHAPTER 8
Scrapers 222
General Information 222
Scraper Types 223
Scraper Operation 228
Scraper Performance Charts 229
Scraper Production Cycle 232
Scraper Production Estimating Format 233
Operational Considerations 247
Scraper Safety 249
  Summary 250
  Problems 250
  References 252
  Website Resources 252

CHAPTER 9
Excavators 253
Hydraulic Excavators 253
Hydraulic Excavator Accidents 255
FRONT SHOVELS 257
General Information 257
Selecting a Front Shovel 258
Calculating Shovel Production 260
Height of Cut Effect on Shovel Production 261
Angle of Swing Effect on Shovel Production 262
HOES 264
General Information 264
Bucket Rating for Hydraulic Hoes 267
Selecting a Hoe 268
Calculating Hoe Production 271
LOADERS 274
General Information 274
Loader Buckets/Attachments 275
Operating Specifications 277
Loader Production Rates 279

Calculating Wheel Loader Production 281
Calculating Track Loader Production 282
Loader Safety 284
SPECIALTY EXCAVATORS 284
Trenching Machines 284
Selecting Equipment for Excavating Trenches 287
Trenching Machine Production 287
Trench Safety 288
Backhoe-Loaders 289
Holland Loaders 290
Vac Excavators 290
Summary 291
Problems 292
References 294
Website Resources 295

CHAPTER 10
Trucks and Hauling Equipment 296
Trucks 296
Rigid-Frame Rear-Dump Trucks 298
Articulated Rear-Dump Trucks 298
Tractors with Bottom-Dump Trailers 300
Capacities of Trucks and Hauling Equipment 301
Truck Size Affects Productivity 303
Calculating Truck Production 305
Production Issues 309
Tires 310
Truck Performance Calculations 312
Truck Safety 317
Summary 317
Problems 318
References 319
Website Resources 319

CHAPTER 11
Finishing Equipment 320
Introduction 320
GRADERS 320
General Information 320
Grader Operations 324
Time Estimates 328
Fine Grading Production 329
Grader Safety 329
GRADALLS 330
General Information 330
Safety 332
TRIMMERS 332
General Information 332
Operation 332
Production 334
Summary 334
Problems 334
References 335
Website Resources 336

CHAPTER 12
Drilling Rock and Earth 337
Introduction 337
Glossary of Drilling Terms 338
Drill Bits 341
Rock Drills 342
Drilling Methods and Production 346
Estimating Drilling Production 350
GPS and Computer Monitoring Systems 358
Drilling Soil 359
Removal of Cuttings 361
Trenchless Technology 362
Safety 367
Summary 368
Problems 368
References 370
Website Resources 371

CHAPTER 13
Blasting Rock 372
Blasting 372
Glossary of Blasting Terms 374
Commercial Explosives 375
Primers and Boosters 379
Initiating Systems 380
Rock Fragmentation 382
Blast Design 383
Powder Factor 395

Trench Rock 397
Breakage Control Techniques 397
Vibration 400
Safety 401
Summary 403
Problems 403
References 405
Website Resources 406

CHAPTER **14**
Aggregate Production 407
Introduction 407
PARTICLE SIZE REDUCTION 409
General Information 409
Jaw Crushers 410
Gyratory Crushers 415
Roll Crushers 419
Impact Crushers 424
Special Aggregate Processing Units 425
Feeders 426
Surge Piles 427
Crushing Equipment Selection 428
SEPARATION INTO PARTICLE SIZE RANGES 431
Scalping Crushed Stone 431
Screening Aggregate 432
OTHER AGGREGATE PROCESSING ISSUES 437
Log Washers 437
Segregation 438
Safety 438
Summary 439
Problems 439
References 441
Website Resources 442

CHAPTER **15**
Asphalt Mix Production and Placement 443
Introduction 443
Glossary of Asphalt Terms 444
Structure of Asphalt Pavements 446
Flexible Pavements 447
Asphalt Concrete 453
ASPHALT PLANTS 454
General Information 454
Batch Plants 455
Drum Mix Plants 460
Dust Collectors 462
Asphalt Storage and Heating 463
Reclaiming and Recycling 464
PAVING EQUIPMENT 465
Sweeper/Broom 466
Haul Trucks 466
Asphalt Distributors 467
Asphalt Pavers 468
Compaction Equipment 474
Safety 479
Summary 479
Problems 480
References 481
Website Resources 482

CHAPTER **16**
Concrete and Concrete Equipment 483
Introduction 483
CONCRETE MIXTURES 485
Proportioning Concrete Mixtures 485
Fresh Concrete 485
Batching Concrete Materials 486
MIXING CONCRETE 490
Concrete Mixing Techniques 490
Ready-Mixed Concrete 496
Central-Mixed Concrete 500
PLACING CONCRETE 502
Buckets 502
Manual or Motor-Propelled Buggies 504
Chutes and Drop Pipes 504
Belt Conveyors 504
Concrete Pumps 505
CONSOLIDATING AND FINISHING 514
Consolidating Concrete 514
Finishing and Curing Concrete 517
CONCRETE PAVEMENTS 519
Slipform Paving 519
ADDITIONAL APPLICATIONS AND CONSIDERATIONS 523

Roller-Compacted Concrete 523
Shotcreting 524
Fly Ash 525
Placing Concrete in Cold Weather 526
Placing Concrete in Hot Weather 527
SAFETY 527
Pumping Concrete 527
Summary 528
Problems 528
References 530
Website Resources 531

CHAPTER **17**
Cranes 533
Major Crane Types 533
MOBILE CRANES 535
Crawler Cranes 535
Telescoping-Boom Truck-Mounted Cranes 538
Lattice-Boom Truck-Mounted Cranes 539
Rough-Terrain Cranes 540
All-Terrain Cranes 541
Modified Cranes for Heavy Lifting 542
Crane Booms 544
Lifting Capacities of Cranes 544
Rated Loads for Lattice- and Telescopic-Boom Cranes 545
Working Ranges of Cranes 548
TOWER CRANES 550
Classification 550
Operation 554
Tower Crane Selection 562
Rated Loads for Tower Cranes 563
RIGGING 567
Rigging Basics 567
Slings 570
SAFETY 572
Crane Accidents 572
Safety Plans and Programs 574
Zones of Responsibility 575
Summary 576
Problems 577
References 578
Website Resources 578

CHAPTER **18**
Draglines and Clamshells 580
Introduction 580
DRAGLINES 581
General Information 581
Description of a Dragline 582
Dragline Production 585
Calculating Dragline Production 588
Factors Affecting Dragline Production 589
CLAMSHELL EXCAVATORS 593
General Information 593
Clamshell Buckets 594
Production Rates for Clamshells 595
Safety 597
Summary 598
Problems 598
References 599
Website Resources 599

CHAPTER **19**
Piles and Pile-Driving Equipment 600
Introduction 600
Glossary of Terms 600
PILE TYPES 602
Classifications of Piles 602
Timber Piles 603
Concrete Piles 604
Steel Piles 610
Composite Piles 611
Sheet Piles 612
DRIVING PILES 618
The Resistance of Piles to Penetration 618
Site Investigation and Test Pile Program 618
Pile Hammers 620
Supporting and Positioning Piles during Driving 630
Jetting Piles 632
Spudding and Preaugering 633
Hammer Selection 633
Pile-Driving Safety 636
Summary 637

Problems 638
References 638
Website Resources 638

CHAPTER 20
Air Compressors and Pumps 639
Support Equipment 639
COMPRESSED AIR 640
Introduction 640
Glossary of Gas Law Terms 641
Gas Laws 642
Glossary of Air Compressor Terms 644
Air Compressors 644
Compressed-Air Distribution System 646
Diversity Factor 652
Safety 653
EQUIPMENT FOR PUMPING WATER 655
Introduction 655
Glossary of Pumping Terms 656
Classification of Pumps 657
Centrifugal Pumps 658
Loss of Head Due to Friction in Pipe 664
Rubber Hose 665
Selecting a Pump 665
Wellpoint Systems 668
Deep Wells 670
Summary 670
Problems 671
References 674
Website Resources 674

CHAPTER 21
Planning for Building Construction 675
Introduction 675
Site Layout 677
Lifting and Support Equipment 683
Delivery of Structural Components 686
Steel Erection 688
Tilt-Up Construction 689
CONTROL OF CONSTRUCTION NUISANCES 692
Construction Noise 692
Noise Mitigation 694
Lighting 697
Dust 698
Vibration 698
Summary 699
Problems 700
References 701
Website Resources 702

CHAPTER 22
Forming Systems 703
Classification 703
Formwork and the Project Engineer 705
Formwork Design 707
Formwork Economics 711
Vertical Systems 718
Horizontal Systems 727
Combined Vertical and Horizontal Systems 733
Shoring Towers 738
Safety 745
Summary 747
Problems 748
References 749
Website Resources 750

APPENDIX A
Alphabetical List of Units with Their SI Names and Conversion Factors 751

APPENDIX B
Selected English-to-SI Conversion Factors 753

APPENDIX C
Selected U.S. Customary (English) Unit Equivalents 754

APPENDIX D
Selected Metric Unit Equivalents 755

Index 756

PREFACE

With the coming of the railroads in the early 1800s there was a need to accomplish sizable grading operations. In Massachusetts, the building of the Western Railroad, which was completed in 1841, required the movement of approximately 6.8 million cubic yards of material. Men wielding pick and shovel accomplished most of the work, with horses and wagons used to move the material from the cuts to the fills. But a young man named William S. Otis, whose firm Carmichael, Fairbanks, and Otis held a grading contract with the Boston & Providence Railroad, developed a machine in 1834 commonly referred to as the "Yankee Geologist" by the English. It was the first steam shovel, and with it the age of mechanically driven construction equipment began.

The Bucyrus Company published a *Handbook of Steam Shovel Work, a Report by the Construction Service Company* in 1911. This publication is a collection of field studies performed in 1909 to analyze shovel production and delay factors. From this early treatise on time and motion applied to the utilization of steam shovels, it is clear that the engineers understood that production was tied to proper planning of the excavation and haulage.

Today, though we have entered the age of the laptop computer and the Internet, and can download data directly for our machines, there is an even greater need to properly plan equipment operations. A machine is only economical if used in the proper manner and in the environment for which it has the mechanical capabilities to engage. Technology improvements greatly enhance our ability to formulate equipment, planning, and construction decisions, but we must first have an understanding of machine capabilities and how to properly apply those capabilities to construction challenges.

This seventh edition follows in the tradition of the first six by providing the reader with fundamentals of machine selection and production estimating in a logical, simple, and concise format. With a grounding in these fundamentals, the constructor is prepared to evaluate those reams of computer-generated data and to develop programs that speed the decision process or that enable easy analysis of multiple options.

Significant changes have been made to this edition. Following a course plotted with the sixth edition, we have introduced more material applicable to building construction. This is particularly true in the chapters addressing cranes and concrete, which have been extensively rewritten, and the two new chapters on "Forming Systems" and "Planning for Building Construction." Today, formwork systems are construction equipment in very much the same manner as cranes and concrete pumps. The new "Forming Systems" chapter focuses on advanced modular and industrialized forming systems.

The chapters on "Compressed Air" and "Equipment for Pumping Water" have been combined because the concept of calculating friction losses is applied to both air and water in designing systems.

We have also found that in the five years since the last edition of *Construction Planning, Equipment, and Methods,* considerably more equipment manufacturers are placing their machine specifications and operation materials on the Internet. Machine data that we originally proposed to present on a CD with the book are now available over the Internet. Therefore Web resource information is provided at the end of every chapter of our text. In addition, Web-based exercises, which in some cases direct the student to specific machine information on the web, have been added to many of the chapters. When you see the website icon in the text margin, visit our website at *www.mhhe.com/peurifoy7e* for additional resources and exercises available on the World Wide Web.

All chapters have undergone revision, ranging from simple clarification to major modifications, depending on the need to improve organization and presentation of concepts. The pictures in all of the chapters have been updated to illustrate the latest equipment and methods, and more pictures of operating equipment have been used in this edition. Drawings have been added beside many of the figures so that the important features under consideration are clearly identified. Safety discussions are now presented in each of the chapters dealing with machine or formwork use.

The world of construction equipment is truly global, and we have tried to search globally for the latest ideas in machine application and technology. We have visited manufacturers and project sites in some 23 countries around the world in gathering the information presented in this edition.

This book enjoys wide use as a practical reference by the profession and as a college textbook. The use of examples to reinforce the concepts through application has been continued. Based on professional practice, we have tried to present standard formats for analyzing production. Many companies use such formats to avoid errors when estimating production during the fast-paced efforts required for bid preparation.

To enhance the value of the book as a college textbook, we have updated and expanded the number of problems at the close of each chapter. We have also included several problems that compel the student to learn using a step-by-step approach: these problems specifically request the solution for each step before moving on to reach a final solution. This approach focuses student learning by clearly defining the critical pieces of information necessary for problem solving. The solutions to some problems are included in the text at the end of the problem statements. Together with the examples, they facilitate learning and give students confidence that they can master the subjects presented.

We are deeply grateful to the many individuals and firms who have supplied information and illustrations. Two individuals are owed a particular debt of gratitude for their support and efforts. Prof. John Zaniewski, Director, Harley O. Staggers National Transportation Center, West Virginia University, has consistently provided assistance with the "Asphalt Mix Production and Placement" chapter, and Mr. R. R. Walker of Tidewater Construction Corporation has done the same for the "Piles and Pile-Driving Equipment" chapter. We would like to express our thanks for many useful comments and suggestions provided by the following reviewers:

David Arditi, *Illinois Institute of Technology*; Ibrahim A. Assakkaf, *University of Maryland*; Frank Atuahene, *South Dakota State University*; Marcia C. Belcher, *University of Akron*; Leonhard E. Bernold, *North Carolina State University*; Keith A. Bisharat, *Sacramento State University*; Carl Bovill, *University of Maryland*; Travis Chapin, *Bowling Green State University*; Jay Christofferson, *Brigham Young University*; Gregg R. Corley, *Clemson University*; Larry G. Crowley, *Auburn University*; Neil N. Eldin, *Texas A&M University*; William C. Epstein, *California Polytechnic State University, San Luis Obispo*; Sean P. Foley, *Northern Kentucky University*; David R. Fritchen, *Kansas State University*; Marvin C. Gabert, *Boise State University*; John Gambatese, *Oregon State University*; Jesus M. de la Garza, *Virginia Tech*; Sanjiv Gokhale, *Vanderbilt University*; Paul M. Goodrum, *University of Kentucky*; F. H. Griffs, *Brooklyn Institute of Technology*; Hanford Gross, *Washington University*; Carl Haas, *The University of Texas at Austin*; Dana Hobson, *Oklahoma State University*; Arpad Horvath, *University of California, Berkeley*; William Ibbs, *University of California, Berkeley*; Saeed Karshenas, *Marquette University*; Henry M. Koffman, *University of Southern California*; Walter Konon, *New Jersey Institute of Technology*; Thomas Mills, *Virginia Tech*; Keith R. Molenaar, *University of Colorado*; Michael L. Nobs, *Washington University in St. Louis*; David Riley, *The Pennsylvania State University*; Jeffrey S. Russell, *University of Wisconsin–Madison*; Richard C. Ryan, *University of Oklahoma*; Robert L. Schmitt, *University of Wisconsin–Platteville*; David N. Sillars, *Oregon State University*; Marlee Walton, *Oregon State University*

However, we take full responsibility for the material. Finally we wish to acknowledge the comments and suggestions for improvement received from persons using the book. We are all aware of how much our students help us to sharpen the subject presentation. Their questions and comments in the classroom have guided us in developing this revised book. For that and much more, we want to thank our students at the Air Force Academy, Arizona State University, Louisiana Tech, Purdue, Technion–Israel Institute of Technology, University of New Mexico, Virginia Tech, the Universidad de Piura, and the Universidad Téchnica Particular de Loja, who have over the years contributed so much helpful advice for clarifying the subject matter.

Most importantly we express our sincere appreciation and love for our wives, Judy and Reuma, who typed chapters, proofread too many manuscripts, kept us healthy, and who otherwise got pushed farther into the exciting world of construction than they probably really wanted. Without their support this text would not be a reality.

We solicit comments on the edition.

Cliff Schexnayder
Del E. Webb School of Construction
Tempe, Arizona

Aviad Shapira
Technion–Israel Institute of Technology
Haifa, Israel

Guided Tour

This book describes the fundamental concepts of machine utilization, which economically match machine capability to specific project construction requirements. The text contains over 300 photos and 300 additional drawings to describe equipment and construction methods. Illustrations and figures have been added to highlight important features.

FIGURE 9.7 Crawler-mounted hydraulic hoe.

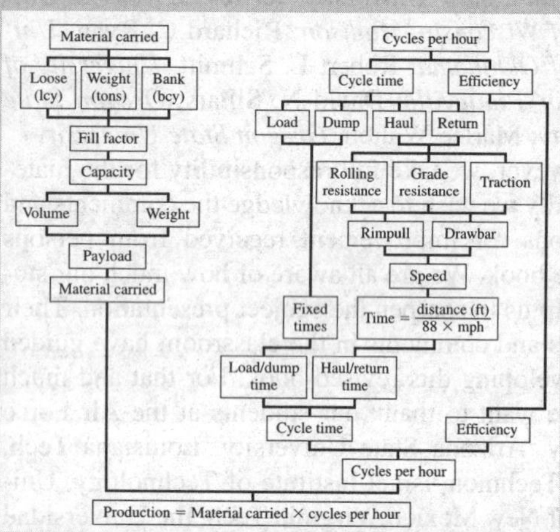

FIGURE 9.3 Excavator production process.

This seventh edition of *Construction Planning, Equipment, and Methods* presents the fundamentals of machine selection and production estimating in a **logical, simple, and concise format.**

Our text features more material applicable to **building construction.** See, for example, the chapters covering **cranes** and **concrete,** which have been extensively rewritten.

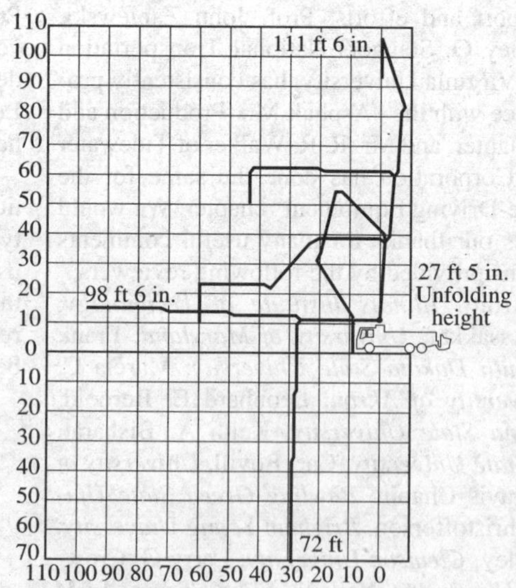

FIGURE 16.16 Truck-mounted boom and pump combination: (a) boom chart.

Guided Tour

Two new chapters, "Forming Systems" and "Planning for Building Construction," are introduced in this edition. The new chapter on forming focuses on advanced modular and industrialized forming systems.

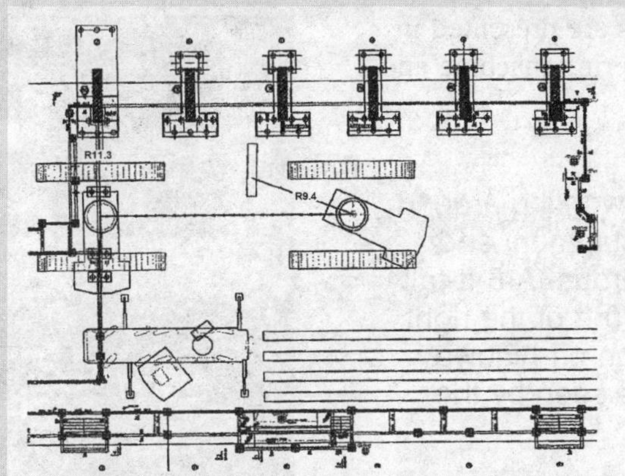

FIGURE 21.10 Lift plan for setting a concrete column.

(1)	Depth of hole:	(a) ______	ft face, (b) ______	ft drill
(2)	Penetration rate:	______	ft/min	
(3)	Drilling time:	______	min	(1b)/(2)
(4)	Change steel:	______	min	
(5)	Blow hole:	______	min	
(6)	Move to next hole:	______	min	
(7)	Align steel:	______	min	
(8)	Change bit:	______	min	
(9)	Total time:	______	min	
(10)	Operating rate:	______	ft/min	(1b)/(9)
(11)	Production efficiency:	______	min/hr	
(12)	Hourly production:	______	ft/hr	(11) × (10)

FIGURE 12.16 Format for estimating drilling production.

We use **examples** to reinforce concepts and applications. Based on professional practice, our text presents **standard formats** for analyzing equipment production. Many companies use such formats to avoid errors when estimating production during the fast-paced efforts required for bid preparation.

Every chapter contains new and expanded **homework problems.** We have also included several problems requiring a **step-by-step approach,** which focuses student learning by clearly defining the critical pieces of information necessary for problem solving.

To purchase a new car it is necessary to borrow $18,550. The bank offers a 5-year loan at an interest rate of 4% compounded annually. If you make only one payment at the end of the loan period repaying the principal and interest, what is the total amount that must be paid back?

(a) What is the number of time periods (n) you should use in solving this problem? **(5)**

(b) What rate of interest (i), per period of time, should be used in solving this problem? **(4%)**

(c) Is the present single amount of money (P) known? (Yes, No) **(Yes)**

Guided Tour

Safety discussions are presented in every chapter covering machine and formwork use.

Operating and working around construction equipment and trucks is dangerous. A 6-ft-tall person within 70 ft of the right side of a 150 ton off-highway truck cannot be seen by the driver.

Web-based exercises have been added to many chapters to draw attention to the expanding volume of information available over the Internet. The computer monitor icon in the text margin will direct you to the text website (www.mhhe.com/peurifoy7e). In addition, extensive **Web resources** are provided at the end of every text chapter.

Visit the Caterpillar website and obtain the flywheel power (hp) rating for both a D6R and a D7R Series II track tractor. For these same tractors, check the blade specifications and determine the width (length) for both an "A" and an "S." Calculate the hp per foot of cutting edge ratio for all four conditions.

For **Instructors,** a comprehensive Solutions Manual and PowerPoint Lectures.

For **Students,** excellent Additional Resources, including video clips, tied directly to the text.

Our **website** contains additional resources for both instructors and students.

CHAPTER 1

Machines Make It Possible

Construction is the ultimate objective of a design, and the transformation of a design by construction into a useful structure is accomplished by men and machines. Men and machines transform a project plan into reality, and as machines evolve there is a continuing transformation of how projects are constructed. This book describes the fundamental concepts of machine utilization, which economically match machine capability to specific project construction requirements. The efforts of contractors and equipment manufacturers, daring to develop new ideas, constantly push machine capabilities forward. As the array of useful equipment expands, the importance of careful planning and execution of construction operations increases.

THE HISTORY OF CONSTRUCTION EQUIPMENT

Machines are a vital resource to the accomplishment of a construction project (see Fig. 1.1). One of the most obvious problems in constructing a project is how to transport heavy building materials. Machines provide the solution to that problem. The proof of how well the planner understands the work that must be accomplished and selects appropriate machines for that purpose is revealed by counting the money when the contract is completed. Did the company make a profit or sustain a loss?

From the time the first man decided to build some type of simple structure for protection from the elements through the construction of the Great Pyramids, the Great Wall of China, the temples at Angkor Wat in Cambodia, and continuing until the middle of the nineteenth century, work was accomplished by the muscle of man and beast. When Ferdinand de Lesseps began excavating the Suez Canal in April 1859, **corvée** laborers, provided by the Egyptian viceroy, did the work of digging that trench in the desert. Human labor assisted only by a very few machines continued the work for the next 4 years.

corvée
Labor required in lieu of taxes.

FIGURE 1.1 Modern hydraulic excavator loading a truck on a dam project in California.

But in 1864 Lesseps and his engineers began turning to machines, and ultimately 300 steam-powered mechanical dredges were at work. Those machines, in the final 3 years of the project, excavated the majority of the main canal's 74 million cubic meters. Mechanization—machines—transformed the project and continue to transform how projects are built.

The Dreams

The development of construction equipment followed the major changes in transportation modes. When travel and commerce were by water systems, builders dreamed of machines that would aid in dredging ports, rivers, and canals. As early as 1420, the Venetian Giovanni Fontana was dreaming and diagramming dredging machines. Leonardo da Vinci designed such a machine in 1503, and at least one of his machines was actually built but the power source was a lonely runner on a treadmill.

On July 4, 1817, at a site near Rome, New York, ground was broken for the 363-mile-long Erie Canal. It was built—excavated—by the efforts of local laborers and Irish immigrants, human labor. However, by 1852 construction in the United States was changing from canal building to railroad construction. The Middlesex Canal, which connected Boston to the Merrimack River at Lowell, had been in service since 1803, but in 1853 the Boston & Lowell Railroad superseded it. Nevertheless, construction, be it building canals or railroads, was still achieved by the brawn of man and beast.

Steam Power Machines

William S. Otis, a civil engineer with the Philadelphia contracting firm of Carmichael & Fairbanks, built the first practical power shovel excavating machine in 1837. The first "Yankee Geologist," as his machines were called, was put to work in 1838 on a railroad project in Massachusetts. The May 10, 1838, issue of the *Springfield Republican* in Massachusetts reported "Upon the road in the eastern part of this town, is a specimen of what the Irishmen call 'digging by stame.' For cutting through a sand hill, this steam digging machine must make a great saving of labor."

Continued development of the steam shovel was driven by a demand for economical mass excavation machines. In the early 1880s, an era of major construction projects began. These projects demanded machines to excavate large quantities of earth and rock. In 1881, Ferdinand de Lessep's French company began work on the Panama Canal. Less than a year earlier, on December 28, 1880, the Bucyrus Foundry and Manufacturing Company, of Bucyrus, Ohio, came into being. Bucyrus became a leading builder of steam shovels (see Fig. 1.2), and 25 years later when the Americans took over the Panama Canal work, the Bucyrus Company was a major supplier of steam shovels for that effort.

Still, the most important driver in excavator development was the railroad. Between 1885 and 1897 approximately 70,000 miles of railway were constructed in the United States. William Otis developed his excavator machine

FIGURE 1.2 An early twentieth-century steam shovel; note this machine is mounted on steel traction wheels.

because the construction company Carmichael & Fairbanks, which he worked for and in which his uncle Daniel Carmichael was a senior partner, was in the business of building railroads.

The Bucyrus Foundry and Manufacturing Company came into being because Dan P. Eells, a bank president in Cleveland, was associated with several railroads. In 1882 the Ohio Central Railroad gave the new company its first order for a steam shovel, and sales to other railroads soon followed.

Internal Combustion Engines

By 1890 courts of law in Europe had ruled that Nikolaus Otto's patented four-cycle gasoline engine was too valuable an improvement to keep restricted. Following the removal of that legal restraint, many companies began experimenting with gasoline-engine-powered carriages. The Best Manufacturing Company (the predecessor to Caterpillar Inc.) demonstrated a gasoline tractor in 1893.

The first application of the internal combustion engine to excavating equipment occurred in 1910 when the Monighan Machine Company of Chicago shipped a dragline powered by an Otto engine to the Mulgrew-Boyce Company of Dubuque, Iowa. Henry Harnischfeger brought out a gasoline-engine-powered shovel in 1914. Following World War I, the diesel engine began to appear in excavators. A self-taught mechanic named C. L. "Clessie" Cummins, working out of an old cereal mill in Columbus, Indiana, developed the Cummins diesel engine in the early 1900s. The Cummins engine soon became popular in power shovels. Warren A. Bechtel, who in 1898 entered the construction profession in Oklahoma Territory and quickly built a reputation for successful railroad grading, pioneered the use of motorized trucks, tractors, and diesel-powered shovels in construction.

In the winter of 1922–1923, the first gas-powered shovel was brought into the state of Connecticut, and in the spring of 1923, it was employed on a federal-aid road construction project. The third phase of transportation construction had begun. Contractors needed equipment for road building. In 1919 Dwight D. Eisenhower, as a young army officer, took an Army convoy cross-country to *experience* the condition of the nation's roads (see Fig. 1.3). But as the country began to improve its road network, World War II intervened, and road building came to a near halt as the war unfolded.

Incubators for Machine Innovation

Los Angeles Aqueduct Large construction projects provide a fertile testing ground for equipment innovation. William Mulholland, as Los Angeles City Engineer, directed an army of 5,000 men for 5 years constructing the Los Angeles Aqueduct that stretches 238 miles from the Owens River to Los Angeles. In 1908 the Holt Manufacturing Company (the other predecessor to Caterpillar Inc.) sold three gas-engine caterpillar tractors to the city of Los Angeles for use in constructing the Los Angeles Aqueduct. Besides crossing several mountain ranges, the aqueduct passed through the Mojave Desert, a severe test site

FIGURE 1.3 Photograph with Eisenhower's description of conditions.
Courtesy Dwight D. Eisenhower Library

for any machine. The desert and mountains presented a challenging test for the Holt machines, but Benjamin Holt viewed the entire project as an experiment and development exercise.

Holt found that cast-iron gears wore out very quickly from sand abrasion, so he replaced them with gears made of steel. The brutal terrain broke suspension springs and burned up the two-speed transmissions in his tractors. The low gear was simply not low enough for climbing the mountains. Holt made modifications to the tractors both at his factory and in the desert. His shop manager, Russell Springer, set up repair facilities in the project work camps. After completion of the project, Mulholland in his final report labeled the Holt tractors as the only unsatisfactory purchase that had been made. But Holt had developed a much better machine because of the experience.

Boulder Dam In the years between the two world wars, one particular construction project stands out because of the equipment contributions that resulted from the undertaking. The Boulder Dam project (later named the Hoover Dam) was an enormous proving ground for construction equipment and techniques.

The use of bolted connections for joining machine pieces together came to an end in the Nevada desert as the project provided the testing ground for R. G. LeTourneau's development of welded equipment and cable-operated attachments. LeTourneau, through his numerous innovations in tractor/scraper design, made possible the machines that later went to build airfields around the world during World War II. Other developments that came from the Boulder

Dam project included sophisticated aggregate production plants, improvements in concrete preparation and placement, and the use of long-flight conveyor systems for material delivery.

Three Significant Developments

torque converter
A fluid-type coupling that enables an engine to be somewhat independent of the transmission.

After World War II, road building surged and in 1956 Eisenhower, now president, signed the legislation that established the interstate highway program. To support the road-building effort, scrapers increased in capacity from 10 to 30 cubic yards (cy). With the development of the **torque converter** and the power shift transmission, the front-end loader began to displace the old "dipper" stick shovels. Concrete batch and mixing plants changed from slow manually controlled contraptions to hydraulically operated and electronically controlled equipment. But the three most important developments were high-strength steels, nylon cord tires, and high-output diesel engines.

1. *High-strength steels.* Up to and through World War II, machine frames had been constructed with steels in the 30,000- to 35,000-psi yield range. After the war, steels in the 40,000- to 45,000-psi range with proportionally better fatigue properties were introduced. The new high-strength steel made possible the production of machines having a greatly reduced overall weight. The weight of a 40-ton off-highway truck body was reduced from 25,000 to 16,000 lb with no change in body reliability.
2. *Nylon cord tires.* The utilization of nylon cord material in tire structures made larger tires with increased load capacity and heat resistance a practical reality. Nylon permitted the actual number of plies to be reduced as much as 30% with the same effective carcass strength, but with far less bulk or carcass thickness. This allowed tires to run cooler and achieve better traction, and improved machine productivity.
3. *High-output diesel engines.* Manufacturers developed new ways to coax greater horsepower from a cubic inch of engine displacement. Compression ratios and engine speeds were raised, and the art of turbo-charging was perfected, resulting in a 10 to 15% increase in flywheel horsepower.

Today there does not appear to be any radically new equipment on the horizon. However, manufacturers are continually refining the inventions of the past, and the development of new attachments will mean improved utility for the contractor's fleet. The future of equipment technology or innovation can be divided into three broad categories:

- *Level of control*: equipment advancements that transfer operational control from the human to the machine.
- *Amplification of human energy*: shift of energy requirements from the man to the machine.
- *Information processing*: gathering and processing of information by the machine.

The Future

A time may come when the base machine is considered only a *mobile counterweight with a hydraulic power plant*. The base machine will perform a variety of tasks through multiple attachments. This trend has started with hydraulic excavators having many attachments such as hammers, compactors, shears, and material-handling equipment. Wheel loaders have seen the introduction of the tool-carrier concept. Wheel loaders are no longer standard bucket machines. There are now other attachments such as brooms, forks, and stingers available so that a loader can perform a multitude of tasks. Other attachments will be developed, offering the contractor more versatility from a base investment. Ultimately, operators sitting in a machine cab may be eliminated altogether.

Safety features and operator station improvements are evolving to compensate for the less experienced workforce available today and in the foreseeable future. Related to workforce quality is the proliferation of supporting machine control technologies. Navigation of equipment is a broad topic, covering a large spectrum of different technologies and applications. It draws on some very ancient techniques, as well as some of the most advanced in space science and engineering.

The new field of geospatial engineering is rapidly expanding and a spectrum of technologies is being developed for the purposes of aeronautic navigation, mobile robot navigation, and geodesy. This technology is rapidly being transferred to construction applications.

The U.S. Army Corps of Engineers conducted a field test of a Computer-Aided Earthmoving System (CAES) developed by Caterpillar Inc. in 2001 (see Fig. 1.4). From that limited test the Corps reported that CAES-equipped CAT 613 scrapers

- Moved 5.4% more earth in the 20-hr test period.
- Reduced preconstruction and restaking time by 28 hrs.
- Reduced manpower requirements by 54%.
- Achieved an accuracy of 2.3 in. vertical, 9.6 in. horizontal.

GPS
A highly precise satellite-based navigation system.

The laser and the global positioning system **(GPS)** guidance will become more common and reduce the need for surveyors. All the grader or dozer operator will need to do is load the digital terrain model into the onboard computer and then guide the machine where the display indicates. Machine position, along with cut or fill information, will be on a screen in front of the operator at all times. This may turn the operator's job into a video game.

Ultimately, operators sitting in a machine cab may be eliminated altogether. Caterpillar is developing and testing automated rock-hauling units for mining. These units are linked by radio to the office and tracked by GPS. The superintendent need only use a laptop to send the start signal and the trucks do the rest. They leave the lineup at set intervals and follow the prescribed course. The superintendent can track the progress of each machine on the computer. If a truck develops a problem, the situation is signaled to the superintendent for corrective action.

(a) Reference station

(b) System mounted on a CAT 613 scraper

(c) CAES-equipped scraper working

(d) CAES-equipped scraper working at night

FIGURE 1.4 Corps of Engineers field test of a Computer-Aided Earthmoving System.

BEING COMPETITIVE

This book introduces the engineering fundamentals for planning, selection, and utilization of construction equipment. It enables one to analyze operational problems and to arrive at practical solutions for completing construction tasks. It is about the application of engineering fundamentals and analysis to construction activities, and the economic comparison of machine choices.

The construction contractor's ability to win contracts and to perform them at a profit is determined by two vital assets: people and equipment. To be economically competitive, a contractor's equipment must be competitive, both mechanically and technologically. Old machines, which require costly repairs, cannot compete successfully with new equipment having lower repair costs and higher production rates.

In most cases, a piece of equipment does not work as a stand-alone unit. Pieces of equipment work in groups. An excavator loads trucks that haul material to a location on the project where it is required. At that point, the material is dumped and a dozer spreads the material. After spreading, a roller compacts the material to the required density. Therefore, a group of machines, in this example an excavator, haul trucks, a dozer, and a roller, constitute what is commonly referred to as an equipment spread.

Optimization in the management of an equipment spread is critical for a contractor, both in achieving a competitive pricing position and in accumulating the corporate operating capital required to finance the expansion of project performance capability. This book describes the basic operational characteristics of the major heavy construction equipment types. More important, however, it explains the fundamental concepts of machine utilization, which economically match machine capability to specific project construction requirements.

There are no unique solutions to the problem of selecting a machine to work on a particular construction project. All machine selection problems are influenced by external environmental conditions. To appreciate how environmental conditions influence the utilization of heavy construction equipment, one must understand the mechanics of how the construction industry operates.

THE CONSTRUCTION INDUSTRY

By the nature of the product, the construction contractor works under a unique set of production conditions that directly affect equipment management. Whereas most manufacturing companies have a permanent factory where raw materials flow in and finished products flow out in a repetitive, assembly-line process, a construction company carries its factory with it from job to job. At each new site, the company proceeds to set up and produce a one-of-a-kind project. If the construction work goes as planned, the job will be completed on time and with a profit.

Equipment-intensive projects present great financial risk. Many projects involving earthwork are bid on a unit-price basis and there can be large variations between estimated and actual quantities. Some projects require an equipment commitment that is greater than the amount that a contractor will be paid for completing the work. Such a situation forces a contractor into a continuing sequence of jobs to support the long-term equipment payments.

Additional risk factors facing contractors in equipment-intensive work include financing structure, construction activity levels (the amount of work being put out for bid), labor legislation and agreements, and safety regulations. Project size and outdoor work that is weather dependent contribute to long project durations. Projects requiring two or more years to complete are not uncommon in the industry.

Government-initiated actions that seriously affect the operating environment of the construction contractor are labor legislation and safety regulation. In each of these areas, many regulations impact on a contractor's operations. These actions can directly influence equipment decisions. Legislative acts that exert direct pressure on equipment questions include the Davis-Bacon Act, which is concerned with wage rates, and the Occupational Safety and Health Act (OSHA), which specifies workplace safety requirements. Over one-half of the dollar volume of work in the equipment-intensive fields of construction is subject to wage determinations under the Davis-Bacon Act, and this strongly influences the labor costs incurred by contractors. OSHA, by its rollover protective structures (ROPS) mandate, substantially increased the cost of those pieces of construction equipment that had to have these structures included as part of the basic machine. That particular regulation had a single-point-in-time effect on equipment decisions, much like that resulting from the introduction of new equipment technology. Similarly, there remains the possibility of additional safety requirements. Sound and emissions are issues that are receiving greater regulatory attention. Some owners, by clauses in the construction contract, are limiting machine noise levels.

> Construction equipment to be certified includes any equipment of the types listed in Table XX brought on-site.
>
> This equipment shall be retested every 6 months while in use on-site. Any equipment used during construction may be subject to confirmatory noise level testing by the contractor at the request of the Engineer.

SAFETY

The rate of personal injury and death resulting from construction work is too high. Of all major industry classifications, construction has one of the poorest safety records. The construction industry employs nearly 6.4 million people. That is about 6% of the American workforce. However, according to the National Safety Council, the industry has about 23% of the deaths and 10.3% of the injury accidents every year. That translates into 1,150 to 2,000 deaths and 400,000 disabling injuries annually. The Construction Industry Institute estimates the direct and indirect costs of construction accidents may be as high as $17 billion annually. The major causes of deaths and injuries are falls from elevations, electrocution, being struck by equipment, being caught in between equipment, and trench excavation cave-ins. As an industry, we are responsible and accountable for those statistics. It is the responsibility of construction managers to create the safety programs that will prevent those accidents. We have both a moral and a business interest in doing so. The key is to provide the leadership, the programs, and the incentives to create a safe industry.

In the late 1960s, Congress began an investigation of construction safety, and in 1970 enacted the Williams-Steiger Act, more commonly referred to as the Occupational Safety and Health Act. The act provided a comprehensive set of safety rules and regulations, inspection procedures, and safety recordkeeping requirements. It imposed nationwide safety standards on the construction industry. It also permitted the states to enact their own OSHA legislation as long as the state legislation is at least as stringent as the federal legislation. Employers are required to provide their employees a safe place to work and to maintain extensive safety records.

The act also established the Occupational Safety and Health Administration (OSHA), with regional offices in several cities throughout the country. OSHA is responsible for the administration of the legislation and the development of rules and regulations to implement the act. The OSHA rules and regulations are published in the *Federal Register. OSHA Safety and Health Standards*, Code of Federal Regulations, Title 29, Part 1910, contains the safety features that must be included in construction projects by the architect or engineer. *Construction and Health Regulations*, Code of Federal Regulations, Part 1926, pertains specifically to construction contractors and construction work. The act provides both civil and criminal penalties for violations of OSHA regulations. The civil penalty for failure to correct a violation is \$7,000 per day with a maximum penalty of \$70,000. Criminal penalties can include both fines and imprisonment. It is OSHA's intent to establish a uniform set of safety standards that apply to construction and to actively enforce those standards. Contractors must maintain a current, up-to-date file of OSHA regulations, and work proactively to comply with OSHA requirements (see Figs. 1.5 and 1.6).

FIGURE 1.5 Proper connections for air hoses.

THE CONTRACTING ENVIRONMENT

Construction contractors work within a unique market situation. The job plans and specifications that are supplied by the client dictate the sales conditions and product, but not the price. Almost all work in the equipment-intensive

FIGURE 1.6 Failure to properly support the crane outriggers.

fields of construction is awarded on a bid basis, through either open or selective tender procedures. Under the design-bid-build method of contracting, the contractor states a price after estimating the cost based on a completed design supplied by the owner. The offered price includes overhead, project risk contingency, and the desired profit.

There is movement toward more design-build contracts, where the contractor also has control of the project design. With a design-build project, the contractor must state a guaranteed price before the design is completed. This adds an additional element of risk, because estimating the quantities of materials required to complete the project becomes very subjective. But the advantage to the contractor is that the design can be matched in the most advantageous way to the contractor's construction skills. In either case, it is tacitly assumed that the winning contractor has been able to underbid the competition because of a more efficient work plan, lower overhead costs, or a willingness to accept a lower profit.

Not infrequently, however, the range between the high and low bids is much greater than these factors would justify. A primary cause of variance in bids is a contractor's inability to estimate costs accurately. The largest portion of estimating variance is probably not caused by the differences between past and future projects but by a lack of accurate cost records. Most contractors have cost-reporting systems, but in numerous cases the systems fail to allocate expenses to the proper sources, and therefore cause false conclusions when used as the historical database for estimating future work.

A construction company owner will frequently use both contract volume and contract turnover to measure the strength of the firm. Contract volume refers to the total dollar value of awarded contracts that a firm has on its books (under contract) *at any given time.* Contract turnover measures the dollar value of work that a firm completes during a specific *time interval.* Contract volume is a guide to the magnitude of resources a firm has committed at any one time, as well as to possible profit if the work is completed as estimated. But contract volume fails to answer any timing questions. A contractor, who, with the same contract volume as the competition, is able to achieve a more rapid project completion, and therefore a higher capital turnover rate while maintaining the revenue-to-expense ratio, will be able to increase the firm's profits. Contractors who finish work ahead of schedule usually make money.

PLANNING EQUIPMENT UTILIZATION

Each piece of construction equipment is specifically designed by the manufacturer to perform certain mechanical operations. The task of the project planner/estimator or the engineer on the job is to match the right machine or combination of machines to the job at hand. Considering individual tasks, the quality of performance is measured by matching the equipment spread's production against its cost. Production is work done; it can be the volume or weight of material moved, the number of pieces of material cut, the distance traveled, or any similar measurement of progress. To estimate the equipment component of project cost it is necessary to first determine machine *productivity*. Productivity is governed by engineering fundamentals and management ability. Chapter 5 covers the principal engineering fundamentals that control machine productivity. Each level of productivity has a corresponding cost associated with the effort expended. The expenses that a firm experiences through machine ownership and use and the method of analyzing such costs are presented in Chapter 2.

Although each major type of equipment has different operational characteristics, it is not always obvious which machine is best for a particular project task. After studying the plans and specifications, visiting the project site, and performing a quantity take-off, the planner must visualize how to best employ specific pieces of equipment to accomplish the work. Is it less expensive to make an excavation with scrapers or to top-load trucks with a dragline? Both methods will yield the required end result, but which is the most economical method of attack for the given project conditions?

To answer that question the planner develops an initial plan for employment of the scrapers and then calculates their production rate and the subsequent cost. The same process is followed for the top-load operation. The type of equipment that has the lowest estimated total cost, including mobilization of the machines to the site, is selected for the job.

To perform such analyses, the planner must consider both machine capability and methods of employment. In developing suitable equipment employment techniques, the planner must have knowledge of the material quantities involved. This book will not cover quantity take-off per se, but that process is strongly influenced by the equipment and methods under consideration. If it is determined that different equipment and methods will be used as an excavation progresses, then it is necessary to divide the quantity take-off in a manner that is compatible with the proposed equipment utilization. The person performing the quantity take-off must calculate the quantities so that groups of similar materials (dry earth, wet earth, rock) are easily accessed. It is not just a question of estimating the total quantity of rock or the total quantity of material to be excavated. All factors, which affect equipment performance and choice of construction method, such as location of the water table, clay or sand seams, site dimensions, depth of excavations, and compaction requirements, must be considered in making the quantity take-off.

The normal operating modes of the particular equipment types are discussed in Chapters 5, 7 to 20, and 22. That presentation, though, should not blind the reader to other possible applications. The most successful construction companies are those that, for each individual project, carefully study all possible approaches to the construction process. These companies use project preplanning, risk identification, and risk quantification techniques in approaching their work. No two projects are exactly alike; therefore, it is important that the planner begins each new project with a completely open mind and reviews all possible options. Additionally, machines are constantly being improved and new equipment being introduced.

Heavy equipment is usually classified or identified by one of two methods: *functional* identification or *operational* identification. A bulldozer, used to push a stockpile of material, could be identified as a support machine for an aggregate production plant, a grouping that could also include front-end loaders. The bulldozer could, however, be *functionally* classified as an excavator. In this book, combinations of functional and operational groupings are used. The basic purpose is to explain the critical performance characteristics of a particular piece of equipment and then to describe the most common applications for that machine.

The efforts of contractors and equipment manufacturers, daring to develop new ideas, constantly push machine capabilities forward. As the array of useful equipment expands, the importance of careful planning and execution of construction operations increases. New machines enable greater economies. It is the job of the estimator and the field personnel to match equipment to project situations, and that is the central focus of this book.

SUMMARY

Civilizations are built by construction efforts. Each civilization had a construction industry that fostered its growth and quality of life. This chapter presented an abridged history of construction equipment, an overview of construction work, and the risk associated with bidding work. Machine production, the amount of earth moved or concrete placed, is only one element of the machine selection process. It is also necessary to know the cost associated with that production. The critical learning objective is

- An understanding of how construction equipment and machines have been developed in response to the demands of the work to be undertaken.

This objective is the basis for the problems that follow.

PROBLEMS

1.1 Research these engineers on the Web and write a one-page paper about their accomplishments.

William Mulholland

Stephen D. Bechtel Sr.

Benjamin Holt

R. G. LeTourneau

William S. Otis

1.2 Research these engineering accomplishments on the Web and write a one-page paper about the equipment used to accomplish their construction.

Hoover Dam

Panama Canal

Interstate highway program

1.3 What is the function of the Occupational Health and Safety Administration? What OSHA office administers your area?

1.4 Why do some construction workers resist the use of safety equipment such as hard hats and fall protection harnesses? Why does the practice of resisting the use of safety equipment persist? What should be done about it?

REFERENCES

1. *Building for Tomorrow: Global Enterprise and the U.S. Construction Industry* (1988). National Research Council, National Academy Press, Washington, DC.
2. *Davis-Bacon Manual on Labor Standards for Federal and Federally Assisted Construction* (1993). The Associated General Contractors (AGC) of America, Alexandria, VA.
3. *OSHA Safety & Health Standards for Construction (OSHA 29 CFR 1926 Construction Industry Standards)* (2003). The Associated General Contractors (AGC) of America, Alexandria, VA.
4. Schexnayder, Cliff J., and Scott A. David (2002). "Past and Future of Construction Equipment," *Journal of Construction Engineering and Management*, ASCE, 128(4), pp. 279–286.

WEBSITE RESOURCES

Significant additional information about the construction industry can be found posted on the following websites.

Associations and Organizations

Sites about construction associations and organizations include

1. www.asce.org The American Society of Civil Engineers (ASCE) is a professional organization of individual members from all disciplines of civil engineering dedicated to developing leadership, advancing technology, advocating lifelong learning, and promoting the profession.
2. www.asme.org The American Society of Mechanical Engineers (ASME) is a nonprofit educational and technical organization that publishes many standards in reference to construction equipment.
3. www.agc.org The Associated General Contractors of America (AGC) is an organization of construction contractors and industry-related companies.
4. construction-institute.org The Construction Industry Institute (CII) is a research organization with the mission of improving the competitiveness of the construction industry. CII is a consortium of owners and contractors who have joined together to find better ways of planning and executing capital construction programs.

Codes and Regulations

Sites that provide information about codes or regulations that impact the construction industry include

1. www.osha.gov The U.S. Department of Labor, Occupational Labor Safety and Health Administration (OSHA) establishes protective standards, enforces those standards, and reaches out to employers and employees through technical

assistance and consultation programs. OSHA's mission is to ensure safe and healthful workplaces in America.

2. www.nist.gov/welcome.html The National Institute of Standards (NIST) is a non-regulatory federal agency within the U.S. Commerce Department's Technology Administration. NIST develops and promotes measurement, standards, and technology.
3. www.ansi.org The American National Standards Institute (ANSI) is a private, nonprofit organization that administers and coordinates the U.S. voluntary standardization and conformity assessment system.
4. www.iso.ch The International Organization for Standardization (ISO) is a non-governmental organization. It is a network of the national standards institutes from 148 countries, with one member per country and a Central Secretariat in Geneva, Switzerland, that coordinates the system.
5. www.astm.org The ASTM International, formerly known as the American Society for Testing and Materials, is a not-for-profit organization that provides a global forum for the development and publication of voluntary consensus standards for materials, products, systems, and services.

Safety

1. www.nsc.org National Safety Council. Excellent library for workplace safety consultants and human resources managers offering resources, member information, services, and publications.
2. www.construction-institute.org Home page of the Construction Industry Institute, University of Texas at Austin. Source of construction management research relating to best practices.
3. www.osha.gov Occupational Safety and Health Administration site. Includes news, statistics, publications, regulations, standards, and reference resources.
4. www.ntsb.gov National Transportation Safety Board. This independent federal agency conducts investigations on significant transportation accidents, and offers synopses and public hearing overviews.
5. www.crmusa.com Contractors Risk Management. Explore details of this company offering manuals, customized plans, and training programs for construction industry safety and health guidelines.
6. www.agc.org Associated General Contractors of America. Deals with contracting and safety issues and construction laws.

CHAPTER 2

Fundamental Concepts of Equipment Economics

A correct and complete understanding of the costs that result from equipment ownership and operation provide companies a market advantage that leads to greater profits. Ownership cost is the cumulative result of those cash flows an owner experiences whether or not the machine is productively employed on a project. Operating cost is the sum of those expenses an owner experiences by working a machine on a project. The process of selecting a particular type of machine for use in constructing a project requires knowledge of the cost associated with operating the machine in the field. There are three basic methods for securing a particular machine to use on a project: (1) buy, (2) rent, or (3) lease.

IMPORTANT QUESTIONS

Equipment cost is often one of a contractor's largest expense categories, and it is a cost fraught with variables and questions. To be successful, equipment owners must carefully analyze and answer two separate cost questions about their machines:

1. How much does it cost to operate the machine on a project?
2. What is the optimum economic life and the optimum manner to secure a machine?

The first question is critical to bidding and operations planning. The only reason for purchasing equipment is to perform work that will generate a profit for the company. This question seeks to identify the expense associated with productive machine work, and is commonly referred to as ownership and operating (O&O) cost. O&O cost is expressed in dollars per machine operating hour (e.g., \$90/hr for a dozer) because it is used in calculating the cost per unit of machine production. If a dozer can push 300 cy per hour and it has a $90/hr O&O cost, production cost is $0.300/cy ($90/hr ÷ 300 cy/hr). The estimator/planner can use the cost per cubic yard figure directly on unit price work. On a lump-sum job, it will be necessary to multiply the cost/unit price by the estimated quantity to obtain the total amount that should be charged.

The second question seeks to identify the optimum point in time to replace a machine and the optimum way to secure a machine. This is important in that it will affect O&O cost and can lower production expense, enabling a contractor to achieve a better pricing position. The process of answering this question is known as replacement analysis. A complete replacement analysis must also investigate the cost of renting or leasing a machine.

The economic analyses that answer these two cost questions require the input of many expense and operational factors. These input factors will be discussed first and a development of the analysis procedures follows.

EQUIPMENT RECORDS

Data on both machine utilization and costs are the keys to making rational equipment decisions, but the collection of individual pieces of data is only the first step. The data must be assembled and presented in usable formats. Many contractors recognize this need and strive to collect and maintain accurate equipment records for evaluating machine performance, establishing operating cost, analyzing replacement questions, and managing projects. Surveys of industrywide practices, however, indicate that such efforts are not universal.

Realizing the advantages to be gained therefrom, owners are directing more attention to accurate record keeping. Advances in computer technology have reduced the effort required to implement record systems. Computer companies offer record-keeping packages specifically designed for contractors. In many cases, the task is simply the retrieval of equipment cost data from existing accounting files.

Automation introduces the ability to handle more data economically and in shorter time frames, but the basic information required to make rational decisions is still the critical item. A commonly used technique in equipment costing and record keeping is the standard rate approach. Under such a system, jobs are charged a standard machine utilization rate for every hour the equipment is employed. Machine expenses are charged either directly to the piece of equipment or to separate equipment cost accounts. This method is sometimes referred to as an internal or company rental system. Such a system usually presents a fairly accurate representation of investment consumption and it properly assigns machines expenses. In the case of a company replacing machines each year and continuing in operation, this system enables a check at the end of each year on estimate rental rates as the internally generated rent should equal the expenses absorbed.

The first piece of information necessary for rational equipment analysis is not an expense but a record of the machine's use. One of the implicit assumptions of a replacement analysis is that there is a continuing need for a machine's production capability. Therefore, before beginning a replacement analysis, the disposal–replacement question must be resolved. Is this machine really necessary? A projection of the ratio between total equipment capacity and utilized capacity provides a quick guide for the dispose–replace question.

The level of detail for reporting equipment use varies. Both independent service vendors and equipment companies (DeereTrax offered by John Deere and Product Link offered by Caterpillar being two examples) offer data collection devices that provide accurate real-time information about machine use. These devices are installed in the machine and transmit data via the most cost-effective wireless network (satellite or cellular networks). As a minimum, data should be collected on a daily basis to record whether a machine worked or was idle. A more sophisticated system will seek to identify use on an hourly basis, accounting for actual production time and categorizing idle time by classifications such as standby, down weather, and down repair. The input for either type of system is easily incorporated into regular personnel timekeeping reports, with machine time and operator time being reported together.

Most of the information required for ownership and operating or replacement analyses is available in the company's accounting records. All owners keep records on a machine's initial purchase expense and final realized salvage value as part of the accounting data required for tax filings. Maintenance expenses can be tracked from mechanics' time sheets, purchase orders for parts, or from shop work orders. Service logs provide information concerning consumption of consumables. Fuel amounts can be recorded at fuel points or with automated systems. Fuel amounts should be cross-checked against the total amount purchased. When detailed and correct reporting procedures are maintained, the accuracy of equipment costs analyses is greatly enhanced.

THE RENT PAID FOR THE USE OF MONEY

What is commonly referred to as the time value of money is the difference—rent—that must be paid if one borrows some money for use today and returns the money at some future date. Many take this charge for granted, as the proliferation of credit cards testifies. This rent or added charge is termed *interest.* It is the profit and risk that the lender applies to the base amount of money that is borrowed. Interest, usually expressed as a percentage of the amount borrowed (owed), becomes due and payable at the close of each billing time period. It is typically stated as a yearly rate. As an example, if $1,000 is borrowed at 8% interest, then $1,000 × 0.08, or $80, in interest plus the original $1,000 is owed after 1 year (yr). Therefore, the borrower would have to repay $1,080 at the end of a 1-yr time period. If this new total amount is not repaid at the end of the 1-yr period, the interest for the second year would be calculated based on the new total amount, $1,080, and thus the interest is *compounded.* Then, after a 2-yr period, the amount owed would be $1,080 + ($1,080 × 0.08), or $1,166.40. If the company's credit is good and it has borrowed the $1,000 from a bank, the banker normally does not care whether repayment is made after 1 yr at \$1,080 or after 2 yr at \$1,166.40. To the bank the three amounts, \$1,000, \$1,080, and \$1,166.40, are equivalent. In other words, \$1,000 today is equivalent to \$1,080 1 yr in the future, which is also equivalent to \$1,166.40 2 yr in the future. The three amounts are obviously not equal; they are *equivalent. Note that the concept of equivalence involves time and a specific rate of interest*. The three amounts are equivalent only for the case of an interest rate of 8%, and then only at the specified points in time. Equivalence means that one sum or series differs from another only by the accrued, accumulated interest at rate i for n periods of time.

Note that in the example the principal amount was multiplied by an interest rate to obtain the amount of interest due. To generalize this concept, the following symbols are used:

P = a present single amount of money
F = a future single amount of money, after n periods of time
i = the rate of interest per period of time (usually 1 yr)
n = the number of time periods

Different situations involving an interest rate and time are presented next, and the appropriate analytical formulas are developed.

Equation for Single Payments

To calculate the future value F of a single payment P after n periods at an interest rate i, these formulations are used:

At the end of the first period, $n = 1$: $F_1 = P + Pi$
At the end of the second period, $n = 2$: $F_2 = P + Pi + (P + Pi)i = P(1 + i)^2$
At the end of the nth period: $F = P(1 + i)^n$

Or the future single amount of a present single amount is

$$F = P(1 + i)^n \tag{2.1}$$

Note that F is related to P by a factor that depends only on i and n. This factor is termed the *single payment compound amount factor (SPCAF)*; it makes *F equivalent to P*.

If a future amount F is given, the present amount P can be calculated by transposing the equation to

$$P = \frac{F}{(1 + i)^n} \tag{2.2}$$

The factor $1/(1 + i)^n$ is known as the *present worth compound amount factor (PWCAF)*.

EXAMPLE 2.1

A constructor wishes to borrow \$12,000 to finance a project. The interest rate is 5% per year. If the borrowed amount and the interest are paid back after 3 yr, what will be the total amount of the repayment?

To solve, use Eq. [2.1]

$$F = \$12{,}000\ (1 + 0.05)^3 = \$12{,}000\ (1.157625)$$

$$= \$13{,}891.50$$

The amount of interest is \$1,891.50

EXAMPLE 2.2

A constructor wants to set aside enough money today in an interest-bearing account to have \$100,000 5 yr from now for the purchase of a replacement piece of equipment. If the company can receive 8% per year on its investment, how much should be set aside now to accrue the \$100,000 5 yr from now?

To solve, use Eq. [2.2]

$$P = \frac{\$100{,}000}{(1 + 0.08)^5} = \frac{\$100{,}000}{(1.469328)}$$

$$= \$68{,}058.32$$

In Examples 2.1 and 2.2 single payments now and in the future were equated. Four parameters were involved: P, F, i, and n. *Given any three parameters, the fourth can easily be calculated.*

Formulas for a Uniform Series of Payments

Often payments or receipts occur at regular intervals, and such uniform values can be handled by use of additional formulas. First, let us define another symbol:

A = uniform *end-of-period* payments or receipts continuing for a duration of n periods

If this uniform amount A is invested at the end of each period for n periods at a rate of interest i per period, then the total equivalent amount F at the end of the n periods will be

$$F = A[(1 + i)^{n-1} + (1 + i)^{n-2} + \cdots + (1 + i) + 1]$$

By multiplying both sides of the equation by $(1 + i)$ we obtain

$$F(1 + i) = A[(1 + i)^n + (1 + i)^{n-1} + (1 + i)^{n-2} + \cdots + (1 + i)]$$

Now by subtracting the original equation from both sides of the new equation, we obtain

$$Fi = A(1 + i)^n - 1$$

which can be rearranged to

$$F = A\left[\frac{(1 + i)^n - 1}{i}\right] \quad [2.3]$$

The relationship $[(1 + i)^n - 1]/i$ is known as the *uniform series, compound amount factor (USCAF)*.

The relationship can be rearranged to yield

$$A = F\left[\frac{i}{(1 + i)^n - 1}\right] \quad [2.4]$$

The relationship $i/[(1 + i)^n - 1]$ is known as the *uniform series sinking fund factor (USSFF)* because it determines the uniform end-of-period investment A that must be made to provide an amount F at the end of n periods.

To determine the equivalent uniform period series required to replace a present value of P, simply substitute Eq. [2.1] for F into Eq. [2.4] and rearrange. The resulting equation is

$$P = A\left[\frac{(1 + i)^n - 1}{i(1 + i)^n}\right] \quad [2.5]$$

This relationship is known as the *uniform series present worth factor (USPWF)*.

By inverting Eq. [2.5] the equivalent uniform series end-of-period value A can be obtained from a present value P. The equation is

$$A = P\left[\frac{i(1 + i)^n}{(1 + i)^n - 1}\right] \quad [2.6]$$

This relationship is known as the *uniform series capital recovery factor (USCRF)*.

As an aid to understanding the six preceding equivalence relationships, appropriate cash flow diagrams can be drawn. *Cash flow diagrams* are drawings where the horizontal line represents time and the vertical arrows represent cash flows at specific times (up positive, down negative). The cash flow diagrams for each relationship are shown in Fig. 2.1. These relationships form the basis for many complicated engineering economy studies involving the time value of money, and many texts specifically address this subject.

Most engineering economy problems are more complicated than the examples we have considered and must be broken down into parts. Example 2.3 illustrates how this is done and demonstrates the use of the other equivalency relationships.

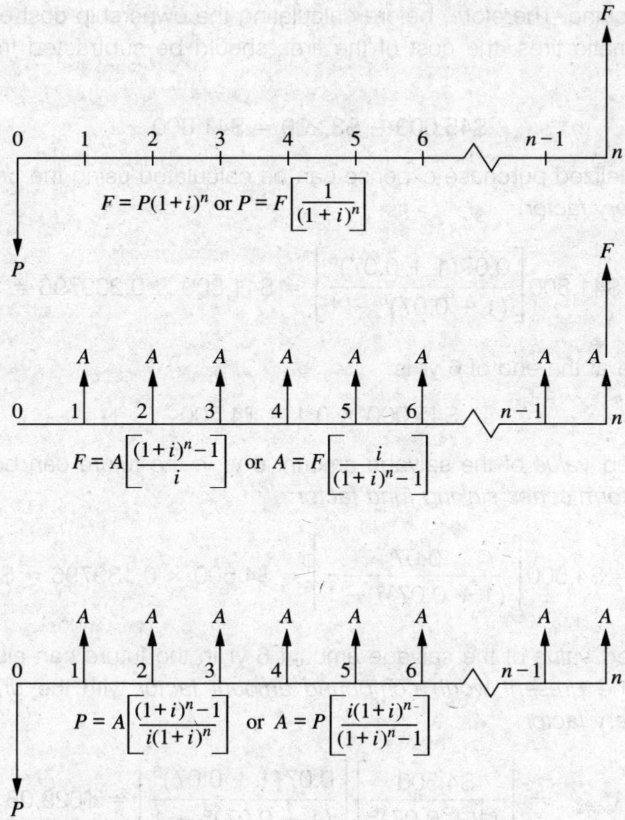

FIGURE 2.1 Cash flow diagrams.

EXAMPLE 2.3

A machine cost \$45,000 to purchase. Fuel, oil, grease (FOG), and minor maintenance are estimated to cost \$12.34 per operating hour (those hours when the engine is operating and the machine is doing work). A set of tires cost \$3,200 to replace, and their estimated life is 2,800 use hours. A \$6,000 major repair will probably be required after 4,200 hr of use. The machine is expected to last for 8,400 hr, after which it will be sold at a price (salvage value) equal to 10% of the original purchase price. A final set of new tires will not be purchased before the sale. How much should the owner of the machine charge per hour of use, if it is expected that the machine will operate 1,400 hr per year? The company's cost of capital rate is 7%. First solve for *n*, the life

$$n = \frac{8{,}400 \text{ hr}}{1{,}400 \text{ hr per yr}} = 6 \text{ yr}$$

Usually ownership and operating cost components are calculated separately. Tires are considered an operating cost element because they wear out much faster than the basic machine. Therefore, before calculating the ownership cost of a machine having pneumatic tires, the cost of the tires should be subtracted from the purchase price.

$$\$45{,}000 - \$3{,}200 = \$41{,}800$$

Now the annualized purchase expense can be calculated using the *uniform series capital recovery factor*:

$$A_{\text{ownership}} = \$41{,}800\left[\frac{0.07(1+0.07)^6}{(1+0.07)^6-1}\right] = \$41{,}800 \times 0.209796 = \$8{,}769.46$$

Salvage value at the end of 6 yr is

$$\$45{,}000 \times 0.1 = \$4{,}500$$

The annualized *value* of the salvage amount 6 yr in the future can be calculated using the *uniform series sinking fund factor*:

$$A_{\text{salvage}} = \$4{,}500\left[\frac{0.07}{(1+0.07)^6-1}\right] = \$4{,}500 \times 0.139796 = \$629.08$$

The annualized *value* of the salvage amount 6 yr in the future can also be calculated using the *present worth compound amount factor* with the *uniform series capital recovery factor*:

$$A_{\text{salvage}} = \left[\frac{\$4{,}500}{(1+0.07)^6}\right]\left[\frac{0.07(1+0.07)^6}{(1+0.07)^6-1}\right] = \$629.08$$

The annual FOG and minor maintenance cost is

$$\$12.34 \text{ per hr} \times 1{,}400 \text{ hr per yr} = \$17{,}276.00$$

In addition to the original set of tires, two sets of replacement tires will have to be purchased, one set 2 yr into the life of the machine and a second set 4 yr into the life. To annualize the tire replacement cost, these future-point-in-time costs must be made equivalent to a present amount at time zero, and then the resulting amount is annualized across the 6-yr life of the machine. To do this, use the *present worth compound amount factor* with the *uniform series capital recovery factor*:

$$A_{\text{tires}} = \left[\$3{,}200 + \frac{\$3{,}200}{(1+0.07)^2} + \frac{\$3{,}200}{(1+0.07)^4}\right]\left[\frac{0.07(1+0.07)^6}{(1+0.07)^6-1}\right]$$
$$= (\$3{,}200.00 + \$2{,}795.00 + 2{,}441.26)(0.209796) = \$1{,}769.89$$

The annualized cost for the major repair 3 yr into the life is

$$A_{\text{major repair}} = \left[\frac{\$6{,}000}{(1+0.07)^3}\right]\left[\frac{0.07(1+0.07)^6}{(1+0.07)^6-1}\right]$$
$$= (\$4{,}897.79)(0.209796) = \$1{,}027.54$$

The resulting total annual cost is

$$A_{\text{total}} = \$8{,}769.46 - \$629.08 + \$17{,}276.00 + \$1{,}769.89 + \$1{,}027.54$$
$$= \$28{,}213.81$$

The total cost per hour is

$$\text{Total cost} = \frac{\$28{,}213.81 \text{ per yr}}{1{,}400 \text{ hr per yr}} = \$20.153 \text{ per hr}$$

COST OF CAPITAL

The interest rate a company experiences is really a weighted average rate resulting from the combined cost associated with all external and internal sources of capital funds—debt (borrowing), equity (sale of stock), and retained earnings (internally generated funds). Furthermore, the cost-of-capital interest rate a company experiences is affected by the risk associated with its business type. The market perceives the risk of the business and applies an after-tax discount rate to the future wealth it expects to derive from the firm. Therefore, the rate that banks charge for borrowed funds cannot be taken alone as the company's cost-of-capital interest rate when evaluating investment alternatives. For a complete treatment of cost of capital, see Modigliani and Miller's classic paper published in *The American Economic Review* [5].

Many discussions of equipment economics include *interest* as a cost of ownership. Sometimes the authors make comparisons with the interest rates that banks charge for borrowed funds or with the rate that could be earned if the funds were invested elsewhere. Such comparisons imply that these are appropriate rates to use in an equipment cost analysis. A few authors appear to have perceived the proper character of interest by realizing that a company requires capital funds for all of its operations. It is not logical to assign different interest costs to machines purchased wholly with retained earnings (cash) as opposed to those purchased with borrowed funds. A single interest rate should be determined by examination of the combined costs associated with all sources of capital funds: debt, equity, and retained earnings.

There are two parts to the common misunderstanding concerning the proper way to account for interest. First, as just discussed, the correct interest rate should reflect the combined effect of the costs associated with all capital funds. The second error comes in trying to recoup *interest costs*. Interest is not a cost to be added together with purchase expense, taxes, and insurance when calculating the total cost of a machine.

This might be easier to understand using an analogy. Consider the situation of a banker trying to decide if a loan should be granted. The question before the banker is one of *risk*: what are the chances the money will be repaid? Based on the perceived risk, a decision is made on how much return must be received to *balance* the risk. If a hundred loans are made, the banker knows some will not be repaid. Those borrowers who repay their loans as promised have to provide the total profit margin for the bank. The good loans must carry the bad loans. A company utilizing equipment should be making a similar analysis every time a decision is made to invest in a piece of equipment.

Based on the risk of the construction business, interest is the fulcrum for determining if the value a machine will create for the company is sufficient. The proper interest rate will ensure that this ratio of gain in value to cost is correctly accounted for in the decision process.

The interest rate at issue is referred to in the economic literature as the *cost of capital* to the company and a market value technique has been developed for its calculation. A complete development of the market value cost-of-capital calculation is beyond the scope of this text, but Reference 4 at the end this chapter is a good presentation of the subject. The resulting cost-of-capital rate is the correct interest rate to use in economic analyses of equipment decisions.

EVALUATING INVESTMENT ALTERNATIVES

The purchase, rental, or replacement of a piece of equipment is a financial investment decision, and as such the real question is how best to use a company's assets. Financial investment decisions are analyzed using time value of money principles. Such analyses require an input interest rate or the calculation of the interest rate that results from the assumed cash flows.

Discounted-Present-Worth Analysis

A discounted-present-worth analysis involves calculating the *equivalent* present worth or present value of all the dollar amounts involved in each of the individual alternates to determine the present worth of the proposed alternates. The present worth is discounted at a predetermined rate of interest i, often termed the minimum attractive rate of return (MARR). The MARR is usually equal to the current *cost-of-capital* rate for the company. Example 2.4 illustrates the use of a discounted-present-worth analysis to evaluate three mutually exclusive investment alternatives.

EXAMPLE 2.4

Ace Builders is considering three different acquisition methods for obtaining pickup trucks. The alternatives are

A. Immediate cash purchase the trucks for $16,800 each, and after 4 yr sell each truck for an estimated $5,000.

B. Lease the trucks for 4 yr for $4,100 per year paid in advance at the beginning of each year. The contractor pays all operating and maintenance costs for the trucks, and the leasing company retains ownership.

C. Purchase the trucks using a time payment plan requiring an immediate down payment of $4,000 and $4,500 per year at the end of each year for 3 yr. Assume the trucks will be sold after 4 yr for $5,000 each.

If the contractor's MARR is 8%, which alternative should be used? To solve, calculate the net present worth (NPW) of each alternative using an 8% interest rate and select the least costly alternative.

For the A alternative, use the present worth compound amount factor to calculate the equivalent salvage value at time zero. Add the purchase price, which is negative because it is a cash outflow, and the equivalent salvage value, which is positive because it is a cash inflow. The result is the net present worth of alternative A:

$$NPW_A = -\$16{,}800 + \frac{\$5{,}000}{PWCAF}$$

Calculating the PWCAF with i equal to 8% and n equal to 4,

$$NPW_A = -\$16{,}800 + \frac{\$5{,}000}{1.360489} = -\$13{,}124.85$$

For alternative B, use the uniform series present worth factor to calculate the time-zero equivalent value of the future lease payments, and add to that result the value of the initial lease payment; both of these are negative because they are cash outflows. There is no salvage in this case, as the leasing company retains ownership of the trucks.

$$NPW_B = -\$4{,}100 - \$4{,}100\ (USPWF)$$

Calculating the USPWF with i equal to 8% and n equal to 4,

$$NPW_B = -\$4{,}100 - \$4{,}100\left[\frac{0.259712}{0.100777}\right] = -\$14{,}666.10$$

For alternative C use the uniform series present worth factor to calculate the time-zero equivalent value of the future payments and the present worth compound amount factor to calculate the equivalent salvage value at time zero. Add the three values, the initial payment and the equivalent annual payment both being negative and the salvage being positive, to arrive at the net present worth of alternate C.

$$NPW_C = -\$4{,}000 - \$4{,}500\ (USPWF) + \left[\frac{\$5{,}000}{PWCAF}\right]$$

$$NPW_C = -\$4{,}000 - \$4{,}500\left[\frac{0.259712}{0.100777}\right] + \left[\frac{\$5{,}000}{1.360489}\right] = -\$11{,}921.79$$

The least costly alternative is C.

Example 2.4 was simplified in two respects. First, the number of calculations required was quite small. Second, all three alternatives involved the same time duration (4-yr lives in the example). Problems involving more data may require more calculations, but the analysis approach is the same as shown in Example 2.4. When the alternatives involve different time durations (machines have different expected durations of usefulness), the analysis must be modified to account for the different time durations. Obviously, if a comparison is made of one alternative having a life of 5 yr and another with a life of 10 yr, the respective discounted present worths are not directly comparable. How is such a situation handled? There are two approaches that are generally used.

Approach 1. Truncate (cut off) the longer-lived alternative(s) to equal the shorter-lived alternative and assume a salvage value for the unused portion of the longer-lived alternative(s). Then make the comparison on the basis of equal lives.

Approach 2. Compute the discounted present worth on the basis of the least common denominator of the different alternatives' lives.

ELEMENTS OF OWNERSHIP COST

Ownership cost is the cumulative result of those cash flows an owner experiences whether or not the machine is productively employed on a job. It is a cost related to finance and accounting exclusively, and it does not include the wrenches, nuts and bolts, and consumables necessary to keep the machine operating.

Most of ownership cash flows are expenses (outflows), but a few are cash inflows. The most significant cash flows affecting *ownership cost* are

1. Purchase expense
2. Salvage valuc
3. Tax saving from depreciation
4. Major repairs and overhauls
5. Property taxes
6. Insurance
7. Storage and miscellaneous

Purchase Expense

The cash outflow the firm experiences in acquiring ownership of a machine is the purchase expense. It is the total delivered cost (drive-away cost), including amounts for all options, shipping, and taxes, less the cost of tires if the machine has pneumatic tires. The machine will show as an asset in the firm's accounting records. The firm has exchanged money (dollars), a liquid asset, for a machine, a fixed asset with which the company hopes to generate profit. As the machine is used on projects, wear takes its toll and the machine can be thought of as being used up or consumed. This consumption reduces the machine's value because the revenue stream it can generate is likewise reduced. Normally, an owner tries to account for the decrease in value by prorating the consumption of the investment over the *service life* of the machine. This prorating is known as depreciation.

It can be argued that the amount that should be prorated is the difference between the initial acquisition expense and the expected future salvage value. Such a statement is correct to the extent of accounting for the amounts involved, but it neglects the timing of the cash flows. Therefore, it is recommended that each cash flow be treated separately to allow for a time value analysis and to allow for ease in changing assumptions during sensitivity analyses.

Salvage Value

Salvage value is the cash inflow a firm receives if a machine still has value at the time of its disposal. This revenue will occur at a future date.

Used equipment prices are difficult to predict. Machine condition (see Fig. 2.2), the movement of new machine prices (see Fig. 2.3), and the machine's possible secondary service applications affect the amount an owner can expect to receive. A machine having a diverse and layered service potential will command a higher resale value. Medium-size dozers, which often exhibit rising salvage values in later years, can have as many as seven different levels of useful life. These may range from an initial assignment as a high-production machine on a dirt spread to an infrequent land-clearing assignment by a farmer.

Historical resale data can provide some guidance in making salvage value predictions and can be fairly easily accessed from auction price books. By studying such historical data and recognizing the effects of the economic environment, the magnitude of salvage value prediction errors can be minimized and the accuracy of an ownership cost analysis improved.

FIGURE 2.2 Salvage value is dependent on machine condition.

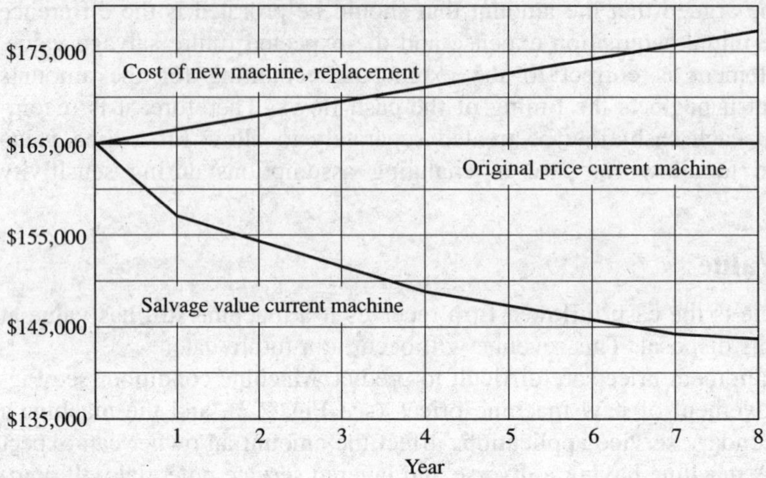

FIGURE 2.3 The movement of new machine prices –cost– is one factor affecting salvage value.

Tax Savings from Depreciation

The tax savings from depreciation are a phenomenon of the tax system in the United States. (This may not be an ownership cost factor under the tax laws in other countries.) Under the tax laws of the United States, depreciating a machine's loss in value with age will lessen the net cost of machine ownership. The cost saving, the prevention of a cash outflow, afforded by tax depreciation is a result of shielding the company from taxes. This is an applicable cash flow factor only if a company is operating at a profit. There are carry-back features in the tax law so that the saving can be preserved even though there is a loss in any one particular year, but the long-term operating position of the company must be at a profit for tax saving from depreciation to come into effect.

The rates at which a company can depreciate a machine are set by the revenue code. These rates usually have no relation to actual consumption of the asset (machine). Therefore, many companies keep several sets of depreciation numbers, one for depreciation tax purposes, one for corporate earnings tax accounting purposes, and one for internal and/or financial statement purposes. The first two are required by the revenue code. The last tries to accurately match the consumption of the asset based on work application and company maintenance policies.

Under the current tax laws, tax depreciation accounting no longer requires the assumption of a machine's future salvage value and useful life. The only piece of information necessary is basis. Basis refers to the cost of the machine for purposes of computing gain or loss. Basis is essential. To compute tax depreciation amounts, fixed percentages are applied to unadjusted basis. The terminology adjusted/unadjusted refers to changing the book value of a machine by depreciation.

The tax law allows the postponement of taxation on gains derived from the exchange of like-kind depreciable property. If there is a gain realized from a like-kind exchange, the depreciation basis of the new machine is reduced by the amount of the gain. However, if the exchange involves a disposal sale to a third party and a separate acquisition of the replacement, the gain from the sale is taxed as ordinary income.

EXAMPLE 2.5

A tractor with an adjusted basis (from depreciation) of $25,000 is traded for a new tractor that has a fair market value of $400,000. A cash payment of $325,000 is made to complete the transaction. Such a transaction is a nontaxable exchange and no gain is recognized on the trade-in. The unadjusted basis of the new tractor is $350,000, even though the cash payment was $325,000 and the apparent gain in value for the traded machine was $50,000 [($400,000 − $325,000) − $25,000].

Cash payment	$325,000
Adjusted basis of the trade-in tractor	25,000
Basis of the new tractor	$350,000

If the owner had sold the old tractor to a third party for $75,000 and then purchased a new tractor for $400,000, the $50,000 profit on the third-party sale would have been taxed as ordinary income and the unadjusted basis of the new tractor would be $400,000.

The current tax depreciation law establishes depreciation percentages that can be used based on the specific year of machine life. These are usually the optimum depreciation rates in terms of tax advantages. However, an owner can still utilize the straight-line method of depreciation or methods that are not expressed in terms of time duration (years). Unit-of-production would be an example of a depreciation system that is not time based.

Straight-Line Depreciation Straight-line depreciation is easy to calculate. The annual amount of depreciation D_n, for any year n, is a constant value, and thus the book value (BV_n) decreases at a uniform rate over the useful life of the machine. The equations are

$$\text{Depreciation rate, } R_n = \frac{1}{N} \quad [2.7]$$

where N = number of years.

$$\text{Annual depreciation amount, } D_n = \text{Unadjusted basis} \times R_n$$

Substituting Eq. [2.7] yields

$$D_n = \frac{\text{Unadjusted basis}}{N} \quad [2.8]$$

$$\text{Book value year } m, \text{BV}_n = \text{Unadjusted basis} - (n \times D_n) \quad [2.9]$$

EXAMPLE 2.6

Consider the new tractor in Example 2.5 and assume it has an estimated useful life of 5 yr. Determine the depreciation and the book value for each of the 5 yr using the straight-line method.

$$\text{Depreciation rate, } R_n = \frac{1}{5} = 0.2$$

Annual depreciation amount, $D_n = \$350{,}000 \times 0.2 = \$70{,}000$

m	BV_{n-1}	D_n	BV_n
0	$ 0	$ 0	$350,000
1	350,000	70,000	280,000
2	280,000	70,000	210,000
3	210,000	70,000	140,000
4	140,000	70,000	70,000
5	70,000	70,000	0

Tax Code Depreciation Schedules Under the tax code, machines are classified as 3-, 5-, 10-, or 15-yr real property. Cars and light-duty trucks (under 13,000 lb unloaded) are classified as 3-yr property. Most other pieces of construction equipment are 5-yr property. The appropriate depreciation rates are given in Table 2.1.

If a machine is disposed of before the depreciation process is completed, no depreciation can be recovered in the year of disposal. Any gain, as measured against the depreciated value or adjusted basis, is treated as ordinary income.

TABLE 2.1 Tax code specified depreciation rates.

Year of life	3-yr property	5-yr property
1	0.33	0.20
2	0.45	0.32
3	0.22	0.24
4	—	0.16
5	—	0.08

EXAMPLE 2.7

A 5-yr life class machine is purchased for $125,000. It is sold in the third year after purchase for $91,000. Using the tax code specified depreciation rates, what are the depreciation amounts and what is the book value of the machine when it is sold? Will there be income tax, if so on what amount?

$125,000 × 0.20 = $25,000 depreciation at end of the first year
$125,000 × 0.32 = $40,000 depreciation at end of the second year
$65,000 total depreciation

Book value when sold = $125,000 − $65,000 = $60,000.

The amount of gain on which there will be a tax is

$91,000 − $60,000 = $31,000

The tax savings from depreciation are influenced by

1. Disposal method for the old machine
2. Value received for the old machine
3. Initial value of the replacement
4. Class life
5. Tax depreciation method

Based on the relationships between these elements, three distinct situations are possible:

1. No gain on the disposal—no income tax on zero gain.
2. A gain on the disposal:
   a. Like-kind exchange—no added income tax, but basis for the new machine is adjusted.
   b. Third-party sale—the gain is taxed as income; the basis of the new machine is fair market value paid.
3. A disposal in which a loss results—the basis of the new machine is the same as the basis of the old machine, decreased by any money received.

Assuming a corporate profit situation, the applicable tax depreciation shield formulas are

1. For a situation where there is no gain on the exchange:

$$\text{Total tax shield} = \sum_{n=1}^{N} t_c D_n \qquad [2.10]$$

where

n = individual yearly time periods within a life assumption of N years
t_c = corporate tax rate
D_n = annual depreciation amount in the nth time period

2. For a situation where a gain results from the exchange
   a. Like-kind exchange—Eq. [2.10] is applicable. It must be realized that the basis of the new machine will be affected.
   b. Third-party sale.

$$\text{Total tax shield} = \left(\sum_{n=1}^{N} t_c D_n\right) - \text{gain} \times t_c \qquad [2.11]$$

Gain is the actual salvage amount received at the time of disposal minus the book value.

The implication of the basis is that in making analysis calculations, the actual salvage derived from the machine directly affects the depreciation saving. To perform a valid analysis, the depreciation accounting practices for tax purposes and the methods of machine disposal and acquisition the company chooses to use must be carefully examined. These dictate the appropriate calculations for the tax effects of depreciation.

Major Repairs and Overhauls

Major repairs and overhauls are included under ownership cost because they result in an extension of a machine's service life. They can be considered as an investment in a new machine. Because a machine commonly works on many different projects, considering major repairs as an ownership cost prorates these expenses to all jobs. These costs should be added to the basis of the machine and depreciated.

Taxes

In this context, taxes refer to those equipment ownership taxes that are charged by any government subdivision. They are commonly assessed at a percentage rate applied against the book value of the machine. Depending on location, property taxes can range up to about 4.5% of assessed machine value. In many locations, there will be no property tax on equipment. Over the service life of the machine, they will decrease in magnitude as the book value decreases.

Insurance

Insurance, as considered here, includes the cost to cover fire, theft, and damage to the equipment. Annual rates can range from 1 to 3%. This cost can be actual premium payments to insurance companies, or it can represent allocations to a self-insurance fund maintained by the equipment owner.

Storage and Miscellaneous

Between jobs or during bad weather, a company will require storage facilities for its equipment. The cost of maintaining storage yards and facilities should be prorated to those machines that require such harborage. Typical expenses include space rental, utilities, and the wages for laborers or watchmen. Usually these expenses are all combined in an overhead account and then allocated on a proportional basis to the individual machines. The rate may range from nothing to perhaps 5%.

ELEMENTS OF OPERATING COST

Operating cost is the sum of those expenses an owner experiences by working a machine on a project. Typical expenses include

1. Fuel
2. Lubricants, filters, and grease

3. Repairs
4. Tires
5. Replacement of high-wear items

Operator wages are sometimes included under operating costs, but because of wage variances between projects, the general practice is to keep operator wages as a separate cost category. Such a procedure aids in estimation of machine cost for bidding purposes as the differing project wage rates can readily be added to the total machine O&O cost. In applying operator cost, all benefits paid by the company must be included—direct wages, fringe benefits, insurance, etc. This is another reason wages are separated. Some benefits are based on an hourly basis, some on a percentage of income, some on a percentage of income to a maximum amount, and some are paid as a fixed amount. The assumptions about project work schedule will therefore affect wage expense.

Fuel

Fuel expense is best determined by measurement on the job. Accurate service records tell the owner how many gallons of fuel a machine consumes over what period of time and under what job conditions. Hourly fuel consumption can then be calculated directly.

When company records are not available, manufacturer's consumption data can be used to construct fuel use estimates. The amount of fuel required to power a piece of equipment for a specific period of time depends on the brake horsepower of the machine and the specific work application. Therefore, most tables of hourly fuel consumption rates are divided according to the machine type and the working conditions. To calculate hourly fuel cost, a consumption rate is found in the tables (see Table 2.2) and then multiplied by the unit price of fuel. The cost of fuel for vehicles used on public highways will include applicable taxes. However, in the case of off-road machines used exclusively on project sites there is usually no fuel tax. Therefore, because of the tax laws, the price of gas or diesel will vary with machine usage.

TABLE 2.2 Average fuel consumption—wheel loaders.

	Type of utilization		
Horsepower (fwhp)	Low (gal/hr)	Medium (gal/hr)	High (gal/hr)
90	1.5	2.4	3.3
140	2.5	4.0	5.3
220	5.0	6.8	9.4
300	6.5	8.8	11.8

Note fwhp is flywheel horsepower.

Fuel consumption can also be calculated on a theoretical basis. The resulting theoretical values must be adjusted by *time and load factors* that account for working conditions. This is because the theoretical formulas are derived assuming that the engine is operating at maximum output. Working conditions that must be considered are the percentage of an hour that the machine is actually working (*time factor*) and at what percentage of rated horsepower (*throttle load factor*). When operating under standard conditions, a *gasoline engine* will consume approximately 0.06 gal of fuel per flywheel horsepower hour (fwhp-hr). A *diesel engine* will consume approximately 0.04 gal per fwhp-hr.

Lubricants—Lube Oils, Filters, and Grease

The cost of lube oils, filters, and grease (see Fig. 2.4) will depend on the maintenance practices of the company and the conditions of the work location. Some companies follow machine manufacturer's guidance concerning time periods between lubricant and filter changes. Other companies have established their own preventive maintenance change period guidelines. In either case, the hourly cost is arrived at by (1) considering the operating hour duration between changes and the quantity required for a complete change plus (2) a small consumption amount representing what is added between changes.

Many manufacturers provide quick cost estimating tables or rules for determining the cost of these items. Whether using manufacturer's data or past experience, notice should be taken about whether the data matches expected field conditions. If the machine is to be operated under adverse conditions, such as deep mud, water, or sever dust, the data values will have to be adjusted.

FIGURE 2.4 Checking the oil on a small loader.

A formula that can be used to estimate the quantity of oil required is

$$\text{Quantity consumed, gph (gal per hour)} = \frac{\text{hp} \times f \times 0.006 \text{ lb/hp} - \text{hr}}{7.4 \text{ lb/gal}} + \frac{c}{t} \qquad [2.12]$$

where

hp = rated horsepower of the engine
c = capacity of the crankcase in gallons
f = operating factor
t = number of hours between oil changes

This formula contains the assumption that the quantity of oil consumed per rated horsepower hour between changes will be 0.006 lb.

Repairs

Repairs, as referred to here, mean normal maintenance-type repairs (see Fig. 2.5). These are the repair expenses incurred on the job site where the machine is operated and would include the costs of parts and labor. Major repairs and overhauls are an ownership cost.

FIGURE 2.5 Normal repairs are included in operating cost.

Repair expenses increase with machine age. The army has found that 35% of its equipment maintenance cost is directly attributable to the oldest 10% of its equipment. Instead of applying a variable rate, an average is usually calculated by dividing the total expected repair cost, for the planned service life of the machine, by the planned operating hours. Such a policy builds up a repair reserve during a machine's early life. That reserve will then be used to cover the higher costs experienced later. As with all costs, company records are the best source of expense information. When such records are not available, manufacturers' published guidelines can be used.

Tires

Tires for wheel-type equipment (see Fig. 2.6) are a major operating cost because they have a short life in relation to the "iron" of a machine. Tire cost will include repair and replacement charges. These costs are very difficult to estimate because of the variability in tire wear with project site conditions and operator skill. Both tire and equipment manufacturers publish tire life guidelines based on tire type and job application. Manufacturers' suggested life periods can be used with local tire prices to obtain an hourly tire cost. It must be remembered, however, that the guidelines are based on good operating practices and do not account for abuses such as overloading haul units.

Replacement of High-Wear Items

The cost of replacing those items that have very short service lives in relation to machine service life can be a critical operating cost. These items will differ depending on the type of machine, but typical items include cutting edges,

FIGURE 2.6 Tires are a major operating cost.

FIGURE 2.7 Bucket teeth are a high-wear item replacement cost.

ripper tips, bucket teeth (see Fig. 2.7), body liners, and cables. By using either past experience or manufacturer life estimates, the cost can be calculated and converted to an hourly basis.

All machine-operating costs should be calculated per working hour. That way it is easy to sum the applicable costs for a particular class of machines and obtain a total operating hour cost.

COST FOR BIDDING

The process of selecting a particular type of machine for use in constructing a project requires knowledge of the cost associated with operating the machine in the field. In selecting the proper machine, a contractor seeks to achieve unit production at the least cost. For project bidding and cost accounting, we are interested in a machine's ownership and operating cost. O&O costs are usually expressed in dollars per equipment operating hour.

Ownership Cost

The expense of purchasing a machine and the inflow of money in the future (when the machine is retired from service) are the two most significant components of ownership cost. The net result of these two cash flows, which defines the machine's decline in value across time, is termed depreciation. As used in this section, depreciation is the measuring system used to account for purchase expense at time zero and salvage value after a defined period of time. Depreciation is expressed on an hourly basis over the service life of a machine. Do not confuse the depreciation discussed here with tax depreciation. Tax depreciation has nothing to do with consumption of the asset; it is simply an artificial calculation for tax code purposes.

> Because tires are a high-wear item that will be replaced many times over a machine's service life, their cost will be not be included in these calculations but will be addressed as a part of operating cost.

The depreciation portion of ownership cost can be calculated by either of two methods: time value or average annual investment.

Depreciation—Time Value Method The time value method will recognize the timing of the cash flows, i.e., the purchase at time zero and the salvage at a future data. The cost of the tires is deducted from the total purchase price, which includes amounts for all options, shipping, and taxes (total cash outflow − cost of tires). A judgment about the expected service life and a corporate cost of capital rate are both necessary input parameters for the analysis. To determine the machine's purchase price equivalent annual cost the uniform series capital recovery factor formula, Eq. [2.6], is used. The input parameters are the purchase price at time zero (P), expected service life (n), and the corporate cost of capital rate (i).

To account for the salvage cash inflow, Eq. [2.4], the uniform series sinking fund factor formula, is utilized. The input parameters are the estimated future salvage amount (F), the expected service life (n), and the corporate cost of capital rate (i).

EXAMPLE 2.8

A company having a cost of capital rate of 8% purchases a $300,000 loader. This machine has an expected service life of 4 yr and will be utilized 2,500 hr per year. The tires on this machine cost $45,000. The estimated salvage value at the end of 4 yr is $50,000. Calculate the depreciation portion of the ownership cost for this machine using the time value method.

Initial cost	$300,000
Cost of tires	−45,000
Purchase price less tires	$225,000

Calculate the equivalent uniform period series required to replace a present value of $255,000. Using Eq. [2.6],

$$A = \$255{,}000\left[\frac{0.08(1+0.08)^4}{(1+0.08)^4-1}\right]$$

$$= \$255{,}000 \times 0.3019208 = \$76{,}990 \text{ per year}$$

Calculate the equivalent uniform end-of-period investments that equal the future salvage value. Using Eq. [2.4],

$$A = \$50{,}000\left[\frac{0.08}{(1+0.08)^4 - 1}\right]$$

$$= \$50{,}000 \times 0.02219208 = \$11{,}096 \text{ per year}$$

Therefore, using the time value method the hourly depreciation portion of the machine's ownership cost is

$$\frac{\$76{,}990/\text{yr} - \$11{,}096/\text{yr}}{2{,}500 \text{ hr/yr}} = \$26.358/\text{hr}$$

Depreciation—Average Annual Investment Method A second approach to calculating the depreciation portion of ownership cost is the average annual investment (AAI) method.

$$\text{AAI} = \frac{P(n+1) + S(n-1)}{2n} \qquad \textbf{[2.13]}$$

where

P = purchase price less the cost of the tires

S = the estimated salvage value

n = expected service life in years

The AAI is multiplied by the corporate cost of capital rate to determine the cost of money portion of depreciation. The straight-line depreciation of the cost of the machine less the salvage and less the cost of tire, if a pneumatic-tired machine, is then added to the cost of money portion (interest) to arrive at the total amount of ownership depreciation.

EXAMPLE 2.9

Using the same machine and company information as in Example 2.8, calculate the ownership depreciation using the AAI method.

$$\text{AAI} = \frac{\$255{,}000(4+1) + \$50{,}000(4-1)}{2 \times 4}$$

$$= \$178{,}125/\text{yr}$$

$$\text{Cost of money portion} = \frac{\$178{,}125/\text{yr} \times 8\%}{2{,}500 \text{ hr/yr}} = \$5.700/\text{hr}$$

Straight-line depreciation portion

Initial cost	\$300,000
Cost of tires	−45,000
Salvage	−50,000
	\$205,000

$$\frac{\$205{,}000}{4 \text{ yr} \times 2{,}500 \text{ hr/yr}} = \$20.500/\text{hr}$$

Total ownership depreciation using the AAI method

$$\$5.700/\text{hr} + \$20.500 = \$26.200/\text{hr}$$

For Examples 2.8 and 2.9, the difference in the calculated depreciation portion of the ownership cost is \$0.158/hr (\$26.358/hr − \$26.200/hr). The choice of which method to use is strictly a company preference, however, all analyses should use the same method. Basically, either method is satisfactory, especially considering the impact of the unknowns concerning service life, operating hours per year, and expected future salvage. There is no single solution to calculating ownership cost. The best approach is to perform several analyses using different assumptions and to be guided by the range of solutions.

Tax Saving from Tax Code Depreciation To calculate the tax saving from depreciation, the government tax code depreciation schedules (Table 2.1) must be used. The resulting depreciation amounts are then multiplied by the company's tax rate to calculate specific savings, using Eq. [2.10] or Eq. [2.11]. The sum of the yearly saving must be divided by the total anticipated operating hours to obtain an hourly cost saving.

EXAMPLE 2.10

Using the same machine and company information as in Example 2.8, calculate the hourly tax saving resulting from tax code depreciation. Assume that under the tax code the machine is a 5-yr property and that there had been no gain on the exchange that procured the machine. The company's tax rate is 37%.

First, calculate the annual depreciation amounts for each of the years. In this case, the tax code depreciation rate must be used to calculate the depreciation.

Year	5-yr property rates	BV_{n-1}	D_n	BV_n
0		\$ 0	\$ 0	\$300,000
1	0.20	300,000	60,000	240,000
2	0.32	240,000	96,000	144,000
3	0.24	144,000	72,000	72,000
4	0.16	72,000	48,000	24,000
5	0.08	24,000	24,000	0

Using Eq. [2.10], the tax shielding effect for the machine's service life would be

Year	D_n	Shielded amount*
1	\$60,000	\$22,200
2	96,000	35,520
3	72,000	26,640
4	48,000	17,760
	Total	\$102,120

* $D_n \times 37\%$

$$\text{Tax saving from depreciation} = \frac{\$102{,}120}{4\text{ yr} \times 2{,}500\text{ hr/yr}} = \$10.21/\text{hr}$$

Major Repairs and Overhauls When a major repair and overhaul takes place, the machine's ownership cost will have to be recalculated. This is done by adding the cost of the overhaul to the book value at that point in time. The resulting new adjusted basis is then used in the depreciation calculation, as already described. If there are separate calculations for true depreciation and for tax code depreciation, both will have to be adjusted.

Taxes, Insurance, and Storage To calculate the taxes, insurance, and storage costs, common practice is to simply apply a percentage value to either the machine's book value or its AAI amount. The expenses incurred for these items are usually accumulated in a corporate overhead account. That value divided by the value of the equipment fleet and multiplied by 100 will provide the percentage rate to be used.

$$\text{Taxes, insurance, and storage portion of ownership cost} = \text{rate}(\%) \times BV_n \text{ (or AAI)} \quad \textbf{[2.14]}$$

EXAMPLE 2.11

Using the same machine and company information as in Examples 2.8 and 2.9 calculate the hourly owning expense associated with taxes, insurance, and storage. Annually, the company pays on average 1% in property taxes for equipment, 2% for insurance, and allocates 0.75% for storage expenses.

Total percentage rate for taxes, insurance, and storage
1% + 2% + 0.75% = 3.75%

From Example 2.9, the average annual investment for the machine is $178,125/yr.

Taxes, insurance, and storage expense

$$\frac{\$178{,}125/\text{yr} \times 3.75\%}{2{,}500 \text{ hr/yr}} = \$2.672/\text{hr}$$

Operating Cost

Figures based on actual company experience should be used to develop operating expenses. Many companies, however, do not keep good equipment operating and maintenance records; therefore, many operating costs are estimated as a percentage of a machine's book value. Even companies that keep records often accumulate expenses in an overhead account and then prorate that total back to individual machines using book value.

Fuel The amount expended on fuel is a product of how a machine is used in the field and the local cost of fuel. In years past, fuel could be purchased on long-term contracts at a fixed price. Today, fuel is usually offered with a *time of delivery price*. A supplier will agree to supply the fuel needs of a project, but the price will not be guaranteed for the duration of the work. Therefore, when bidding a long-duration project, the contractor must make an assessment of future fuel prices.

To calculate hourly fuel expense, a consumption rate is multiplied by the unit price of fuel. Service records are important for estimating fuel consumption.

EXAMPLE 2.12

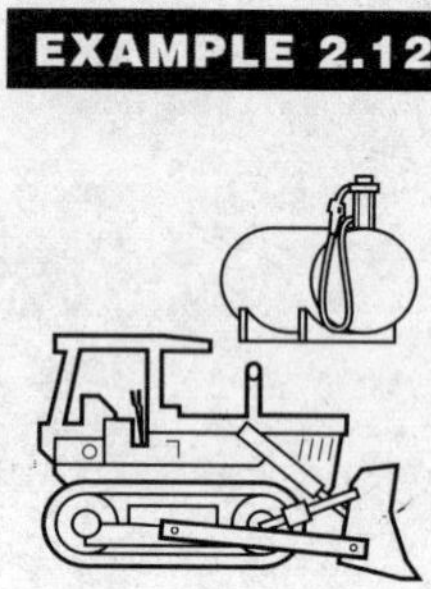

A 220-fwhp dozer will be used to push aggregate in a stockpile. This dozer is diesel powered. It is estimated that the work will be steady at an efficiency equal to a 50-min hour. The engine will work at full throttle while pushing the load (30% of the time) and at three-quarter throttle to reverse travel and position. Calculate the fuel consumption using the engine consumption averages. If diesel cost $1.07/gal, what is the expected fuel expense?

Fuel consumption diesel engine 0.04 gal per fwhp-hr.

Throttle load factor (operating power):

Push load	1.00 (power) × 0.30 (% of the time) = 0.30
Travel and position	0.75 (power) × 0.70 (% of the time) = 0.53
	0.83

Time factor (operating efficiency): 50-min hour: 50/60 = 0.83

Combined factor: 0.83 × 0.83 = 0.69

Fuel consumption = 0.69 × 0.04 gal/fwhp-hr × 220 fwhp = 6.1 gal/hr

Lubricants The quantity of lubricants used by an engine will vary with the size of the engine, the capacity of the crankcase, the condition of the piston rings, and the number of hours between oil changes. For extremely dusty conditions, it may be desirable to change oil every 50 hr, but this is an unusual condition. It is common practice to change oil every 100 or 200 hr. The quantity of the oil consumed by an engine per change will include the amount added during the change plus the makeup oil between changes.

EXAMPLE 2.13

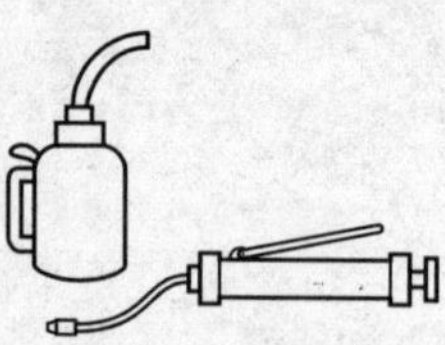

Calculate the oil required, on a per hour basis, for the 220-fwhp dozer in Example 2.12. The operating factor will be 0.69, as calculated in that example. The crankcase capacity is 8 gal and the company has a policy to change oil every 150 hr.

Quantity consumed, gph (gal per hour) is

$$\frac{220 \text{ fwhp} \times 0.69 \times 0.006 \text{ lb/hp-hr}}{7.4 \text{ lb/gal}} + \frac{8 \text{ gal}}{150 \text{ hr}} = 0.18 \text{ gal/hr}$$

The cost of hydraulic oils, filters, and grease will be added to the expense of engine oil. The hourly cost of filters is simply the actual expense to purchase the filters divided by the hours between changes. If a company does not keep good machine servicing data, it is difficult to accurately estimate the cost of hydraulic oil and grease. The usual solution is to refer to manufacturers' published tables of average usage or expense.

Repairs The cost of repairs is normally the largest single component of machine cost (see Table 2.3). Some general guidelines published, in the past, by the Power Crane and Shovel Association (PCSA) estimated repair and maintenance expenses at 80 to 95% of depreciation for crawler-mounted excavators, 80 to 85% for wheel-mounted excavators, 55% for crawler cranes, and 50% for wheel-mounted cranes. The lower figures for cranes reflect the work they perform and the intermittent nature of their use. The data assumed that half of the cost was materials and parts, and half was labor, in the case of mechanical machines. For hydraulic machines, two-thirds of the cost is for materials and parts, and one-third for labor.

Equipment manufacturers supply tables of average repair costs based on machine type and work application. Repair expenses will increase with machine usage (age). The repair cost to establish a machine rate for bidding should be an average rate.

Tires Tire expenses include both tire repair and tire replacement. Tire maintenance is commonly handled as a percentage of straight-line tire depreciation. Tire hourly cost can be derived simply by dividing the cost of a set of tires by their expected life, and this is how many companies prorate this expense. A more sophisticated approach is to use a time-value calculation recognizing that tire replacement expenses are single point in time outlays that take place over the life of a wheel-type machine.

TABLE 2.3 Breakdown of machine cost over its service life.

Cost category	Percentage of total cost (%)
Repair	37
Depreciation	25
Operating	23
Overhead	15

EXAMPLE 2.14

Calculate the hourly tire cost that should be part of machine operating cost if a set of tires can be expected to last 5,000 hr. Tires cost \$38,580 per set of four. Tire repair cost is estimated to average 16% of the straight-line tire depreciation. The machine has a service life of 4 yr and operates 2,500 hr per year. The company's cost of capital rate is 8%.

Not considering the time value of money:

$$\text{Tire repair cost} = \frac{\$38{,}580}{5{,}000\ \text{hr}} \times 16\% = \$1.235/\text{hr}$$

$$\text{Tire use cost} = \frac{\$38{,}580}{5{,}000\ \text{hr}} = \$7.716/\text{hr}$$

Therefore, tire operating cost is \$8.951/hr (\$1.235/hr + \$7.716/hr).

Considering the time value of money:
Tire repair cost is the same $1.235/hr.
Calculate the number of times the tires will have to be replaced.

$$\left(\frac{4\text{ yr} \times 2{,}500\text{ hr/yr}}{5{,}000\text{ hr per set of tires}}\right) = 2\text{ sets}$$

Will have to purchase a second set at the end of the second year.

First set: Calculate the uniform series required to replace a present value of $38,580. Using Eq. [2.6],

$$A = \$38{,}580\left[\frac{0.08(1+0.08)^4}{(1+0.08)^4 - 1}\right]$$

$$\frac{\$38{,}580 \times 0.301921}{2{,}500\text{ hr/yr}} = \$4.659\text{/hr}$$

Second set: The second set will be purchased 2 yr in the future. Therefore, what amount at time zero is equivalent to $38,580 2 yr in the future? Using the present worth compound amount factor (Eq. [2.2]), the equivalent time-zero amount is calculated:

$$P = \frac{\$38{,}580}{(1+0.08)^2} = \$33{,}076$$

Calculate the uniform series required to replace a present value of $33,076.

$$A = \$33{,}076\left[\frac{0.08(1+0.08)^4}{(1+0.08)^4 - 1}\right]$$

$$\frac{\$33{,}076 \times 0.301921}{2{,}500\text{ hr/yr}} = \$3.995\text{/hr}$$

Therefore, considering the time value of money, tire operating cost is $9.889/hr ($1.235/hr + $4.659/hr + $3.995/hr).

High-Wear Items Because the cost of high-wear items is dependent on job conditions and machine application, the cost of these items is usually accounted for separate from general repairs.

EXAMPLE 2.15

A dozer equipped with a three-shank ripper will be used in a loading and ripping application. Actual ripping will take place only about 20% of total dozer operating time. A ripper shank consists of the shank itself, a ripper tip, and a shank protector. The estimated operating life for the ripper tip is 30 hr. The estimated operating life of the ripper shank protector is 3 times tip life. The local price for a tip is $40 and $60 for shank protectors. What hourly high-wear item charge should be added to the operating cost of a dozer in this application?

Tips: $\frac{30\text{ hr}}{0.2} = 150$ hr of dozer operating time

$$\frac{3 \times \$40}{150\text{ hr}} = \$0.800\text{/hr for tips}$$

Shank protectors: 3 times tip life × 150 hr = 450 hours of dozer operating time

$$\frac{3 \times \$60}{450\text{ hr}} = \$0.400\text{/hr for shank protectors}$$

Therefore, cost of high-wear items is \$1.200/hr (\$0.800/hr tips + \$0.400/hr shank protectors).

REPLACEMENT DECISIONS

A piece of equipment has two lives: (1) a physically limited working life and (2) a cost-limited economic life. Because equipment owners are in business to make money, the economic life of their equipment is of critical importance. A machine in good mechanical condition and working productively enjoys a strong bias in favor of its retention in the equipment inventory. The equipment manager may look only at the high initial cash outflow associated with the purchase of a replacement machine and consequently ignore the other cost factors involved. All cost factors must be examined when considering a replacement decision. A simple example will help to illustrate the concept.

EXAMPLE 2.16

A small dozer is purchased for \$106,000. A forecast of expected operating hours, salvage values, and maintenance expense is presented in the table.

Year	Operating hours	Salvage (\$)	Maintenance expense (\$)
1	1,850	79,500	3,340
2	1,600	63,600	3,900
3	1,400	76,320	4,460
4	1,200	74,200	5,000
5	800	63,600	6,600

A replacement analysis might look like this:

Year	1	2	3	4	5
Purchase	\$106,000	\$106,000	\$106,000	\$106,000	\$106,000
Salvage	\$79,500	\$69,500	\$76,320	\$73,000	\$70,000
Cost	\$26,500	\$36,500	\$29,680	\$33,800	\$36,000
Cumulative operating hours	1,850	3,450	4,850	6,050	6,850
Ownership cost \$/hr	\$14.32	\$10.58	\$6.12	\$5.45	\$5.26
Cumulative maintenance expense	\$3,340	\$7,240	\$11,700	\$16,700	\$23,300
Operating cost \$/hr	\$1.81	\$2.10	\$2.41	\$2.76	\$3.40
Total \$/hr	\$16.13	\$12.68	\$8.53	\$8.21	\$8.66

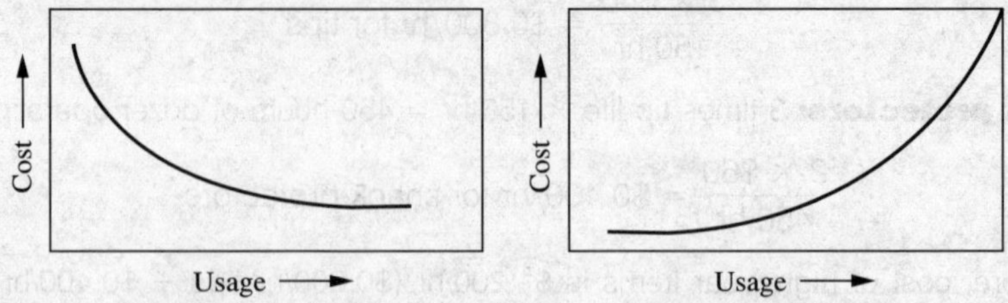

(a) Ownership cost (b) Operating cost

FIGURE 2.8 Effect of cumulative usage on cost.

If an owner considers only purchase price and expected salvage, the numbers argue (ownership cost $/hr) that the machine should not be traded (see Fig. 2.8a). However, if only operating cost is examined, the owner would want to trade the machine after the first year, as operating expenses are continually rising with usage (see Fig. 2.8b). A correct analysis of the situation requires that total cost be considered. So in the case of Example 2.16, the most economical service life of this machine is 4 yr, as $8.21/operating hour is the minimum total cost.

The analysis is based on *cumulative* hours. This is an important point that is often missed. If the owner chooses to keep the machine 5 yr, the effective loss is $0.45 ($8.66 − $8.21) on *every operating hour*, not just the 800 hr of the last year. When the total operating hours are large, the significance of this cumulative effect can become much greater than it would appear by simply looking at the *combined cost per hour* values.

The replacement analysis should present all the cost and timing information affecting a machine or class of machines in a usable format. The format should be such that it is easy to perform sensitivity analyses to determine the correctness of the results. As described here, the model is a cost-minimization model. With such a model, the optimum economic life of a machine is that ownership time duration that results in a minimum hourly cost.

The cash flows being studied in a replacement analysis take place at different points in time, therefore the model should consider the timing effects by use of present-value techniques. The company's cost-of-capital rate is the appropriate interest rate to use in the present-value equations.

RENT AND LEASE CONSIDERATIONS

There are three basic methods for securing a particular machine to use on a project: (1) *buy* (direct ownership), (2) *rent*, or (3) *lease*. Each method has inherent advantages and disadvantages. Ownership guarantees control of machine availability and mechanical condition, but it requires a continuing sequence of projects to pay for the machine. Additionally, ownership may force a company into using obsolete equipment. The calculations applicable for determining the cost of direct ownership have been developed.

Rental

The rental of a machine is a short-term alternative to direct equipment ownership. With a rental, a company can pick the machine that is exactly suited for the job at hand. This is particularly advantageous if the job is of short duration or if the company does not foresee a continuing need for the particular type of machine in question. Rentals are very beneficial to a company in such situations, even though the rental charges are higher than *normal* direct ownership expense. The advantage lies in the fact that direct ownership costing assumes a continuing need and utilization of the machine. If that assumption is not valid, a rental should be considered. Another important point to consider is the fact that with a rental, the company loses the tax depreciation shield of machine ownership but gains a tax deduction because rental payments are treated as an expense.

It must be remembered that rental companies only have a limited number of machines and, during the peak work season, all types are not always available. Furthermore, many specialized or custom machines cannot be rented.

Firms many times use rentals as a way to test a machine prior to a purchase decision. A rental provides the opportunity for a company to operate a specific make or model machine under actual project conditions. The profitability of the machine, based on the company's normal operating procedures, can then be evaluated before a major capital expenditure is approved to purchase the machine.

The general practice of the industry is to price rental rates for equipment on either a daily (8 hr), weekly (40 hr), or monthly (176 hr) basis. In the case of larger pieces of equipment, rentals may be available only on a monthly basis. Cost per hour usually is less for a longer-term rental (i.e., the monthly rate figured on a per hour basis would be less than the daily rate on an hourly basis).

Responsibility for repair cost is stated in the rental contract. Normally, on tractor-type equipment, the renter is responsible for all repairs. If it is a pneumatic-tired machine (on rubber), the renting company will measure tread wear and charge the renter for tire wear. In the case of cranes and shovels, the renting company usually bears the cost of normal wear and tear. The user must provide servicing of the machine while it is being used. The renter is almost always responsible for fuel and lubrication expenses. Industry practice is that rentals are payable in advance. The renting company will require that the user furnish certificates of insurance before the machine is shipped to the job site.

Equipment cost is very sensitive to changes in use hours. Fluctuations in maintenance expenses or purchase price barely affect cost per hour. But a decrease in use hours per year can make the difference between cost-effective machine ownership and renting. The basic cost considerations that need to be

TABLE 2.4 Rental versus ownership, operating hour breakeven points.

Rental Duration	Rate ($)	Hours	Rental rate ($/hr)	Operating hour breakeven point ($26,304 /$/hr)	Operating hour breakeven point ($3,558 /$/hr)
Monthly	3,558	176	20.22	1,300	—
Weekly	1,182	40	29.55	890	120
Daily	369	8	46.13	570	77

examined when considering a possible rental can be illustrated by a simple set of circumstances. Consider a small wheel loader with an ownership cost of $10.96 per hr. Assume that the cost is based on the assumption that the machine will work 2,400 hr each year of its service life. If $10.96/hr is multiplied by 2,400 hr/yr the yearly ownership cost is found to be $26,304.

Checking with the local rental company, the construction firm receives rental quotes of $3,558 per month, $1,182 per week, and $369 per day for this size loader. By dividing with the appropriate number of hours, these rates can be expressed as hourly costs. Likewise, by dividing the calculated hourly rental rates into the construction firm's yearly ownership cost figure ($26,304) the operating hour breakeven points can be determined (see Table 2.4).

If the loader will be used for less than 1,300 hr but more than 890 hr, the construction company should consider a monthly rental instead of ownership. When the projected usage is less than 1,300 hr but more than 120 hr, a weekly rental would be appropriate. In the case of very limited usage, that is, less than 26 hr, the daily rate is optimal.

The point is that when a company rents, it pays for the equipment only when project requirements dictate a need. The company that owns a machine must continue to make the equipment payments even when the machine sits idle. When investigating a rental the critical question is usually expected *hours of usage*.

Lease

A lease is a long-term agreement for the use of an asset. It provides an alternative to direct ownership. During the lease term, the leasing company (lessor) always owns the equipment and the user (lessee) pays the owner to use the equipment. The lessor must retain ownership rights for the contract to be considered a true lease by the Internal Revenue Service. The lessor will receive lease payments in return for providing the machine. The lease payments do not have to be uniform across the lease period. The payments can be structured in the agreement to best fit the situation of the lessee or the lessor. In the lessee's case, cash flow at the beginning may be low, so the lessee wants payments that are initially low. Because of tax considerations, the lessor may agree to such a payment schedule. Lease contracts are binding legal documents, and most equipment leases are noncancellable by either party.

A lease pays for the use of a machine during the most reliable years of a machine's service life. Sometimes the advantage of a lease is that the lessor provides the equipment management and servicing. This frees the contractor from hiring mechanics and service personnel and enables the company to concentrate on the task of building.

Long-term, when used in reference to lease agreements, is a period of time that is long relative to the life of the machine in question. An agreement that is for a very short period of time, as measured against the expected machine life, is a rental. A conventional—true—lease will have one of three different end-of-lease options: (1) buy the machine at fair market value, (2) renew the lease, or (3) return the equipment to the leasing company.

As in the case of a rental, a lessee loses the tax depreciation shield of machine ownership but gains a tax deduction because lease payments are treated as an expense. The most important factor contributing to a decision to lease is reduced cost. Under specific conditions, the actual cost of a leased machine can be less than the ownership cost of a purchased machine. This is caused by the different tax treatments for owning and leasing an asset. An equipment user must make a careful examination of the cash flows associated with each option to determine which results in the lowest total cost.

Working capital is the cash that a firm has available to support its day-to-day operations. This *cash asset* is necessary to meet the payroll on Friday, to pay the electric bill, and to purchase fuel to keep the machines running. To be a viable business, working capital assets must be greater than the inflow of bills. A machine is an asset to the company, but it is not what the electric company will accept as payment for their bill.

A commonly cited advantage of leasing is that working capital is not tied up in equipment. This statement is only partly true. It is true that when a company borrows funds to purchase a machine, the lender normally requires that the company establish an equity position in the machine, a *down payment*. Additionally, the costs of delivery and initial servicing are not included in the loan and must be paid by the new owner. Corporate funds are therefore tied up in these up-front costs of a purchase. Leasing does not require these cash outflows and is often considered as 100% financing. However, most leases require an advance lease payment. Some even require security deposits and charge other up-front costs.

Still another argument is that because borrowed funds are not used, credit capacity is preserved. Leasing is often referred to as off-balance-sheet financing. A lease is considered an operating expense, not a liability, as is the case with a bank loan. With an operating lease (used when the lessee does not ultimately want to purchase the equipment), leased assets are expensed. Therefore, such assets do not appear on the balance sheet. Standards of accounting, however, require disclosure of lease obligations. It is hard to believe that *lenders* would be so naive as to not consider all of a company's fixed obligations, including both loans and leases. But the off-balance-sheet lease typically will not hurt *bonding capacity*, which is important to a company's ability to bid work.

Before entering into a contract with a construction company, most owners require that the company post a bond guaranteeing that it will complete the project. A third-party surety company secures this bond. The surety closely examines the construction company's financial position before issuing the bond. Based on the financial strength of the construction company, the surety typically restricts the total volume of work that the construction company can have under contract at any one time. This restriction is known as bonding capacity. It is the total dollar value of work under contract that a surety company will guarantee for a construction company.

Owners should make a careful examination of the advantages of a lease situation. The cash flows, which should be considered when evaluating the cost of a lease, include

1. Inflow initially of the equivalent value of the machine.
2. Outflow of the periodic lease payments.
3. Tax shielding provided by the lease payments. (This is allowed only if the agreement is a true lease. Some "lease" agreements are essentially installment sale arrangements.)
4. Loss of salvage value when the machine is returned to the lessor.

These costs all occur at different points in time, so present-value computations must be made before the costs can be summed. The total present value of the lease option should be compared to the minimum ownership costs, as determined by a time-value replacement analysis. In most lease agreements, the lessee is responsible for maintenance. If, for the lease in question, maintenance expense is the same as for the case of direct ownership, then the maintenance expense factor can be dropped from the analysis. A leased machine would exhibit the same aging and resulting reduced availability as a purchased machine.

SUMMARY

Equipment owners must carefully calculate machine ownership and operating cost. This cost is usually expressed in dollars per operating hour. The most significant cash flows affecting *ownership cost* are (1) purchase expense; (2) salvage value; (3) tax saving from depreciation; (4) major repairs and overhauls; and (5) property taxes, insurance, storage, and miscellaneous expenses. *Operating cost* is the sum of those expenses an owner experiences by working a machine on a project: (1) fuel; (2) lubricants, filters, and grease; (3) repairs; (4) tires; and (5) replacement of high-wear items. *Operator wages* are sometimes included under operating costs, but because of wage variance between jobs, the general practice is to keep operator wages as a separate cost category. Critical learning objectives include:

- An ability to calculate ownership cost.
- An ability to calculate operating cost.

- An understanding of the advantages and disadvantages associated with direct ownership, renting, and leasing machines.

These objectives are the basis for the problems that follow.

PROBLEMS

2.1 To purchase a new car it is necessary to borrow $18,550. The bank offers a 5-yr loan at an interest rate of $4\frac{1}{4}\%$ compounded annually. If you make only one payment at the end of the loan period, repaying the principal and interest, what is the total amount that must be paid back?

a. What is the number of time periods (n) you should use in solving this problem?

b. What rate of interest (*i*), per period of time, should be used in solving this problem?

c. Is the present single amount of money (*P*) known? (Yes, No)

d. Which time value factor should be used to solve this problem?

e. What is the total amount that must be paid back?

f. How much of the total amount repaid represents interest?

2.2 To purchase a new car it is necessary to borrow $18,550. A car dealer offers a 6-yr loan at an interest rate of 4% compounded annually. If you make only one payment at the end of the loan period, repaying the principal and interest, what is the total amount that must be paid back?

a. What is the number of time periods (*n*) you should use in solving this problem?

b. What rate of interest (*i*), per period of time, should be used in solving this problem?

c. Is the present single amount of money (*P*) known? (Yes, No)

d. Which time value factor should be used to solve this problem?

e. What is the total amount that must be paid back?

f. How much of the total amount repaid represents interest?

2.3 What amount must be invested today at an annual interest rate of $4\frac{1}{2}\%$, if you want to purchase a $450,000 machine 4 yr in the future?

a. What is the number of time periods (*n*) you should use in solving this problem?

b. What rate of interest (*i*), per period of time, should be used in solving this problem?

c. Is the present single amount of money (*P*) known? (Yes, No)

d. Which time value factor should be used to solve this problem?

e. What is the total amount that must be invested today?

2.4 What amount must be invested today at an interest rate of $4\frac{1}{2}\%$, compounded monthly, if you want to purchase a $450,000 machine 4 yr in the future?

a. What is the number of time periods (n) you should use in solving this problem?
b. What rate of interest (i), per period of time, should be used in solving this problem?
c. Is the present single amount of money (P) known? (Yes, No)
d. Which time value factor should be used to solve this problem?
e. What is the total amount that must be invested today?

2.5 To purchase a new truck it is necessary to borrow \$18,550. The bank offers a 5-yr loan at an interest rate of $4\frac{1}{4}\%$ compounded monthly. You will be making monthly payments on the loan. What is the total amount that must be paid back?
a. What is the number of time periods (n) you should use in solving this problem?
b. What rate of interest (i), per period of time, should be used in solving this problem?
c. Is the present single amount of money (P) known? (Yes, No)
d. Which time value factor should be used to solve this problem?
e. What amount must be paid back each month?
f. What is the total amount that will be paid back over the life of the loan?
g. How much of the total amount repaid represents interest?

2.6 Use an interest rate equal to 7% compounded annually to solve this problem.
a. If \$20,000 is borrowed for 5 yr, what total amount must be paid back?
b. How much of the total amount repaid represents interest?

2.7 A company's interest rate for acquiring outside capital is 5.5% compounded annually. If \$40,000 must be borrowed for 4 yr, what is the total amount of interest that will be accrued?

2.8 A contractor is saving to purchase a \$200,000 machine. How much will the company have to bank today if the interest rate is 10% and they would like to purchase the machine in 4 yr?

2.9 What amount must be invested today at an annual interest rate of $4\frac{1}{2}\%$, if you want to purchase a \$450,000 machine 4 yr in the future?

2.10 What amount must be invested today at an interest rate of $4\frac{1}{2}\%$, compounded monthly, if you want to purchase a \$450,000 machine 4 yr in the future?

2.11 What amount will a company have to place in savings today in order to purchase a \$300,000 machine 5 yr in the future? The expected interest rate is 7%.

2.12 A track dozer cost \$163,000 to purchase. Fuel, oil, grease, and minor maintenance are estimated to cost \$32.14 per operating hour. A major engine repair costing \$12,000 will probably be required after 7,200 hr of use. The expected resale price (salvage value) is 21% of the original purchase price. The machine should last 10,800 hr. How much should the owner of the machine charge per hour of use, if it is expected that the machine will operate 1,800 hr per year? The company's cost of capital rate is 7.3%.

2.13 A machine cost $245,000 to purchase. Fuel, oil, grease, and minor maintenance are estimated to cost $47.64 per operating hour. A set of tires cost $13,700 to replace, and their estimated life is 3,100 use hours. A $15,000 major repair will probably be required after 6,200 hr of use. The machine is expected to last for 9,300 hr, after which it will be sold at a price (salvage value) equal to 17% of the original purchase price. A final set of new tires will not be purchased before the sale. How much should the owner of the machine charge per hour of use, if it is expected that the machine will operate 3,100 hr per year? The company's cost of capital rate is 8%.

2.14 To purchase a new car it is necessary to borrow $18,550. The bank offers a 5-yr loan at an interest rate of $4\frac{1}{4}$% compounded annually while the car dealer offers a 6-yr loan at an interest rate of 4% compounded annually. If you make only one payment at the end of the loan period repaying the principal and interest:

a. For each case, what is the total amount that must be paid back?

b. For each case, how much of the total amount repaid represents interest?

2.15 To purchase a new truck it is necessary to borrow $18,550. The bank offers a 5-yr loan at an interest rate of $4\frac{1}{4}$% compounded monthly while the car dealer offers a 6-yr loan at an interest rate of 2% compounded monthly. If you will be making monthly payments:

a. For each case, what is the total amount that must be paid back?

b. For each case, how much of the total amount repaid represents interest?

2.16 A contractor is considering the following three alternatives:

a. Purchase a new microcomputer system for $15,000. The system is expected to last 6 yr with a salvage value of $1,000.

b. Lease a new microcomputer system for $3,000 per year, payable in advance. The system should last 6 yr.

c. Purchase a used microcomputer system for $8,200. It is expected to last 3 yr with no salvage value.

Use a common-multiple-of-lives approach. If a MARR of 8% is used, which alternative should be selected using a discounted present worth analysis? If the MARR is 12%, which alternate should be selected?

2.17 What is the largest single equipment cost?

2.18 Regardless of how much a machine is used, the owner must pay owning cost. (True : False).

2.19 A machine's owning cost includes:

a. Tires
b. Storage expenses
c. Taxes
d. General repair

2.20 A tractor with an adjusted basis (from depreciation) of $55,000 is sold for $60,000 and a new tractor purchased with a cash payment of $325,000. These are two separate transactions. What is the tax depreciation basis of the new tractor?

2.21 A tractor with an adjusted basis (from depreciation) of $65,000 is traded for a new tractor that has a fair market value of $300,000. A cash payment of $225,000 is made to complete the transaction. What is the tax depreciation basis of the new tractor?

2.22 Asphalt Pavers, Inc., purchases a loader to use at its asphalt plant. The purchase price delivered is $235,000. Tires for this machine cost $24,000. The company believes it can sell the loader after 7 yr (3,000 hr/yr) of service for $79,000. There will be no major overhauls. The company's cost of capital is 6.3%. What is the depreciation part of this machine's ownership cost? Use the time value method to calculate depreciation. ($9.236/hr)

2.23 For the machine described in Problem 2.22 what is the depreciation part of machine ownership cost? Use the AAI method of calculation.

2.24 Pushem Down clearing contractors purchases a dozer with a delivered price of $275,000. The company believes it can sell the used dozer after 4 yr (2,000 hr/yr) of service for $56,000. There will be no major overhauls. The company's cost of capital is 9.2%, and its tax rate is 33%. Property taxes, insurance, and storage will run 4%. What is the owning cost for the dozer? Use the time-value method to calculate the depreciation portion of the ownership cost. ($29.943/hr)

2.25 Earthmovers, Inc., purchases a grader to maintain haul roads. The purchase price delivered is $165,000. Tires for this machine cost $24,000. The company believes it can sell the grader after 6 yr (15,000 hr) of service for $26,000. There will be no major overhauls. The company's cost of capital is 7.3%, and its tax rate is 35%. There are no property taxes, but insurance and storage will run 3%. What is the owning cost for the grader? Use the time-value method to calculate the depreciation portion of the ownership cost.

2.26 Lifters, Inc., purchases a mobile crane. The purchase price delivered is $863,000. Tires for this machine cost $60,000. The company believes it can sell the crane after 10 yr (1,500 hr/yr) of service for $240,000. There will be no major overhauls. The company's cost of capital is 3.9%. What is the depreciation part of this machine's ownership cost? Use the time-value method to calculate depreciation.

2.27 Using the AAI method to calculate depreciation and the Problem 2.26 information, what is the depreciation part of the machine's ownership cost?

2.28 A 140-fwhp diesel-powered wheel loader will be used at an asphalt plant to move aggregate from a stockpile to the feed hoppers. The work will be steady at an efficiency equal to a 55-min hour. The engine will work at full throttle while loading the bucket (32% of the time) and a three-quarter throttle to travel and dump. Calculate the fuel consumption using the engine consumption averages and compare the result to a medium rating in Table 2.2. (4.3 gal/hr, 4.0 gal/hr)

2.29 A 60-fwhp gasoline-powered pump will be used to dewater an excavation. The work will be steady at an efficiency equal to a 60-min hour. The engine will work at half throttle. Calculate the theoretical fuel consumption.

2.30 A 260-fwhp diesel-powered wheel loader will be used to load shot rock. This loader was purchased for $330,000. The estimated salvage value at the end of 4 yr is $85,000. The company's cost of capital is 8.7%. A set of tires costs $32,000. The work efficiency will be equal to a 45-min hour. The engine will work at full throttle while loading the bucket (33% of the time) and at three-quarter throttle to travel and dump. The crankcase capacity is 10 gal and the company has a policy to change oil every 100 hr on this job. The annual cost of repairs equals 70% of the straight-line machine depreciation. Fuel cost $1.07/gal, and oil is $2.50/gal. The cost of other lubricants and filters is $0.45/hr. Tire repair is 17% of tire depreciation. The tires should give 3,000 hr of service. The loader will work 1,500 hr/yr. In this usage the estimated life for bucket teeth is 120 hr. The local price for a set of teeth is $640. What is the operating cost for the loader in this application? ($52.007/hr)

2.31 A 400-fwhp diesel-powered dozer will be used to support a scraper fleet. This dozer was purchased for $395,000. The estimated salvage value at the end of 4 yr is $105,000. The company's cost of capital is 7.6%. The work efficiency will be equal to a 50-min hour. The engine will work at full throttle while push loading the scrapers (59% of the time) and at three-quarter throttle to travel and position. The crankcase capacity is 16 gal, and the company has a policy to change oil every 150 hr. The annual cost of repairs equals 68% of the straight-line machine depreciation. Fuel cost $1.03/gal. and oil is $2.53/gal. The cost of other lubricants and filters is $0.65/hr. The dozer will work 1,800 hr/yr. In this usage, the estimated life for cutting edges is 410 hr. The local price for a set of cutting edges is $1,300. What is the operating cost for the dozer in this application?

2.32 A 265-fwhp diesel-powered wheel loader will be used to load shot rock. This loader was purchased for $365,000. The estimated salvage value at the end of 4 yr is $105,000. The company's cost of capital is 6.4%. A set of tires costs $35,000. The work will be at an efficiency equal to a 50-

min hour. The engine will work at full throttle while loading the bucket (30% of the time) and at three-quarter throttle to travel and dump. The crankcase capacity is 12 gal and the company has a policy to change oil every 150 hr on this job. The annual cost of repairs equals 65% of the straight-line machine depreciation. Fuel cost $1.15/gal and oil is $3.00/gal. The cost of other lubricants and filters is $0.65/hr. Tire repair is 20% of tire depreciation. The tires should give 4,000 hr of service. The loader will work 2,000 hr/yr. In this usage, the estimated life for bucket teeth is 110 hr. The local price for a set of teeth is $900. What is the operating cost for the loader in this application?

REFERENCES

1. *Caterpillar Performance Handbook*, Caterpillar Inc., Peoria, IL, issued annually.
2. Collier, Courtland A., and W. B. Ledbetter (1988). *Engineering and Economic Cost Analysis*, 2nd ed., Harper & Row, New York.
3. Johnson, Robert W. (1977). *Capital Budgeting*, Kendall/Hunt Publishing Co., Dubuque, IA.
4. Lewellen, Wilbur G. (1976). *The Cost of Capital*, Kendall/Hunt Publishing Co., Dubuque, IA.
5. Modigliani, Franco, and Merton H. Miller (1958). "The Cost of Capital, Corporate Finance and the Theory of Investment," *The American Economic Review*, Vol. XLVIII, No. 3, June.
6. Schexnayder, C. J., and Donn E. Hancher (1981). "Contractor Equipment Management Practices," *Journal of the Construction Division, Proceedings, ASCE*, Vol. 107, No. CO4, December.
7. Schexnayder, C. J., and Donn E. Hancher (1981). "Interest Factor in Equipment Economics," *Journal of the Construction Division, Proceedings, ASCE*, Vol. 107, No. CO4, December.
8. Schexnayder, C. J., and Donn E. Hancher (1982). "Inflation and Equipment Replacement Economics," *Journal of the Construction Division, Proceedings, ASCE*, Vol. 108, No. CO2, June.

WEBSITE RESOURCES

1. www.aednet.org The Associated Equipment Distributors, Inc. (AED) is an international trade association representing construction equipment distributors, manufacturers, and industry-service firms.
2. www.aem.org The Association of Equipment Manufacturers (AEM) is a trade and business development resource for companies that manufacture equipment. AEM was formed on January 1, 2002, from the consolidation of the Construction Industry Manufacturers Association (CIMA) and the Equipment Manufacturers Institute (EMI).
3. www.caterpillar.com Caterpillar Inc. is the world's largest manufacturer of construction and mining equipment.

4. www.constructionequipment.com *Construction Equipment* magazine online.
5. www.deere.com Deere and Company, "John Deere's" Construction and Forestry Equipment Division.
6. www.elaonline.com Equipment Leasing Association (ELA) is a national organization comprised of member companies within the equipment leasing and finance industry.
7. www.equipmentworld.com Equipmentworld.com is an online magazine featuring construction industry and equipment news.
8. www.goodyearotr.com The Goodyear Tire & Rubber Company off-the-road tire information.
9. www.hcmac.com Hitachi Construction Machinery Co., Ltd. is an earthmoving construction machinery manufacturer.
10. www.machinerytrader.com Machinery Trader is a marketplace for buying and selling heavy construction equipment.
11. earthmover.webmichelin.com/na_eng Michelin's tire company's construction equipment tire information site.

CHAPTER 3

Planning for Earthwork Construction

Projects are executed under conditions that vary immensely from one project to another. Therefore, before a project is undertaken it is necessary to systematically analyze project conditions and develop alternatives that potentially provide success. When the engineer prepares a plan and cost estimate for an earthwork project, the critical attributes that must be determined are (1) the quantities involved, basically volume or weight; (2) the haul distances; and (3) the grades for all segments of the hauls. An earthwork volume sheet allows for the systematic recording of information and making the necessary earthwork calculations. The mass diagram is an analysis tool for selecting the appropriate equipment for excavating and hauling material.

PLANNING

Every construction project is a unique undertaking. Although similar work may have been performed previously, no two projects will have identical job conditions. The pace, complexity, and cost of modern construction are incompatible with trial-and-error corrections as the work proceeds. Therefore, planning is undertaken to understand the problems and to develop courses of action.

The goal of planning is to minimize resource expenditures required to successfully complete the project and to ensure that the work is accomplished in a safe manner. Planning is necessary in order to

1. Understand project objectives and requirements.
2. Define work elements.
3. Develop safe construction methods and avoid hazards.
4. Improve efficiency.
5. Coordinate and integrate activities.
6. Develop accurate schedules.

7. Respond to future changes.
8. Provide a yardstick for monitoring and controlling execution of project activities.

It is a decision-making process, an anticipatory decision-making process that chooses courses of action that will be performed in the future or when certain events occur. Systematic planning requires screening of variables and alternatives that potentially affect the success of the project.

There is no single best way to conduct a construction activity. Performance of a work task depends on variables related to the work itself: (1) the quantities involved, basically volume or weight; (2) the haul distances; (3) the grades for all segments of the hauls; (4) the work hazards; and (5) the various contractual and legal constraints. Typical constraints are

1. Contract requirements described in the drawings and technical specifications, including time duration or requirements as to completion dates.
2. Legal requirements (OSHA, licensing, environmental control) that must be satisfied.
3. Physical and/or environmental limits of the job, which may necessitate off-site fabrication and material storage, or sequencing of construction operations (traffic control).
4. Climatic conditions that limit when certain activities can be performed, such as paving or stabilization operations, or that limit earthwork operations because of moisture content and the inability to dry the material.

The earlier you plan before execution the more you will be able to influence execution by having time to carefully consider the impacts of all constraints and to devise efficient strategies for dealing with project requirements.

For planning purposes, tasks should be divided into smaller independent executable subtasks. This enables us to structure unwieldy and complex problems into smaller digestible tasks. The more time and resources allocated for planning at each stage of planning, the greater the opportunity to develop optimal solutions rather than something that is "good enough."

Safety

In planning any construction operation it is necessary to give considerable attention to safety. Engineers have both a moral and a legal responsibility to the public, to construction workers, and to the end-users of our projects to make sure that workplaces, work processes, and the general environment are safe. The machines that are used in heavy construction greatly extend human capabilities and are very sophisticated, but they can also present a variety of hazards (Fig. 3.1). If a large earthmoving truck strikes a normal size vehicle the results will most likely be fatal. In 2002 there were 1,181 highway construction work zone fatalities in the United States.* Safety planning will identify hazards and develop ways to protect both the construction worker and the public.

*Work Zone Safety Fact Sheet, United States Department of Transportation, Federal Highway Administration, February 23, 2004.

FIGURE 3.1 The driver of an off-highway truck has a limited field of vision.

Accidents are nearly always the result of multiple mistakes. Safety planning should institute procedures that "break the chain of mistakes" that lead to accidents and disasters. Mr. Howard I. Shapiro, P.E., of Howard Shapiro Associates, New York, discussed this idea of breaking the chain of mistakes at ASCE's 150th Anniversary conference in 2003.

> Most of the time crane accidents are blamed on the crane operator; you'll often hear terms such as—"operator error"—"he reached out too far with the load"—"the outrigger sank into the earth and the crane turned over." Those phrases all refer to single errors that led to failure. But, my 30 years of experience with crane operations, and my investigation of many accidents, has taught me that cranes are robust, and the people who operate them are both talented and resourceful—enough so to overcome the effects of most single errors. I've found that it usually takes a string of independent and often unconnected errors to cause crane failures.
>
> We can't hope to eliminate all accidents or all errors, but by breaking the string—eliminating any one error in the chain—we can prevent many accidents.

In the case of the fatal crane accident at Milwaukee's Miller Park Stadium in July 1999, the chain of errors that occured during the lifting of a large piece of the stadium's retractable roof included:

1. The lifting sequence was on the project's critical path, placing pressure on management to proceed in spite of weather conditions.

2. Wind speed limits were not established.
3. Field supervision did not adhere to the rating chart warnings concerning wind.

All equipment operations entail risk. By making an assessment of risk, you are able to focus your resources. Money and management time are always limited; therefore, they should be invested in proportion to the risks and the potential consequences of an accident. The level of safety planning that is needed should be based on the specific risks associated with work conditions. When risks are identified in advance, and steps are taken to control, reduce, or remove them, links are being removed from the chain of potential errors. The overall level of risk is thereby diminished. What Howard Shapiro advised concerning crane safety is applicable to all construction operations—"Don't assume anything. Ask questions and check for yourself." Because those simple pieces of advice were not followed at Milwaukee, three ironworkers in the crane basket were killed and the crane operator and four others were injured.

Earthwork Planning

When the engineer prepares a plan and cost estimate for an earthwork project the decision process is often not a chain of sequential activities. The process takes the form of recurrent chains with feedback. As decisions are proposed, further investigation—collection of more information—is usually necessary to diminish uncertainty. The process begins with the project documents. The information and requirements as set forth in the drawings and specifications are analyzed.

Project work elements are defined in physical terms: volume of stripping, soil excavation, rock excavation, embankment, waste material, etc. This is a project material take-off or **quantity survey**. The take-off effort must calculate not just the total quantity of material to be handled but must segregate the total quantity of material based on factors affecting productivity.

quantity survey *The process of calculating the quantity of materials required to build a project.*

Mass excavation involves moving a substantial volume of material and the excavation work is a primary part of the project. On the Eastside project in California, the contractor building the West Dam moved over 68 million cy of material. Mass excavations are typically operations of considerable excavation depth and horizontal extent, and may include requirements to drill and blast rock (e.g., the movement of consolidated materials). **Structural excavation** is a different type of undertaking. The excavation work is performed to support the construction of other structural elements. This work is usually done in a confined area, it is typically vertical in extent, and the banks of the work may require support systems. The volume of excavated material is not a decisive factor as much as dealing with limited work space and vertical movement of the material. The project plan sheets provide the graphical information necessary to calculate work quantities.

mass excavation *The requirement to excavate substantial volumes of material, usually at considerable depth or over a large area.*

structural excavation *Excavation undertaken in support of structural element construction, usually involves removing materials from a limited area.*

Risks associated with subsurface conditions, specifically those inherent with the types of materials encountered and their behavior during the construction processes, are identified. "You bid it—you build it," is not quite correct.

Contractors have a right to rely on owner-provided information. Additionally, many contracts will contain a *differing site condition* clause. Material differences in conditions are applicable in either of two cases. A Type 1 differing site condition exists when actual conditions differ materially from those "indicated in the contract." A Type 2 differing site condition arises when actual conditions differ from reasonable expectations. These clauses provide the constructor some protection from geotechnical risk. They do not, however, eliminate the contractor's responsibility for performing a thorough examination of project conditions. After these first two steps have been accomplished for the first time, it is almost always necessary to visit the field. This is one of those recurrent chains with feedback.

Field investigations, geologic and soil studies, and analysis of meteorological data enable the planner to better quantify what has been presented in the project documents. The contract documents will usually include geotechnical data and information that was gathered during the design phase of the project. If not included directly in the documents, this material is typically available to bidders as supplemental information. The geotechnical data are gathered to support the design effort; interpretation for design and interpretation for construction are two very different things. The designer is interested in structural capacities and the constructor is interested in how the *material will handle* during the construction processes. Supplemental field studies by the constructor are almost always necessary to fully appreciate work opportunities and limitations. Field investigations are also necessary to locate material sources and/or sites to use in disposing of excess or unsatisfactory materials.

Detailed studies of machine production can be performed only after the questions of material uncertainty have been answered or at least reduced. At that point, the planner is ready to undertake production analyses based on different types of equipment spreads. This chapter is devoted to the first two elements in the list; the performance of production studies for individual machine types is covered in Chapters 5, 7 to 12, 14 to 20, and 22.

Planning is a thought process where complex idea arrays are arranged in different combinations and the possible resulting effects of the combinations are visualized. Because it is difficult to interrelate incremental decisions made at different times, planning should be carried out without interruption during shielded blocks of time and the planner must carefully document the process and reasons for decisions.

GRAPHICAL PRESENTATION OF EARTHWORK

station
A horizontal distance of 100 feet.

Horizontal distances along a project are referenced in stations. The term **station** refers to locations on a base-100 numbering system. Therefore, the distance between two adjacent stations is 100 ft. Station 1 is written 1 + 00. The plus sign is used in this system of referencing points. The term refers to the surveyor notation for laying out a project in the field and is used on the plans to denote locations along the length of the project.

Three kinds of views are presented in the contract documents to show earthwork construction features:

1. *Plan view*. The **plan view** is drawn looking down on the proposed work and presents the horizontal alignment of features. Figure 3.2 is a plan view of a highway project; it shows the project centerline with stationing noted and the project limits, the two dark exterior lines.
2. *Profile view*. The **profile view** is a cut view, typically along the centerline of the work. It presents the vertical alignment of features. Figure 3.3 is a profile view. The bottom horizontal scale shows the centerline stationing, the vertical scale gives elevation, the dashed line is the existing ground line, and the solid line is the proposed final grade of the work.
3. *Cross section view*. The **cross section view** is formed by a plane cutting the work vertically and at right angles to its long axis. Figures 3.4 and 3.5 present fill and cut cross sections, respectively. The continuous line denotes the final grade of the work and the dashed line shows the existing ground.

plan view
A construction drawing representing the horizontal alignment of the work.

profile view
A construction drawing depicting a vertical plane cut through the centerline of the work. It shows the vertical relationship of the ground surface and the finished work.

cross section view
A construction drawing depicting a vertical section of earthwork at right angles to the centerline of the work.

Cross Sections

In the case of a project that is linear in extent, material volumes are usually determined from **cross sections**. Cross sections are pictorial drawings produced from a combination of the designed project layout and measurements taken in the field at right angles to the project centerline or the centerline of a project feature, such as a drainage ditch. When the ground surface is regular, field measurements are typically taken at every full station (100 ft). When the ground is irregular, measurements must be taken at closer intervals and particularly at points of change. Typical cross sections are shown in Figures 3.4 and 3.5.

cross sections
Earthwork drawings created by combining the project design with field measurements of existing conditions.

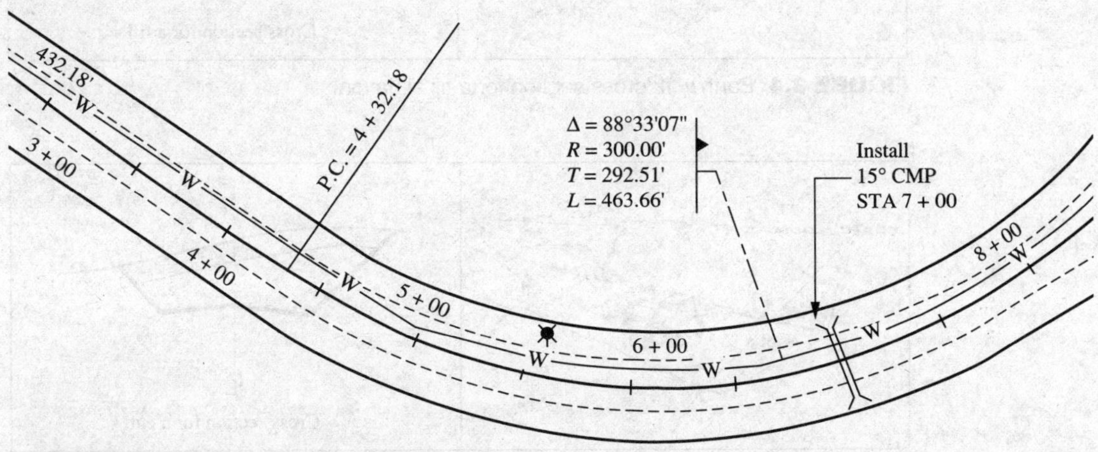

FIGURE 3.2 Plan view of a highway project.

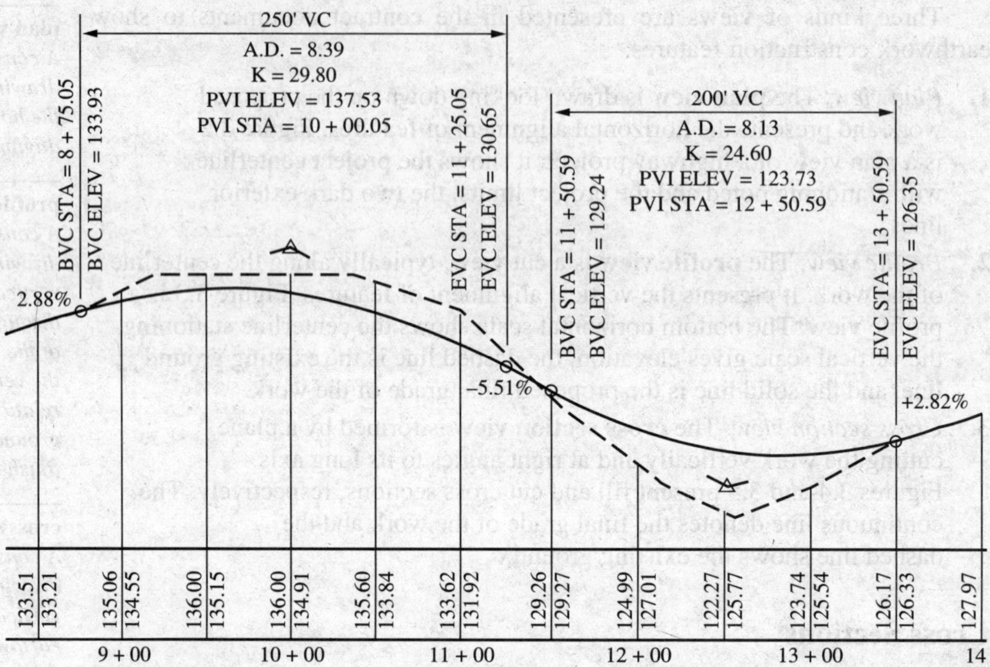

FIGURE 3.3 Profile view of a highway project.

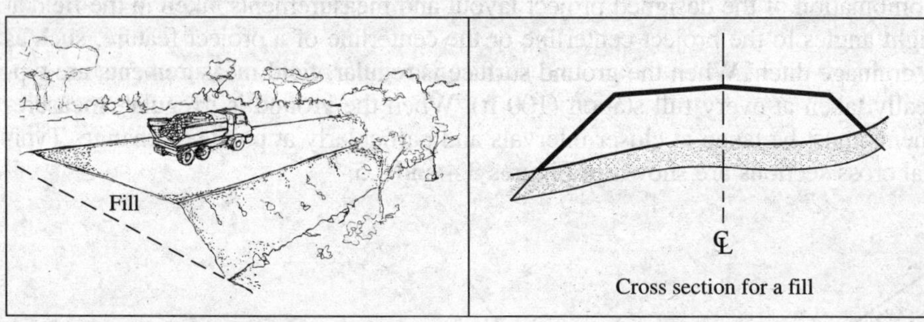

FIGURE 3.4 Earthwork cross section for a fill situation.

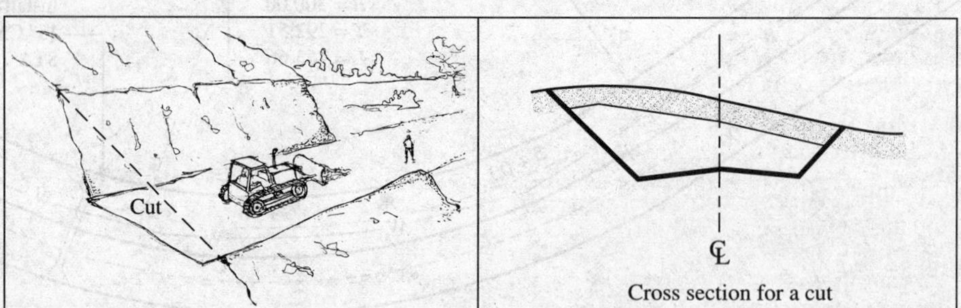

FIGURE 3.5 Earthwork cross section for a cut situation.

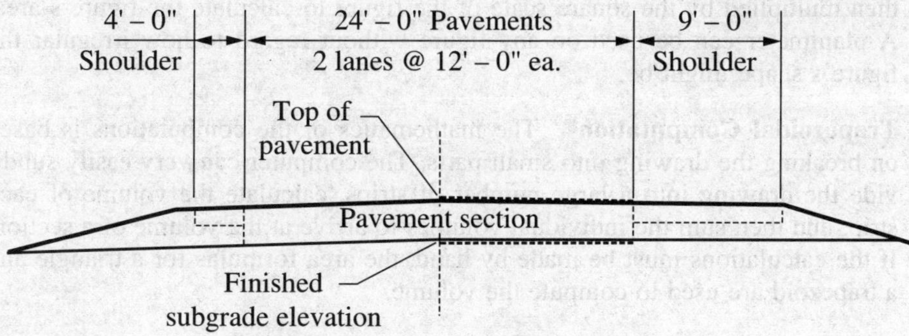

FIGURE 3.6 Cross section view of a pavement section.

The cross sections usually depict the finished subgrade elevations; this fact should always be verified (Fig. 3.6). If instead, the grades depicted are top of pavement, the thickness of the pavement section must be used to adjust the computations. From the cross section drawings, *end areas* can be computed using any of several methods.

EARTHWORK QUANTITIES

Earthwork computations involve the calculation of earthwork volumes, balancing cuts and fills, and planning the most economical material hauls. The first step in planning an earthmoving operation is estimating the quantities involved in the project. The exactness with which earthwork computations can be made is dependent on the extent and accuracy of field measurements portrayed on the drawings.

End-Area Determination

The method chosen to compute a cross section end area will depend on the time available and the aids at hand. Most companies use commercial computer software and digitizing tablets (see references at the end of the chapter) to arrive at cross section end areas. Other methods include the use of a planimeter, subdivision of the area into geometric figures with definite formulas for areas (rectangles, triangles, parallelograms, and trapezoids), and the use of the trapezoidal formula.

Digitizing Tablet A digitizing tablet is a board with a wire mesh grid imbedded into it. When the cursor is traced over the board, a current is picked up by the board's grid, and the coordinates of the tracing device are passed on to the computer. When a plan is taped to the board and a scale is entered, the traced plan is converted to actual measurements, and then a software program computes the area, length, and volume of the traced data.

Planimeter A planimeter is a drafting instrument that is used to move a tracing point around the perimeter of the plotted area. It provides a value that is then multiplied by the square scale of the figure to calculate the figure's area. A planimeter can be used on any figure without regard to how irregular the figure's shape might be.

Trapezoidal Computations The mathematics of the computations is based on breaking the drawing into small parts. The computer can very easily subdivide the drawing into a large number of strips, calculate the volume of each strip, and then sum the individual volumes to arrive at the volume of a section. If the calculations must be made by hand, the area formulas for a triangle and a trapezoid are used to compute the volume.

$$\text{Area of a triangle} = \frac{1}{2} hw \qquad \textbf{[3.1]}$$

where h = height of the triangle
w = base of the triangle

$$\text{Area of a trapezoid} = \frac{(h_1 + h_2)}{2} \times w \qquad \textbf{[3.2]}$$

where (see Fig. 3.7) w = distance between the parallel sides
h_1 and h_2 = the lengths of the two parallel sides

The general trapezoidal formula for calculating area is

$$\text{Area} = \left(\frac{h_0}{2} + h_1 + h_2 + \cdots + h_{(n-1)} + \frac{h_n}{2}\right) \times w \qquad \textbf{[3.3]}$$

where (see Fig. 3.7) w = distance between the parallel sides
$h_0 \ldots h_n$ = the lengths of the individual adjacent parallel sides

The precision achieved using this formula depends on the number of strips but is about ±0.5%.

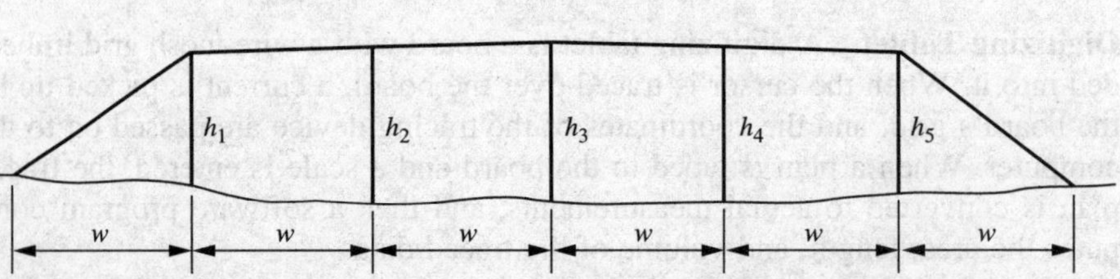

FIGURE 3.7 Division of a cross section drawing into triangles and trapezoids.

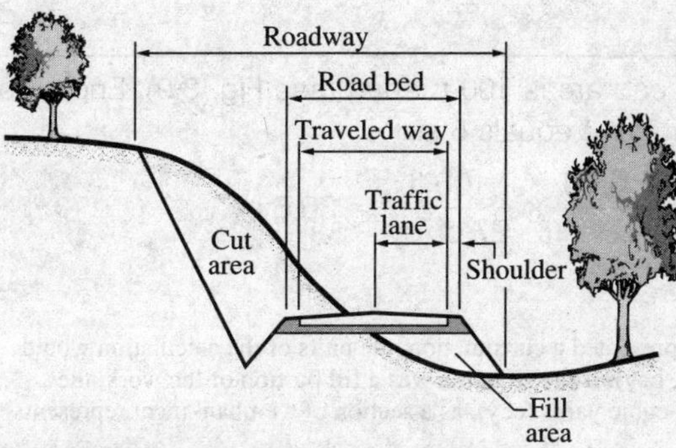

FIGURE 3.8 Cut-and-fill areas in the same section.

In the case of sidehill construction, there can be both a cut area and a fill area in the same cross section (Fig. 3.8). When making area computations, it is necessary to always calculate cut and fill areas separately.

Average End Area

average end area *A calculation method for determining the volume of material bounded by two cross sections or end areas.*

The **average-end-area** method is commonly used to determine the volume bounded by two cross sections or end areas. The principle is that the volume of the solid bounded by two parallel, or nearly parallel, cross sections is equal to the average of the two end areas times the distance between the cross sections along their centerline (see Fig. 3.9). The average-end-area formula is

$$\text{Volume (net cy)} = \frac{(A_1 + A_2)}{2} \times \frac{L}{27} \qquad [3.4]$$

where (see Fig. 3.9) A_1 and A_2 = area in square feet of the respective end areas
L = the length in ft between the end areas

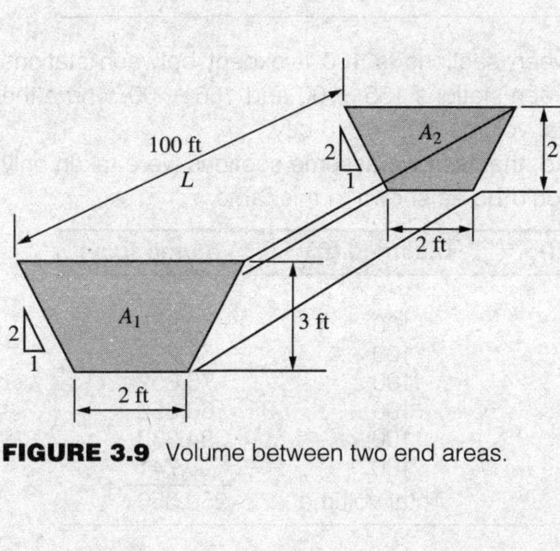

FIGURE 3.9 Volume between two end areas.

EXAMPLE 3.1

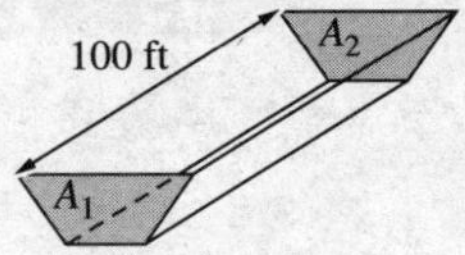

Calculate the volume between two end areas 100 ft apart (see Fig. 3.9). End area 1 (A_1) equals 10.5 sf and end area 2 (A_2) equals 6 sf.

$$\text{Volume} = \frac{(10.5\text{ sf} + 6\text{ sf})}{2} \times \frac{100\text{ ft}}{27\text{ cf/cy}} = 30.6\text{ cy}^*$$

where cf means cubic feet.

* If the cross sections (the two end areas) represented a cut situation, the units of the calculation would be bank cubic yards (1 cy in its natural state, bcy). If the situation was a fill portion of the work, the units of the calculation would be compacted cubic yards (ccy), as a section of a embankment represents a compacted condition.

The average-end-area principle is not altogether true because the average of the two end areas is not the arithmetic mean of many intermediate areas. The method gives volumes generally slightly in excess of the actual volumes. The precision is about ±1.0%.

EXAMPLE 3.2

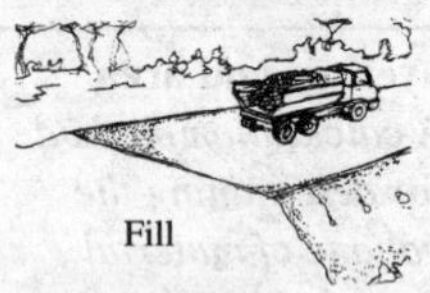

The volume between the embankment (fill) end areas given in the table is calculated using the average-end-area method. The total volume of the section (the sum of the individual volumes) is also given.

Station	End area (sf)	Distance (ft)	Volume (ccy)
150 + 00	360	—	—
150 + 50	3,700	50	3,759
151 + 00	10,200	50	12,870
152 + 00	18,000	100	52,222
153 + 00	23,500	100	76,852
154 + 00	12,600	100	66,852
155 + 00	5,940	100	34,333
155 + 50	2,300	50	7,630
156 + 00	400	50	2,500
		Total volume	257,018

In this example, the interval between sections is 100 ft except between stations 150 + 00 and 151 + 00 and between stations 155 + 00 and 156 + 00, where the interval is 50 ft. The calculated total volume is 257,018 ccy.

If the 50-ft intervals are omitted, that is, if we assume sections were taken only on full stations, the calculations would be as shown in this table.

Station	End area (sf)	Distance (ft)	Volume (ccy)
150 + 00	360	—	—
151 + 00	10,200	100	19,556
152 + 00	18,000	100	52,222
153 + 00	23,500	100	76,852
154 + 00	12,600	100	66,852
155 + 00	5,940	100	34,333
156 + 00	400	100	741
		Total volume	250,556

The difference is 6,462 ccy. This example emphasizes the importance of cross section spacing and the possible introduction of volume calculation error. The error in this case is 2.5%, based on an actual 257,018 ccy volume, and using more data.

Although cross sections can be taken at any conservative interval along the centerline, judgment should be exercised, depending particularly on the irregularity of the ground and the tightness of curves. In the case of tight curves, a spacing of 25 ft is often appropriate.

Stripping

The upper layer of material encountered in an excavation is often topsoil (organic material), resulting from the decomposition of vegetative matter. Such organic material is commonly referred to as **stripping**. This material is unsuitable for use in an embankment and must usually be handled in a separate excavation operation. It can be collected and wasted, or stockpiled for later use on the project to plate slopes. If the embankments are shallow, the organic material below the footprint of the fill sections must be stripped before embankment placement can commence (see Fig. 3.10). In the case of embankments over 5 ft in height, most specifications allow the organic material to remain if its thickness is only a few inches. When calculating the volume of cut sections, this stripping quantity must be subtracted from the net volume (see Fig. 3.11). In the case of fill sections, the quantity must be added to the calculated fill volume (see Fig. 3.11).

stripping
The upper layer of organic material that must be removed before beginning an excavation or embankment.

Net Volume

The computed volumes from the cross sections represent two different material states. The volumes from the fill cross sections represent compacted volume. If the volume is expressed in cubic yards, the notation is compacted cubic yards (ccy).

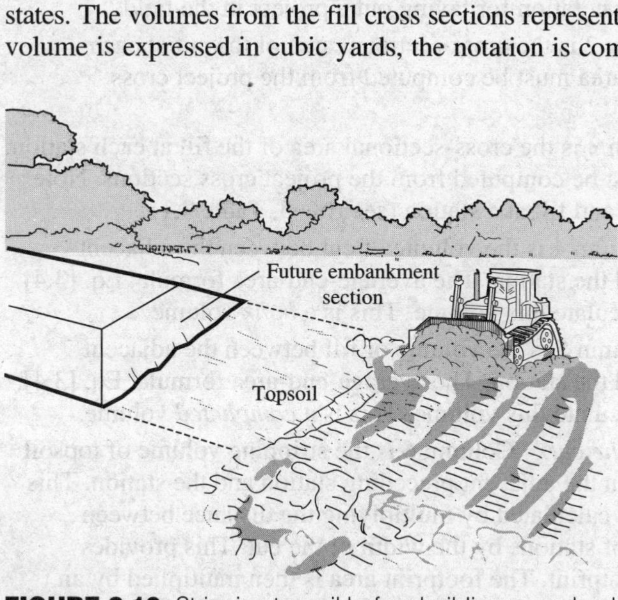

FIGURE 3.10 Stripping topsoil before building an embankment.

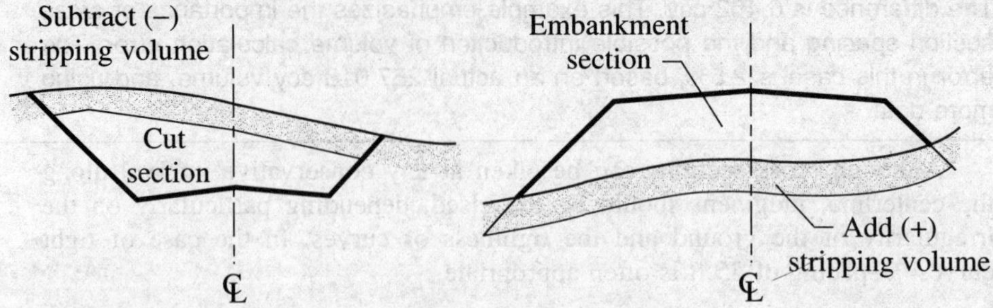

FIGURE 3.11 Effect of stripping on embankment and cut volume calculations.

In the case of cut sections, the volume is a natural in situ volume. The term bank volume is used to denote this in situ volume; if the volume is expressed in cubic yards the notation is bank cubic yards (bcy). If the cut and fill volumes are to be combined, they must be converted into compatible volumes. In Table 3.1, the conversion from compacted to bank cubic yards is made by dividing the compacted volume by 0.90. The theoretical basis for making soil volume conversions is explained in Chapter 4.

Earthwork Volume Sheet

An earthwork volume sheet, which can easily be constructed using a spreadsheet program, allows for the systematic recording of information and making the necessary earthwork calculations (see Table 3.1).

Stations. Column 1 is a listing of all stations at which cross-sectional areas have been recorded. A full station is a 100-ft interval. The term refers to the surveyor notation for laying out a project in the field.

Area of cut. Column 2 is the cross-sectional area of the cut at each station. Usually this area must be computed from the project cross sections.

Area of fill. Column 3 is the cross-sectional area of the fill at each station. Usually this area must be computed from the project cross sections. Note there can be both cut and fill at a station (see row 5, Table 3.1).

Volume of cut. Column 4 is the volume of cut between the adjacent preceding station and the station. The average-end-area formula, Eq. [3.4], is usually used to calculate this volume. This is a *bank* volume.

Volume of fill. Column 5 is the volume of fill between the adjacent preceding station and the station. The average-end-area formula, Eq. [3.4], is usually used to calculate this volume. This is a *compacted* volume.

Stripping volume in the cut. Column 6 is the stripping volume of topsoil above the cut between the adjacent preceding station and the station. This volume is commonly calculated by multiplying the distance between stations or fractions of stations by the width of the cut. This provides the area of the cut footprint. The footprint area is then multiplied by an

TABLE 3.1 Earthwork volume calculation sheet.

	Station (1)	End-area cut (sf) (2)	End-area fill (sf) (3)	Volume of cut (bcy) (4)	Volume of fill (ccy) (5)	Stripping cut (bcy) (6)	Stripping fill (ccy) (7)	Total cut (bcy) (8)	Total fill (ccy) (9)	Adj. fill (bcy) (10)	Algebraic sum (bcy) (11)	Mass ordinate (12)
(1)	0 + 00	0	0									
(2)	0 + 50	0	115	0	106	0	18	0	124	138	−138	−138
(3)	1 + 00	0	112	0	210	0	30	0	240	267	−267	−405
(4)	2 + 00	0	54	0	307	0	44	0	351	390	−390	−796
(5)	2 + 50	64	30	59	78	0	22	59	100	111	−52	−847
(6)	3 + 00	120	0	170	28	26	0	144	28	31	114	−734
(7)	4 + 00	160	0	519	0	76	0	443	0	0	443	−291
(8)	5 + 00	317	0	883	0	74	0	809	0	0	809	518
(9)	6 + 00	51	0	681	0	60	0	621	0	0	621	1,140
(10)	6 + 50	46	6	90	6	21	0	69	6	6	63	1,202
(11)	7 + 00	0	125	43	121	0	25	43	146	163	−120	1,082
(12)	8 + 00	0	186	0	576	0	81	0	657	730	−730	352
(13)	8 + 50	0	332	0	480	0	69	0	549	610	−160	−257

average depth of topsoil to derive the stripping volume. This represents a bank volume of cut material. Usually topsoil material is not suitable for use in the embankment. The average depth of topsoil must be determined by field investigation.

Stripping volume in the fill. Column 7 is the stripping volume of topsoil under the fill between the adjacent preceding station and the station. This volume is commonly calculated by multiplying the distance between stations or fractions of stations by the width of the fill. This provides the area of the fill footprint. To derive the stripping volume, the area of the embankment footprint is multiplied by an average depth of topsoil. The stripping is a *bank* volume but it also represents an additional requirement for fill material, compacted volume of fill.

Total volume of cut. Column 8 is the volume of cut material available for use in embankment construction. It is derived by subtracting the cut stripping (column 6) from the cut volume (column 4). Columns 4 and 6 are both bank volume quantities.

Total volume of fill. Column 9 is the total volume of fill required. It is derived by adding the fill stripping (column 7) to the fill volume (column 5). In this case, columns 5 and 7 both characterize compacted volume quantities.

Adjusted fill. Column 10 is the total fill volume converted from compacted volume to bank volume.

Algebraic sum. Column 11 is the difference between column 10 and column 8. This indicates the volume of material that is available (cut is positive) or required (fill is negative) within station increments after intrastation balancing.

Mass ordinate. Column 12 is the running total of column 11 values from some point of beginning on the project profile. When the stations being summed are excavation sections, the value of this column will increase. Although summing a fill section will result in a decrease of the column 12 value, note that any material that could be used within a *station length* is not accounted for in the mass ordinate and therefore it is not accounted for in the **mass diagram**.

mass diagram
In earthwork calculations, a graphical representation of the algebraic cumulative quantities of cut and fill along the centerline, where cut is positive and fill is negative. Used to calculate haul in terms of station yards.

The mass diagram only accounts for material that must be transported beyond the limits of the two cross sections that define the volume of material. Where there is both cut and fill between a set of stations, only the excess of one over the other is used in computing the mass ordinate.

> Cut material between two successive stations is first used to satisfy fill requirements between those same two successive stations before there is a contribution to the mass ordinate value. Likewise, if there is a greater fill requirement between two successive stations than there is cut available, the cut contribution is accounted for first. Only after all of the cut material is utilized will there be a fill contribution to the mass ordinate value.

FIGURE 3.12 The mass diagram provides the information for deciding in which direction material should be hauled on the project.

The material used between the two successive stations is considered to move at right angles to the centerline of the project and therefore is often termed *crosshaul*. The remaining material in either case represents a longitudinal haul along the length of the project, and the mass diagram enables us to decide in which direction the material should be hauled (Fig. 3.12).

MASS DIAGRAM

Earthmoving is basically an operation where material is removed from high spots and deposited in low spots, with the "making up" of any deficit with borrow or the wasting of excess cut material. The mass diagram is an excellent method of analyzing linear earthmoving operations. It is a graphical means for measuring haul distance (stations) in terms of earthwork volume. A station yard (specifically a cubic yard) is a measure of work, the movement of 1 cy through a distance of one station.

On a mass diagram graph, the horizontal dimension represents the stations of a project (column 1, Table 3.1) and the vertical dimension (column 12, Table 3.1) represents the cumulative sum of excavation and embankment from some point of beginning on the project profile. The diagram provides information concerning

1. Quantities of materials,
2. Average haul distances,
3. Types of equipment that should be considered for accomplishing the work.

When combined with a ground profile, the average slope of haul segments can be estimated. The mass diagram is one of the most effective tools for planning the movement of material on any project of linear extent.

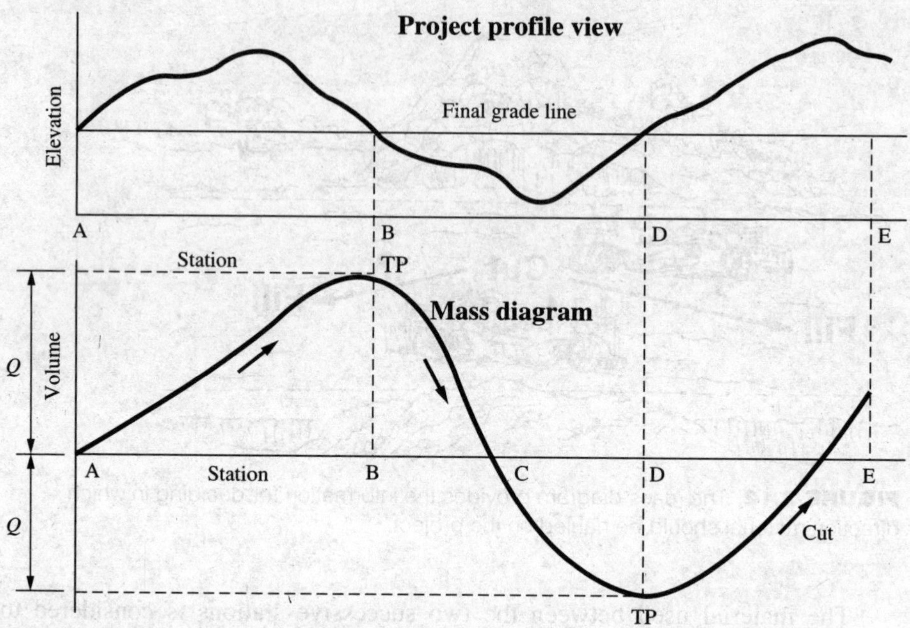

FIGURE 3.13 Properties of a mass diagram.

A mass diagram is plotted by using the column 1 data of the earthwork volume sheet as the horizontal scale location of a point and the column 12 data as the corresponding vertical scale location of that same point (see the bottom portion of Fig. 3.13). Positive mass ordinate values are plotted above the zero datum line and negative values below. The top portion of Fig. 3.13 is the profile view of the same project.

Mass Diagram Properties

A mass diagram is a running total of the quantity of material that is surplus or deficient along the project profile. An excavation operation produces an *ascending* mass diagram curve; the excavation quantity exceeds the embankment quantity requirements. Excavation is occurring between stations A and B, and stations D and E in Fig. 3.13. The total volume of excavation between A and B is obtained by projecting horizontally to the vertical axis *the mass diagram points* at stations A and B and reading the difference of the two volumes. Conversely, if the operation is a fill situation, there is a deficiency of material and a *descending* curve is produced; the embankment requirements exceed the excavation quantity being generated. Filling is occurring between stations B and D. The volume of fill can be calculated in a similar manner as the excavation calculation by a projection of the mass diagram line points to the vertical scale.

The *maximum* or *minimum points* on the mass diagram, where the curve transitions from rising to falling or from falling to rising, indicate a change from an excavation to fill situation or vice versa. These points are referred to as *transition points*. On the ground profile, the project grade line is crossing the natural ground line (see Fig. 3.13 at stations B and D).

When the mass diagram curve crosses the datum (or zero volume) line (as at station C) exactly as much material is being excavated (between stations A and C) as is required for fill between stations A and B. There is no excess or deficit of material at point C in the project. The final position of the mass diagram curve above or below the datum line indicates whether the project has surplus material that must be wasted or if there is a deficiency that must be made up by borrowing material from outside the project limits. Figure 3.13 station E indicates a waste situation, and excess material will have to be hauled off of the project.

USING THE MASS DIAGRAM

The mass diagram is an analysis tool for selecting the appropriate equipment for excavating and hauling material. The analysis is accomplished using balance lines and calculating average hauls.

Balance Lines

A **balance line** is a horizontal line of specific length that intersects the mass diagram in two places. The balance line can be constructed so that its length is the maximum haul distance for different types of equipment. The maximum haul distance is the limiting economical haul distance for a particular type of equipment (see Table 3.2).

balance line
A horizontal line of specific length that intersects the mass diagram in two places.

Figure 3.14 shows a balance line drawn on a portion of a mass diagram. If this was constructed for a large push-loaded scraper, the distance between stations A and C would be 5,000 ft. Between the ends of the balance line, the cut volume generated equals the fill volume required. Between stations A and C,

TABLE 3.2 Economical haul distances based on basic machine types.

Machine type		Economical haul distance
Large dozers, pushing material		Up to 300 ft*
Push-loaded scrapers		300 to 5,000 ft*
Trucks		Hauls greater than 5,000 ft

*The specific distance will depend on the size of the dozer or scraper.

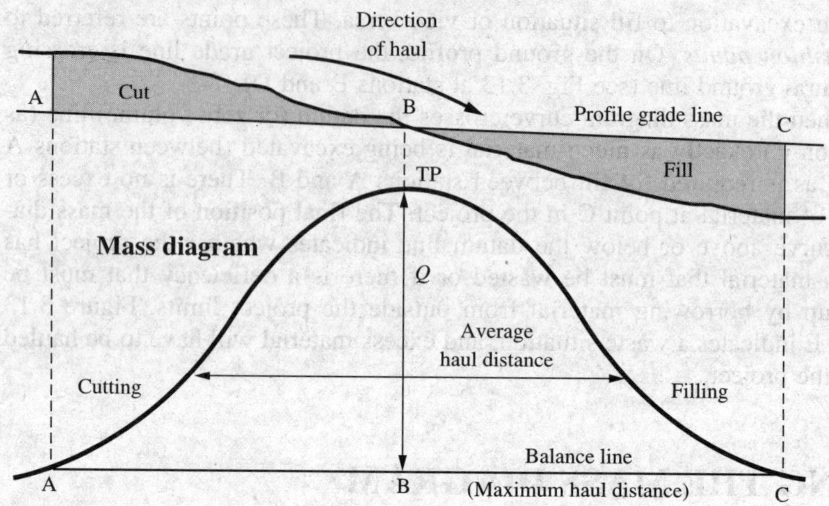

FIGURE 3.14 Mass diagram with a balance line.

the amount of material the scrapers will haul is dimensioned on the vertical scale depicted by the vertical line Q. By examining either the profile view or the mass diagram, it is easy to determine the direction of haul for cut material. The mass diagram establishes to which fill location each portion of the cut should be hauled. Note the arrow on the profile view Fig. 3.14.

In accomplishing the balanced earthwork operation between stations A and C, some of the hauls will be short, while some will approach the maximum haul distance. The *average haul distance* can be calculated by first computing the area enclosed by the balance line and the mass diagram curve and then dividing that area by the total quantity of material to be hauled (the maximum vertical distance between the balance line and the mass diagram curve).

If the curve is above the balance line, the direction of haul is from left to right, that is, up stationing. When the curve is below the balance line the haul is from right to left, that is, down stationing.

Because the lengths of balance lines on a mass diagram are equal to the maximum or minimum haul distances for the balanced earthmoving operation, they should be drawn to conform to the capabilities of the particular equipment that will be utilized. The equipment will therefore operate at haul distances that are within its range of efficiency. Figure 3.15 illustrates a portion of a mass diagram on which two balance lines have been drawn. In this situation, it is planned that dozers will be used to push the short-haul material. Using dozers, the excavation between stations C and D will be placed between stations D and E. Then there will be a scraper operation to excavate the material between stations A and C and haul it to fill between stations E and G.

Average Grade

When the mass diagram and the project profile are plotted one on top of the other, as shown in Figure 3.14 and 3.16, the average haul grades of the earth-

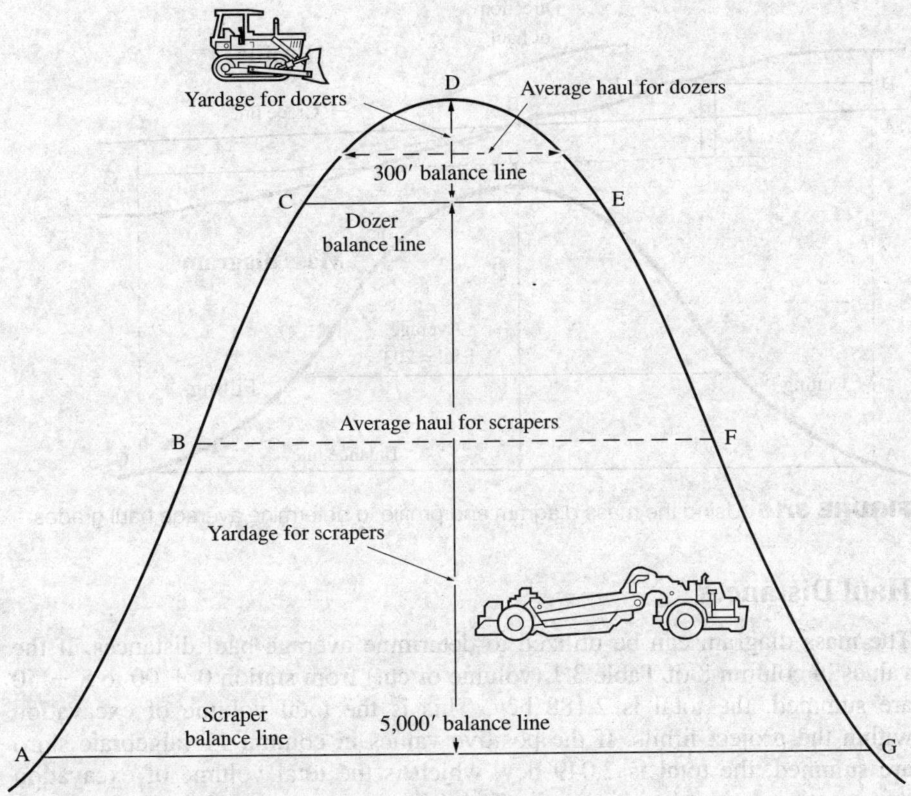

FIGURE 3.15 Two balance lines on a mass diagram.

moving operations can be approximated. On the profile view, draw a horizontal line that roughly divides the cut area in half in the vertical dimension (see Fig. 3.16). Do the same for the fill area. This is a division of only that portion of the cut or fill area defined by the balance line in question. The difference in elevation between these two lines provides the vertical distance to use in calculating the average grade for the haul involving the material in the balance. The average haul distance, as determined by construction of a horizontal line on the mass diagram, is the denominator in the grade calculation.

$$\text{Average grade percent} = \frac{\text{Change in elevation}}{\text{Average haul distance}} \times 100 \qquad [3.5]$$

EXAMPLE 3.3

Calculate the average grade for the balance haul shown in Fig. 3.16.

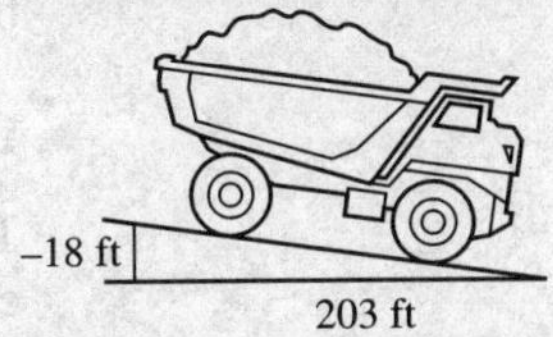

$$\text{Average grade going from the cut to the fill} = \frac{-18 \text{ ft}}{203 \text{ ft}} \times 100 = -8.9\%$$

The return trip will be up an 8.9% grade.

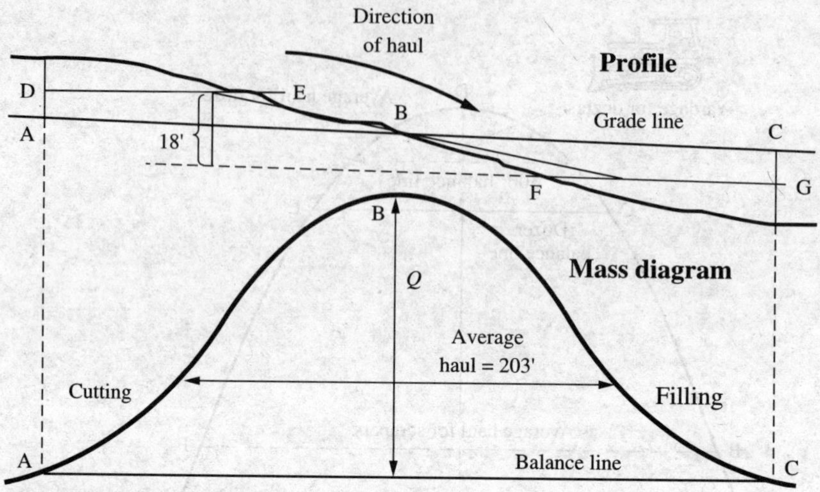

FIGURE 3.16 Using the mass diagram and profile to determine average haul grades.

Haul Distances

The mass diagram can be utilized to determine average haul distances. If the values in column 8 of Table 3.1 (volume of cut) from station 0 + 00 to 8 + 50 are summed, the total is 2,188 bcy. This is the total volume of excavation within the project limits. If the positive values in column 11 (algebraic sum) are summed, the total is 2,049 bcy, which is the total volume of excavation that must be moved longitudinally. The difference between these two values is the crosshaul for the project, 139 bcy. Many contractors treat the crosshaul as dozer work.

Reviewing the values in column 12 of Table 3.1, we see there is a low point or valley of −847 bcy at station 2 + 50 and a high point or crest of 1,202 bcy at 6 + 50. The sum of the absolute values of the peaks and low points equals the total excavation that must be moved longitudinally (847 + 1,202 = 2,049 bcy). The curve can have intermediate peaks and low points (see Fig. 3.17), and all must be accounted for when calculating the amount of material that must be moved longitudinally.

Reviewing the values in column 12 of Table 3.3, we see there is a low point or valley of −28,539 bcy at station 5 + 00, a high point or crest of −17,080 bcy at 8 + 00, and a second low point of −22,670 bcy at station 10 + 00. The sum of the absolute values of the peaks and low points equals the total excavation that must be moved longitudinally [(28,539 − 17,080) + (22,670 − 17,080) + (17,080 − 0) = 34,120 bcy)]. Haul 1 involves 11,459 bcy, haul 2 is 5,590 bcy, and haul 3 is 17,080 bcy.

Calculating Haul Distance

The average haul distance for any of the individual hauls can be determined by calculation. Dividing the area (in Fig. 3.17 the units are stations-cubic yards)

TABLE 3.3 Earthwork volume calculation sheet for the Mass Diagram in Fig. 3.17.

Station (1)	End-area cut (sf) (2)	End-area fill (sf) (3)	Volume of cut (bcy) (4)	Volume of fill (ccy) (5)	Stripping cut (bcy) (6)	Stripping fill (ccy) (7)	Total cut (bcy) (8)	Total fill (ccy) (9)	Adj. fill (bcy) (10)	Algebraic sum (bcy) (11)	Mass ordinate (12)
0 + 00	0	0		0					0.90		
1 + 00	0	1,700	0	3,148	0	120	0	3,268	3,631	−3,631	−3,631
2 + 00	0	3,100	0	8,889	0	120	0	9,009	10,010	−10,010	−13,641
3 + 00	0	1,500	0	8,519	0	120	0	8,639	9,598	−9,598	−23,240
4 + 00	60	600	111	3,889	60	80	51	3,969	4,410	−4,359	−27,598
5 + 00	400	200	852	1,481	80	60	772	1,541	1,713	−941	**−28,539** ←
6 + 00	1,300	30	3,148	426	110	10	3,038	436	484	2,554	−25,985
7 + 00	2,400	400	6,852	796	120	85	6,732	881	979	5,753	−20,223
8 + 00	800	850	5,926	2,315	90	100	5,836	2,415	2,683	3,153	**−17,080** ←
9 + 00	50	1,250	1,574	3,889	5	120	1,569	4,009	4,454	−2,885	−19,965
10 + 00	95	180	269	2,648	20	10	249	2,658	2,953	−2,705	**−22,670** ←
11 + 00	200	8	546	348	60	0	486	348	387	99	−22,571
12 + 00	560	0	1,407	15	65	0	1,342	15	16	1,326	−21,245
13 + 00	1,430	0	3,685	0	100	0	3,585	0	0	3,585	−17,660
14 + 00	3,580	0	9,278	0	120	0	9,158	0	0	9,158	−8,502
15 + 00	2,600	0	11,444	0	110	0	11,334	0	0	11,334	2,833

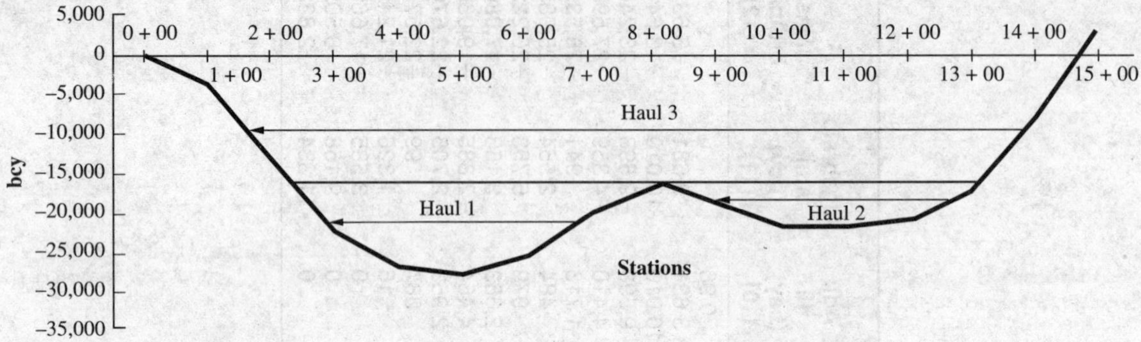

FIGURE 3.17 Mass diagram plotted from the data in Table 3.3.

that is enclosed by the balance line and the mass diagram curve by the amount of material hauled (in Fig. 3.17 the volume units are cubic yards) will yield the average haul distance (stations).

The area (stations-cubic yards, usually referred to as station-yards) that is enclosed by the balance line, the mass diagram curve, and the zero datum line that is haul 3 can be calculated using the trapezoidal formula [Eq. 3.3]. The vertical at h_0 (station 0 + 00) is 0 bcy, at h_1 is 3,631 bcy, at h_2 is 13,641 bcy, at h_3 through h_{13} is 17,080 bcy, at h_{14} is 8,502 bcy, and at h_{15} is 0 bcy. For stations 0 + 00 and 15 + 00, the beginning and ending stations in the equation (h_0 and h_n), the value is divided by 2. But 0 divided by 2 is still 0, so the area equals the sum of the individual verticals times the distance between verticals, which is 1 (one station). The sum of the verticals is 213,654 bcy, and multiplying by 1 sta., produces an area of 213,645 station-yards (sta.-yd).

The amount of material hauled is 17,080 bcy. Therefore, the sum of verticals' calculated average haul for haul 3 is 12.51 stations or 12,510 ft.

$$\left(\frac{213{,}654 \text{ station} - \text{yards}}{17{,}080 \text{ bcy}} = 12.51 \text{ stations}\right)$$

There is a slight error in the computation because the descending mass diagram curve actually (1) reaches −17,080 somewhere between stations 2 + 00 and 3 + 00, (2) the ascending curve reaches −17,080 at a point located between stations 13 + 00 and 14 + 00, and (3) reaches the zero balance line at a point between stations 14 + 00 and 15 + 00. There are no data to determine exactly where the curve reaches −17,080 or begins to ascend from that level, or where it reaches 0. If we assume a linear slope between the values at the given stations, the points can be calculated. In the case of the Table 3.3 data and the mass (Fig. 3.17) diagram, the calculated locations are (1) descending curve reaches −17,080 bcy at station 2 + 64.2, (2) begins to ascend from −17,080 bcy at station 13 + 06.3, and (3) reaches 0 bcy at station 14 + 75.0. When those points are used in the calculation, the area enclosed by the balance line, the mass diagram curve, and the zero datum line is −213,966 sta.-yd and the average haul is 12.53 sta. Table 3.4 presents a summary of the average haul values using both calculation approaches.

TABLE 3.4 Calculated haul distances.

Haul	Volume	Sum of verticals (sta.)	Calculated points (sta.)
1	11,459 bcy	3 + 51	3 + 41
2	5,590 bcy	3 + 35	3 + 30
3	17,080 bcy	12 + 51	12 + 53

Consolidated Average Hauls

Using the individual average hauls and the quantity associated with each, a project average haul can be calculated. The calculation process is similar to that used to calculate an average haul for the individual hauls. Consider the three hauls depicted in Fig. 3.17 and their sum of verticals' average haul distances. Haul 1 is 11,459 bcy with an average haul distance of 351 ft or 3 + 51 stations. Haul 2 is 5,590 bcy with an average haul of 335 ft, and haul 3 is 17,080 bcy with an average haul of 12,510 ft. By multiplying each haul quantity by its respective haul distance a station-yard value can be determined.

Haul 1	11,459 bcy	3 + 51 stations	40,221 sta.-yd
Haul 2	5,590 bcy	3 + 35 stations	18,727 sta.-yd
Haul 3	17,080 bcy	12 + 51 stations	213,671 sta.-yd
	34,129 bcy		272,619 sta.-yd

If the individual station-yard values are summed and that value divided by the total quantity moved, an average haul for the project is the result. In this case, the project average haul is 8.0 stations (272,619 sta.-yd/34,129 yd).

If all the hauls are of about the same length, the estimator can consolidate the production calculations by using an averaging process, as described. In the case of the data presented in Table 3.3 and Fig. 3.17, it is clear that there are two very distinct haul situations on this project. There are two short-haul sections and one long-haul section. These two situations will most likely each require a different number of haul units. The short-haul situation will have shorter haul times and therefore will require fewer units to achieve continuous production. The long haul will require more units. Assuming the haul grades are about the same, the estimator will not calculate production for every individual part of the mass diagram. This is also driven by the practical situation of not mobilizing and demobilizing various numbers of machines for small differences in hauls.

Considering the data in Table 3.3, the estimator would most likely develop two production scenarios. The first scenario would be for the short hauls of hauls 1 and 2.

Haul 1	11,459 bcy	3.51 stations	40,221 sta.-yd
Haul 2	5,590 bcy	3.35 stations	18,727 sta.-yd
	17,049 bcy		58,948 sta.-yd
Haul 3	17,080 bcy	12.5 stations	

The average haul that results from combining these two is 3.5 stations. The second scenario would be for the long-haul situation of haul 3.

There are other mass diagram applications that can be investigated. On some projects, it can prove economical to borrow material rather than undertake extremely long hauls. The same holds true for waste material. The mass diagram provides the engineer the ability to analyze such situations. An example mass diagram analysis including calculation of equipment production is provided at **W3.1**.

PRICING EARTHWORK OPERATIONS

The cost of earthwork operations will vary with the kind of soil or rock encountered and the methods used to excavate, haul, and place the material in its final deposition. It is usually not too difficult to compute the volume of earth or rock to be moved, but estimating the cost of actually performing the work depends on both a careful study of the project plans and a diligent site investigation. The site investigation should seek to identify the characteristics of the subsurface soils and rock that will be encountered.

The earthwork quantities and average movement distances can be determined using the techniques described in the Earthwork Quantities and Mass Diagram sections of this chapter. Proper equipment and estimated production rates are determined by (1) selecting an appropriate type of machine and (2) using machine performance data (as discussed in Chapter 6). The costs to be matched with the selected machines are derived as described in Chapter 2.

Spread Production

To accomplish a task, machines usually work together and are supported by auxiliary machines. To accomplish a loading, hauling, and compacting task would involve an excavator, several haul units, and auxiliary machines to distribute the material on the embankment (Fig. 3.18), and achieve compaction (Fig. 3.19).

spread
A group of construction machines that work together to accomplish a specific construction task such as excavating, hauling, and compacting material.

Such groups of equipment are referred to as an equipment **spread**. An excavator and a fleet of trucks can be thought of as a linked system, one link of which will control the spread production. If spreading and compaction of the hauled material is required, a two-linked system is created. Because the systems are linked, the capabilities of the individual components of the spread must be compatible in terms of overall production (Fig. 3.20). The number of machines and specific types of machines in a spread will vary with the proposed task.

The production capacity of the total system is dictated by the lesser of the production capacities of individual systems. Our objective is to predict the spread production rate (*linked-system* production rate) and the cost per unit of production. In the case of the Figure 3.17 data, the estimator would develop two production spreads. The first spread would be for the short 3.5-station haul and the second for the 12.5-station haul.

FIGURE 3.18 Spreading dumped embankment material with a dozer.

FIGURE 3.19 Water truck on the fill with the scrapers.

FIGURE 3.20 Two-link earthwork system.

> Always ensure that a consistent set of units is used when estimating spread production. If the mass diagram quantities are expressed in bank cubic yards, excavator capacity, hauling capacity, and compaction should all be converted to bcy. The units used in the estimate are usually chosen to match those used in the owner's bid documents.

SUMMARY

The goal of planning is to minimize resource expenditures required to successfully complete the project. Three kinds of views are presented in the contract documents to show earthwork construction features: (1) plan view, (2) profile view, and (3) cross-section view. Earthwork computations involve the calculation of earthwork volumes, the balancing of cuts and fills, and the planning of the most economical material hauls. Critical learning objectives that support earthwork planning include

- Ability to calculate earthwork volume.
- Ability to adjust quantities for stripping requirements.
- Ability to construct an earthwork calculation sheet.
- Ability to construct and understand a mass diagram.

These objectives are the basis for the problems that follow.

PROBLEMS

3.1 Using the average-end-area method, calculate the cut-and-fill volumes for stations 125 + 00 through 131 + 00.

Station	End-area cut (sf)	End-area fill (sf)	Volume of cut (bcy)	Volume of fill (ccy)
125 + 00	0	785	—	—
126 + 00	652	0		
127 + 00	2,150	0		
128 + 00	3,210	0		
129 + 00	1,255	147		
130 + 00	95	780		
131 + 00	0	3,666		

3.2 Using the average-end-area method, calculate the cut-and-fill volumes for stations 19 + 00 through 24 + 00.

Station	End-area cut (sf)	End-area fill (sf)	Volume of cut (bcy)	Volume of fill (ccy)
19 + 00	326	0	—	—
20 + 00	157	0		
21 + 00	44	0		
21 + 50	0	0		
22 + 00	0	147		
23 + 00	0	165		
24 + 00	0	133		

3.3 Using the average-end-area method, calculate the cut-and-fill volumes for stations 25 + 00 through 31 + 00.

Station	End-area cut (sf)	End-area fill (sf)	Volume of cut (bcy)	Volume of fill (ccy)
25 + 00	0	3,525	—	—
26 + 00	355	985		
27 + 00	786	125		
28 + 00	2,515	55		
29 + 00	1,255	23		
29 + 25	620	0		
29 + 50	25	845		
30 + 00	0	3,655		
31 + 00	0	8,560		

3.4 Complete the earthwork calculation sheet here and plot the resulting mass diagram. Divide ccy by 0.9 to convert to bcy.

Station	End-area cut (sf)	End-area fill (sf)	Vol. of cut (bcy)	Vol. of fill (ccy)	Stripping cut (bcy)	Stripping fill (ccy)	Total cut (bcy)	Total fill (ccy)	Adj. fill (bcy)	Algebraic sum (bcy)	Mass ordinate
									0.90		
0 + 00	0	0									
1 + 00	0	90			0	24					
2 + 00	0	154			0	40					
3 + 00	147	0			15	19					
4 + 00	192	0			50	0					
4 + 50	205	0			57	0					
5 + 00	179	0			21	0					
6 + 00	121	0			43	0					
6 + 50	100	0			19	0					
7 + 00	52	10			8	0					
8 + 00	0	180			10	20					
9 + 00	0	231			0	69					
10 + 00	0	285			0	18					

3.5 Complete the earthwork calculation sheet here and plot the resulting mass diagram. Divide ccy by 0.9 to convert to bcy. Calculate the area under the mass diagram curve. What quantity of material must be hauled along the length of the project? What is the average haul distance, in stations, for this material? (2,616,667 cy-ft; 4,190 cy; 6 + 24 stations)

Station	End-area cut (sf)	End-area fill (sf)	Vol. of cut (bcy)	Vol. of fill (ccy)	Stripping cut (bcy)	Stripping fill (ccy)	Total cut (bcy)	Total fill (ccy)	Adj. fill (bcy)	Algebraic sum (bcy)	Mass ordinate
									0.90		
0 + 00	0	0			0	0					
1 + 00	200	0			0	0					
2 + 00	310	0			0	0					
3 + 00	230	0			0	0					
4 + 00	230	0			0	0					
5 + 00	185	30			0	0					
6 + 00	115	86			0	0					
7 + 00	120	150			0	0					
8 + 00	20	250			0	0					
9 + 00	0	380			0	0					
10 + 00	0	340			0	0					
11 + 00	0	33			0	0					
12 + 00	0	00			0	0					

3.6 Complete the earthwork calculation sheet here and plot the resulting mass diagram. Divide ccy by 0.88 to convert to bcy. Calculate the area under the mass diagram curve. What quantity of material must be hauled along the length of the project? What is the average haul distance, in stations, for this material?

Station	End-area cut (sf)	End-area fill (sf)	Vol. of cut (bcy)	Vol. of fill (ccy)	Stripping cut (bcy)	Stripping fill (ccy)	Total cut (bcy)	Total fill (ccy)	Adj. fill (bcy)	Algebraic sum (bcy)	Mass ordinate
									0.88		
0 + 00	0	0			0	0					
1 + 00	0	95	0	0	0	0					
2 + 00	0	285	0	0	0	0					
3 + 00	3	270	0	0	0	0					
4 + 00	6	220	0	0	0	0					
5 + 00	12	180	30	0	0	0					
6 + 00	50	42	86	210	0	0					
7 + 00	120	0	150	437	0	0					
8 + 00	250	0	250	800	0	0					
9 + 00	290	0	380	1,167	0	0					
10 + 00	310	0	340	1,333	0	0					
11 + 00	200	0	33	691	0	0					
12 + 00	0	0	0	10	0	0					

3.7 Complete the earthwork calculation sheet here and plot the resulting mass diagram. Divide ccy by 0.9 to convert to bcy. Calculate the average haul (trapezoidal formula) for the balances on this project. Is this a waste or borrow project?

Station	End-area cut (sf)	End-area fill (sf)	Vol. of cut (bcy)	Vol. of fill (ccy)	Stripping cut (bcy)	Stripping fill (ccy)	Total cut (bcy)	Total fill (ccy)	Adj. fill (bcy)	Algebraic sum (bcy)	Mass ordinate
									0.90		
10 + 00	0	0									
11 + 00	580	0			80	0					
12 + 00	2,100	0			90	0					
13 + 00	4,650	0			100	0					
14 + 00	6,000	0			100	0					
15 + 00	3,250	560			80	60					
16 + 00	1,300	1,620			80	80					
17 + 00	700	2,450			80	85					
18 + 00	0	7,800			0	100					
19 + 00	0	3,620			0	90					
20 + 00	0	1,980			0	80					
21 + 00	0	1,310			0	80					
22 + 00	580	860			80	10					
23 + 00	1,620	250			100	10					
24 + 00	3,850	0			100	0					
25 + 00	2,600	0			100	0					

REFERENCES

1. *Construction Estimating & Bidding Theory Principles Process* (1999). Publication No. 3505, Associated General Contractors of America, 333 John Caryle Street, Suite 200, Alexandria, VA.
2. Ringwald, Richard C. (1993). *Means Heavy Construction Handbook*. R. S. Means Company, Inc., Kingston, MA.
3. *RSMeans Heavy Construction Cost Data*. R. S. Means Company, Inc., Kingston, MA (published annually).
4. Schexnayder, Cliff (2003). "Construction Forum," *Practice Periodical on Structural Design and Construction*, American Society of Civil Engineers, Reston, VA, Vol. 8, No. 2, May.
5. Smith, Francis E. (1976). "Earthwork Volumes by Contour Method," *Journal of the Construction Division*, American Society of Civil Engineers, New York, Vol. 102, CO1, March.

WEBSITE RESOURCES

1. www.usbr.gov/main/index.html Bureau of Reclamation, Construction Cost Trends. The Bureau of Reclamation's construction cost trends (CCT) were developed to track construction relevant to the primary types of projects being constructed by the organization.
2. www.hcss.com Heavy Construction Systems Specialists Inc. (HCSS). HCSS specializes in construction estimating, bidding, and job cost software for contractors.
3. www.trimble.com Trimble is a leading innovator of global positioning system (GPS) technology.
4. www.trakware1.com TRAKWARE provides excavation-estimating software for the construction industry.
5. www.agtek.com AGTEK Development Company, Inc. supplies construction computer products including earthwork quantity take-off and graphical grade positioning systems.

CHAPTER 4

Soil and Rock

Knowledge of the properties, characteristics, and behavior of different soil types and aggregates is important to both design and construction. The constructor is interested in how the material will handle during the construction process. Density is the most commonly used parameter for specifying construction operations because there is a direct correlation between a soil's properties and its density. The effectiveness of different compaction methods is dependent on the individual soil type being manipulated.

INTRODUCTION

Soil and rock are the principal components of many construction projects. They are used to support structures (static load); to support pavements for highways and airport runways (dynamic loads); and in dams and levees, as impoundments, to resist the passage of water. Most soils must be excavated, processed, and compacted to meet the engineering requirements of a project. Additionally, either natural aggregate deposits or quarried rock constitute approximately 95% by weight of asphalt concrete and 75% of Portland cement concrete.

Knowledge of the properties, characteristics, and behavior of different soil and rock types and crushed rock—aggregates—is important to those persons who are associated with the design or construction of projects involving these materials. Some soil and rock types are suitable for structural purposes in their natural state. In many cases, however, the locally available materials do not meet engineering requirements, and thus it is necessary to cost-effectively modify them to meet project demands. Processing can be as simple as adjusting the moisture content or mixing and blending. Because there is a direct relationship between increased density and increased strength and bearing capacity, the engineering properties of many soils can be improved simply by compaction.

GLOSSARY OF TERMS

The following glossary defines important terms that are used in discussing soil and rock materials, and compaction.

Aggregate, coarse. Crushed rock or gravel, generally greater than $\frac{1}{4}$ in. in size.

Aggregate, fine. The sand or fine-crushed stone used for filling voids in coarse aggregate. Generally less than $\frac{1}{4}$ in. and greater than a No. 200 sieve in size.

Cohesion. The quality of some soil particles to be attracted to like particles, manifested in a tendency to stick together, as in clay.

Cohesive materials. A soil having properties of cohesion.

Grain-size curve. A graph showing the percentage of soil sizes by weight contained in a sample.

Optimum moisture content. The water content, for a given compactive effort, at which the greatest density of a soil can be obtained.

Pavement. A layer, above the base, of rigid surfacing material that provides high bending resistance and distributes loads to the base. Pavements are usually constructed of asphalt or concrete.

Plasticity. The capability of being molded. Plastic materials do not assume their original shape after the force causing deformation is removed.

Rock. The hard, mineral matter of the earth's crust, occurring in masses and often requiring blasting to cause breakage before excavation can be accomplished.

Shrinkage. A soil volume reduction usually occurring in fine-grained soils when they are subjected to moisture.

Soil. The loose surface material of the earth's crust, created naturally from the disintegration of rocks or decay of vegetation. Soil can be excavated easily using power equipment in the field.

SOIL AND ROCK PROPERTIES

Before discussing earth- and rock-handling techniques or analyzing problems involving these materials, it is necessary to first become familiar with some of their physical properties. These properties have a direct effect on the ease or difficulty of handling the material, on the selection of equipment, and on equipment production rates.

Types of Geotechnical Materials

Steel and concrete are construction materials that are basically homogeneous and uniform in composition. As such, their behavior can be predicted. Soil and rock are just the opposite. By nature they are heterogeneous. In their natural state, they are rarely uniform and work processes are developed by comparison to a similar type material with which previous experience has been gained. To accomplish this, soil and rock types must be classified. Soils can be classified according to the sizes of the particles of which they are composed, by their physical properties, or by their behavior when the moisture content varies.

A constructor is concerned primarily with five types of soils: gravel, sand, silt, clay, and organic matter, or with combinations of these types. Different agencies and specification groups denote the sizes of these types of soil differently, causing some confusion. The following size limits represent those set forth by the American Society for Testing and Materials (ASTM):

Gravel is rounded or semiround particles of rock that will pass a 3-in. and be retained on a 2.0-mm No. 10 sieve. Sizes larger than 10 in. are usually called boulders.

Sand is disintegrated rock whose particles vary in size from the lower limit of gravel 2.0 mm down to 0.074 mm (No. 200 sieve). It can be classified as coarse or fine sand, depending on the sizes of the grains. Sand is a granular noncohesive material whose particles have a bulky shape.

Silt is a material finer than sand, and thus its particles are smaller than 0.074 mm but larger than 0.005 mm. It is a noncohesive material and it has little or no strength. Silt compacts very poorly.

Clay is a cohesive material whose particles are less than 0.005 mm. The cohesion between the particles gives clays a high strength when air-dried. Clays can be subject to considerable changes in volume with variations in moisture content. They will exhibit plasticity within a range of "water contents." Clay particles are shaped like thin wafers.

Organic matter is partly decomposed vegetation. It has a spongy, unstable structure that will continue to decompose and is chemically reactive. If present in soil that is used for construction purposes, organic matter should be removed and replaced with a more suitable soil.

Generally, the soil types are found in nature in some mixed proportions. Table 4.1 presents a classification system based on combinations of soil types.

Soils existing under natural conditions may not contain the relative amounts of desired material types necessary to produce the properties required for construction purposes. For this reason, it may be necessary to obtain soils from several sources and then blend them for use in a fill.

borrow pit
A pit from which fill material is mined.

If the material in a **borrow pit** consists of layers of different types of soils, the specifications for the project may require the use of equipment that will excavate vertically through the layers in order to mix the soil.

Rock was formed by one of three different means:

Igneous rocks solidified from molten masses.

Sedimentary rocks formed in layers, settling out of water solutions.

Metamorphic rocks were transformed from material of the first two by heat and pressure.

Their respective formation processes will affect how rocks can be excavated and handled.

TABLE 4.1 Unified soil classification system.

Symbol	Primary	Secondary	Supplementary
GW	Coarse-grained soils	Well-graded gravels, gravel-sand mixtures, little or no fines	Wide range of grain size
GP	Coarse-grained soils	Poorly graded gravels, gravel-sand mixtures, little or no fines	Predominantly one size or a range of intermediate sizes missing
GM	Gravel mixed with fines	Silty gravels and gravel-sand-silt mixtures—may be poorly graded	Predominantly one size or a range of intermediate sizes missing
GC	Gravel mixed with fines	Clayey gravels, gravel-sand-clay mixtures, which may be poorly graded	Plastic fines
SW	Clean sands	Well-graded sands, gravelly sands, little or no fines	Wide range in grain sizes
SP	Clean sands	Poorly graded sands, gravelly sands, little or no fines	Predominantly one size or a range of sizes with some intermediate sizes missing
SM	Sands with fines	Silty sands and sand-silt mixtures, which may be poorly graded	Nonplastic fines or fines of low plasticity
SC	Sands with fines	Clayey sands, sand-clay mixtures, which may be poorly graded	Plastic fines
ML	Fine-grained soils	Inorganic silts, clayey silts, rock flour, silty very fine sands	Plastic fines
CL	Fine-grained soils	Inorganic clays of low to medium plasticity, silty sandy or gravelly clays	Plastic fines
OL	Fine-grained soils	Organic silts and organic silt-clay of low plasticity	
MH	Fine-grained soils	Inorganic silts, clayey silts, elastic silts	
CH	Fine-grained soils	Inorganic clays of high plasticity, fat clays	
OH	Fine-grained soils	Organic clays and silty clays of medium to high plasticity	

Symbol classification

COARSE-GRAINED MATERIAL Symbol	Subdivision
G—Gravel grain size from 3" to No. 4 sieve size	W—Well graded, little or no fines
S—Sand grain size from No. 4 to 200 sieve size	P—Poorly graded, little or no fines
	M—Concentration of silty or nonplastic fines
	C—Concentration of clay or plastic fines

FINE-GRAINED MATERIAL Symbol	Subdivision
M—Silt very fine grain size, floury appearance	L—Low plastic material, lean soil
C—Clay finest grain size, high dry strength—plastic	H—High plastic material, fat soil
O—Organic matter partly decomposed, appears fibrous, spongy and dark in color	

Categorization of Materials

In contract documents, excavation is typically categorized as common, rock, muck, or unclassified. *Common* refers to ordinary earth excavation, while the term *unclassified* reflects the lack of clear distinction between soil and rock. The removal of common excavation will not require the use of explosives, although tractors equipped with rippers may be used to loosen consolidated formations. The specific engineering properties of the soil—plasticity, grain size distribution, and so on—will influence the selection of the appropriate equipment and construction methods.

In construction, *rock* is a material that cannot be removed by ordinary earth-handling equipment.* Drilling and blasting or some comparable method must be used to remove rock. This normally results in considerably greater expense than earth excavation. Rock excavation involves the study of the rock type, faulting, dip and strike, and explosive characteristics as the basis for selecting material removal and aggregate production equipment.

Muck includes materials that will decay or produce subsidence in embankments. It is usually a soft organic material having a high water content. Typically, it would include such things as decaying stumps, roots, logs, and humus. These materials are hard to handle and can present special construction problems both at their point of excavation, and in transportation and disposal.

> You should never price an earth- or rock-handling project without first making a thorough study of the materials.

In many cases, the contract documents include geotechnical information. This owner-furnished information provides a starting point for your *independent* investigation. Other good sources of preliminary information are topographic maps, agriculture maps, geologic maps, well logs, and aerial photographs. The investigation is not complete, however, until you make an on-site visit and conduct either drilling or test pit exploration. In the case of a project involving rock, it is often necessary to have private seismic surveys performed. For one large dam project in California, 13 private seismic studies were made of the spillway rock excavation for the bidding contractors.

Even though many owners provide good geotechnical data that has been put together by qualified engineers, the design engineer's primary concern is with how well the material will perform structurally. The constructor is interested in how the material will handle during the construction process and what volume or quantity of material is to be processed to yield the desired final structure.

Soil Weight-Volume Relationships

The primary relationships (see Fig. 4.1) are expressed by Eqs. [4.1] through [4.6].

*Note that this definition of rock will be affected by equipment developments. Larger and heavier machines are continually changing the limits of this definition for rock.

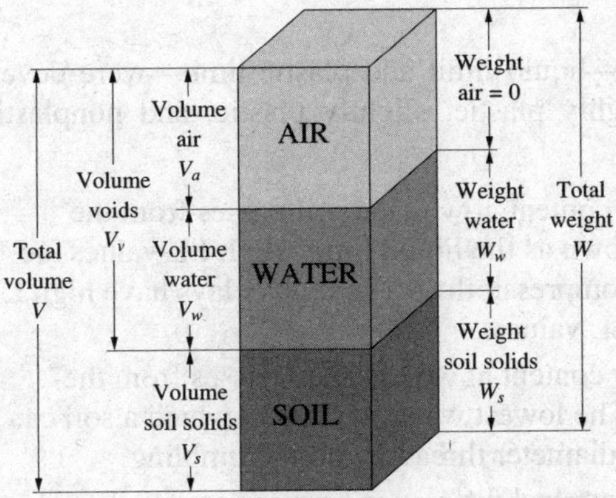

FIGURE 4.1 Soil mass weight and volume relationships.

$$\text{Unit weight } (\gamma) = \frac{\text{total weight of soil}}{\text{total soil volume}} = \frac{W}{V} \qquad [4.1]$$

$$\text{Dry unit weight } (\gamma_d) = \frac{\text{weight of soil solids}}{\text{total soil volume}} = \frac{W_s}{V} \qquad [4.2]$$

$$\text{Water content } (\omega) = \frac{\text{weight of water in soil}}{\text{weight of soil solids}} = \frac{W_w}{W_s} \qquad [4.3]$$

$$\text{Void ratio } (e) = \frac{\text{volume of voids}}{\text{volume of soil solids}} = \frac{V_v}{V_s} \qquad [4.4]$$

$$\text{Porosity } (n) = \frac{\text{volume of voids}}{\text{total soil volume}} = \frac{V_v}{V} \qquad [4.5]$$

$$\text{Specific gravity (SG)} = \frac{\text{weight of soil solids/volume of solids}}{\text{unit weight of water}} = \frac{W_s/V_s}{\gamma_w} \qquad [4.6]$$

Many more formulas can be derived from these basic relationships. Two such formulas, useful in analyzing compaction specifications, are

$$\text{Total soil volume } (V) = \text{ volume voids } (V_v) + \text{ volume solids } (V_s) \qquad [4.7]$$

$$\text{Weight of solids } (W_s) = \frac{\text{weight of soil } (W)}{1 + \text{ water content}(\omega)} \qquad [4.8]$$

If unit weights are known, which is the usual case, then Eq. [4.8] becomes

$$\gamma_d = \frac{\gamma}{1 + \omega} \qquad [4.9]$$

Soil Limits

Certain limits of soil consistency—liquid limit and plastic limit—were developed to differentiate between highly plastic, slightly plastic, and nonplastic materials.

Liquid limit (LL). The water content at which a soil passes from the plastic to the liquid state is known as the liquid limit. High LL values are associated with soils of high compressibility. Typically, clays have high LL values; sandy soils have low LL values.

Plastic limit (PL). The water content at which a soil passes from the plastic to the semisolid state. The lowest water content at which a soil can be rolled into a $\frac{1}{8}$ in. (3.2 mm) diameter thread without crumbling.

Plasticity index (PI). The numerical difference between a soil's liquid limit and its plastic limit is its plasticity index (PI = LL − PL). Soils having high PI values are quite compressible and have high cohesion.

On many projects, the specifications will specify a certain material gradation, a maximum LL and a maximum PI. The American Association of State Highway and Transportation Officials (AASHTO) system of soil classification, which is the most widely used for highway construction, illustrates this point (see Table 4.2).

Volumetric Measure

For bulk materials, volumetric measure varies with the material's position in the construction process (see Fig. 4.2). The same weight of a material will occupy different volumes as the material is handled on the project. In general, most cohesive soils will shrink 10 to 30% from bank to compacted state. Solid rock will swell 20 to 40% from bank to placement in an embankment. Between the bank and loose states, cohesive soils swell about 40% and solid rock swells as much as 65%.

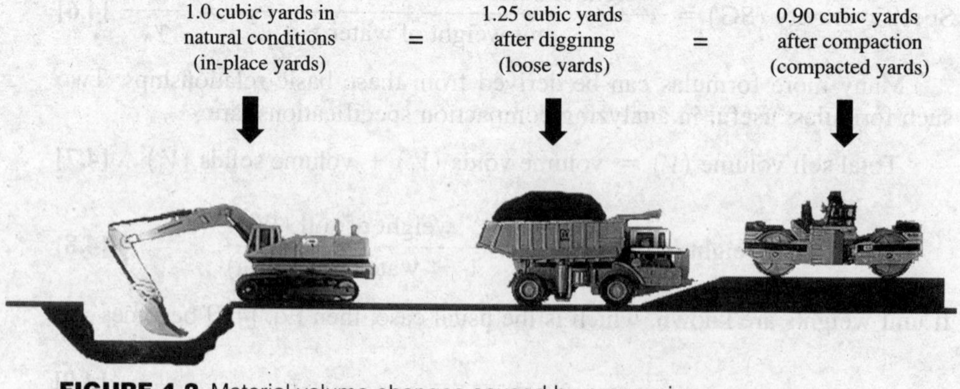

FIGURE 4.2 Material volume changes caused by processing.

TABLE 4.2 AASHTO soil classification system.*

General classification	Granular materials (35% or less of total sample passing No. 200)							Silt-clay materials (more than 35% of total sample passing No. 200)			
	A-1		**A-3**	**A-2**				**A-4**	**A-5**	**A-6**	**A-7**
Group classification	A-1-a	A-1-b		A-2-4	A-2-5	A-2-6	A-2-7				A-7-5, A-7-6
Sieve analysis, percentage passing											
No. 10	50 max.										
No. 40	30 max.	50 max.	51 min.								
No. 200	15 max.	25 max.	10 max.	35 max.	35 max.	35 max.	35 max.	36 min.	36 min.	36 min.	36 min.
Characteristics of fraction passing No. 40											
Liquid limit				40 max.	41 min.	40 max.	41 min.	40 max.	41 min.	40 max.	41 min.
Plasticity index	6 max.		NP	10 max.	10 max.	11 min.	11 min.	10 max.	10 max.	11 min.	11 min.
Group index	0		0	0		4 max.		8 max.	12 max.	16 max.	20 max.

*A group index based on a formula that considers particle size, LL, and PI is given at the bottom of the table. The group index indicates the suitability of a given soil for embankment construction. A group index number of "0" indicates a good material while an index of "20" indicates a poor material.

Soil volume is measured in one of three states:

Bank cubic yard

1 cubic yard (cy) of material as it lies in the *natural* state, bcy

Loose cubic yard

1 cy of material after it has been disturbed by a loading process, lcy

Compacted cubic yard

1 cy of material in the compacted state, also referred to as a net in-place cubic yard, ccy

In planning or estimating a job, the engineer must use a consistent volumetric measure in any set of calculations. The necessary consistency of units is achieved by use of shrinkage and swell factors. The *shrinkage factor* is the ratio of the compacted dry weight per unit volume to the bank dry weight per unit volume:

$$\text{Shrinkage factor} = \frac{\text{compacted dry unit weight}}{\text{bank dry unit weight}} \qquad \textbf{[4.10]}$$

The weight shrinkage due to compacting a fill can be expressed as a percent of the original bank measure weight:

$$\begin{array}{c}\text{Shrinkage}\\\text{percent}\end{array} = \frac{(\text{compacted dry unit weight}) - (\text{bank unit weight})}{\text{compacted unit weight}} \times 100 \qquad \textbf{[4.11]}$$

The *swell factor* is the ratio of the loose dry weight per unit volume to the bank dry weight per unit volume:

$$\text{Swell factor} = \frac{\text{loose dry unit weight}}{\text{bank dry unit weight}} \qquad \textbf{[4.12]}$$

The percent swell, expressed on a gravimetric basis, is

$$\text{Swell percent} = \left(\frac{\text{bank dry unit weight}}{\text{loose dry unit weight}} - 1\right) \times 100 \qquad \textbf{[4.13]}$$

TABLE 4.3 Representative properties of earth and rock.

Material	Bank weight lb/cy	Bank weight kg/m³	Loose weight lb/cy	Loose weight kg/m³	Percent swell	Swell factor*
Clay, dry	2,700	1,600	2,000	1,185	35	0.74
Clay, wet	3,000	1,780	2,200	1,305	35	0.74
Earth, dry	2,800	1,660	2,240	1,325	25	0.80
Earth, wet	3,200	1,895	2,580	1,528	25	0.80
Earth and gravel	3,200	1,895	2,600	1,575	20	0.83
Gravel, dry	2,800	1,660	2,490	1,475	12	0.89
Gravel, wet	3,400	2,020	2,980	1,765	14	0.88
Limestone	4,400	2,610	2,750	1,630	60	0.63
Rock, well blasted	4,200	2,490	2,640	1,565	60	0.63
Sand, dry	2,600	1,542	2,260	1,340	15	0.87
Sand, wet	2,700	1,600	2,360	1,400	15	0.87
Shale	3,500	2,075	2,480	1,470	40	0.71

*The swell factor is equal to the loose weight divided by the bank weight per unit volume.

Table 4.3 gives representative swell values for different classes of earth. These values will vary with the extent of loosening and compaction. If more accurate values are desired for a specific project, tests should be made on several samples of the earth taken from different depths and different locations within the proposed cut. The test can be made by weighing a given volume of undisturbed, loose, and compacted earth.

EXAMPLE 4.1

An earth fill, when completed, will occupy a net volume of 187,000 cy. The borrow material that will be used to construct this fill is a stiff clay. In its "bank" condition, the borrow material has a wet unit weight of 129 lb per cubic foot (cf) (γ), a water content (ω%) of 16.5%, and an in-place void ratio (e) of 0.620. The fill will be constructed in layers of 8-in. depth, loose measure, and compacted to a dry unit weight (γ_d) of 114 lb per cf at a moisture content of 18.3%. Compute the required volume of borrow pit excavation.

$$\text{Borrow } \gamma_d = \frac{129}{1 + 0.165} = 111 \text{ lb/cf}$$

$$\text{Fill } \gamma_d = 114 \text{ lb/cf}$$

$$\overbrace{187{,}000 \text{ cy} \times \frac{27 \text{ cf}}{\text{cy}} \times \frac{114 \text{ lb}}{\text{cf}}}^{\text{Fill}} = \overbrace{\chi \times \frac{27 \text{ cf}}{\text{cy}} \times \frac{111 \text{ lb}}{\text{cf}}}^{\text{Borrow}}$$

$$187{,}000 \text{ cy} \times \frac{114}{111} = 192{,}054 \text{ cy, } \textit{borrow required}$$

Note that the element 114/111 is the shrinkage factor 1.03.

The key to solving this type of problem is unit weight of the solid particles (dry weight) that make up the soil mass. In the construction process, the specifications may demand that water be either expelled or added to the soil mass. In Example 4.1, the contractor would be required to add water to the borrow to increase the moisture content from 16.5 to 18.3%. Adjusting for the extra borrow cubic yards required to make one fill cubic yard, note the water difference:

Fill		Borrow
$\gamma = 114 \times 1.183 =$	135 lb/cf	129 lb/cf
$\gamma_d =$	114 lb/cf	− 111 lb/cf
Water	21 lb/cf	18 lb/cf
		× 1.03 (shrinkage factor)
		19 lb/cf

To achieve the desired fill density and water content, the contractor will have to add water. This water must be hauled in by water wagon and is not part of the in-place borrow unit weight. The quantity of water that must be added is

Fill

Water content = 0.183

$$187{,}000 \text{ cy} \times \frac{114 \text{ lb}}{\text{cf}} \times 27\,\frac{\text{cf}}{\text{cy}} \times 0.183 = 105{,}332{,}238 \text{ lb} \quad \text{water}$$

Borrow

Water content = 0.165

$$192{,}054 \text{ cy} \times \frac{111 \text{ lb}}{\text{cf}} \times 27\,\frac{\text{cf}}{\text{cy}} \times 0.165 = 94{,}971{,}663 \text{ lb}$$

$$10{,}360{,}575 \text{ lb} \quad \text{water}$$

which is 1,241,941 gal or approximately 6.5 gal per cy of borrow.

The method of soil preparation prior to compaction is not sufficiently appreciated as an important factor influencing achievement of successful results. This includes adding water or conversely drying the soil. The blending of the excavation material to achieve a homogeneous composition and uniform water content within a placed layer is especially important.

Constructors commonly apply what is referred to as a *swell factor* when estimating jobs. This rule-of-thumb factor should not be confused with the previously defined factors. The term swell factor is used in this case because of how the number is applied. The embankment yardage of the job is multiplied by this factor; that is, it is swelled in order to put it in the same reference units as the borrow. The job is then figured in *borrow yards*. This swell factor is strictly a guess—based on past experience with similar materials. It may, also, reflect consideration of the project design. A case in point would be when the embankment is less than 3 ft in total height, in which case more embankment material will be required to compensate for the compaction of the natural ground below the fill. The constructor would therefore apply a higher swell factor when calculating the required borrow material for fills of minimum height.

COMPACTION SPECIFICATION AND CONTROL

INTRODUCTION

Prior to preparing the specifications for a project, representative soil samples are usually collected and tested in the laboratory to determine material properties. Normal testing would include grain-size analysis, because the size of the grains and the distribution of those sizes are important properties that affect a soil's suitability.

Maximum Dry Density/Optimum Moisture

Another critical test is the laboratory-established compaction curve. From such a curve, the maximum dry unit weight (density) and the percentage of water required to achieve maximum density can be determined. This percentage of water, which corresponds to the maximum dry density (for a given compactive effort), is known as the *optimum water content*. It is the amount of water required for a given soil to reach maximum density.

Figure 4.3 shows two compaction curves based on different input energy levels. The curves are plotted in dry weight (in pounds per cubic foot) against water content (percent by dry weight). Each illustrates the effect of varying amounts of moisture on the density of a soil subjected to given compactive effort (energy input level). The two energy levels depicted are known as standard and modified Proctor tests. Note that the modified Proctor (higher input energy) gives a higher density at a lower water content than the standard Proctor. For the material depicted by the curves in Fig. 4.3, the optimum moisture for the standard Proctor is 16% versus 12% for the modified Proctor.

The difference in optimum water content is a result of mechanical energy replacing the lubricating action of the water during the densification process. The contractor working to a modified Proctor specification (higher input energy) will have to plan to either make more passes with the compaction equipment or use heavier compaction equipment on the project. But at the same time there will be less of a requirement to haul water and to mix water into the material.

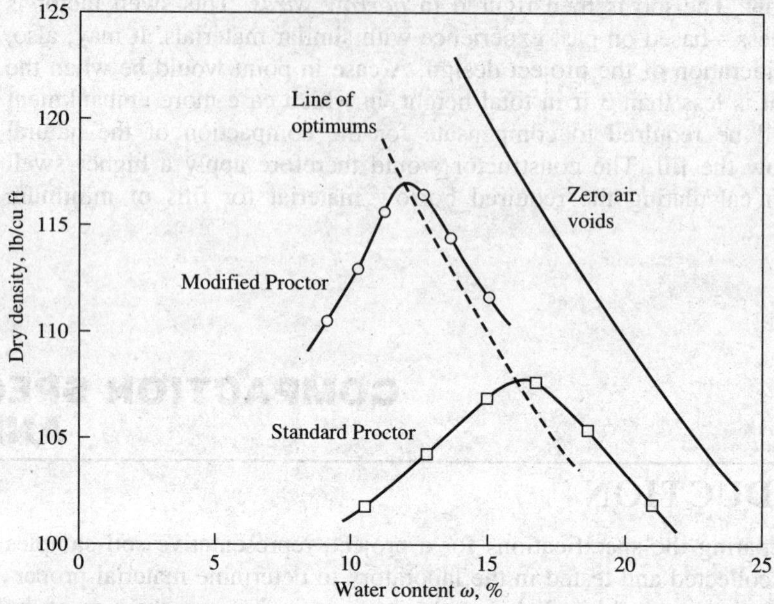

FIGURE 4.3 Standard and modified compaction curves.

COMPACTION TESTS

Proctor test
A method developed by R. R. Proctor for determining the moisture-density relationship in soils subjected to compaction.

The laboratory compaction test that is accepted by highway departments and other agencies is the **Proctor test**. For this test, a sample of soil consisting of $\frac{1}{4}$ in. and finer material is used. The sample is placed in a steel mold in three equal layers. The cylindrical steel mold has an inside diameter of 4.0 in. and a height of 4.59 in. In the standard test, each of three equal layers are compacted by dropping a 5.5-lb rammer, with a 2-in. circular base, 25 times from a height of 12 in. above the specimen (see Fig. 4.4). The specimen is removed from the mold and the entire specimen is immediately weighed. Then a sample of the specimen is taken and weighed. That sample is dried to a constant weight to remove all moisture and weighed again so that the water content can be determined. With the water content information, the dry weight of the specimen can be determined. The test is repeated, using varying water content specimens, until the water content that produces the maximum density is determined. This test is designated as ASTM D-698 or AASHTO T 99.

The modified Proctor test, designated as ASTM D-1557 or AASHTO T 180, is performed in a similar manner, except the applied energy is approximately five times greater because a 10-lb rammer is dropped 18 in. on each of five equal layers (see Fig. 4.4).

Compaction Control

The specifications for a project may require a contractor to compact the soil to 100% of a relative density, based on the standard Proctor test or a laboratory

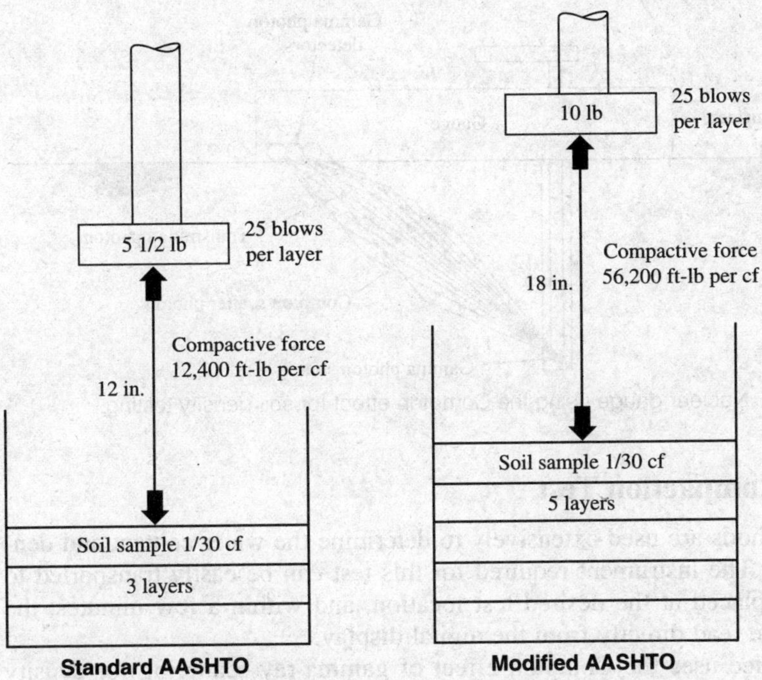

FIGURE 4.4 Standard and modified compaction test.

test at some other energy level. If the maximum laboratory dry density of the soil is determined to be 120 lb per cf, the contractor must compact the soil in the field to a density of 120 lb per cf.

Field verification tests of achieved compaction can be conducted by any of several accepted methods: sandcone, balloon, or nuclear. The first two methods are destructive tests. They involve

- Excavating a hole in the compacted fill and weighing the excavated material,
- Determining the water content of the excavated material,
- Measuring the volume of the resulting hole by use of the sandcone or a water-filled balloon, and
- Computing the density based on the obtained total weight of the excavated material and hole volume.

The dry density conversion can be made because the water content is known. Difficulties associated with such methods are (1) they are too time-consuming to conduct sufficient tests for statistical analysis, (2) there are problems with oversized particles, and (3) there is a time delay in determining the water content. Because the tests are usually conducted on each placement **lift**, delays in testing and acceptance can delay construction operations.

lift
A layer of soil placed on top of previously placed embankment material. The term can be used in reference to material as spread or as compacted.

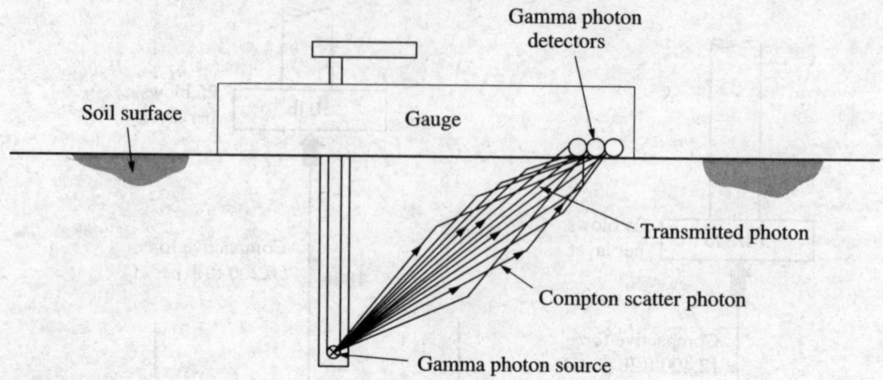

FIGURE 4.5 Nuclear gauge using the Compton effect for soil density testing.

Nuclear Compaction Test

Nuclear methods are used extensively to determine the water content and density of soils. The instrument required for this test can be easily transported to the fill and placed at the desired test location, and within a few minutes, the results can be read directly from the digital display.

The device uses the Compton effect of gamma-ray scattering for density determinations (Fig. 4.5) and hydrogenous thermalization of fast neutrons for moisture determinations. The emitted rays enter the ground, where they are partially absorbed and partially reflected. Reflected rays pass through Geiger-Müller tubes in the surface gauge. Counts per minute are read directly on a reflected-ray-counter gauge and are related to moisture and density calibration curves.

Advantages of the nuclear method when compared with other methods include the fact that it

1. Decreases the time required for a test from as much as a day to a few minutes, thereby eliminating potentially excessive construction delays, and because more samples can be taken per unit of time, the engineer is better able to characterize the achieved density.
2. Is nondestructive in that it does not require the removal of soil samples from the site of the tests.
3. Provides a means of performing density tests on soils containing large-sized aggregates.
4. Reduces or eliminates, when properly calibrated and used correctly, the effect of the personal element, and possible errors. Erratic results can be easily and quickly rechecked.

Because nuclear tests are conducted with instruments that present a potential source of radiation, an operator needs to be certified and should exercise reasonable care to ensure that no harm can result from the use of the instruments. By following the instructions furnished with the instruments and by exercising proper care, exposure can be kept well below the limits set by the Nuclear Regulatory Commission (NRC). In the United States and most individual states, a license is required to own, possess, or use nuclear-type instruments.

GeoGauge

Another nondestructive device that does not require the removal of soil samples from the site of the tests is the GeoGauge. This device is new to the field. In 1994 the Minnesota Department of Transportation tested the first prototype models in a program sponsored by the Federal Highway Administration (FHWA). Today production models are available and each year more agencies are conducting independent field evaluations [8].

The GeoGauge is a portable instrument that provides a simple, rapid, and precise means of directly measuring lift stiffness and soil modulus. It also provides an alternative means of measuring soil density. This device imparts very small displacements to the soil ($< 1.27 \times 10^{-6}$ m or < 0.00005 in.) at 25 steady state frequencies between 100 and 196 **hertz** (Hz). Stiffness is determined at each frequency, and the average is displayed. The entire process takes about 1 min. If a Poisson's ratio is assumed and the gauge's physical dimensions are known, the shear and Young's modulus of the soil can be derived. The gauges weigh about 10 kg ($\sim$22 lb) and rest on the soil surface on a ring-shaped footing.

hertz
The number of complete cycles of the vibrating mechanism per second.

Laboratory versus Field

Maximum dry density is only a maxim for a specific compaction effort (input energy level) and method by which that effort is applied. If more energy is applied in the field, a density greater than 100% of the laboratory value can be achieved. Dissimilar materials have individual curves and maximum values for the same input energy (Fig. 4.6). Well-graded sands have a higher dry density than uniform soils. As plasticity increases, the dry density of clay soils decreases.

It should be noted that at some point, higher moisture contents result in decreased density. This is because initially the water serves to "lubricate" the soil grains and helps the mechanical compaction operation move them into a compact physical arrangement. But the density of water is less than that of the soil solids, and at water contents above optimum, water is replacing soil grains in the matrix. If compaction is attempted at a water content that is much above optimum, no amount of effort will overcome these physical facts. Under such conditions, extra compactive effort will be wasted work. In fact, soils can be "overcompacted." Shear planes are established and there is a large reduction in strength.

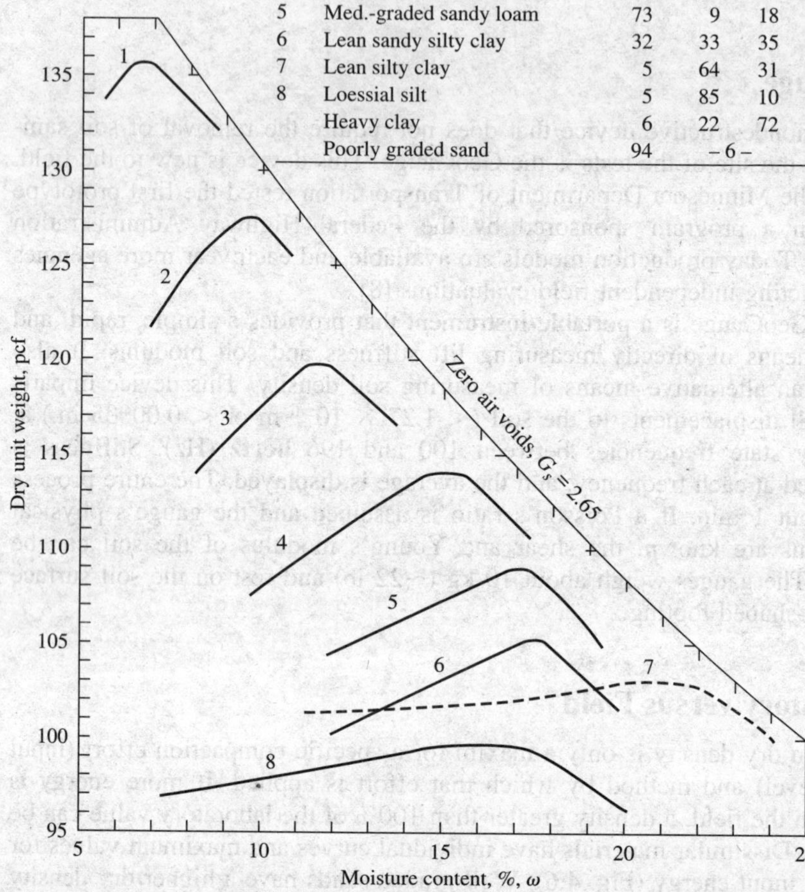

SOIL TEXTURE AND PLASTICITY DATA

No.	Description	Sand	Silt	Clay	LL	PI
3	Well-graded loamy sand	88	10	2	16	NP
4	Well-graded sandy loam	78	15	13	16	NP
5	Med.-graded sandy loam	73	9	18	22	4
6	Lean sandy silty clay	32	33	35	28	9
7	Lean silty clay	5	64	31	36	15
8	Loessial silt	5	85	10	26	2
	Heavy clay	6	22	72	67	40
	Poorly graded sand	94	– 6 –		NP	—

FIGURE 4.6 Compaction curves for eight soils compacted according to AASHTO T99.
Source: The Highway Research Board.

SOIL PROCESSING

The optimum water content for compaction varies from about 12 to 25% for fine-grained soils and from 7 to 12% for well-graded granular soils. Because it is difficult to attain and maintain the exact optimum water content, normal practice is to work within an acceptable moisture range. This range, which is usually ±2% of optimum, is based on attaining the maximum density with the minimum compactive effort.

FIGURE 4.7 Adding water prior to excavating the soil at the pit.

Adding Water to Soil

If the moisture content of a soil is below the optimum moisture range, water must be added to the soil prior to compaction. When it is necessary to add water consider the

- Amount of water required.
- Rate of water application.
- Method of application.
- Effects of the climate and weather.

Water can be added to the soil at the borrow pit (Fig. 4.7) or in-place (at the construction site). When processing **granular materials**, best results are usually obtained by adding water in-place. After water is added, it must be thoroughly and uniformly mixed with the soil.

granular material *A soil whose particle sizes and shapes are such that they do not stick together.*

Amount of Water Required

It is essential to determine the amount of water required to achieve a soil water content within the acceptable moisture range for compaction. The amount of water that must be added or removed is normally computed in gallons per station (100 ft of length); therefore, the volume in the following formula would normally be that for one station length. The computation is based on the dry weight of the soil and compacted volume. This formula can be used to compute the amount of water to be added or removed from the soil:

$$\text{Gallons} = \text{desired dry density pounds per cf (pcf)} \times \frac{(\text{desired water content \%}) - (\text{water content borrow \%})}{100} \times \frac{\text{compacted vol. of soil (cf)}}{8.33 \text{ lb per gal}} \quad [4.14]$$

When having to add water, it is good practice to adjust the desired moisture content 2% above optimum, but this depends on the environmental conditions (temperature and wind) and the soil type. A negative answer indicates that water must be removed from the borrow material before it is compacted on the fill. The 8.33 lb per gal is the weight of a gallon of water.

EXAMPLE 4.2

Job specifications require placement of the embankment fill soil in 6-in. (compacted) lifts. The desired dry unit weight of the embankment is 120 pcf. The laboratory compaction curve indicates that the optimum water content, sometimes referred to as optimum moisture content (OMC), of the soil is 12%. Soil tests indicated that the water content of the borrow material is 5%. The roadway lift to be placed is 40 ft wide. Compute the amount of water in gallons to add on a per station basis for each lift of material.

$$\text{Gallons per station} = 120\text{ pcf} \times \frac{12\%(\text{OMC}) - 5\%}{100} \times \frac{40\text{ ft} \times 100\text{ ft} \times 0.5\text{ ft}}{8.33\text{ lb/gal}}$$

$$= 120\text{ pcf} \times 0.07 \times = \frac{2{,}000\text{ cf}}{8.33\text{ lb/gal}}$$

$$= 2{,}017\text{ gallons per station}$$

Application Rate

Once the total amount of water has been calculated, the application rate can be calculated. The water application rate is normally calculated in gallons per square yard, using the following formula:

$$\text{Gallons per square yard} = \text{desired dry density of soil (pcf)} \times \frac{\%\text{ moisture added or removed}}{100} \times \text{lift thickness (ft) (compacted)} \times \frac{9\text{ sf/sy}}{8.33\text{ lb/gal}} \quad \textbf{[4.15]}$$

EXAMPLE 4.3

Using the data from Example 4.2, determine the required application rate in gallons per square yard:

$$\text{Gallons per square yard} = 120\text{ pcf} \times 0.07 \times 0.5\text{ ft} \times \frac{9\text{ sf/sy}}{8.33\text{ lb/gal}}$$

$$= 4.5\text{ gal per sy}$$

FIGURE 4.8 Water truck working on the project.

Application Methods

Once the application rate has been calculated, the method of application must be determined. Regardless of which method of application is used, it is important to ensure that the proper application rate is achieved and that the water is uniformly distributed.

Water Distributor On construction projects the most common method of adding water to a soil is with a water distributor. Water distributors are designed to evenly distribute the correct amount of water over the fill. These truck-mounted (see Fig. 4.8) or towed water distributors are designed to distribute water under various pressures, or by gravity feed. Many distributors are equipped with rear-mounted spray bars. The operator can maintain the water application rate by controlling the forward speed of the vehicle.

Ponding If time is available, water can be added to a soil by ponding or prewetting the area (see Fig. 4.7) until the desired depth of water penetration is achieved. With this method, it is difficult to control the application rate. Ponding usually requires several days to achieve a uniform moisture distribution.

Reducing the Moisture Content

As previously stated, soil that contains more water than desired (above the optimum moisture range) is correspondingly difficult to compact. Excess water makes achieving the desired density very difficult. In these cases, steps must be taken to reduce the moisture content to within the required moisture range. Drying actions may be as simple as aerating the soil. They may, however, be as complicated as adding a soil stabilization agent that actually changes the physical properties of the soil. Lime or fly ash is the typical stabilization agent for fine-grained soils. If a high water table is causing the excess moisture, some form of subsurface drainage may be required before the soil's moisture content can be reduced.

The most common method of reducing the moisture is to scarify the soil prior to compaction. This can be accomplished with either the scarifying teeth or rippers on a motor grader (see Fig. 4.9), or by disking (see Fig. 4.10) the soil. A motor grader can also use its blade to toe the soil over into furrows in order to expose more material for drying (see Fig. 4.11).

Effects of Weather

Weather conditions substantially affect soil moisture content. Cold, rainy, and cloudy weather will allow a soil to retain water. Hot, dry, sunny, and windy weather is conducive to drying the soil. In a desert climate, evaporation claims a large amount of water intended for the soil lift. Thus, for a desert project, the engineer might have to go as high as 6% above the optimum moisture content

FIGURE 4.9 A motor grader using rear rippers to scarify material.

FIGURE 4.10 Disk harrow used to scarify soil.

FIGURE 4.11 A motor grader using its blade to mix a soil for drying.

as a target for all water application calculations so that the actual moisture content will fall very near to the desired content when the material is placed and compacted.

Mixing and Blending

Whether adding water to a soil to increase the moisture content or adding a drying agent to reduce it, it is essential to mix the water or drying agent thoroughly and uniformly with the soil. Even if additional water is not needed, mixing may still be necessary to ensure a uniform distribution of the existing moisture. Mixing can be accomplished using motor graders, farm disks, or rotary cultivators.

Conventional motor graders can be used to mix or blend a soil additive (water or stabilizing agent) by windrowing the material from one side of the working lane to the other (see Fig. 4.11).

SUMMARY

Soils and rock are the principal components of many construction projects. These materials are by nature heterogeneous. In their natural state, they are rarely uniform and can be comprehended only by comparison to a similar type material with which previous experience has been gained. You should never price an earth- or rock-handling project without first making a thorough study of the materials. For bulk materials, volumetric measure varies with the material's position in the construction process. Soil volume is measured in one of three states: bank cubic yard, loose cubic yard, or compacted cubic yard. The engineer must use a consistent volumetric state in any set of calculations. A compaction curve graphically presents the maximum dry unit weight (density) and the percentage of water required to achieve maximum density based on a standard compactive effort (input energy).

Critical learning objectives include:

- An understanding of liquid limit, plastic limit, and plasticity index as physical properties of a soil.
- An ability to calculate soil volumetric changes.
- An understanding of compaction tests and specifications, and compaction curves.

These objectives are the basis for the problems that follow.

PROBLEMS

4.1 A dragline is excavating a ditch having a cross-sectional area of 120 sf. The material swells 28% from the bank to loose state. The loose material has a 37° angle of repose. What is the height and width of the spoil pile?

4.2 A soil borrow material in its natural state in the ground has a unit weight of 120 pcf. When a sample of this soil is dried in the laboratory, its dry unit weight is 103 pcf. What is the water content of this borrow material?

4.3 A soil borrow material in its natural state in the ground has a unit weight of 122 pcf. When a sample of this soil is dried in the laboratory, its dry unit weight is 105 pcf. What is the water content of this borrow material?

4.4 A soil borrow material in its natural state in the ground has a unit weight of 120 pcf. Its natural water content is 8%. What is the dry unit weight of this material?

4.5 The borrow material to construct an embankment has a dry unit weight of 86.2 pcf and a water content of 5%. The specific gravity of the solids is 2.64. The contract specifications require that the soil be placed in the fill at a γ_d of 115 pcf and a water content of 8%. How many cubic yards of borrow are required to construct an embankment having a 65,000-cy net volume? (86,717 bcy)

4.6 The soil borrow material to be used to construct a highway embankment has a mass unit weight of 96.0 pcf, a water content of 8%, and the specific gravity of the soil solids is 2.66. The specifications require that the soil be compacted to a dry unit weight of 112 lb per cf and the water content be held to 13%. (315,000 bcy, 14.4 gal/bcy borrow, 132.3 lb/cf)

   a. How many cubic yards of borrow are required to construct an embankment having a 250,000-cy net section volume?
   b. How many gallons of water must be added to the borrow material, on a per cubic yard basis, assuming no loss by evaporation?
   c. If the compacted fill becomes saturated at constant volume, what will be the water content and mass unit weight of the soil?

4.7 The natural material in a borrow pit has a mass unit weight of 97.0 pcf, a water content of 7%, and the specific gravity of the soil solids is 2.66. The specifications require that the soil be compacted to a dry unit weight of 115 lb per cf and the water content be held to 8%.

a. How many cubic yards of borrow are required to construct an embankment having a net section volume of 250,000 cy?

b. How many gallons of water must be added per cubic yard of borrow material assuming no loss by evaporation?

4.8 The soil borrow material to be used to construct a highway embankment has a mass unit weight of 98.0 lb/cf, a water content of 9%, and the specific gravity of the soil solids is 2.67. The specifications require that the soil be placed in the fill so that the dry unit weight is 114 lb/cf and the water content be held to 12%.

a. How many cubic yards of borrow are required to construct an embankment having a 800,000-cy net section volume?

b. How many gallons of water must be added per cubic yard of borrow material assuming no loss evaporation?

c. If the compacted fill becomes saturated at constant volume, what will be the water content and unit weight?

4.9 An embankment is to be constructed at a 12% moisture content. Material will be placed at the rate of 270 ccy/hr. The specified dry weight of the compacted fill is 2,900 lb/cy. How many gallons of water must be supplied each hour to increase the moisture content of the material from 7 to 12% by weight?

4.10 The borrow material to construct an embankment has a mass unit weight of 96.5 pcf and a water content of 8%. The specific gravity of the solids is 2.66. The contract specifications require that the soil be placed in the fill at a γ_d of 114 pcf and a water content of 10%.

a. How many cubic yards of borrow are required to construct an embankment having a 455,000-cy net volume?

b. How many gallons of water must be added per cubic yard of borrow material assuming no loss by evaporation?

4.11 Earth is placed in fill at the rate of 190 cy/hr, compacted measure. The placement moisture content is 10% and the dry weight of the compacted earth is 2,890 lb/cy. How many gallons of water must be supplied each hour to increase the moisture content of the earth from 4 to 10% by weight?

REFERENCES

1. *ASTM Standards on Soil Compaction* (1992). American Society for Testing and Materials, Philadelphia, PA.
2. *ASTM Standards on Soil Stabilization with Admixtures* (1990). American Society for Testing and Materials, Philadelphia, PA.
3. *Construction and Controlling Compaction of Earth Fills* (2000). ASTM Special Technical Publication, 1384, D. W. Shanklin Ed., American Society for Testing and Materials, Philadelphia, April.
4. *Guide to Earthwork Construction* (1990). State of the Art Report 8, TRB, National Research Council, Washington, DC.

5. Holtz, R. D. (1989). *NCHRP, Synthesis of Highway Practice 147: Treatment of Problem Foundations for Highway Embankments*, TRB, National Research Council, Washington, DC.
6. Rollings, M. P., and R. S. Rollings (1996). *Geotechnical Materials in Construction*, McGraw-Hill, New York.
7. Welsh, Joseph P. (1987). *Soil Improvement—A Ten Year Update, Geotechnical Special Publication No. 12*, American Society of Civil Engineers, Washington, DC.

WEBSITE RESOURCES

1. www.fhwa.dot.gov/bridge/geo.htm Federal Highway Administration. Contains information on FHWA geotechnical programs, including publications, software, and training.
2. www.geosynthetic-institute.org Geosynthetic Institute (GSI). GSI's mission is to transfer knowledge about geosynthetics.
3. www.nrcs.gov/TechRes.html National Resources Conservation Service. Serves as the gateway to the soils database, including GIS-based polygon data.
4. www4.national-academies.org/trb/onlinepubs.nsf Transportation Research Board (TRB). TRB's website provides a searchable index of the board's publications and articles, which cover all modes and aspects of transportation, including geotechnical engineering.
5. www.erdc.usace.army.mil U.S. Army Corps of Engineers. Links are provided to the geotechnical engineering laboratory at the Waterways Experiment Station and the Cold Regions Research and Engineering Laboratory.
6. www.usgs.gov U.S. Geological Survey. Provides links to geological, water, and mapping information.
7. www.usucger.org U.S. Universities Council on Geotechnical Engineering Research (USUCGER). Pertinent information about geomedia research at 96 member universities in the United States is provided.

CHAPTER 5

Compaction and Stabilization Equipment

Compaction equipment must be matched to the type of material being manipulated. Equipment manufacturers have developed a variety of compactors that incorporate at least one of the compaction methods and in some cases more than one into machine performance capabilities. In engineering construction, stabilization refers to when compaction is preceded by the addition and mixing of an inexpensive admixture, termed a "stabilization agent," that alters the chemical makeup of the soil, resulting in a more stable material.

COMPACTION OF SOIL AND ROCK

With time, material will settle or compact itself naturally, but the objective of compaction is to achieve the required density quickly. The earliest recorded use of compaction can be found in the Roman Empire records of their road construction projects. The Romans realized that compaction would improve the engineering properties of soils; therefore, they used large cylindrical stone rollers to achieve mechanical densification of their road bases, (see Fig. 5.1).

Obtaining a greater soil unit weight is not the direct objective of compaction. The reason for compaction is to improve soil properties to

1. Reduce or prevent settlements.
2. Increase strength.
3. Improve bearing capacity.
4. Control volume changes.
5. Lower permeability.

Density, however, is the most commonly used parameter for specifying construction operations because there is a direct correlation between these desired

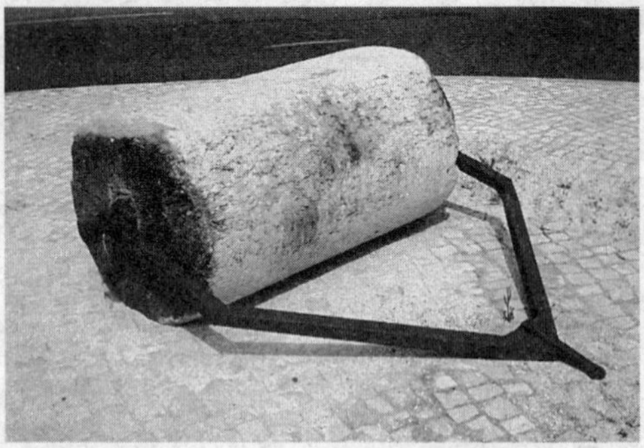

FIGURE 5.1 A cylindrical stone roller.

properties and a soil's density. Construction contract documents usually call for achieving a specified density, even though one of the other soil properties is the crucial objective.

There may be other methods whereby the desired properties could be attained, but by far the most widely used method of soil strengthening is compaction of the soil at optimum moisture. The benefits of proper compaction are enormous, far outweighing their costs. Typically, a uniform layer, or lift, of soil from 4 to 12 in. thick is compacted by means of several passes of mechanized compaction equipment.

GLOSSARY OF TERMS

The following glossary defines important terms that are used in discussing compaction equipment.

Backfill. Material used in refilling a cut or other excavation.

Binder. Fine aggregate or other materials that fill voids and hold coarse aggregate together.

Subbase. A constructed layer of select material installed to furnish strength to the base under a road or airfield pavement. In areas where the construction goes through marshy, swampy, unstable land it is often necessary to excavate the natural materials in the area under the roadway and replace them with more stable materials. The material used to replace the unsuitable natural soils is generally called subbase material, and when compacted it is known as the subbase.

Subgrade. The surface produced by grading native earth, or imported materials that serve as the foundation layer for a paving structural section.

TYPES OF COMPACTING EQUIPMENT

Applying energy to a soil by one or more of the following methods will cause compaction:

1. Impact—sharp blow
2. Pressure—static weight
3. Vibration—shaking
4. Kneading—manipulation or rearranging

The effectiveness of different compaction methods is dependent on the individual soil type being manipulated. Appropriate compaction methods based on soil type are identified in Table 5.1.

Equipment manufacturers have developed a variety of compactors that incorporate at least one of the compaction methods, and in some cases more

TABLE 5.1 Soil types versus the method of compaction.

Material	Impact	Pressure	Vibration	Kneading
Gravel	Poor	No	Good	Very good
Sand	Poor	No	Excellent	Good
Silt	Good	Good	Poor	Excellent
Clay	Excellent with confinement	Very good	No	Good

than one, into their performance capabilities. Many types of compacting equipment are available, including:

1. Sheepsfoot rollers

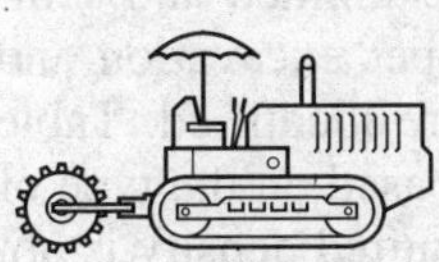

2. Tamping rollers

3. Smooth-drum vibratory soil compactors

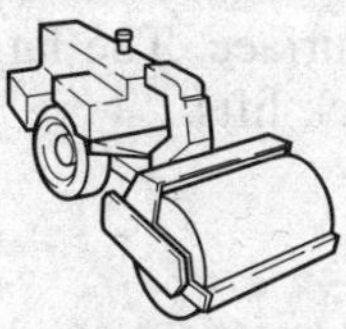

4. Pad-drum vibratory soil compactors

5. Pneumatic-tired rollers

Table 5.2 summarizes the principal methods of compaction for the various types of compactors. There are also static, smooth steel-wheel rollers. These generally consist of two tandem drums, one in front and one behind. Steel three-wheel rollers are another version of this type of compactor, but they are not as common today as they were in the past. Vibratory rollers are more efficient than static steel-wheel rollers for earthwork and have largely replaced them.

> On some projects, it may be desirable to use more than one type of compaction equipment to attain the desired results and to achieve the greatest economy.

TABLE 5.2 Principal method of compaction used by various compactors.

Compactor type	Impact	Pressure	Vibration	Kneading
Sheepsfoot		X		
Tamping foot	X	X		
Vibrating smooth	X		X	
Vibrating padfoot	X		X	
Pneumatic		X		X

The ultimate goal is to construct a quality embankment in the shortest time at the least cost, and that means the compaction equipment must be matched to the material. Therefore, the job should always be closely examined and samples taken of the excavation or borrow materials. The proper excavation and compaction equipment cannot be selected until the soils are identified. Table 5.3 provides guidance for selecting compaction equipment based on the type of material that must be compacted. As seen in the table, if required density is not achieved within four to eight coverages, a different type compactor should be considered.

Rock fills are usually spread in 18- to 48-in. lifts. Attention to spreading the material in a uniform lift is vital to achieving density during the compaction process. Consistent spreading helps to fill voids and orients the rocks so as to provide the compaction equipment with an even surface. The largest possible smooth-drum vibratory rollers are used for deep rock lifts.

TABLE 5.3 Appropriate project compaction equipment based on material type.

Material	Lift thickness (in.)	Number of passes	Compactor type	Comments
Gravel	8–12	3–5	Vib. padfoot	Foot psi 150–200
			Vib. smooth	—
			Pneumatic	Tire psi 35–130
			Sheepsfoot	Foot psi 150–200
Sand	8–10	3–5	Vib. padfoot	—
			Vib. smooth	—
			Pneumatic	Tire psi 35–65
			Smooth static	Tandem 10–15 ton
Silt	6–8	4–8	Vib. padfoot	Foot psi 200–400
			Tamping foot	—
			Pneumatic	Tire psi 35–50
			Sheepsfoot	Foot psi 200–400
Clay	4–6	4–6	Vib. padfoot	Foot psi 250–500
			Tamping foot	—
			Sheepsfoot	Foot psi 250–500

FIGURE 5.2 A two-drum towed sheepsfoot roller.
Source: Kokosing Fru-Con.

Sheepsfoot Roller

These rollers are usually found as towed drum models (see Fig. 5.2). The sheepsfoot roller is suitable for compacting all fined-grained materials, but is generally not suitable for use on cohesionless granular materials. These rollers have steel wheels equipped with cylindrical pads (or feet), normally less than 10 in. in length. Varying the weight of the roller by the use of ballast in the drum will vary the foot-contact pressure.

The pads on a sheepsfoot drum penetrate through the top lift and actually compact the lift below. When the drum rotates the pads out of the soil, they kick up or fluff the material because of their shape. Sheepfoot rollers can work only at speeds from 4 to 6 mph. Usually 6 to 10 passes will be needed to compact an 8-in. clay lift.

Since the sheepsfoot roller tends to aerate the soil as it compacts, it is ideally suited for working soils that have moisture contents above the acceptable moisture range. The sheepsfoot roller does not adequately compact the upper 2 to 3 in. of a lift, and it should, therefore, be followed by a lighter pneumatic-tired or steel-wheeled roller if no succeeding lift is to be placed.

Tamping Rollers

Tamping foot compactors (Fig. 5.3) are high-speed, self-propelled, nonvibratory rollers. These rollers usually have four steel-padded wheels and can be equipped with a small blade to help level the lift. The pads are tapered with an oval or rectangular face. The pad face is smaller than the base of the pad at the drum. As a tamping roller moves over the surface, the feet penetrate the soil to produce a kneading action and a pressure to mix and compact the soil from the bottom to the top of the layer. With repeated passages of the roller over the surface, the penetration of the feet decreases until the roller is said to walk out

FIGURE 5.3 Two self-propelled tamping rollers equipped with leveling blades.

of the fill. Because the pads are tapered, a tamping foot roller can walk out of the lift without fluffing the soil. If it does not walk out, the roller is too heavy or the soil is too wet and the roller is shearing the soil.

The working speed for these rollers is in the 8- to 12-mph range. Generally two to three passes over an 8- to 12-in. lift will achieve density, but this is dependent on the size of the roller. Four passes may be necessary in poorly graded plastic silt or very fine clays. A tamping foot roller is effective on all soils except pure sand. To realize their true economical compaction potential, they need long uninterrupted passes so the roller can build up speed, which generates high production.

Like the sheepsfoot roller, tamping-foot compactors do not adequately compact the upper 2 to 3 in. of a lift. Therefore, if a succeeding lift is not going to be placed, follow up with a pneumatic-tired or smooth-drum roller to complete the compaction or to seal the surface.

Vibrating Compactors

Vibration creates impact forces, and these forces result in greater compacting energy than an equivalent static load. This fact is the economics behind a vibratory compactor. The impact forces are higher than the static forces because the vibrating drum converts potential energy into kinetic energy. Vibratory compactors may have one or more drums. Typically on two-drum models, one drum is powered to transmit unit propulsion. Single-drum models usually have two rubber-tired drive wheels. There are also towed vibratory compactors.

Certain types of soils such as sand, gravel, and relatively large shot rock respond quite well to compaction produced by a combination of pressure and vibration. When these materials are vibrated, the particles shift their positions and nestle more closely with adjacent particles to increase the density of the mass.

Vibrating drum rollers are actuated by an eccentric shaft that produces the vibratory action. The eccentric shaft need be only a body that rotates about an axis other than the one through the center of mass. The vibrating mass (drum) is always isolated from the main frame of the roller. Vibrations normally vary from 1,000 to 5,000 per min.

Vibration has two measurements—**amplitude**, which is the measurement of the movement, or throw, and frequency, which is the rate of the movement, or number of vibrations (oscillations) per second or minute (vpm). The amplitude controls the effective area, or depth to which the vibration is transmitted into the soil, while the frequency determines the number of blows or oscillations that are transmitted in a period of time.

amplitude
The vertical distance the vibrating drum or plate is displaced from the rest position by an eccentric moment.

The impacts imparted by the vibrations produce pressure waves that set the soil particles in motion, producing compaction. In compacting granular material, frequency (the number of blows in a given period) is usually the critical parameter as opposed to amplitude.

Compaction results are a function of the frequency of the blows, the force of the blows, and the time period over which the blows are applied. The frequency/time relationship accounts for the slower working speed requirement when using vibratory compactors. Working speed is important as it dictates how long a particular part of the fill is compacted. A working speed of 2 to 4 mph provides the best results when using vibratory compactors.

Smooth-Drum Vibratory Soil Compactors

The smooth-drum compactors, whether single- or dual-drum models, generate three compactive forces: (1) pressure, (2) impact, and (3) vibration. These rollers (see Fig. 5.4) are most effective on granular materials, with particle sizes ranging from large rocks to fine sand. They can be used on semicohesive soils with up to about 10% of the material having a PI of 5 or greater. Large steel-drum vibratory rollers can be effective on rock lifts as thick as 3 ft.

FIGURE 5.4 Smooth-drum vibratory soil compactor.

FIGURE 5.5 Padded-drum vibratory soil compactor with a leveling blade.

Padded-Drum Vibratory Soil Compactors

These rollers (see Fig. 5.5) are effective on soils with up to 50% of the material having a PI of 5 or greater. The edges of the pads are rolled inward enabling them to walk out of the lift without fluffing the soil. The typical lift thickness for padded-drum units on cohesive soil is 12 to 18 in. These units are sometimes equipped with a leveling blade.

Small walk-behind and/or remotely controlled vibratory rollers having widths in the range of 24 to 38 in. are available (see Figs. 5.6 and 5.7). These units are designed specifically for trench work or for working in confined areas. The drums of the roller extend beyond the sides of the roller body, so the compaction can be accomplished adjacent to the trench walls. Many of these small compactors can be equipped with remote control systems so that the operator can control the roller without having to enter the trench. Nearly all of the remote control systems use a digitized radio frequency; this eliminates the need to have control cables dragged around the construction site.

base materials
The layer of material, in a roadway or runway section, on which the pavement is placed. It may be of different types of materials, ranging from selected soils to crushed stone or gravel.

Pneumatic-Tired Rollers

These are surface rollers that apply the principle of kneading action to affect compaction below the surface. They may be self-propelled (see Fig. 5.8) or towed. Pneumatics are used on small- to medium-size soil compaction jobs, primarily on bladed granular **base materials**. Small pneumatics are not suited for high-production, thick lift embankment compaction projects. Pneumatic-tired rollers are also used in compacting asphalt, chip seals, recycled pavement, and base and subbase materials. Because of their relatively gentle

FIGURE 5.6 Padded-drum walk-behind vibratory roller.

FIGURE 5.7 Remotely controlled padded-drum walk-behind vibratory roller.

FIGURE 5.8 Self-propelled pneumatic rollers.

kneading action they are well suited for intermediate and breakdown compaction of Superpave and stone mastic asphalt mixes. The flexible tire surface permits conformance of the tire to slightly irregular surfaces. This helps to maintain uniform density and bearing capacity whereas a steel-drum roller would bridge over low spots while applying more pressure to high spots.

FIGURE 5.9 Fifty-ton pneumatic roller being used to proof roll a roadway subgrade.

The small-tired units usually have two tandem axles with four to five wheels on each axle. The wheels oscillate, enabling them to follow the surface contour and reach into low areas for uniform compaction. The rear tires are spaced to track over the uncompacted surface left by the passage of the front tires. This path tracking produces complete coverage of the surface. The wheels may be mounted slightly out of line with the axle, giving them a weaving action (the name "wobbly wheel") to increase the kneading action of the soil. By adding ballast, the weight of a unit may be varied to suit the material being compacted.

Large-tired rollers are available in sizes varying from 15 to 200 tons gross weight (Fig. 5.9). They utilize two or more big earth-moving tires on a single axle. The air pressure in the tires may vary from 80 to 150 psi (pounds per square inch). Because of the heavy loads and high tire pressures, they are capable of compacting all types of soils to greater depths. The expense is in propelling these large units over the lift, as they require tractors having considerable drawbar pull and traction.

These units are frequently used to proof roll roadway subgrades and airfields bases, and on earth-fill dams. On the Painted Rock Dam in Arizona, 50-ton rubber-tired rollers were used to compact sandy gravel embankment material. The pervious material was prewetted in the borrow area and then placed in the embankment in 24-in. lifts. To achieve density, four passes of the 50-ton roller were required.

Because the area of contact between a tire and the ground surface over which it passes varies with the air pressure in the tire, specifying the total weight or the weight per wheel is not necessarily a satisfactory method of indicating the compacting ability of a pneumatic roller. Four parameters must be known to determine the compacting ability of pneumatic rollers:

1. Wheel load
2. Tire size
3. Tire ply
4. Inflation pressure

Pneumatic-Tired Rollers with Variable Inflation Pressures

When a pneumatic-tired roller is used to compact soil through all stages of density, the first passes over a lift should be made with relatively low tire pressures to increase flotation and ground coverage. However, as the soil is compacted, the air pressure in the tires should be increased up to the maximum specified value for the final pass. Prior to the development of rollers having the capability of varying their tire pressure while in operation, it was necessary to stop the rolling and either (1) adjust the pressure in the tires, (2) vary the weight of the ballast on the roller, or (3) keep rollers of different weights and tire pressures on a project to provide units to fit the particular needs of a given compaction condition.

Several manufacturers produce rollers that are equipped to enable the operator to vary the tire pressure without stopping the machine. The first passes are made with relatively low tire pressures. As the soil is compacted, the tire pressure is increased to suit the particular conditions of the soil. The use of this type of roller usually enables adequate compaction with fewer passes than are required by constant pressure rollers.

Towed Impact Compactors

Beginning in 1949, engineers in South Africa began experimenting with impact compactors or "square wheels." These compactors have used three-, four- (see Fig. 5.10), and five-sided drums. As the compactor is towed, the drum rotates, lifting itself up on edge, and then falls back to earth. The impact of the drum striking the ground provides the compactive force. Much of the impetus for this compactor design was the need to develop a high-energy

FIGURE 5.10 Four-sided impact compactor.

FIGURE 5.11 Compaction wheel mounted on a hydraulic excavator boom.

compactive device that could be used for densifying materials at low moisture content in arid regions. There was also the desire to have a device that could help collapse the unstable structure of certain soils often found in arid regions.

These compactors can be used on a wide range of materials: rock, sand, gravel, silt, and clay. They will handle lifts up to 3 ft, and because they impart such high energy to the ground, density can be achieved over a wider range of moisture contents.

Compaction Wheels

To avoid the hazards of having men work in excavations of limited dimensions, a compaction wheel attached to an excavator boom is often used to achieve compaction when backfilling utility trenches (see Fig. 5.11). The feet on these wheels can be of either the sheepsfoot or tamping shape. The wheels are designed to compact all types of soil. Changing from an excavator bucket to a compaction wheel can be accomplished quickly. Wheels are manufactured in sizes to fit 7- to 45-ton excavators.

FIGURE 5.12 Self-propelled vibratory-plate compactors.

Manually Operated Vibratory-Plate Compactors

Figure 5.12 illustrates self-propelled vibratory-plate compactors, which are used for compacting granular soils, crushed aggregate, and asphalt concrete in locations where large compactors could not operate. These gasoline- or diesel-powered units are rated by centrifugal force, exciter revolutions per minute, depth of vibration penetration (lift), foot-per-minute travel, and area of coverage per hour. Many of these compactors can be operated either manually as a walk-behind unit or by remote control.

Manually Operated Rammer Compactors

Gasoline-engine-driven rammers (see Fig. 5.13) are used for compacting cohesive or mixed soils in confined areas. These units range in impact from 300 to 900 foot-pounds (ft-lb) per sec at an impact rate up to 850 per min, depending on the specific model. Performance criteria include pounds per blow, area covered per hour, and depth of compaction (lift) in inches. Rammers are self-propelled in that each blow moves them ahead slightly to contact a new area.

Small compactors such as the self-propelled vibratory-plate or the rammer will provide adequate compaction if

1. Lift thickness is minimal (usually 3 to 4 in.),
2. Moisture content is carefully controlled, and
3. Coverages are sufficient.

FIGURE 5.13 Manually operated rammers.

The primary causes of density problems when backfilling utility trenches are (1) an inadequate number of coverages with the small equipment that must be used in the confined space, (2) lifts that are too thick, and (3) inconsistent control of moisture.

ROLLER PRODUCTION ESTIMATING

The compaction equipment used on a project must have a production capability matched to that of the excavation, hauling, and spreading equipment. Usually, excavation or hauling capability will set the expected maximum production for the job. The production formula for a compactor is

$$\text{Compacted cubic yards per hour} = \frac{16.3 \times W \times S \times L \times \text{efficiency}}{n} \quad \textbf{[5.1]}$$

where

W = compacted width per roller pass in feet

S = average roller speed in miles per hour

L = compacted lift thickness in inches

n = number of roller passes required to achieve the required density

The computed production is in compacted cubic yards (ccy), so it will be necessary to apply a shrinkage factor to convert the production to bank cubic yards (bcy), as that is how the excavation and hauling production is usually expressed.

EXAMPLE 5.1

A self-propelled tamping foot compactor will be used to compact a fill being constructed of clay material. Field tests have shown that the required density can be achieved with four passes of the roller operating at an average speed of 3 mph. The compacted lift will have a thickness of 6 in. The compacting width of this machine is 7 ft. One bank cubic yard equals 0.83 ccy. The scraper production, estimated for the project, is 510 bcy/hr. How many rollers will be required to maintain this production? Assume a 50-min hour efficiency.

$$\text{Compacted cubic yards per hour} = \frac{16.3 \times 7 \times 3 \times 6 \times 50/60}{4} = 428 \text{ ccy/hr}$$

$$\frac{428 \text{ ccy per hour}}{0.83} = 516 \text{ bcy/hr}$$

$$\frac{510 \text{ bcy/hr required}}{516 \text{ bcy/hr}} = 0.99$$

Therefore, only one roller will be required.

DYNAMIC COMPACTION

The densification technique of repeatedly dropping a heavy weight onto the ground surface is commonly referred to as "dynamic compaction." This process has also been described as heavy tamping, impact densification, dynamic consolidation, pounding, and dynamic precompression. For either a natural soil deposit or a placed fill, the method can produce densification to depths of greater than 35 ft. Most projects have used drop weights weighing from 6 to 30 tons, and typical drop heights range from 30 to 75 ft. However, on the Jackson Lake Dam project in Grand Teton National Park, a 32-ton weight from a height of 103 ft was employed.

Conventional cranes are used for drop weights of up to 20 tons and drop heights below 100 ft. The weight is attached to a single hoist line. During the drop, the hoist line is released and the hoist drum is allowed to free spool. When heavier weights are used, specially designed dropping machines are required (see Fig. 5.14). With this densification technique, the **in situ** strata are compacted from the ground surface at their prevailing water contents. A possible disadvantage is that ground vibrations can be produced that travel significant distances from the impact point.

in situ
Soil in its original or undisturbed position.

The most successful projects have been those where coarse-grained pervious soils were present. The position of the water table will have a major influence on dynamic compaction project success. It is best to be at least 6.5 ft above the water table. Operations on saturated **impervious** deposits have resulted in only minor improvement at high cost and should be considered ineffective.

impervious
A material that resists the flow of water through it is termed impervious.

FIGURE 5.14 Dynamic compaction using a Lampson Thumper.

The depth of improvement that can be achieved is a function of the weight of the tamper and the drop height:

$$D = n(W \times H)^{1/2} \qquad [5.2]$$

where

D = depth of improvement in meters (m)

n = an empirical coefficient, which is less than 1.0

W = weight of tamper in metric tons

H = drop height in meters

An n value of 0.5 has been suggested for many soil deposits. That value is a reasonable starting point; however, the coefficient is affected by the

- Type and characteristics of the material being compacted.
- Applied energy.
- Contact pressure of the tamper.
- Influence of cable drag.
- Presence of energy-absorbing layers.

SOIL STABILIZATION

GENERAL INFORMATION

Many soils are subject to differential expansion and shrinkage when they undergo changes in moisture content. Many soils also shift and rut when subjected to moving wheel loads. If pavements are to be constructed on such soils, it is usually necessary to stabilize them to reduce the volume changes and to strengthen them to the point where they can carry the imposed load, even under adverse weather and climatic conditions. In the broadest sense, *stabilization* refers to any treatment of the soil that increases its natural strength. There are two kinds of stabilization—(1) mechanical and (2) chemical. In engineering construction, however, stabilization most often refers to when compaction is preceded by the addition and mixing of an inexpensive admixture, termed a "stabilization agent," which alters the chemical makeup of the soil, resulting in a more stable material.

Stabilization may be applied in-place to a soil in its natural position, or mixing can take place on the fill. Also, stabilization can be applied in a plant, and then the blended material is transported to the job site for placement and compaction.

The two primary methods of stabilizing soils are

1. Incorporating lime or lime–fly ash into soils that have a high clay content and
2. Incorporating Portland cement (with or without fly ash) with soils that are largely granular in nature

Fly ash is a by-product in the production of electricity from burning coal. As such, it can be a highly variable product, and its engineering usefulness can

FIGURE 5.15 A rotary stabilizer.

range from superior to extremely poor. However, good quality fly ash can replace a *portion* of the lime needed to stabilize a clay-type soil. Because lime is relatively expensive and fly ash often quite inexpensive, lime–fly ash stabilization of soils is often utilized.

Stabilizers

Rotary stabilizers (Fig. 5.15) are extremely versatile pieces of equipment ideally suited for mixing, blending, and aerating soil. A stabilizer consists of a rear-mounted, removable tine, rotating tiller blade, which is covered by a removable hood. In place, the hood creates an enclosed mixing chamber that enhances thorough blending of the soil. The tiller blade lifts the material and throws it against the hood. The material, deflecting off of the hood, falls back onto the tiller blades for thorough blending. As the stabilizer moves forward, the material is ejected from the rear of the mixing chamber. As the material is ejected, it is struck off by the trailing edge of the hood, resulting in a fairly level working surface. With the hood removed, the blades churn the soil, exposing it to the drying action of the sun and wind. Some models are equipped with a forward-mounted spray bar that can be used to add water or stabilizing agents to the soil during the blending process. The stabilizer's use is limited to material less than 4 in. in diameter. The tines are designed to penetrate up to 10 in. below the existing surface so the unit can be used for scarifying and blending in-place (in situ) material as well as fill material.

STABILIZING SOILS WITH LIME

In general, lime reacts readily with postplastic soils containing clay, either the fine-grained clays or clay-gravel types. Such soils range in PI from 10 to 50 or more. Unless stabilized, these soils usually become very soft when water is introduced. The only exception would be organic soils containing more than 20% organic matter.

Soil-Lime Chemistry

In combination with compaction, soil stabilization with lime involves a chemical process whereby the soil is improved with the addition of lime. Lime, in its hydrated form [$Ca(OH)_2$], will rapidly cause cation exchange and flocculation/agglomeration, provided it is intimately mixed with the soil. A high-PI clay soil will then behave much like a material having a lower PI. This reaction begins to occur within an hour after mixing, and significant changes are realized within a very few days, depending on the PI of the soil, the temperature, and the amount of lime used. The observed effect in the field is a drying action.

Following this rapid soil improvement, a longer, slower soil improvement takes place, termed a "pozzolanic reaction." In this reaction, the lime chemically combines with siliceous and aluminous constituents in the soil to cement the soil together. Here some confusion exists. Some people refer to this as a "cementatious reaction," which is a term normally associated with the hydraulic action occurring between Portland cement and water, in which the two constituents chemically combine to form a hard, strong product. The confusion is increased by the fact that almost two-thirds of Portland cement is lime (CaO). But the lime in Portland cement starts out already chemically combined during manufacture with silicates and aluminates, and thus is not in an available or "free" state to combine with the clay.

The cementing reaction of the lime, as $Ca(OH)_2$, with the clay is a very slow process, quite different from the reaction of Portland cement and water, and the final form of the products is thought to be somewhat different. The slow strength with time experienced with lime stabilization of clay provides flexibility in manipulation of the soil. Lime can be added and the soil mixed and compacted, initially drying the soil and causing flocculation. Several days later, the soil can be remixed and compacted to form a dense stabilized layer that will continue to gain strength for many years. The resulting stabilized soils have been shown to be extremely durable.

Lime Stabilization Construction Procedures

Lime treatments can be characterized into three classes:

1. *Subgrade (or subbase) stabilization* includes stabilizing fine-grained soils in place or borrow materials, which are employed as subbases.
2. *Base stabilization* includes plastic materials, such as clay-gravels that contain at least 50% coarse material retained on a No. 40 mesh screen.

3. *Lime modification* includes the upgrading of fine-grained soils with small amounts of lime, that is, $\frac{1}{2}$ to 3% by weight.

The distinction between modification and stabilization is that generally no credit is accorded the lime-modified layer in the structural design. It is used usually as a contractor technique to dry wet areas, to help "bridge" across underlying spongy subsoil, or to provide a working table for subsequent construction.

The basic steps in lime stabilization construction are

1. *Scarification and pulverization*. To accomplish complete stabilization, adequate pulverization of the clay fraction is essential. This is best accomplished with a rotary stabilizer (Fig. 5.15).
2. *Lime spreading*. Dry lime should not be spread under windy conditions. Lime slurry can be prepared in a central mixing tank and spread on the grade using standard water distributor trucks.
3. *Preliminary mixing and addition of water*. During rotary mixing of the lime with the soil material, the water content should be raised to at least 5% above optimum. This may require the addition of water.
4. *Preliminary curing*. The lime-soil mixture should cure for 24 to 48 hr to permit the lime and water to break down (or mellow) the clay clods. In the case of extremely heavy clays, the curing period may extend to 7 days.
5. *Final mixing and pulverization*. During final mixing, pulverization should continue until all clods are broken down to pass a 1-in. screen and at least 60% to pass a No. 4 sieve.
6. *Compaction*. The soil-lime mixture should be compacted as required by specification.
7. *Final curing*. The compacted material should be allowed to cure for 3 to 7 days prior to placing subsequent layers. Curing can be accomplished by light sprinkling with water so that the surface is maintained in a moist condition for the desired period of time. Membrane curing is another acceptable curing method. It involves sealing the compacted layer with a bituminous material.

CEMENT-SOIL STABILIZATION

Stabilizing soils with Portland cement is an effective method of strengthening certain soils. As long as the soils are predominately granular with only minor amounts of clay particles, the use of Portland cement has been found to be effective. A rule of thumb is that soils with a PI of less than 10 are likely candidates for this type of stabilization. Soils with higher amounts of clay particles are very difficult to manipulate and thoroughly mix with the cement before the cement sets. The terms "soil cement" and "cement-treated base" are often used interchangeably, and generally describe this type of stabilization. However, in some areas, the term "soil cement" refers strictly to mixing and treatment of in-place soils on the grade. The term "cement-treated base" is then used to describe an aggregate/cement blend produced in a pugmill plant and hauled onto the grade. The amount of cement mixed with the soil is usually 3 to 7% by dry weight of the soil.

As discussed in connection with lime stabilization, fly ash is plentiful in many areas of the world, and it can be effectively used to replace a portion of the Portland cement in a soil cement treatment. Replacement percentages on an equal weight basis or on a 1.25:1.0 fly ash/Portland cement replacement ratio have been used.

Soil Cement Construction Procedures

Cement stabilization involves scarifying the grade; spreading the Portland cement uniformly over the surface of the soil (see Fig. 5.16); mixing the cement into the soil to the specified depth, preferably with a pulverizer-type machine (Fig. 5.17); compaction; fine grading; and curing. If the moisture content of the soil is low, it will be necessary to add water during the mixing operation. The material should be compacted within 30 min after it is mixed, using either tamping foot or pneumatic-tired rollers, followed by final rolling with a smooth-wheel roller. A seal of asphalt or another acceptable material may have to be applied to the surface to retain the moisture in the mix.

The nature of mix in-place soil cement does not allow a 7 A.M. to 3 P.M. work schedule. When the cement is applied, the material cannot be left overnight, you must complete the operation even if overtime work is necessary.

FIGURE 5.16 Bulker with rear Flynn spreader being used to uniformly apply cement during a "soil cement" stabilization project.

FIGURE 5.17 Stabilizer for soil mixing operation.
Source: Kokosing Fru-Con.

The Project Site When estimating soil cement, the configuration of the areas to be treated must be considered. There are many aspects of the process that can be affected by configuration. It may be necessary to travel over treated areas to obtain water that is required to treat other areas of the project. In some instances, sites are very small and it is very difficult to load and unload equipment in a safe manner. Cement dust is another issue that should be taken into consideration. Personnel working on the grade need to have proper protective equipment, including eye protection. Cement dust can also cause damage to nearby vehicles. Always work with favorable wind conditions. Of major concern in a soil-cement operation is the amount of rock in excess of softball size. Rocks can damage the pulverizing equipment and make grading difficult. Removal of oversize rock is labor intensive, making it a critical cost consideration.

Scarifying Loosening the soil or scarifying can help identify problem areas. If rocks are an unforeseen problem, scarifying can bring the unacceptable rocks to the surface. A good effort at rock removal will save equipment from damage. If large amounts of rock are removed, the grade must be reconfirmed after the removal process is completed. Scarifying can also reveal soft, yielding, or wet areas and hidden organic matter.

Spreading During the spreading operations, the bulk application (spread rate) is checked and necessary adjustments are made. The calibration of equipment is a crucial point in federal, state, and airport work. The spread pattern of the cement is determined by site configuration. Bulker types vary somewhat (see Fig. 5.16) and the size of the site will determine the bulk spreader to be used. The bulker should move at a continuous rate without stopping while the product is being discharged.

Mixing Mixing should be accomplished immediately after application of the cement (see Fig. 5.17). The mixed soil and cement should be checked to ensure the uniformity of blend. It is important that the cement be blended thoroughly with the soil to achieve the desired finished product. The normal soil-cement mix procedure is in a down cut motion using tine-type mixer teeth (stabilization). In cohesive soils, an up cut machine with conical type bits is better.

In the case of very dry or windy conditions, prewetting the grade can be helpful to provide an adequate material moisture content and to control dust. In extreme conditions, prewetting prior to scarifying the material may be a necessity. During the application of water, the water truck drivers should take care not to allow the discharge of water to puddle or pool. Puddles and wet spots can occur when water trucks are parked or allowed to stand idle on the grade. Such situations can result in soft or yielding grade conditions. These soft spots may, in turn, become areas where compaction cannot be achieved. Strict moisture control should be adhered to at all times.

Compaction Initial compaction with a vibratory-pad foot roller immediately follows the mixing operation. A roller pattern should be established at the beginning of the initial compaction effort. Normally, two to three complete roller passes will be required. The depth of mix, type of soil, and required density are factors that control the number of roller passes required. The compactor should keep pace with the mixing operation.

Following the initial compaction, the treated material should be shaped to the approximate line and grade. During grading special attention should always be given to maintaining proper drainage. It is hard to make grade corrections to cured material. The compaction effort should continue until required density is achieved.

The fine-grading operation should follow after acceptable density is achieved and the material has been rolled with a smooth-drum vibratory or pneumatic roller. The treated material should be kept from drying at all times during rolling, shaping, and fine grading. Again, it is important that the water truck drivers exercise care not to allow puddling or pooling of water to occur.

Curing There are several acceptable methods of curing soil cement. Some projects will specify the use of liquid asphalt curing. That method has possible environmental impacts and material can be tracked onto adjacent roadways, requiring a cleanup operation. In many areas, the effect of runoff is of great concern, particularly for projects located near waterways. The use of white curing compound has proven effective. It has to be applied at a heavy rate (0.25 to 0.30 gal/square yard) to achieve complete coverage. This can be a time-consuming operation and requires special equipment. Therefore there is a preference for the wet cure method.

Keep the finish grade wet/damp for a period of 7 days, with as little traffic as possible being allowed to travel on the newly treated material. The water truck should be careful not to make sharp turns during the initial curing. Construction equipment, particularly track machines, should be strictly prohibited from entering the area.

During the curing operation, the treated area must be kept from freezing for 7 days. Frost should not be allowed to form on the grade during the first 48 hr of the curing process. The use of straw (4 in. minimum) and/or poly can help reduce possible freezing, depending on the anticipated low temperature. Temperature should always be taken into consideration at the beginning of a project. Do not begin a project without a weather projection that will allow for seven continuous days of temperatures above freezing after completion of mixing and grading.

SUMMARY

Applying energy to a soil by one or more of the following methods will cause compaction: impact, pressure, vibration, or kneading. The effectiveness of different compaction methods is dependent on the individual soil type being manipulated. The compaction equipment used on a project must have a production capability matched to that of the excavation, hauling, and spreading equipment. Usually, excavation or hauling capability will set the expected maximum production for the job.

Stabilization refers to any treatment of the soil that increases its natural strength: (1) mechanical and (2) chemical. In engineering construction, however, stabilization most often refers to when compaction is preceded by the addition and mixing of an inexpensive admixture, termed a "stabilization agent," which alters the chemical makeup of the soil, resulting in a more stable material.

Critical learning objectives include:

- An understanding of when to use which compaction method.
- An ability to calculate estimated compaction production.
- An understanding of the construction procedures involved in either lime or cement stabilization.

These objectives are the basis for the problems that follow.

PROBLEMS

5.1 Earth—whose in situ weight is 112 lb/cf, whose loose weight is 95 lb/cf, and whose compacted weight is 120 lb/cf—is placed in a fill at the rate of 240 cy/hr, measured as compacted earth. The thickness of the compacted layers is 6 in. A tractor pulls a towed sheepsfoot roller having drums, 5 ft wide, at a speed of 2 mph. Assume a 45-min hour operating efficiency. Determine the number of drums required to provide the necessary compaction if eight drum passes are specified for each layer of earth.

5.2 Estimate the production in compacted cubic yards per hour for a roller if its average speed will be 8 mph and it will cover 7.33 ft in one pass. The job specifications limit the compacted lift thickness to 6 in. and require seven passes per lift. The company usually figures production based on a 50-min hour.

5.3 Earth whose in situ weight is 110 lb/cf and whose compacted weight is 124 lb/cf is placed in a fill at the rate of 280 cy/hr, measured as compacted earth. The thickness of the compacted layers is 8 in. A tractor pulls a towed sheepsfoot roller having drums, 6 ft wide, at a speed of 1.4 mph. Assume a 50-min hour operating efficiency. Determine the number of drums required to provide the necessary compaction if four drum passes are specified for each layer of earth.

REFERENCES

1. *ASTM Standards on Soil Compaction* (1992). American Society for Testing and Materials: Philadelphia, PA.
2. *ASTM Standards on Soil Stabilization with Admixtures* (1990). American Society for Testing and Materials: Philadelphia, PA.
3. *Construction and Controlling Compaction of Earth Fills* (2000). ASTM Special Technical Publication, 1384, D. W. Shanklin Ed., American Society for Testing and Materials, Philadelphia, April.
4. *Guide to Earthwork Construction* (1990). State of the Art Report 8, TRB, National Research Council, Washington, D.C.
5. Lukas, Robert G. (1986). *Dynamic Compaction for Highway Construction Volume I: Design and Construction Guidelines*, U.S. Department of Transportation, Federal Highway Administration, July.

WEBSITE RESOURCES

1. www.cat.com Caterpillar is a large manufacturer of construction and mining equipment. In the Products, Equipment section of the website can be found the specifications for Caterpillar manufactured rollers.
2. www.deere.com Deere & Company is a worldwide corporation with a construction and forestry products division.
3. www.equipmentworld.com Equipmentworld.com is a news and e-commerce website for construction contractors, equipment manufacturers and dealers, and providers of services and supplies to the construction industry.
4. www.fhwa.dot.gov/pavement/fafacts.pdf *Fly Ash Facts for Engineers.* This document provides basic technical information about the various uses of fly ash in highway construction.
5. www.lime.org/Construct104.pdf *Lime-Treated Soil Construction Manual, Lime Stabilization & Lime Modification,* published by the National Lime Association.
6. www.multiquip.com/Multiquip/pages-products/compaction/compaction.html Multiquip, Inc., is a manufacturer of rammers, plate compactors, and small rollers.

CHAPTER 6

Mobile Equipment Power Requirements

The constructor must select the proper equipment to relocate and/or process materials economically. The analysis procedure for matching the best possible machine to the project task requires inquiry into a machine's mechanical capability. The engineer must first calculate the power required to propel the machine and its load. This power requirement is established by two factors: (1) rolling resistance and (2) grade resistance. Equipment manufacturers publish performance charts for individual machine models. These charts enable the equipment planner to analyze a machine's ability to perform under a given set of job and load conditions.

GENERAL INFORMATION

Heavy/highway construction projects require the handling and processing of large quantities of bulk materials. The constructor of such projects must select the proper equipment to relocate and/or process bulk materials economically. The decision process for matching the best possible *machine* to the project task requires that the engineer take into account both the properties of the material to be handled and the mechanical capabilities of the machines.

When the engineer considers the material-handling problem of a project, there are three crucial *material* considerations: (1) total quantity of material, (2) rate at which it must be moved, and (3) size of the individual pieces. The quantity of material to be handled and the time constraints resulting from the project contract specifications or from expected weather conditions influence the selection of machines as to type, size, and number to be employed. Larger units generally have lower unit-production cost, but there is a trade-off in higher mobilization and fixed costs. The size of the individual material pieces will affect the machine size alternatives that can be considered. A loader used in a quarry to move shot rock must be capable of handling the largest rock sizes produced.

Payload

The payload capacity of construction excavation and hauling equipment can be expressed either *volumetrically* or *gravimetrically*. Volumetric capacity can be stated as struck or heaped volume and either volume can be expressed in terms of loose cubic yard, bank cubic yard, or compacted cubic yard.

The payload capacity of excavation buckets and hauling units is often stated by the manufacturer in terms of the volume of loose material, assuming that the material is heaped in some specified angle or repose. A gravimetric capacity represents the safe operational weight that the axles or structural frame of the machine is designed to handle.

From an economic standpoint, overloading a truck or any other haul unit to improve production looks attractive and overloading by 20% might increase the haulage rate 15%, allowing for slight increases in time to load and haul. The cost per ton hauled should show a corresponding decrease, since direct labor costs will not change and fuel costs will increase only slightly. This apparently favorable situation is only temporary, for the advantage is being bought at the cost of premature aging of the truck and a corresponding increased replacement capital expense.

Machine Performance

Cycle time and payload determine a machine's production rate, and machine travel speed directly affects cycle time. "Why does the machine only travel at 12 mph when its top speed is listed as 35 mph?" To answer the travel speed question, it is necessary to examine three power questions:

1. Required power
2. Available power
3. Usable power

REQUIRED POWER

Power required is the power needed to overcome resisting forces and cause machine motion. The magnitude of resisting forces establishes this power requirement. The forces resisting the movement of mobile equipment are (1) **rolling resistance** and (2) **grade resistance**. Therefore, power required is the power necessary to overcome the *total resistance* to machine movement, which is the sum of rolling and grade resistance.

rolling resistance *The resistance of a level surface to constant-velocity motion across it.*

grade resistance *The force-opposing movement of a machine up a frictionless slope.*

Total resistance (TR) = Rolling resistance (RR) + Grade resistance (GR) **[6.1]**

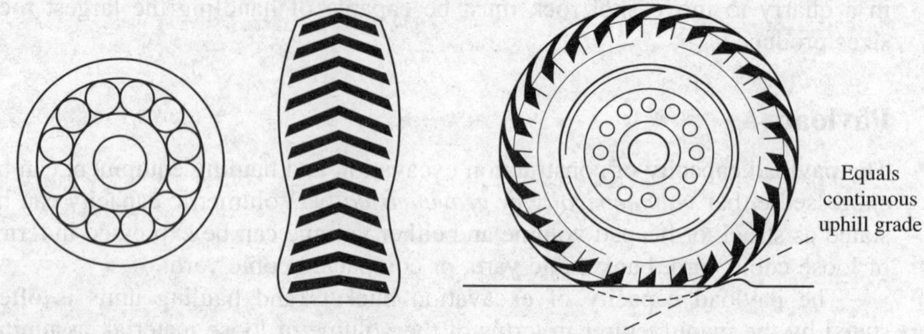

FIGURE 6.1 Mechanisms of rolling resistance.

Rolling Resistance

Rolling resistance is the resistance of a level surface to constant-velocity motion across it. This is sometimes referred to as *wheel resistance* or *track resistance*. Rolling resistance results from friction of the driving mechanism, tire flexing, and the force required to shear through or ride over the supporting surface (Fig. 6.1).

This resistance varies considerably with the type and condition of the surface over which a mobile machine moves (see Fig. 6.2). Soft earth offers a higher resistance than hard-surfaced roads such as those constructed of concrete or asphalt. For machines that move on rubber tires, the rolling resistance varies with the size of, pressure on, and the tread design of the tires. For equipment that moves on crawler tracks, such as tractors, the resistance varies primarily with the type and condition of the road surface.

FIGURE 6.2 Rolling resistance varies with the condition of the surface over which a mobile machine moves.

A narrow-tread, high-pressure tire gives lower rolling resistance than a broad-tread, low-pressure tire on a hard-surfaced road. This is the result of the small area of contact between the tire and the road surface. If the road surface is soft and the tire tends to sink into the earth, a broad-tread, low-pressure tire will offer a lower rolling resistance than a narrow-tread, high-pressure tire. The reason for this condition is that the narrow tire sinks farther into the earth than the broad tire and thus is always having to climb out of a deeper rut—a situation equivalent to climbing a steeper grade.

The rolling resistance of an earthen-haul road probably will not remain constant under varying climatic conditions or for varying types of soil that exist along the road. If the earth is stable, highly compacted, and well maintained by a grader, and if the moisture content is kept near optimum, it is possible to provide a surface with a rolling resistance about as low as that of concrete or asphalt. Moisture can be added, but following an extended period of rain, it may be difficult to remove the excess moisture and the haul road will become soft and rutted; when this happens, rolling resistance will increase. Providing good surface drainage will speed the removal of the water and should permit the road to be reconditioned quickly. For a major earthwork project, it is good economy to provide graders, water trucks, and even rollers to keep the haul road in good condition.

> The maintenance of low-rolling resistance haul roads is one of the best financial investments an earthmoving contractor can make. The cost of having a grader to maintain the haul road is repaid in increased production.

A tire sinks into the soil until the product of bearing area and bearing capacity is sufficient to sustain the load; then the tire is always attempting to climb out of the resulting rut. The rolling resistance will increase about 30 lb/ton for each inch of tire penetration. Total rolling resistance is a function of the riding gear characteristics (independent of speed), the total weight of the vehicle, and torque. It is usually expressed as pounds of resistance per ton of vehicle weight, or as an equivalent grade resistance. Consider a loaded truck that has a gross weight of 20 tons and is moving over a level road having a rolling resistance of 100 lb/ton. The force required to overcome rolling resistance and keep the truck moving at a uniform speed will be 2,000 lb (20 tons × 100 lb/ton).

The estimation of off-road rolling resistance is based largely on empirical information, which may include experience with similar soils. Rarely are rolling resistance values based on actual field tests of haul roads. Much of the actual test data that is available comes from aircraft tire research performed at the U.S. Army Waterways Experiment Station. Although it is impossible to give complete accurate values for the rolling resistances for all types of haul roads and wheels, the values given in Table 6.1 are reasonable estimates.

If desired, one can determine the rolling resistance of a surface by towing a truck or other vehicle whose gross weight is known along a level section at a uniform speed. The tow cable must be equipped with a dynamometer or some

TABLE 6.1 Representative rolling resistances for various types of wheels and crawler tracks versus various surfaces.*

	Steel tires, plain bearings		Crawler type track and wheel		Rubber tires, antifriction bearings			
					High pressure		Low pressure	
Type of surface	lb/ton	kg/m ton	lb/ton	kg/m ton	lb/ton	kg/m ton	lb/ton	kg/m ton
Smooth concrete	40	20	55	27	35	18	45	23
Good asphalt	50–70	25–35	60–70	30–35	40–65	20–33	50–60	25–30
Earth, compacted and maintained	60–100	30–50	60–80	30–40	40–70	20–35	50–70	25–35
Earth, poorly maintained	100–150	50–75	80–110	40–55	100–140	50–70	70–100	35–50
Earth, rutted, muddy, no maintenance	200–250	100–125	140–180	70–90	180–220	90–110	150–200	75–100
Loose sand and gravel	280–320	140–160	160–200	80–100	260–290	130–145	220–260	110–130
Earth, very muddy, rutted, soft	350–400	175–200	200–240	100–120	300–400	150–200	280–340	140–170

*In pounds per ton or kilograms per metric ton of gross vehicle weight.

other device that will enable determination of the average tension in the cable. This tension is the total rolling resistance of the gross weight for the truck. The rolling resistance in pounds per gross ton will be

$$R = \frac{P}{W} \tag{6.2}$$

where

R = rolling resistance in pounds per ton
P = total tension in tow cable in pounds
W = gross weight of mobile vehicle in tons

When tire penetration is known, an approximate rolling resistance value for a wheeled vehicle can be calculated using Eq. [6.3]:

$$\text{RR} = [40 + (30 \times \text{TP})] \times \text{GVW} \tag{6.3}$$

where

RR = rolling resistance in pounds
TP = tire penetration in inches
GVW = gross vehicle weight in tons

FIGURE 6.3 Articulated truck moving up an adverse slope.

Grade Resistance

The force-opposing movement of a machine up a frictionless slope is known as grade resistance. It acts against the total weight of the machine, whether track type or wheel type. When a machine moves up an adverse slope (see Fig. 6.3), the power required to keep it moving increases approximately in proportion to the slope of the road. If a machine moves down a sloping road, the power required to keep it moving is reduced in proportion to the slope of the road. This is known as **grade assistance**.

grade assistance *The effect of gravitational force in aiding movement of a vehicle down a slope.*

The most common method of expressing a slope is by gradient in percent. A 1% slope is one where the surface rises or drops 1 ft vertically in a horizontal distance of 100 ft. If the slope is 5%, the surface rises or drops 5 ft per 100 ft of horizontal distance. If the surface rises, the slope is defined as plus, whereas if it drops, the slope is defined as minus. This is a physical property not affected by the type of machine or the condition of the road, but in respect to analyzing forces its effect is dependent upon the machine's direction of travel.

For slopes of less than 10%, the effect of grade is to increase, for a plus slope, or decrease, for a minus slope, the required tractive effort by 20 lb per gross ton of machine weight for each 1% of grade. This can be derived from elementary mechanics by calculating the required driving force.

From Fig. 6.4, the following relationships can be developed:

$$F = W \sin \alpha \qquad [6.4]$$

$$N = W \cos \alpha \qquad [6.5]$$

For angles less than 10°, $\sin \alpha \approx \tan \alpha$ (the small-angle assumption); with that substitution:

$$F = W \tan \alpha \qquad [6.6]$$

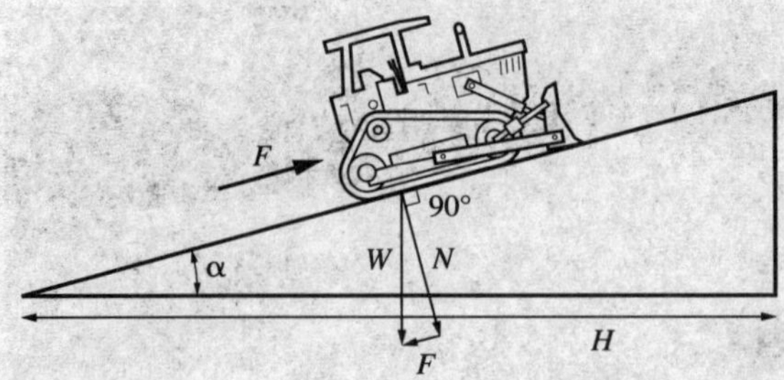

FIGURE 6.4 Frictionless slope-force relationships.

But

$$\tan \alpha = \frac{V}{H} = \frac{G\%}{100}$$

where $G\%$ is the gradient. Hence,

$$F = W \times \frac{G\%}{100} \qquad [6.7]$$

If we substitute $W = 2{,}000$ lb/ton, the formula reduces to

$$F = 20 \text{ lb/ton} \times G\% \qquad [6.8]$$

This formula is valid for a G up to about 10%, that is, the small angle assumption ($\sin \alpha \approx \tan \alpha$).

Total Resistance

Total resistance equals rolling resistance plus grade resistance or rolling resistance minus grade assistance, Eq. [6.1]. It can also be expressed as an *effective grade*.

Using the relationship expressed in Eq. [6.8], a rolling resistance can be equated to an equivalent gradient.

$$\frac{\text{Rolling resistance expressed in lb/ton}}{20 \text{ lb/ton}} = G\% \qquad [6.9]$$

Table 6.2 gives values for the effect of slope, expressed in pounds per gross ton or kilograms per metric ton (m ton) of vehicle weight.

By combining the rolling resistance, expressed as an equivalent grade, and the grade resistance, expressed as a gradient in percent, one can express the total resistance as an effective grade. The three terms—power required, total resistance, and effective grade—all denote the same thing. Power required is expressed in pounds. Total resistance is expressed in pounds or pounds per ton of machine weight, and effective grade is expressed in percent.

TABLE 6.2 The effect of grade on the tractive effort of vehicles.

Slope (%)	lb/ton*	kg/m ton*	Slope (%)	lb/ton*	kg/m ton*
1	20.0	10.0	12	238.4	119.2
2	40.0	20.0	13	257.8	128.9
3	60.0	30.0	14	277.4	138.7
4	80.0	40.0	15	296.6	148.3
5	100.0	50.0	20	392.3	196.1
6	119.8	59.9	25	485.2	242.6
7	139.8	69.9	30	574.7	287.3
8	159.2	79.6	35	660.6	330.3
9	179.2	89.6	40	742.8	371.4
10	199.0	99.5	45	820.8	410.4
11	218.0	109.0	50	894.4	447.2

*Ton or metric ton of gross vehicle weight.

EXAMPLE 6.1

The haul road from the borrow pit to the fill has an adverse grade of 4%. Wheel-type hauling units will be used on the job, and it is expected that the haul-road rolling resistance will be 100 lb per ton. What will be the effective grade for the haul? Will the units experience the same effective grade for the return trip?

Using Eq. [6.9], we obtain

$$\text{Equivalent grade (RR)} = \frac{100 \text{ lb/ton rolling resistance}}{20 \text{ lb/ton}} = 5\%$$

$$\text{Effective grade } (\text{TR}_{\text{haul}}) = 5\% \text{ RR} + 4\% \text{ GR} = 9\%$$

$$\text{Effective grade } (\text{TR}_{\text{return}}) = 5\% \text{ RR} - 4\% \text{ GR} = 1\%$$

where

RR = rolling resistance
GR = grade resistance

Note that the effective grade is not the same for the two cases. During the haul, the unit must overcome the uphill grade; on the return, the unit is aided by the downhill grade.

Haul Routes During the life of a project the haul-route grades (and, therefore, grade resistance) may remain constant. One example of this is the trucking of aggregate from a rail-yard off-load point to the concrete batch plant. In most cases, however, the haul-route grades change as the work progresses. On a liner highway or airfield projects, the tops of the hills are excavated and hauled into the valleys. Early in the work, the grades are steep and reflect the existing natural ground. Over the life of the project, the grades begin to assume the design profile. Therefore, the engineer must first study the project's mass diagram to determine the direction that the material has to be moved. Then the natural ground and the design profiles depicted on the plans must be checked to determine the grades that the equipment will encounter during haul and return cycles.

Site work projects are usually not linear in extent; therefore a mass diagram is not very useful. The engineer in that case must look at the cut-and-fill areas, lay out probable haul routes, and then check the natural and finish grade contours to determine the haul-route grades.

This process of laying out haul routes is critical to machine productivity. If a route can be found that results in less grade resistance, machine travel speed can be increased and production will likewise increase. In planning a project, a constructor should always check several haul-route options before deciding on a final construction plan.

> Hauling efficiency is achieved by careful planning of haul routes.

Equipment selection is affected by travel distance because of the time factor distance introduces into the production cycle. All other factors being equal, increased travel distances will favor the use of high-speed large-capacity machines. The difference between the self-loading scraper and a push-loaded scraper can be used as an illustration. The self-loading scraper will load, haul, and spread without any assisting equipment, but the extra weight of the loading mechanism reduces the scrapers' maximum travel speed and load capacity. A scraper, which requires a push tractor to help it load, does not have to expend power to haul a loading mechanism with it on every cycle. It will be more efficient in long-haul situations as it does not have to expend fuel transporting extra machine weight.

AVAILABLE POWER

Internal combustion engines power most construction equipment (see Fig. 6.5). Because diesel engines perform better under heavy-duty applications than gasoline engines, diesel-powered machines are the workhorses of the construction industry. Additionally, diesel engines have longer service lives and lower fuel consumption, and diesel fuel presents less of a fire hazard. No matter which type of engine serves as the power source, the mechanics of energy transmission are the same.

Work and Power

Work is defined as force through a distance, Eq. [6.10]. Work is achieved when a force causes an object to move.

$$\text{Work} = \text{Force} \times \text{Distance} \qquad \textbf{[6.10]}$$

FIGURE 6.5 Cutaway of an internal combustion engine.

EXAMPLE 6.2

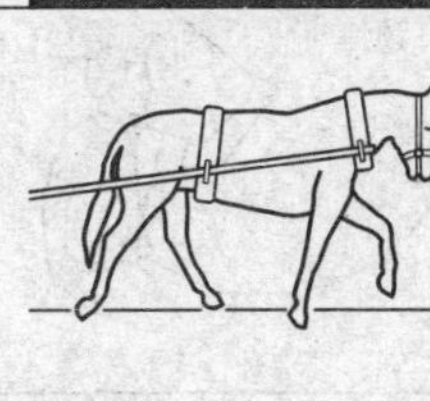

A 180-lb horse walks in a circular path operating a pump that raises water from a mine. The horse is hitched to a 12-ft lever arm that is connected to the pumping mechanism. The horse makes 144 revolutions (rev) per hour. How much work does the horse do in 1 hour?

The 180-lb horse is the moving force.
The distance moved is the circumference of the circle 144 times.

$$\text{Circumference} = 2 \times \pi \times \text{radius}$$

$$\text{Circumference (12-ft radius)} = 2 \times \pi \times 12 \text{ ft} = 75.4 \text{ ft}$$

$$\text{Total distance moved} = 75.4 \text{ ft} \times 144 \text{ rev} = 10{,}857 \text{ ft}$$

$$\text{Work} = 180 \text{ lb} \times 10{,}857 \text{ ft} = 1{,}954{,}320 \text{ lb-ft/hr}$$

When James Watt developed the first practical steam engine and wanted to express the work his engine could do, he related it to the horse in Example 6.2. This was caused by the fact that the purpose of his engine was to replace the horses that were used to power pumping apparatuses used in the mines across England. Watt defined power as the amount of work that can be done in a certain amount of time, as

$$\text{Power} = \frac{\text{Work}}{\text{Time}} \qquad \textbf{[6.11]}$$

Therefore, the power of a horse would be

$$\text{Power of a horse} = \frac{1{,}954{,}320 \text{ lb-ft/hour}}{60 \text{ min/hr}} = 32{,}572 \text{ lb-ft/min}$$

Watt rounded this value to 33,000 lb-ft/min (or 550 lb-ft/sec), which is the definition of one horsepower. Horsepower is a unit of power.

Torque

An internal combustion engine by the combustion of fuel in a piston develops a mechanical force that acts on a crankshaft having a radius r. The crankshaft in turn drives the flywheel and gears that power the other components of the machine. The force from a rotating object, such as crankshafts (a "twisting" force), is termed *torque*. A pound-foot of torque is the twisting force necessary to support a 1-lb weight on a weightless horizontal bar, 1 ft from the fulcrum.

EXAMPLE 6.3

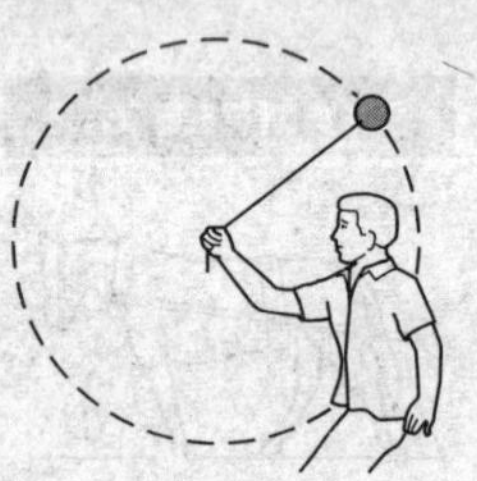

Calculate the work accomplished by moving a 1-lb weight one revolution if the radius of the fulcrum is one foot.

One revolution equals the circumference of a circle having a one-foot radius.

$$\text{One revolution (circumference)} = 2 \times \pi \times \text{radius}$$

$$\text{One revolution (1-ft radius)} = 2 \times \pi \times 1 \text{ ft} = 6.2832 \text{ ft}$$

$$\text{Work} = 1 \text{ lb} \times 6.2832 \text{ ft} = 6.2832 \text{ lb-ft}$$

The torque represented by one revolution of work is 6.2832 lb-ft. The power exerted by a rotating object is the torque it exerts multiplied by the speed at which it rotates (revolutions per minute—rpm). The relationship between horsepower and torque (T) at a specified rpm can therefore be established:

$$\text{Horsepower (hp)} = \frac{6.2832 \times \text{rpm} \times T}{33{,}000} = \frac{\text{rpm}}{5{,}252} \times T \qquad \textbf{[6.12]}$$

Conversely, to calculate torque

$$\text{Torque } (T) = \frac{5{,}252}{\text{rpm}} \times \text{hp} \qquad \textbf{[6.13]}$$

Horsepower Rating

Manufacturers rate machine horsepower as either gross or flywheel (sometimes listed as net horsepower). Gross horsepower is the actual power generated by the engine prior to load losses for auxiliary systems, such as the alternator, air conditioner compressors, and water pump. Flywheel horsepower (fwhp) can be considered as *usable* horsepower. It is the power available to operate a machine—power the driveline—after deducting for power losses in the engine. This horsepower is sometimes listed as brake horsepower (bhp). Prior to electronic bench testing, horsepower was quantified as the amount of resistance against a flywheel brake. Although the method is no longer used, the term remains in the industry.

Flywheel power	209 kW (280 hp)
Rated power SAE (net)	198 kW (265 hp)

The Society of Automotive Engineers (SAE) standardized engine-rating procedure (J1349) measures horsepower at the flywheel, using an engine dynamometer. The engine is tested with all accessories installed, including a full exhaust system, all pumps, the alternator, the starter, and emissions controls. So today, equipment manufacturers measure torque on a dynamometer and then calculate horsepower by converting the radial force of torque into work units of horsepower.

Power Output and Torque

Figure 6.6 shows the typical curves for brake horsepower and torque as an engine increases its crankshaft speed to the governed rpm value. The important feature of this plot is the shape of the torque curve. Maximum torque is not obtained at maximum rpm. Remember Eq. [6.13], where 5,252 is divided by the rpm (5,252 ÷ rpm). This tells us that when the engine turns at an rpm less than 5,252, the effect is to produce a ratio value greater than one—to increase torque for the same hp value. Once the engine rpm increases past 5,252, the effect is to produce a ratio value less than one—to reduce the torque for the same hp value. This effect provides the engine with a power reserve. When a machine is subjected to a momentary overload, the rpm drops and the torque goes up, keeping the engine from stalling. This is commonly referred to as "lugging" the engine.

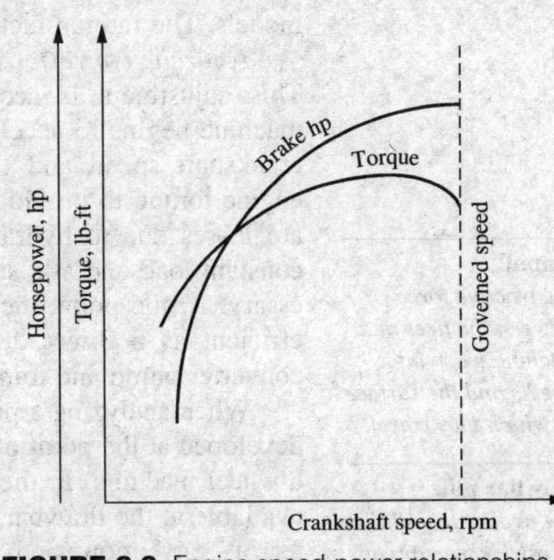

FIGURE 6.6 Engine speed-power relationships.

The power output from the engine, fwhp, becomes the power input to the transmission system. This system consists of the drive shaft, a

FIGURE 6.7 Cutaway of the gearing between the drive shaft and drive axles.

transmission, planetary gears, drive axles (see Fig. 6.7), and drive wheels. Machines can be purchased with either a direct drive (standard) or torque-converter drives. With a direct-drive machine, the operator must manually shift gears to match engine output to the resisting load. The difference in power available when considering maximum torque and torque at the governed speed is the machine's operating range for a given gear. In those applications where load is constantly changing, the operation of a direct-drive machine requires a skilled operator. Operator skill is a significant factor controlling the amount of wear and tear a direct-drive machine will experience. Operators of direct-drive machines will be subjected to more operator fatigue than those on power-shift models. The fatigue factor will affect machine productivity.

A *torque converter* is a device that adjusts power output to match the load. This adjustment is accomplished hydraulically by a fluid coupling. As a machine begins to accelerate, the engine rpm will quickly reach the governed crankshaft speed and the torque converter will automatically multiply the engine torque to provide the required acceleration force. In this process, there are losses due to hydraulic inefficiencies. If the machine is operating under constant load and at a steady whole-body speed, no torque multiplying is necessary. At the point, the transmission of engine torque can be made nearly as efficient as a direct drive transmission by locking ("lock-up") the torque-converter pump and transmission together.

rimpull
The tractive force between the tires of a machine's driving wheels and the surface on which they travel.

drawbar pull
The available pull that a crawler tractor can exert on a towed load.

When analyzing a piece of equipment, we are interested in the usable force developed at the point of contact between the tire and the ground **(rimpull)** for a wheel machine. In the case of a track machine, the force in question is that available at the drawbar **(drawbar pull)**. The difference in the name is a matter of convention; both rimpull and drawbar pull are measured in the same units, pounds pull.

Rimpull

Rimpull is a term that is used to designate the tractive force between the tires of a machine's driving wheels and the surface on which they travel. If the **coefficient of traction** is sufficiently high there will be no tire slippage, in which case maximum rimpull is a function of the power of the engine and the gear ratios between the engine and the driving wheels. If the driving wheels slip on the supporting surface, the maximum effective rimpull will be equal to the total pressure the tires exert on the surface multiplied by the coefficient of traction. Rimpull is expressed in pounds.

coefficient of traction *The factor that determines the maximum possible tractive force between the powered running gear of a machine and the surface on which it travels.*

If the rimpull of a machine is not known, it can be determined from the equation

$$\text{Rimpull} = \frac{375 \times \text{hp} \times \text{efficiency}}{\text{speed (mph)}} \text{(lb)} \qquad \textbf{[6.14]}$$

This is the formulation of the available power accounting for the machine's available horsepower and operating speed. The efficiency of most tractors and trucks will range from 0.80 to 0.85. SAE recommends using 0.85 when the efficiency is not known.

EXAMPLE 6.4

Assuming that the coefficient of traction is sufficient such that the machine's total power can be developed, calculate the rimpull of a pneumatic-tired truck with a 140-hp engine. The tractor is operating at a speed of 3.3 mph in first gear. Use the SAE recommended efficiency.

$$\text{Rimpull} = \frac{375 \times 140 \times 0.85}{3.3} = 13{,}523 \text{ lb}$$

The maximum rimpull in all gear ranges for the truck in Example 6.4 is shown in Table 6.3.

In computing the pull that a tractor can exert on a towed load (speed the machine can attain), it is necessary to deduct from the rimpull of the tractor the force required to overcome the total resistance—the combination of rolling and

TABLE 6.3 Maximum rimpulls for the Example 6.4 tractor.

Gear	Speed(mph)	Rimpull(lb)
First	3.3	13,523
Second	7.1	6,285
Third	12.9	3,542
Fourth	21.5	2,076
Fifth	33.9	1,316

grade. Consider a tractor that weighs 12.4 tons and whose maximum rimpull in the first gear is 13,523 lb. If it is operated on a haul road with a positive slope of 2% and a rolling resistance of 100 lb/ton, the full 13,523 lb of rimpull will not be available for towing, as a portion will be required for overcoming the total resistance arising from the haul conditions. Given the stated conditions, the pull available for towing a load will be

Maximum rimpull = 13,523 lb

Force required to overcome grade resistance,
$12.4 \text{ ton} \times (20 \text{ lb/ton} \times 2\%) = 496 \text{ lb}$

Force required to overcome rolling resistance,
$12.4 \text{ ton} \times 100 \text{ lb/ton} = 1{,}240 \text{ lb}$

Total resistance, 496 lb + 1,240 lb = 1,736 lb

Power available for towing a load (13,523 lb − 1,736 lb) = 11,787 lb

Drawbar Pull

The towing force a crawler tractor can exert on a load is referred to as drawbar pull. Drawbar pull is typically expressed in pounds. To determine the drawbar pull available for towing a load it is necessary to subtract from the total pulling force available at the engine the force required to overcome the total resistance imposed by the haul conditions. If a crawler tractor tows a load up a slope, its drawbar pull will be reduced by 20 lb for each ton of weight of the tractor for each 1% slope.

The performance of crawler tractors, as reported in the specifications supplied by the manufacturer, is usually based on the Nebraska tests. In testing a tractor to determine its maximum drawbar pull at each of the available speeds, the haul road is calculated to have a rolling resistance of 110 lb/ton. If a tractor is used on a haul road whose rolling resistance is higher or lower than 110 lb/ton, the drawbar pull will be reduced or increased, respectively, by an amount equal to the weight of the tractor in tons multiplied by the variation of the haul road from 110 lb/ton.

EXAMPLE 6.5

A track-type tractor whose weight is 15 tons has a drawbar pull of 5,685 lb in sixth gear when operated on a level road having a rolling resistance of 110 lb/ton. If the tractor is operated on a level road having a rolling resistance of 180 lb/ton, the drawbar pull will be reduced by what amount?

$$15 \text{ tons} \times (180 \text{ lb/ton} - 110 \text{ lb/ton}) = 1{,}050 \text{ lb}$$

Thus, the effective drawbar pull will be 5,685 − 1,050 = 4,635 lb.

The drawbar pull of a crawler tractor will vary indirectly with the speed of each gear. It is highest in the first gear and lowest in the top gear. The specifications supplied by the manufacturer should give the maximum speed and drawbar pull for each of the gears.

USABLE POWER

Usable power depends on project conditions: primarily, haul-road surface condition, altitude, and temperature. Underfoot conditions determine how much of the available power can be transferred to the surface to propel the machine. As altitude increases, the air becomes less dense. Above 3,000 ft, the decrease in air density may cause a reduction in horsepower output of some engines. Manufacturers provide charts detailing appropriate altitude power reductions (see Table 6.5 later in the chapter). Temperature will also affect engine output.

Coefficient of Traction

The total energy of an engine in any machine designed primarily for pulling a load can be converted into tractive effort only if sufficient traction can be developed between the driving wheels or tracks and the haul surface. If there is insufficient traction, the full power of the engine will not be available to do work, as the wheels or tracks will slip on the surface.

The coefficient of traction can be defined as the factor by which the total weight on the drive wheels or tracks should be multiplied to determine the maximum possible tractive force between the wheels or tracks and the surface just before slipping will occur.

$$\text{Usable force} = \text{Coefficient of traction} \times \text{Weight on powered running gear} \quad \textbf{[6.15]}$$

The power that can be developed to do work is often limited by traction. The factors controlling usable horsepower are the weight on the powered running gear (drive wheels for wheel type, total weight for track type—see Fig. 6.8), the characteristics of the running gear, and the characteristics of the travel surface.

The coefficient of traction between rubber tires and travel surfaces will vary with the type of tread on the tires and with the travel surface. For crawler tracts, it will vary with the design of the grosser and the travel surface. These variations are such that exact values cannot be given. Table 6.4 gives approximate values for the coefficient of traction between rubber tires and crawler tracks, and surface materials and conditions. These coefficients are sufficiently accurate for most estimating purposes.

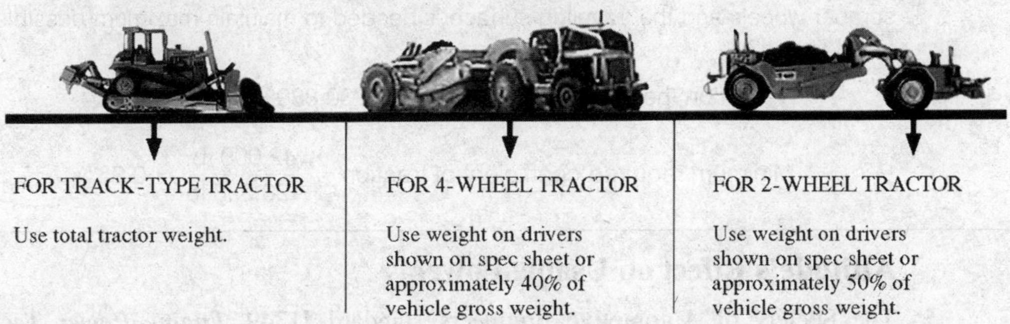

FIGURE 6.8 Weight distribution on powered running gear.

TABLE 6.4 Coefficients of traction for various road surfaces.

Surface	Rubber tires	Crawler tracks
Dry, rough concrete	0.80–1.00	0.45
Dry, clay loam	0.50–0.70	0.90
Wet, clay loam	0.40–0.50	0.70
Wet sand and gravel	0.30–0.40	0.35
Loose, dry sand	0.20–0.30	0.30
Dry snow	0.20	0.15–0.35
Ice	0.10	0.10–0.25

EXAMPLE 6.6

Assume that a rubber-tired tractor has a total weight of 18,000 lb on its driving wheels. The maximum rimpull in low gear is 9,000 lb. If the tractor is operating in wet sand, with a coefficient of traction of 0.30, what is the the maximum possible rimpull prior to slippage of the tires?

$$0.30 \times 18{,}000 \text{ lb} = 5{,}400 \text{ lb}$$

Regardless of the engine horsepower, because of wheel slippage not more than 5,400 lb of force (power) is available to do work. If the same tractor is operating on dry clay, with a coefficient of traction of 0.60, what is the maximum possible rimpull prior to slippage of the wheels?

$$0.60 \times 18{,}000 \text{ lb} = 10{,}800 \text{ lb}$$

For this surface, the engine will not be able to cause the tires to slip. Thus, the full 9,000 lb of rimpull is available to do work.

EXAMPLE 6.7

A wheel tractor scraper is used on a road project. When the project begins, the scraper will experience high rolling and grade resistance at one work area. The rimpull required to maneuver in this work area is 42,000 lb. In the fully loaded condition, 52% of the total vehicle weight is on the drive wheels. The fully loaded vehicle weight is 230,880 lb. What minimum value of coefficient of traction between the scraper wheels and the traveling surface is needed to maintain maximum possible travel speed?

$$\text{Weight on the drive wheels} = 0.52 \times 230{,}880 \text{ lb} = 120{,}058 \text{ lb}$$

$$\text{Minimum required coefficient of traction} = \frac{42{,}000 \text{ lb}}{120{,}058 \text{ lb}} = 0.35$$

Altitude's Effect on Usable Power

The Society of Automotive Engineers standard J1349, *Engine Power Test Code—Spark Ignition and Compression Ignition—Net Power Rating Standard,* specifies a basis for a net engine power rating. The standard conditions for the SAE rating are a temperature of 60°F (15.5°C) and sea-level barometric pressure of 29.92 in. mercury (Hg) [103.3 kilopascals (kPa)].

The important point here is that the ratings are based on a specific barometric pressure. For naturally aspirated engines, operation at altitudes above sea level will cause a significant decrease in available engine power as the barometric pressure decreases. A decrease in barometric pressure causes a corresponding decrease in air density, and to operate at peak efficiency the engine must have the proper amount of air. A reduction in air density affects the combustion fuel-to-air ratio in the engine's pistons. For specific machine applications, the manufacturer's performance data should be consulted. Table 6.5 presents data for selected Caterpillar Inc. machines.

The effect of the loss in power due to altitude can be eliminated by the installation of a turbocharger or a supercharger. These are mechanical forced-induction systems that compress the air flowing into the combustion chamber of the engine, thus permitting sea-level performance at higher altitudes. The

TABLE 6.5 Percent flywheel horsepower available for select Caterpillar machines at specified altitudes.

Model	0–2,500 ft (0–760 m)	2,500–5,000 ft (760–1,500 m)	5,000–7,500 ft (1,500–2,300 m)	7,500–10,000 ft (2,300–3,000 m)	10,000–12,500 ft (3,000–3,800 m)
Tractors					
D6D, D6E	100	100	100	100	94
D7G	100	100	100	94	86
D8L	100	100	100	100	93
D8N	100	100	100	100	98
D9N	100	100	100	96	89
D10N	100	100	100	94	87
Graders					
120G	100	100	100	100	96
12G	100	100	96	90	84
140G	100	100	100	100	94
14G	100	100	100	94	87
16G	100	100	100	100	100
Excavators					
214B	100	100	100	100	92
235D	100	100	100	98	91
245D	100	100	100	94	87
Scrapers					
615C	100	100	95	88	81
621E	100	100	94	87	80
623E	100	100	94	87	80
631E	100	100	96	88	82
Trucks					
769C	100	100	100	97	89
773B	100	100	100	100	96
Loaders					
966E	100	100	100	93	86
988B	100	100	100	100	93

Source: Reprinted courtesy of Caterpillar Inc.

fundamental difference between a turbocharger and a supercharger is the unit's source of power. With a turbocharger, the exhaust stream powers a turbine, which in turn spins the compressor. The power source for a supercharger is a belt connected directly to the engine. If equipment is to be used at high altitudes for long periods of time, the increased performance will probably pay for installing one of these two devices.

PERFORMANCE CHARTS

performance charts *Graphical presentation of power and corresponding speed that an engine and transmission of a mobile machine can deliver.*

Equipment manufacturers publish **performance charts** for individual machine models. These charts enable the equipment estimator/planner to analyze a machine's ability to perform under a given set of project-imposed load conditions. The performance chart is a graphical representation of the power and corresponding speed the engine and transmission can deliver. The load condition is stated as either rimpull or drawbar pull. It should be noted that the drawbar pull/rimpull-speed relationship is inverse because as vehicle speed increases pull decreases.

Drawbar Pull Performance Chart

In the case of the track machine whose drawbar pull performance chart is shown in Fig. 6.9, the available power ranges from 0 to 56,000 lb (the vertical scale) and the speed ranges from 0 to 6.5 mph (the horizontal scale).

Assuming the power required for a certain application is 25,000 lb, this machine would travel efficiently at a speed of approximately 1.4 mph in first gear. This is found by first locating the 25,000-lb mark on the left vertical scale and then moving horizontally across the chart. At the intersect point of this horizontal projection and the gear curve, project a vertical line downward to the speed scale at the bottom of the chart. The horizontal projection at 25,000 lb in this case also intersects the second gear curve. If the tractor was operated in second gear, it could only obtain a speed of 1 mph.

Rimpull Performance Charts

Each manufacturer has a slightly different graphical layout for presenting performance chart information. However, the procedures for reading a performance chart are basically the same. The steps described here are based on the chart shown in Fig. 6.10.

Required Power—Total Resistance The graphical arrangement of a rimpull chart enables a determination of machine speed using total resistance expressed either in terms of force (rimpull) or percent effective grade. The procedures to determine machine speed from a rimpull performance chart are

1. Ensure that the proposed machine has the same engine, gear ratios, and tire size as those identified for the machine on the chart. If the gear ratios or rolling radius of a machine's tires are changed, the performance curve will shift along both the rimpull and speed axes.

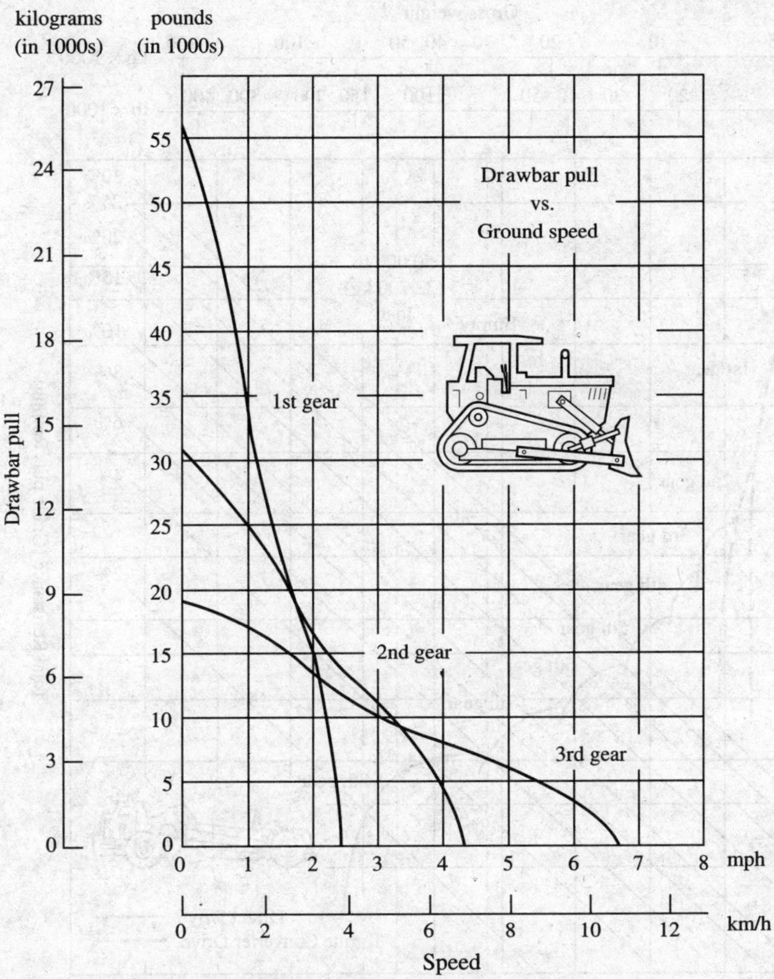

FIGURE 6.9 Drawbar pull performance chart for a crawler tractor.
Reprinted courtesy of Caterpillar Inc.

2. Estimate the rimpull (power) required—total resistance (rolling resistance plus grade resistance)—based on the probable job conditions.
3. Locate the power requirement value on the left vertical scale and project a line horizontally to the right intersecting a gear curve. The point of intersection of the projected horizontal line with a gear curve defines the operating relationship between horsepower and speed.
4. From the point at which the horizontal line intersects the gear curve, project a line vertically to the bottom *x* axes, which indicates the speed in mph and km/h. Sometimes the horizontal line from the power requirement (rimpull) will intersect the gear range curve at two different points (see Fig. 6.11). In such a case, the speed can be interpreted in two ways.

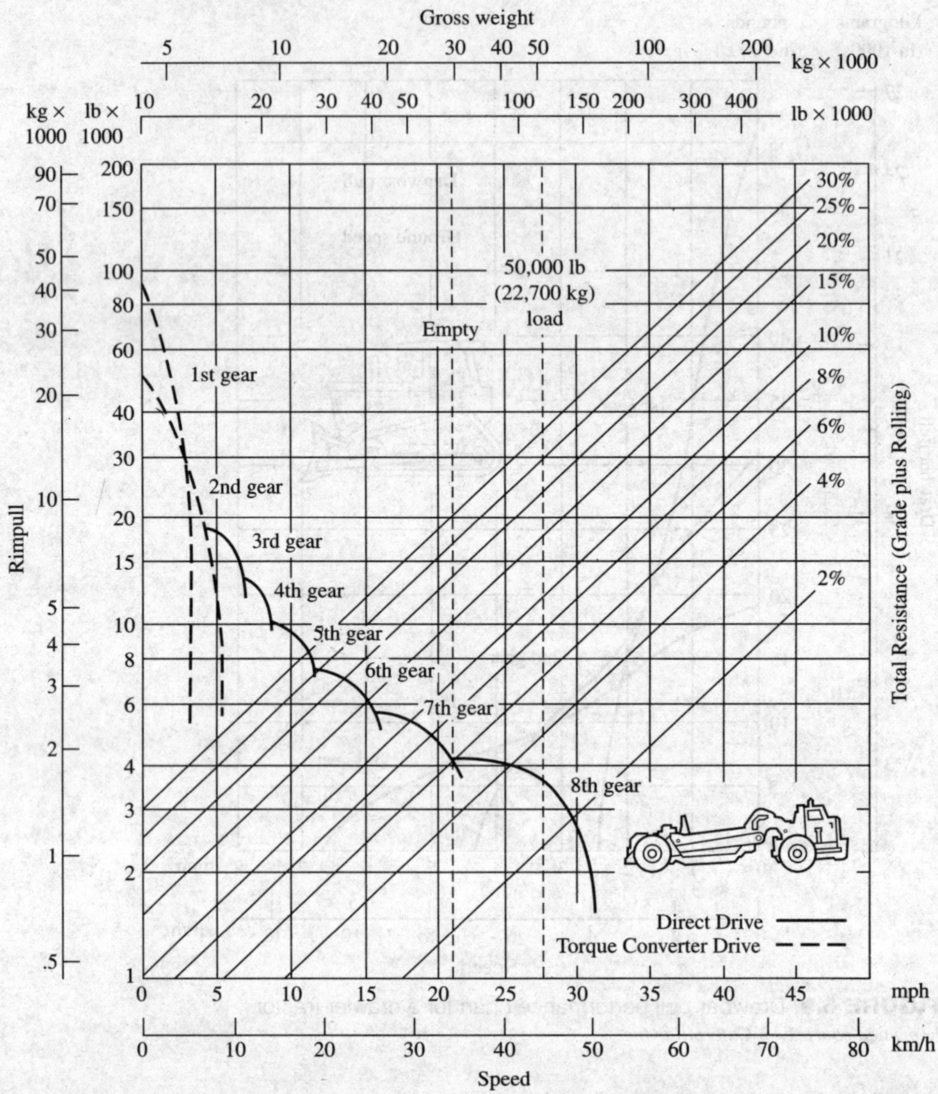

FIGURE 6.10 Rimpull performance chart for a wheel tractor scraper.
Reprinted courtesy of Caterpillar Inc.

A guide in determining the appropriate speed is

- If the required rimpull is *less* than that required on the previous stretch of haul road, use the higher gear and speed.
- If the required rimpull is *greater* than that required on the previous stretch of haul road, use the lower gear and speed.

Effective Grade—Total Resistance Assuming that the total resistance has been expressed as an effective grade, the procedures to determine speed are

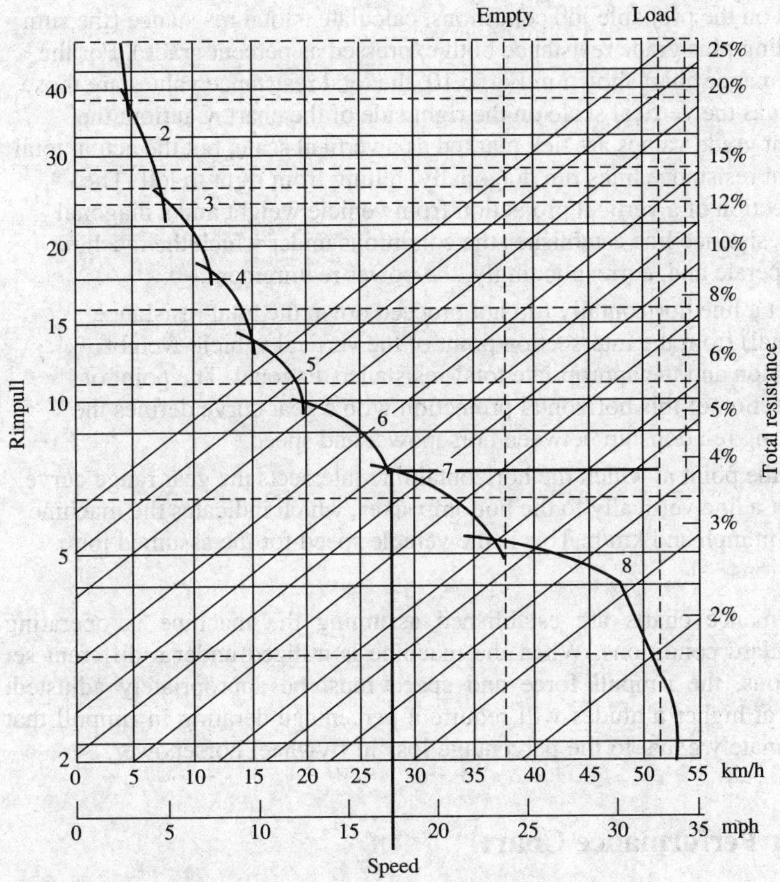

FIGURE 6.11 Rimpull performance chart—gear effect detail.
Reprinted courtesy of Caterpillar Inc.

1. Ensure that the proposed machine has the same engine, gear ratios, and tire size as those identified for the machine on the chart.
2. Determine the machine weight both when the machine is empty and loaded. The empty weight is the operating weight and should include coolants, lubricants, full fuel tanks, and operator. Loaded weight depends on the density of the carried material and the load size (volume). These two weights, empty and loaded, are often referred to as the *net vehicle weight* (NVW) and the *gross vehicle weight* (GVW), respectively. The NVW (empty weight) is usually marked on the performance chart. Likewise the GVW (gross weight), based on the gravimetric capacity of the machine, will also usually be indicated on the chart. Vehicle weights are depicted on the upper horizontal scale of the chart. Note that this horizontal weight data is presented as a log scale.

3. Based on the probable job conditions, calculate a total resistance (the sum of rolling plus grade resistance both expressed as percent grade). For the performance chart shown in Fig. 6.10, the total resistance values are those shown as the vertical scale on the right side of the chart. Caution: the percent grade values are tick marked as a vertical scale, but the actual total percent resistance lines run diagonally, falling from right to left. The intersection of a vertical projection from vehicle weight and a diagonal total resistance line establishes the conditions under which the machine will operate and, correspondingly, the power requirement.
4. Project a line horizontally (do not proceed down the total resistance diagonal) from the intersection point of the vertical vehicle-weight projection and the appropriate total-resistance diagonal. The point of intersection of this horizontal projection with a gear curve defines the operating relationship between horsepower and speed.
5. From the point at which the horizontal line intersects the gear range curve, project a line vertically to the bottom x axes, which indicates the machine speed in mph and km/h. This is the vehicle speed for the assumed job conditions.

Performance charts are established assuming the machine is operating under standard conditions. When the machine is utilized under a different set of conditions, the rimpull force and speed must be appropriately adjusted. Operation at higher altitudes will require a percentage derating in rimpull that is approximately equal to the percentage loss in flywheel horsepower.

Retarder Performance Chart

When operating on steep downgrades, a machine's speed may have to be limited for safety reasons. A retarder is a dynamic speed control device. By the use of an oil-filled chamber between the torque converter and the transmission, machine speed is retarded. The retarder will not stop the machine; rather it provides speed control for long downhill hauls, reducing wear on the service brake. Figure 6.12 presents a **retarder chart** for a wheel tractor scraper.

retarder chart *A graphical presentation that identifies the controlled speed of a machine descending a slope when the magnitude of the grade assistance is greater than the rolling resistance.*

The retarder performance chart (see Fig. 6.12) identifies the speed that can be maintained when a vehicle is descending a grade having a slope such that the magnitude of the grade assistance is greater than the resisting rolling resistance. This retarder-controlled speed is steady state, and use of the service brake will not be necessary to prevent acceleration.

A retarder performance chart is read in a manner similar to that already described for performance charts, remembering that the total resistance (effective grade) values in this case are actually negative values. As with the rimpull chart, the horizontal line can intersect more than one gear. In a particular gear, the vertical portion of the retarder curve indicates maximum retarder effort and resulting speed. If haul conditions dictate, the operator will shift into a lower gear and a lower speed would be applicable. Many times the decision as to

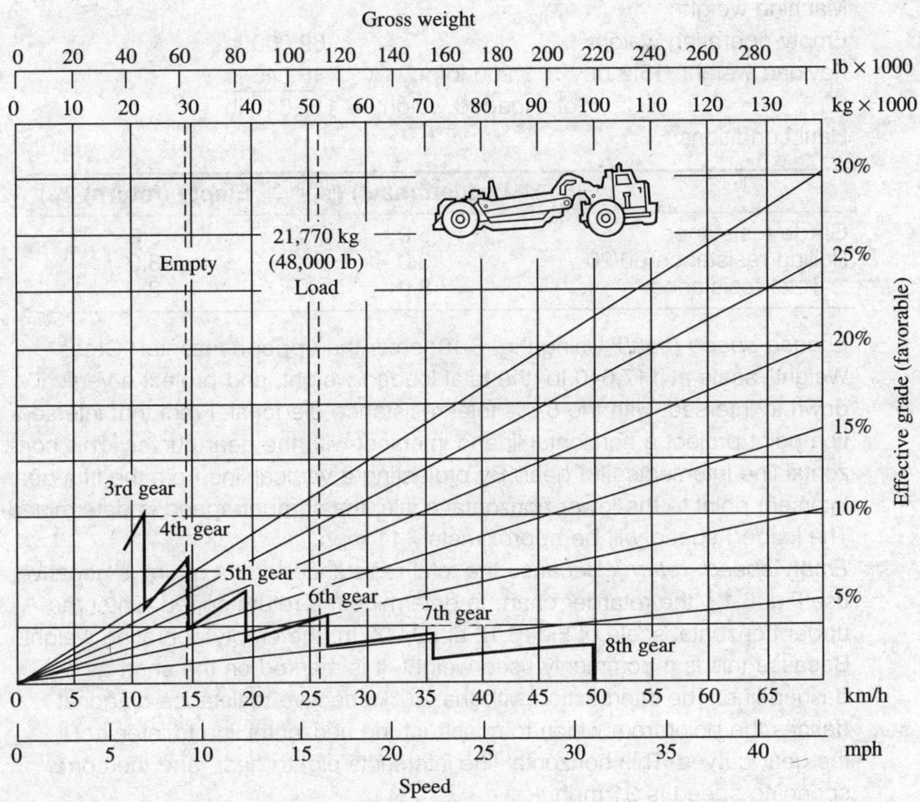

FIGURE 6.12 Retarder performance chart for a wheel tractor scraper.
Reprinted courtesy of Caterpillar Inc.

which speed to select is answered by the question: "How much effort will be expended in haul-route maintenance?" Route smoothness is often the controlling factor affecting higher operating speeds.

EXAMPLE 6.8

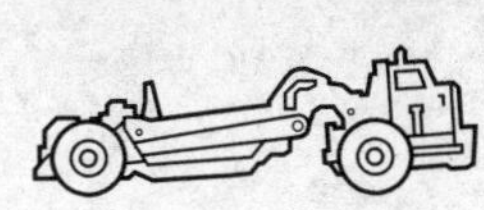

A contractor proposes to use scrapers on an embankment job. The performance characteristics of the machines are shown in Figs. 6.10 and 6.12. The scrapers have a rated capacity of 14-cy struck. Operating weight empty is 69,000 lb. The weight distribution of the scraper when loaded is 53% on the drive wheels.

The contractor believes that the average scraper load will be 15.2 bcy. The haul from the excavation area is a uniform adverse gradient of 5.0% with a rolling resistance of 60 lb/ton. The material to be excavated and transported is a common earth with a bank unit weight of 3,200 lb/bcy.

a. Calculate the maximum travel speeds that can be expected.

Machine weight:

Empty operating weight	69,000 lb
Payload weight, 15.2 bcy × 3,200 lb/bcy	48,640 lb
Total loaded weight =	117,640 lb

Haul conditions:

	Loaded (haul) (%)	Empty (return) (%)
Grade resistance	5.0	−5.0
Rolling resistance 60/20	3.0	3.0
Total resistance	8.0	−2.0

Loaded speed (haul): Using Fig. 6.10, enter the upper horizontal "Gross Weight" scale at 117,640 lb, the total loaded weight, and project a vertical down to intersect with the 8.0% total resistance diagonal. From that intersection point project a horizontal line to intersect with the gear curves. This horizontal line intersects fifth gear. By projecting a vertical line from the fifth-gear intersect point to the lower horizontal scale, the scraper speed is determined. The loaded speed will be approximately 11 mph.

Empty speed (return): Because the total resistance for the return is negative, use Fig. 6.12, the retarder chart, to determine the return speed. Enter the upper horizontal scale of Fig. 6.12 at 69,000 lb, the empty operating weight. Because this is a commonly used weight, it is marked on the chart as a dashed line. The intersection with the 2.0% effective resistance diagonal defines the point from which to construct the horizontal line to intersect the gear curves. This horizontal line intersects eighth gear, and the corresponding speed is 31 mph.

b. If the job is at elevation 12,500 ft, what will be the operating speeds when consideration of the nonstandard barometric pressure is included in the analysis? For operating at an altitude of 12,500 ft, the manufacturer has reported that these scrapers, which have turbocharger engines, can deliver 82% of rated flywheel horsepower.

At an altitude of 12,500 ft, the rimpull that is necessary to overcome a total resistance of 8% must be adjusted for the altitude derating.

$$\text{Altitude-adjusted effective resistance: } \frac{8.0}{0.82} = 9.8\%$$

Now proceed as in part a by locating the intersection of the total loaded weight line and the altitude-adjusted total resistance line. Projecting a horizontal line from the 9.8% altitude-adjusted effective resistance diagonal yields an intersection with the fourth-gear curve and, by projecting a vertical line at that point, a speed of approximately 8 mph is determined.

In this problem, the nonstandard altitude would not affect the empty (return) speed because the machine does not require rimpull force for downslope motion. The retarder controls the machine's downhill momentum. Therefore, the empty (return) speed is still 31 mph at 12,500-ft altitude.

SUMMARY

The payload of hauling equipment can be expressed either *volumetrically* or *gravimetrically*. Power required is the power necessary to overcome the *total resistance* to machine movement, which is the sum of the rolling and grade resistances. Rolling resistance is the resistance of a level surface to constant-velocity motion across it. The force-opposing movement of a machine up a frictionless slope is known as *grade resistance*.

The *coefficient of traction* is the factor by which the total weight on the drive wheels or tracks of a machine should be multiplied in order to determine the maximum possible tractive force between the wheels or tracks and the supporting surface just before slipping occurs. A performance chart is a graphical representation of power and the corresponding speed the engine and transmission can deliver. Critical learning objectives include:

- An ability to calculate vehicle weight.
- An ability to determine rolling resistance based on anticipated haul-road conditions.
- An ability to calculate grade resistance.
- An ability to use performance charts to determine machine speed.

These objectives are the basis for the problems that follow.

PROBLEMS

6.1 A four-wheel tractor whose operating weight is 48,000 lb is pulled up a road whose slope is +4% at a uniform speed. If the average tension in the towing cable is 4,680 lb, what is the rolling resistance of the road? (115 lb/ton)

6.2 The new Ferrari has a V-12 engine rated at 540 hp. The Ferrari has a maximum speed of 196 mph in sixth gear. Determine the maximum rimpull of a Ferrari operated in sixth gear if its engine efficiency is 90%. The Ferrari weighs 4,055 lb and is operated over a concrete road with rolling resistance of 40 lb/ton and a slope of 3%. Determine the maximum external pull the Ferrari has when operated in sixth gear at maximum speed.

6.3 A CAT 834G four-wheel tractor (rubber tires, high pressure), whose operating weight can be found at www.cat.com, is pulled up a road whose slope is +3.2% at a uniform speed. If the average tension in the towing cable is 6,600 lb, what is the rolling resistance of the road? This would be what type of road surface?

6.4 What is the rated flywheel power for a CAT 834G four-wheel tractor equipped with a converter drive? What are the maximum speeds, for this machine, in each of its four forward gears? Determine the maximum rimpull of the tractor in each of the indicated gears if the efficiency is 90%. If the tractor is operated over a haul road whose slope is +2%

with rolling resistance of 80 lb/ton, determine the maximum external pull by the tractor in first gear. Can the tractor climb the slope in fourth gear? The required data can be found at www.cat.com/cda/layout?m=37840 &x=7.

6.5 A fully loaded CAT 621G scraper is operating in fifth gear at its full rated rpm. Ambient air temperature is 60°F and the altitude is approximately sea level. The scraper is climbing a uniform 4.0% slope with a rolling resistance of 70 lb/ton, and the pneumatic-tired scraper bowl is fully loaded with fill material.

a. What is the operating weight of the empty scraper in pounds (lb)?
b. What is the rated load in pounds for this scraper?
c. What is the total operating weight of the fully loaded scraper?
d. What portion, in pounds and percent, of the fully loaded weight is carried by the front axle?
e. What portion, in pounds and percent, of the fully loaded weight is carried by the rear axle (the axle of the bowl)?
f. What is the speed of the scraper in mph when operated in fifth gear (forward)?
g. At what net horsepower does the manufacturer rate the tractive effort of the scraper, for the environmental conditions described and operating in fifth gear?
h. What percentage of this rated rimpull does the tractor actually develop? In performing your calculations, it will be acceptable to assume that the component of weight normal to the traveling surface is equal to the weight itself (i.e., cos 0 taken as 1.00, where 0° is the angle of total resistance). The "20 lb of rimpull required ton of weight per % of slope" approximation will be acceptable. You may also disregard the power required to overcome wind resistance and to provide acceleration. Assume traction is not a limiting factor.
i. What is the value of the coefficient of traction if the drive wheels of the tractor are at the point of incipient slippage for the conditions just described?

The required data can be found at: www.cat.com/cda/layout?m=37840&x=7.

6.6 A wheel-type tractor-pulled scraper having a combined weight of 172,000 is push-loaded down a 6% slope by a crawler tractor whose weight is 70,000 lb. What is the equivalent gain in loading force for the tractor and scraper resulting from loading the scraper downslope instead of upslope?

6.7 A pneumatic-tired tractor with a 300-hp engine has a maximum speed of 3.6 mph in first gear.

a. Determine the maximum rimpull of the tractor in each of the indicated gears if the engine efficiency is 92%.

Gear	Speed (mph)
First	3.6
Second	6.5
Third	11.5
Fourth	20.5

b. The tractor weighs 29.2 tons. It is operated over a haul road having a slope of +2.0% and a rolling resistance of 84 lb/ton. Determine the maximum external pull by the tractor in each of five gears.

6.8 If the tractor of Problem 6.7 is operated down a 4% slope and the rolling resistance is 80 lb/ton, determine the maximum external pull by the tractor in each of the five gears.

6.9 A wheeled tractor with high-pressure tires and weighing 81,000 lb is pulled up a 5% slope at a uniform speed. If the tension in the tow cable is 15,200 lb, what is the rolling resistance of the road? What type of surface would this be?

6.10 A wheel tractor-scraper is operating on a 4% adverse grade. Assume that no power derating is required for equipment condition, altitude, and temperature. Use equipment data from Fig. 6.10. Disregarding traction limitations, what is the maximum value of rolling resistance (in pounds per ton) over which the empty unit can maintain a speed of 14 mph?

6.11 A wheel tractor-scraper is operating on a level grade. Assume no power derating is required for equipment condition, altitude, and temperature. Use equipment data from Fig. 6.10.

a. Disregarding traction limitations, what is the maximum value of rolling resistance (in lb/ton) over which the fully loaded unit can maintain a speed of 20 mph?

b. What minimum value of coefficient of traction between the tractor wheels and the traveling surface is needed to satisfy the requirements of part a? For the fully loaded condition 67% of the weight is distributed to the drive axle. Operating weight of the empty scraper is 70,000 lb.

6.12 A wheel-type tractor unit, operating in its fourth-gear range and at its full rated rpm, is observed to maintain a steady speed of 7.50 mph, when operating under the conditions described herein. Ambient air temperature is 60°F. Altitude is sea level. The tractor is climbing a uniform 5.5% slope with a rolling resistance of 55 lb/ton, and it is towing a pneumatic-tired trailer loaded with fill material. The single-axle tractor has an operating weight of 66,000 lb. The loaded trailer has a weight of 48,000 lb. The weight distribution for the combined tractor-trailer unit is 55% to the drive axle and 45% to the rear axle.

a. The manufacturer, for the environmental conditions just described, rates the tractive effort of the new tractor at 330 rimpull horsepower. What percentage of this rated rimpull hp does the tractor actually develop? The "20 lb of rimpull required per ton of weight per % of slope" approximation will be acceptable. Assume traction is not a limiting factor.

b. What is the value of the coefficient of traction if the drive wheels of the tractor are at the point of incipient slippage for the conditions just described?

6.13 A wheel-type tractor with a 210-hp engine has a maximum speed of 4.65 mph in first gear. Determine the maximum rimpull of the tractor in each of the indicated gears if the engine efficiency is 90%.

Gear	Speed (mph)
First	4.65
Second	7.60
Third	11.50
Fourth	17.50
Fifth	26.80

6.14 Determine the maximum external pull, in each of five gears, for the tractor in Problem 6.13 when it is operated over a haul road whose slope is +3%. The rolling resistance of the haul road is 80 lb/ton. The tractor weighs 21.4 tons. (12,246 lb; 6,330 lb; 3,167 lb; 1,077 lb; none)

6.15 If the tractor of Problems 6.13 and 6.14 is operated down a 4% slope and the rolling resistance is 80 lb/ton, what is its the maximum external pull in each of the five gears?

6.16 A wheel-type tractor, operating in its second-gear range and at its full rated rpm, is observed to maintain a steady speed of 1.50 mph when operating under the conditions described herein. Ambient air temperature is 60°F, altitude is sea level. The tractor is climbing a uniform 6.5% slope with a rolling resistance of 65 lb/ton, and it is towing a pneumatic-tired trailer loaded with fill material. The two-axle tractor has a total weight of 60,000 lb, 55% of which is distributed to the power axle. The loaded trailer has a weight of 75,000 lb.

a. For the environmental conditions just described, a manufacturer rates the tractive effort of the new tractor at 60 rimpull horsepower. What percentage of this rated rimpull does the tractor actually develop? The "20 lb of rimpull required ton of weight per % of slope" approximation will be acceptable. Assume traction is not a limiting factor. (87.7%)

b. What is the value of the coefficient of traction if the drive wheels of the tractor are at the point of incipient slippage for the conditions just described? (0.4)

6.17 A wheel-type tractor unit, operating in its fourth-gear range and at its full rated rpm, is observed to maintain a steady speed of 7.50 mph when operating under the conditions described herein. Ambient air temperature is 60°F. Altitude is sea level. The tractor is climbing a uniform 5.5%

slope with a rolling resistance of 55 lb/ton, and it is towing a pneumatic-tired trailer loaded with fill material. The single-axle tractor has an operating weight of 66,000 lb. The loaded trailer has a weight of 48,000 lb. The weight distribution for the combined tractor-trailer unit is 55% to the drive axle and 45% to the rear axle.

a. The manufacturer, for the environmental conditions described, rates the tractive effort of the new tractor at 330 rimpull horsepower. What percentage of this rated rimpull hp does the tractor actually develop? The "20 lb of rimpull required per ton of weight per % of slope" approximation will be acceptable. Assume traction is not a limiting factor.

b. What is the value of the coefficient of traction if the drive wheels of the tractor are at the point of incipient slippage for the conditions just described?

6.18 A wheel tractor-scraper is operating on a level grade. Assume no power derating is required for equipment condition, altitude, and temperature. Use equipment data from Fig. 6.10.

a. Disregarding traction limitations, what is the maximum value of rolling resistance (in lb per ton) over which the fully loaded unit can maintain a speed of 20 mph?

b. What minimum value of coefficient of traction between the tractor wheels and the traveling surface is needed to satisfy the requirements of part a? For the fully loaded condition, 67% of the weight is distributed to the drive axle. Operating weight of the empty scraper is 70,000 lb.

REFERENCES

1. *Caterpillar Performance Handbook*, Caterpillar Inc., Peoria, Ill. (Published annually.) Machine data can also be found at: www.cat.com/cda/layout?m=37840&x=7.
2. Schexnayder, Cliff, Sandra L. Weber, and Brentwood T. Brooks, "Effect of Truck Payload Weight on Production," *Journal of Construction Engineering and Management*, ASCE, 125(1), pp. 1–7.

WEBSITE RESOURCES

1. www.aem.org The Association of Equipment Manufacturers (AEM) is the international trade and business development resource for companies that manufacture equipment for the construction, mining, forestry, and utility industries.
2. www.cat.com Caterpillar is the world's largest manufacturer of construction and mining equipment. In the Products, Equipment section of the website can be found the specifications for Caterpillar-manufactured scrapers.

3. www.constructionequipment.com *Construction Equipment* magazine online.
4. www.equipmentworld.com Equipmentworld.com is a news and e-commerce website for construction industry and equipment news.
5. www.hcmac.com Hitachi Construction Machinery Co., Ltd., manufactures construction and transportation equipment.
6. www.terex.com Terex Corporation is a diversified manufacturer of construction equipment. The specifications for their scrapers can be found in the "Construction" equipment section of the website.
7. www.wes.army.mil The Waterways Experiment Station (WES) is headquarters for the U.S. Army Engineer Research and Development Center (ERDC).

CHAPTER 7

Dozers

A dozer is a tractor unit that has a blade attached to the machine's front. It is designed to provide tractive power for drawbar work. A dozer has no set volumetric capacity. The amount of material the dozer moves is dependent on the quantity that will remain in front of the blade during the push. Crawler dozers equipped with special clearing blades are excellent machines for land clearing. Heavy ripping of rock is accomplished by crawler dozers equipped with rear-mounted rippers because of the power and tractive force that they can develop.

INTRODUCTION

Dozers may be either crawler (tracklaying) or wheel-type machines. They are tractor units equipped with a front blade for pushing materials (dozing). These machines are designed to provide tractive power for drawbar work. Consistent with their purpose, as a unit for drawbar work, they are low-center-of-gravity machines. This is a prerequisite of an effective dozer. The larger the difference between the line-of-force transmission from the machine and the line-of-resisting force, the less effective the utilization of developed power. Besides dozing, these machines are used for land clearing, ripping, assisting scrapers in loading, and towing other pieces of construction equipment. They can be equipped with either a rear-mounted winch or a ripper. For long moves between projects or within a project, the track dozer should be transported. Moving them under their own power, even at slow speeds, increases track wear and shortens the machine's operational life.

PERFORMANCE CHARACTERISTICS OF DOZERS

Dozers are classified on the basis of running gear:

1. Crawler type (see Fig. 7.1)
2. Wheel type (see Fig. 7.2)

Crawler dozers are tracklaying machines. They have a continuous track of linked shoes (see Fig. 7.3) that moves in the horizontal plane across fixed rollers. At the rear of the machine, the track passes over a sprocket drive wheel

FIGURE 7.1 Crawler dozer.

FIGURE 7.2 Wheel dozer.

FIGURE 7.3 Track shoes.

that is mounted vertically. As the sprocket turns, it forces the track forward or back, imparting motion to the dozer. In the front of the machine, the track passes over a vertically mounted idler wheel that is connected to a recoil device having adjustable tension. The idler wheel maintains the proper tension in the track and enables it to absorb heavy shocks. The linked shoes are made of heat-treated steel designed to resist wear and abrasion. There are several companies that now offer tracks having rubber-covered steel shoes.

As discussed in Chapter 4, the usable force that a machine has available to perform work is often limited by traction. This limitation is dependent on two factors:

1. Coefficient of traction for the surface being traversed.
2. Weight carried by the drive wheels of the tractor.

Sometimes users weight the tires of wheel-type dozers to overcome tractive-power limitations. A mixture of calcium chloride and water is recommended as tire ballast. Care must be taken to ensure that the new weight distribution is equal between all drive wheels.

Traction or flotation requirements can be met by proper undercarriage or tire selection. A standard crawler dozer undercarriage is appropriate for general work in rock to moderately soft ground. Typical ground pressure for a crawler dozer with a standard undercarriage is about 6 to 9 psi (41–62 kPa). There are low-ground-pressure (LGP) undercarriage configurations for dozers operating in soft ground conditions. The ground pressure exerted by a crawler dozer with an LGP undercarriage is about 3 to 4 psi (21–28 kPa). LGP machines should not be used in hard ground or in rocky conditions; as such a practice will reduce undercarriage life. There are extralong (XL) undercarriages available for machines dedicated to **finish work**.

finish work
Shaping the material to the final earthwork grade required by the specifications.

In the case of wheel machines, wider tires provide greater contact area and increase flotation. It must be remembered, however, that rimpull charts are based on standard equipment, including tires. Larger tires will reduce the developed rimpull.

The crawler-type tractor is designed for those jobs requiring high tractive effort. No other piece of equipment can provide the power, traction, and flotation needed in such a variety of working conditions. A crawler dozer can operate on slopes as steep as 45°.

Both track- and wheel-type dozers are rated by flywheel power (fwhp) and weight. Normally the weight is an operating weight and includes lubricants, coolants, a full fuel tank, a blade, hydraulic fluid, the OSHA rollover protective standards canopy (ROPS), and an operator. Dozer weight is important on many projects because the maximum tractive effort that a tractor can provide is limited to the product of the weight times the coefficient of traction for the unit and the particular ground surface, regardless of the power supplied by the engine. Table 6.4 gives the coefficients of traction for various surfaces.

An advantage of a wheel-type dozer as compared with a crawler dozer is the higher speed possible with the former machines—in excess of 30 mph for some models. To attain a higher speed, however, a wheel dozer must sacrifice pulling effort. Also, because of the lower coefficient of traction between rubber tires and some ground surfaces, the wheel dozer may slip its wheels before developing its rated pulling effort. Table 7.1 provides a comparison of crawler dozer and wheel dozer utilization.

Internal combustion engines are used to power most dozers, with diesel engines being the most common primary power units. Gasoline engines are

TABLE 7.1 Dozer-type utilization comparison.

Wheel dozer	Crawler dozer
Good on firm soils and concrete and abrasive soils that have no sharp-edged pieces	Can work on a variety of soils; sharp-edged pieces not as destructive to dozer, though fine sand will increase running gear wear
Best for level and downhill work	Can work over almost any terrain
Wet weather, causing soft and slick surface conditions, will slow or stop operation	Can work on soft ground and over mud-slick surfaces; will exert very low ground pressures with special low ground pressure undercarriage and track configuration
The concentrated wheel load will provide compaction and kneading action to ground surface	
Good for long travel distances	Good for short work distances
Best in handling loose soils	Can handle tight soils
Fast return speeds, 8–26 mph	Slow return speeds, 5–10 mph
Can handle only moderate blade loads	Can push large blade loads

used in some smaller machines. There are electric- and air-powered dozers available for tunnel work.

Because the crankshaft rotation derived from the engine is usually too fast and does not have sufficient force (torque), machines have transmissions that reduce the rotational speed of the crankshaft and increase the force available to do work. Transmissions provide the operator with the ability to change the machine's speed-power ratio so that it matches the work requirements (force necessary to do work). Manufacturers provide dozers with a variety of transmissions, but primarily the options are

- Direct drive
- Torque converter and power-shift transmission

Some less-than-100-hp dozers are available with hydrostatic powertrains. The small to medium, less-than-300-hp, diesel-powered machines are commonly available with either direct- or power-shift-type transmissions. Larger dozers are always equipped with power-shift transmissions.

Crawler Dozers with Direct Drive

The term direct drive means the power is transmitted straight through the transmission as if there was a single shaft. This is usually what happens when the transmission is in its highest gear. In all other gears, mechanical elements match speed and torque. Direct-drive dozers are superior when the work involves constant load conditions. A job where full blade loads must be pushed long distances would be an appropriate application of a direct drive machine.

The specifications provided by some manufacturers list two sets of drawbar pulls—rated and maximum for direct-drive dozers. The rated value is the drawbar pull that can be sustained for continuous operation. The maximum drawbar pull is the pull that the dozer can exert for a short period while lugging the engine, such as when passing over a soft spot in the ground that requires a temporary higher tractive effort. Thus, the rated pull should be used for continuous operation. Available drawbar pull is subject to the limitation imposed by the traction that can be developed between the tracts and the ground.

Crawler Dozers with Torque Converter and Power-Shift Transmissions

Transmissions that can be shifted while transmitting full engine power are known as power shifts. These transmissions are teamed with torque converters to absorb drive train shock loads caused by changes in gear ratios. A power-shift transmission provides an efficient flow of power from the engine to the tracks and gives superior performance in applications involving variable load conditions. Figure 7.4 illustrates the performance curves for a track-type dozer equipped with a power-shift transmission.

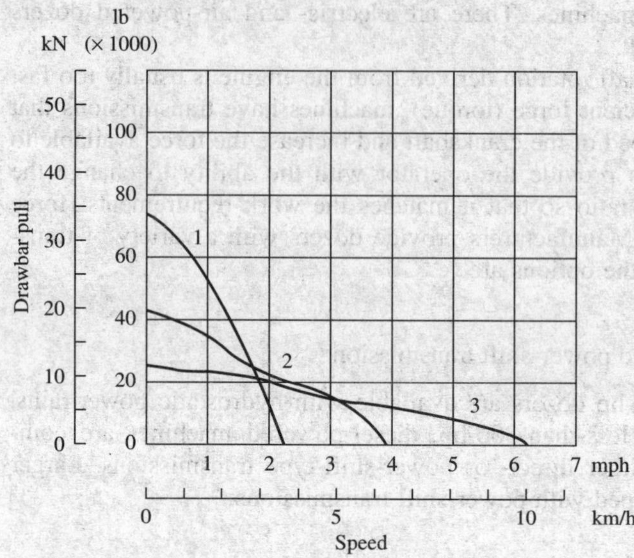

FIGURE 7.4 Performance chart for a 200-hp 45,560-lb track-type dozer with a power shift.
Reprinted courtesy of Caterpillar Inc.

Crawler Dozers with Hydrostatic Powertrains

Confined oil under pressure is an effective means of power transmission. A hydrostatic powertrain offers an infinitely variable speed range with constant power to both of the tractor's tracks. This type of powertrain improves machine controllability and increases operational efficiency. Hydrostatic powertrain transmissions are available on some lower-horsepower dozers.

Wheel Dozers

Most wheel dozers are equipped with torque converters and power-shift transmissions. Figure 7.5 illustrates the performance curves for a wheel dozer equipped with a power-shift transmission. Wheel dozers exert comparatively high ground pressures, 25 to 35 psi (172–241 kPa).

Comparison of Performance

Heed the cautionary note on Fig. 7.5 that usable pull/rimpull will depend on the weight and traction of the fully equipped dozer. This is a warning that, even though the engine can develop a certain drawbar pull or rimpull force, all of that pull may not be available to do work. The caution is a restatement of Eq. [6.15]. If the project-working surface is dry clay loam, Table 6.4 provides the following coefficient of traction factors:

Rubber tires	0.50–0.70
Track	0.90

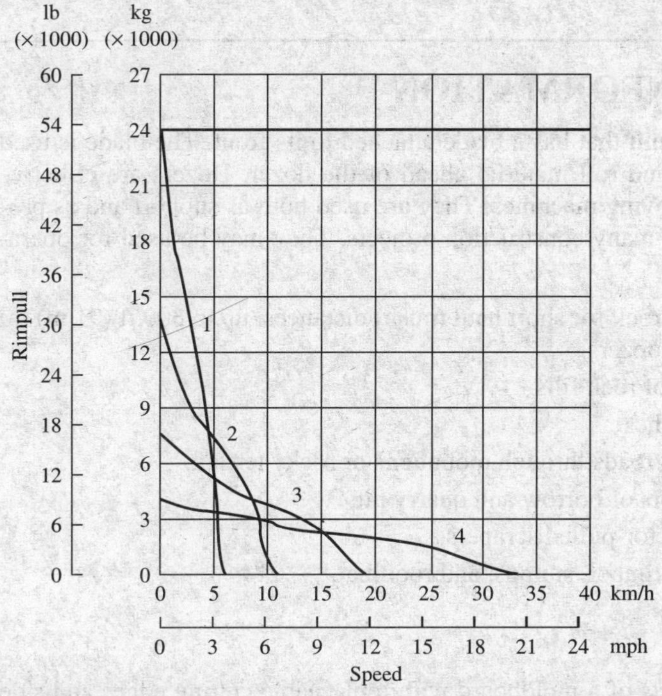

FIGURE 7.5 Performance chart for a 216-hp 45,370-lb wheel dozer with a power-shift transmission. Usable rimpull depends on traction and weight of dozer.
Reprinted courtesy of Caterpillar Inc.

Using the factor for tracks, 0.90 and considering a track-type dozer with a power shift (see Fig. 7.4), the usable drawbar pull is found to be

$$45{,}560 \text{ lb} \times 0.90 = 41{,}004 \text{ lb}$$

Now consider a wheel-type dozer (Fig. 7.5):

$$45{,}370 \text{ lb} \times 0.60 = 27{,}222 \text{ lb}$$

The two machines have approximately the same operating weight and flywheel power; yet, because of the effect of traction, the track machine can supply one and a half times the *usable* power.

In the case of most soil conditions, the coefficient of traction for wheels is less than that of tracks. Therefore, a wheel-type dozer must be considerably heavier (approximately 50%) than a crawler dozer to develop the same amount of usable force.

As the weight of a wheel-type dozer is increased, a larger engine will be required to maintain the weight-to-horsepower ratio. There is a limit to the weight that can be added to the wheel-type dozer and still have a machine with a speed and mobility advantage over track dozers.

PUSHING MATERIAL

GENERAL INFORMATION

A *dozer* is a tractor unit that has a blade attached to its front. The blade is used to push, shear, cut, and roll material ahead of the dozer. Dozers are effective and versatile earthmoving machines. They are used both as support and as production machines on many construction projects. They may be used for operations such as

1. Moving earth or rock for short haul (push) distances, up to 300 ft (91 m) in the case of large dozers.
2. Spreading earth or rock fills.
3. Backfilling trenches.
4. Opening up pilot roads through mountains or rocky terrain.
5. Clearing the floors of borrow and quarry pits.
6. Helping load tractor-pulled scrapers.
7. Clearing land of timber, stumps, and root mat.

BLADES

A dozer blade consists of a moldboard with replaceable cutting edges and side bits. Push arms and tilt cylinders or a C-frame connect the blade to the dozer (Fig. 7.6). Blades vary in size and design based on specific work applications. The hardened-steel cutting edges and side bits are bolted on because they receive most of the abrasion and wear out rapidly. The bolted connection enables easy replacement. The design of some machines enables either end of the blade to be raised or lowered in the vertical plane of the blade, *tilt*. The top of the blade can be pitched forward or backward varying the angle of attack of

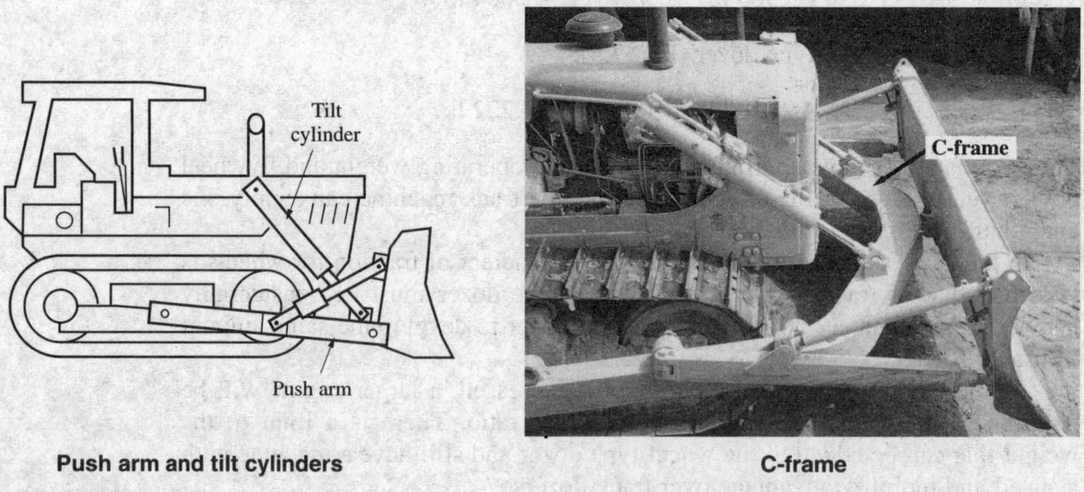

Push arm and tilt cylinders **C-frame**

FIGURE 7.6 Dozer blade mounting arrangements.

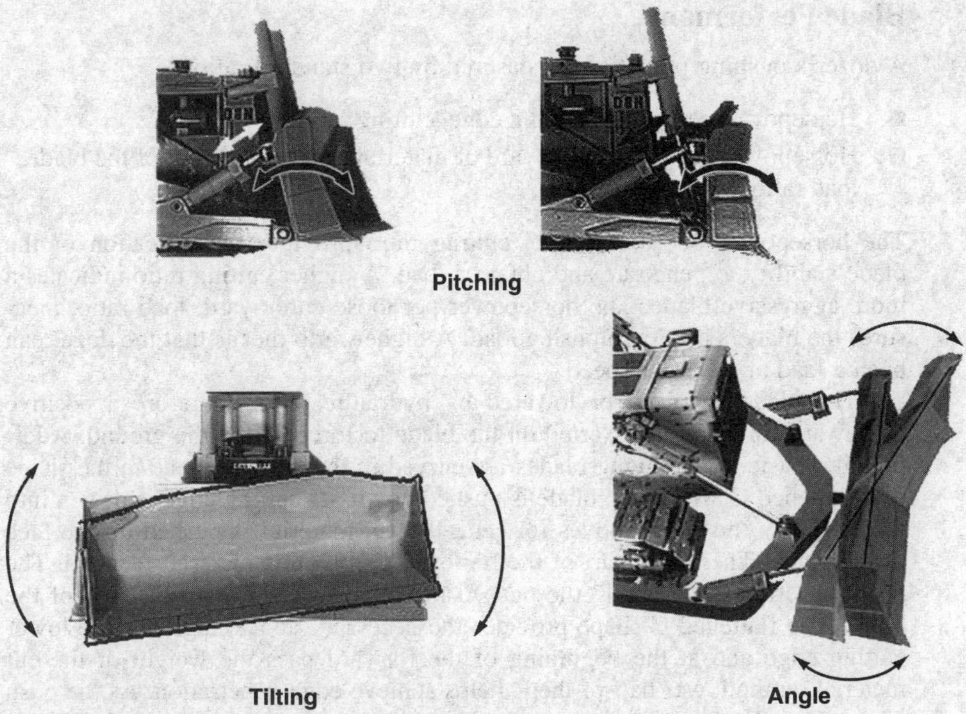

FIGURE 7.7 Bulldozer blade adjustments—tilt, pitch, and angle.

the cutting edge, *pitch*. Blades mounted on a C-frame can be turned from the direction of travel, *angling*. These features are not applicable to all blades, but any two of these may be incorporated in a single mount. Figure 7.7 illustrates tilt, pitch, and angling.

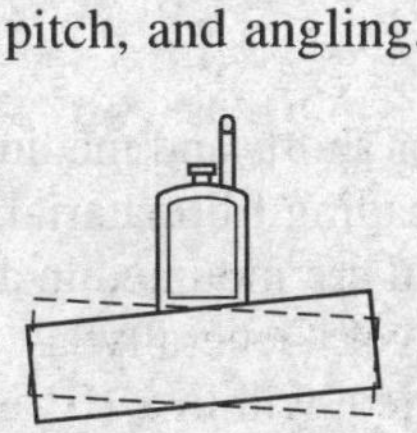

Tilt. This movement is within the vertical plane of the blade. Tilting permits concentration of dozer driving power on a limited portion of the blade's length.

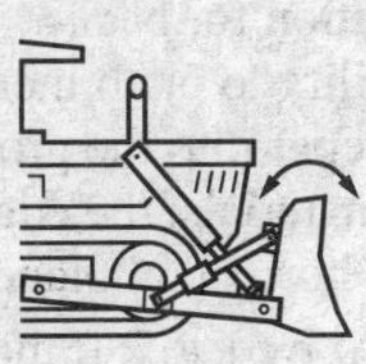

Pitch. This is a pivotal movement about the point of connection between the dozer and blade. When the top of the blade is pitched forward, the bottom edge moves back; this increases the angle of cutting edge attack.

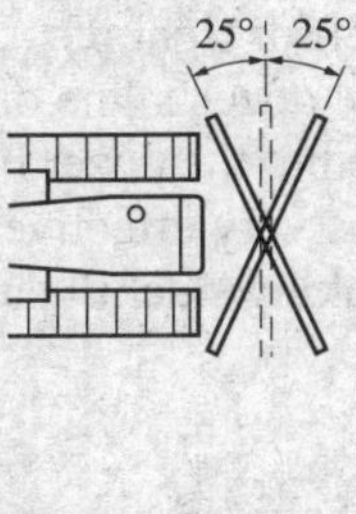

Angling. Turning the blade so that it is not perpendicular to the direction of the dozer's travel is known as angling. Angling causes the pushed material to roll off the trailing end of the blade. This procedure of rolling material off one end of the blade is called *side casting*.

Blade Performance

A dozer's pushing potential is measured by two standard ratios:

- Horsepower per foot of cutting edge, cutting ratio
- Horsepower per loose cubic yard of material retained in front of the blade, load ratio

The horsepower per foot (hp/ft) cutting ratio provides an indication of the blade's ability to penetrate and obtain a load. A higher cutting ratio indicates a more aggressive blade. The horsepower per loose cubic yard, load ratio, measures the blade's ability to push a load. A higher ratio means that the dozer can push a load at a greater speed.

The blade is raised or lowered by hydraulic rams, therefore a positive downward force can be exerted on the blade to force it into the ground. Additionally, basic earthmoving blades are curved in the vertical plane in the shape of a flattened C. When the blade is pushed down, its cutting edge is driven into the earth. As the dozer moves forward, the cut material is pushed up the face of the blade. The upper part of the flattened C rolls this material forward. The resulting effect is to "boil" the pushed material over and over in front of the blade. The flattened C shape provides the necessary attack angle for the lower cutting edge and at the beginning of the forward pass the weight of the cut material on the lower half of the C helps achieve edge penetration. As the push progresses, the load in front of the blade passes the midpoint of the C and begins to exert an upward force on the blade. This "floats" the blade, reducing the penetration of the cutting edge, and helps the operator control the load.

Many different special application blades can be attached to a tractor (see Fig. 7.8), but basically only five blades are common to earthwork: (1) the *straight* "S" blade, (2) the *angle* "A" blade, (3) the *universal* "U" blade, (4) the *semi-U* "SU" blade, and (5) the *cushion* "C" blade.

Straight blades "S." The straight blade is designed for short- and medium-distance passes, such as backfilling, grading, and spreading fill material. These blades have no curvature across their length and are mounted in a fixed position, perpendicular to the dozer's line of travel. Generally, a straight blade is heavy-duty and normally it can be tilted, within a 10° arc, increasing penetration for cutting or decreasing penetration for back-dragging material. It may be equipped to pitch. The ability to pitch means that the operator can set one end of the cutting edge deeper into the ground to dig or pry hard materials. For easy drifting of light materials, the edge's are brought to the same level—the blade is level in the horizontal plane.

Angle blades "A." An angle blade is wider (face length) by 1 to 2 ft than an S blade. It can be angled up to a maximum of 25° left or right of perpendicular to the tractor or held perpendicular to the dozer's line of travel. The blade can be tilted, but because it is attached to the dozer by a C-frame mount, it cannot be pitched. The angle blade is very effective for side-casting material, particularly for backfilling or making sidehill cuts.

FIGURE 7.8 Common earthmoving dozer blades.

Universal blades "U." This blade is wider than a straight blade and the long-dimension outside edges are canted forward about 25°. This canting of the edges reduces the spillage of loose material making the U blade efficient for moving large loads over long distances. The cutting ratio is lower for the U than the S blade mounted on a similar dozer. Penetration is not a prime objective of the blade's design (shape) as the lower cutting ratio relationship indicates. The U blade's load ratio is lower than that of a similar S blade. This denotes that the blade is best suited for lighter materials. Typical usages are working stockpiles and drifting loose or noncohesive materials.

Semi-U blades "SU." This blade combines the characteristics of the S and U-blade designs. By the addition of short wings it has increased capacity compared to an S blade.

Cushion blades "C." Cushion blades are mounted on large dozers that are used primarily for push-loading scrapers. The C blade is shorter than the S blade so as to avoid pushing the blade into and cutting the rear tires of the scraper while push-loading. The shorter length also facilitates maneuvering into position behind the scrapers. Rubber cushions and springs in the mounting enable the dozer to absorb the impact of contacting the scraper push block. By using a cushion blade instead of a "pusher block" to push scrapers, the dozer has the ability to clean up the cut area and increase the total fleet production. It is a blade of limited utility in pushing material and should not be used for production dozing. It cannot be tilted, pitched, or angled.

PROJECT EMPLOYMENT

The first bulldozers were adapted from farm tractors that were used to plow the fields. In fact, it has been claimed that the name comes from the fact that these machines replaced farming machines powered by bulls and now the bulls get to sleep, or doze, so hence the name. No matter the name, these machines are the mainstay for major civil construction projects and are capable of performing a variety of work tasks.

Stripping

Dozers are excellent machines for *stripping*, which is the removal of a thin layer of covering material. On most projects, this is a term used to describe the removal of topsoil. Dozers are economical machines for moving material a maximum of 300 ft in the case of large machines. The economical push distance decreases as dozer size decreases, but economical push distance also depends on the material being handled. A material exhibiting cohesion is easier to push than a granular material (sand), which tends to run in front of the blade.

Backfilling

A dozer can efficiently accomplish backfilling by drifting material sideways with an angle blade. This enables forward motion parallel to the excavation. If a straight blade is used, the dozer will approach the excavation at a slight angle and then, at the end of the pass, turn in toward the excavation. No part of the tracks should hang over the edge of the excavation.

Caution must be exercised in making the initial pass completely across pipes and culverts. As a minimum, 12 in. of material should cover the pipe or structure before accomplishing a crossing. The diameter of the pipe, the pipe type, the distance between the sidewalls of the excavation, and the number of lines of pipe in the excavation dictate the minimum required cover. Larger-diameter pipe, larger excavation widths, and multiple lines of pipe are factors that all dictate more cover before crossing the structure.

Spreading

The spreading of material dumped by trucks or scrapers is a common dozer task. Ordinarily, project specifications state a maximum loose lift thickness. Even when lift thickness limits are not stated in the contract specifications, density requirements and proposed compaction equipment will force the contractor to control the thickness of each lift. A dozer accomplishes uniform spreading by keeping the blade straight and at the desired height above the previously placed fill surface. The dumped material is forced directly under the blade's cutting edge. Today laser blade controls are available for this type of work (see Fig. 7.9).

FIGURE 7.9 Spreading using a laser blade control.

Slot Dozing

Slot dozing is the technique whereby the blade end spillage from the first pass or the sidewalls from previous cuts are used to hold material in front of the dozer blade on subsequent passes. When employing this method to increase production, align cuts parallel, leaving a narrow uncut section between slots. Then, remove the uncut sections by normal dozing. The technique prevents spillage at each end of the blade and usually increases production by about 20%. The production increase is highly dependent on the slope of the push and the type of material being pushed.

Blade-to-Blade Dozing

Another technique used to increase bulldozer production is blade-to-blade dozing (see Fig. 7.10). The technique is sometimes referred to as *side-by-side dozing*. As the names imply, two machines maneuver so that their blades are right next to each other during the pushing phase of the production cycle. This reduces the side spillage of each machine by 50%. The extra time necessary to position the machines together increases that phase of the cycle. Therefore, the technique is not effective on pushes of less than 50 ft because of the excess maneuver time required. When machines operate simultaneously, delay to one machine is in effect a double delay. The combination of less spillage but increased maneuver time tends to make the total increase in production for this technique somewhere between 15 and 25%.

FIGURE 7.10 Blade-to-blade dozing used to increase production by minimizing spillage.

DOZER PRODUCTION ESTIMATING

A dozer has no set volumetric capacity. There is no hopper or bowl to load; instead, the amount of material the dozer moves is dependent on the quantity that will remain in front of the blade during the push. The factors that control dozer production rates are

1. Blade type
2. Type and condition of material
3. Cycle time

Blade Type

By design, straight blades roll material in front of the blade, and universal and semi-U blades control side spillage by holding the material within the blade. Because the U and SU blades force the material to move to the center, there is a greater degree of material volumetric swell. The U or SU blade's quantity of loose material will be greater than that of the S blade. But the ratio of this difference is not the same when considering bank cubic yards. This is because the factor to convert loose cubic yards to bank cubic yards for the universal-type blades is not the same as that for a straight blade. The U or SU blade's boiling effect causes the difference.

The same type of blade comes in different sizes to fit different size dozers. Blade capacity then is a function of blade type and physical size. Manufacturers' specification sheets will provide information concerning blade dimensions.

Type and Condition of Material

The type and condition of the material being handled affects the shape of the pushed mass in front of the blade. Cohesive materials (clays) will "boil" and heap. Materials that exhibit a slippery quality or those that have a high mica content will ride over the ground and swell out. Cohesionless materials (sands) are known as "dead" materials because they do not exhibit heap or swell properties. Figure 7.11 illustrates these material attributes.

Blade Volumetric Load

The volumetric load a blade will carry can be estimated by several methods:

1. Manufacturer's blade rating
2. Previous experience (similar material, equipment, and work conditions)
3. Field measurements

Manufacturer's Blade Ratings Manufacturers provide blade ratings based on SAE Standard J1265. The purpose of SAE Standard J1265 is to provide a uniform method for calculating blade capacity. It is for making relative comparisons of dozer blade capacity and not for predicting productivity in the field.

Clay material boiling in front of the blade.

Cohesionless, sandy loam in front of the blade.

FIGURE 7.11 Bulking attributes of materials when being pushed.

$$V_s = 0.8WH^2 \quad \textbf{[7.1]}$$

$$V_u = V_s + ZH(W-Z)\tan x^\circ \quad \textbf{[7.2]}$$

where

V_s = capacity of straight or angle blade, in lcy
V_u = capacity of universal blade, in lcy
W = the blade width, in yards, exclusive of end bits
H = the effective blade height, in yards
Z = the wing length measured parallel to the blade width, in yards
x = the wing angle

Previous Experience Properly documented past experience is an excellent blade load estimating method. Documentation requires that the excavated area be cross-sectioned to determine the total volume of material moved and that the number of dozer cycles be recorded. Production studies can also be made based on the weight of the material moved. In the case of dozers, the mechanics of weighing the material is normally harder to accomplish than surveying the volume.

Field Measurement A procedure for measuring blade loads follows:

1. Obtain a normal blade load:
   a. The dozer pushes a normal blade load onto a level area.
   b. Stop the dozer's forward motion. While raising the blade move forward slightly to create a symmetrical pile.
   c. Reverse and move away from the pile.

2. Measurement (see Fig. 7.12):
   a. Measure the height (H) of the pile at the inside edge of each track.

b. Measure the width (W) of the pile at the inside edge of each track.
c. Measure the greatest length (L) of the pile. This will not necessarily be at the middle.

3. Computation: Average both the two height and the two width measurements. If the measurements are in feet, the blade load in lcy is calculated by the formula

$$\text{Blade load (lcy)} = 0.0139HWL \qquad [7.3]$$

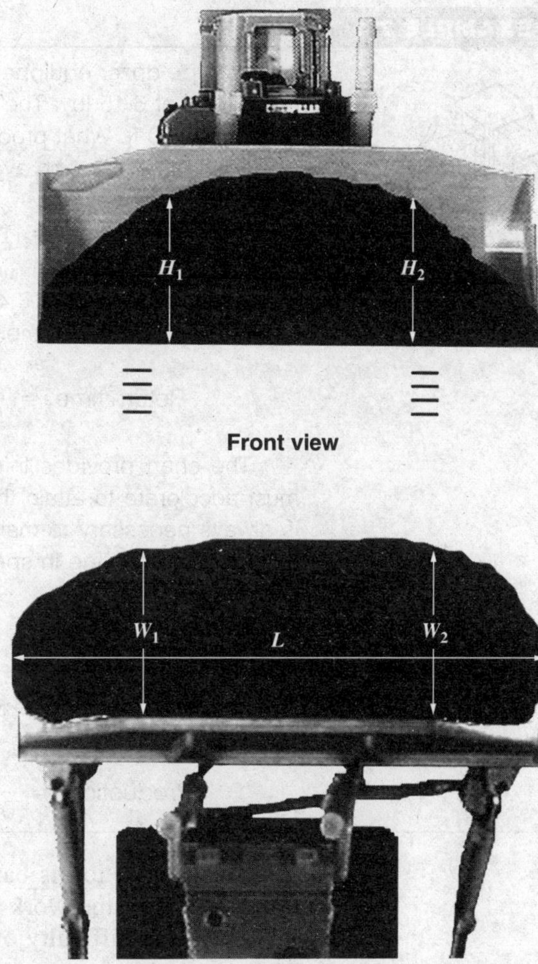

FIGURE 7.12 Measurements for calculating blade loads.

Cycle Time

The sum of the time required to push a load, backtrack, and maneuver into position to push again represents one dozer production cycle. The time required to push and backtrack can be calculated for each dozing situation considering the travel distance and obtaining a speed from the machine's performance chart.

Dozing, however, is generally performed at slow speed, 1.5 to 2 mph. The lower figure is appropriate for very heavy cohesive materials. Return speed is usually the maximum that can be attained in the distance available. When using performance charts to determine possible speeds, remember the chart identifies instantaneous speeds. In calculating cycle duration, the estimator must use an average speed that accounts for the time required to accelerate to the attainable speed as indicated by the chart. Usually the operator cannot shift the machine past second gear in the case of distances that are less than 100 ft. If the distance is greater than 100 ft and the ground conditions are relatively smooth and level, maximum machine speed may be obtained. Maneuver time for power-shift dozers used in pushing material is about 0.05 min.

Production

The formula to calculate dozer pushing production in loose cubic yards per a 60-min hour is

$$\begin{matrix}\text{Production} \\ \text{(lcy per hour)}\end{matrix} = \frac{60 \text{ min} \times \text{blade load}}{\begin{matrix}\text{push time} \\ \text{(min)}\end{matrix} + \begin{matrix}\text{return time} \\ \text{(min)}\end{matrix} + \begin{matrix}\text{maneuver time} \\ \text{(min)}\end{matrix}} \qquad [7.4]$$

EXAMPLE 7.1

A track-type dozer equipped with a power shift (see Fig. 7.4) can push an average blade load of 6.15 lcy. The material being pushed is silty sand. The average push distance is 90 ft. What production, in loose cubic yards, can be expected?

Push time: 2 mph average speed (sandy material):

$$\text{Push time} = \frac{90 \text{ ft}}{5{,}290 \text{ ft/mi}} \times \frac{1}{2 \text{ mph}} \times 60 \text{ min/hr} = 0.51 \text{ min}$$

Return time: Figure 7.4, second gear because less than 100 ft
Maximum speed 4 mph:

$$\text{Return time} = \frac{90 \text{ ft}}{5{,}290 \text{ ft/mi}} \times \frac{1}{4 \text{ mph}} \times 60 \text{ min/hr} = 0.26 \text{ min}$$

The chart provides information based on a *steady state velocity*. The dozer must accelerate to attain that velocity. Therefore, when using such speed data, it is always necessary to make an allowance for acceleration time. Because, in this example, the change in speed is very small, an allowance of 0.05 min is made for acceleration time.

$$\text{Return time} = (0.26 + 0.05) = 0.31 \text{ min}$$

$$\text{Maneuver time} = 0.05 \text{ min}$$

$$\text{Production} = \frac{60 \text{ min} \times 6.15 \text{ lcy}}{0.51 \text{ min} + 0.31 \text{ min} + 0.05 \text{ min}} = 424 \text{ lcy/hr}$$

This production is based on working 60 min per hour, or an ideal condition. How well the work is managed in the field, the condition of the equipment, and the difficulty of the work are factors that will affect job efficiency. The efficiency of an operation is usually accounted for by reducing the number of minutes worked per hour. The efficiency factor is then expressed as working minutes per hour, for example, a 50-min hour or a 0.83 efficiency factor.

If it is necessary to calculate the production, in terms of bank cubic yards (bcy), the swell factor, Eq. [4.12], or the percent swell, Eq. [4.13], can be used to make the conversion.

EXAMPLE 7.2

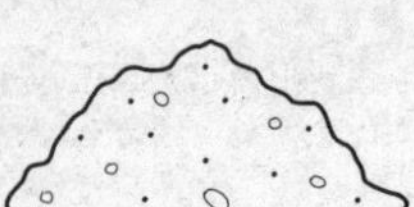

Assume a percent swell of 0.25 for the silty sand of Example 7.1 and a job efficiency equal to a 50-min hour. What is the actual production that can be expected in bank cubic yards?

$$\text{Production} = \frac{424 \text{ lcy}}{1.25} \times \frac{50 \text{ min}}{60 \text{ min}} = 283 \text{ bcy/hr} \approx 280 \text{ bcy/hr}$$

The final step is to compute the unit cost for pushing the material. A large machine should be able to push more material per hour than a small one. However, the cost to operate a large machine will be greater than the cost to operate a small one. The ratio of the *total cost to operate (O&O plus operator) to the amount of material moved* determines the most economical machine for the job. That ratio is the cost figure that is used in bidding unit price work.

EXAMPLE 7.3

The machine in Example 7.2 has an owning and operating cost of \$40.50 per hour. Operators in the area where the proposed work will be performed are paid a wage of \$15.50 per hour. What is the unit cost for pushing the silty sand?

$$\text{Unit cost} = \frac{\$40.50 \text{ per hour} + \$15.50 \text{ per hour}}{280 \text{ bcy/hr}} = \$0.20 \text{ per bcy}$$

Manufacture Production Estimation Guidance

Equipment manufacturers provide dozer production guidance through formulas and factor information. Equation [7.5] is a *rule-of-thumb* formula proposed by International Harvester (IH). This formula equates the net horsepower of a power-shift crawler dozer to an lcy production value.

$$\text{Production (lcy per 60-min hr)} = \frac{\text{net hp} \times 330}{D + 50} \qquad [7.5]$$

where

net hp = net horsepower at the flywheel for a power-shift crawler dozer

D = one-way push distance, in feet

EXAMPLE 7.4

The power-shift dozer whose characteristics are shown in Fig. 7.4 will be used to push material 90 ft. Use the IH formula to calculate the lcy production that can be expected for this operation.

From Fig. 7.4, net hp = 200.

$$\frac{200 \times 330}{(90 + 50)} = 471 \text{ lcy per 60-min hr}$$

Again, the actual production will be less, as a 60-min hour represents an ideal situation.

Caterpillar-Developed Production Factors

Figures 7.13 and 7.14 are production estimating curves published by Caterpillar Inc. in the *Caterpillar Performance Handbook*. The data is specific to that manufacturer's dozers. A production number taken from the Caterpillar curves is a maximum value in lcy per hour based on a set of "ideal" conditions:

1. A 60-min hour (100% efficiency).
2. Power-shift machines with 0.05-min fixed time.

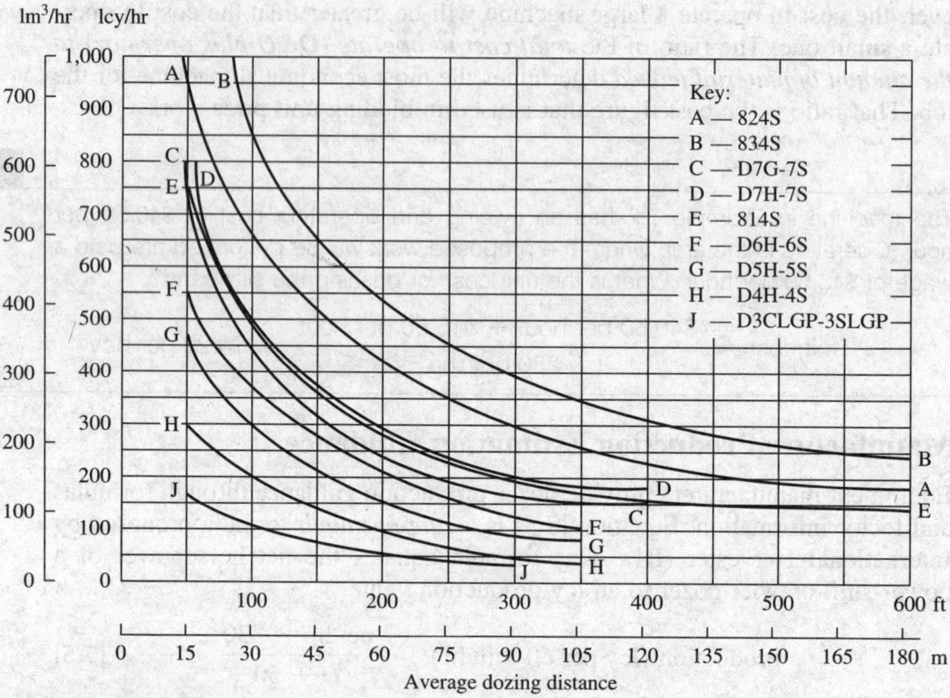

FIGURE 7.13 Dozing production estimating curves for straight blade Caterpillar D3, D4, D5, D6, D7, 814, 824, and 834 dozers equipped with straight blades.
Reprinted courtesy of Caterpillar Inc.

3. The machine cuts for 50 ft, then drifts the blade load to dump over a high wall.
4. A soil density of 2,300 lb per lcy.
5. Coefficient of traction:
   a. Track machines—0.5 or better
   b. Wheel machines—0.4 or better*
6. The use of hydraulic-controlled blades.

To calculate field production rates, the curve values must be adjusted by the expected job conditions. Table 7.2 lists the correction factors to use with the Caterpillar dozer production curves. The formula for production calculation is

$$\underset{\text{(lcy per hour)}}{\text{Production}} = \underset{\text{from curve}}{\text{ideal production}} \times \underset{\text{correction factors}}{\text{product of the}} \qquad \textbf{[7.6]}$$

* Poor traction affects both track and wheel machines, causing smaller blade loads. Wheel dozers, however, are affected more severely. There are no fixed rules to predict the resulting production loss caused by poor traction. A rough rule of thumb for wheel-dozer production loss is a 4% decrease for each 0.01 decrease in the coefficient of traction below 0.40.

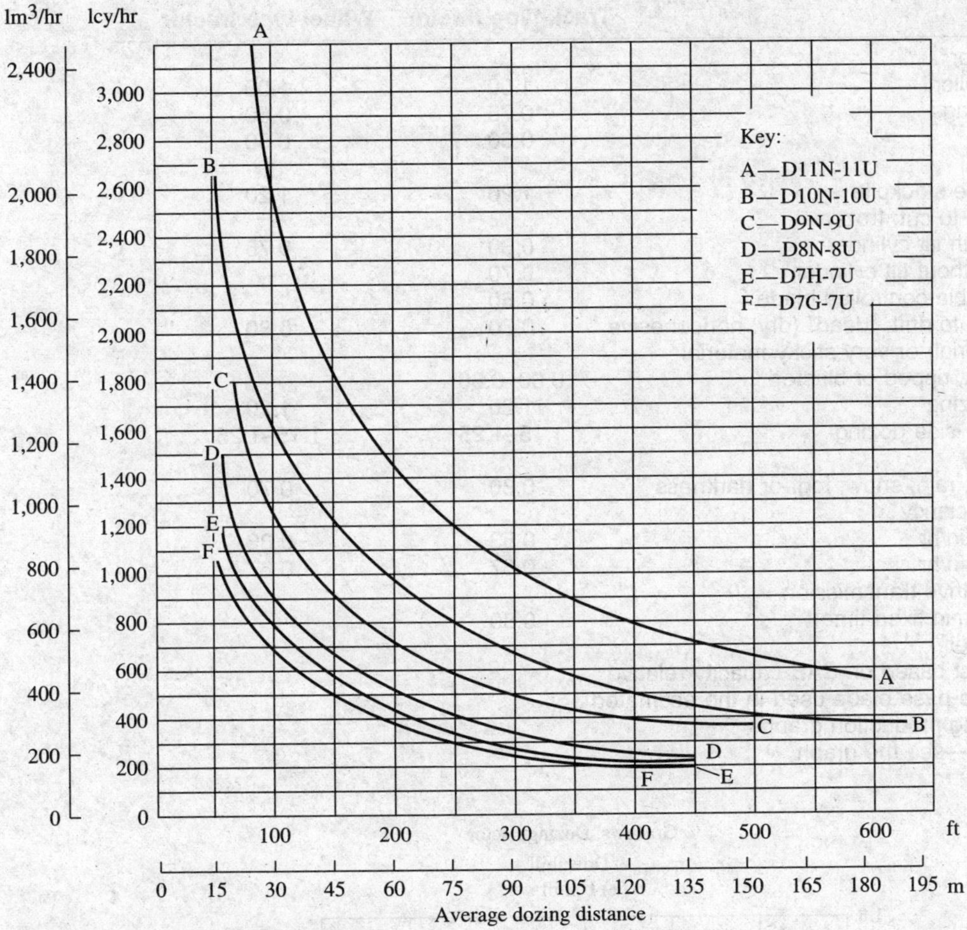

FIGURE 7.14 Dozing production estimating curves for Caterpillar D7 through D11 dozers equipped with universal blades.
Reprinted courtesy of Caterpillar Inc.

DOZER PRODUCTION ESTIMATING FORMAT

Here is a format that can be used to analyze dozer production. The calculations are based on using a Caterpillar D7G crawler dozer with a straight blade for which the specification information is given in Table 7.2 and the performance chart is shown in Fig. 7.13. This will be a slot-dozing operation. The material is a dry, noncohesive silty sand, and it is to be moved a distance of 300 ft from the beginning of the cut. Dozing is downhill on a 10% grade. The operator will have average skill, the dozer will have a power-shift transmission, and both visibility and traction should be satisfactory. The material weighs 108 pcf (lb per cf) in the bank state and is estimated to swell 12% when excavated (bank to loose state). Job efficiency is assumed to be equivalent to a 50-min hour. Calculate the direct cost of the proposed earthmoving operation in dollars per

TABLE 7.2 Caterpillar job condition correction factors for estimating dozer production.

	Track-type tractor	Wheel-type tractor
Operator		
Excellent	1.00	1.00
Average	0.75	0.60
Poor	0.60	0.50
Material		
Loose stockpile	1.20	1.20
Hard to cut; frozen		
with tilt cylinder	0.80	0.75
without tilt cylinder	0.70	—
cable-controlled blade	0.60	—
Hard to drift; "dead" (dry, noncohesive material) or very sticky material	0.80	0.80
Rock, ripped or blasted	0.60–0.80	—
Slot dozing	1.20	1.20
Side-by-side dozing	1.15–1.25	1.15–1.25
Visibility		
Dust, rain, snow, fog, or darkness	0.80	0.70
Job efficiency		
50 min/hr	0.83	0.83
40 min/hr	0.67	0.67
Direct-drive transmission		
(0.1-min fixed time)	0.80	—
Bulldozer*		
Adjust based on SAE capacity relative to the base blade used in the estimated dozing production graphs		
Grades—see the graph		

% Grade vs. Dozing Factor

(−) Downhill

(+) Uphill

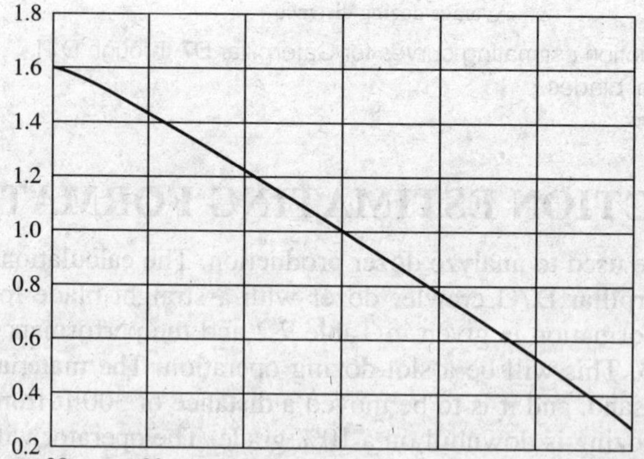

*Note: Angling blades and cushion blades are not considered production-dozing tools. Depending on job conditions, the A blade and C blade will average 50–75% of straight-blade production.

Reprinted courtesy of Caterpillar Inc.

bcy. Assume that the owning and operating (O&O) cost for the dozer is \$32.50 per hour and the operator's wage is \$14.85 per hour.

Step 1: Ideal Maximum Production

Determine the ideal maximum production from the appropriate curve based on the particular dozer model and type of blade. Find the dozing distance on the bottom horizontal scale of the proper figure. Draw a vertical line upward until it intersects the production curve for the dozer under consideration, then construct a line horizontally to the vertical scale on the left side of the figure. Read from the intersect point on vertical scale the maximum production in lcy per hr.

D7 with straight blade: From Fig. 7.13 ideal production for 300-ft push is 170 lcy/hr.

Step 2: Material-Weight Correction Factor

If the actual unit weight of the material to be pushed is not available from soil investigations, the average values found in Table 4.3 can be used. Divide 2,300 lb/lcy by the lcy weight of the material being pushed in order to determine the correction factor.

Bank weight for this project material is given as 108 pcf; therefore,

$$108 \text{ lb/cf} \times 27 \text{ cf/cy} = 2{,}916 \text{ lb/bcy}$$

Swell is 12%, therefore the loose weight of the material being pushed is

$$\frac{2{,}916}{1.12} = 2{,}604 \text{ lb/lcy}$$

Standard condition is 2,300 lb/lcy

$$\text{Material weight correction} = \frac{2{,}300 \text{ lb/lcy}}{2{,}604 \text{ lb/lcy}} = 0.88$$

Step 3: Operator Correction Factor

Table 7.2 presents correction factors for operator skill.

Operator 0.75 (average skill, track-type tractor)

Step 4: Material-Type Correction Factor

Dozer blades are designed to cut material and to roll the cut material in front of the blade. The normal condition is a production factor of 1, however, some materials do not behave in the ideal manner and a correction factor must be applied (see Table 7.2).

Material (type) 0.80 (dry, noncohesive)

Step 5: Operating-Technique Correction Factor

In the case of a dozer operating alone the factor is 1, see Table 7.2. In the case of slot or side-by-side dozing:

Operating technique 1.20 (Slot dozing)

Step 6: Visibility Correction Factor

In the case of good visibility, use 1; see Table 7.2 for factors based on other conditions.

Visibility 1.00

Step 7: Efficiency Factor

See Table 7.2 or use the assumed number of operating minutes per hour divided by 60 min.

Job efficiency 0.83 (50-min hour)

Step 8: Machine Transmission Factor

Table 7.2 presents correction factors for different tractor transmissions.

Transmission 1.00 (the D7G is a power-shift tractor)

Step 9: Blade Adjustment Factor

See the note at the bottom of Table 7.2.

Blade 1.00

Step 10: Grade Correction Factor: (−) Favorable or (+) Unfavorable

Find the percent grade on the bottom of the horizontal scale (Table 7.2 graph). Move up vertically and intersect the grade correction curve, then move left horizontally and locate the grade correction factor on the vertical scale.

Grade 1.24 (−10% grade)

Step 11: Determine the Product of the Correction Factors

Product, correction factors

$$= 0.88 \times 0.75 \times 0.80 \times 1.20 \times 1.0 \times 0.83 \times 1.0 \times 1.0 \times 1.24 = 0.652$$

Step 12: Determine the Dozer Production

$$\text{Production} = 170 \text{ lcy/hr} \times 0.652 = 111 \text{ lcy/hr}$$

Step 13: Determine the Material Conversion, if required

The average values that are found in Table 4.3 can be used if no project-specific data is available. Note this conversion does not change the dozer production, only the units of how the production is stated.

$$\frac{111 \text{ lcy/hr}}{1.12} = 99 \text{ bcy/hr}$$

Step 14: Determine the Total Cost to Operate the Dozer

Cost:	
O&O	$32.50 per hour
Operator	$14.85
Total	$47.35 per hour

Step 15: Determine the Direct Unit Production Cost

$$\text{Direct production cost} = \frac{\$47.35 \text{ per hour}}{99 \text{ bcy/hr}} = \$0.478 \text{ per bcy}$$

> Dozer production estimating using average data such as that presented in the Caterpillar production curves of Figs. 7.13 and 7.14 and the appropriate correction factors should not be considered exact. The first complication is the difficulty of reading the curves, but more important, the work in the field is never a fixed set of conditions. Therefore, when making such production estimates it is good practice to investigate the cost effects for a range of production values. With such knowledge, management can evaluate the risk resulting from estimate assumptions. Good production record keeping will greatly aid management in evaluating the reliability of production estimates.

Step 16: Sensitivity Check

$$\text{Direct production cost} = \frac{\$47.35 \text{ per hour}}{110 \text{ bcy/hr}} = \$0.431 \text{ per bcy}$$

$$\text{Direct production cost} = \frac{\$47.35 \text{ per hour}}{99 \text{ bcy/hr}} = \$0.478 \text{ per bcy}$$

$$\text{Direct production cost} = \frac{\$47.35 \text{ per hour}}{90 \text{ bcy/hr}} = \$0.526 \text{ per bcy}$$

Therefore, the effect of achieving a production of only 90 bcy/hr would be a $0.048 increase in cost for every bcy moved.

DOZER SAFETY

On May 14, 2003, an operating engineer in the state of Washington was killed when the bulldozer he was operating slid on ice, overturned, and came to rest at the bottom of a 150-ft embankment. The victim, *who was not wearing a seat belt*, was thrown from the bulldozer and crushed beneath it when it came to rest. The bulldozer's rollover protective structure (ROPS) was not significantly damaged.

Project engineers are not expected to operate heavy equipment, but as part of management, they have both a moral responsibility and business reasons to ensure the equipment operators receive proper training and can recognize potential hazards associated with their machines. OSHA Construction Standards [29 CFR Part 1926] specifies many safety rules. Section 1926.602(a)(2)(i) specifically states, "Seat belts shall be provided on all equipment covered by this section and shall meet the requirements of the Society of Automotive Engineers, J386-1969, Seat Belts for Construction Equipment." Yet a quick search of the Web will yield several accident reports of fatalities that resulted from failure to use seat belts while operating a dozer (see website resource 4).

Morally (and legally) companies have to provide employees a safe working place. From a purely business point of view, an effective safety program has the potential to drastically reduce the cost of doing business. Companies that maintain extremely low accident rates reap the benefits of doing so. Practicing and demanding good safety will pay dividends throughout one's career.

LAND CLEARING

LAND-CLEARING OPERATIONS

Crawler dozers equipped with special clearing blades are excellent machines for land clearing. Clearing of vegetation and trees is usually necessary before undertaking earthmoving operations. Trees, brush, and even grass and weeds make material handling very difficult. If these organic materials are allowed to become mixed into an embankment fill, their decay over time will cause settlement of the fill.

Clearing land can be divided into several operations, depending on the type of vegetation, the condition of the soil and topography, the amount of clearing required, and the purpose for which the clearing is performed:

1. Removing all trees and stumps, including roots
2. Removing all vegetation above the surface of the ground only, leaving stumps and roots in the ground
3. Disposing of vegetation by stacking and burning

TYPES OF EQUIPMENT USED

Several types of equipment are used, with varying degrees of success, for clearing land:

1. Crawler tractors with earthmoving blades
2. Crawler tractors with special clearing blades
3. Crawler tractors with clearing rakes

Crawler Tractors with Earthmoving Blades

Crawler tractors with earthmoving blades were once used extensively to clear land. There are at least two valid objections to the use of tractors with earthmoving blades. Prior to felling large trees, they must excavate earth from around the trees and cut the main roots, which leaves objectionable holes in the ground and requires considerable time. Also, when stacking the felled trees and other vegetation, a considerable amount of earth is transported to the piles, which makes burning difficult. Therefore, the use of such equipment is not recommended.

Crawler Tractors with Special Clearing Blades

There are blades specially designed for use in felling trees. Maybe the most common clearing blade is the angled blade with projecting stinger. This blade is often referred to as a *"K/G" blade* (Fig. 7.15). This name comes from the highly successful *Rome K/G clearing blade*, manufactured by the Rome Plow Company, Cedartown, Georgia.

The major components of a single-angle clearing blade are the stinger, web, cutting edge, and guide bar. The exterior edges of the stinger and web are sharpened like a knife blade by machining with a grinder. The stinger is in effect a protruding vertical knife. It is designed to be used as a knife to cut and split trees, stumps, and roots. The stinger is a vertical knife and the web is a horizontal knife; together the stinger and web cut the tree in both planes simultaneously (see Fig. 7.15). The blade can be tilted and is mounted at a 30° angle with the stinger forward. The guide bar serves to push the cut material forward and to the side of the tractor.

FIGURE 7.15 The Rome K/G clearing blade.

Clearing blades are most efficient when the tractor is operating on level ground and the cutting edges can maintain good contact with the ground surface. It is easier to work with soil types that hold the vegetation's root structure while the trunks are sheared. Large rocks will slow production by damaging the cutting edges of the blade.

Crawler Tractors with Clearing Rakes

A *rake* is a frame with multiple vertical teeth or tines mounted in the place of a solid-faced blade. Clearing rakes are used to grub and to pile trees after the clearing blades have worked an area (see Fig. 7.16). Like earthmoving blades, the teeth of a rake are curved in the vertical plane to form a flattened "C." With this shape, the teeth can easily be driven under roots, rocks, and boulders. As the tractor moves forward, it forces the teeth of the rake below the ground surface. The teeth will catch the belowground roots and surface brush left from the felling operation, while allowing the soil to pass through. Rakes have an upward extension brush guard that can be either solid plate or opened spaced ribs. Some rakes have a steel center plate to protect the tractor's radiator.

The size, weight, and spacing of the rake's teeth depend on the intended application. Rakes used to grub out stumps and heavy roots must have teeth of sufficient strength so that a single tooth can take the push of the tractor at full power. Lighter rakes having smaller and closer-spaced teeth are used for finish raking and to clear light root systems and small branches from the ground.

Rakes are also used to push, shake, and turn piles of trees and vegetation before and during burning operations. These tasks shake the dirt out of the piles and improve burning.

FIGURE 7.16 Crawler dozer with land-clearing rake.

Disposal of Trees, Brush, Stumps, and Root Debris

When debris is to be disposed of by burning, it should be piled into stacks or rows with a minimum amount of soil (these piled-up or created rows are referred to as windrows). Shaking a rake while it is moving the debris will reduce the amount of soil in the pile.

If the debris is burned while the moisture content is high, it may be necessary to provide an external source of fuel, such as diesel, to start combustion. A burner, consisting of a gasoline-engine-driven pump and a propeller, is capable of maintaining a fire even under adverse conditions. The liquid fuel is blown as a stream into the pile of material while the propeller furnishes a supply of air to ensure vigorous burning. Once combustion is started, the fuel is turned off.

In many urban areas, the burning of debris is restricted. Therefore, it is common to chip brush and trees. Mixed chips, bark, and clean wood can be sold as mulch or used as boiler fuel. Clean wood chips without bark are used to manufacture pressed boards.

LAND-CLEARING PRODUCTION ESTIMATING

Typically, clearing of timber is performed with 160- to 460-hp crawler tractors. The speed at which the tractor can move through the vegetation will depend on the nature of the growth and the size of the machine. It is best to estimate land clearing by the use of historical data from similar projects. When data from past projects are not available, the estimator can utilize the formula presented in this section as a rough guide for probable production rates. Production rates calculated strictly from the formula, however, should be used with caution.

The estimator should always walk the project site prior to preparing a production estimate. This is necessary to obtain information needed to properly evaluate the impacting factors, such as soil condition, and to develop a complete understanding of the project requirements and possible variations from the clearing formula assumptions.

Constant Speed Clearing

When there is only small-size vegetation and it is possible to clear at a constant speed, production can be estimated by the tractor speed and the width of the pass:

$$\underset{(\text{acre/hr})}{\text{Production}} = \frac{\underset{\text{of cut (ft)}}{\text{width}} \times \text{speed (mph)} \times 5{,}280\ \text{ft/mi} \times \text{efficiency}}{43{,}560\ \text{sf/acre}} \qquad [7.7]$$

The American Society of Agricultural Engineers' formula for estimating land-clearing production at constant speed is based on a 49.5-min hour, which is a 0.825 efficiency. When this efficiency is used, formula [7.7] reduces to

$$\text{Production (acre/hr)} = \frac{\text{width of cut (ft)} \times \text{speed (mph)}}{10} \qquad \textbf{[7.8]}$$

The width of cut is the resulting cleared width, measured perpendicular to the direction of dozer travel. With an angled blade, it is obvious that this is not equal to the width (length) of the blade. Even when working with a straight blade it may not be the same as the blade width. The width of cut should be determined by field measurement.

EXAMPLE 7.5

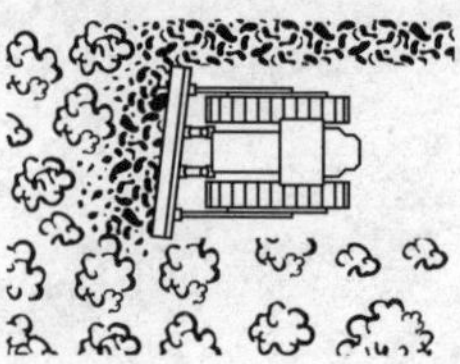

A 200-hp crawler dozer will be used to clear small trees and brush from a 10-acre site. By operating in first gear, the dozer should be able to maintain a continuous forward speed of 0.9 mph. An angled clearing blade will be used on the project; the width of this K/G blade is 12 ft 9 in. From past experience the average resulting clear width will be 8 ft. Assuming normal efficiency, how long will it take to knock down the vegetation?

Using Eq. [7.8], we have

$$\frac{8\text{ ft} \times 0.9\text{ mph}}{10} = 0.72\text{ acre/hr}$$

$$\frac{10\text{ acres}}{0.72\text{ acre/hr}} = 13.9\text{ hr} \approx 14\text{ hr}$$

Tree Count Method—Cutting and Piling-Up Production

The Rome Plow Company developed formulas for estimating cutting and piling-up production (refer to the *Caterpillar Performance Handbook*). The Rome formula and tables of constants provide guidance for variable-speed clearing operations, but use of the results should be tempered with field experience.

To develop the necessary input data for the Rome formula, the estimator must make a field survey of the area to be cleared and collect information on the following items:

1. Density of vegetation *less than 12 in.* in diameter:

Dense—600 trees per acre.

Medium—400 to 600 trees per acre.

Light—less than 400 trees per acre.

2. Presence of hardwoods expressed in percent.

3. Presence of heavy vines.

TABLE 7.3 Production factors for felling with Rome K/G blades.*

Tractor (hp)	Base minutes per acre* (*B*)	Diameter range 1–2 ft (M_1)	2–3 ft (M_2)	3–4 ft (M_3)	4–6 ft (M_4)	Above 6 ft (F)
165	34.41	0.7	3.4	6.8	—	—
215	23.48	0.5	1.7	3.6	10.2	3.3
335	18.22	0.2	1.3	2.2	6.0	1.8
460	15.79	0.1	0.4	1.3	3.0	1.0

*Based on power-shift tractors working on reasonably level terrain (10% maximum grade) with good footing and no stones, and an average mix of soft- and hardwoods.

Reprinted courtesy of Caterpillar Inc.

4. Average number of trees per acre in each of the following size ranges:

Less than 1 ft in diameter.

1 to 2 ft in diameter.

2 to 3 ft in diameter.

3 to 4 ft in diameter.

4 to 6 ft in diameter.

The diameter of the tree is taken at breast-height or 4.5 ft above the ground. In the case of a tree having a large buttress, the measurement should be taken where the trunk begins to run straight and true.

5. Sum of diameter of all trees per acre above 6 ft in diameter at ground level.

Once the field information is collected, the estimator can enter the table of production factors for cutting (Table 7.3) to determine the time factors that should be used in the Rome cutting formula, Eq. [7.9]. The formula is based on the assumptions that a power shift tractor is being used, the ground is reasonably level (less than 10% grades), and the machine has good footing.

Time (min) per acre for cutting

$$\text{Time (min) per acre for cutting} = H[A(B) + M_1N_1 + M_2N_2 + M_3N_3 + M_4N_4 + DF] \quad \textbf{[7.9]}$$

where H = hardwood factor affecting total time.

Hardwoods affect overall time as follows:

75 to 100% hardwoods; add 30% to total time ($H = 1.3$)

25 to 75% hardwoods; no change ($H = 1.0$)

0 to 25% hardwoods; reduce total time 30% ($H = 0.7$)

A = tree density and the effect the presence of vines have on base time

The density of undergrowth material less than 1 ft in diameter and the presence of vines affect base time.

Dense: greater than 600 trees per acre; add 100% to base time ($A = 2.0$)

Medium: 400 to 600 trees per acre; no change ($A = 1.0$)

TABLE 7.4 Production factors for piling-up in windrows.*

Tractor (hp)	Base minutes per acre* (*B*)	Diameter range				
		1–2 ft (M_1)	2–3 ft (M_2)	3–4 ft (M_3)	4–6 ft (M_4)	Above 6 ft (*F*)
165	63.56	0.5	1.0	4.2	—	—
215	50.61	0.4	0.7	2.5	5.0	—
335	44.94	0.1	0.5	1.8	3.6	0.9
460	39.27	0.08	0.1	1.2	2.1	0.3

*May be used with most types of raking tools and angled shearing blades. Windrows to be spaced approximately 200 ft apart.

Reprinted courtesy of Caterpillar Inc.

Light: less than 400 trees per acre; reduce base time 30% ($A = 0.7$)

Presence of heavy vines; add 100% to base time ($A = 2.0$)

B = base time per acre for each dozer size

M = minutes per tree in each diameter range

N = number of trees per acre in each diameter range, from field survey

D = sum of diameters in foot increments of all trees per acre above 6 ft in diameter at ground level, from field survey

F = minutes per foot of diameter for trees above 6 ft in diameter

If the removal of the trees and grubbing of roots and stumps greater than 1 ft in diameter is to be accomplished in one operation, increase total time per acre by 25%. If the removal of stumps is accomplished in a separate operation, increase the time per acre by 50%. Note that this tree count method has no correction for efficiency. The time values in Tables 7.3 and 7.4 are based on normal efficiency.

EXAMPLE 7.6

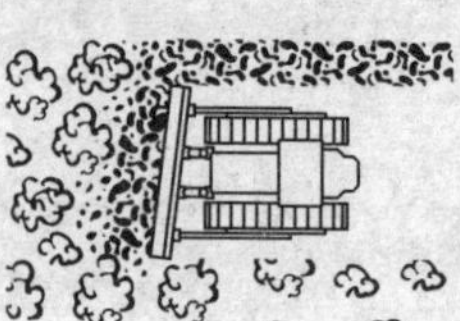

Estimate the rate at which a 215-hp tractor equipped with a K/G blade can fell the vegetation on a highway project. Project specifications require grubbing of stumps resulting from trees greater than 12 in. in diameter. Felling and grubbing will be performed in one operation.

The site is reasonably level terrain with firm ground and less than 25% hardwood. The field survey gathered the following tree counts:

Average number of trees per acre, 700
1 to 2 ft in diameter, 100 trees
2 to 3 ft in diameter, 10 trees
3 to 4 ft in diameter, 2 trees
4 to 6 ft in diameter, 0 trees
Sum of diameter increments above 6 ft, none
The necessary input values for Eq. [7.9] are
$H = 0.7$, less than 25% hardwoods

$A = 2.0$, dense, >600 trees per acre

From Table 7.3 for a 215-hp dozer:

$B = 23.48$, $M_1 = 0.5$, $M_2 = 1.7$, $M_3 = 3.6$, $M_4 = 10.2$, and $F = 3.3$

Time per acre = 0.7[2.0(23.48) + 0.5(100) + 1.7(10) + 3.6(2) + 10.2(0) + 0(3.3)]

Time per acre = 84.8 min per acre

Because the operation will include grubbing, the time must be increased by 25%.

Time per acre = 84.8 min per acre × 1.25 = 106.0 min per acre

Clearing rates are often expressed in acres per hour, so for this example the rate would be 0.57 acres per hour (60 min per hr/106 min per acre).

Table 7.3 and Eq. [7.9] are used to calculate the time for a cutting operation. On many projects, the cut material must be piled for burning or so that it can be easily picked up and hauled away. The Rome Plow Company developed a separate formula and set of constants for estimating piling-up production rates:

Time (min) per acre for piling-up

$$\text{Time (min) per acre for piling-up} = B + M_1N_1 + M_2N_2 + M_3N_3 + M_4N_4 + DF \quad [7.10]$$

The factors have the same definitions as when used previously in Eq. [7.9], but their values must be determined from Table 7.4 for input into the piling-up Eq. [7.10]. In the case of piling-up grubbed vegetation increase the total piling-up time by 25%.

EXAMPLE 7.7

Consider that the vegetation cut in Example 7.8 must be piled. What is the estimated rate at which piling-up can be accomplished?

Time per acre = 50.61 + 0.4(100) + 0.7(10) + 2.5(2) + 5.0(0) + 0(−)

= 103 min/acre

Because the operation will include grubbing, the time must be increased by 25%.

103 min/acre × 1.25 = 129 min/acre or 0.47 acre piled-up per hr

---

Safety during Clearing Operations

During clearing operations that involve multiple dozers, it is necessary to maintain considerable clear distance between machines. This is because falling trees can easily strike a neighboring machine if they are operated close together. Operators must also exercise care when pushing trees over. If the dozer follows a falling tree too closely, the stump and root mass can catch under the front of the machine. But the greatest hazard is from fire. The dozer's belly pan must be cleaned often during a clearing operation as accumulated debris in the engine compartment can easily ignite.

RIPPING ROCK

RIPPERS

Crawler tractors can be fitted with rear-mounted rippers specifically designed by the manufacturer to match tractor characteristics. Rippers are made for a range of tractor sizes with various ripper configurations and linkage designs for depth control and adjustment of the tip's attack angle. Because of the power and tractive force available with large tractors, penetration depth of a rear ripper on such machines (see Fig. 7.17) can be as great as 4 to 5 ft (1.2 to 1.5 m) but decreases significantly with smaller, lighter tractors. A ripper is a relatively narrow-profile implement. It penetrates the earth and is pulled by the crawler tractor to loosen and split hard ground, weak rock, or old pavements and bases. Motor graders can also be equipped with rippers for light-duty applications.

Although rock has been ripped with varying degrees of success for many years, developments in methods, equipment, and knowledge have greatly increased the range of materials that can be ripped economically. Rock that was once considered to be unrippable is now ripped with relative ease, and at cost reductions—including ripping and hauling with scrappers—amounting to as much as 50% when compared with the cost of drilling, blasting, loading with loaders, and hauling with trucks.

The major developments responsible for increased ripping capability include:

1. Heavier and more powerful tractors
2. Improvements in the sizes and performance of rippers, to include development of impact rippers

FIGURE 7.17 Dozer-mounted hydraulically operated double-shank ripper.

3. Better instruments for determining the rippability of rocks
4. Improved techniques in using instruments and equipment

DETERMINING THE RIPPABILITY OF ROCK

The first step in investigating a project that involves rock excavation is determining if the rock can be ripped (plowed). Prior to selecting the method of excavating and hauling rock, the engineer should ask the question "Can I rip?" not "Do I have to drill and blast?" Evaluating the rippability of a rock formation involves study of the rock type and a determination of the rock's density. Igneous rocks, such as granites and basaltic types, are normally impossible to rip, because they lack stratification and cleavage planes, and are very hard. Sedimentary rocks have layered structures caused by the manner in which they were formed. This characteristic makes them easier to rip. The metamorphic rocks, such as gneiss, quartzite, schist, and slate, being a changed form from either igneous or sedimentary, vary in rippability with their degree of lamination or cleavage.

Physical characteristics that favor ripping are

1. Fractures, faults, and joints; these act as planes of weakness facilitating ripping.
2. Weathering; the greater the degree of weathering the more easily the rock is ripped.
3. Brittleness and crystalline structure.
4. High degree of stratification or lamination offers good possibilities for ripping.
5. Large grain size, coarse-grained rocks rip more easily than fine-grained rocks.

Because the rippability of most rock types is related to the speed at which seismic (sound) waves travel through the rock, it is possible to use refraction seismographic methods to determine with reasonable accuracy if a rock can be ripped. Refraction seismographic methods are based on Snell's law,[†] which defines how a wave bends when it crosses the boundary between two different materials. In the earth, shear and compression waves travel with different velocities in different types of rocks. Compression waves (p-waves) travel faster than shear waves (s-waves). When a wave intersects a boundary at an angle other than 90°, the wave will reflect and/or refract across that boundary, depending on the velocity of each material. A refracted wave then travels along the boundary between the two layers and is not transmitted into the lower layer. Measuring these waves allows the seismic velocity of the rock to be determined.

†Willebrord Snell (1580–1626) was a Dutch scientist who studied the behavior of light as it passed through a medium. He found that there is a direct relationship between the sine of the angle of incidence and the sine of the angle of refraction.

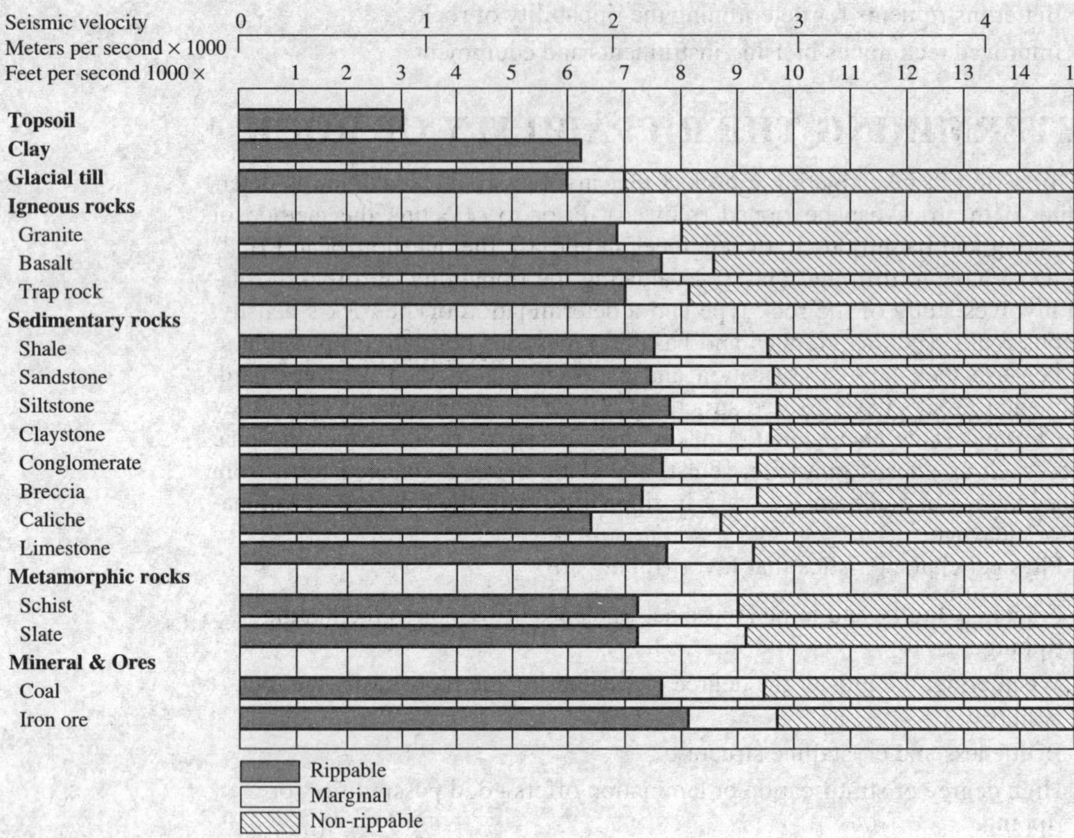

FIGURE 7.18 Ripper performance for Caterpillar 370-hp crawler tractor with multi- or single-shank rippers. Estimated by seismic wave velocities.
Reprinted courtesy of Caterpillar Inc.

Rocks that propagate seismic waves at low velocities, less than 7,000 ft/sec, are rippable, whereas rocks that propagate waves at high velocities, 10,000 ft/sec or greater, are not rippable. Rocks having intermediate velocities are classified as marginal. Figure 7.18 indicates, for a specific size dozer, rippability based on velocity ranges for various types of soil and rocks. The indication that a rock may be rippable, marginal, or nonrippable is based on using multi- or single-shank rippers mounted on a crawler tractor. The information appearing in the figure should be used only as a guide and should be supplemented with other data such as boring logs or core samples. The decision to rip or not to rip rock should be based on the estimated costs as compared to the costs for other excavating methods. Field tests may be necessary to determine if a given rock can be ripped economically.

DETERMINING THE THICKNESS AND STRENGTH OF ROCK LAYERS

A refraction seismograph can be used to determine the top of bedrock, and the thickness and strength of rock layers at or near the ground surface. The paths followed by seismic (sound) waves from a wave-generating source through a formation to detecting instruments (a seismograph) are illustrated in Fig. 7.19.

A geophone, which is a sound sensor, is driven into the ground at station 0. Equally spaced points 1, 2, 3, and so on are located along a line, as indicated in Fig. 7.19. Due to the geometry of refraction it is necessary for the length of the seismic "spread" to be approximately 3 to 5 times the depth to the rock layers. A wire is connected from the geophone to the seismic timer (see Fig. 7.20), and another wire is connected from the timer to a sledgehammer, or other seismic energy source. A steel plate is placed on the ground at the sensing stations, in successive order. When an impulsive seismic source creates a seismic wave (the hammer strikes the steel plate), a switch closes instantly to send an electric signal, which starts the timer. At the same instant, the blow from the hammer sends seismic waves into the formation, and these waves travel to the geophone. On receipt of the first wave, the geophone signals the timer to stop recording elapsed time. With the distance and time known, the velocity of a wave can be determined.

As the distance from the geophone to the wave source, namely, the striking hammer, is increased, waves will enter the lower and denser formation, through which they will travel at a higher speed than through the topsoil. These waves that travel through the denser formation will reach the geophone before the waves traveling through the topsoil arrive. The velocity through the

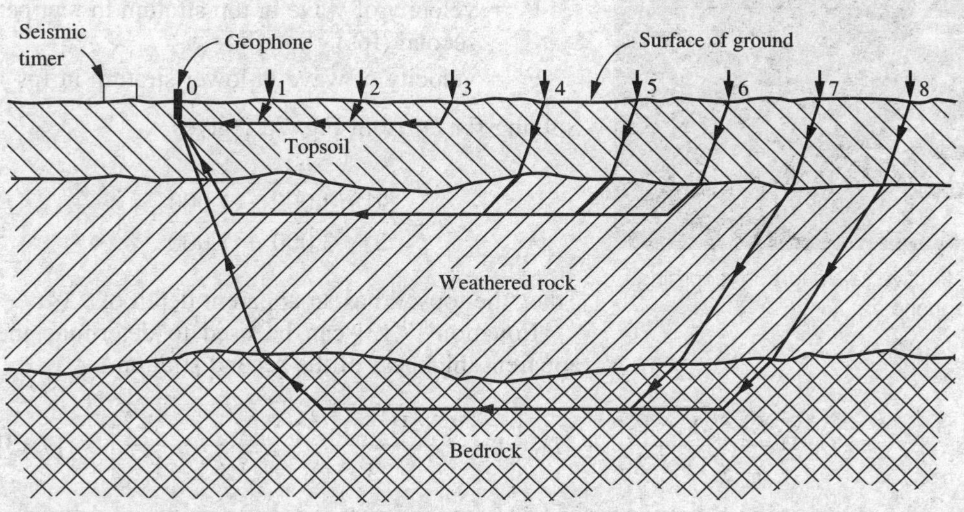

FIGURE 7.19 Path of seismic waves through earth layers having increased density with depth.

FIGURE 7.20 Refractive seismograph recording instrument and geophones.

denser formation is a higher value and it will remain approximately constant as long as the waves travel through a formation of uniform density.

A plot such as that shown in Fig. 7.21 yields wave velocity by measuring the distance and time associated with each layer and dividing the distance by the corresponding time:

$$\text{Velocity, } V_i = \frac{\text{horizontal distance, } L_i}{\text{time to travel distance } L_i} \quad \textbf{[7.11]}$$

The velocity remains constant as the wave passes through material of uniform density. Breakpoints in the slope of the lines indicate changes in velocity and therefore material density.

The depth to the surface separating the two strata depends on the critical distance and the velocities in the two materials. It can be computed from the equation

$$D_1 = \frac{L_1}{2}\sqrt{\frac{V_2 - V_1}{V_2 + V_1}} \quad \textbf{[7.12]}$$

where

D_1 = depth in feet of the first layer

L_1 = critical distance in feet

V_1 = velocity of wave in top stratum in feet per second (fps)

V_2 = velocity of wave in lower stratum in fps

Solving for D_1 from Fig. 7.21 gives

$$D_1 = \frac{36}{2}\sqrt{\frac{3{,}000 - 1{,}000}{3{,}000 + 1{,}000}} = 13 \text{ ft}$$

Thus, the topsoil has an apparent depth of 13 ft.

Equation [7.13] can be used to determine the apparent thickness of the second stratum.

$$D_2 = \frac{L_2}{2}\sqrt{\frac{V_3 - V_2}{V_3 + V_2}} + D_1\left[1 - \frac{V_2\sqrt{V_3^2 - V_1^2} - V_3\sqrt{V_2^2 - V_1^2}}{V_1\sqrt{V_3^2 - V_2^2}}\right] \quad \textbf{[7.13]}$$

where L_2 is 70 ft, D_1 is 13 ft, V_1 is 1,000 fps, V_2 is 3,000 fps, and V_3 is 6,000 fps. Solving, we obtain

$$D_2 = 31 \text{ ft}$$

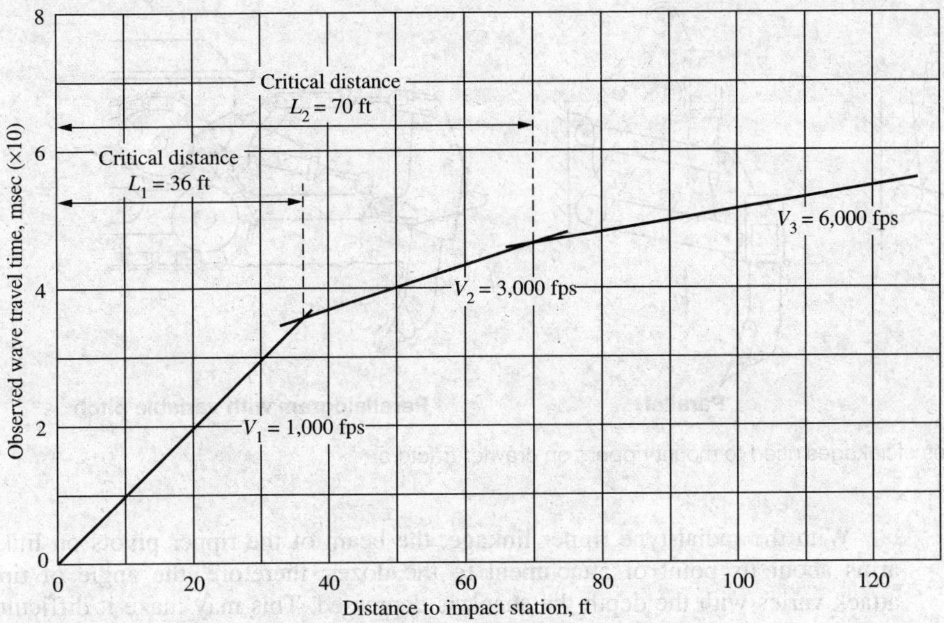

FIGURE 7.21 Plot of seismic wave travel time versus distance between source and the geophone.

Ripping is normally an operation involving the top strata of a formation. Determination of the depth (thickness) of the upper two rock layers and the velocity of the upper three layers will provide the necessary data to estimate most construction jobs. The procedure can become very complicated, however, when dipping strata are encountered. The investigation of complicated formations may require, in addition to seismic studies, the core drilling of samples and the excavation of test pits.

RIPPER ATTACHMENTS

Ripper attachments for crawler tractors are generally rear mounted. The mounting may be fixed radial, fixed parallelogram, or a parallelogram linkage with hydraulically variable pitch (see Fig. 7.22). The vertical member of the ripper that is forced down into the material to be ripped is known as a *shank*. A ripper tip (a tooth, point, or tap) is fixed to the lower, cutting end of the shank. The tip is detachable for easy replacement, as it constitutes the real working part of the ripper and receives all the abrasive action of the rock—it is a high-wear surface. A tip can have a service life from only 30 min to 1,000 hr, depending on the abrasive characteristics of the material being ripped.

Both straight and curved shanks are available. Straight shanks are used for massive or blocky formations. Curved shanks are used for bedded or laminated rocks, or for pavements, where a lifting action will help shatter the material.

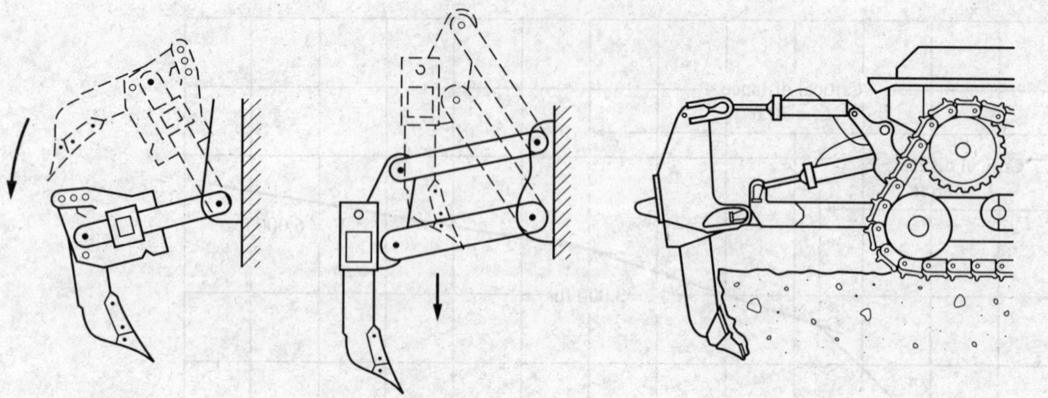

FIGURE 7.22 Types of linkages used to mount rippers on crawler tractors.

With the radial-type ripper linkage, the beam of the ripper pivots on link arms about its point of attachment to the dozer; therefore, the angle of tip attack varies with the depth the shank is depressed. This may make it difficult to achieve penetration in tough materials. The shank may also tend to "dig itself in" like an anchor.

The parallelogram-type ripper maintains the shank in a vertical position and keeps the tip at a constant angle. With adjustable parallelogram-type rippers the operator can control the tip angle.

Heavy-duty rippers are available in either single-shank or multishank models. The multishank models are available with up to five shanks in the case of smaller tractors and up to three shanks for larger tractors. The shanks are inserted into slots on the ripper frame and pin connected to the frame. This arrangement facilitates shank removal so that the number of engaged shanks can be matched to material and project requirements. In the case of very heavy work (i.e., high seismic velocity material) a single center-mounted shank will maximize production. The use of multiple shanks will produce more uniform breakage if full penetration can be obtained. A single shank often rolls individual oversized pieces to the side.

The effectiveness of a ripper depends on

1. Down pressure at the ripper tip
2. The tractor's usable power to advance the tip; a function of power available, tractor weight, and coefficient of traction
3. Properties of the material being ripped; laminated, faulted, weathered, and crystalline structure

A rule of thumb in sizing a tractor for a ripping operation is that it must have 1 fwhp per 100 lb of down pressure on the ripper and it must have 3 lb of machine weight per lb of down pressure to ensure adequate traction.

The number of shanks used depends on the size of the dozer, the depth of penetration desired, the resistance of the material being ripped, and the degree of breakage of the material desired. If the material is to be excavated by scrapers, it should be broken into particles that can be loaded into scrapers, usually of not more than 24 to 30 in. maximum sizes. Two shanks can be effective in softer, easily fractured materials that are to be scraper-loaded. Three shanks should be used only in very easy-to-rip material such as hardpan or some weak shales. Only a field test conducted at the project site will demonstrate which method, depth, and degree of breakage is appropriate and economical.

RIPPING PRODUCTION ESTIMATES

Although the cost of excavating rock by ripping and scraper loading is considerably higher than for earth that requires no ripping, it may be much less expensive than using an alternative method, such as drilling, blasting, excavator loading, and truck hauling. Estimating ripping production is best accomplished by working a test section and conducting a study of the operational methods. This enables a production determination based on weight of ripper material. However, the opportunity to conduct such field tests is often nonexistent; therefore, most initial estimates are based on production charts supplied by equipment manufacturers.

Quick Method

By field-timing several passes of a ripper over a measured distance, an approximate production rate can be determined. The timed duration should include the turnaround time at the end of each pass. An average cycle time can be determined from the timed cycles. The quantity (volume) is determined by measuring the length, width, and depth of the ripped area. Experience has shown that the production rate calculated by this method is about 20% higher than an accurate cross-section study. Therefore, the quick-estimating formula is

$$\text{Ripping production (bcy/hr)} = \frac{\text{measured volume (bcy)}}{1.2 \times \text{average time (hr)}} \qquad [7.14]$$

Seismic-Velocity Method

Manufacturers have developed relationships between the seismic wave velocities of different rock types and rippability (see Fig. 7.18). Ripping performance charts, such as those in Fig. 7.18, enable the estimator to make an initial determination of equipment that may be suitable for ripping a particular material based on general rock-type classifications. After the initial determination of applicable machines is made, production rates for the machines are calculated from production charts (see Fig. 7.23).

Ripping production charts developed from field tests conducted in a variety of materials have been published. However, because of the extreme variations possible among materials of a specific classification, a great deal of judgment is necessary when using such charts. The production rates obtained from the charts must be adjusted to reflect the actual field conditions of the project.

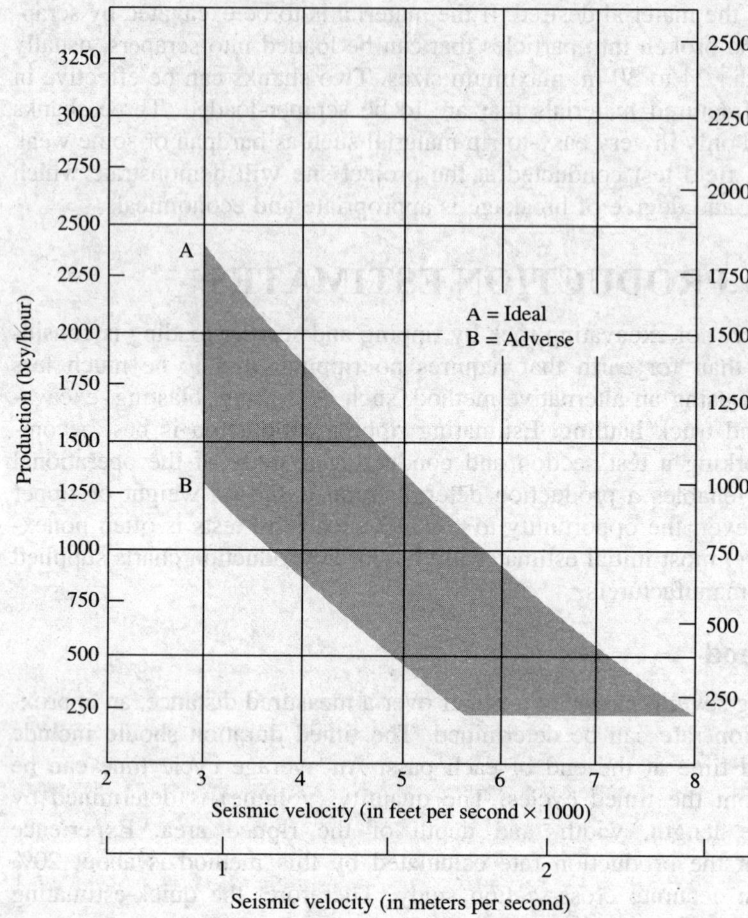

FIGURE 7.23 Ripper production Caterpillar 370-hp crawler tractor with a single shank.
Reprinted courtesy of Caterpillar Inc.

Owning and operating (O&O) costs of a tractor will increase for machines used regularly in ripping operations. Caterpillar warns that normal O&O costs must be increased by 30 to 40% if the machine is used in heavy ripping applications.

EXAMPLE 7.8

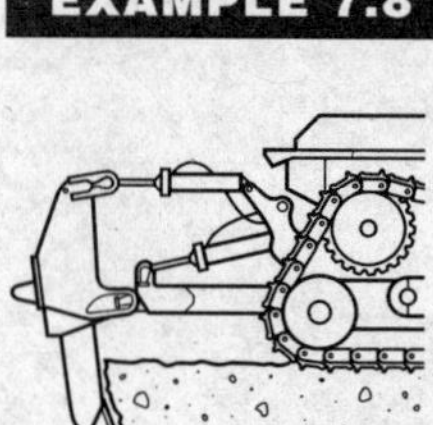

A contractor encounters a shale formation at a shallow depth in a cut section of a project. Seismographic tests indicate a seismic velocity of 7,000 fps for the shale. On this basis, it is proposed to rip the material with a 370-hp crawler tractor. Estimate the production in bcy for full-time ripping, with efficiency based on a 45-min hour (a typical factor for ripping operations). Assume that the ripper is equipped with a single shank and that ripping conditions are ideal. The "normal" O&O cost excluding operator for the tractor-ripper combination is $86 per hour. Operator wages are $9.50 per hour. What is the estimated ripping production cost in dollars per bcy?

Step 1. From Fig. 7.18, ripper performance is 370 hp—crawler tractor can rip shale having a seismic velocity up to 7,500 fps

The tractor is applicable for this situation according to the chart, but it is close to the limit of its capability. Therefore, the contractor might want to consider using a larger machine.

Step 2. Using the 370-hp tractor ripper production chart (see Fig. 7.23) for a seismic velocity of 7,000 fps and ideal conditions,

$$\text{Ideal production 370-hp tractor} = 560 \text{ bcy/hr}$$

$$\text{Adjusted production} = 560 \times \frac{45}{60} = 421 \text{ bcy/hr}$$

Step 3. Increase normal O&O cost because of ripping application:

$$\$86.50 \text{ per hour} \times 1.35 = \$116.10 \text{ per hour}$$

$$\text{Total cost with operator: } \$116.10 + \$9.50 = \$125.60 \text{ per hour}$$

$$\text{Production cost} = \frac{\$125.60 \text{ per hour}}{420 \text{ bcy/hr}} = \$0.299 \text{ per bcy}$$

Operating Techniques

Ripping should be done at the maximum penetration depth that traction will allow, but it should also be accomplished at a uniform depth. For the most economical production, ripping is performed in low gear at low speed, typically 1 to $1\frac{1}{2}$ mph. At speeds only slightly higher, there can be a dramatic increase in operating cost from undercarriage and ripper tip wear. When ripping to load scrapers, it is best to rip in the same direction that the scrapers load. When removing the ripped material, always leave a "cushion" 4 to 6 in. in depth of loose material. The "cushion" will create better underfoot conditions for the tractor doing the ripping and reduce track wear.

Take advantage of gravity and rip downhill when possible. Cross-ripping will make the pit rougher, increase scraper tire wear, and requires twice as many passes. Cross-ripping does, however, help to break up "hard spots" or material that comes loose in large slabs.

SUMMARY

Dozers are used for dozing (pushing materials), land clearing, ripping, assisting scrapers in loading, and towing other pieces of construction equipment. The factors that control dozer production rates are (1) blade type, (2) type and condition of material, and (3) cycle time. Manufacturers provide production curves for estimating the amount of material their dozers can push. The curves provide a maximum value in lcy per hour based on a set of "ideal" conditions.

Crawler dozers equipped with special clearing blades are excellent machines for land clearing. Typically, land clearing of timber is performed with crawler dozers having between 160 and 460 hp. It is best to estimate land-clearing production by the use of historical data from similar projects. If historical data are not available, the Rome Plow Company has developed a formula for estimating cutting and piling-up production.

Heavy ripping of rock is accomplished by crawler tractors equipped with rear-mounted rippers because of the power and tractive force available from such machines. A refraction seismograph can be used to determine the thickness and degree of consolidation of rock layers. Equipment manufacturers have developed relationships between seismic-wave velocities and rippability. Additionally, there are ripping production charts that manufacturers have developed from field tests. Critical learning objectives include:

- An ability to calculate dozer pushing production
- An ability to calculate dozer production for clearing operations
- An ability to calculate tractor production for ripping operations

These objectives are the basis for the problems that follow.

PROBLEMS

7.1 The width of a D6 blade is 11 ft. Its effective height is 4 ft 10 in. Considering SAE J1265, calculate the capacity of the blade in lcy.

7.2 Visit the Caterpillar website and obtain the flywheel power (hp) rating for both a D6R and a D7R Series II track tractor. For these same tractors check the blade specifications and determine the width (length) for both an "A" and a "S." Calculate the hp per foot of cutting edge ratio for all four conditions.

7.3 When material is to be moved long distances semi-U (SU) blades are a good choice as the short wings increase capacity compared to straight blades. A measure of a dozer blade's effectiveness in moving material is the ratio of horsepower to the cubic yards of blade load. A measure of the blade's ability to cut hard material is the ratio of horsepower to foot of blade length. Considering the use of a semi-U blade, plot these ratios for CAT D6R Series II, CAT D7R Series II, CAT D8R Series II, and D9R dozers. Horsepower (use the flywheel horsepower values) and blade data can be found at www.cat.com/.

a. Is it better to use a D8R or a D9R if we want better volumetric pushing ability? (D8R 27.2 compared to 23.2 hp/cy)

b. Which is the best dozer to use if we need cutting ability? (D9R highest hp/ft ratio, 28.9 hp/ft)

7.4 There is a report posted at the National Institute for Occupational Safety and Health (NIOSH) homepage, www.cdc.gov/niosh/homepage.html, under Publications and Products (Publications list 2001, NIOSH Pub. #2001-126) that describes serious injuries occurring to bulldozer operators working at U.S. coal, metal, and nonmetal mines. The Mine Safety and Health Administration (MSHA) collected the data. A total of 873 injury records are examined. These injuries resulted in 18 fatalities and 31,866 lost workdays. All of these injuries occurred to dozer operators while they were doing common production tasks.

a. How many fatalities occurred while the operator was getting on or off the dozer?
b. How many fatalities occurred while operating the dozer?
c. Did any dozer fatalities occur at U.S. coal, metal, and nonmetal mines during 1997?
d. In 12% (14 out of 116 incidents) of the incidents involving a rollover or a dozer falling over an edge, the operator died. Not wearing seatbelts (or older equipment without a ROPS or seatbelt) contributed to five of the fatalities. (True, False)

7.5 A CAT D7H (power-shift) dozer is used in a pushing operation. The dozer is equipped with a straight blade. The material (dry and noncohesive) weighs 106 pcf in the bank state. It is estimated that the material will swell 7%, from bank to loose state. The center-of-mass–to–center-of-mass pushing distance is 300 ft downhill on a 3% grade. The operators have average skill and the job will be performed in dusty conditions. Job efficiency can be assumed to be equivalent to a 45-min hour. Calculate the direct cost of the proposed earthmoving operation in dollars per bcy. The company's normal O&O cost for these machines is $70 per hour and the operator's wage is $12.18 per hour plus 41% for fringes, insurance, worker's compensation, and so on.

7.6 A CAT 834 with a straight blade is to be used to push a material that weighs 110 pcf in the bank state. It is estimated that the material will swell 9%, from the bank to loose state. This cohesive soil is to be moved an average distance of 250 ft on a 0% grade. The operator is a new hire with average skill. Job efficiency is estimated to be equivalent to a 50-min hour. Assume that the dozer will have a 0.38 coefficient of traction. Calculate the direct cost of the proposed earthmoving operation in dollars per bcy. The (O&O) cost for the dozer is $68 per hour and the operator's wage is $8.76 per hour plus 34% for fringes, insurance, worker's compensation, and so on.

7.7 A contractor wants to investigate using a 6S blade on the D6H dozer. The material is a dry, noncohesive silty sand, and is to be moved a distance of 200 ft from the beginning of the cut. The dozing is uphill on a 2% grade. The operator will have average skill, the dozer has a power-shift transmission, and both visibility and traction are assumed to be satisfactory. The material weighs 110 pcf in the bank state. Job efficiency is assumed to be equivalent to a 50-min hour. Calculate the direct cost of the proposed earthmoving operation in dollars per bcy. Assume that the O&O cost for the dozer is $54.00 per hour and the operator's wage is $12.00 per hour plus 35% for fringes, worker's compensation, and so on. ($0.900 per bcy)

7.8 A contractor wants to investigate the production and cost differences of using a 7S blade on a D7H dozer or an S blade on an 824 dozer. The material is a dry clay, and it is to be moved a distance of 200 ft from the beginning of the cut. The dozing is uphill on a 5% grade. The operator will have average skill, the dozers have power-shift transmissions, and traction is assumed to be satisfactory. The site will be dusty. The material weighs 98 pcf in the bank state. Job efficiency is assumed to be equivalent to a 50-min hour. Calculate the direct cost of the proposed earthmoving operation in dollars per bcy. Assume that the O&O cost for the D7 dozer is $76.00 per hour and $73.00 per hour for the 824. The operator's wage is $16.00 per hour plus 35% for fringes, worker's compensation, and other benefits.

a. Which machine would you use on this job?

b. It rains and the coefficient of traction for the 824 is now 0.37. The coefficient of traction for the D7 is 0.55. Which machine would you use on this job?

7.9 A crawler dozer with a universal blade is to be used in a slot-dozing operation. The material is a dry, noncohesive silty sand, and is to be moved a distance of 350 ft from the beginning of the cut. The dozing is downhill on a 10% grade. The operator will have average skill, the dozer has a power-shift transmission, and both visibility and traction are assumed to be satisfactory. The material weighs 108 pcf in the bank state, and will swell 10% when excavated by dozing. Job efficiency is assumed to be equivalent to a 50-min hour.

a. Assuming a Caterpillar D8 is used, calculate the direct cost of the proposed earthmoving operation in dollars per bcy. The O&O cost for the dozer is $64.80 per hour and the operator's wage is $10.85 per hour plus 36% for fringes.

b. Assuming that a CAT 834 wheel-type dozer with a straight blade is substituted for the D8, calculate the direct cost of the proposed earthmoving operation in dollars per bcy. (The 834 has 58% more flywheel horsepower than the D8.) The O&O cost for the 834 is $69.95 per hour, all other conditions of the job are unchanged.

c. Assume that the 834 dozer, when operating on this particular soil, will have a coefficient of traction of 0.34. Estimate the unit-production cost in dollars per bcy, with all other job conditions remaining unchanged.

(a, unit production cost = $0.479/bcy; b, unit production cost = $0.522/bcy; c, 834 unit production cost with traction of 0.34% = $0.687 per bcy)

7.10 Two CAT D-10 (power-shift) dozers are to be used in a side-by-side pushing operation. The dozers are equipped with universal blades. The material, dry and noncohesive, weighs 102 pcf in the bank state. It is estimated that the material will swell 8%, from bank to loose state. The center-of-mass–to–center of mass pushing distance is 380 ft downhill on a 5% grade. The operators have average skill and the job will be performed in dusty conditions. Job efficiency can be assumed to be equivalent to a 50-min hour.

Calculate the direct cost of the proposed earthmoving operation in dollars per bcy. The company's normal O&O cost for these machines is $93.82 per hour and the operator's wage is $12.18 per hour plus 42% for fringes.

7.11 A CAT 824C with a straight blade is to be used to push a material that weighs 125 pcf in the bank state. It is estimated that the material will swell 20%, from the bank to loose state. This cohesive soil is to be moved an average distance of 100 ft up a 12% grade. The operator is a new hire who has poor skills. Job efficiency is estimated to be equivalent to a 45-min hour. Assume that the dozer will have a 0.36 coefficient of traction. Calculate the direct cost of the proposed earthmoving operation in dollars per bcy. The O&O cost for the dozer is $47.30 per hour and the operator's wage is $8.76 per hour plus 34% for fringes.

7.12 A contract is being bid to clear and grub 154 acres. You are considering the cost associated with using a 335-hp dozer. The dozer's O&O cost is $63.96 per hour. The wage determination for operators is $10.85 per hour with fringes. Project overhead will run $357 per workday. Assume a 10-hr workday.

A field engineer has made a site visit and provided you with the following information. The site is reasonably level and the machines will have no problem with bearing or traction. About 10% of the trees are pines (softwood) and the rest are oak (hardwood). The total number of trees per acre is about 300; of that number

210 trees per acre are 1 to 2 ft in diameter.

9 trees per acre are 2 to 3 ft in diameter.

1 tree per acre is 3 to 4 ft in diameter.

You plan to clear and grub in one operation, and then to follow with piling-up in windrows and burning. Burning will require that the dozer be employed twice the time expected for piling-up alone.

a. What is the estimated clearing and grubbing production rate?

b. What is the cost associated with the total operation, including overhead, when only one 335-hp dozer is used?

c. Is it cheaper to employ two 335-hp dozers?

7.13 A contract is being bid to clear and grub 193 acres. You are considering the cost associated with using a 460-hp dozer. The dozer's O&O cost is $96.96 per hour. The wage determination for heavy-duty operators is $13.85 per hour, plus 30% for fringes. Project overhead will run $957 per workday. Assume a 10-hr workday.

A field engineer has made a site visit and provided you with the following information. The site is practically level and the machines will have no problem with bearing or traction. About 80% of the trees are pines (softwood) and the rest are oak (hardwood). The total number of trees per acre is about 500; of these

200 trees per acre are 1 to 2 ft in diameter.

50 trees per acre are 2 to 3 ft in diameter.

20 trees per acre are 3 to 4 ft in diameter.

You plan to clear and grub in one operation, and then to follow with piling-up in windrows and burning. Burning will require that the dozer be employed twice the time expected for piling-up alone.

a. What is the estimated production rate for clearing, grubbing, and piling and burning?
b. What is the cost associated with the total operation, including overhead, when only one 460-hp dozer is used?
c. Is it cheaper to employ two, three, or four of the 460-hp dozers?

7.14 A contract is being bid to clear and grub 147 acres. You are estimating the cost associated with using either of two different power-shift dozers. One dozer is a 215-hp model and the other is rated at 335 hp. The 215-hp has an O&O cost, without operator, of $43 per hour and the 335-hp's O&O is $63.96 per hour. The wage determination for heavy-duty operators is $10.85 per hour. Project overhead will run $357 per workday. Assume a 10-hr workday.

A field engineer has made a site visit and provided you with the following information. The site is reasonably level and the machines will have no problem with bearing or traction. About 10% of the trees are pines (softwood) and the rest are oak (hardwood). There are about

210 trees per acre 1 to 2 ft in diameter.

9 trees per acre 2 to 3 ft in diameter.

1 tree per acre 3 to 4 ft in diameter.

You plan to clear and grub in one operation, and then to follow with piling-up in windrows and burning. Burning will require that the dozer be employed twice the time expected for piling-up alone.

a. What is the estimated clearing and grubbing production rate for each dozer?
b. What is the cost associated with the total operation, including overhead, when only one 215-hp dozer is used?
c. What is the cost associated with the total operation, including overhead, when only one 335-hp dozer is used?
d. Is it cheaper to employ three 215-hp dozers or two 335-hp dozers?

7.15 Given the geophone data shown here, calculate the seismic-wave velocity in each layer. Find the bottom depth of each layer except the last.

Distance (ft)	15	30	45	60	75	90	105	125	145
Time (msec)	7.5	15.0	22.5	26.3	30.0	33.8	37.5	40.4	43.2

7.16 An earthmoving contractor encounters a trap-rock formation at shallow depth in a cut. Seismographic tests indicate a seismic velocity of 6,500 fps for the material. On this basis, it is proposed to rip the rock with a 370-hp crawler tractor. Estimate the production in bcy, for full-time ripping, with efficiency based on a 50-min hour. Assume that the ripper is equipped with a single shank and that ripping conditions are *average* (i.e., intermediate between extreme conditions). The *normal* O&O cost per hour, including operator, for a 370-hp tractor with ripper is $95.82. This is classified as *heavy ripping*. Using your estimate of hourly production, estimate the ripping unit-production cost in dollars per bcy.

7.17 A limestone formation with an average depth of 5 ft is exposed over the full 80-ft width and 23,700-ft length of a highway cut. Preliminary seismic investigations indicate that the rock layer has a seismic velocity of 5,500 fps. The contractor proposes to loosen this rock by using a single-shank ripper on a power shift, track-type tractor. The job conditions relative to lamination and other rock characteristics are *average*, or midway between *adverse* and *ideal*. A 40-min hour working efficiency is possible. The contractor wishes to explore the use of two 370-hp dozers. O&O cost for the tractors in this ripping application is $126.00 per hour including rippers. The operator's wage is $14.30 per hour. What is the unit cost to rip this rock? What will be the total cost of the work and how many hours will the project require? (unit cost $0.439 per bcy, total cost $153,997, time with two dozers 405 hr)

7.18 A limestone formation, average depth of 2.4 ft, is exposed over the full 80-ft width and 23,700-ft length of a highway cut. Seismic investigations indicate that the rock layer has a seismic velocity of 5,500 fps. The contractor proposes to loosen this rock by using a single-shank ripper on a power shift, track-type tractor. It is estimated that job conditions relative to the rock characteristics are *average* or midway between *adverse* and *ideal*. A 40-min hour working efficiency should be obtainable while the ripper is actually in production, but it is estimated that the ripper will be out of service for repairs for 5% of the total hours that the ripper could be used on the job (i.e., availability of 0.95). It is further estimated that the ripper will be out of service due to adverse weather for 20% of the same total hours.

The contractor wishes to explore the use of two 370-hp tractors for the ripping operations. O&O cost, including rippers, is estimated at $79.82 per working hour for each tractor. Each machine is to be charged against the job at 60% of its normal O&O rate during working hours when the equipment is not used because of adverse weather or scheduling problems, but will not be charged against the job when the equipment is down for repair.

The operator's wage, paid during hours worked and hours when the equipment is being repaired, is $14.30 per hour. Associated charges against the ripping operation for the foreman and mechanic are estimated to be $25 per nominal (i.e., regularly scheduled, including downtime for repairs and adverse weather) working hours. Move-in and move-out costs, combined, are estimated to be $200 per tractor. Calculate the total time in weeks, based on a normal 40-hr workweek, to loosen the rock cut, and calculate the total loosening cost, in dollars per bcy. (production 0.205 acre/hr, total cost $109,870, cheaper to use two dozers, $96,661)

7.19 An earthmoving contractor encounters a shale formation at shallow depth in a rock cut. He performs seismographic tests, which indicate a seismic velocity of 5,000 fps for the material. On this basis, he proposes to rip the material with a 370-hp dozer. Estimate the production in bcy, for full-time ripping, with efficiency based on a 45-min hour. Assume that the dozer is equipped with a single shank and that ripping conditions are *adverse*. The *normal* O&O cost per hour, including operator for a 370-hp dozer with ripper is $111. This work would be classified as a *heavy ripping* application. Using your estimate of hourly production, estimate the ripping unit-production cost in dollars per bcy.

REFERENCES

1. *Caterpillar Performance Handbook*, Caterpillar Inc., Peoria, Ill (published annually).
2. *Handbook of Ripping*, 7th ed. (January 1983). Caterpillar Tractor Co., Peoria, Ill.
3. *Land Clearing*, Caterpillar Tractor Co., Peoria, Ill.
4. Schexnayder, Cliff (1998). Discussion of "Rational Equipment Selection Method Based on Soil Conditions," by Ali Touran, Thomas C. Sheahan, and Emre Ozcan, *Journal of Construction Engineering and Management, ASCE*, 124 (6).
5. *The Rome K/G Clearing Blade, Operator's Handbook*, Rome Plow Company, Cedartown, Ga.
6. Touran, Ali, Thomas C. Sheahan, and Emre Ozcan (1997)."Rational Equipment Selection Method Based on Soil Conditions," *Journal of Construction Engineering and Management*, ASCE, 123(1).
7. Burger, Henry Robert, Douglas C. Burger, and Robert H. Burger (1992). *Exploration Geophysics of the Shallow Subsurface*, Prentice Hall PTR.

WEBSITE RESOURCES

1. www.cat.com Caterpillar is the world's largest manufacturer of construction and mining equipment. In the Products, Equipment section of the website can be found the specifications for Caterpillar manufactured dozers.
2. www.deere.com Deere & Company is a worldwide corporation with a construction products division.
3. www.globalsecurity.org/military/library/policy/army/fm/5-430-00-1/CH4.htm Chapter 4, "Clearing, Grubbing, and Stripping," of Department of the Army, Field Manual, No. 5-430-00-1, Air Force Joint Pamphlet, No. 32-8013, *Planning and Design of Roads, Airfields, and Heliports in the Theater of Operations—Road Design* (August 1994).

4. www.imac.ca/technofocus/ratingblades.htm IMAC Design Group Ltd., 7622 18th Street, Edmonton, Alberta, Canada T6P 1Y6, provides at this website a detailed explanation of the SAE Standard J1265 method for calculating the capacity of bulldozer blades. The site also has detailed drawings explaining dozer blade terminology.
5. tirupati-international.com/product/dozer/faq/faq.htm Liebherr-International AG, Case postale 272,CH-1630 Bulle/Fribourg, Germany. This part of Liebherr's website provides detailed information about dozer blade selection, track pads, drawbar pull, and travel speed performance of Liebherr dozers, formulas for calculating blade capacities, and formulas for calculating dozing production.
6. www.cdc.gov/niosh/pdfs/2001-126.pdf National Institute for Occupational Safety and Health, Pittsburgh Research Laboratory, Pittsburgh, Pa. Information Circular 9455, *An Analysis of Serious Injuries to Dozer Operators in the U.S. Mining Industry* (April 2001).

CHAPTER 8

Scrapers

The key to a pusher-scraper spread's economy is that both the push tractor and the scraper share in the work of obtaining the load. The production cycle for a scraper consists of six operations: (1) loading, (2) haul travel, (3) dumping and spreading, (4) turning, (5) return travel, and (6) turning and positioning to pick up another load. A systematic analysis of the individual elements that make up the scraper production cycle is the fundamental approach for identifying the most economical employment of these machines. Scrapers are best suited for haul distances greater than 500 ft but less that 3,000 ft, although with very large units, the maximum distance can approach a mile.

GENERAL INFORMATION

shot rock
Rock that has been blasted from an excavation.

Tractor-pulled scrapers are designed to load, haul, and dump loose material in controlled lifts. The greatest advantage of tractor-scraper combinations is their versatility. The key to a pusher-scraper spread's economy is that both the pusher and the scraper share in the work of obtaining the load. Scrapers can be used to load and haul a wide range of material types, including **shot rock**, and are economical over a wide range of haul lengths and haul conditions. To the extent that they can self-load, they are not dependent on other equipment. If one machine in the spread experiences a temporary breakdown, it will not shut down the job, as would be the case for a machine that is used exclusively for loading. If the loader breaks down, the entire job must stop until repairs can be made. Scrapers are available with loose-heaped capacities up to 44 cy, although in the past a few machines as large as 100 cy were offered.

Since scrapers are a compromise between machines designed exclusively for either loading or hauling, they are not superior to function-specific equipment in either hauling or loading. Excavators, such as hydraulic hoes and front shovels or loaders, will usually surpass scrapers in loading efficiency. Trucks can have faster travel speeds and therefore surpass them in hauling, especially over long distances. However, for off-highway situations having hauls of less than a mile, a scraper's ability to both load and haul gives it an advantage. Additionally, the ability of these machines to deposit their loads in layers of uniform thickness facilitates compaction operations.

SCRAPER TYPES

There are several types of scrapers, primarily classified according to the number of powered axles or by the method of loading. These are wheel-tractor machines that tow a hydraulically operated wheeled trailer—bowl. In the past, **crawler-tractor "towed" two-axle scraper** bowls were manufactured. These proved effective in short-haul situations, less than 600 ft one way. Another machine of the past is the two-axle pulling tractor that can now be found only in much older spreads. Machines currently available include:

crawler-tractor "towed" two-axle scraper

1. Pusher-loaded (conventional)
   - **a.** Single-powered axle

   - **b.** Tandem-powered axles

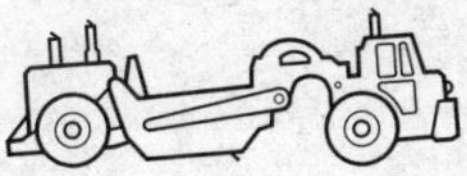

2. Self-loading
   - **a.** Push-pull, tandem-powered axles

   - **b.** Elevating

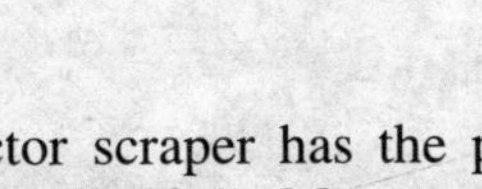

   - **c.** Auger

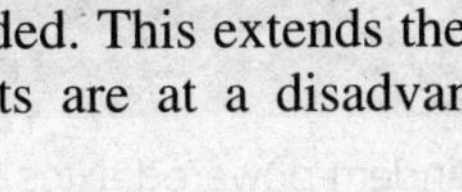

Pusher-Loaded Scrapers

The push-loaded wheel-type tractor scraper has the potential for high travel speeds on favorable haul roads (see Fig. 8.1). Many models can achieve speeds as great as 35 mph when fully loaded. This extends the economic haul distance of the units. However, these units are at a disadvantage when it comes to individually providing the high tractive effort required for economical loading. For the single-powered axle scraper (see Fig. 8.1) only a portion, on the order of 50 to 55% of the total loaded weight, bears on the drive axle.

Additionally, in most materials, the coefficient of traction for rubber tires is less than for tracks. Therefore, it is necessary to supplement the loading power of these scrapers. The external source of loading power is usually a crawler-tractor pusher. Loading costs are still relatively low because both the scraper and the pusher share in providing the total power required to obtain a full load. Even tandem-powered scrapers normally require help loading (see Fig. 8.2). Tandem-powered scrapers have an initial cost that is about 25% more than that for a single-powered axle scraper. For that reason, they are commonly considered a specialized machine good for opening up a job, working extremely adverse grades, or working in soft ground conditions.

Single-powered axle pusher-loaded wheel-tractor scrapers are suited for jobs where haul-road rolling resistance is low and grades are minimal. They become uneconomical when

- Haul grades are greater the 5% and
- Return grades are greater than 12%.

FIGURE 8.1 Single-powered axle wheel-tractor scraper being push-loaded.

FIGURE 8.2 Push-pull, tandem-powered axles, wheel-tractor scrapers.

Tandem-powered axle wheel-tractor scrapers have separate engines for the wheel-tractor unit and for the scraper (bowl) unit. This twin-engine arrangement produces extra power for overcoming high rolling resistance and/or steep grades.

Self-Loading Scrapers

Self-loading scrapers, although heavier and more costly to purchase than comparable conventional scrapers, can be economical in certain applications, particularly in isolated work and for stripping materials.

Push-Pull Scrapers These are basically tandem-powered axle scrapers having a cushioned-push block and bail mounted on the front (see Figs. 8.2 and 8.3) and a hook on the rear above the usual back push block. These features enable two scrapers to assist one another during loading by pushing and pulling one another. The trailing scraper pushes the lead scraper as it loads. Then the lead scraper pulls the trailing scraper to assist it in loading. This feature enables two scrapers to work without assistance from a push tractor. They can also function individually with a pusher.

When used in rock or abrasive materials, tire wear on these scrapers will increase because of more slippage from the four-wheel-drive action.

Elevating Scrapers This is a completely self-contained loading and hauling scraper (see Fig. 8.4). A chain elevator (see Fig. 8.5), mounted vertically on the front of the bowl, serves as the loading mechanism. The disadvantage of this machine is that the weight of the elevator loading assembly is deadweight during the haul cycle. Such scrapers are economical in short-haul situations where the ratio of haul time to load time remains low. Elevating scrapers are used for utility work, dressing-up behind high-production spreads, or shifting material during fine-grading operations. They are very good in small-quantity

FIGURE 8.3 Hook on the rear and cushioned-push block and bail on the front of two push-pull wheel-tractor scrapers.

FIGURE 8.4 Elevating wheel-tractor scraper.

situations. No pusher is required, so there is never a mismatch between pusher and the number of scrapers. Because of the elevator mechanism, they cannot handle rock or material containing rocks.

Auger Scrapers This is another completely self-contained loading and hauling scraper. Auger scrapers (see Fig. 8.6) can self-load in difficult conditions, such as laminated rock, granular materials, or frozen material. In a soft limestone mining operation, 44-cy-capacity tandem-power auger scrapers making 3- to 5-in.-deep cuts achieved an average load time of about $1\frac{1}{2}$ min.

An independent hydrostatic system powers the auger that is located in the center of the bowl. The rotating auger lifts the material off the scraper cutting edge and carries it to the top of the load, creating a void that enables new material to easily enter the bowl. This action reduces the cutting edge resistance, permitting the wheel-tractor scraper to continue moving through the cut. Auger scrapers are available in both single- and tandem-powered configurations. As with an elevating scraper, the auger adds nonload weight to the

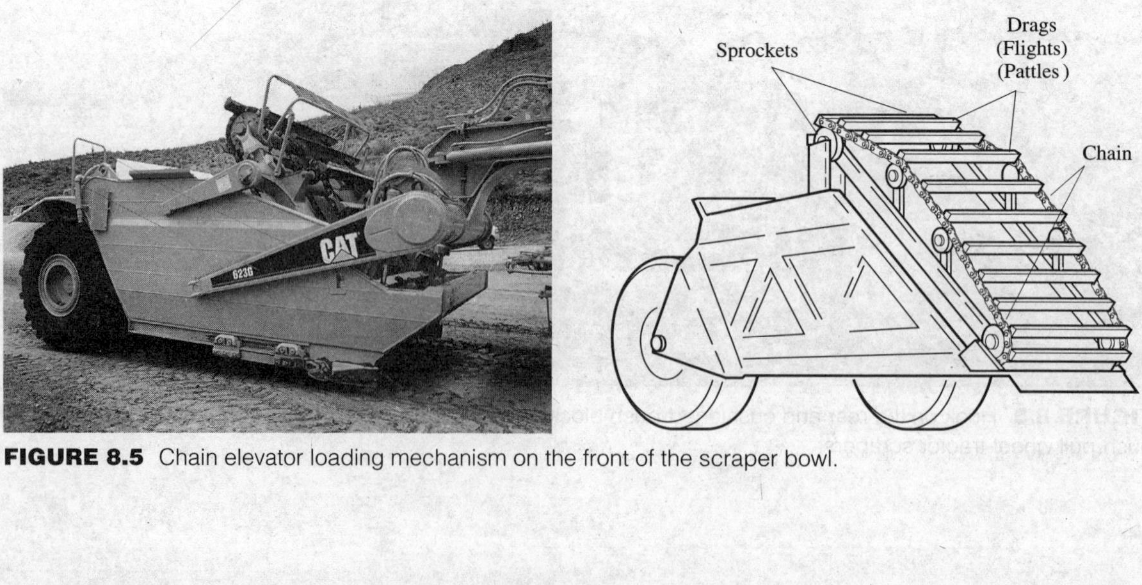

FIGURE 8.5 Chain elevator loading mechanism on the front of the scraper bowl.

FIGURE 8.6 Auger wheel-tractor scraper.

scraper during the travel cycles. These scrapers are more costly to own and operate than conventional single- or tandem-power machines.

Volume of a Scraper

The volumetric load of a scraper may be specified as either the struck or heaped capacity of the bowl expressed in cubic yards. The struck capacity is the volume that a scraper would hold if the top of the material was struck off even at the top of the bowl. In specifying the heaped capacity of a scraper, manufacturers usually specify the slope of the material above the sides of the bowl with the designation SAE. The Society of Automotive Engineers (SAE) specifies a repose slope of 1:1 for scrapers. The SAE standard for other haul units and loader buckets is 2:1, as will be discussed in Chapters 9 and 11. Remember that the actual slope will vary with the type of material handled. In practice, both of these volumes represent loose cubic yards (lcy) of material because of how a scraper loads.

The capacity of a scraper, expressed in cubic yards bank measure (bcy), can be approximated by multiplying the loose volume in the scraper bowl by an appropriate swell factor (see Table 4.3). Because of the compacting effect on the material in a push-loaded scraper, resulting from the pressure required to force additional material into the bowl, the swell is usually less than for material dropped into a truck by a hydraulic excavator. Tests indicate that the swell factors in Table 4.3 should be increased by approximately 10% for material push-loaded into a scraper. When computing the bank measure volume for an elevating scraper, no correction is required for the factors in Table 4.3.

EXAMPLE 8.1

If a push-loaded scraper hauls a heaped load measuring 22.5 cy and the appropriate swell factor from Table 4.3 is 0.8, what is the calculated bank measure volume?

$$22.5 \text{ cy} \times 0.8 \times 1.1 = 19.8 \text{ bcy}$$

FIGURE 8.7 Weighing loaded scrapers in the field.

The use of rated volumetric capacity and swell values from tables provides an estimate of the bank measure scraper load. This is satisfactory for small jobs or if the estimator has developed an accurate set of swell values for the materials commonly encountered in the work area. However, over time, actual field weights of loaded scrapers should be obtained for a variety of materials (see Fig. 8.7). From such weights, the average carrying capacity of those scrapers actually in the fleet can be calculated directly by the methods outlined in Chapter 4. This can be important on large jobs where the significance of small differences between table values and actual measured values can have large cost effects.

SCRAPER OPERATION

The basic operating parts of a scraper (see Fig. 8.8) are

- *Bowl.* The bowl is the loading and carrying component of a scraper. It has a cutting edge that extends horizontally across its front bottom edge. The bowl is lowered for loading and raised during travel.
- *Apron.* The apron is the front wall of the bowl. It is independent of the bowl. It is raised during the loading and dumping operations to enable the material to flow into or out of the bowl. The apron is lowered during hauling to prevent material spillage.
- *Ejector.* The ejector is the rear vertical wall of the bowl. The ejector is in the rear position during loading and hauling. During spreading, the ejector is activated and moves forward, providing positive discharge of the material in the bowl.

A scraper is loaded by lowering the front end of the bowl until the cutting edge that is attached to and extends across the width of the bowl enters the ground. At the same time, the front apron is raised to provide an open slot

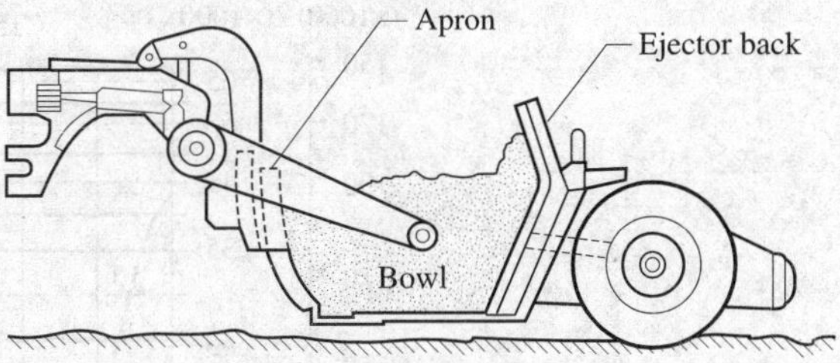

FIGURE 8.8 Scraper bowl with apron and ejector identified.

through which the earth can flow into the bowl. As the scraper moves forward, making a horizontal cut, a ribbon of material is forced into the bowl. This is continued until the bowl is filled, at which point the bowl (cutting edge) is raised and the apron lowered to prevent spillage during the haul. During the haul, the bowl should be kept just high enough to clear the ground. Keeping the bowl low during travel enhances stability at high speeds and on uneven haul roads.

The dumping operation consists of lowering the cutting edge to the desired height above the fill, raising the apron, and then forcing the material out by means of the movable ejector mounted at the rear of the bowl.

The elevating scraper is equipped with horizontal flights (Fig. 8.5) operated by two endless elevator chains, to which the ends of the flights are connected. As the scraper moves forward with its cutting edge scraping loose a ribbon of material, the flights rake the material upward into the bowl. The pulverizing action of the flights enables a complete filling of the bowl and enhances uniform spreading of the fill. Like a push-loaded scraper, an elevating scraper has an ejector for discharging the material. Additionally, the front portion of the bowl's floor can be retracted, and the elevator reverses to aid ejection of material.

SCRAPER PERFORMANCE CHARTS

Manufacturers of scrapers provide specifications and performance charts for each of their machines. These charts contain information that can be used to analyze the performance of a scraper under various operating conditions. Table 8.1 gives the specifications for a Caterpillar 631E single-powered axle scraper. Figures 8.9 and 8.10 present the performance charts for this particular scraper.

It should be noted that owners sometimes add sideboards to the bowls of their scrapers. This practice will increase the volumetric load the machine can carry. If the boards cause the volumetric load of the scraper, whose specifications

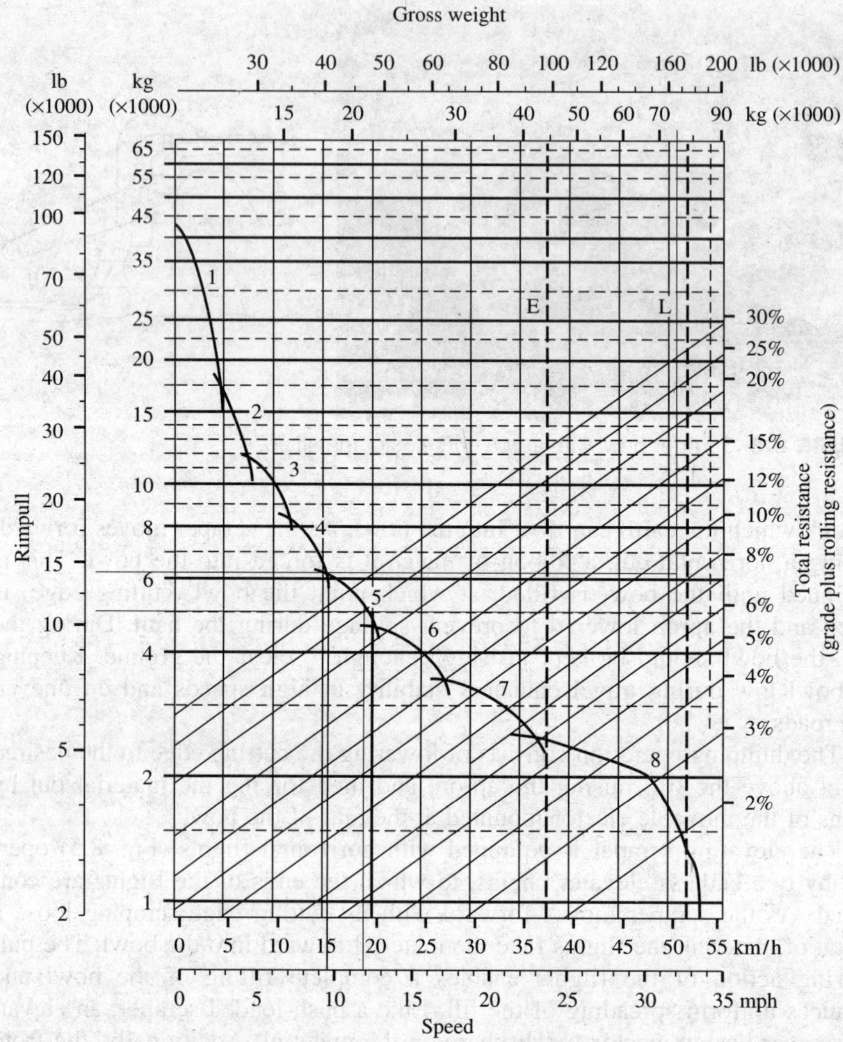

1. 1st gear torque converter drive
2. 2nd gear torque converter drive
3. 3rd gear direct drive
4. 4th gear direct drive
5. 5th gear direct drive
6. 6th gear direct drive
7. 7th gear direct drive
8. 8th gear direct drive

E Empty 43,945 kg (96,880 lb)
L Loaded 77,965 kg (171,880 lb)

FIGURE 8.9 Rimpull performance chart for a Caterpillar 631E scraper.
Reprinted courtesy of Caterpillar Inc.

are given in Table 8.1, to be increased to 25.8 bcy, the weight the scraper is carrying will also increase. This scraper has a rated payload of 75,000 lb. In this sideboarding case, if the unit weight of the material being hauled is 3,000 lb per bcy, the weight of the load would be 77,400 lb (25.8 bcy × 3,000 lb per bcy). This is greater than the rated load. The effect will be a few more yards

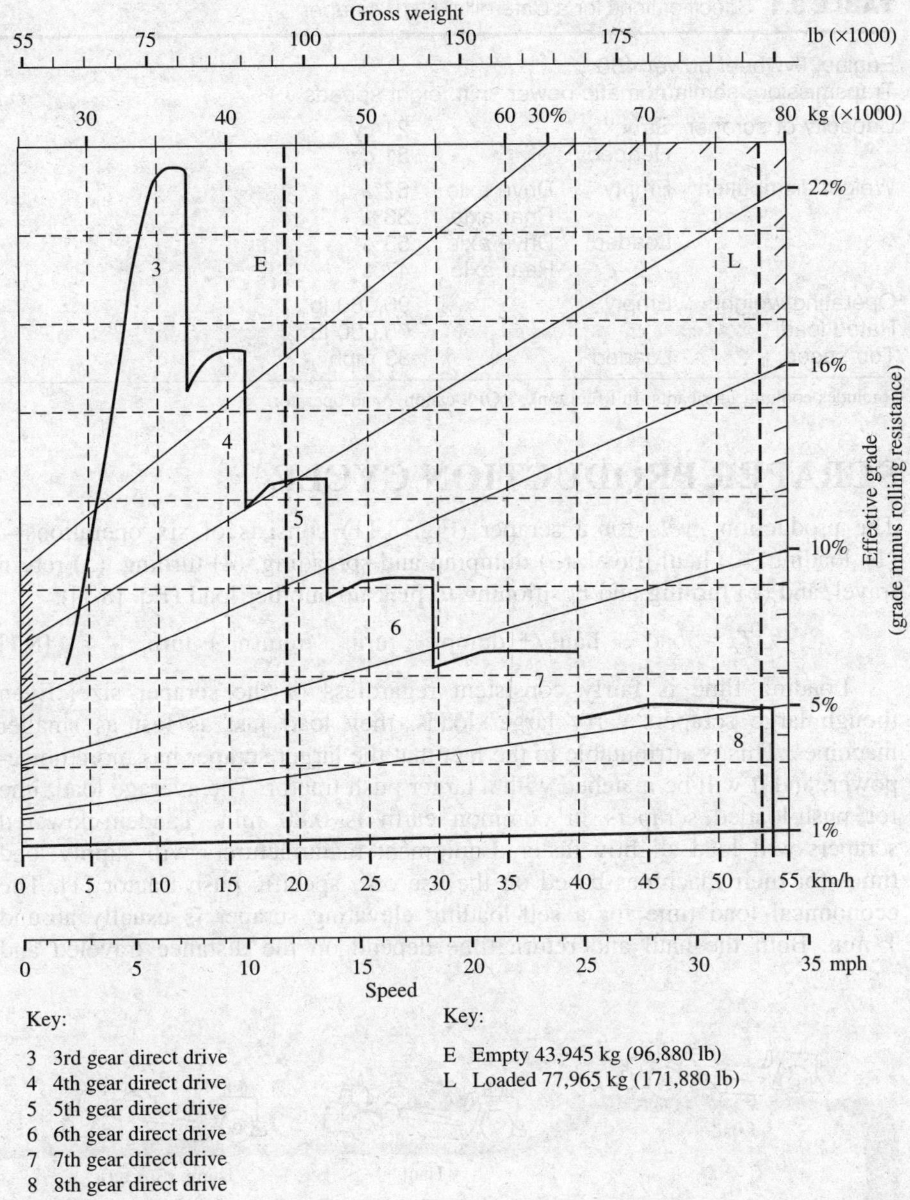

FIGURE 8.10 Retarding performance chart for a Caterpillar 631E scraper.
Reprinted courtesy of Caterpillar Inc.

per load, but in time, maintenance costs will increase because of overloading the machine. A material having a higher unit weight would aggravate this effect. However, if the machine was to be operated in lightweight materials, sideboards would be a reasonable consideration.

TABLE 8.1 Specifications for a Caterpillar 631E scraper.

Engine: flywheel power 450			
Transmission: semiautomatic power shift, eight speeds			
Capacity of scraper:	Struck		21 cy
	Heaped		31 cy
Weight distribution:	Empty	Drive axle	67%
		Rear axle	33%
	Loaded	Drive axle	53%
		Rear axle	47%
Operating weight:	Empty		96,880 lb*
Rated load:			75,000 lb
Top speed:	Loaded		33 mph

*Includes coolant, lubricants, full fuel tank, ROPS canopy, and operator.

SCRAPER PRODUCTION CYCLE

The production *cycle* for a scraper (Fig. 8.11) consists of six operations—(1) loading, (2) haul travel, (3) dumping and spreading, (4) turning, (5) return travel, and (6) turning and positioning to pick up another load (Eq. [8.1]):

$$T_s = \text{load}_t + \text{haul}_t + \text{dump}_t + \text{turn}_t + \text{return}_t + \text{turn}_t \qquad \textbf{[8.1]}$$

Loading time is fairly consistent regardless of the scraper size. Even though large scrapers carry larger loads, they load just as fast as smaller machines. This is attributable to the fact that the larger scraper has more horsepower, and it will be matched with a larger push tractor. The average load time for push-loaded scrapers in common earth is 0.80 min. Tandem-powered scrapers will load slightly faster. Equipment manufacturers will supply load times for their machines based on the use of a specific push tractor [1]. The economical load time for a self-loading elevating scraper is usually around 1 min. Both the haul and return time depend on the distance traveled and

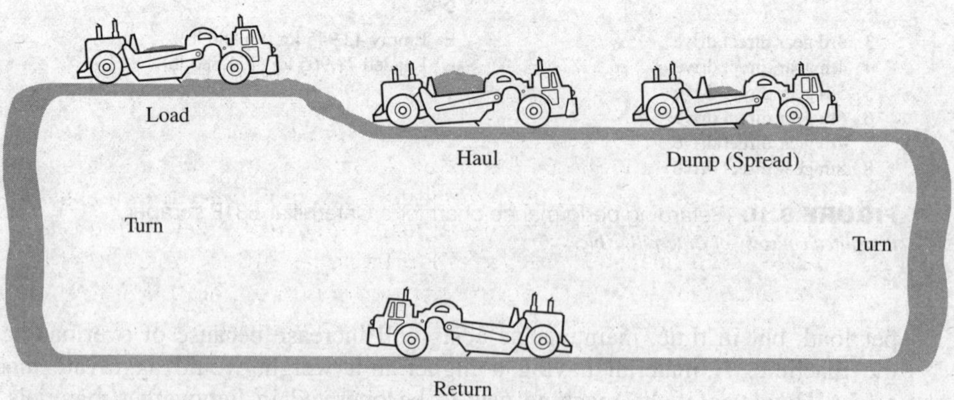

FIGURE 8.11 Scraper production cycle.

scraper speed. Hauling and returning are usually at different speed ranges. Therefore, it is necessary to determine the time for each separately. If the haul road has multiple grade or rolling resistance conditions, a speed should be calculated for each segment of the route. In the calculations, it is appropriate to use a short distance at a slower speed when coming out of the pit (cut), approaching the dump area, leaving the dump area, and again when entering the pit, to account for acceleration/deceleration time. A distance on the order of 200 ft at a speed of about 5 mph in each case is usually appropriate. Extremely steep downhill (favorable) grades can result in longer travel times than those calculated. This is caused by the fact that operators tend to downshift to keep speeds from becoming excessive. When using the charts, always consider the human element; do not blindly accept chart speeds.

SCRAPER PRODUCTION ESTIMATING FORMAT

The most profitable methods and equipment to be used on any job can only be determined by a careful project investigation. The basic approach is a systematic analysis of the scraper cycle with a determination of the cost per cubic yard under the conditions existing or anticipated on the project. Such an analysis is helpful when making decisions for estimating purposes and for job control.

Here is a format that can be used to analyze scraper production. The calculations are based on using a CAT 631E scraper, for which the specification information is given in Table 8.1 and the performance charts are shown in Figs. 8.8 and 8.9. The total length of haul when moving from the cut to the fill is 4,000 ft in three segments, each having a different grade:

1,200 ft +4% grade
1,400 ft +2% grade
1,400 ft −2% grade

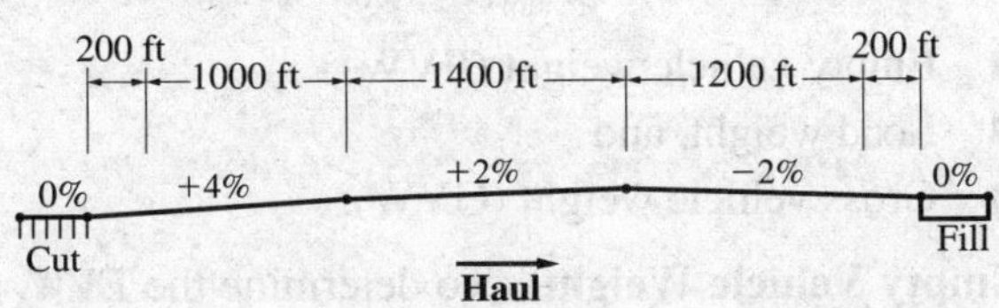

The soil being moved is a clay having a unit weight of 3,000 lb per bcy. The earth haul road will be well maintained; therefore the assumed rolling resistance is 80 lb per ton or 4%. Assume an average load time of 0.85 min. If it is also assumed that the shape of the load-growth curve (Fig. 8.12) is valid for this example, the expected load will be 96% of heaped capacity. A 50 min per hour efficiency factor will be used.

> A load-growth curve for analyzing push-loaded scrapers only describes the loadability of a given scraper for a specific set of conditions including the use of a particular size push tractor.

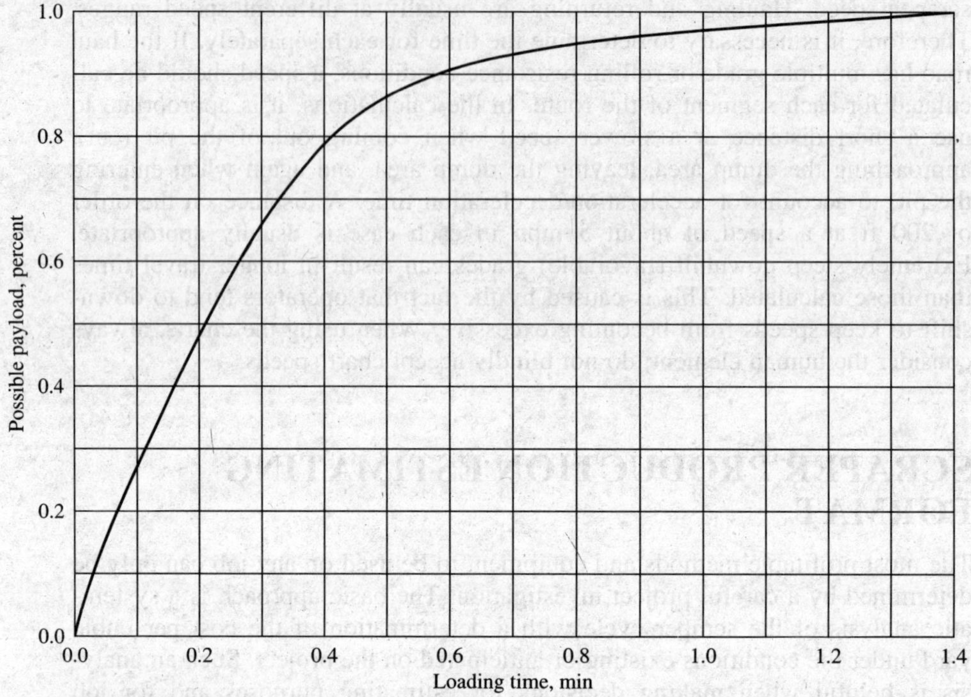

FIGURE 8.12 A typical load-growth curve for scraper loading.

Step 1: Weight

The first step in calculating scraper production is to determine the

- Empty vehicle weight (EVW),
- Load weight, and
- Gross vehicle weight (GVW).

Empty Vehicle Weight To determine the *EVW*, reference the manufacturer's data for the specific make and model of scraper being considered. When referring to the manufacturer's data be careful about notes concerning what is include in the stated weight. Typically, empty operating weight includes a full fuel tank, coolants, lubricants, a rollover protective system (ROPS) canopy, and the operator.

Load Weight The *load weight* is a function of scraper load volume and the unit weight of the material being hauled. The load volume is a loose material volume, so the unit weight needs to be an lcy unit weight or the loose volume can be converted to match the units of the unit weight.

Gross Vehicle Weight The *GVW* is the sum of the EVW and the load weight.

Example Data and Calculations

Empty weight (EVW)	Table 8.1	96,880 lb
Load volume: (based on load time, Fig. 8.10 for this example)		
	0.96×31 cy $= 29.8$ lcy	
	From Table 4.3: swell factor clay $= 0.74$	
Load volume bank measure:	29.8 lcy $\times$ 0.74 $\times$ 1.1 = 24.3 bcy	
Weight of load:	24.3 bcy $\times$ 3,000 lb/bcy =	72,900 lb
	Gross weight (GVW)	169,780 lb

Step 2: Rolling Resistance

Rolling resistance (RR) is the result of a conscious management decision as to how much effort (money) will be expended maintaining the haul road. As the effort (and money) expended for maintenance of haul roads increases, production will also increase. The reverse is also true—if no effort is made to improve the haul road, production suffers. Graders and sometimes dozers are necessary to eliminate ruts and rough (washboard) surfaces. Water trucks will be required to provide moisture for compacting the haul road and for dust control. Visibility is improved by keeping roads moist. This simple action lessens the chance of accidents. Dust control also helps to alleviate mechanical wear.

Rolling Resistance and Scraper Production Haul-road rolling resistance is a job condition that is sometimes neglected. The effect of haul-road conditions on rolling resistance, and thereby on scraper production and the cost of hauling earth, should always be carefully analyzed. A well-maintained haul road enables faster travel speeds and reduces the costs of scraper maintenance and repair. Figure 8.13 is from a field study of scraper haul times. The shaded area represents the range of average travel times on numerous projects. The lower boundary indicates travel times on projects having well-maintained haul roads. The upper boundary indicates poor haul roads. Considering a haul distance of 4,000 ft, the total time for the haul and return combined on a well-maintained haul road was 5.20 min. Under poor conditions, this could be as high as 7.55 min. A difference of 2.35 min in cycle time is a 4.7% production loss in a 50-min hour.

Representative rolling resistance values are given in Table 6.1. Rolling resistance can be expressed as either a percentage or as pounds per ton of vehicle weight.

For this example it is anticipated that the earth haul road will be well maintained, and have a rolling resistance of 80 lb per ton or 4%.

Step 3: Grade Resistance/Assistance

Grade resistance (GR) or assistance (GA) is usually a function of the project topography. From where the material must be excavated and to where it must be hauled is a physical condition imposed by the project requirements. Plan the

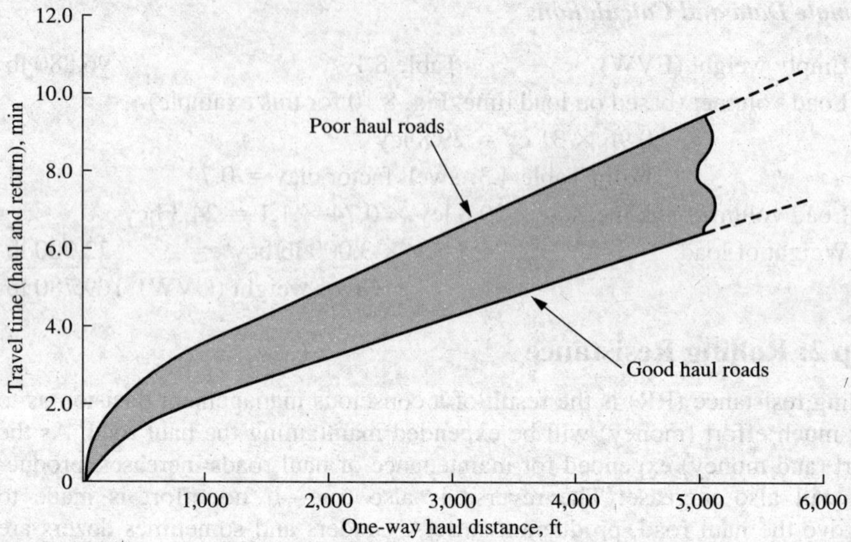

FIGURE 8.13 Average travel times single-powered axle scrapers, capacity <25 cy, negligible grade.

Source: U.S. Department of Transportation, FHWA.

job to avoid adverse grades that could drastically reduce production. There can sometimes be a selection of haul-route placement that enables flatter grades, but those possibilities must usually be considered in terms of increased haul-route length. Such alternatives may mean that several estimates of scraper production will have to be produced.

Grade resistance can be expressed as either a percentage or as pounds per ton of vehicle weight.

The assumed haul-route length and grades from the cut to the fill for this analysis are

1,200 ft	80 lb/ton	+4% grade
1,400 ft	40 lb/ton	+2% grade
1,400 ft	−40 lb/ton	−2% grade

Step 4: Total Resistance/Assistance

The total resistance (TR) or assistance (TA) for each segment of the haul and return must be determined. This is the sum of the rolling resistance and the grade resistance/assistance for each segment. It is recommended practice to use a table format when developing an estimate of scraper production. Such a format is illustrated in this analysis. In such a table format, a separate row can be used for each segment of both the haul and return routes.

To account for acceleration, 200 ft at the beginning of the first segment on both the haul and return routes will be at a reduced speed. Likewise, to account for deceleration, 200 ft at the end of the last segment on both the haul and return routes will be at a reduced speed. These do not represent additional

TABLE 8.2 Scraper production estimating format for the haul cycle.

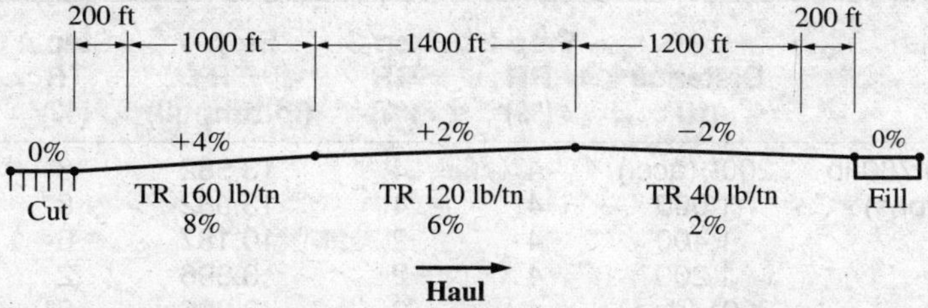

	Distance (ft)	Step 2 RR (%)	Step 3 GR (%)	Step 4 TR (lb per ton)	Step 4 TR (%)
Haul (169,780 lb or 84.89 tons)	200 (acc.)	4	4	160	8
	1,000	4	4	160	8
	1,400	4	2	120	6
	1,200	4	−2	40	2
	200 (dec.)	4	−2	40	2

Total resistance can be calculated either as lb per ton or as a percent. Both are included in the chart to illustrate this fact. However, it is not necessary to use both when making a production calculation, one or the other will suffice. The choice does affect how the performance charts are used, but either method will yield the same speed.

TABLE 8.3 Scraper production estimating format for the return.

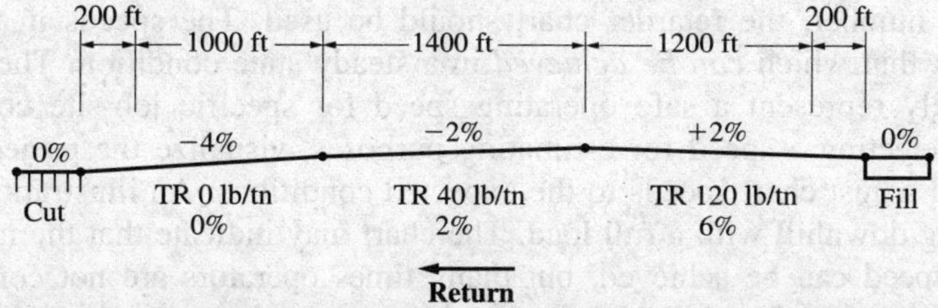

	Distance (ft)	Step 2 RR (%)	Step 3 GR (%)	Step 4 TR (lb per ton)	Step 4 TR (%)
Return (96,880 lb or 48.44 tons)	200 (acc.)	4	2	120	6
	1,200	4	2	120	6
	1,400	4	−2	40	2
	1,000	4	−4	0	0
	200 (dec.)	4	−4	0	0

Total resistance can be calculated either as lb per ton or as a percent. Both are included in the chart to illustrate this fact. However, it is not necessary to use both when making a production calculation, one or the other will suffice. The choice does affect how the performance charts are used, but either method will yield the same speed.

distance, they are part of their respective segments, but instead of using a speed appropriate to the weight of the scraper and the total resistance, a reduced speed will be used in calculating travel time for these segments. This is illustrated in Tables 8.2 and 8.3.

TABLE 8.4 Scraper production estimating format including Step 5—speed.

	Distance (ft)	Step 2 RR (%)	Step 3 GR (%)	Step 4 TR (lb rimpull)	Step 4 TR (%)	Step 5 speed (mph)
Haul (169,780 lb	200* (acc.)	4	4	13,582	8	5†
or 84.89 tons)	1,000	4	4	13,582	8	10
	1,400	4	2	10,187	6	14
	1,200	4	−2	3,396	2	32
	200* (dec.)	4	−2	3,396	2	16†
Return (96,880 lb	200* (acc.)	4	2	5,813	6	11†
or 48.44 tons)	1,200	4	2	5,813	6	23
	1,400	4	−2	1,938	2	33
	1,000	4	−4	0	0	33
	200† (dec.)	4	−4	0	0	16†

*Assumed acceleration and deceleration distances.

†Reduced speed to account for acceleration or deceleration.

Step 5: Travel Speed

Steady state travel speeds based on total vehicle weight and total resistance can be determined from the manufacturer's performance or retarder charts for the specific scrapers under consideration. If the total resistance is a positive number, the performance chart should be used, and if the total resistance is a negative number, the retarder chart should be used. The speeds in the chart represent that which *can be achieved* in a steady state condition. They do not necessarily represent a safe operating speed for specific job-site conditions. Before selecting a speed for estimating purposes, visualize the project conditions and adjust chart speeds to the expected conditions. An illustration of this is hauling downhill with a full load. The chart may indicate that the maximum vehicle speed can be achieved, but many times operators are not comfortable traveling that fast downhill (they realize it is an unsafe condition) and will shift into a lower gear. The speeds used in this analysis are shown in Table 8.4.

Step 6: Travel Time

Travel time is the sum of the times the scraper requires to traverse each segment of the haul and return routes. Based on the speeds determined from the performance or retarder charts, or the assumed speeds because of job conditions, travel time can be calculated using Eq. [8.2] (Table 8.5).

$$\text{Travel time per segment (min)} = \frac{\text{Segment distance, ft}}{88 \times \text{travel speed, mph}} \quad \textbf{[8.2]}$$

Step 7: Load Time

Load time is a management decision that should be made after a careful evaluation of the production and cost effects. The tendency in the field is to take too long in loading scrapers [4].

TABLE 8.5 Scraper production estimating format including Step 6—time.

	Distance (ft)	Step 2 RR (%)	Step 3 GR (%)	Step 4 TR (%)	Step 5 speed (mph)	Step 6 time (min)
Haul (172,210 lb)	200* (acc.)	4	4	8	5†	0.45
	1,000	4	4	8	10	1.14
	1,400	4	2	6	14	1.14
	1,200	4	−2	2	32	0.43
	200* (dec.)	4	−2	2	16†	0.14
Return (96,880 lb)	200* (acc.)	4	2	6	11†	0.21
	1,200	4	2	6	23	0.56
	1,400	4	−2	2	33	0.48
	1,000	4	−4	0	33	0.34
	200* (dec.)	4	−4	0	16†	0.14
					Total Travel time	5.03

*Assumed acceleration and deceleration distances.
†Reduced speed to account for acceleration or deceleration.

Scraper Load-Growth Curve Without a critical evaluation of available information, it may appear that the lowest cost for moving earth with scrapers is to load every scraper to its maximum capacity before it leaves the cut. However, numerous studies of loading practices have revealed that loading scrapers to their maximum capacities will usually reduce, rather than increase, production (see Table 8.6).

When a scraper starts loading, the earth flows into it rapidly and easily, but as the quantity of earth in the bowl increases, the incoming earth encounters

TABLE 8.6 Variations in the rates of production of scrapers with loading times (a 2,500-ft one-way haul distance).

Loading time (min)	Other time (min)	Cycle time (min)	Number trips per hr	Payload*		Production per hour†	
				(cy)	(cu m)	(cy)	(cu m)
0.5	5.7	6.2	8.07	17.4	(13.3)	140	(107)
0.6	5.7	6.3	7.93	18.3	(14.0)	145	(111)
0.7	5.7	6.4	7.81	18.9	(14.5)	147	(112)
0.8	5.7	6.5	7.70	19.2	(14.7)	148‡	(113)
0.9	5.7	6.6	7.57	19.5	(14.9)	147	(112)
1.0	5.7	6.7	7.46	19.6	(15.0)	146	(112)
1.1	5.7	6.8	7.35	19.7	(15.1)	145	(111)
1.2	5.7	6.9	7.25	19.8	(15.2)	143	(109)
1.3	5.7	7.0	7.15	19.9	(15.2)	142	(109)
1.4	5.7	7.1	7.05	20.0	(15.3)	141	(108)

*Determined from measured performance.
†For a 50-min hour.
‡The economical time is 0.8 min.

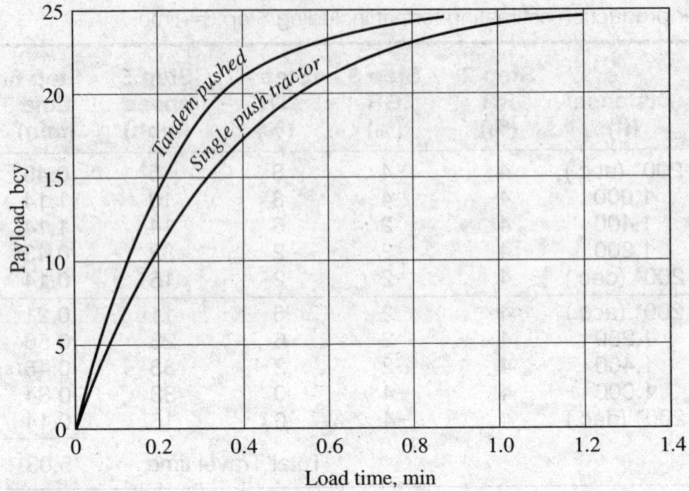

FIGURE 8.14 Load-growth curves for a particular scraper when tandem pushed and when pushed by a single tractor.

greater resistance and the rate of loading decreases quite rapidly, as illustrated in Figs. 8.12 and 8.14. These figures are load-growth curves for specific scraper and pusher combinations used to excavate a particular type of material; they show the relation between the load in the scraper and the loading time. An examination of the Fig. 8.12 curve reveals that during the first 0.5 min, the scraper loads about 85% of its maximum possible payload. During the next 0.5 min it loads only an additional 12%, and if loading is continued to 1.4 min, the gain in volume during the last 0.4 min is only 3%—an instructive example of the law of diminishing returns.

Figure 8.14 presents two load-growth curves. While both curves are for the same scraper they show the effect of using different push tractor combinations. The steeper curve on the left is produced when the scraper is tandem pushed by two tractors.

Economical loading time is a function of haul distance. As haul distance increases, economical load time will increase. Figure 8.15 illustrates the relationship between loading time, production, and haul distance; it is presented only to demonstrate graphically those interrelations. The economical haul distance for scrapers is usually much less than a mile. This can be recognized from the rapid rate at which production falls with increasing haul distance.

For this analysis an average load time 0.85 min has been assumed.

Step 7: Load time 0.85 min

Step 8: Dump Time

Dump times vary with scraper size, but project conditions will affect the dumping duration. Average values for dump time are presented in Table 8.7. Physical constraints in the dump area may dictate scraper-dumping techniques.

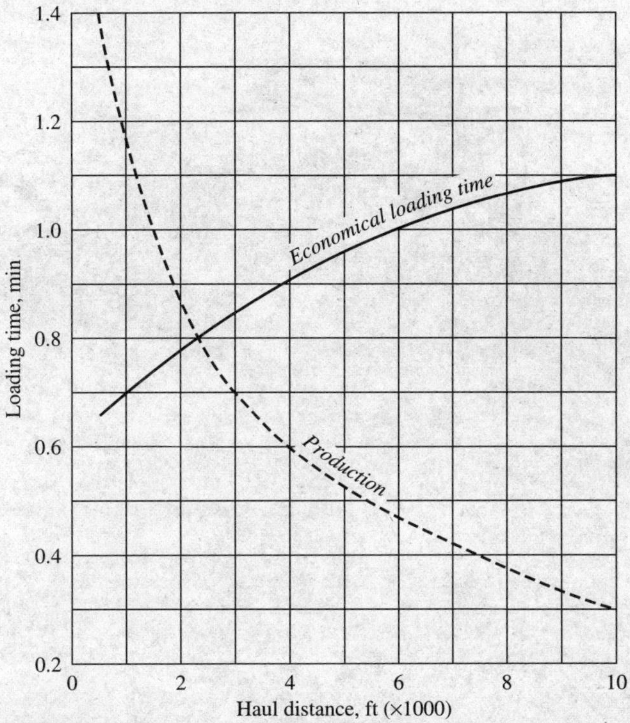

FIGURE 8.15 The effect of haul distance on the economical loading time of a scraper.

TABLE 8.7 Scraper dump cycle times.

	Scraper type	
Scraper size (cy) heaped	**Single engine (min)**	**Tandem-powered (min)**
<25	0.30	—
25–34	0.37	0.26
35–44	0.44	0.28

Source: U.S. Department of Transportation, FHWA.

The most common method is for the scraper to dump before turning. This uses the haul-speed momentum to carry the scraper bowl over and through the dumped material. It reduces the possibility of the scraper becoming stuck in the newly dumped material and yields a fairly even spreading of material, thereby reducing later spreading and compaction equipment effort.

Sometimes it is necessary to negotiate the turn before dumping. This generally increases dump time and the chances of the scraper becoming stuck while dumping. The last method is to dump during the turning maneuver. This definitely increases dumping time, results in uneven spreading, and can cause

FIGURE 8.16 A scraper dumping at a bridge abutment.

scrapers to become bogged down. When access is limited, such as the situation at bridge abutments (see Fig. 8.16) or when backfilling culverts, a turning/dumping maneuver may be necessary. Wet material is difficult to eject and will increase the dump time.

Considering the condition of this analysis, a 631 scraper, that is, a single-engine machine, with a rated heaped capacity of 31 cy (Table 8.1); the dump time will be 0.37 min (Table 8.7).

Step 8: Dump time 0.37 min

Step 9: Turning Times

Turning time is not significantly affected by either the type or size of scraper. Based on FHWA studies, the average turn time in the cut is 0.30 min, and on the fill, the average turn time is 0.21 min. The slightly slower turn in the cut is primarily caused by congestion in that area and the necessity to spot the scraper for the pusher.

Step 9: Turn time fill 0.21 min
Turn time cut 0.30 min

Step 10: Total Cycle Time

Total scraper cycle time is the sum of the times for the six operations that have been examined in the previous steps 1 to 9—haul travel and return travel, which have been combined; loading; dumping and spreading; turning on the fill; and turning and positioning to pick up another load at the cut.

Step 6: Travel time	5.03 min
Step 7: Load time	0.85 min
Step 8: Dump time	0.37 min
Step 9: Turn time fill	0.21 min
Turn time cut	0.30 min
Step 10: Total cycle time scraper	6.76 min

Step 11: Pusher Cycle Time

If push-loaded tractor scrapers are to attain their volumetric capacities, they need the assistance of a push tractor during the loading operation. Such assistance will reduce the loading time duration and thereby reduce total cycle time. When using push tractors, the number of such tractors must be matched with the number of scrapers available at a given time. If either the pusher or scraper must wait for the other, operating efficiency is lowered and production costs are increased.

The pusher cycle time includes the time required to push-load the scraper (the time duration of pusher-scraper contact—contact time) and the time required for the pusher to move into position to push-load the next scraper. The cycle time for a push tractor will vary with the conditions in the loading area, the relative size of the tractor and the scraper, and the loading method. Figure 8.17 shows three loading methods. *Backtrack loading* is the most common method employed. It offers the advantage of always being able to load in the direction of the haul. *Chain loading* can be used when the excavation is conducted in a long cut. *Shuttle loading* is used infrequently. However, if one pusher can serve scrapers hauling in opposite directions from the cut, it is a viable method. Loading patterns that reduce pusher and scraper moves in the pit gain valuable seconds and increase production.

Caterpillar recommended calculating backtrack push tractor cycle time, T_p, by the formula

$$T_p = 1.4L_t + 0.25 \quad \textbf{[8.3]}$$

where L_t = scraper load time (pusher contact time).

The formula is based on the concept that pusher cycle time is a function of four components:

1. Load time of the scraper
2. Boost time, time assisting scraper out of the cut, 0.15 min

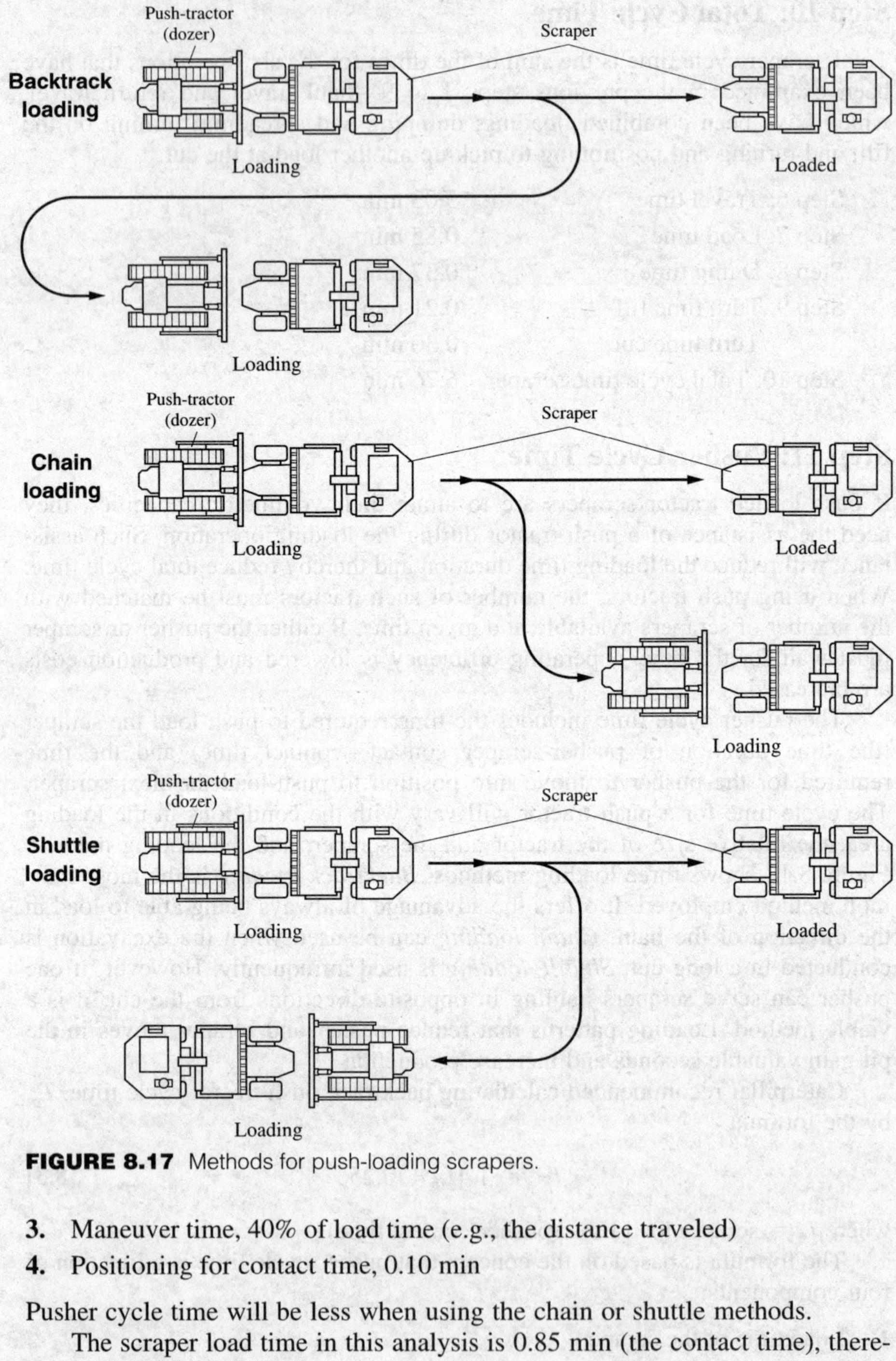

FIGURE 8.17 Methods for push-loading scrapers.

3. Maneuver time, 40% of load time (e.g., the distance traveled)
4. Positioning for contact time, 0.10 min

Pusher cycle time will be less when using the chain or shuttle methods.

The scraper load time in this analysis is 0.85 min (the contact time), therefore applying Eq. [8.3]:

$$T_p = 1.4(0.85) + 0.25 = 1.44 \text{ min}$$

Favorable loading conditions will reduce loading time and increase the number of scrapers that a pusher can serve. A large pit or cut, ripping hard soil prior to loading, loading downgrade, and using a push tractor whose power is matched with the size of the scraper are all factors that can improve loading times. Likewise, tight soils with no prior ripping, very large scrapers, and loading rock are factors that may create a situation where multiple pushers are required to load a scraper effectively.

Step 12: Balance Fleet

The number of scrapers a push tractor can serve is simply the ratio of the scraper cycle time to the pusher cycle time:

$$N = \frac{T_s}{T_p} \quad \textbf{[8.4]}$$

where N = number of scrapers per one pusher.

Rarely, if ever, will the value of N be an integer. This means that either the pusher or the scrapers will be idle for some portion of time during a cycle.

In this analysis, the total cycle time for the scraper was 6.76 min and the pusher required 1.44 min, therefore applying Eq. [8.4]:

$$N = \frac{6.76 \text{ min}}{1.44 \text{ min}} = 4.7$$

Consequently, the economics of using either 4 or 5 scrapers should be investigated.

Step 13: Efficiency

The term *efficiency* or *operating efficiency* is used to account for the actual productive operations in terms of an average number of minutes per hour that the machines will operate. A 50-min hour average would yield a 0.83 (50/60) efficiency factor. The 50-min hour is a reasonable starting point if no company- and/or equipment-specific efficiency data are available. The estimator should always try to visualize the work site and how the work will be performed in the field before applying an efficiency factor. If the pit will not be congested and if the dump area is wide open, a 55-min hour may be appropriate. But if the cut involves a confined area, such as a ditch, or if the embankment area is a narrow bridge header, the estimator should consider a reduced efficiency, maybe a 45-min hour.

For the analysis, a 50-min per hour efficiency factor has been assumed.

Step 14: Production

If the number of scrapers placed on the job is less than the balance number from Eq. [8.4], the scrapers will control production and the push tractor will experience idle time.

$$\begin{array}{c}\text{Production}\\ \text{(scrapers controlling)}\end{array} = \frac{\text{efficiency, min/hr}}{\begin{array}{c}\text{total cyc. time}\\ \text{scraper, min}\end{array}} \times \begin{array}{c}\text{number}\\ \text{of scrapers}\end{array} \times \begin{array}{c}\text{volume}\\ \text{per load}\end{array} \qquad \textbf{[8.5]}$$

If the number of scrapers placed on the job is greater than the balance from Eq. [8.4], the pusher will control production and the scrapers will experience idle time.

$$\begin{array}{c}\text{Production}\\ \text{(pusher controlling)}\end{array} = \frac{\text{efficiency, min/hr}}{\text{total cyc. time pusher, min}} \times \text{volume per load} \qquad \textbf{[8.6]}$$

In this analysis, if only four scrapers are placed on the job, production (Eq. [8.5]) would be

$$\begin{array}{c}\text{Production}\\ \text{(scrapers controlling)}\end{array} = \frac{50 \text{ min/hr}}{\begin{array}{c}6.76 \text{ min}\\ \text{(scraper cycle time)}\end{array}} \times 4 \times 24.3 \text{ bcy} = 719 \text{ bcy/hr}$$

If five scrapers were used on the job, production (Eq. [8.6]) would be

$$\begin{array}{c}\text{Production}\\ \text{(pusher controlling)}\end{array} = \frac{50 \text{ min/hr}}{\begin{array}{c}1.44 \text{ min}\\ \text{(pusher cycle time)}\end{array}} \times 24.3 \text{ bcy} = 844 \text{ bcy/hr}$$

Step 15: Cost

If project schedule is more important than unit cost, it is obvious in this example that five scrapers should be used for the work because of the increased production when compared to using four scrapers. But usually the decision concerning the number of scrapers to employ is a question of unit production cost. Let us assume that each of these scrapers has an O&O cost of $89 per hour and that the O&O cost for the push tractor is $105 per hour. Furthermore, assume that the scraper operators are paid $12 per hour and the pusher operator $20 per hour. With this cost information available it is possible to determine the unit cost for moving the material.

4 scrapers @ $89/hr	+ 1 pusher @ $105/hr	= $461/hr
4 operators @ $12/hr	+ 1 operator @ $20/hr	= 68/hr
	Cost for a four-scraper spread	$529/hr
5 scrapers @ $89/hr	+ 1 pusher @ $105/hr	= $550/hr
5 operators @ $12/hr	+ 1 operator @ $20/hr	= 80/hr
	Cost for a five-scraper spread	$630/hr

Unit cost to move the material using a four-scraper spread:

$$\frac{\$529/\text{hr}}{719\ \text{bcy/hr}} = \$0.736/\text{bcy}$$

Unit cost to move the material using a five-scraper spread:

$$\frac{\$630/\text{hr}}{844\ \text{bcy/hr}} = \$0.746/\text{bcy}$$

The unit costs are very close. If this were for a project having a large quantity of material to move, it would most likely be best to use a five-scraper spread of equipment. However, if it is for a project having a very limited amount of material to move, the mobilization cost of the fifth scraper might amount to more of an expense than the extra unit production cost. Many scraper projects involve moving large quantities of material, therefore the difference of only a few pennies is important. For this same reason, when moving large quantities, many companies carry the cost figures to three places past the decimal.

The final decision concerning the size of spread (number of scrapers) to use must include consideration of the costs to mobilize the equipment and the costs of daily overhead for the project. When daily overhead expenses are great, higher production is usually justified.

OPERATIONAL CONSIDERATIONS

Scrapers are best suited for medium-haul-distance earthmoving operations. Haul distances greater than 500 ft but less that 3,000 ft are typical, although with larger scrapers, the maximum distance can approach a mile. The selection of a particular type of scraper for a project should take into consideration the type of material being loaded and transported. Figure 8.18 provides a summary of applicable scraper type based on project material type.

To obtain a higher profit from earthwork, a contractor must organize and operate the spread in a manner that will ensure maximum production at the lowest cost. There are several methods whereby this objective can be attained.

Ripping

Most types of tight soils will load faster if they are ripped ahead of the scraper. Additionally, delays pertaining to equipment repairs will be reduced substantially as the scraper will not be operated under as much strain. If the value of the increased production resulting from ripping exceeds the ripping cost, the material should be ripped.

Hauling Rock When rock is ripped for scraper loading, the depth ripped should always exceed the depth to be excavated. This is done to leave a loose layer of material under the tires to provide better traction and to reduce the wear on the tracks and tires. The cost of hauling ripped or shot rock with

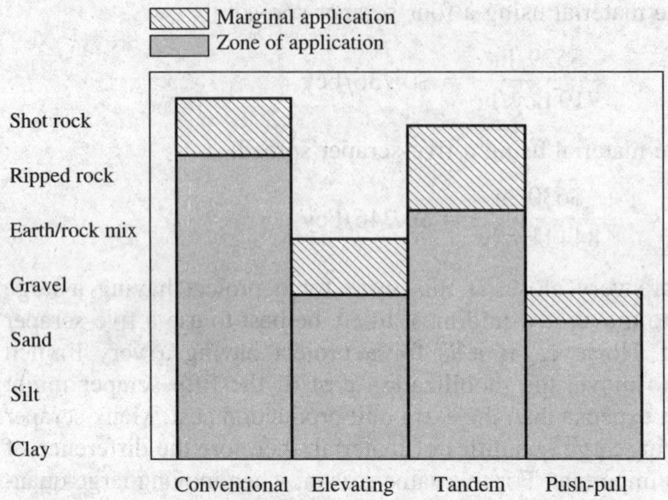

FIGURE 8.18 Zones of application for different type scrapers.

scrapers will be higher than that for moving common earth, but it can still be more economical than an excavator-truck operation. The volume of material carried will be about 70% of normal payload. Repair costs will be about 150% of normal, and tire life will only be about 30 to 40% of normal.

Prewetting the Soil

Some soils will load more easily if they are reasonably moist. To achieve uniform moisture conditioning of the soil, prewetting can be performed in conjunction with ripping and ahead of loading. Prewetting the soil in the cut can reduce or eliminate the use of water trucks on the fill, thereby reducing equipment congestion on the fill. The elimination of excess moisture on the surface of the fill may facilitate the movement of the scrapers on the fill.

Loading Downgrade

When it is practicable to do so, scrapers should be loaded downgrade and in the direction of haul. Downgrade loading results in faster loading times, whereas loading in the direction of haul both shortens the length of haul and eliminates the need to turn in the cut with a loaded scraper. Each 1% of favorable grade is the equivalent of increasing the loading force by 20 lb/ton of gross weight of the push tractor and scraper unit.

Dumping Operations

It is easier to compact materials that are dumped in thin lifts. Thick lifts require greater spreading effort and impede travel on the fill. Compaction equipment must work in patterns to be effective. An orderly pattern of dumping will therefore make it easier to intermesh material placement and compaction operations.

Supervision

Full-time supervisory control should be provided in the cut. A more efficient operation will result through the elimination of confusion and traffic congestion. A spotter should always control the fill operations, being responsible for coordinating the scrapers with the spreading and compacting equipment. The spotter's job is to maintain the dumping pattern of the scrapers. Typically, the spotter will direct each scraper operator to dump the load at the end of the preceding coverage until the end of the fill area is reached. Then the next coverage is started parallel to and adjacent the first coverage. This enables compaction equipment to work the freshly dumped material without interfering with the scrapers.

SCRAPER SAFETY

To achieve production, scrapers should travel in the highest gear that is safe for haul-road conditions. They should never be operated, however, at unsafe speeds. Operators must always wear seat belts as uneven terrain and ruts in the haul road can cause violent pitching and bouncing of the operator. Such violent movement can cause the operator to be thrown from the scraper if not secured by the seat belt. Examples of injuries resulting from not wearing a seat belt can be found in many OSHA reports posted on the Web: ". . . entered rough, rutted terrain that caused scraper to become uncontrollable. The driver was ejected from the operating position and fell to the ground. Scraper continued on and stalled out about 50 yards further on. Scraper driver sustained multiple serious injuries from the fall."

In other cases, the operator violently turns the steering wheel, causing an accident. The fill should be constructed as shown in Fig. 8.19. Make the fill higher on the outside edges. By keeping the edges higher, scrapers are not prone to slide to the outside and over the edge.

Large machines like scrapers have large blind spots. Being "struck by" or "caught in between" are two of the leading causes of injuries and fatalities on construction projects. Personnel being "struck by" equipment or vehicles accounts for 22% of the accidents experienced on construction sites and "caught in between" accounts for 18%. Therefore, it is evident that access to the fill area and haul roads should be restricted. Nonessential personnel should not be allowed in areas were scrapers are working. All employees who are at risk—working on the ground—must receive basic safety indoctrination about hazardous conditions and must be provided with proper reflective vests.

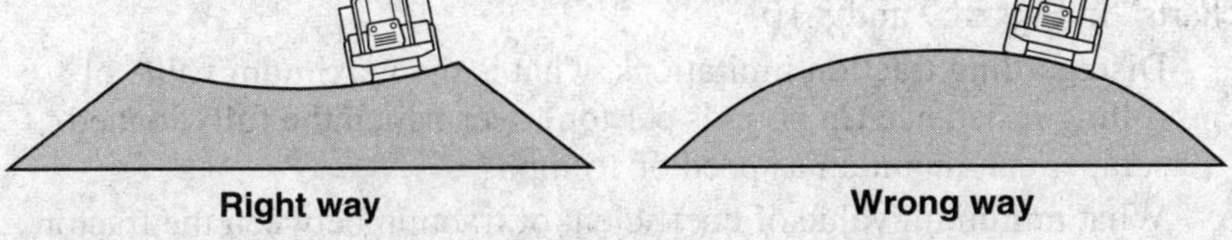

FIGURE 8.19 Proper and safe construction of an embankment.

SUMMARY

Profitable scraper employment on any job can only be determined by a careful project investigation. Equipment manufacturers provide scraper performance charts that are used to analyze the performance of a scraper under various operating conditions. The first part of the analysis is to assume a scraper load time and volume of load. With the load information and haul-route data generated from a study of the project layout, and possibly a mass diagram, scraper cycle times can be calculated. The ratio of the pusher cycle time to the scraper cycle time provides guidance as to the number of scrapers to employ for the work. However, the final decision concerning the number of scrapers should only be made after the spread production and unit hauling cost have been determined for the applicable number of scrapers. Critical learning objectives include:

- An ability to properly use scraper performance charts
- An ability to calculate the time required to complete each of the six operations of the scraper production cycle
- An ability to calculate pusher cycle time
- An ability to calculate scraper spread production

These objectives are the basis for the problems that follow.

PROBLEMS

8.1 What is the travel time for a scraper moving at 18 mph over a distance of 2,300 ft? (1.45 min)

8.2 What is the travel time, in minutes, for a scraper moving at 18 mph over a distance of 0.8 miles?

8.3 What is the travel time for a scraper moving at 16 mph over a distance of 3,200 ft?

8.4 What is the travel time for a scraper moving at 23 mph over a distance of 1,300 ft?

8.5 A wheel-tractor scraper is operating on a level grade. Assume no power derating is required for equipment condition, altitude, temperature, and so on. Disregarding traction limitations, what is the maximum value of rolling resistance (in pounds per ton) over which an empty scraper can maintain a speed of 8 mph? Use the scraper specifications in Table 8.1 and the "performance charts" in Figs. 8.9 and 8.10.

8.6 A wheel-tractor scraper is operating on a level grade. Assume no power derating is required for equipment condition, altitude, temperature, and so on. Use the scraper specifications in Table 8.1 and the "performance charts" in Figs. 8.9 and 8.10.

a. Disregarding traction limitations, what is the maximum value of rolling resistance (in pounds per ton) over which the fully loaded scraper can maintain a speed of 15 mph?

b. What minimum value of coefficient of traction between the tractor wheels and the traveling surface is needed to satisfy the requirements of part a of the question?

8.7 Determine the cycle time for a single-engine scraper rated at 20 cy heaped that is used to haul material from a pit to a fill 1,700 ft away under severe conditions. The average haul speed will be 10 mph and the average return speed will be 16 mph. Assume, at an average speed of 5 mph, that 200 ft is required to both accelerate and decelerate. The operating efficiency will be equal to a 50-min hour. It will take 0.85 min to load this scraper.

8.8 Based on the scraper specifications in Table 8.1 and on the "perform ance charts" in Figs. 8.9 and 8.10, and for haul conditions as stated here, analyze the probable scraper production. How many scrapers should be used and what will be the production in bcy per hr? The material to be hauled is a sandy clay (dry earth), 2,800 lb/bcy. The expected rolling resistance for the well-maintained haul road is 40 lb/ton. Use a 0.85-min load time. This will result in an average load of 91% heaped capacity. To account for both acceleration and deceleration use an average speed of 5 mph over a distance of 200 ft. Use a 50-min hour efficiency factor. The total length of haul is 3,200 ft and has the following individual segments when moving from cut to fill:

600 ft	+3% grade
2,200 ft	0% grade
400 ft	+4% grade

8.9 Based on the scraper specifications in Table 8.1 and the "performance charts" in Figs. 8.9 and 8.10, and for haul conditions as stated here, analyze the probable scraper production. How many scrapers should be used and what will be the production in bcy per hr? The material to be hauled is cohesive. It has a swell factor of 0.76 and a unit weight of 2,900 lb/bcy. The expected rolling resistance for the well-maintained haul road is +3%. Assume a 0.80-min load time and that average load will be 90% of heaped capacity. To account for both acceleration and deceleration use an average speed of 5 mph over a distance of 200 ft. Use a 50-min hour efficiency factor. The total length of haul is 2,600 ft and has the following individual segments when moving from cut to fill:

600 ft	+5% grade
1,800 ft	−2% grade
200 ft	−4% grade

8.10 Determine the maximum hauling production given the following conditions. As many scrapers as required can be used, but only one push tractor will be available. The material is a sandy clay (dry earth), 2,900 lb/bcy. The expected haul-road rolling resistance is 80 lb/ton. The average scraper load will be 28.2 lcy. There are three segments to the haul route: 600 ft at a grade of +3%; 2,200 ft at 0% grade; and 400 ft at +4% grade (moving from the cut to the fill). To account for acceleration and deceleration use an average speed of 4 mph for 200 feet at each end of the haul and return. Use the scraper specifications in Table 8.1 and the "performance charts" in Figs. 8.9 and 8.10. Assume a 50-min hour efficiency.

8.11 From a production analysis for using scrapers to move earth on a project, the following has been determined:

Number of scrapers	6 scrapers, RR = 4%	5 scrapers, RR = 4%	5 scrapers, RR = 3%	4 scrapers, RR = 3%
Production	850 bcy	830 bcy	850 bcy	700 bcy

The cost to operate a scraper is $120 per hour and to operate the push tractor is $100 per hour. To achieve the 3% RR it will be necessary to add a grader to the equipment spread. A grader costs $60 per hour. Calculate the cost to move a bcy of material for all of the conditions.

a. For the 4% RR condition, is it better to use five or six scrapers?
b. For the 3% RR condition, is it better to use four or five scrapers?
c. Is it economical to invest in the grader and improve production?

8.12 Based on the scraper specifications in Table 8.1 and the "performance charts" in Figs. 8.9 and 8.10, and for haul conditions as stated here, analyze the probable scraper production. How many scrapers should be used and what will be the production in bcy per hour? The material to be hauled is dry clay. It has a unit weight of 2,800 lb per bcy. The expected rolling resistance for the haul road is +5%. Assume a 0.80-min load time and an average load of 86% heaped capacity. Accelerate and decelerate at an average speed of 5 mph over a distance of 200 ft. Use a 50-min hour efficiency factor. The total length of haul is 3,300 ft and has the following individual segments when moving from cut to fill:

600 ft	+3% grade
800 ft	+2% grade
1,900 ft	0% grade

REFERENCES

1. *Caterpillar Performance Handbook,* Caterpillar Inc., Peoria, IL, published annually.
2. *Making the Most of Scraper Potential* (1993). Caterpillar Inc., Peoria, IL.
3. "Danish Quarry Uses Auger Scrapers in Lime Production," *Rock Products*, July 1993.
4. *Optimum Scraper Load Time* (1994). Caterpillar Inc., Peoria, IL.

WEBSITE RESOURCES

1. www.cat.com Caterpillar is a large manufacturer of construction and mining equipment. In the "Products, Equipment" section of the website can be found the specifications for Caterpillar manufactured scrapers.
2. www.ehso.com/oshaConstruction_N_O.htm OSHA Health & Safety Construction-related Regulations, Subpart N—Cranes, Derricks, Hoists, Elevators, and Conveyors and Subpart O—Motor Vehicles, Mechanized Equipment, and Marine Operations. In section 1926.602—Material Handling Equipment—the mandatory safety features for scrapers are detailed.
3. www.terex.com Terex Corporation is a diversified manufacturer of equipment for the construction industry. The specifications for their scrapers can be found in the "Construction" equipment section of the website.

CHAPTER 9

Excavators

Hydraulic power is the key to the versatility of many excavators. Hydraulic front shovels are used predominantly for hard digging above track level and for loading haul units. Hydraulic hoe-type excavators are used primarily to excavate below the natural surface of the ground on which the machine rests. The loader is a versatile piece of equipment designed to excavate at or above wheel/track level. Unlike a shovel or hoe, a loader must maneuver and travel to position the bucket to load or dump. Additionally, there are a variety of excavators available for specialty applications.

HYDRAULIC EXCAVATORS

Hydraulic excavators (Fig. 9.1) may be either crawler or pneumatic-tire-carrier-mounted, and many different specialized attachments are available for individual job applications. With the options in types, attachments, and sizes of machines, there are machines for almost any application, but each offers variations in economical advantage. This chapter takes into account the important machine operating features and calls attention to production consequences of specific machine applications.

Hydraulic power is the key to the advantages offered by these machines. The hydraulic control of machine components provides

- Faster cycle times.
- Positive control of attachments.
- Precise control of attachments.
- High overall efficiency.
- Smoothness and ease of operation.

Boom and stick hydraulic excavators are classified by the digging motion of the bucket (see Fig. 9.2). An upward motion machine is known as a "front shovel." A shovel develops breakout force by crowding material away from

FIGURE 9.1 A hydraulic-operated front shovel loading shot rock into a truck.

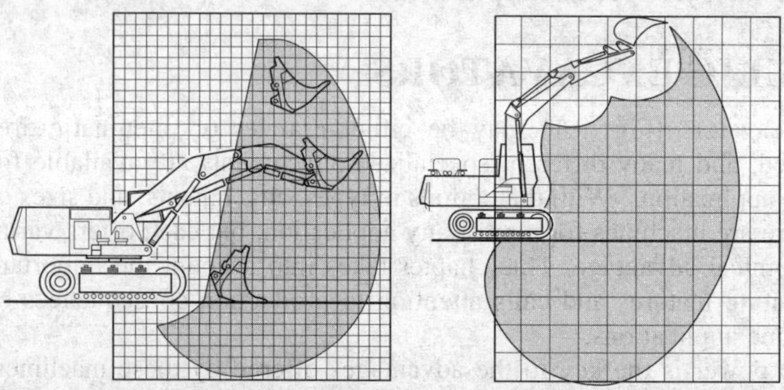

FIGURE 9.2 Digging motion of hydraulic excavators.

the machine. The boom of a shovel swings in an upward arc to load; therefore, the machine requires a material face above the running gear to work against. A downward arc machine is classified as a "hoe." It develops excavation break-out force by pulling the bucket toward the machine and curling the bucket inward. The downward swing of a hoe dictates usage for excavating below the running gear.

If an excavator is considered as an independent machine (a one-link system), its production rate can be estimated using the following steps:

Step 1. Obtain the heaped bucket load volume from the manufacturers' data sheet. This would be a loose volume (lcy) value.

Step 2. Apply a bucket **fill factor** based on the type of machine and the class of material being excavated.

fill factor
A numerical value used to adjust rated-heaped excavator bucket capacity based on the type of material being handled and the type of excavator.

Step 3. Estimate a peak cycle time. This is a function of machine type and job conditions to include angle of swing, depth or height of cut, and in the case of loaders, travel distance.

Step 4. Apply an efficiency factor.

Step 5. Conform the production units to the desired volume or weight units (lcy to bcy or tons).

Step 6. Calculate the production rate.

The basic production formula is: Material carried per load × cycles per hour. In the case of excavators, this formula can be refined and written as

$$\text{Production} = \frac{3{,}600 \text{ sec} \times Q \times F \times (\text{AS:D})}{t} \times \frac{E}{60\text{-min hr}} \times \frac{1}{\text{volume correction}} \quad [9.1]$$

where

Q = heaped bucket capacity (lcy)
F = bucket fill factor
AS:D = angle of swing and depth (height) of cut correction
t = cycle time in seconds
E = efficiency (min per hour)
volume correction for loose volume to bank volume, $\frac{1}{1 + \text{swell factor}}$;
for loose volume to tons, $\frac{\text{loose unit weight, lb}}{2{,}000 \text{ lb/ton}}$.

This production analysis process is shown in Fig. 9.3.

HYDRAULIC EXCAVATOR ACCIDENTS

A 26-year-old construction worker was killed while working in an 8-ft-deep trench, trying to remove a concrete sewer casing. Because it was impossible for the excavator operator to see the bottom of the trench where the casing was located, the victim was standing inside a trench box, giving hand signals to the operator above him. While pulling off the encasement, the bucket teeth slipped off the edge of the concrete and the excavator arm and bucket swung toward the victim, crushing him against the side of the trench box (Iowa NIOSH Fatality Assessment and Control Evaluation Investigation 96IA0).

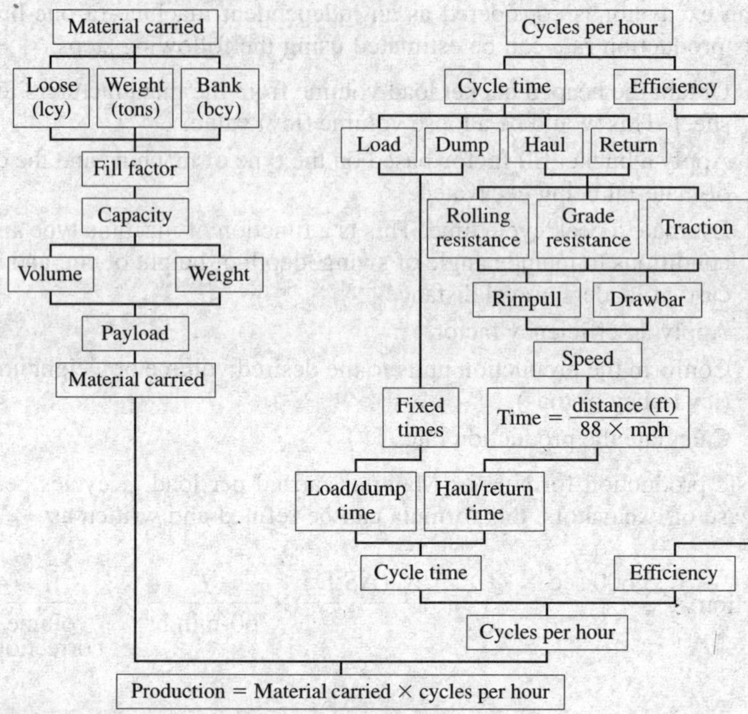

FIGURE 9.3 Excavator production process.

Bureau of Labor Statistics (BLS) data identified 346 deaths associated with excavators or backhoe loaders during the period 1992 to 2000. A review of the data by the National Institute for Occupational Safety and Health (NIOSH) identified two common causes of injury:

- Being struck by the moving machine, swinging booms, or other machine components.
- Being struck by quick-disconnect excavator buckets that unexpectedly detach from the stick [1].

Other leading causes of fatalities are rollovers, electrocutions, and machines sliding into trenches after cave-ins.

Managers have a responsibility to ensure that excavator operators and those working around such equipment always follow safe practices. Safe practice dictates that

- *Operators* keep machine attachments at a safe distance from workers at all times.
- *Workers* are trained regarding safe practices when working in close proximity to heavy equipment.
- *Supervisors* consider alternative working methods that eliminate the need to place workers in close proximity to heavy equipment.

GENERAL INFORMATION

Front shovels are used predominantly for hard digging above track level and for loading haul units. Loading of shot rock would be a typical application (see Fig. 9.4). Shovels are capable of developing high breakout force with their buckets, but the material being excavated should be such that it will stand as a vertical bank, i.e., a wall of material that stands perpendicular to the ground. Most shovels are crawler-mounted and have very slow travel speeds, less than 3 mph. The parts of the shovel are designed for machine balance; each element of the front-end attachment—shovel—is designed for the anticipated load. The front-end attachment weighs about one-third as much as the superstructure with its power parts and cab.

Size Rating of Front Shovels

The size of a shovel is indicated by the size of the bucket, expressed in cubic yards. There are three different bucket-rating standards, Power Crane and Shovel Association (PCSA) Standard No. 3, Society of Automotive Engineers (SAE) Standard J67, and the Committee on European Construction Equipment (CECE) method. All of these methods are based only on the physical dimensions of the bucket and do not address the "bucket loading motion" of a specific machine. For buckets greater than 3-cy capacity, ratings are on $\frac{1}{4}$-cy intervals and on $\frac{1}{8}$-cy intervals for buckets less than 3 cy in size.

FIGURE 9.4 A hydraulic-operated front shovel loading shot rock into a truck.

TABLE 9.1 Fill factors for front shovel buckets.

Material	Fill factor* (%)
Bank clay; earth	100–110
Rock-earth mixture	105–115
Rock—poorly blasted	85–100
Rock—well blasted	100–110
Shale; sandstone—standing bank	85–100

*Percent of heaped bucket capacity.

Reprinted courtesy of Caterpillar Inc.

Struck capacity The industry definition for struck capacity is that volume actually enclosed by the bucket with no allowance for the bucket teeth.

Heaped capacity Both PCSA and SAE use a 1:1 angle of repose for evaluating heaped bucket capacity. CECE specifies a 2:1 angle of repose.

Rated-heaped capacities represent a net section bucket volume; therefore rated-heaped capacities must be amended to "average" bucket payload based on the characteristics of the material being handled. Manufacturers usually suggest factors, commonly named "fill factors," for making such corrections. Fill factors account for the void spaces between individual material particles of a particular type of material when it is loaded into an excavator bucket. Materials that can be described as flowing (sand, gravel, or loose earth) should easily fill the bucket to capacity with a minimum of void space. At the other extreme are the bulky-shaped rock particles. If all the particles are of the same general size, void spaces can be significant especially with large-size pieces. If the material dug has a significant amount of oversize chunks or is extremely sticky, significant voids can occur.

Fill factors are percentages that, when multiplied by a rated-heaped capacity, adjust the volume by accounting for how the specific material will load into the bucket (see Table 9.1). To validate fill factors, it is best, when possible, to conduct field tests based on weight of material per bucket load.

Basic Parts and Operation

The basic parts of a front shovel include the mounting (substructure), cab, boom, stick, and bucket (see Fig. 9.5). With a shovel in the correct position, near the face of the material to be excavated, the bucket is lowered to the floor of the pit, with the teeth pointing into the face. A crowding force is applied by hydraulic pressure to the stick cylinder at the same time the bucket cylinder rotates the bucket through the face.

SELECTING A FRONT SHOVEL

The two fundamental factors that should be taken into account when selecting a front shovel for project work are (1) cost per cubic yard of material excavated and the (2) job conditions under which the shovel will operate.

FIGURE 9.5 Basic parts of a hydraulic front shovel.

In estimating the cost per cubic yard, one should consider the following factors:

1. The size of the job; a job involving a large quantity of material may justify the higher O&O and mobilization costs of a larger shovel.
2. The cost of mobilizing (transporting) the machine to the project (a large shovel will involve more cost than a small one).
3. The combined cost of drilling, blasting, and excavating; for a large shovel, these costs may be less than for a small shovel, as a large machine will handle more massive rocks than a small one. Large shovels may enable savings in drilling and blasting cost.

These job conditions should be considered in selecting a shovel:

1. If the material is hard to excavate, the bucket of the large shovel that exerts higher digging pressures will handle the material more easily.
2. If blasted rock is to be excavated, the large-size bucket will handle larger individual pieces.
3. The size of available hauling units should be considered in selecting the size of a shovel (see Fig. 9.6). If small hauling units must be used, the size of the shovel should be small, whereas if large hauling units are available, a large shovel should be used. The haul-unit capacity should be approximately five times excavator bucket size. This ratio provides for an efficient balance between excavator product capability and haul-unit product capability. Proper matching of excavator bucket size to haul-unit capacity eliminates wasted cycle time caused by mismatches that create the necessity for partial buckets loads to properly fill the haul units.

FIGURE 9.6 Size of haul truck matched to hydraulic front shovel bucket size.

Maximum bucket dumping height is especially important when the shovel is loading haul-units. It is necessary to select a shovel that has the physical dimensions that enable placement of the bucket for dumping at a height above the haul-unit's structure. Manufacturers' specifications should always be consulted for exact values of machine dimensions and clearances.

CALCULATING SHOVEL PRODUCTION

There are four elements in the production cycle of a shovel:

1. Load bucket
2. Swing with load
3. Dump load
4. Return swing

A shovel does not travel during the digging and loading cycle. Travel is limited to moving into or along the face as the excavation progresses. One study of shovel travel found that on the average it was necessary to move after about 20 bucket loads. This movement into the excavation took an average of 36 sec.

Typical cycle element times under average conditions, for 3- to 5-cy-size shovels, are

1. Load bucket	7–9 sec
2. Swing with load	4–6 sec
3. Dump load	2–4 sec
4. Return swing	4–5 sec

The actual production of a shovel is affected by numerous factors, including the

1. Class of material
2. Height of cut
3. Angle of swing
4. Operator skill
5. Condition of the shovel
6. Haul-unit exchange
7. Size of hauling units
8. Handling of oversize material
9. Cleanup of loading area

Haul-unit exchange refers to the total time required for a loaded truck to clear its loading position under the excavator and for the next empty truck to be positioned for loading.

When handling shot rock, carefully evaluate the amount of oversize material to be moved. A machine with a bucket whose bite width and pocket are satisfactory for the average-size pieces may spend too much time trying to handle individual oversize pieces. A larger bucket, or a larger machine, or changing the blasting pattern should be considered when there is a large percentage of oversize material.

The use of auxiliary equipment in the loading area, such as a dozer, can reduce cleanup delays. Control of haul units and operator breaks are within the control of field management.

The capacity of a bucket is based on its heaped volume. This would be loose cubic yards (lcy). To obtain the bank-measure volume of a bucket when considering a particular material, the average loose volume should be divided by 1 plus the material's swell. For example, if a 2-cy bucket, excavating material whose swell is 25%, will handle an average loose volume of 2.25 lcy, the bank-measure volume will be $\frac{2.25}{1.25} = 1.8$ bcy. If this shovel can make 2.5 cycles/min, which includes no allowance for lost time, the output will be $2.5 \times 1.8 = 4.5$ bcy/min, or 270 bcy/hour. This is an ideal production. Ideal production is based on digging at optimum height with a 90° swing and no delays.

HEIGHT OF CUT EFFECT ON SHOVEL PRODUCTION

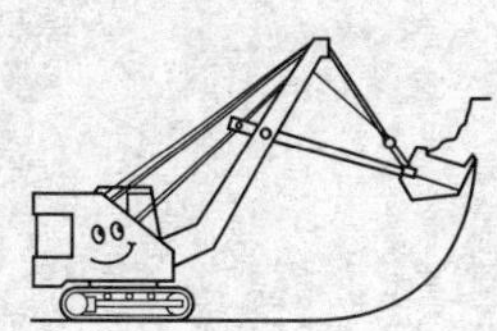

A loose, flowing material will fill a digging bucket in a shorter sweep up the embankment than will chunky material. If the face against which a shovel is excavating material does not have sufficient height, it will be difficult or impossible to fill the bucket in one pass up the face. The operator will have a choice of making more than one pass to fill the bucket, a process that increases the cycle time, or with each cycle the partly filled bucket can be carried to the hauling unit. In either case, the effect will be to reduce shovel production.

If the height of the face is greater than the minimum required for filling the bucket, the operator is presented with three options. The depth of bucket

TABLE 9.2 Factors for height of cut and angle of swing effect on shovel production.

Percentage of optimum depth	Angle of swing (degrees)						
	45	60	75	90	120	150	180
40	0.93	0.89	0.85	0.80	0.72	0.65	0.59
60	1.10	1.03	0.96	0.91	0.81	0.73	0.66
80	1.22	1.12	1.04	0.98	0.86	0.77	0.69
100	1.26	1.16	1.07	1.00	0.88	0.79	0.71
120	1.20	1.11	1.03	0.97	0.86	0.77	0.70
140	1.12	1.04	0.97	0.91	0.81	0.73	0.66
160	1.03	0.96	0.90	0.85	0.75	0.67	0.62

penetration into the face may be reduced in order to fill the bucket in one full pass up the face. This will increase the time for a cycle. The operator may maneuver the bucket so as to begin digging above the base of the face, and then remove the lower portion of the face later. Or the bucket may be run up the full height of the face and the excess material allowed to spill onto the working area of the bench. This spillage will have to be picked up later. The choice of any one of the procedures will result in lost time, compared to the time required to fill the bucket when digging at optimum height.

The PCSA has published findings on the optimum height of cut based on data from studies of small cable-operated shovels (see Table 9.2). In the table, the percentage of optimum height of cut is obtained by dividing the actual height of cut by the optimum height for the given material and bucket and multiplying the result by 100. Thus, if the actual height of cut is 6 ft and the optimum height is 10 ft, the percentage of optimum height of cut is $\frac{6}{10} \times 100 = 60\%$. In most cases, other types of excavators, track or rubber-tire loaders, have replaced the small shovels of the PCSA studies. But some general guidelines can still be gleaned from the data.

Optimum height of cut ranges from 30 to 50% of maximum digging height, with the lower percentage being representative of easy-to-load materials, such as loam, sand, or gravel. Hard-to-load materials, sticky clay or blasted rock, necessitate a greater optimum height, in the range of 50% of the maximum digging height value. Common earth would require slightly less than 40% of the maximum digging height.

ANGLE OF SWING EFFECT ON SHOVEL PRODUCTION

The angle of swing of a shovel is the horizontal angle, expressed in degrees, between the position of the bucket when it is excavating and the position where it discharges the load. The total time in a cycle includes digging, swinging to the dumping position, dumping, and returning to the digging position. If the angle of swing is increased, the time for a cycle will be increased, whereas if the angle of swing is decreased, the time for a cycle will be decreased. Ideal production of a shovel is based on operating at a 90° swing and optimum height of cut. The effect of the angle of swing on the production of a shovel is illustrated in Table 9.2. The ideal production should be multiplied by the proper correction factor to adjust the production for any given height and swing angle.

Proper excavation planning can reduce the angle of swing. For example, if a shovel, which is digging at optimum depth, has the angle of swing reduced from 90° to 60°, the production will be increased by 16%.

EXAMPLE 9.1

A shovel with a 5-cy heaped capacity bucket is loading poorly blasted rock, a situation similar to that shown in Fig. 9.2. It is working a 12-ft-high face. The shovel has a maximum rated digging height of 34 ft. The haul units can be positioned so the swing angle is only 60°. What is a conservative ideal loose cubic yard production if the ideal cycle time is 21 sec?

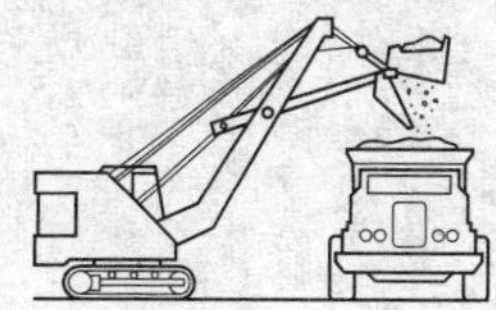

Step 1. Size of bucket, 5 cy

Step 2. Bucket fill factor (Table 9.1) for poorly blasted rock: 85 to 100%, use 85%, conservative estimate

Step 3. Cycle time given 21 sec

Average height of excavation 12 ft

Optimum height for this machine and material (poorly blasted rock):

$$0.50 \times 34 \text{ ft (max. height)} = 17 \text{ ft}$$

Percent optimum height, $\frac{12 \text{ ft}}{17 \text{ ft}} = 0.71$

Correcting for height and swing from Table 9.2, by interpolation, 1.08

Step 4. Efficiency factor—ideal production, 60-min hour

Step 5. Production will be in lcy

Step 6. Ideal production per 60-min hour

$$\frac{3{,}600 \text{ sec/hr} \times 5 \text{ cy} \times 0.85 \text{ (fill factor)} \times \begin{matrix}1.08 \text{ (height-}\\ \text{swing factor}\end{matrix}}{21 \text{ sec/cycle}} = 787 \text{ lcy/hr}$$

Although the information given in the text and in Tables 9.1 and 9.2 is based on extensive field studies, the reader is cautioned against using it too literally without adjusting for conditions that will probably exist on a particular project.

The estimator must consider all factors and decide on an efficiency factor with which to adjust peak production. Experience and good judgment are essential to selecting the appropriate efficiency factor. Transportation Research Board (TRB) studies have shown that actual production times for shovels used in highway construction excavation operations are 50 to 75% of available working time. Therefore, production efficiency is only 30 to 45 min-hr. The best estimating method is to develop specific historical data by type of machine and project factors. But information such as the TRB study that presents data from thousands of shovel cycles provides a good benchmark for selecting an efficiency factor.

EXAMPLE 9.2

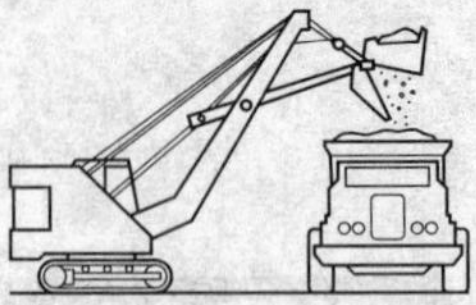

A shovel with a 3-cy heaped capacity bucket is loading well-blasted rock on a highway project. The average face height is expected to be 22 ft. The shovel has a maximum rated digging height of 30 ft. Most of the cut will require a 140° swing of the shovel to load the haul units. What is a conservative production estimate in bank cubic yards?

Step 1. Size of bucket, 3 cy

Step 2. Bucket fill factor (Table 9.1) for well-blasted rock 100 to 110%, use 100%, conservative estimate

Step 3. Cycle element times

Load	9 sec	(because of material, rock)
Swing loaded	4 sec	(small machine, 3 cy)
Dump	4 sec	(into haul units)
Swing empty	4 sec	(small machine, 3 cy)
Total time	21 sec	

Average height of excavation 22 ft

Optimum height: 50% of max.: $0.5 \times 30\text{ ft} = 15\text{ ft}$

Percent optimum height: $\dfrac{22\text{ ft}}{15\text{ ft}} \times 100 = 147\%$

Height and swing factor: From Table 9.2, for 147%, by interpolation, 0.73

Step 4. Efficiency factor: If the TRB information were used, the efficiency would be 30 to 45 working minutes per hour. Assume 30 min for a conservative estimate.

Step 5. Class of material, well-blasted rock, swell 60% (Table 4.3)

Step 6. Production:

$$\frac{3{,}600\text{ sec/hr} \times 3\text{ cy} \times 1.0 \times 0.73}{21\text{ sec/cycle}} \times \frac{30\text{ min}}{60\text{ min}} \times \frac{1}{(1 + 0.6)} = 117\text{ bcy/hr}$$

HOES

GENERAL INFORMATION

Hoes are used primarily to excavate below the natural surface of the ground on which the machine rests (see Fig. 9.7). A hoe is sometimes referred to by other names, such as backhoe or back shovel. Hoes are adept at excavating trenches and pits for basements, and the smaller machines can handle general grading work. Because of their positive bucket control, they are superior to draglines in operating on close-range work and loading into haul units.

FIGURE 9.7 Crawler-mounted hydraulic hoe.
Source: Kokosing Fru-Con.

Wheel-mounted hydraulic hoes (see Fig. 9.8) are available with buckets up to $1\frac{1}{2}$ cy. Maximum digging depth for the larger machines is about 25 ft. With all four outriggers down, the large machines can handle 10,000-lb loads at a 20-ft radius. These are not production excavation machines. They are designed for mobility and general-purpose work.

FIGURE 9.8 Wheel-mounted hydraulic hoes.

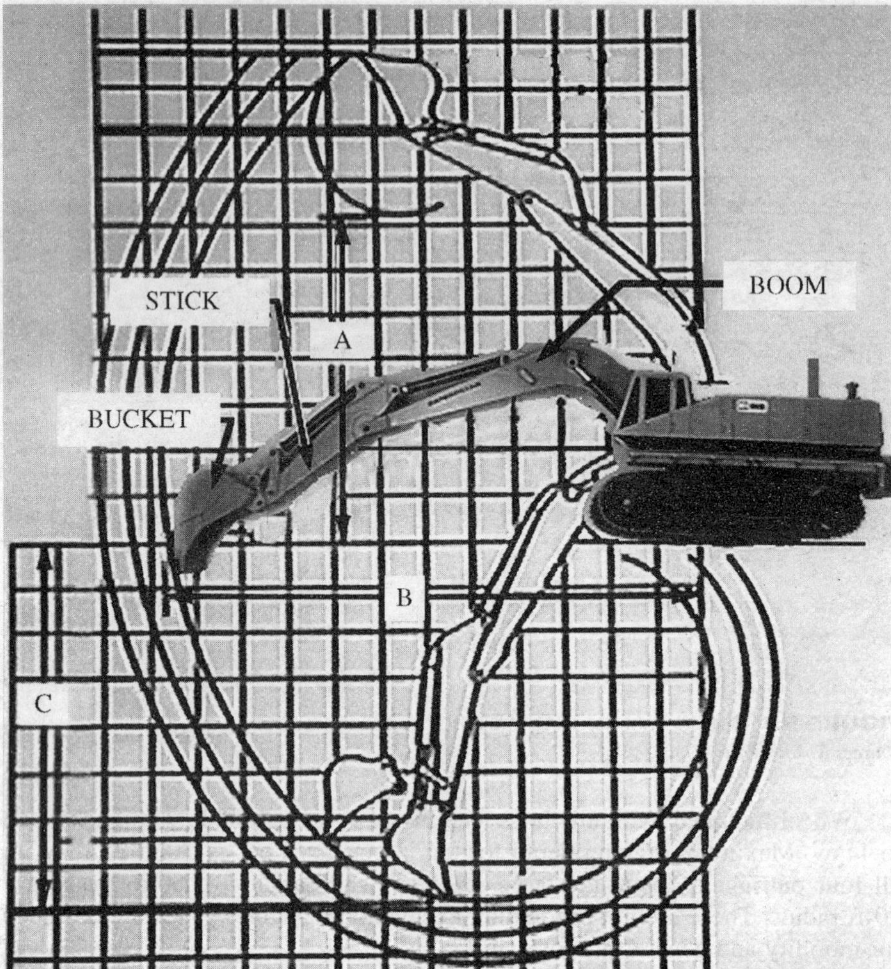

FIGURE 9.9 Basic parts and operating ranges of a hydraulic hoe. A, dumping height; B, digging reach; C, maximum digging depth.

Basic Parts and Operation of a Hoe

The basic parts and operating ranges of a hoe are illustrated in Fig. 9.9. Table 9.3 gives representative dimensions and clearances for hydraulic crawler-mounted hoes. Buckets are available in varying widths to suit the job requirements.

Penetration force into the material being excavated is achieved by the stick and bucket cylinders. Maximum crowd force is developed when the stick cylinder operates perpendicular to the stick (see Fig. 9.10). The ability to break material loose is best at the bottom of the arc because of the geometry of the boom, stick, and bucket and the fact that at that point, the hydraulic cylinders exert the maximum force drawing the stick in and curling the bucket.

TABLE 9.3 Representative dimensions, loading clearance, and lifting capacity hydraulic crawler hoes.

Size bucket (cy)	Stick length (ft)	Maximum reach @ ground level (ft)	Maximum digging depth (ft)	Maximum loading height (ft)	Lifting capacity at 15 ft			
					Short stick		Long stick	
					Front (lb)	Side (lb)	Front (lb)	Side (lb)
$\frac{3}{8}$	5–7	19–22	12–15	14–16	2,900	2,600	2,900	2,600
$\frac{3}{4}$	6–9	24–27	16–18	17–19	7,100	5,300	7,200	5,300
1	5–13	26–33	16–23	17–25	12,800	9,000	9,300	9,200
$1\frac{1}{2}$	6–13	27–35	17–21	18–23	17,100	10,100	17,700	11,100
2	7–14	29–38	18–27	19–24	21,400	14,500	21,600	14,200
$2\frac{1}{2}$	7–16	32–40	20–29	20–26	32,600	21,400	31,500	24,400
3	10–11	38–42	25–30	24–26	32,900*	24,600*	30,700*	26,200*
$3\frac{1}{2}$	8–12	36–39	23–27	21–22	33,200*	21,900*	32,400*	22,000*
4	11	44	29	27	47,900*	33,500*		
5	8–15	40–46	26–32	25–26	34,100†	27,500†	31,600†	27,600†

*Lifting capacity @ 20 ft.
† Lifting capacity @ 25 ft.

FIGURE 9.10 Arrangement of a hoe's hydraulic cylinders to develop digging forces.

BUCKET RATING FOR HYDRAULIC HOES

Hoe buckets are rated like shovel buckets by PCSA and SAE standards using a 1:1 angle of repose for evaluating heaped capacity (see Fig. 9.11). Buckets should be selected based on the material being excavated. The hoe can develop high penetration forces. By matching bucket width and bucket tip radius to the resistance of the material, full advantage can be taken of the hoe's potential. For easily excavated materials, wide buckets should be used. When excavating

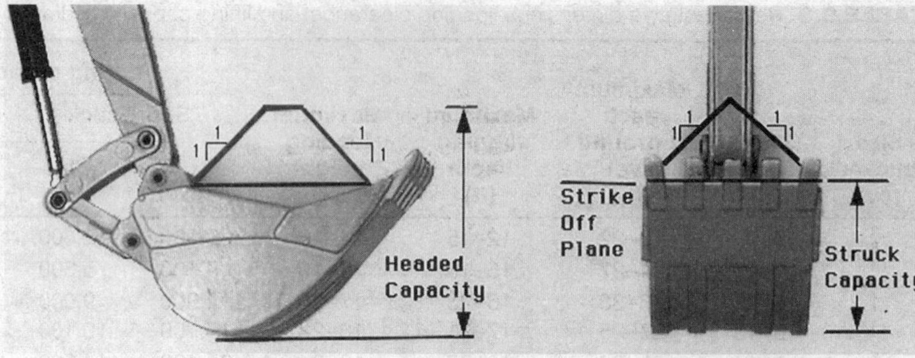

FIGURE 9.11 Hydraulic hoe bucket capacity rating dimensions.

TABLE 9.4 Fill factors for hydraulic hoe buckets.

Material	Fill factor* (%)
Moist loam/sandy clay	100–110
Sand and gravel	95–110
Rock—poorly blasted	40–50
Rock—well blasted	60–75
Hard, tough clay	80–90

*Percent of heaped bucket capacity.
Reprinted courtesy of Caterpillar Inc.

rocky material or blasted rock, a narrow bucket with a short tip radius is best. In utility work, the width of the required trench may be the critical consideration. Fill factors for hydraulic hoe buckets are presented in Table 9.4.

SELECTING A HOE

In the selection of a hoe for use on a project the following must be considered:

1. Maximum excavation depth required
2. Maximum working radius required for digging and dumping
3. Maximum dumping height required
4. Hoisting capability required (where applicable, i.e., handling pipe and trench boxes)

Multipurpose Tool Platform

It must also be recognized that the hydraulic hoe has evolved from a single-purpose excavating machine into a versatile, multipurpose tool (see Fig. 9.12). It is a platform designed for literally hundreds of applications.

A quick-coupler enables the hoe to change attachments and perform a variety of tasks in rapid succession. There are rock drills, earth augers, grapples for land

(a) Hydraulic hoe with land-clearing grapple

(b) Hydraulic ram attachment

(c) Special round-bottom hoe bucket

(d) Wheel-mounted hydraulic hoe with clamshell bucket

FIGURE 9.12 The hydraulic hoe as a multipurpose tool platform.

clearing (see Fig. 9.12a), impact hammers (see Fig. 9.12b), demolition jaws, and vibratory-plate compactors, all of which can easily be attached to the stick in place of the normal excavation bucket. Additionally, there is a broad range of special purpose buckets such as trapezoidal buckets for digging and cleaning irrigation ditches, round-bottom buckets (see Fig. 9.12c) for cast-in-place pipe operations, and clamshell buckets (see Fig. 9.12d) for vertical excavation of footings that enhance this machine's versatility.

Rated Lifting Capacity

In storm drain and utility work, the hoe can perform the trench excavation and handle the pipe, eliminating a second machine (see Fig. 9.13). Manufacturers provide machine-lifting capacities (rated hoist load) based on (1) distance from the center of gravity of the load to the axis of rotation of the machine's superstructure and (2) height of the bucket lift point above the bottom of the tracks or wheels (see Fig. 9.14). Typical lifting data at only one specific distance are provided in Table 9.3. To evaluate lifting ability it is necessary to plot the data for multiple distances as capability varies with positioning of the boom and stick (see Fig. 9.15), because to reach a required position the boom and stick may have to be manipulated through various distances and depths.

FIGURE 9.13 Crawler-mounted hydraulic hoe handling a trench box.

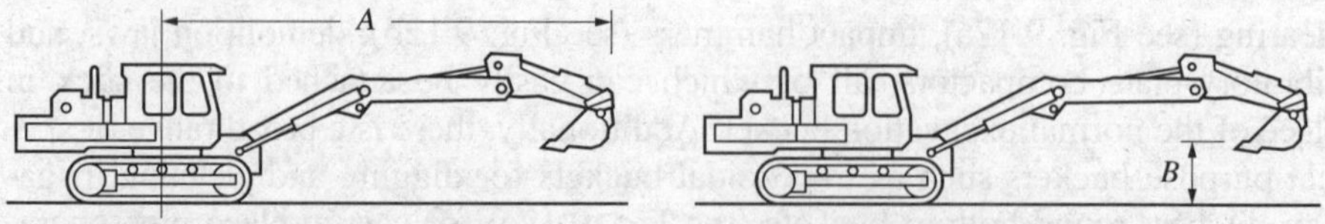

Load radius is the horizontal distance *A*. **Load height is the vertical distance *B*.**

FIGURE 9.14 Hoe lifting capacity position definitions.

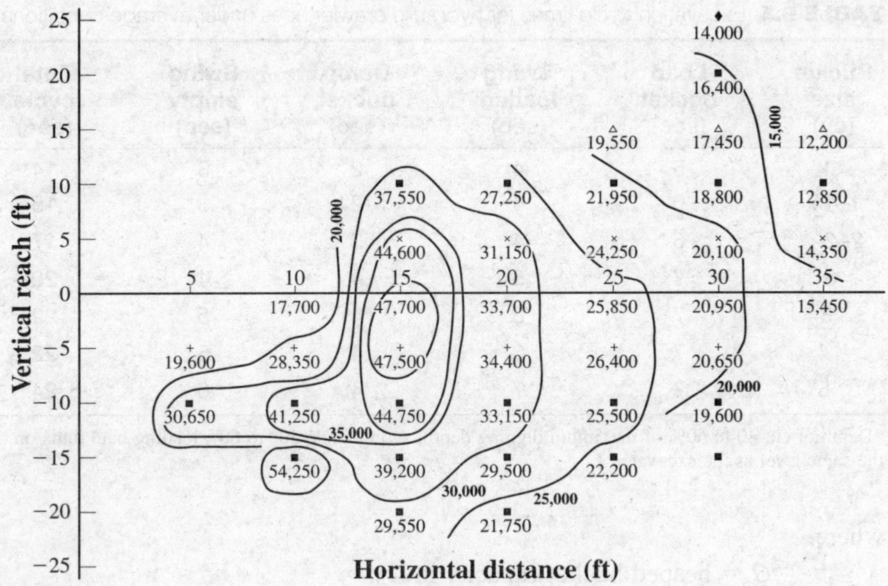

FIGURE 9.15 Crawler-mounted hydraulic hoe over the front-lifting capacity contours, in lb.

Rated hoist load is typically established based on the following guidelines:

1. Rated hoist load shall not exceed 75% of the tipping load.
2. Rated hoist load shall not exceed 87% of the excavator's hydraulic capacity.
3. Rated hoist load shall not exceed the machine's structural capabilities.

CALCULATING HOE PRODUCTION

The same elements that affect shovel excavation production are applicable to hoe excavation operations. Hoe cycle times are approximately 20% longer in duration than those of a similar-size shovel because the hoisting distance is greater as the boom and stick must be fully extended to dump the bucket.

Optimum depth of cut for a hoe will depend on the type of material being excavated and bucket size and type. As a rule, the optimum depth of cut for a hoe is usually in the range of 30 to 60% of the machine's maximum digging depth. Table 9.5 presents cycle times for hydraulic track hoes based on bucket size and average conditions. There are no tables available that relate average hoe cycle time to variations in depth of cut and horizontal swing. Therefore, when using Table 9.5, consideration must be given to those two factors when deciding on a load bucket time and the two swing times.

The basic production formula for a hoe used as an excavator is

$$\text{Hoe (excavation) production} = \frac{3{,}600\text{ sec} \times Q \times F}{t} \times \frac{E}{60\text{-min hour}} \times \frac{1}{\text{volume correction}} \qquad [9.2]$$

TABLE 9.5 Excavation cycle times for hydraulic crawler hoes under average conditions.*

Bucket size (cy)	Load bucket (sec)	Swing loaded (sec)	Dump bucket (sec)	Swing empty (sec)	Total cycle (sec)
<1	5	4	2	3	14
$1–1\frac{1}{2}$	6	4	2	3	15
$2–2\frac{1}{2}$	6	4	3	4	17
3	7	5	4	4	20
$3\frac{1}{2}$	7	6	4	5	22
4	7	6	4	5	22
5	7	7	4	6	24

*Depth of cut 40 to 60% of maximum digging depth; swing angle 30° to 60°; loading haul units on the same level as the excavator.

where

Q = heaped bucket capacity in lcy

F = bucket fill factor for hoe buckets

t = cycle time in seconds

E = efficiency in minutes per hour

volume correction = for loose volume to bank volume, $\frac{1}{1 + \text{swell factor}}$;

for loose volume to tons, $\frac{\text{loose unit weight, lb}}{2{,}000\ \text{lb/ton}}$

EXAMPLE 9.3

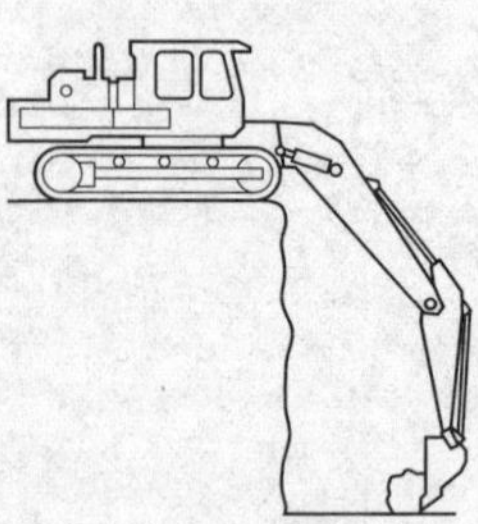

A crawler hoe having a $3\frac{1}{2}$-cy bucket is being considered for use on a project to excavate very hard clay from a borrow pit. The clay will be loaded into trucks having a loading height of 9 ft 9 in. Soil-boring information indicates that below 8 ft, the material changes to an unacceptable silt material. What is the estimated production of the hoe in cubic yards bank measure, if the efficiency factor is equal to a 50-min hour?

Step 1. Size of bucket, $3\frac{1}{2}$ cy

Step 2. Bucket fill factor (Table 9.4), hard clay 80 to 90%; use average 85%

Step 3. Typical cycle element times

Optimum depth of cut is 30 to 60% of maximum digging depth. From Table 9.3 for a $3\frac{1}{2}$-cy size hoe, maximum digging depth is 23 to 27 ft

Depth of excavation, 8 ft

$$\frac{8\ \text{ft}}{23\ \text{ft}} \times 100 = 34\% \geq 30\%;\ \text{okay}$$

$$\frac{8\ \text{ft}}{27\ \text{ft}} \times 100 = 30\% \geq 30\%;\ \text{okay}$$

Therefore, under average conditions and for a $3\frac{1}{2}$-cy size hoe, cycle times from Table 9.5 would be:

1.	Load bucket	7 sec	very hard clay
2.	Swing with load	6 sec	load trucks
3.	Dump load	4 sec	load trucks
4.	Return swing	5 sec	
	Cycle time	22 sec	

Step 4. Efficiency factor, 50-min hour

Step 5. Class of material, hard clay, swell 35% (Table 4.3)

Step 6. Probable production:

$$\frac{3{,}600 \text{ sec/hr} \times 3\frac{1}{2} \text{ cy} \times 0.85}{22 \text{ sec/cycle}} \times \frac{50 \text{ min}}{60 \text{ min}} \times \frac{1}{(1 + 0.35)} = 300 \text{ bcy/hr}$$

Check maximum loading height to ensure the hoe can service the trucks, from Table 9.3, 21 to 22 ft

$$21 \text{ ft} > 9 \text{ ft } 9 \text{ in. okay}$$

As stated, hoe cycle times are usually of greater duration than shovels times. Part of the reason for this is that after making the cut, the hoe bucket must be raised above the ground level to load a haul unit or to position above a spoil pile. If the haul units can be spotted on the floor of the pit, the bucket will be above the haul unit when the cut is completed (see Fig. 9.16). Then it would not be necessary to raise the bucket any higher before swinging and dumping the load. Every movement of the bucket equals increased cycle time. The spotting of haul units below the level of the hoe will increase production. One study found a 12.6% total cycle-time savings between loading at the same level and working the hoe from a bench above the haul units [7].

Often in trenching operations, the volume of material moved is not the primary concern. The critical issue is matching the hoe's ability to excavate linear feet of trench per unit of time with the pipe laying production.

FIGURE 9.16 Location of haul units affects hoe production.

LOADERS

GENERAL INFORMATION

Loaders are used extensively in construction work to handle and transport bulk material, such as earth and rock; to load trucks; to excavate earth; and to charge aggregate bins at asphalt and concrete plants. The loader is a versatile piece of equipment designed to excavate at or above wheel/track level. The hydraulic-activated lifting system exerts maximum breakout force with an upward motion of the bucket. It does not require other equipment to level, smooth, or clean up the area in which it is working.

Types and Sizes

Classified on the basis of running gear, there are two types of loaders: (1) the crawler-tractor-mounted type (see Fig. 9.17) and (2) the wheel-tractor-mounted type (see Fig. 9.18). They may be further grouped by the capacities of their buckets or the weights that the buckets can lift. Wheel loaders may be steered by the rear wheels, or they may be articulated. To increase stability during load lifting, the tracks of crawler loaders are usually longer and wider than those found on comparable-size tractors.

FIGURE 9.17 Track-type loader.

FIGURE 9.18 Wheel-tractor loader.

LOADER BUCKETS/ATTACHMENTS

The most common loader attachments are a shovel-type bucket and a forklift. The loader's hydraulic system provides the control necessary for operating these attachments.

Buckets

Common loader buckets are the one-piece conventional type, *general-purpose*; the hinged-jaw, *multipurpose*; and the heavy-duty, *rock bucket;* but there are also many special-purpose buckets and attachments. Buckets are attached to the tractor by a push frame and lift arms.

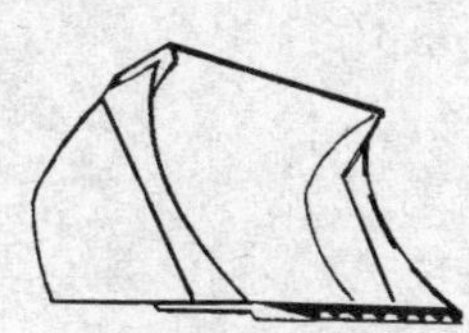

General Purpose The general-purpose (one-piece) bucket is made of heavy-duty, all-welded steel. The replaceable cutting edges are bolted onto the bucket proper. These buckets are usually equipped with replaceable teeth that bolt onto the cutting edge, but they also come with a straight lip (edge) and no teeth.

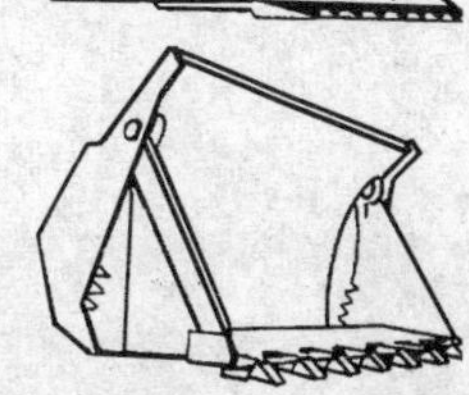

Multipurpose The multipurpose segmented (two-piece) hinged-jawed bucket is made of heavy-duty, all-welded steel. Multipurpose buckets are often referred to as four-in-one buckets because they can be used to dig like a normal bucket, blade, clam, and grapple. These buckets have bolted-on replaceable cutting edges. Bolt-on-type replaceable teeth are also common on these buckets.

FIGURE 9.19 Wheel-tractor loader with V-shaped rock bucket.

Rock The rock bucket is of one piece, heavy-duty construction, having a protruding V-shaped cutting edge (see Fig. 9.19). This protruding edge can be used for prying up and loosening shot rock.

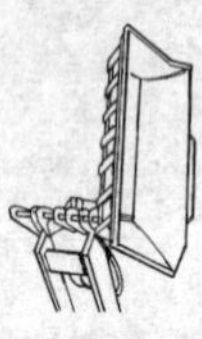

Side-Dump A side-dump bucket allows versatility for working in confined areas, along roads in traffic, and for filling trucks. There are both left- and right-hand dumping buckets available.

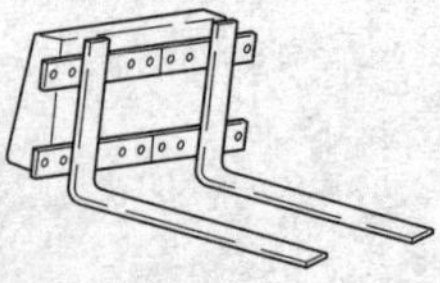

Forklift The forklift can be attached to the loader in place of a bucket.

Other There are many other specialty bucket types and attachments available including demolition buckets, plow blades for snow removal, brush rakes for clearing applications, heavy-duty sweeper brooms, and front booms designed for lifting and moving sling loads.

Fill Factors for Loader Buckets

The heaped capacity of a loader bucket is based on SAE Standard J742b—Front End Loader Bucket Rating. That standard specifies a 2:1 angle of repose for the material above the struck load. This repose angle (2:1) is different from that specified by both SAE and PCSA for shovel and hoe buckets, (1:1). The rated capacity of loader buckets is expressed in cubic yards for all sizes $\frac{3}{4}$ cy or over, and in cubic feet for all sizes under $\frac{3}{4}$ cy. Rated capacities are stated in intervals of 1 cf for buckets under $\frac{3}{4}$ cy, $\frac{1}{8}$ cy for buckets from $\frac{3}{4}$ to 3 cy, and $\frac{1}{4}$ cy for buckets over 3 cy.

TABLE 9.6 Bucket fill factors for wheel and track loaders.

Material	Wheel loader fill factor (%)	Track loader fill factor (%)
Loose material		
Mixed moist aggregates	95–100	95–100
Uniform aggregates		
up to $\frac{1}{8}$ in.	95–100	95–110
$\frac{1}{8}-\frac{3}{8}$ in.	90–95	90–110
$\frac{1}{2}-\frac{3}{4}$ in.	85–90	90–110
1 in. and over	85–90	90–110
Blasted rock		
Well blasted	80–95	80–95
Average	75–90	75–90
Poor	60–75	60–75
Other		
Rock dirt mixtures	100–120	100–120
Moist loam	100–110	100–120
Soil	80–100	80–100
Cemented materials	85–95	85–100

Reprinted courtesy of Caterpillar Inc.

The fill factor correction for a loader bucket (see Table 9.6) adjusts heaped capacity based on the type of material being handled and the type of loader, wheel or track. Mainly because of the relationship between traction and developed breakout force, the bucket fill factors for the two types of loaders are different.

Operating Loads

Once the bucket volumetric load is determined, a check must be made of payload weight. Unlike a shovel or hoe, to position the bucket to dump, a loader must maneuver and travel with the load. A shovel or hoe simply swings about its center pin and does not require travel movement when moving the bucket from loading to dump position. SAE has established operating load weight limits for loaders. A wheel loader is limited to an operating load, by weight, that is less than 50% of rated full-turn static tipping load considering the combined weight of the bucket and the load, measured from the center of gravity of the extended bucket at its maximum reach, with standard counterweights and nonballasted tires. In the case of track loaders, the operating load is limited to less than 35% of static tipping load. The term "operating capacity" is sometimes used interchangeably for operating load. Most buckets are sized based on 3,000 lb/lcy material.

OPERATING SPECIFICATIONS

Representative operating specifications for a wheel loader furnish information such as that listed in Table 9.7. Tables 9.8 and 9.9 present the operating specifications across the ranges of commonly available wheel and track loaders.

TABLE 9.7 Representative specifications for a 119-hp wheel loader.

Speeds, forward and reverse	
Low	0– 3.9 mph
Intermediate	0–11.1 mph
High	0–29.5 mph
Operating load (SAE)	6,800 lb
Tipping load, straight ahead	17,400 lb
Tipping load, full turn	16,800 lb
Lifting capacity	18,600 lb
Breakout force, maximum	30,000 lb

TABLE 9.8 Representative specifications for wheel loaders.

Size, heaped bucket capacity (cy)	Bucket dump clearance (ft)	Static tipping load, @ full turn (lb)	Maximum forward speed				Maximum reverse speed				Raise/ dump/ lower cycle (sec)
			First (mph)	Second (mph)	Third (mph)	Fourth (mph)	First (mph)	Second (mph)	Third (mph)	Fourth (mph)	
1.25	8.4	9,600	4.1	7.7	13.9	21	4.1	7.7	13.9	—	9.8
2.00	8.7	12,700	4.2	8.1	15.4	—	4.2	8.3	15.5	—	10.7
2.25	9.0	13,000	4.1	7.5	13.3	21	4.4	8.1	14.3	23	11.3
3.00	9.3	17,000	5.0	9.0	15.7	26	5.6	10.0	17.4	29	11.6
3.75	9.3	21,000	4.6	8.3	14.4	24	5.0	9.0	15.8	26	11.8
4.00	9.6	25,000	4.3	7.7	13.3	21	4.9	8.6	14.9	24	11.6
4.75	9.7	27,000	4.4	7.8	13.6	23	5.0	8.9	15.4	26	11.5
5.50	10.7	37,000	4.0	7.1	12.4	21	4.6	8.1	14.2	24	12.7
7.00	10.4	50,000	4.0	7.1	12.7	22	4.6	8.2	14.5	25	16.9
14.00	13.6	98,000	4.3	7.6	13.0	—	4.7	8.3	14.2	—	18.5
23.00	19.1	222,000	4.3	7.9	13.8	—	4.8	8.7	15.2	—	20.1

TABLE 9.9 Representative specifications for track loaders.

Size, heaped bucket capacity (cy)	Bucket dump clearance (ft)	Static tipping load (lb)	Maximum forward speed (mph)	Maximum reverse speed (mph)	Raise/ dump/ lower cycle (sec)
1.00	8.5	10,500	6.5	6.9	11.8
1.30	8.5	12,700	6.5	6.9	11.8
1.50	8.6	17,000	5.9*	5.9*	11.0
2.00	9.5	19,000	6.4*	6.4*	11.9
2.60	10.2	26,000	6.0*	6.0*	9.8
3.75	10.9	36,000	6.4*	6.4*	11.4

*Hydrostatic drive.

LOADER PRODUCTION RATES

The two critical factors to be considered in choosing a loader are (1) the type of material and (2) the volume of material to be handled. Wheel loaders are excellent machines for soft to medium-hard material. However, wheel loader production rates decrease rapidly when used in medium to hard material. Another factor to consider is the height that the material must be lifted. When loading trucks, the loader must be able to reach over the side of the truck's cargo container. A wheel loader attains its highest production rate when working on a flat smooth surface with sufficient space to maneuver. In poor underfoot conditions or when there is a lack of space to maneuver efficiently, other types of equipment may be more effective.

Wheel loaders work in repetitive cycles, constantly reversing direction, loading, turning, and dumping. The production rate for a wheel loader will depend on the

1. Fixed cycle time required to load the bucket, maneuver with four reversals of direction, and dump the load
2. Time required to travel from the loading to the dumping position
3. Time required to return to the loading position
4. Volume of material hauled each cycle

Table 9.10 gives fixed cycle times for both wheel and track loaders. Figure 9.20 illustrates a typical loading situation. Because wheel loaders are more maneuverable and can travel faster on smooth haul surfaces, their production rates should be higher than those of track units under favorable conditions requiring longer maneuver distances.

When travel distance is more than minimum, it will be necessary to add travel time to the fixed cycle time. For travel distances of less than 100 ft, a wheel loader should be able to travel, with a loaded bucket, at about 80% of its maximum speed in low gear and return empty at about 60% of its maximum speed in second gear. In the case of distances over 100 ft, return travel should be at about 80% of its maximum speed in second gear. If the haul surface is not well maintained, or is rough, these speeds should be reduced accordingly.

TABLE 9.10 Fixed cycle times for loaders.

Loader size, heaped bucket capacity (cy)	Wheel loader cycle time* (sec)	Track loader cycle time* (sec)
1.00–3.75	27–30	15–21
4.00–5.50	30–33	—
6.00–7.00	33–36	—
14.00–23.00	36–42	—

*Includes load, maneuver with four reversals of direction (minimum travel), and dump.

FIGURE 9.20 Loading travel cycle for a loader.

Consider a wheel loader with a $2\frac{1}{2}$-cy-heaped capacity bucket, handling well-blasted rock weighing 2,700 lb/lcy, for which the swell is 25%. The speed ranges for this loader, equipped with a torque converter and a power-shift transmission, are given in Table 9.7.

The average speeds [in feet per minute (fpm)] should therefore be about

Hauling, all distances	0.8 × 3.9 mph × 88 fpm per mph = 274 fpm
Returning, 0–100 ft	0.6 × 11.1 mph × 88 fpm per mph = 586 fpm
Returning, over 100 ft	0.8 × 11.1 mph × 88 fpm per mph = 781 fpm

The effect of increased haul distance on production is shown by the Table 9.11 calculations.

TABLE 9.11 Effect of haul distance on production.

Haul distance (ft)	25	50	100	150	200
Fixed time	0.45	0.45	0.45	0.45	0.45
Haul time	0.09	0.18	0.36	0.55	0.73
Return time	0.04	0.09	0.13	0.19	0.26
Cycle time (min)	0.58	0.72	0.94	1.19	1.44
Trips per 50-min hour	86.2	69.4	53.2	42.0	34.7
Production (tons)*	262	210	161	127	105

*0.9 bucket fill factor.

CALCULATING WHEEL LOADER PRODUCTION

The following example demonstrates the process for estimating loader production.

EXAMPLE 9.4

A 4-cy wheel loader will be used to load trucks from a quarry stockpile of processed aggregate having a maximum size of $1\frac{1}{4}$ in. The haul distance will be negligible. The aggregate has a loose unit weight of 3,100 lb/cy. Estimate the loader production in tons based on a 50-min hour efficiency factor. Use a conservative fill factor.

Step 1. Size of bucket, 4 cy

Step 2. Bucket fill factor (Table 9.6), aggregate over 1 in., 85–90%; use 85% conservative estimate

Check tipping:

Load weight: 4 cy × 0.85 = 3.4 lcy

3.4 lcy × 3,100 lb/lcy (loose unit weight of material) = 10,540 lb

From Table 9.8: 4-cy machine static tipping load at full turn is 25,000 lb

Therefore, operating load (50% static tipping at full turn) is

$$0.5 \times 25{,}000 \text{ lb} = 12{,}500 \text{ lb}$$

10,540 lb actual load < 12,500 lb operating load; therefore okay

Step 3. Typical fixed cycle time (Table 9.10) 4-cy wheel loader, 30 to 33 sec; use 30 sec.

Step 4. Efficiency factor, 50-min hour

Step 5. Class of material, aggregate 3,100 lb per lcy

Step 6. Probable production:

$$\frac{3{,}600 \text{ sec/hr} \times 4 \text{ cy} \times 0.85}{30 \text{ sec/cycle}} \times \frac{50 \text{ min}}{60 \text{ min}} \times \frac{3{,}100 \text{ lb/lcy}}{2{,}000 \text{ lb/ton}} = 527 \text{ ton/hr}$$

EXAMPLE 9.5

The loader in Example 9.4 will also be used to charge the aggregate bins of an asphalt plant that is located at the quarry. The one-way haul distance from the $1\frac{1}{4}$ in. aggregate stockpile to the cold bins of the plant is 220 ft. The asphalt plant uses 105 tons per hour of $1\frac{1}{4}$-in. aggregate. Can the loader meet this requirement?

Step 3. Typical fixed cycle time (Table 9.10) 4-cy wheel loader, 30 to 33 sec; use 30 sec.

From Table 9.8: Travel speeds forward

First, 4.3 mph; second, 7.7 mph; third, 13.3 mph

Travel speeds reverse

First, 4.9 mph; second, 8.6 mph; third, 14.9 mph

Travel loaded: 220 ft, because of short distance and time required to accelerate and brake, use 80% maximum first gear maximum speed.

$$\frac{4.3 \text{ mph} \times 80\% \times 88 \text{ fpm/mph}}{60 \text{ sec/min}} = 5.0 \text{ ft/sec}$$

Return empty: 220 ft; because of short distance and time required to accelerate and brake, use 80% of second gear maximum speed.

$$\frac{7.7 \text{ mph} \times 80\% \times 88 \text{ fpm/mph}}{60 \text{ sec/min}} = 9.0 \text{ ft/sec}$$

1.	Fixed time	30 sec	4-cy wheel loader
2.	Travel with load	44 sec	220 ft, 80% first gear
3.	Return travel	24 sec	220 ft, 80% second gear
	Cycle time	98 sec	

Step 6. Probable production:

$$\frac{3{,}600 \text{ sec/hr} \times 4 \text{ cy} \times 0.85}{98 \text{ sec/cycle}} \times \frac{50 \text{ min}}{60 \text{ min}} \times \frac{3{,}100 \text{ lb/lcy}}{2{,}000 \text{ lb/ton}} = 161 \text{ ton/hr}$$

161 tons/hr > 105 tons/hr requirement

The loader will meet the requirement.

CALCULATING TRACK LOADER PRODUCTION

The production rates for track loaders are determined in the same manner as for wheel loaders.

EXAMPLE 9.6

A 2-cy track loader having the following specifications is used to load trucks from a bank of moist loam. This operation will require that the loader travel 30 ft for both the haul and return. Estimate the loader production in bcy based on a 50-min hour efficiency factor. Use a conservative fixed cycle time.

Travel speeds by gear for 2-cy track loader

Gear	mph	fpm
Forward		
First	1.9	167
Second	2.9	255
Third	4.0	352
Reverse		
First	2.3	202
Second	3.6	317
Third	5.0	440

Assume that the loader will travel at an average of 80% of the specified speeds in second gear, forward and reverse. The fixed time should be based on time studies for the particular equipment and job; Table 9.10 provides average times by loader size.

Step 1. Size of bucket, 2 cy

Step 2. Bucket fill factor (Table 9.6), moist loam, 100 to 120%; use average 110%. Check tipping:
Load weight:

$$2 \text{ cy} \times 1.10 = 2.2 \text{ lcy}$$

Unit weight moist loam (earth, wet) (Table 4.3) 2,580 lb lcy

$$2.2 \text{ lcy} \times 2{,}580 \text{ lb/lcy} = 5{,}676 \text{ lb}$$

From Table 9.9: 2-cy track machine static tipping load is 19,000 lb. Therefore, operating load (35% static tipping) is

$$0.35 \times 19{,}000 \text{ lb} = 6{,}650 \text{ lb}$$

5,676-lb actual load < 6,650-lb operating load, therefore okay.

Step 3. Typical fixed cycle time (Table 9.10) 2-cy track loader, 15 to 21 sec; use 21 sec, conservative.

Travel loaded: 30 ft, use 80% first gear maximum speed.

$$\frac{1.9 \text{ mph} \times 80\% \times 88 \text{ fpm/mph}}{60 \text{ sec/min}} = 2.2 \text{ ft/sec}$$

Return empty: 30 ft, use 60% (less than 100 ft) of second gear maximum speed.

$$\frac{2.9 \text{ mph} \times 60\% \times 88 \text{ fpm/mph}}{60 \text{ sec/min}} = 2.6 \text{ ft/sec}$$

1.	Fixed time	30 sec	2-cy track loader
2.	Travel with load	13 sec	30 ft, 80% first gear
3.	Return travel	12 sec	30 ft, 60% second gear
	Cycle time	55 sec	

Step 4. Efficiency factor, 50-min hour

Step 5. Class of material, moist loam, swell 25 (Table 4.3)

Step 6. Probable production:

$$\frac{3{,}600 \text{ sec/hr} \times 2 \text{ cy} \times 1.1}{55 \text{ sec/cycle}} \times \frac{50 \text{ min}}{60 \text{ min}} \times \frac{1}{1.25} = 96 \text{ bcy/hr}$$

LOADER SAFETY

Employees working in areas where loaders are operating are exposed to the hazard of being struck or run over.

Incident from an OSHA Report

The operator of the front-end loader after dumping a load into the dump truck proceeded to back up while looking over his left shoulder toward the back left side of the loader. While he was backing up, the supervisor was walking into the work area, approaching the right rear of the loader. A laborer, who witnessed the incident, stated that although he saw the supervisor walk into the work area, he lost sight of him momentarily as the front-end loader was backing up but noticed that it ran over something. The laborer immediately motioned to the operator to stop backing up. The supervisor had been fatally struck and run over by the right front tire of the front-end loader.

OSHA standard 29 CFR 1926.602(a)(9)(ii) states:

> No employer shall permit earthmoving or compacting equipment which has an obstructed view to the rear to be used in reverse gear unless the equipment has in operation a reverse signal alarm distinguishable from the surrounding noise level or an employee signals that it is safe to do so.

SPECIALTY EXCAVATORS

There are a variety of excavators available for specialty applications. There are machines designed strictly for one work application such as trenchers, but there are also general utility machines such as backhoe-loaders. If there is a requirement to excavate larger quantities of material from an extensive borrow area, the Holland Loader is designed for such mass excavation tasks. However, if the excavation is in a busy city street, the contractor should consider a vac excavator.

TRENCHING MACHINES

Trenching machines have been around for over 100 years. Supposedly, the first mechanical trenching machine was assembled in 1893. The early models were steam-driven monsters, but today there are many small utility trenchers available including two-wheel-drive walk-behind trenchers for trenching in soft ground conditions.

Trenching or ditching machines are designed for excavating trenches or ditches of considerable length and having a variety of widths and depths. The term "trenching machine," as used in this book, applies to the wheel-and-ladder-type machines. Most general construction work requires trenches that are no wider than 2 ft or deeper than 7 ft, so 95% of the machines are in the

40- to 150-hp range. Machines in this horsepower range are satisfactory for digging utility trenches for water and gas and shoulder drains on highways. Larger trenchers, having gross horsepower rating of over 340, are used to excavate the much wider and deeper trenches necessary to lay large-diameter cross-country pipelines or water and sewage system main-distribution lines.

There are fully hydrostatic trenchers having self-leveling tracks. A hydrostatic trencher delivers horsepower to the cutting chain and wheels/tracks more efficiently and tends to be smoother to operate. With self-leveling systems, a vertical trench can be maintained on uneven terrain having up to an 18.5% slope.

Trenchers provide relatively fast digging, with positive control of trench depth and width, thereby reducing expensive finishing operations. Most large trenchers are crawler-mounted to increase their stability and to distribute the weight over a greater area.

Wheel-Type Trenching Machines

Figure 9.21 illustrates a wheel-type trenching machine. These machines are available with maximum cutting depths exceeding 8 ft, with trench widths from 12 in. to approximately 60 in. Many are available with 25 or more digging speeds to enable the selection of the most suitable speed for almost any job condition.

The excavating part of the machine consists of a power-driven wheel on which are mounted a number of removable buckets, equipped with cutter teeth. Buckets are available in varying widths to which side cutters can be attached when it is necessary to increase the width of a trench. The machine is operated

FIGURE 9.21 Wheel-type trenching machine.

by lowering the rotating wheel to the desired depth, while the unit moves forward slowly. The earth is picked up by the buckets and deposited onto an endless belt conveyor that can be adjusted to discharge the earth on either side of the trench.

Wheel-type machines are especially suited to excavating trenches for utility services that are placed in relatively shallow trenches.

Ladder-Type Trenching Machines

Figure 9.22 illustrates a ladder-type trenching machine. By installing extensions to the ladders or booms, and by adding more buckets and chain links, it is possible to dig trenches in excess of 30 ft with large machines. Trench widths in excess of 12 ft can be dug. Most of these machines have booms whose lengths can be varied, thereby permitting a single machine to be used on trenches varying considerably in depth. This eliminates the need to own a different machine for each depth range. A machine may have 30 or more digging speeds to suit the needs of a given job.

The excavating part of the machine consists of two endless chains that travel along the boom, to which are attached cutter buckets equipped with teeth. In addition, shaft-mounted side cutters can be installed on each side of the boom to increase the width of a trench. As the buckets travel up the underside of the boom, they bring out earth and deposit it on a belt conveyor that discharges it along either side of the trench. As a machine moves over uneven ground, it is possible to vary the depth of cut, by adjusting the position, but not the length of the boom.

FIGURE 9.22 Ladder-type trenching machine.

Ladder-type trenching machines have considerable flexibility with regard to trench depths and widths. However, these machines are not suitable for excavating trenches in rock or where large quantities of groundwater, combined with unstable soil, prevent the walls of a trench from remaining in place. If the soil, such as loose sand or mud, tends to flow into the trench, it may be desirable to adopt some other method of excavating the trench.

SELECTING EQUIPMENT FOR EXCAVATING TRENCHES

The choice of equipment to be used in excavating a trench will depend on

1. The job conditions
2. The depth and width of the trench
3. The class of soil
4. The extent to which groundwater is present
5. The width of the right-of-way for disposal of excavated earth

If a relatively shallow and narrow trench is to be excavated in firm soil, the wheel-type machine is probably the most suitable. However, if the soil is rock that requires blasting, the most suitable excavator will be a hoe. If the soil is an unstable, water-saturated material, it may be necessary to use a hoe or a clamshell and let the walls establish a stable slope. If it is necessary to install solid sheeting to hold the walls in place, either a hoe or a clamshell that can excavate between the trench braces that hold the sheeting in place will probably be the best equipment for the job.

TRENCHING MACHINE PRODUCTION

Many factors influence the production rates of trenching machines. These include the class of soil; depth and width of the trench; extent of shoring required; topography; climatic conditions; extent of vegetation such as trees, stumps, and roots; physical obstructions such as buried pipes, sidewalks, paved streets, and buildings; and the speed with which the pipe can be placed in the trench. Trencher performance defined in terms of forward speed and volume of excavation is significantly affected by the resistance of the ground being cut. The design of these machines is such that irrespective of machine power, the cutting or excavation power remains constant. An increase in ground strength (rock) reduces cutting rates exponentially [5].

If a trench is dug for the installation of sewer pipe under favorable conditions, it is possible that the machine could dig 300 ft of trench per hour. However, an experienced pipe-laying crew may not be able to lay more than 25 joints of small-diameter pipe, 3 ft long, in an hour. Thus the speed of the machine will be limited to about 75 ft/hr regardless of its ability to excavate more trench. In estimating the probable rate of digging a trench, one must apply an appropriate operating factor to the speed at which the machine could dig if there were no interferences.

TRENCH SAFETY

Time and again noncompliance with trench safety guidelines and common sense results in the loss of life from cave-ins and entrapments. The death rate for trench-related accidents is nearly double that for any other type of construction accident. The first line of defense against cave-ins is a basic knowledge of soil mechanics combined with knowledge as to the type of material that will be excavated.

A critical question, often neglected, is previous disturbance of the material being excavated. In this regard, OSHA standard 1926.651(b)(1) states: "The estimated location of utility installations, such as sewer, telephone, fuel, electric, water lines, or any other underground installations that reasonably may be expected to be encountered during excavation work, shall be determined prior to opening an excavation."

Any trench measuring 5 ft or more in depth must be sloped, shored, or shielded (see Fig. 9.23).

> *Sloping* is the most common method employed to protect workers in trenches. The trench walls are excavated in a V-shaped manner so that the angle of repose prevents the cave-in. Required slope angles vary depending on the specific soil type and soil moisture.
>
> *Benching* is a subsidiary class of sloping that involves the formation of "steplike" horizontal levels. Both sloping and benching require ample right-of-way space. When the work area is restricted, shoring, sheathing, or shielding is necessary.

Shoring is a structural system that applies pressure against the walls of a trench to prevent collapse of the soil. Sheathing is a barrier driven into the ground to provide support to the vertical sides of an excavation. **Shields** or trench boxes (Fig. 9.24) are designed to protect the workers, not the excavation, from collapse.

shields
A structural system designed to protect the workers should a trench in which they are working collapse.

Means of Egress

OSHA standard 1926.651(c)(2) states: "A stairway, ladder, ramp or other safe means of egress shall be located in trench excavations that are 4 feet (1.22 m) or more in depth so as to require no more than 25 feet (7.62 m) of lateral travel for employees."

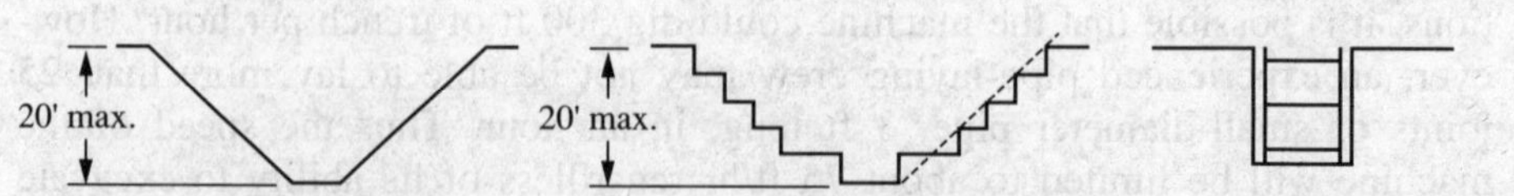

FIGURE 9.23 Methods for protecting workers in trenches.

FIGURE 9.24 Using a trench box to protect workers in an excavation.

Other Requirements

OSHA is serious about trenching accident prevention and every construction manager should likewise be concerned. In 2003, OSHA proposed $99,400 in penalties against a Louisiana contractor for failure to protect employees from potential trenching and excavation hazards. Three alleged repeat violations were issued for failing to provide a ladder for employees to get into and out of the trench, and placing soil from the trench within 2 ft from the trench's edge. The regulations clearly state and common sense dictates that "Spoil piles, tools, equipment, and materials must be kept at least 2 ft from the excavation's edge (1926.651(j)(2))." Another regulation states that in the case of trenches 20 ft or more in depth, a registered professional engineer must design the excavation protection [1926.651(i)(2)(iii) and 1926.652(b)(4)].

Visit www.osha.gov/SLTC/trenchingexcavation/index.html for excellent tools that aid in identifying and controlling trenching hazards.

BACKHOE-LOADERS

The success of the backhoe-loader (see Fig. 9.25) comes from its versatility. This is not a high-production machine for any one task, but it provides flexibility to accomplish a variety of work tasks. 9.2 It is three construction machines combined into one unit—a tractor, a loader, and a hoe. Because of its four-wheel-drive tractor capability, the backhoe-loader can work in unstable ground conditions. It is an excellent excavator for digging loosely packed moist clay or sandy clay. It is not really suitable for continuous high-impact diggings such as with hard clay or caliche.

FIGURE 9.25 Backhoe-loader using the rear hoe to excavate.

Loader functions A loader bucket attached to the front of the tractor enables this machine to excavate above wheel level.

Backhoe functions A hoe attached to the rear of the tractor enables it to excavate below wheel level. The hoe bucket can be replaced with a breaker or hammer, turning the unit into a demolition machine.

HOLLAND LOADERS

A Holland loader is an excavating unit mounted between two crawler tractors that operate in tandem from a single operator in the lead tractor. In continuous passes through excavation areas, the loader carves material from the ground and belt-loads it into hauling units (see Fig. 9.26). There are both vertical- and horizontal-cut Holland loaders. Working with large-bottom dump trailers, these loaders have had reported production rates between 2,000 and 3,300 bcy/hr.

VAC EXCAVATORS

Vac excavation units make use of powerful vacuum systems to safely excavate around buried utilities (Fig. 9.27). There are similar hydrovac units that excavate using a combination of high-pressure water jets and vacuum systems.

FIGURE 9.26 Side-cut Holland loader loading a bottom dump trailer.

FIGURE 9.27 Truck-mounted vacuum excavating system.

These units are excellent tools when it is necessary to work between previously buried electrical and communication conducts and utility pipelines. They are very good for city work where there is no space to dispose of excavated material on site. The larger units have debris body tanks capable of holding up to 15 cy. The excavated material can then be easily transported away for disposal without the need for rehandling.

SUMMARY

The same elements govern the production of all types of excavators. The first issue is how much material is actually loaded into the bucket. This is a function of the bucket size—volume—and the type of material being excavated.

The effect of material type is accounted for by the use of a bucket fill factor applied to the rated-heaped capacity of the bucket. The second issue is cycle time. In the case of shovels and hoes, cycle time is a function of height or depth of cut and swing angle. When travel distance is more than minimum for a loader, the travel cycle time is influenced by travel speed. Critical learning objectives include:

- An ability to adjust bucket volume based on an appropriate fill factor
- An ability to ascertain bucket cycle time as a function of machine type and job conditions to include angle of swing, depth or height of cut, and in the case of loaders, travel distance
- An ability to select an efficiency factor to use in the production calculation by considering both project-specific physical conditions and management ability

These objectives are the basis for the problems that follow.

PROBLEMS

9.1 Visit the Japan International Center for Occupational Safety and Health (JICOSH) website (www.jicosh.gr.jp/english/cases/sacl/saigai01e.htm) and report on two of the hydraulic excavator accidents that are presented.

9.2 What is a good efficiency factor for a front shovel working on a highway construction project?

9.3 A contractor has both a 3-cy and a 5-cy shovel in the equipment fleet. Select the minimum-size shovel that will excavate 400,000 bcy of common earth in a minimum of 130 working days of 8 hr each. The average height of excavation will be 15 ft, and the average angle of swing will be 120°. The 3-cy shovel has a maximum digging height of 30 ft and the 5-cy machine's maximum digging height is 34 ft. The efficiency factor will be a 45-min hour. Appropriate-size haul units can be used with either shovel. How many days will it require to complete the work? (5-cy machine, 121 days to complete)

9.4 For each of the stated conditions determine the probable production expressed in cubic yards per hour bank measure for a shovel equipped with a 3-cy bucket. The shovel has a maximum digging height of 32 ft. Use a 30-min-hour efficiency factor.

	Class of material			
Condition	Common earth	Common earth	Rock-earth/ earth-gravel	Shale
Height of excavation (ft)	12	8	12	19
Angle of swing (degrees)	90	120	60	130
Loading haul units	no	yes	no	no

9.5 A shovel having a 5-cy bucket whose cost per hour, including the wages to an operator, is \$96 will excavate well-blasted rock and load haul units under each of the stated conditions. The maximum digging height of the machine is 35 ft. Determine the cost per cubic yard for each condition.

Condition	(1)	(2)	(3)	(4)
Height of excavation (ft)	10	18	24	28
Angle of swing (degrees)	60	90	120	150
Efficiency factor (min per hr)	45	40	30	45

9.6 A crawler hoe having a $2\frac{1}{4}$-cy bucket and whose cost per hour, including the wages to an operator, is \$67 will excavate and load haul units under each of the stated conditions. The maximum digging depth of the machine is 20 ft. Determine the cost per bank cubic yard for each condition. (Condition 1. 300 bcy/hr, \$0.223/bcy; condition 2. 319 bcy/hr, \$0.210/bcy; condition 3. 128 bcy/hr, \$0.523/bcy)

Condition	(1)	(2)	(3)
Material	Moist loam, earth	Sand and gravel	Rock, well blasted
Depth of excavation (ft)	12	10	14
Angle of swing (degrees)	60	80	120
Percent swell		14	
Efficiency factor (min per hr)	45	50	45

9.7 A crawler hoe having a 3-cy bucket and whose cost per hour, including the wages to an operator, is \$86 will excavate and load haul units under each of the stated conditions. The maximum digging height of the machine is 25 ft. Determine the cost per cubic yard for each condition.

Condition	(1)	(2)	(3)	(4)
Material	Sandy clay	Hard clay	Rock, well blasted	Sand and gravel
Depth of excavation (ft)	10	18	15	20
Angle of swing (degrees)	60	90	90	150
Percent swell	20			13
Efficiency factor (min per hr)	45	50	50	55

9.8 A 3-cy wheel loader will be used to load trucks from a quarry stockpile of processed aggregates having a maximum size of $\frac{1}{4}$ in. The haul distance will be negligible. The aggregate has a loose unit weight of 2,950 lb/cy. Estimate the loader production in tons based on a 50-min-hour efficiency factor. Use an aggressive cycle time and fill factor. (311 ton/hr)

9.9 A 7-cy wheel loader will be used to load a crusher from a quarry stockpile of blasted rock (average breakage) 180 ft away. The rock has a loose unit weight of 2,700 lb/cy. Estimate the loader production in tons based on a 50-min-hour efficiency factor.

9.10 A 4-cy wheel loader will be used to load trucks from a quarry stockpile of uniform aggregates 1 in. and over in size. The haul distance will be negligible. The aggregate has a loose unit weight of 2,750 lb/cy. Estimate the loader production in tons based on a 50-min-hour efficiency factor. Use a conservative cycle time and fill factor.

9.11 For each of the stated conditions, determine the probable production expressed in cubic yards per hour bank measure for a shovel equipped with a 6.8-cy bucket. The shovel has a maximum digging height of 35 ft. Use a 50-min-hour efficiency factor but be aggressive with the assumed fill factor. In the case of poorly blasted rock, use a swell factor of 65%.

	Class of material	
Condition	**Shot rock, poorly blasted**	**Shot rock, well blasted**
Height of excavation (ft)	24.5	20
Angle of swing (degrees)	60	130
Loading haul units	yes	yes

9.12 A crawler hoe having a $2\frac{1}{2}$-cy bucket and whose cost per hour is $81 (without operator) will excavate and load haul units under each of the stated conditions. The maximum digging depth of the machine is 24 ft. Determine the cost per cubic yard for each condition. Be aggressive with the assumed fill factor.

Condition	**(1)**	**(2)**
Material	Moist loam (earth)	Sand and gravel (gravel, wet)
Depth of excavation (ft)	12	10
Angle of swing (degrees)	60	80
Percent swell		14
Efficiency factor (min per hr)	45	50

REFERENCES

1. Barnes, Jonathan (2005). "OSHA may update warning on quick excavator attachments," ENR, McGraw-Hill Construction, New York, NY, February 28, p.12.
2. Boom, Jim (1999). "Trenching Is a Dangerous and Dirty Business!" *Job Safety & Health Quarterly,* Vol. 11, No. 1, Fall. U.S. Department of Labor, Occupational Safety and Health Administration, 200 Constitution Avenue, NW, Washington, D.C. 20210, available in a pdf file at www.osha.gov/Publications/JSHQ/jshq-v11-1-fall1999.pdf.
3. *Caterpillar Performance Handbook,* Caterpillar Inc., Peoria, Ill. (published annually). www.cat.com.
4. *Construction Standards for Excavations,* AGC publication No. 126, promulgated by the Occupational Safety and Health Administration, Associated General Contractors of America, Washington, D.C.
5. *Excavations, OSHA 2226* (2002). U.S. Department of Labor, Occupational Safety and Health Avenue, NW, Washington, D.C., available in a pdf file at www.osha.gov/Publications/osha2226.pdf.
6. Farmer, Ian W. (1996). "Performance of Chain Trenchers in Mixed Ground," *Journal of Construction Engineering and Management, ASCE,* Vol. 122, No. 2, June, pp. 115–118.
7. Lewis, Chris R., and Cliff J. Schexnayder (1986). "Production Analysis of the CAT 245 Hydraulic Hoe," in *Proceedings Earthmoving and Heavy Equipment*

Specialty Conference, American Society of Civil Engineers, February, pp. 88–94.

8. Nichols, Herbert L., Jr. and David A. Day (1998). *Moving the Earth, the Workbook of Excavation,* 4th ed., McGraw-Hill, New York.
9. O'Brien, James J., John A. Havers, and Frank W. Stubbs, Jr. (1996). *Standard Handbook of Heavy Construction,* 3rd ed., McGraw-Hill, New York.
10. *OSHA Technical Manual* (OTM), Section V: Chapter 2, "Excavations: Hazard Recognition in Trenching and Shoring." U.S. Department of Labor, Occupational Safety and Health Administration, 200 Constitution Avenue, NW, Washington, D.C.
11. Schexnayder, Cliff, Sandra L. Weber, and Brentwood T. Brooks (1999). "Effect of Truck Payload Weight on Production," *Journal of Construction Engineering and Management, ASCE,* Vol. 125, No. 1, January/February, pp. 1–7.

WEBSITE RESOURCES

1. www.case.com CNH Global N.V. is a company organized under the laws of The Netherlands and includes a family—Case, Kobelco, and New Holland—of construction equipment brands.
2. www.cat.com Caterpillar is a large manufacturer of construction and mining equipment. In the Products, Equipment section of the website can be found the specifications for Caterpillar manufactured excavators.
3. www.deere.com Deere & Company is a worldwide corporation with a construction products division.www.osha.gov/SLTC/index.html OSHA Technical Links to Safety and Health Topics, U.S. Department of Labor, Occupational Safety and Health Administration, 200 Constitution Avenue, NW, Washington, D.C.
4. www.hitachiconstruction.com Hitachi Construction Machinery Co. produces equipment for the mining and construction industry, offerings include hydraulic excavators, shovels and rigid haul trucks.
5. www.howstuffworks.com/backhoe-loader.htm "How Caterpillar Backhoe Loaders Work," Howstuffworks Inc.
6. www.kobelcoamerica.com CNH Global N.V. is a company organized under the laws of The Netherlands and includes a family—Case, Kobelco, and New Holland—of construction equipment brands.
7. www.komatsuamerica.com Komatsu America Corp manufactures Komatsu, Dressta and Galion lines of hydraulic excavators, wheel loaders, crawler dozers, off-highway trucks and motor graders.
8. www.liebherr.com Liebherr is, a Germany based, global manufacturer of construction equipment. For the earthmoving machinery sector Liebherr produces an extensive range of hydraulic excavators, hydraulic rope excavators, crawler tractors and loaders, wheel loaders and dumper trucks.
9. www.newhollandconstruction.com CNH Global N.V. is a company organized under the laws of The Netherlands and includes a family—Case, Kobelco, and New Holland—of construction equipment brands.
10. www.orenstein-koppel.com O&K Orenstein & Koppel is a brand of CNH Baumaschinen GmbH, Amtsgericht Charlottenburg of Berlin.
11. www.terexca.com Terex Corporation is a diversified manufacturer of equipment for the construction industry.
12. www.volvoce.com Volvo is a worldwide producer of construction equipment to include the excavators and Mack trucks.

CHAPTER 10

Trucks and Hauling Equipment

Trucks are hauling units that provide relatively low hauling costs because of their high travel speeds. The weight capacity of a truck may limit the volume of the load that a unit can haul. The productive capacity of a truck depends on the size of its load and the number of trips it can make in an hour. The number of trips completed per hour is a function of cycle time. Truck cycle time has four components: (1) load time, (2) haul time, (3) dump time, and (4) return time. Tires for trucks and all other haul units should be suitably matched to the job requirements.

TRUCKS

In transporting excavated material, processed aggregates, and construction materials, and for moving other pieces of construction equipment (see Fig. 10.1), trucks serve one purpose: they are hauling units that, because of their high travel speeds, provide relatively low hauling costs. The use of trucks as the primary hauling unit provides a high degree of flexibility, as the number in service can usually be increased or decreased easily to permit modifications in the total hauling capacity of a fleet. Most trucks can be operated over any haul road for which the surface is sufficiently firm and smooth, and on which the grades are not excessively steep. Some units are designated as off-highway trucks because their size and weight are greater than that permitted on public highways (see Fig. 10.2). Off-highway trucks are used for hauling materials in quarries and on large projects involving the movement of substantial amounts of earth and rock. On such projects, the size and costs of these large trucks are easily justified because of the increased production capability they provide.

Trucks can be classified by many factors, including

1. The method of dumping the load—rear-dump, bottom-dump, side-dump
2. The type of frame—rigid-frame or articulated
3. The size and type of engine—gasoline, diesel, butane, or propane

FIGURE 10.1 Truck tractor unit towing a low-profile trailer hauling a mobile crane.

FIGURE 10.2 Off-highway truck on a dam project being loaded by a shovel.

4. The kind of drive—two-wheel, four-wheel, or six-wheel
5. The number of wheels and axles, and the arrangement of driving wheels
6. The class of material hauled—earth, rock, coal, or ore
7. The capacity—gravimetric (tons) or volumetric (cubic yards)

If trucks are to be purchased for general material hauling, the purchaser should select units adaptable to the multipurposes for which they will be employed. On the other hand, if trucks are to be used on a given project for a single purpose, they should be selected specifically to fit the requirements of the project.

FIGURE 10.3 Highway rigid-frame rear-dump truck.

RIGID-FRAME REAR-DUMP TRUCKS

Rigid-frame rear-dump trucks are suitable for use in hauling many types of materials (see Fig. 10.3). The shape of the body, such as the extent of sharp angles and corners, and the contour of the rear, through which the materials must flow during dumping, will affect the ease or difficulty of loading and dumping. The bodies of trucks that will be used to haul wet clay and similar materials should be free of sharp angles and corners. Dry sand and gravel will flow easily from almost any shape of body (see Fig. 10.3). When hauling rock, the impact of loading on the truck body is extremely severe. Continuous use under such conditions will require a heavy-duty rock body made of high-tensile strength steel. Even with the special body, the loader operator must use care in placing material in the truck.

Off-highway dump trucks do not have tailgates, therefore the body floor slopes upward at a slight angle toward the rear, typically less than 15° (see Fig. 10.2). The floor shape perpendicular to the length of the body for some models is flat, while other models utilize a "V" bottom to reduce the shock of loading and to help center the load. Low sides and longer, wider bodies are a better target for the excavator operator. The result of such a configuration is quicker loading cycles.

ARTICULATED REAR-DUMP TRUCKS

The articulated dump truck (ADT) is specifically designed to operate through high-rolling-resistance material and in confined working locations where a rigid-frame truck would have problems (see Fig. 10.4). An articulated joint and

FIGURE 10.4 An articulated dump truck.

oscillating ring between the tractor and dump body enable all of the truck's wheels to maintain contact with the ground at all times. The articulation, all-wheel drive, high clearance, and low-pressure radial tires combine to produce a truck capable of moving through soft or sticky ground.

When haul-route grades are an operating factor, articulated trucks can typically climb steeper grades than rigid-frame trucks. Articulated trucks can operate on grades up to about 35%, whereas rigid-frame trucks can only navigate grades of 20% for short distances, and for continuous grades, 8 to 10% is a more reasonable limit.

The most common ADTs are the 4 × 4 models, but there are larger 6 × 6 models. Articulated dump trucks usually have high hydraulic system pressure, which means the dumping cylinder hoists the bed faster. The bed also achieves a steeper dump angle. One model can attain a 72° dump angle in 15 sec. The combination of these two attributes, hoist speed and a steep angle, translates into quick discharge times. To solve the problem of unloading sticky materials, one manufacturer is equipping its truck bed with an ejector.

Rear-dumps, be they rigid-frame or articulated, should be considered when

1. The material to be hauled is free flowing or has bulky components.
2. The hauling unit must dump into restricted locations or over the edge of a bank or fill.
3. There is ample maneuver space in the loading or dumping area.

TRACTORS WITH BOTTOM-DUMP TRAILERS

Tractors towing bottom-dump trailers are economic haulers when the material to be moved is free flowing, such as sand, gravel, and reasonably dry earth. The use of bottom-dump trailers will reduce the time required to unload the material. To take full advantage of this time saving, there must be a large, clear dumping area where the load can be spread into windrows. Bottom-dump units are also very good for unloading into a drive-over hopper. The rapid rate of discharging the load gives the bottom-dump wagons a time advantage over rear-dump trucks. There are both large off-highway units (see Fig. 10.5) and highway-sized units (see Fig. 10.6). With either the large off-highway or the highway units, relatively flat haul roads are required if maximum travel speed is to be obtained.

FIGURE 10.5 Off-highway tractor towing a bottom-dump trailer.

FIGURE 10.6 Highway bottom-dump hauling hot mix asphalt to the paver.

The clamshell doors through which these units discharge their loads have a limited opening width. Difficulties may be experienced in discharging materials such as wet, sticky clay, especially if the material is in large lumps.

Tractor-towed bottom-dump trailers are economical hauling units on projects where large quantities of materials are to be transported and haul roads can be kept in reasonable condition. The bottom-dump trailer can have a single axle, tandem axles, or even triaxles. Hydraulic excavators, loaders, draglines, or belt loaders, such as Holland loaders, can be used to load these units.

Bottom-dumps should be considered when

1. The material to be hauled is free flowing.
2. There are unrestricted loading and dump sites.
3. The haul-route grades are less than about 5%.

Because of its unfavorable power-weight ratio and the fact that there is less weight on the drive wheels of the tractor unit, thereby limiting traction, bottom-dump units have limited ability to climb steep grades

CAPACITIES OF TRUCKS AND HAULING EQUIPMENT

There are at least three methods of rating the capacities of trucks and wagons:

1. *Gravimetric*—the load it will carry, expressed as a weight

2. *Struck volume*—the volumetric amount it will carry, if the load is water level in the body (bowl or dump box)

3. *Heaped volume*—the volumetric amount it will carry, if the load is heaped on a 2:1 slope above the body (bowl or dump box)

The gravimetric rating is usually expressed in pounds or kilograms and the latter two ratings in cubic yards or cubic meters (Table 10.1).

The struck capacity of a truck is the volume of material that it will haul when it is filled level to the top of the body sides (see Fig. 10.7). The heaped capacity is the volume of material that it will haul when the load is heaped

TABLE 10.1 Example specifications for a large off-highway truck.

Axle	Empty weight	Gross weight
Front drive	229,000 lb (103,855 kg)	398,000 lb (180,499 kg)
Drive axle	249,000 lb (112,925 kg)	800,000 lb (362,812 kg)
Total	478,000 lb (216,780 kg)	1,189,000 lb (543,311 kg)
Volume		
Struck (SAE)	207 cy	158 cubic m
Heaped (SAE 2:1)	285 cy	218 cubic m

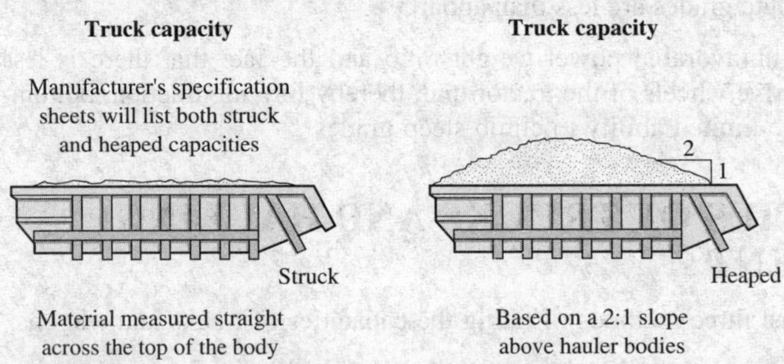

FIGURE 10.7 Measurement of volumetric capacity.

above the sides. The standard for rated-heaped capacity (SAE J1363) uses an assumed 2:1 slope (see Fig. 10.7). The actual heaped capacity will vary with the material that is being hauled. Wet earth or sandy clay can be hauled with a slope of about 1:1, while dry sand or gravel may not permit a slope greater than about 3:1. To determine the actual heaped capacity of a unit, it is necessary to know the struck capacity, the length and width of the body, and the slope at which the material will remain stable while the unit is moving. Smooth haul roads will permit a larger heaped capacity than rough haul roads.

The truck's weight capacity may limit the volumetric load a unit can carry. This happens when hauling a material having a high unit weight, such as iron or even wet sand. However, when the unit weight of the materials is such that the safe load is not exceeded, a unit can be filled to its heaped capacity. Always check to ensure that the volumetric load does not cause a condition where the load weight exceeds the gravimetric capacity of the truck or trailer. Overloading will cause the unit's tires to flex too much, and flexing produces excessive internal tire temperature. Such a condition will cause permanent tire damage and increase operating costs.

In some instances, it is possible to add sideboards to increase the depth of the truck or wagon's cargo body, thereby enabling it to haul a larger load. When this is done, the weight of the new volumes must be checked against the vehicle's gravimetric load capacity. If the weight is greater than the rated gravimetric capacity, the practice will probably increase the hourly cost of operating the haul unit, because of higher fuel consumption, reduced tire life, more frequent failures of parts (such as axles, gears, brakes, and clutches), and higher maintenance costs. If the value of the extra material hauled is greater than the total increase in the cost of operating the vehicle, overloading is justified. In considering the option of sideboarding and hauling larger volumes of materials, the maximum safe loads on the tires must be checked to prevent excessive loading, which might result in considerable lost time due to tire failures.

> Tires are about 35% of a truck's operating cost. Overloading a truck abuses the tires.

The Rubber Manufacturers Association publication *Care and Service of Off-the-Highway Tires* addresses overloading and provides load and inflation tables.

TRUCK SIZE AFFECTS PRODUCTIVITY

The productivity of a truck depends on the size of its load and the number of trips it can make in a unit of time. The number of trips completed per hour is a function of cycle time. Truck cycle time has four components: (1) load time, (2) haul time, (3) dump time, and (4) return time. Examining a match between truck cargo body size and excavator bucket size yields the size of the load and the load time. The haul and return cycle times will depend on the weight of the truck, the horsepower of the engine, the haul and return distances, and the condition of the roads traversed. Dump time is a function of the type of equipment and conditions in the dump area.

When an excavator is used to load material into trucks, the size of the truck cargo body introduces several factors, which affect the production rate and the cost of handling the material.

Small Trucks—Advantages

1. Maneuvering flexibility, which may be an advantage on restricted work sites
2. Speed, can achieve higher haul and return speeds
3. Production, little impact if one truck breaks down
4. Balance of fleet, easy to match number of trucks to excavator production

Small Trucks—Disadvantages

1. Number, more trucks increases operational dangers in the pit, along the haul road, and at the dump
2. More drivers required, more needed for a given output
3. Loading impediment, small target for excavator bucket
4. Positioning time, total spotting time greater because of the number required

Large Trucks—Advantages

1. Number, fewer needed for a given output
2. Drivers required, fewer needed for a given output
3. Loading advantage, larger target for the excavator bucket
4. Positioning time, frequency of spotting trucks is reduced

Large Trucks—Disadvantages

1. Cost of truck time at loading greater, especially with small excavators
2. Loads heavier, possible damage to the haul roads thus increasing the cost for maintenance of the haul road
3. Balance of fleet, difficult to match number of trucks to excavator production
4. Size, may not be permitted to haul on highways

Balancing the capacities of hauling units with the excavator bucket size and production capability is important. When loading with excavators such as hydraulic hoes or shovels, draglines, or loaders, it is desirable to use haul units whose cargo body volume is balanced with the excavator bucket volume. If this is not done, operating difficulties will develop and the combined cost of excavating and hauling material will be higher than when a balance between trucks and excavators is achieved.

A practical rule of thumb frequently used in selecting the size of trucks is to use trucks with a capacity of four to five times the capacity of the excavator bucket. The dependability of this practice is discussed in the following analysis.

EXAMPLE 10.1

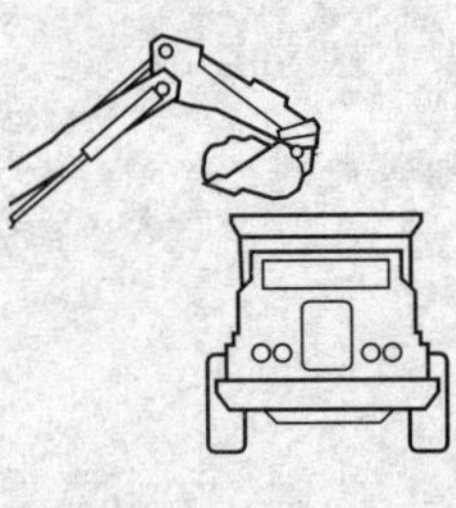

Consider a 3-cy shovel excavating good common earth with a 90° swing, with no delays waiting for hauling units, and with a 20-sec cycle time. Assume for this example that if the bucket and the trucks are operated at their heaped capacities, the swelling effect of the earth will permit each truck to carry its rated struck capacity, expressed in bank cubic yards (bcy). Assume that the number of buckets required to fill a truck will equal the capacity of the truck divided by the size of the shovel bucket, both expressed in cubic yards. Assume further that the time for the travel and dump cycle, *excluding the time for loading*, will be the same for the several sizes of trucks considered. The time for a travel cycle, which includes traveling to the dump, dumping, and returning to the shovel, will be 6 min.

12-cy trucks: If 12-cy trucks are used, it will require four buckets (12 ÷ 3) to fill a truck. The time that is required to load a truck would be 80 sec (4 × 20 sec), or 1.33 min. The round-trip cycle for a truck will be 7.33 min. The minimum number of trucks required to keep the shovel busy will be 7.33 ÷ 1.33 = 5.51.

Five trucks: The time required to load five trucks will be 5 × 1.33 = 6.65 min. Thus the shovel will lose 7.33 − 6.65 = 0.68 min when only five trucks are used. The percentage of time lost will be (0.68 ÷ 6.65) × 100 = 10.2%.

Six trucks: If six trucks are used, the total loading time required will be 6 × 1.33 = 7.98 min. As this will increase the total round-trip cycle of each truck from 7.33 to 7.98 min, the lost time per truck cycle will be 0.65 min per truck. This will result in a loss of

$$\frac{0.65}{7.98} \times 100 = 8.2\% \text{ for each truck}$$

which is equivalent to an operating factor of 91.8% for the trucks.

24-cy trucks: If 24-cy trucks are used, it will require eight buckets to fill a truck. The time that is required to load a truck would be 160 sec, or 2.66 min. The round-trip cycle for a truck will be 8.66 min. The minimum number of trucks required to keep the shovel busy will be 8.66 ÷ 2.66 = 3.26.

Four trucks: Using four trucks, the time required to load will be 4 × 2.66 = 10.64 min, the lost time per truck cycle will be 10.64 − 8.66 = 1.98 min per truck. This will produce an operating factor of 8.66/10.64 × 100 = 81.4% for the trucks.

In Example 10.1, note that the production of the shovel is based on a 60-min hour. This policy should be followed when balancing an excavator with hauling units because at times both will operate at maximum capacity if the number of units is properly balanced. However, the average production of a unit, excavator or truck, for a sustained period of time, should be based on applying an appropriate efficiency factor to the maximum productive capacity. Attention must be called to the fact that in this example we have chosen truck sizes that exactly matched the loader, i.e., to load a 12-cy truck with a 3-cy shovel results in an integer number of bucket loads. In practice, this is not always the case, but physically only an integer number of bucket loads can be used in loading the truck.

CALCULATING TRUCK PRODUCTION

The most important consideration when matching excavators and trucks is finding equipment having compatible capacities. Matched capacities yield maximum loading efficiency. The following is a format that can be used to calculate truck production.

Number of Bucket Loads

The first step in analyzing truck production is to determine the number of excavator bucket loads it takes to load the truck.

$$\text{Balanced number of bucket loads} = \frac{\text{Truck capacity (lcy)}}{\text{Bucket capacity (lcy)}} \qquad [10.1]$$

Load Time and Truckload Volume

The actual number of bucket loads placed on the truck must be an integer number. It is possible to not completely fill the bucket (light load) to match the bucket volume to the truck volume, but that practice is usually inefficient as it results in wasted loading time.

If one less bucket load is placed on the truck, the loading time will be reduced; but the load on the truck is also reduced. Sometimes job conditions will dictate that a lesser number of bucket loads be placed on the truck, i.e., the load size is adjusted if haul roads are in poor condition or if the trucks must traverse steep grades. The truckload in such cases will equal the bucket volume multiplied by the number of bucket loads.

Next Lower Integer For the case where the number of bucket loads is *rounded down to an integer lower* than the balance number of loads or reduced because of job conditions:

$$\text{Load time} = \text{Number of bucket loads} \times \text{Bucket cycle time} \qquad \textbf{[10.2]}$$

$$\text{Truckload}_{\text{LI}}\ (\text{volumetric}) = \text{Number of bucket loads} \times \text{Bucket volume} \qquad \textbf{[10.3]}$$

Next Higher Integer If the division of truck cargo body volume by the bucket volume is *rounded to the next higher integer* and that higher number of bucket loads is placed on the truck, excess material will spill off the truck. In such a case, the loading duration equals the bucket cycle time multiplied by the number of bucket swings. But the volume of the load on the truck equals the truck capacity, not the number of bucket swings multiplied by the bucket volume.

$$\text{Load time} = \text{Number of bucket loads} \times \text{Bucket cycle time} \qquad \textbf{[10.2]}$$

$$\text{Truckload}_{\text{HI}}\ (\text{volumetric}) = \text{Truck volumetric capacity} \qquad \textbf{[10.4]}$$

Gravimetric Check Always check the load weight against the gravimetric capacity of the truck.

$$\begin{matrix}\text{Truckload}\\ (\text{gravimetric})\end{matrix} = \text{Volumetric load (lcy)} \times \begin{matrix}\text{Unit weight}\\ (\text{loose vol. lb/lcy})\end{matrix} \qquad \textbf{[10.5]}$$

$$\text{Truckload (gravimetric)} < \text{Rated gravimetric payload?} \qquad \textbf{[10.6]}$$

Haul Time

Hauling should be at the highest safe speed and in the proper gear. To increase efficiency, use one-way traffic patterns.

$$\text{Haul time (min)} = \frac{\text{Haul distance (ft)}}{88\ \text{fpm/mph} \times \text{Haul speed (mph)}} \qquad \textbf{[10.7]}$$

Based on the gross weight of the truck with the load, and considering the rolling and grade resistance from the loading area to the dump point, haul travel speeds can be estimated using the truck manufacturer's performance chart (see Fig. 10.8).

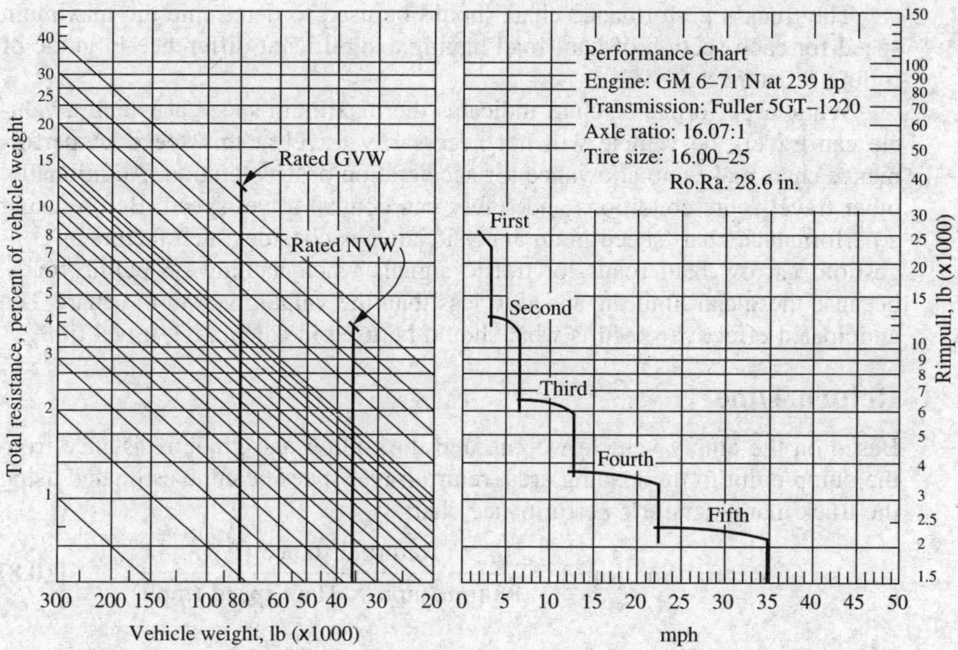

FIGURE 10.8 Performance chart for a 22-ton rear-dump truck.

EXAMPLE 10.2

Determine the speed of a truck when it is hauling a load of 22 tons up a 6% grade on a haul road having a rolling resistance of 60 lb per ton, equivalent to a 3% adverse grade. The Fig. 10.8 performance chart applies, and the truck specifications are as follows:

Capacity
  Stuck, 14.7 cy
  Heaped, 2:1, 18.3 cy
Net weight empty = 36,860 lb
Payload = 44,000 lb
Gross vehicle weight = 80,860 lb

It is necessary to combine the grade and rolling resistance, which gives an equivalent total resistance equal to 9% (6 + 3) of the vehicle weight.

The procedures for using the Fig. 10.8 chart are

1. Find the vehicle weight on the lower left horizontal scale.
2. Read up the vehicle weight line to the intersection with the slanted total resistance line.
3. From this intersection, read horizontally to the right to the intersection with the performance curve.
4. From this intersection, read down to find the vehicle speed.

Following these four steps, it is found that the truck will operate in the second-gear range, and its maximum speed will be 6.5 mph.

The truck's performance chart should be used to determine the maximum speed for each section of haul road having a significant difference in grade or rolling resistance.

While a performance chart indicates the maximum speed at which a vehicle can travel, the vehicle will not necessarily travel at this speed. A performance chart makes no allowance for acceleration or deceleration. Additionally, other travel route conditions and safety can control travel speed. Before using a performance chart speed in an analysis, always consider such factors as congestion, narrow haul roads, or traffic signals when hauling on public roads, because these can limit the speed to less than the value given in the chart. The anticipated effective speed is what should be used in calculating travel time.

Return Time

Based on the empty vehicle weight, and the rolling and grade resistance from the dump point to the loading area, return travel speeds can be estimated using the truck manufacturer's performance chart.

$$\text{Return time (min)} = \frac{\text{Return distance (ft)}}{88\ \text{fpm/mph} \times \text{Haul speed (mph)}} \qquad [10.8]$$

Dump Time

Dump time will depend on the type of hauling unit and congestion in the dump area. Consider that the dumping area is usually crowded with support equipment. Dozers are spreading the dumped material, and multiple pieces of compaction equipment may be working in the area. Rear-dumps must be spotted before dumping. This usually means that the truck must come to a complete stop and then back up some distance. Total dumping time in such cases can exceed 2 min. Bottom-dumps will customarily dump while moving. After dumping, the truck normally turns and returns to the loading area. Under favorable conditions, a rear-dump can dump and turn in 0.7 min but an average unfavorable time is about 1.5 min. Bottom-dumps can dump in 0.3 min under favorable conditions, but they too may average 1.5 min when conditions are unfavorable. Always try to visualize the conditions in the dump area when estimating dump time.

Truck Cycle Time

The cycle time of a truck is the sum of the load time, the haul time, the dump time, and the return time:

$$\text{Truck cycle time} = \text{Load}_{\text{time}} + \text{Haul}_{\text{time}} + \text{Dump}_{\text{time}} + \text{Return}_{\text{time}} \qquad [10.9]$$

Number of Trucks Required

The number of trucks required to keep the loading equipment working at capacity is

$$\text{Balanced number of trucks} = \frac{\text{Truck cycle time (min)}}{\text{Excavator cycle time (min)}} \quad \textbf{[10.10]}$$

Production

The number of trucks must be an integer number.

Integer Lower Than Balance Number If an integer number of trucks *lower* than the result of Eq. [10.10] is chosen, then the trucks will control production.

$$\begin{matrix}\text{Production} \\ \text{(lcy/hr)}\end{matrix} = \begin{matrix}\text{Truck} \\ \text{load (lcy)}\end{matrix} \times \begin{matrix}\text{Number} \\ \text{of trucks}\end{matrix} \times \frac{60 \text{ min}}{\text{Truck cycle time (min)}} \quad \textbf{[10.11]}$$

Integer Greater Than Balance Number When an integer number of trucks *greater* than the result of Eq. [10.10] is selected, production is controlled by the loading equipment.

$$\begin{matrix}\text{Production} \\ \text{(lcy/hr)}\end{matrix} = \begin{matrix}\text{Truck} \\ \text{load (lcy)}\end{matrix} \times \frac{60 \text{ min}}{\text{Excavator cycle time (min)}} \quad \textbf{[10.12]}$$

As a rule, it is better to never keep the loading equipment waiting. If there is not a sufficient number of haul trucks, there will be a loss in production. Truck bunching or queuing will reduce production 10 to 20% even when there is a perfect match between excavator capability and the number of trucks. If there are extra haul units, this queuing effect is reduced. Therefore, it is usually best to have more trucks; that is, with Eq. [10.10] round up to the next integer.

Efficiency

The production calculated with either Eq. [10.11] or Eq. [10.12] is based on a 60-min working hour. That production should be adjusted by an efficiency factor. Longer hauling distances usually result in better driver efficiency. Driver efficiency increases as haul distances increase out to about 8,000 ft, after which efficiency remains constant. Other critical elements affecting efficiency are bunching, discussed above, and equipment condition.

$$\text{Adjusted production} = \text{Production} \times \frac{\text{Working time (min/hour)}}{60 \text{ min}} \quad \textbf{[10.13]}$$

PRODUCTION ISSUES

A number of other factors must be considered when matching excavators and hauling units.

- Reach of the excavator
- Dumping height of the bucket
- Width of the bucket

Reach of the Excavator

The excavator must be able to physically reach—extend its bucket—from its digging position to the dumping point over the cargo body of the truck. In the case of a shovel or hoe, this involves the reach of the stick and boom when extended at dump height. A loader's reach is measured from the front of its front tires to the tip of the bucket's cutting edge when the boom is fully raised and the bucket is dumped at 45°.

Bucket Dump Height

Compare the excavator's bucket dump height to that of the sides on the cargo body. This comparison must consider the actual configuration of the excavator and its bucket. The teeth on a rock bucket can reduce dump height by as much as a foot. Sideboarding the cargo body or the use of larger tires on the truck can increase its height significantly.

Width of the Bucket

Compare the width of the excavator bucket to the length of the truck's cargo body. It is recommended that bucket-width-to-cargo body length should fall within a ratio of 1:1.4 or 1.5. The bucket should not be so wide that it will be difficult for the excavator operator to avoid striking the back of the truck cab or dumping material too close to the end of the cargo body. Bulky shaped material pieces—rocks—that land too close to the end of the dump body are likely to fall off the truck the first time it goes up a grade. Rocks spilled on the haul road can damage tires, and there will be an additional requirement for road maintenance. Even if the load does not roll off the truck, it places an undesirable weight distribution on the rear axle, which can increase axle and tire wear.

TIRES

Tires for trucks and all other haul units should be suitably matched to the job requirements. The selection of proper tire sizes and the practice of maintaining correct air pressure in the tires will reduce that portion of the rolling resistance due to the tires. A tire supports its load by deforming where it contacts the road surface until the area in contact with the road will produce a total force on the road equal to the load on the tire. Neglecting any supporting resistance furnished by the sidewalls of the tire, if the load on a tire is 5,000 lb and the air pressure is 50 psi, the area of contact will be 100 square inches (sq. in.). If, for the same tire, the air pressure is permitted to drop to 40 psi, the contact area will increase to 125 sq. in. The additional area of contact will be produced by additional deformation of the tire. This will increase the rolling resistance because the tire will be continually climbing a steeper grade as it rotates.

The tire size selected and the inflated pressure should be based on the resistance, which the surface of the road offers to penetration, by the tire. For rigid road surfaces, such as concrete, small-diameter high-pressure tires will give lower rolling resistance, while, for yielding road surfaces, large-diameter low-pressure tires will give lower rolling resistance because the larger areas of contact will reduce the tire penetration depth.

Many tire failures can be traced to constant overload, excessive speed, incorrect tire selection, and poorly maintained haul roads. Underinflating tires can cause radial sidewall cracking and ply separation. Overinflation subjects the tire to excessive wear in the center of the tread. Mismatched duals will cause unequal weight distribution, overloading the larger tire.

Tires generate heat as they roll and flex. As a tire's operating temperature increases, the rubber and textiles within significantly lose strength. The manufacturers of earthmover tires provide a ton-miles-per-hour (TMPH) limit for their tires. The TMPH is a numerical expression of the working capacity of a tire. It is good practice to calculate a job TMPH value and make a comparison with the TMPH for the tires on the equipment.

$$\begin{array}{c}\text{TMPH}\\ \text{job rate}\end{array} = \text{Average tire load} \times \begin{array}{c}\text{Average speed during}\\ \text{a day's operation}\end{array} \quad [10.14]$$

$$\begin{array}{c}\text{Average tire}\\ \text{load (tons)}\end{array} = \frac{\text{“Empty” tire load (tons)} + \text{“Loaded” tire load (tons)}}{2} \quad [10.15]$$

$$\begin{array}{c}\text{Average}\\ \text{speed (mph)}\end{array} = \frac{\text{Round trip distance (miles)} \times \text{number of trips}}{\text{Total hours worked}} \quad [10.16]$$

When calculating the job TMPH value, always select the tire that carries the highest average load. If the tires being used on the trucks have a TMPH rating of less than the job TMPH, either the speed or the load or both must be reduced, or the trucks must be equipped with tires having a higher rating.

EXAMPLE 10.3

An off-highway truck weights 70,000 lb empty and 150,000 lb when loaded. The weight distribution empty is 50% front and 50% rear. The weight distribution loaded is 33% front and 67% rear. The truck has two front tires and four rear tires. The truck works an 8-hr shift hauling rock to a crusher. The one-way haul distance is 5.5 miles. The truck can make 14 trips per day. Calculate the job TMPH value for the truck.

$$\text{Total weight on two front tires (empty)} = 70{,}000 \text{ lb} \times 50\% = 35{,}000 \text{ lb}$$

$$\text{Total weight on two front tires (loaded)}$$

$$= 150{,}000 \text{ lb} \times 33\% = 50{,}000 \text{ lb}$$

$$\text{Weight on individual front tire (empty)} = \frac{35{,}000 \text{ lb}}{2} = 17{,}500 \text{ lb}$$

$$\text{Weight on individual front tire (loaded)} = \frac{50{,}000 \text{ lb}}{2} = 25{,}000 \text{ lb}$$

$$\text{Average front tire load} = \frac{17{,}500 \text{ lb} + 25{,}000 \text{ lb}}{2} = 42{,}500 \text{ lb or 10.6 to}$$

$$\text{Total weight on four rear tires (empty)} = 70{,}000 \text{ lb} \times 50\% = 35{,}000 \text{ lb}$$

$$\text{Total weight on four rear tires (loaded)} = 150{,}000 \text{ lb} \times 67\%$$

$$= 100{,}000 \text{ lb}$$

$$\text{Weight on individual rear tire (empty)} = \frac{35{,}000 \text{ lb}}{4} = 8{,}750 \text{ lb}$$

$$\text{Weight on individual rear tire (loaded)} = \frac{100{,}000 \text{ lb}}{4} = 25{,}000 \text{ lb}$$

$$\text{Average rear tire load} = \frac{8{,}750 \text{ lb} + 25{,}000 \text{ lb}}{2} = 16{,}875 \text{ lb or 8.4 ton}$$

The front tire carries the highest average load.

$$\text{Average speed} = \frac{(2 \times 5.5 \text{ mile}) \times 14 \text{ trips}}{8 \text{ hr}}$$

$$\text{Average speed} = 19.25 \text{ mph}$$

$$\text{Job TMPH value} = 10.6 \text{ tons} \times 19.25 \text{ mph}$$

$$\text{Job TMPH} = 204$$

This means that a tire with a TMPH rating of 204 or higher must be used under these job conditions.

TRUCK PERFORMANCE CALCULATIONS

Example 10.4 analyzes the performance of a fleet of 22-ton rear-dump trucks being loaded by hydraulic hoe having a 3-cy bucket.

EXAMPLE 10.4

Rear-dump trucks with specifications as follows are used to haul sandy clay waste material. The performance chart shown in Fig. 10.8 is valid for these trucks.

Capacity
- Stuck, 14.7 cy
- Heaped, 2:1, 18.3 cy

Net weight empty	= 36,860 lb
Payload	= 44,000 lb
Gross vehicle weight	= 80,860 lb

FIGURE 10.9 Rear-dump truck being loaded by a hoe.

The trucks will be loaded by a hydraulic hoe having a 3-cy bucket (Fig. 10.9). The haul route from the loading point to the waste site is a 3-mile downhill grade of 1%. After turning off the city streets, the haul route will be on earth, poorly maintained. Use the rolling resistance for that condition as most of the travel is off of the city streets. Dump time will average 2 min because of expected congestion on the waste site. The hoe should be able to cycle in 20 sec. The sandy clay has a loose unit weight of 2,150 lb/cy. A realistic efficiency estimate for this work is a 50-min hour.

Step 1. Number of bucket loads. The bucket fill factor for the hoe handling sandy clay has been determined to be 110%. The hoe bucket volume will therefore be 3.3 lcy (3 × 1.1). The heaped capacity of the truck is 18.3 lcy.

$$\text{Balanced number of bucket loads} = \frac{18.3 \text{ lcy}}{3.3 \text{ lcy}} = 5.5$$

The actual number of buckets should be an integer number; therefore, two cases, placing either five or six bucket loads on the truck, should be investigated.

Step 2. Load time. Check production based on both situations, 5 or 6 bucket loads to fill the truck.

$$\text{Load time (five buckets) } 5 \times \frac{20 \text{ sec}}{60 \text{ sec/min}} = 1.66 \text{ min}$$

Load volume (five buckets) 5×3.3 lcy/bucket load = 16.5 lcy

Check load weight 16.5 lcy × 2,150 lb/lcy = 35,045 lb

Okay 35,045 lb < 44,000 lb rated payload

$$\text{Load time (six buckets) } 6 \times \frac{20 \text{ sec}}{60 \text{ sec/min}} = 2.00 \text{ min}$$

Load volume (six buckets) equals truck capacity 18.3 lcy, excess spills off.

Check load weight 18.3 lcy × 2,150 lb/lcy = 39,345 lb

Okay 39,345 lb < 44,000 lb rated payload

Step 3. Haul time.

Rolling resistance (Table 6.1); earth, poorly maintained, 100 to 140 lb/ton

Using an average value 120 lb/ton or 6.0%

Grade resistance −1%

Total resistance 5.0% [6.0% + (−1%)]

	Five buckets	Six buckets
Empty truck net weight	36,860 lb	36,860 lb
Load weight	35,045 lb	39,345 lb
Gross weight	71,905 lb	76,205 lb
Speed (Fig. 10.8)	16 mph	13 mph

In this particular case, the effect of total resistance and the difference in gross weight for the two load scenarios result in a speed difference for the conditions. If the total resistance had been only 4%, the achievable speed would have been 22 mph for both load conditions.

$$\text{Haul time (five buckets)} = \frac{3 \text{ mile} \times 5{,}280 \text{ ft/mile}}{88 \text{ fpm/mph} \times 16 \text{ mph}} = 11.25 \text{ min}$$

$$\text{Haul time (six buckets)} = \frac{3 \text{ mile} \times 5{,}280 \text{ ft/mile}}{88 \text{ fpm/mph} \times 13 \text{ mph}} = 13.85 \text{ min}$$

Step 4. Return time.

Rolling resistance 120 lb per ton or 6.0%

Grade resistance 1%

Total resistance 7.0% [6.0% + (+1%)]

Empty truck weight 36,860 lb

Speed (Fig. 10.8) 22 mph

$$\text{return time } \frac{3 \text{ mile} \times 5{,}280 \text{ ft/mile}}{88 \text{ fpm/mph} \times 22 \text{ mph}} = 8.18 \text{ min}$$

Step 5. Dump time. Congestion on the fill, dump time expected to be 2 min.

Step 6. Truck cycle time.

	Five bucket loads on the truck (min)	Six bucket loads on the truck (min)
Load time	1.66	2.00
Haul time	11.25	13.85
Dump time	2.00	2.00
Return time	8.18	8.18
Truck cycle time	23.09	26.03

Step 7. Number of trucks required.

	Five bucket loads on the truck	Six bucket loads on the truck
Truck cycle time	23.09 min	26.03 min
Loader cycle time	1.66 min	2.00 min
Number of trucks	13.9	13.0

Step 8. Production. The number of trucks must be an integer number. For the case of five buckets to load the truck, consider using 13 or 14 trucks. If 13 trucks are used, the loader will have to wait for trucks; therefore, the truck cycle will control production.

Production (five buckets and 13 trucks)

$$16.5 \text{ lcy} \times 13 \text{ trucks} \times \frac{60 \text{ min}}{23.09 \text{ min}} = 557 \text{ lcy/hr}$$

If 14 trucks are used, the loader will control production and the trucks will sometimes wait to be loaded.

Production (five buckets and 14 trucks)

$$16.5 \text{ lcy} \times \frac{60 \text{ min}}{1.66 \text{ min}} = 596 \text{ lcy/hr}$$

Considering the case of six bucket loads and using 13 trucks

Production (six buckets and 13 trucks)

$$18.3 \text{ lcy} \times 13 \text{ trucks} \times \frac{60 \text{ min}}{26.03 \text{ min}} = 548 \text{ lcy/hr}$$

In this particular case, considering only production, it is best to load light and use 14 trucks for an hourly production of 596 lcy. If an effort was made to improve the haul road, the rolling resistance would decrease, and the travel speed advantage to the lighter load would disappear.

TABLE 10.2 Production comparison with various numbers of bucket loads to fill the truck and different numbers of trucks.

Total resistance ≈ 4%	Five bucket loads on the truck	Six bucket loads on the truck
Truck cycle time	20.02 min	20.36 min
Load time	1.66 min	2.00 min
Balance no. of trucks	12.06	10.18
Production 10 trucks	495 lcy/hr	539 lcy/hr
Production 11 trucks	544 lcy/hr	549 lcy/hr
Production 12 trucks	593 lcy/hr	549 lcy/hr
Production 13 trucks	596 lcy/hr	549 lcy/hr

For the case of an earth, compacted and maintained haul road the rolling resistance would be between 40 and 70 lb/ton considering rubber high-pressure tires. Using an average condition the total resistance for the haul would be just over 4% and the haul speed for both loading conditions (five or six buckets) would be 22 mph. This would reduce the number of trucks required because of the faster cycle time. Note in Table 10.2 that if the haul road is improved the same 596 lcy/hr production can be achieved with one less truck. The same production with fewer trucks will cause production cost to be reduced. The analysis also illustrates the fact that there is a mismatch between the size of the hoe bucket and the size of the truck cargo capacity. It is very inefficient to be loading so light or to be wasting so much spillage with the last bucket.

Step 9. Efficiency.

$$\text{Adjusted Production: } 596 \text{ lcy/hr} \times \frac{50 \text{ min}}{60 \text{ min}} = 497 \text{ lcy/hr}$$

Step 10. Production in desired units (volume or weight).

Finally, this production can be converted as necessary into bank cubic yards or tons by using material property information specific to the job or average values as found in Table 4.3.

$$\text{Adjusted production} \quad 497 \text{ lcy/hr} \times 0.74 = 367 \text{ bcy/hr}$$

$$\text{Adjusted production} \quad 497 \text{ lcy/hr} \times \frac{2{,}100 \text{ lb/lcy}}{2{,}000 \text{ lb/ton}} = 522 \text{ ton/hr}$$

Because the cost of the excavation equipment is usually greater than the cost of a haul truck, it is common practice to use a greater number of trucks than the balance number derived from the ratio of the loader and truck cycle times. When considering this decision the mechanical condition of the trucks should be considered. Another consideration is the availability of standby trucks. These are not necessarily idle units but could be trucks assigned to lower priority tasks from which they can easily be diverted.

After the job has started, the number of trucks required may vary because of changes in haul-road conditions, reductions or increases in the length of hauls, or changes in conditions at either the loading or dumping areas. Management should always continue to monitor hauling operations for changes in assumed conditions.

TRUCK SAFETY

The driver of a dump truck was killed in 2002 when he backed the truck too close to the edge of an embankment, which gave way. A signalperson told the driver to stop approximately 8 ft before the edge and had given the sign to dump the load. The driver, however, continued to back the truck. The ground under the rear wheels gave way, and the truck slid backward down the embankment and then rolled over, landing on its roof. The embankment berm was insufficient to stop the truck when it got close to the edge. The truck was supplied with a lap-type safety belt, which the victim was wearing.

In 2001 a traffic control person died when a dump truck backed over her. The truck was delivering asphalt to a highway crew. The victim was directly behind the truck when it began backing.

Operating trucks and working around trucks can be hazardous. The employers of the individuals killed in these two truck accidents had well-established safety programs and written Accident Prevention Plans that contained all the required elements. Still, two fatal accidents occurred. Operating and working around construction equipment and trucks is dangerous. A 6-ft-tall person standing closer than 70 ft from the right side of a 150-ton off-highway truck cannot be seen by the driver. Employees must be mentored daily about the dangers, and management must be exceedingly proactive if accidents are to be prevented.

The investigators of these two accidents made the following recommendations:

- Conduct a hazard assessment of the work site every day, and ensure operators and drivers are aware of the hazards.
- Ensure berms are adequately built to prevent trucks from over traveling at dump locations and on haul roads.
- Ensure that heavy equipment operators follow all operating signals.
- Use a spotter when backing heavy equipment with blind spots.
- Ensure that employees stay out of vehicle travel paths and in clear view of those who are operating equipment.
- Wear high-visibility safety clothing.

SUMMARY

The use of trucks as the primary hauling unit provides a high degree of flexibility, as the number in service can usually be increased or decreased easily to permit modifications in hauling capacity. When estimating what a truck will carry, both the rated *gravimetric* load and the rated-heaped *volume* must be examined. The heaped capacity is the volume of material that the truck will haul when the load is heaped above the sides. The actual heaped capacity will vary with the material that is being carried. Critical learning objectives include:

- An understanding of the necessity to achieve balance between excavator bucket volume and truck load volume
- An ability to use performance charts to calculate truck speed
- An understanding of the job-site constraints that affect dump times
- An ability to calculate the number of trucks required to keep the excavating equipment working at capacity

These objectives are the basis for the problems that follow.

PROBLEMS

10.1 Visit the California Fatality Assessment and Control Evaluation (FACE) Program website (www.dhs.ca.gov/ohb/OHSEP/FACE/) and report on the causes of two construction trucking accidents that resulted in fatalities. What could have been done to prevent the accidents you selected?

10.2 How many reports of fatal "backed over" construction site dump truck accidents can you locate by searching the Web for 10 min?

10.3 A truck for which the information in Fig. 10.8 applies operates over a haul road with a +3% slope and a rolling resistance of 140 lb/ton. If the gross vehicle weight is 90,000 lb, determine the maximum speed of the truck. (rimpull 9,000 lb, speed 6.5 mph)

10.4 A truck for which the information in Fig. 10.8 applies operates over a haul road with a +4% slope and a rolling resistance of 90 lb/ton. If the gross vehicle weight is 70,000 lb, determine the truck's maximum speed.

10.5 A truck for which the information in Fig. 10.8 applies operates over a haul road with a −4% slope and a rolling resistance of 200 lb/ton. If the gross vehicle weight is 80,000 lb, determine the maximum speed of the truck. (rimpull 4,800 lb, speed 12 mph)

10.6 The truck of Problem 10.4 operates on a haul road having a −4% slope. Determine the truck's maximum speed.

10.7 An articulated truck weighs 46,300 lb empty and 96,300 lb when loaded. This truck has one front axle and two rear axles. The axle weight distribution empty is 58% front, 21% center, and 21% rear. The weight distribution loaded is 32% front, 34% center, and 34% rear. The truck has two tires on each axle. The truck works a 10-hr shift hauling sand. The one-way haul distance is 3.5 miles. The truck can make 24 trips per day. The truck is equipped with tires having a 110 TMPH rating. Are these tires satisfactory under the specified job conditions?

10.8 Prepare a table similar to Table 10.2, using a $3\frac{1}{2}$-cy shovel loading poorly blasted rock (2,600 lb/lcy). Consider loading 15- and 20-cy-size trucks. Assume that both size trucks can handle the gravimetric load and the shovel bucket swing cycle time is 26 sec. Use a bucket fill factor of 100%. Using the information in Fig. 10.8, the net empty weight of the 15-cy truck is 44,000 lb and that of the 20-cy truck is 50,000 lb. Dump time will be 1.5 min. The haul distance is 4 miles from the excavation area to the fill up a 2% grade. The rolling resistance of the haul road will be maintained at 4%.

10.9 Evaluate cost and production using a 3-cy hoe to load wet gravel into 14-cy-size trucks. Assume a bucket fill factor of 105% and that the hoe bucket swing cycle time is 24 sec. Use the information in Fig. 10.8, a truck net empty weight of 36,860 lb, and a rated payload of 40,000 lb. Dump time will be 1.3 min. The haul distance is 2.5 miles from the pit to the plant up a 3% grade. The rolling resistance of the haul road will be maintained at 3%. The hoe and operator cost $97 per hour, and trucks cost $49 per hour. What is the most cost-effective mix of bucket loads to use in loading the trucks and number of trucks to place on this job?

REFERENCES

1. *Care and Service of Off-the-Highway Tires*, Manual Pub No. OHM-882, Rubber Manufacturers Association, Washington, DC, www.rma.org/publications.
2. *Caterpillar Performance Handbook*, Caterpillar Inc., Peoria, IL (published annually). www.cat.com.
3. Gove, D., and W. Morgan (1994). "Optimizing Truck-Loader Matching," *Mining Engineering*, Vol. 46, October, pp. 1179–1185.
4. Hull, Paul E. (1999). "Moving Materials," *World Highways/Routes Du Monde*, November–December, pp. 79–82.
5. Schexnayder, Cliff, Sandra L. Weber, and Brentwood T. Brooks (1999). "Effect of Truck Payload Weight on Production," *Journal of Construction Engineering and Management*, ASCE, Vol. 125, No. 1, pp. 1–7, January–February.
6. *Capacity Rating—Dumper Body and Trailer Body*, J1363, SAE Standards, Society of Automotive Engineers International, 400 Commonwealth Drive, Warrendale, PA.

WEBSITE RESOURCES

1. www.cat.com Caterpillar Inc., Peoria, IL.
2. www.osha.gov/SLTC/index.html OSHA Technical Links to Safety and Health Topics, U.S. Department of Labor, Occupational Safety and Health Administration, 200 Constitution Avenue, NW, Washington, DC.

CHAPTER 11

Finishing Equipment

Finishing operations follow closely behind excavation operations or compaction of embankments. Graders are multipurpose machines used for finishing and shaping. The gradall is a utility machine that combines the operating features of a hoe, dragline, and motor grader. It is designed as a versatile machine for both excavation and finishing work. There are also a variety of highly specialized trimming machines for finish grading. These automatic trimmers use an automatic control system.

INTRODUCTION

Finishing, finish grading, and fine grading are all terms used to refer to the process of shaping materials to the required line and grade specified in the contract documents. Finishing operations follow closely behind excavation (rough grading) operations or compaction of embankments. These operations include finishing to prescribed grade those sections supporting structural members, and the smoothing and shaping of slopes. On many projects, graders are used as the finishing machine. In the case of long linear projects, such as roads and airfields, there are special trimming machines to accomplish the finishing under the pavement sections.

GRADERS

GENERAL INFORMATION

Graders (see Fig. 11.1) are multipurpose machines used for finishing, shaping, bank sloping (see Fig. 11.2), and ditching. They are also used for mixing, spreading, side casting, leveling and crowning, light stripping operations,

FIGURE 11.1 Grader fine grading a building parking lot.

FIGURE 11.2 Grader working a slope.

general construction, and dirt road maintenance. A grader's primary purpose is cutting and moving material with the moldboard. These machines are restricted to making shallow cuts in medium-hard materials; they should not be used for heavy excavation. A grader can move small amounts of material but cannot perform dozer-type work because of the structural strength and location of its moldboard.

Graders are capable of progressively cutting ditches to a depth of 3 ft and for working on slopes as steep as 3:1. However, it is not advisable to run graders parallel with such steep slopes because they have a comparatively high

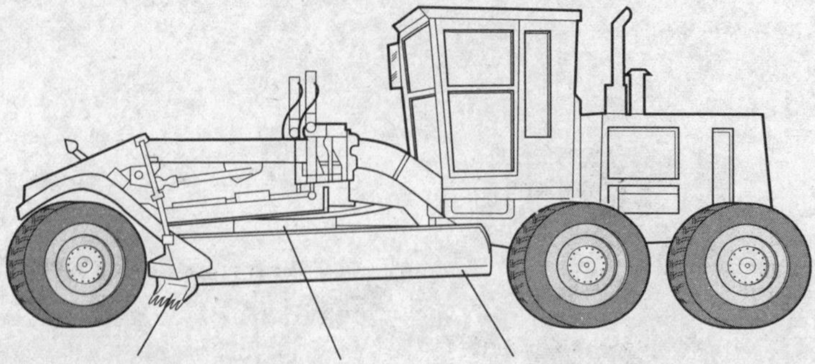

FIGURE 11.3 The components of a grader.

FIGURE 11.4 Grader with lightweight rear-mounted rippers.

center of gravity, and the right pressure at a critical point on the moldboard could cause the machine to roll over. It is more economical to use other types of equipment to cut ditches deeper than 3 ft.

The components of the grader that actually do the work are the moldboard (blade) and the scarifier (see Fig. 11.3). Graders may also be equipped with lightweight rear-mounted rippers (see Fig. 11.4).

Moldboard

The moldboard, commonly referred to as the blade, is the working member of the grader. A rotating circle carries the moldboard. Through intricate hydraulics,

Moldboard shifted to right side **Moldboard raised into vertical position**

FIGURE 11.5 Moldboard positions.

the moldboard can be placed into many positions, either under the grader or to the side (see Fig. 11.5). It can be side-shifted horizontally for increased reach outside of the tires.

The moldboard is used to side cast the material it encounters. The ends of the moldboard can be raised or lowered together or independently of one another. By convention, the toe of the moldboard is the foremost end of the moldboard in the direction of travel and the heel is the discharge end.

Moldboard Angle

The moldboard can be angled (positioned) at almost any angle to the line of travel, parallel to the direction of travel, shifted to either side, or raised into vertical position (see Figs. 11.5 and 11.6).

Moldboard Pitch

For normal work, the moldboard is kept near the center of the pitch adjustment, which keeps the top of the moldboard directly over the cutting edge. However, the top of the moldboard can be pitched (leaned) forward or backward (see Fig. 11.7). When leaned forward, the cutting ability of the moldboard is decreased, and it has more dragging action. It will tend to ride over material rather than cut and push, and it has less chance of catching on solid obstructions. A forward pitch is used to make light, rapid cuts and to blend materials. When leaned to the rear, the moldboard cuts readily but tends to let material spill back over the moldboard.

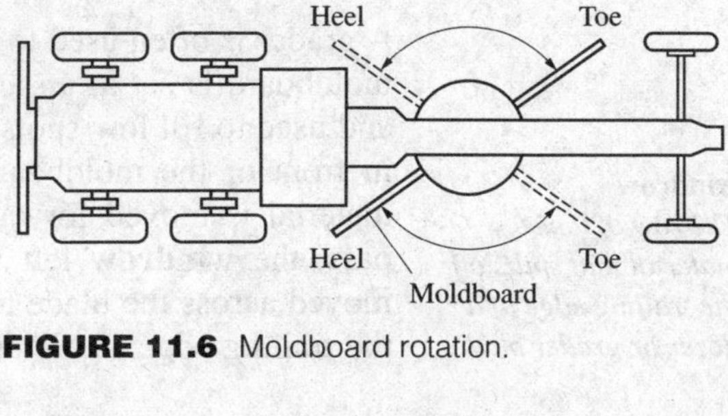

FIGURE 11.6 Moldboard rotation.

Scarifier

Material too hard to cut with the moldboard should be broken up with the scarifier. A scarifier is an attachment hung between the front axle and the moldboard. It is composed of a scarifier log with

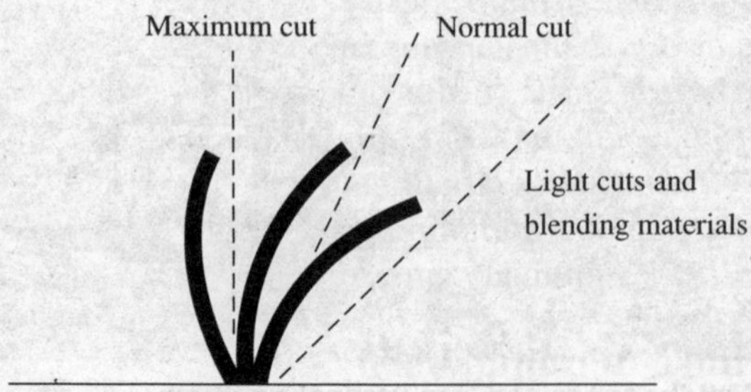

FIGURE 11.7 Grader moldboard pitch.

removable teeth. The teeth can be adjusted to cut to a depth of 12 in. When operating in hard material, it may be necessary to remove some of the teeth from the scarifier log. A maximum of five teeth may be removed from the log. If more than five teeth are removed, the force against the remaining teeth could shear them off. When removing teeth, take the center one out first and then alternately remove the other four teeth. This balances the scarifier and distributes the load evenly. With the top of the scarifier pitched to the rear, the teeth lift and tear the material being loosened. This position is also used for breaking up asphalt pavement. The pitch of the scarifier log can be adjusted for the particular material being manipulated.

GRADER OPERATIONS

When the moldboard is set at an angle, the load being pushed tends to drift off to the trailing end of the moldboard. Rolling action caused by the moldboard curve assists this side movement. As the moldboard is angled more sharply, the speed of the side drift increases, so that material is not carried forward as far and deeper cuts can be made. To shape and maintain most roads, set the moldboard at a 25° to 30° angle. The angle should be decreased for spreading windrows and increased for hard cuts and ditching.

Planing Surfaces

A grader is often used to plane or smooth off fill or cut surfaces. To do this the moldboard is set at an angle to surface with high material being scrapped off and used to fill low spots. The operator tries to keep enough of the cut material in front of the moldboard to accomplish the necessary filling. The loosened material is moved forward and sideways to distribute it evenly. On the next pass, the **windrow** left at the trailing edge of the moldboard is caught and moved across the blade to the heel. On the final pass, a lighter cut is made and the trailing edge (heel) of the moldboard lifted to allow the surplus material to

windrow
The row of loose material that spills off the trailing edge of a dozer or grader blade.

pass under rather than off the end. This avoids leaving a ridge. Windrows should not be piled in front of the rear wheels because they will adversely affect traction and cutting accuracy.

Road and embankment finish work and shallow ditch cuts are basic grader operations. These operations are normally performed as follows.

Ditch Cuts

Normally, ditching cuts are done in second gear at full throttle. For better grader control and straighter ditches, a 3- to 4-in.-deep marking cut is made at the outer edge of the bank slope (usually identified by slope stakes) on the first pass. The toe of the moldboard should be in line with the outside edge of the lead tire. The marking cut provides a guide for subsequent operations. Cuts are made as deep as possible without stalling or losing control of the grader. Each successive cut is brought in from the edge of the bank slope so the toe of the moldboard will be in line with the ditch bottom on the final cut.

Moving Windrows

When a grader makes a cut, a windrow will form between the heel of the moldboard and rear wheel. This windrow will impart a **side draft** force. The front wheels of a grader can be leaned both left and right. The grader wheels are inclined against the direction of side draft (see Fig. 11.8) and the windrow can then be shifted across the grade with successive cuts. Sometimes cuts produce more material than needed for the roadbed and shoulders. This excess material can be used as fill at other locations on the project. In that case, the

side draft
The force at the moldboard that tends to pull the front of the grader to one side.

FIGURE 11.8 Leaning the wheels of the grader against the direction of side draft.

excess material is drifted into a windrow and picked up by an elevating scraper. The scraper can haul the material to the appropriate location on the project.

Haul-Road Maintenance

Haul roads should always be kept in good condition as that improves haul-unit production. Graders are the best machines for maintaining haul roads. The most efficient method of road maintenance is to use sufficient graders to complete one side of a road with one pass of each grader (tandem operation). With this method, one side of the road is completed, while the other side is left open to traffic.

Working the material across the road from one side to the other provides leveling and maintenance of the surface. However, to maintain a satisfactory surface in dry weather, traffic-eroded material should be worked from the edges and shoulders of the road toward the center. The surface is easier to work if it is damp; therefore after a rain is a good time to perform surface maintenance. A water truck may be necessary to dampen material that is too dry to be worked.

Smoothing Pitted Surfaces When binder (fine material) is present and moisture content is appropriate, rough or badly pitted surfaces can be cut smooth. The material cut from the surface is then respread over the smooth base. Again, the best time to reshape earth and gravel roads is after a rain. Dry roads should be watered using a water distributor. This ensures that the material will have sufficient moisture content to recompact readily.

Correcting Corrugated Roads When correcting corrugated roads, care should be exercised so as not to make the situation worse. Deep cuts on a washboard surface will set up moldboard *chatter*, which emphasizes rather than corrects corrugations. Scarifying may be required if the surface is too badly corrugated. With proper moisture content, the surface can be leveled by cutting across the corrugations. The operator should alternate the moldboard so the cutting edge will not follow the rough surface and cut the surface to the bottom of the corrugations. Then reshape the road surface by spreading the windrows in an even layer across the road. Compaction after shaping gives longer-lasting results.

Spreading

Graders are often used to spread and mix dumped loads. Because of their mechanical structure and operating characteristics, graders can only be effective spreading and mixing free-flowing materials. A general formula for figuring grader spreading and mixing production is

$$\text{Production (bcy) per hr} = 3.0 \times \text{hp} \qquad \textbf{[11.1]}$$

where hp is the engine flywheel-brake horsepower of the grader and efficiency is assumed to be a 50-min working hour.

EXAMPLE 11.1

A large grader is rated at 220 hp. What is its expected production, in bcy, when used to spread sand dumped from haul trucks?

Sand is a free-flowing material, therefore Eq. [11.1] can be used to calculate the production.

$$\text{Production (bcy) per hr} = 3.0 \times 220 = 660 \text{ bcy/hr}$$

Proper Working Speeds

Always operate as fast as the skill of the operator and the condition of the grade permit. Graders should be operated at full throttle in each gear. If less speed is required, it is better to use a lower gear, rather than run at less than full throttle. Correct gear ranges for various grader operations, under normal conditions, are listed in Table 11.1.

Turns

When making a number of passes over a short distance (less than 1,000 ft), backing the grader to the starting point is normally more efficient than turning it around and continuing the work from the far end. Turns should never be make on newly laid asphalt surfaces.

Number of Passes

Grader efficiency is in direct proportion to the number of passes made. Operator skill, coupled with planning, is most important in eliminating unnecessary passes. For example, if four passes will complete a job, every additional pass increases the time and cost of the job.

Tire Inflation

Overinflated tires cause less contact between the tires and the road surface, resulting in a loss of traction. Air pressure differences in the rear tires cause wheel slippage and grader bucking. To achieve good results, it is necessary to always keep tires properly inflated.

TABLE 11.1 Proper gear ranges for grader operations.

Operation	Gear
Road maintenance	Second to third
Spreading	Third to fourth
Mixing	Fourth to sixth
Bank sloping	First
Ditching	First to second
Finishing	Second to fourth

TIME ESTIMATES

The following formula can be used to prepare estimates of the total time (in hours or minutes) required to complete a grader operation:

$$\text{Total time} = \frac{P \times D}{S \times E} \qquad \textbf{[11.2]}$$

where

P = number of passes required

D = distance traveled in each pass, in miles or feet

S = speed of grader, in mph or feet per minute (fpm) (use 88 fpm to change mph to fpm)

E = grader efficiency factor

Factors in Formula

The number of passes depends on project requirements and is estimated before construction begins. For example, five passes may be needed to clean out a ditch and reshape a road. Travel distance per pass depends on the length of the work.

Speed Grader travel speed is the most difficult factor in the formula to estimate accurately. As work progresses, conditions may require that speed estimates be increased or decreased. Work output should be computed for each rate of speed used in an operation. The speed depends largely on the skill of the operator and the type of material being handled.

Efficiency Factor A reasonable efficiency factor for grader operations is 60%.

EXAMPLE 11.2

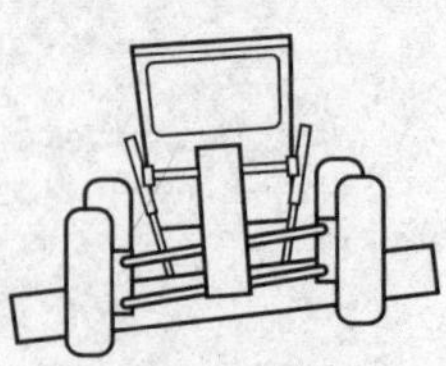

Application of Eq. [11.2] with distance expressed in miles

Maintenance of 5 miles of haul road requires cleaning the ditches, and leveling and reshaping the roadway. Use an efficiency factor of 0.60. Cleaning the ditches requires two passes in first gear (2.3 mph); leveling the road requires two passes in second gear (3.7 mph); and final shaping of the road requires three passes in fourth gear (9.7 mph).

$$\text{Total time} = \frac{2 \times 5 \text{ miles}}{2.3 \text{ mph} \times 0.60} + \frac{2 \times 5 \text{ miles}}{3.7 \text{ mph} \times 0.60} + \frac{3 \times 5 \text{ miles}}{9.7 \text{ mph} \times 0.60}$$

$$= 7.3 \text{ hr} + 4.5 \text{ hr} + 2.6 \text{ hr} = 14.4 \text{ hr}$$

EXAMPLE 11.3

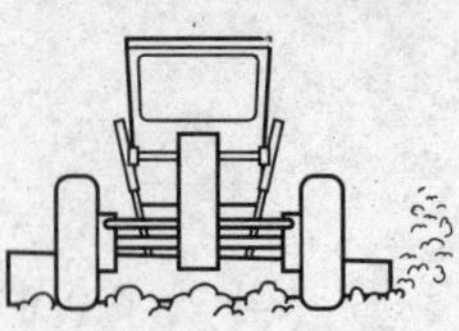

Application of Eq. [11.2] with distance expressed in feet

A haul road of 1,500 ft requires leveling and reshaping. Use an efficiency factor of 0.60. The work requires two passes in second gear (3.7 mph) and three passes in third gear (5.9 mph).

Total time

$$= \frac{2 \times 1{,}500 \text{ ft}}{88 \text{ fpm/mph} \times 3.7 \text{ mph} \times 0.60} + \frac{3 \times 1{,}500 \text{ ft}}{88 \text{ fpm/mph} \times 5.9 \text{ mph} \times 0.60}$$

$$= 15.4 \text{ min} + 14.4 \text{ min} = 29.8 \text{ min}$$

FINE GRADING PRODUCTION

When used for finishing and fine grading work such as the final shaping of a surface layer, production on a square yard per hour basis can be calculated using Eq. [11.3].

$$\text{Production (sy/hr)} = \frac{5{,}280 \times S \times W \times E}{9} \qquad \textbf{[11.3]}$$

where

S = speed of grader, in mph

W = effective width per grader pass, in feet

E = grader efficiency factor

EXAMPLE 11.4

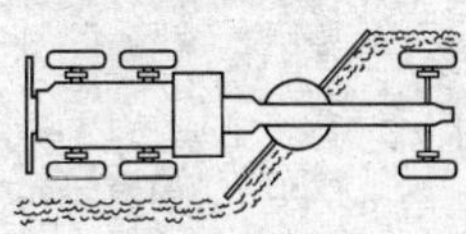

It is required to fine grade the subgrade of a roadway before proceeding to construct the subbase. Use an efficiency factor of 0.60. The grader will be operated in second gear (3.5 mph) for this work. The effective blade width per pass is 9 ft. Estimate the production rate for these conditions.

Using Eq. [11.3],

$$\text{Production (sy/hr)} = \frac{5{,}280 \times 3.5 \times 9 \times 0.6}{9}$$

$$\text{Production} = 11{,}088 \text{ sy/hr}$$

Note in this example, it was assumed that the fine grading could be accomplished with only one grader pass. This is not usually the case. Typically, multiple passes are required to meet the specified grade tolerance, which can be $\frac{1}{8}$ in. in 10 ft (10 mm/m) or less.

GRADER SAFETY

Work-zone safety efforts usually concentrate on keeping workers and construction equipment, and public separated. The National Institute of Occupational Safety and Health (NIOSH) notes that over half of work-zone fatalities are inside the work area and do not involve the public. Many work-zone fatalities are the result of works on foot in the work zone being hit by construction vehicles moving in reverse.

Sight Lines

"A 32-year-old construction surveyor died when backed over by a motor grader" [1]. Operators of large equipment including motor graders cannot see what is directly behind their machines. Therefore fleet managers are beginning to employ rear-vision cameras [3] on graders and other machines. These systems enable the operator to view on a small in-cab monitor what the wideview camera picks up behind the vehicle. One fleet manager reported that prior to employing the cameras they had accidents where motor graders actually backed up over the hood of vehicles.

Adjoining Activities

Many construction projects involve work along or work that crosses railroad tracks. Maybe the worst motor grader accident on record occurred in November 1961 when a grader was driven onto a rail-highway grade crossing immediately in front of an approaching train. The resulting collision and derailment of the Chicago, Rock Island, and Pacific passenger train caused injuries to 110 people, including 82 passengers and the driver of the motor grader.

Safety Rules

Listed next are specific safety rules for grader operations.

- All graders should be equipped with accident prevention signs and tags as per the OSHA standards 29 CFR 1926.145 "Specifications for accident prevention signs and tags," and should have slow-moving vehicle emblems per 1910.145(d)(10).
- When operating a grader slowly on a highway or roadway, display a red flag or flashing light on a staff at least 6 ft above the left rear wheel.
- Never allow other personnel to ride on the tandem, moldboard, or rear of the grader.
- Keep the moldboard angled well under the machine when not in use.

GRADALLS

GENERAL INFORMATION

The *gradall* is a utility machine that combines the operating features of the hoe, dragline, and motor grader (see Figs. 11.9 and 11.10). The full revolving superstructure of the unit can be mounted on either crawler tracks or wheels. The unit is designed as a versatile machine for both excavation and finishing work. Being designed as a multiuse machine affects production efficiency in respect to individual applications, when compared to a unit designed specifically for a particular application. The gradall will have lower production capability than those single-purpose machines.

FIGURE 11.9 A gradall on display at an equipment show.

FIGURE 11.10 Gradall working on a city street project.

The bucket of a gradall can be rotated (that is, the gradall's arm can rotate) 90° or more, enabling it to be effective in reaching restricted working areas and where special shaping of slopes is required. The three-part telescoping boom can be hydraulically extended or retracted to vary digging or shaping reach. It can exert breakout force either above or below ground level.

When used in a hoe application to excavate below the running gear, its production rate will be less than a hoe equipped with an equal-size bucket. Similarly, it can perform dragline-type tasks, but it has limited reach compared to a dragline. Because the machine provides the operator with positive hydraulic control of the bucket, it can be used as a finishing tool for fine-grading slopes and confined areas, tasks that would normally be grader work if there were no space constraints.

SAFETY

As with other excavators that rely on a rear counterweight to counterbalance the excavating arm, there is the danger of personnel being crushed between the counterweight and fixed worksite objects. Even with properly installed rear-view mirrors, the operator cannot always see personnel working behind the machine. Good safety training that makes operators aware of this danger is a management responsibility.

TRIMMERS

GENERAL INFORMATION

In the case of linear projects, there are a variety of large, highly specialized, but extremely versatile machines to do fine grade trimming. The result is better accuracy and greater production compared to fine grading with a grader. It has been reported that the production from one dual-lane trimmer is equal to that achievable with four to six graders. Another benefit is that automatic trimmers enable grade control to closer tolerances.

OPERATION

There are large four-track multilane trimmers and smaller three-track machines (see Fig. 11.11) for bridge approach work and parking lot applications. Both the large and small machines work using the same general control and mechanical systems.

Grade Control

Automatic trimmers use a sensor control system that establishes the elevation and cross-slope of its cutting teeth, based on input from either an arm riding on a grade wire or a ski riding on an established grade. A tightly stretched wire set

FIGURE 11.11 Small trimmer working off a wire grade line and loading a bottom-dump truck.

at a known elevation above the specified project grade and having a known offset distance from the alignment establishes a reference for the trimmer. A horizontal movable arm, mounted on the trimming machine and spring-tensioned, rides on the wire. This arm is connected to an electric switch and activates control signals for vertical movement of the trimmer. A similar vertical arm travels along the wire, steering the trimmer.

On projects where it is necessary to match an existing grade, a ski-type runner travels over the existing surface and controls the vertical elevation of the trimmer. In this case, the operator usually controls the alignment manually.

The frames of four-track multilane trimmers can provide a trimming width of 40 ft or more. These machines typically have four crawler-tracks, one at each corner. Each crawler mount has a vertical member that is hydraulically adjustable either manually or by the automatic control system.

Trimmer Assembly

The actual cutting of the grade is accomplished by a series of cutting teeth mounted on a full-width *cutter mandrel*. The cutter assembly is usually fixed at the ends, but it can break, allowing for the programming of a crown in the grade. An adjustable moldboard assembly follows the cutter mandrel and strikes off excess material. An *auger* for directing the spoil follows the moldboard. Typically, the spoil can be cast to either or both sides of the machine. On many large machines, the right and left augers are independently driven.

There are typically *wastegates* located at each end of the auger and adjacent to the centerline of the machine. These are for depositing excess material in a windrow on the grade.

A gathering and discharge conveyor system can be added to many trimmers. These attach to the rear for removing and reclaiming material from the finished grade. The material can be directly loaded into haul units with these systems (see Fig. 11.11) or deposited on the shoulder of the finished grade.

PRODUCTION

The large, full-width trimmers have operating speeds of about 30 fpm but this is dependent on the amount (depth of cut) of material being handled. In the case of the smaller single-lane trimmers, operating speed increases significantly. Some of these machines are rated at 128 fpm, but again speed is controlled by the amount of material being cut. As operating speed is increased there is usually a decrease in quality.

SUMMARY

The speed at which a grader can accomplish quality work depends largely on the skill of the operator and the type of material being handled. Consider the specific operation and the proper gear range for the task when calculating a grader production estimate. The number of passes, the anticipated speed, the distance of the operation, and an efficiency factor are the input variables for the production calculation.

If a gradall is used as an excavator, production can be estimated by using the general excavator production Eq. [9.1]. Trimmer production is a function of forward operating speed. Critical learning objectives include:

- An ability to estimate grader speed based on operational task
- An ability to calculate grader production

These objectives are the basis for the problems that follow.

PROBLEMS

11.1 Perform a Web search of accidents involving gradalls. Report on one of the accidents that discussed personnel being crush by the counterweight.

11.2 A grader will be used to maintain 3 miles of haul road. It is estimated that this work will require two passes in first gear, leveling the road requires two passes in second gear, and final shaping of the road requires three passes in fourth gear. Use an efficiency factor of 60%. Grader speeds are given in the chart. What is the time requirement for this task?

Forward gears	Maximum travel speed (mph)
First	2.2
Second	3.0
Third	4.3
Fourth	6.0

11.3 A grader will be used to maintain 1,800 ft of haul road. It is estimated that this work will require two passes in second gear and three passes in third gear. Use an efficiency factor of 60%. Grader speeds are given in the chart in Problem 11.2. What is the time requirement for this task?

11.4 A grader will be used to maintain 2.5 miles of haul road. It is estimated that this work will require three passes in first gear, leveling the road requires two passes in second gear, and final shaping of the road requires two passes in fourth gear. Use an efficiency factor of 60%. Grader speeds are given in the next chart. What is the time requirement for this task?

Forward gears	Maximum travel speed (mph)
First	2.3
Second	3.2
Third	4.4
Fourth	6.4

11.5 A grader will be used to maintain 1,400 ft of haul road. It is estimated that this work will require three passes in second gear and four passes in third gear. Use an efficiency factor of 60%. Grader speeds are given in the chart in Problem 11.4. What is the time requirement for this task?

11.6 It is required to fine grade the base of a roadway before proceeding to construct the pavement section. Use an efficiency factor of 0.65. The grader will be operated in second gear (3.4 mph) for this work. The effective blade width per pass is 8 ft. To meet the 10 mm/m specification it will be necessary to make four passes on an area. Estimate the production rate for these conditions.

REFERENCES

1. "A Construction Surveyor is Run Over by a Motor Grader That Was Backing Up" (2001). California FACE Report #01CA008, California FACE Program, California Department of Health Services, Occupational Health Branch, 1515 Clay St. Suite 1901, Oakland, CA.
2. *Caterpillar Performance Handbook*, Caterpillar Inc., Peoria, Ill. (published annually).
3. MacDonald, Chyck (2004). "NAPA Members Do the Extraordinary and the Ordinary to Improve Safety," *HMAT—Hot Mix Asphalt Technology*, May–June, pp. 16–17.

4. *Regulations* (*Standards—29 CFR*), Occupational Safety & Health Administration, 200 Constitution Avenue, NW, Washington, D.C.

WEBSITE RESOURCES

1. www.cat.com Caterpillar is a large manufacturer of construction and mining equipment. In the Products, Equipment section of the website can be found the specifications for Caterpillar manufactured graders.
2. www.cmicorp.com CMI Terex Corporation, P.O. Box 1985, Oklahoma City, OK 73101, is a major supplier of automated grading and paving machines.
3. www.deere.com Deere & Company is a worldwide corporation with a construction and forestry products division.
4. www.gradall.com Gradall, New Philadelphia, Ohio 44663 (in 1999, JLG Industries, Inc. acquired Gradall Industries, Inc.).
5. www.casece.com/products/products.asp?RL=NAE&ID=190&industryID=16 Case motor graders and Case construction equipment are marketed by CNH Global (in 1999, Case merged with New Holland to become CNH Global).
6. www.osha.gov/comp-links.html Occupational Safety & Health Administration, 200 Constitution Avenue, NW, Washington, D.C.

CHAPTER 12

Drilling Rock and Earth

The purposes for which drilling is performed vary a great deal from general applications to highly specialized work. The rates of drilling rock will vary with a number of factors such as the type of drill and bit size, hardness of the rock, depth of holes, drilling pattern, terrain, and time lost because of sequencing other operations. The first step in estimating drilling production is to make an assumption about the type of equipment that will be used and then to account for the total depth to be drilled, penetration rate, time for changing drill rods and bits, and time to clean the hole.

INTRODUCTION

The manner of achieving a hole in hard materials did not change from ancient times until about the middle of the nineteenth century. Across that long span of time, drilling was accomplished by manpower—a man swinging a hammer against a pointed drill. In 1861 the first practical mechanical drilling machine was employed in the Alps on the Mount Cenis tunnel work. The first pneumatic drill employed in the United States was at the eastern header of the Hoosac tunnel in western Massachusetts in June of 1866.

This chapter deals with the equipment and methods used by the construction and mining industries to drill holes in both rock and earth. Drilling may be performed to explore the types of materials to be encountered on a project (exploratory drilling), or it may involve production work such as drilling holes for loading explosive charges to blast rock. Other purposes would include holes for grouting or rock bolt stabilization work and borings for the placement of utility lines. Some jobs also require seep holes for drainage to reduce hydrostatic pressure. Rock (see Fig. 12.1) and earth drilling (see Fig. 12.2) will be treated separately in this chapter, although in some instances the same or similar equipment may be used for drilling both materials.

FIGURE 12.1 Drilling rock in a tunnel to load explosives.

Because the purposes for which drilling is performed vary a great deal from production work to highly specialized applications, it is necessary to select the equipment and methods best suited to the specific service. A contractor engaged in highway construction must usually drill rock under varying conditions; therefore, equipment that is suitable for variable applications would be selected. However, if equipment is needed to drill rock in a quarry where the material and conditions will not change, specialized equipment should be considered. In some instances, custom-made equipment designed for use on a single project may be justified.

GLOSSARY OF DRILLING TERMS

The following glossary defines the important terms used in describing drilling equipment and procedures. The dimension terminology frequently used for drilling is illustrated in Fig. 12.3.

Bit. The portion of a drilling tool that actually cuts through the rock or soil by a combination of crushing and shearing actions. There are many types of bits.

Burden. The horizontal distance from a rock face to the first row of drill holes or the distance between rows of drill holes.

Burden distance. The distance between the rock face and a row of holes or between adjacent rows of holes (see Fig. 12.3).

FIGURE 12.2 Drilling earth for a foundation pier.

FIGURE 12.3 Drilling and blasthole dimensional terminology.

Cuttings. These are the disintegrated rock particles caused by the action of the drill bit against the rock.

Drills

Abrasion. A drill that grinds rock into small particles through the abrasive effect of a bit that rotates in the hole.

Blasthole. A rotary drill consisting of a steel-pipe drill stem on the bottom of which is a roller bit. As the bit rotates it grinds the rock.

Churn. A percussion-type drill consisting of a long steel bit that is mechanically lifted and dropped to disintegrate the rock. It is used to drill deep holes, usually 6 in. in diameter or larger.

Core. A drill designed for obtaining samples of rock from a hole, usually for exploratory purposes. Diamond and shot drills are used for core drilling.

Diamond. A rotary abrasive-type drill whose bit consists of a metal matrix in which are embedded a large number of diamonds. As the drill rotates, the diamonds disintegrate the rock.

Downhole drill or downhole bit. The bit and the power system providing rotation and percussion are one unit suspended at the bottom of the drill steel.

Percussion. A drill that breaks rock into small particles by the impact from repeated blows. Compressed air or hydraulic fluids can power percussion drills.

Shot. A rotary abrasive-type drill whose bit consists of a section of steel pipe with a roughened surface at the bottom. As the bit is rotated under pressure, chilled-steel shot is supplied under the bit to accomplish the disintegration of the rock.

Shank or striker bar. A short piece of steel that attaches to the percussion drill piston for receiving the blow and transferring the energy to the drill steel (see Fig. 12.4).

Spacing. The distance between adjacent holes in the same row (see Fig. 12.3).

Subdrilling. The depth to which a blasthole will be drilled below the proposed final grade. This extra depth is necessary to ensure that rock breakage will occur completely to the required elevation.

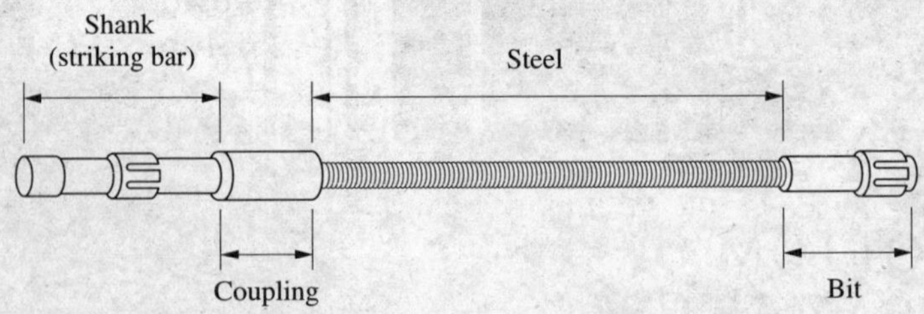

FIGURE 12.4 High-wear drilling items: shank, couplings, steel, and bit.

DRILL BITS

The *bit* is the essential part of a drill, as it is the part that must engage and disintegrate the rock. The success of a drilling operation depends on the ability of the bit to remain sharp under the impact of the drill. Many types and sizes of bits are available. Most bits are replaceable units that screw to the **drill steel** (rod). Bits are available in various sizes, shapes, and hardnesses.

drill steel (rod)
Steel rods that transmit the blow energy and drill rotation from the shank to the bit.

Carbide-Insert Bits

The actual drilling edges of the bit consist of a very hard metal, tungsten carbide, which is embedded in steel (Fig. 12.5). Although these bits are considerably more expensive than steel bits, the increased **drilling rate** and depth of hole obtained per bit provide an overall economy in drilling hard rock. Typical sizes for carbide-insert bits are from $1\frac{3}{8}$ to 5 in. in diameter.

drilling rate
The total feet of hole drilled per hour per drill.

Carbide-insert bits are available in four grades in order of increasing hardness (see Table 12.1). Susceptibility to breakage increases with hardness. However, abrasion resistance also increases. If excessive bit breakage occurs using a specific grade, a softer grade should be tried.

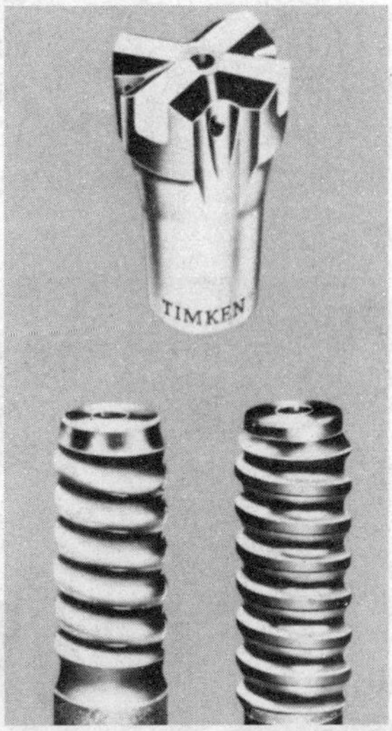

FIGURE 12.5 Carbide insert rock bit.
Source: The Timken Company.

Button Bits

Button bits can yield faster penetration rates in a wide range of drilling applications. Figure 12.6 illustrates several button bits. These bits are available in different cutting-face designs with a choice of insert grades. Most button bits are run to destruction and never reconditioned.

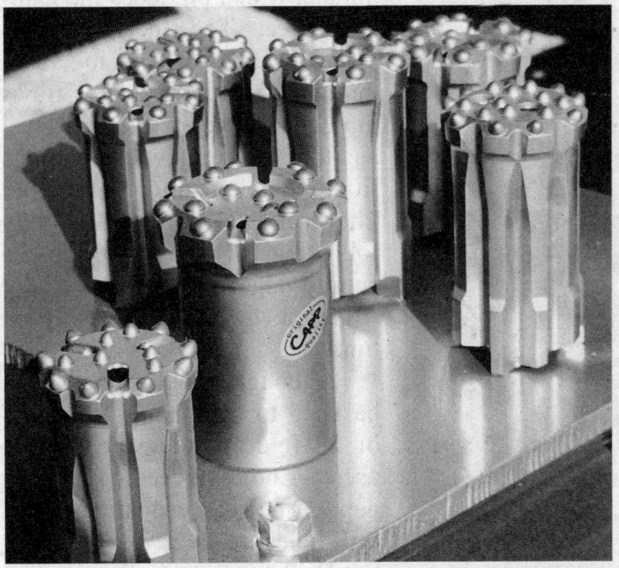

FIGURE 12.6 Button bits.

TABLE 12.1 Carbide-insert bit grades.

Grade	Abrasion resistance
Shock	Fair
Intermediate	Good
Wear	Excellent
Extra wear	Outstanding

ROCK DRILLS

There is an extensive range of rock drills designed for a variety of construction and mining applications. Drills use three methods to cause fragmentation of rock: (1) percussion, (2) rotary grinding, and (3) abrasion. Percussion and rotary drills are the primary production tools of construction work. Abrasion is used for special drilling applications.

Percussion Drills

Percussion drilling accomplishes disintegration of the rock by hammer impacts to the bit at the same time a rotating motion is applied to the bit. The percussion drill bit is literally hammered into the rock, it smashes the rock to pieces. These drills may vary in size from handheld units such as jackhammers to large crawler-mounted rigs.

Jackhammers The term jackhammer comes from the days when two men—a driver (John Henry being maybe the most famous) and a shaker, the poor fellow who held the steel rod, which was called a "Jack," between his legs—manually accomplished the drilling of rock. Jackhammers are handheld, air-operated percussion-type drills used primarily for drilling in a downward direction (Fig. 12.7). They are classified according to their weight, such as 45 or 55 lb. A complete drilling unit consists of a hammer, drill steel, and bit. As the compressed air flows through a hammer, it causes a piston to reciprocate at a speed up to 2,200 blows per minute, which produces the hammer effect. The energy of this piston is transmitted to a bit through the drill steel. Air flows through a hole in the drill steel and the bit to remove the cuttings from the hole and to cool the bit. The drill steel is rotated slightly following each blow so that the cutting edges of the bit will not strike at the same place each time.

FIGURE 12.7 Handheld air-operated jackhammer.

Although jackhammers can be used to drill holes in excess of 20 ft deep, they are seldom used for holes exceeding 10 ft in depth. The heavier hammers will drill holes up to $2\frac{1}{2}$ in. in diameter. Drill steel is usually supplied in 2-, 4-, 6-, and 8-ft lengths.

Drifters The larger and heavier percussion drills are mounted either on a traveling carriage (tractor) (see Fig. 12.1) or a frame. The combination drill and mount is known as a drifter. These tools can drill in the downward, horizontal, or upward directions. These drills are used extensively in rock excavation, mining, and tunneling. Either air or water can be used to remove the cuttings. Drifter drills are similar to jackhammers in operation, but they are larger and are used as mounted tools for downward, horizontal, or upward drilling.

The drifter's weight is usually sufficient to supply the necessary feed pressure for downward drilling. But when used for horizontal or upward drilling, a hand-operated screw, or a pneumatic or hydraulic piston, supplies the feed pressure.

Track-Mounted Drills To provide mobility, drills are commonly mounted on track carriages (Fig. 12.8). This special type of drifter is the workhorse drill on heavy/highway construction projects. Because of the mounting, these drills are often referred to as "air-track" drills, as originally these were drills that used pneumatic power for both driving the drill steel and tramming the machine. Typical air compressor requirements were around 750 cubic feet per minute (cfm). Today almost all of these small drills are hydraulically powered. Hydraulic motors tram the machine, and power the hammer, the rotation, and the feed of the

FIGURE 12.8 Track-mounted percussion drills.

drill. Hydraulic drills have a small, hydraulic-motor-driven air compressor onboard (175–250 cfm) for blowing the hole.

Hydraulic drills can normally achieve better penetration rates than air-driven models. Additionally, hydraulic drills use less fuel and are "gentler" on the drill steel and striking bar.

These small, self-contained drills are very productive tools because of their ability to move quickly between locations and because the hydraulically operated boom enables easy positioning of the drill. Holes can be drilled at any angle from under 15° back from the vertical to above the horizontal, ahead, or on either side of the unit. All operation, including tramming (travel), can be powered by compressed air. There are also hydraulic-powered models, but compressed air is still used for hole cleaning.

Figure 12.9 illustrates a drill mounted on a small hydraulic excavator boom. This is another special type of drifter. These drills are similar in sizes and capacities to the track-mounted drills.

FIGURE 12.9 Air-operated drill mounted on a hydraulic excavator boom.

Rotary Drills

With rotary drilling, the rock is ground away by applying a down pressure on the drill steel and bit and at the same time continuously rotating the bit in the hole (Fig. 12.10). To remove the rock cuttings and cool the bit, compressed air is constantly forced down the drill steel and through the bit during this process. Rotary or blasthole drills are self-propelled drills that can be mounted on a truck or on crawler tracks (Fig. 12.11). Rigs are available to drill holes to different diameters and to depths up to approximately 300 ft. These drills are suitable for drilling soft to medium rock, such as hard dolomite and limestone, but are not suitable for drilling the harder igneous rocks.

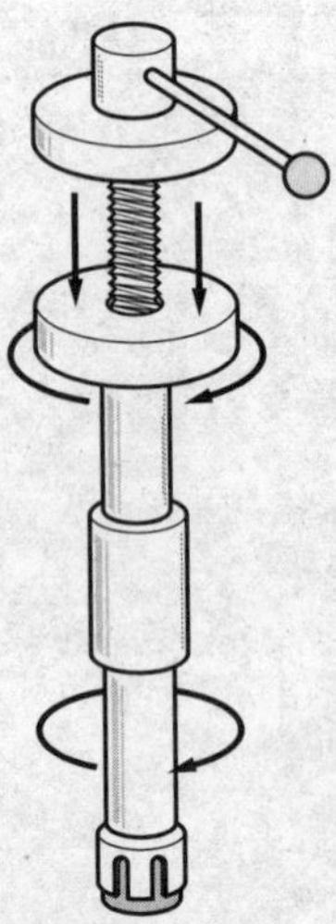

FIGURE 12.10 Rotary drilling mechanism.
Source: Ingersoll-Rand Co.

Reported project drilling speeds have varied from $1\frac{1}{2}$ ft/hr in dense, hard dolomite to 50 ft/hr in limestone. The speed of drilling is regulated by pressure delivered through a hydraulic feed.

FIGURE 12.11 Rotary (blasthole) drill.

Rotary-Percussion Drills

Rotary-percussion drilling combines the hard-hitting reciprocal action of the percussion drill with the turning-under-pressure action of the rotary drill (Fig. 12.12). Whereas the percussion drill only has a rotary action to reposition the bit's cutting edges, the rotation of this combination drill, with the bit under constant pressure, has demonstrated its ability to drill much faster than the regular percussion drill. Rotary-percussion drills require special carbide bits, with the carbide inserts set at a different angle than those used with standard carbide bits.

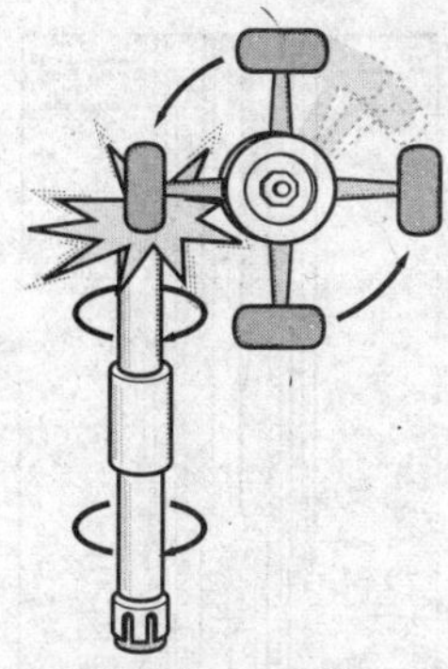

FIGURE 12.12 Rotary-percussion drilling mechanism. *Source: Ingersoll-Rand Co.*

Downhole Drills

When drilling deep holes, it can often be more efficient to place the driving mechanism in the hole with the bit (Fig. 12.13). This eliminates having to transmit the rotation and percussion forces through the drill steel. Typically these units are air-operated hammers (Fig 12.14). They can be operated underwater by maintaining a higher air pressure in the hammer than the pressure outside. Air or water can be used to clear cuttings from the hole. Standard sizes are available from 4 in. (102 mm) to 30 in. (762 mm).

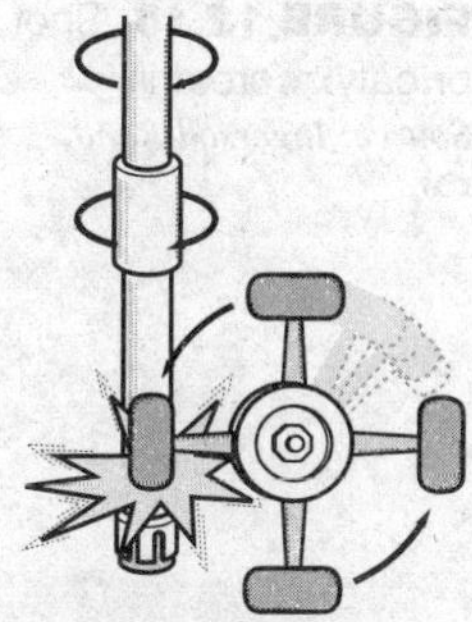

FIGURE 12.13 Downhole drilling mechanism. *Source: Ingersoll-Rand Co.*

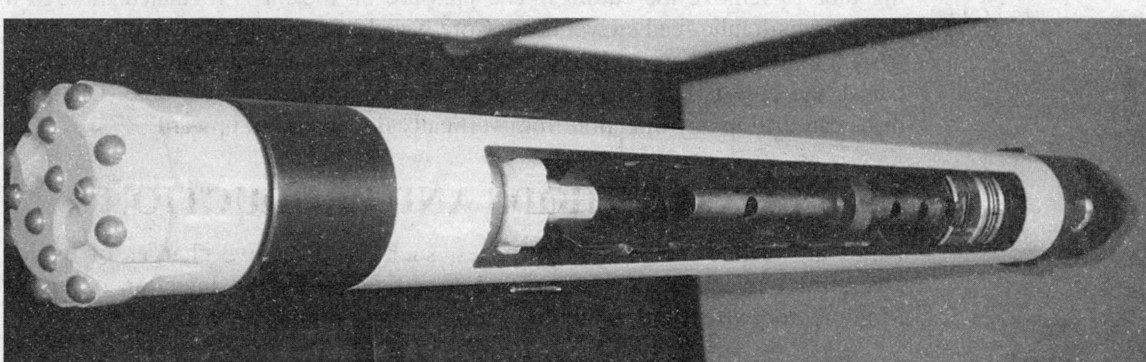

FIGURE 12.14 Cutaway of a downhole drill.

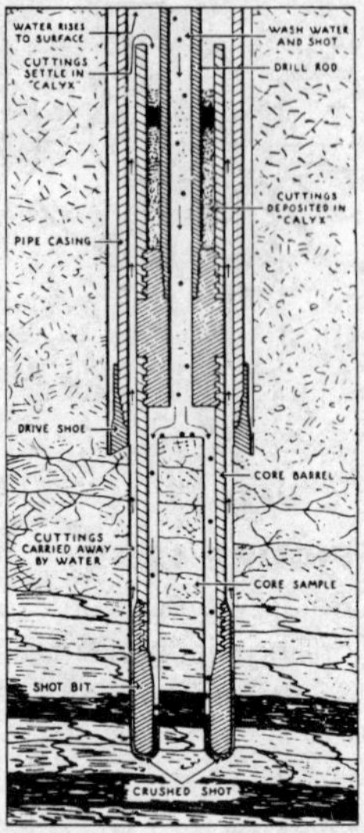

FIGURE 12.15 Shot or calyx core drill.
Source: Ingersoll-Rand Co.

Abrasion Drills

Rock can be drilled by the mechanical wearing away of its surface by way of frictional contact with a harder material. Two of the most common types of abrasion drills are shot drills and diamond drills.

Shot Drills A shot drill (Fig. 12.15) depends on the abrasive effect of chilled steel shot to penetrate the rock. The essential parts include a shot bit, core barrel, sludge barrel, drill rod, water pump, and power-driven rotation unit. The bit consists of a section of steel pipe, with a serrated lower end extending through the drill rod. The cutting medium, chilled steel shot, is fed with wash water. It lodges around and is partially embedded in a coring bit of slot steel. To be effective, the shot must be crushed during the coring. Precrushed shot, known under the trade name Calyxite, is often used in coring relatively soft rock. Water that is supplied through the drill rod forces the rock cuttings up around the outside of the drill, where they settle in a sludge barrel, to be removed when the entire unit is pulled from the hole. The flow of water must be carefully regulated so that it removes the cuttings but not the chilled shot. Periodically, it is necessary to break the core off and remove it from the hole so that drilling may proceed.

A primary purpose of small-hole shot drilling is to provide continuous cores for examination of structural information, as rock of any hardness may be drilled. The rate of drilling with a shot drill is relatively slow, sometimes less than 1 ft/hr, depending on the size of the drill and the hardness of the rock. Shot drilling can only be used for downward boring and is best suited for vertical holes.

Diamond Drills Used primarily for exploration drilling, diamond drilling offers the advantage that it can bore in any desired direction from vertically downward to upward. The Diamond Core Drill Manufacturers' Association lists four sizes as standard—$1\frac{1}{2}$, $1\frac{7}{8}$, $2\frac{3}{8}$, and 3 in. Larger sizes are available, but the investment in diamonds increases rapidly with an increase in size.

diamond bit
A bit whose cutting elements consist of diamonds embedded in a metal matrix.

A drilling rig consists of a **diamond bit**, a core barrel, a jointed driving tube, and a rotary head to supply the driving torque. Water is pumped through the driving tube to remove the cuttings. The pressure on the bit is regulated through a screw or hydraulic-feed swivel head. Core barrels are available in lengths varying from 5 to 15 ft. When the bit advances to a depth equal to the length of the core barrel, the core is broken off and the drill is removed from the hole. Diamond drills can drill in any direction from vertically downward to upward.

DRILLING METHODS AND PRODUCTION

Holes are drilled for various purposes, such as to receive charges of explosives, for exploration, or for ground modification by the injection of grout. Many factors affect the selection of equipment. Among these are

12.2

1. The purpose of the holes, such as blasting, exploration, or grout injection.
2. The nature of the terrain. Rough terrain may dictate track-mounted drills.

3. The required depth of holes.
4. The hardness of the rock.
5. The extent to which the rock formation is broken or fractured.
6. The size of the project (total linear amount of drilling).
7. The availability of water for drilling purposes. Lack of water favors dry drilling.
8. The size cores required for exploration. Small cores permit the use of diamond drills, whereas large cores suggest shot drills.

For small-diameter, shallow blastholes, especially on rough terrain where larger drills cannot operate, it is usually necessary to use track-mounted drills or even jackhammers, though the production rates will be low and the costs higher. For blastholes up to about 6 in. in diameter and up to about 50 ft deep, where machines can operate, the choice may be track-mounted, rotary-percussion drills. With each addition of drill rod, necessary to reach a greater depth, penetration production is reduced. There is approximately a 20% reduction in penetration production with the addition of the second rod and another 20% loss with the addition of the third rod. With the addition of a fourth rod the reduction is roughly another 10%.

For drilling holes from 6 to 12 in. in diameter, from 50 to 300 ft deep, the rotary or blasthole drill is usually the best choice, but the type of rock affects the drilling method and bit selection.

If cores up to 3 in. are desired, the diamond coring drill is satisfactory.

If intermediate-size cores (3 to 8 in. outside diameter) are desired, the choice will be between diamond and shot drills. A diamond drill will usually drill faster than a shot drill. Also, a diamond drill can drill holes in any direction, whereas a shot drill is limited to holes vertically down, or nearly so.

The Drilling Pattern

In the case of holes to be loaded with explosives, the blast design sets the drilling pattern. The pattern is the repeated distance between the drill holes in both directions, usually stated as "burden distance × spacing." This pattern will vary with the type of rock, the maximum rock breakage size permissible, and the depth of the blasted rock **face**. The blast design and drilling pattern in turn set the diameter of hole, depth of hole, and total linear footage of drilling requirements.

face
The approximately vertical surface extending upward from the floor of a pit to the level at which drilling is accomplished.

Drilling operations for rock excavation where the material will be used in an embankment fill must consider the project specifications concerning the maximum physical size of individual pieces placed in the fill. The blast design will be developed so as to produce rock sizes small enough to enable most of the blasted material to be handled by the excavator and to pass into the crusher opening without secondary blasting. While meeting either condition is possible, the cost of excess drilling and greater amounts of explosives to produce such material may be so high that the production of some oversized rocks will be cost effective. The oversized rocks will still have to be handled on an individual basis, possibly with a headache ball.

If small-diameter holes are spaced close together, the better distribution of the explosives will result in a more uniform rock breakage. However, if the added cost of drilling more holes (i.e., more drilling footage) exceeds the value of the benefits resulting from better breakage, the close spacing is not justified.

Large-diameter holes enable greater explosive loading per hole, making it possible to increase the spacing between holes, and thereby reducing the number of holes and the cost of drilling.

Drill Penetration Rates

The drill penetration rate will vary with a number of factors such as the type of drill and bit size, hardness of the rock, depth of holes, drilling pattern, terrain, and time lost because of sequencing other operations.

> If not determined by actual field tests, the prediction of a penetration rate for estimating purposes is guided by technical knowledge, as explained here, but that knowledge must be tempered by practical field experience—it is art as much as science.

The critical rock properties that affect penetration rate are

- Hardness
- Texture
- Tenacity
- Formation

Hardness A scientific definition of hardness is a measure of a material's resistance to localized plastic deformation. Many hardness tests involve indentation, with hardness reported as resistance to scratching. The main factors controlling rock hardness are porosity, grain size, and grain shape. In drilling practice, the term hardness is usually used in reference to the crystalline solid.

The *Moh* hardness classifications are based on the resistance of a smooth surface to abrasion—the ability of one mineral to scratch another—and rates hardness by a 10-point scale with talc rated as 1, the softest, and diamond rated as 10, the hardest, (see Table 12.2). Moh's scale does not represent an actual measured hardness. The relationship of rock hardness, using Moh's scale as an indicator, and drill penetration rate is presented in Table 12.3.

TABLE 12.2 Moh's scale for rock hardness.

Rock	Moh number	Scratch test
Diamond	10	Will scratch glass
Schist	5	Knife
Granite	4	Knife
Limestone	3	Copper coin
Potash	2	Fingernail
Gypsum	2	Fingernail

TABLE 12.3 Relationship between Moh's hardness rating and drill penetration rate.

Hardness	Drilling speed
1–2	Fast
3–4	Fast–medium
5	Medium
6–7	Slow–medium
8–9	Slow

The *Vickers* test, which yields a Vickers Hardness Number (VHN) provides a more scientific method of stating hardness (see Table 12.4). With the Vickers test, a pyramid-shaped diamond indenter with a 136° angle between opposite faces is pressed into the specimen with a force of from 1 to 50 kg. Both diagonals of the indent are measured with a microscope. The advantage of this test is that on homogeneous materials, the hardness value is considered not to be load-dependent.

Rocks are composed of mineral combinations. Therefore, a Vickers Hardness Number Rock (VHNR) that accounts for the hardness of individual minerals within the rock has been developed. The VHNR is a composite value determined by percentage weighing each mineral's hardness contribution to arrive at a single hardness value (Table 12.5). VHNR is also a good parameter for measuring the life of drill bits.

TABLE 12.4 Comparison of hardness classifications.

Mineral	Moh's number	Vickers Hardness Number
Diamond	10	1,600
Corundum	9	400
Corundum	9	400
Topaz	8	200
Quartz	7	100
Apatite	5	48
Fluorite	4	21
Calcite	3	9
Gypsum	2	3
Talc	1	1

TABLE 12.5 VHNR for a sample of gneiss.

Mineral	Percentage	VHN	Contribution to total hardness
Quartz	30	1,060	318
Plagioclase	63	800	504
Amphibole	2	600	12
Biotite	5	110	6
		VHNR	840

TABLE 12.6 Effect of rock tenacity on drilling speed.

Breaking characteristics	Drilling speed
Shatters	Fast
Brittle	Fast–medium
Shaving	Medium
Strong	Slow–medium
Malleable	Slow

Texture The term texture in relation to rock refers to the grain structure—degree of crystallinity, grain size, and shape and the geometric relationships between the grains. A loose-grained, structured rock (porous, cavities) drills fast. If the grains are large enough to be seen individually (granite), the rock will drill medium. Fine-grained rocks drill slowly.

Tenacity The term tenacity refers to the ability of a substance to resist breakage. Terms such as brittle and malleable are used to describe rock tenacity or breaking characteristics. The impact of rock tenacity on drilling rates is shown in Table 12.6.

Formation The structure of the rock mass, its formation, affects drilling speed. Solid rock masses tend to drill fast. If there are horizontal strata (layers), the rock should drill between medium and fast. Rock with dipping planes drills slow to medium. Dipping planes also make it difficult to maintain the drill hole alignment. All of these factors should be carefully considered when estimating drilling penetration rate without the benefit of actual field tests.

Historical drill penetration rates based on very broad rock-type classification are shown in Table 12.7. These rates should be used only as an order-of-magnitude guide. Actual project estimates need to be based on the results of drilling tests on the specific rock that will be encountered.

ESTIMATING DRILLING PRODUCTION

The first step in estimating drilling production is to make an assumption about the type of equipment that will be used. The type of rock to be drilled will guide that first assumption. Information useful in making such a decision is presented in Table 12.7. However, it must again be emphasized, the final decision on type of equipment should only be made after test drilling the specific formation. The drilling test should yield data on penetration rate based on bit size and type. Once a drill type and bit are selected, the format given in Fig. 12.16 can be used to estimate production.

Total Depth of Hole

Usually when drilling for blasting, it is necessary to subdrill below the desired finish grade of the excavation. This is because when explosives are fired in blastholes, rock breakage is not normally achieved to the full bottom-depth of

TABLE 12.7 Order-of-magnitude drilling production rates.

Bit size	Drill type Compressed air	Direct penetration rate		Estimated* production rate—good conditions	
		Granite (ft/hr)	Dolomite (ft/hr)	Granite (ft/hr)	Dolomite (ft/hr)
	Rotary-percussion				
$3\frac{1}{2}$	750 cfm @ 100 psi	65	125	35	55
$3\frac{1}{2}$	900 cfm @ 100 psi	85	175	40	65
	Downhole				
$4\frac{1}{2}$	600 cfm @ 250 psi	70	110	45	75
$6\frac{1}{2}$	900 cfm @ 350 psi	100	185	65	90
	Rotary				
$6\frac{1}{4}$	30,000 pulldown	NR	100	NR	65
$6\frac{3}{4}$	40,000 pulldown	75	120	30	75
$7\frac{7}{8}$	50,000 pulldown	95	150	45	85

NR—Not recommended.
*Estimated production rates are for ideal conditions, but they do account for all delays including blasting.

(1) Depth of hole:	(a) ________ ft face,	(b) ________ ft drill		
(2) Penetration rate:		________ ft/min		
(3) Drilling time:		________ min	(1b)/(2)	
(4) Change steel:		________ min		
(5) Blow hole:		________ min		
(6) Move to next hole:		________ min		
(7) Align steel:		________ min		
(8) Change bit:		________ min		
(9) Total time:		________ min		
(10) Operating rate:		________ ft/min	(1b)/(9)	
(11) Production efficiency:		________ min/hr		
(12) Hourly production:		________ ft/hr	(11)×(10)	

FIGURE 12.16 Format for estimating drilling production.

the hole. This extra drilling depth is dependent on the blasting design. Factors that impact blasthole drilling include hole diameter, hole spacing, pounds of explosive per cubic yard of rock, and firing sequence. Normally 2 or 3 ft of extra depth is required. For example, even though the depth to finish grade is 25 ft (Fig. 12.16, item 1a), it may be necessary to actually drill 28 ft (item 1b).

Penetrations Rate

Penetration rate (item 2 in Fig. 12.16) is the rate at which the drill penetrates the rock. This is usually developed by drilling test holes and is based on a specific bit size and type. In an attempt to make drilling penetration rate estimating more scientific, a drilling rate index (DRI) has been developed in Europe. DRI is an indirect method of predicting drillability. It is based on two laboratory tests, the brittleness value (S_{20}) and a Siever J-value (SJ). The S_{20} is the percentage by weight of rock, from the original sample, that passes through an 11.2-mm screen after pounding with a 14-kg impactor 20 times. The original sample is crushed, and screened rock that passes the 16-mm screen is retained on the 11.2-mm screen. The SJ is determined by miniature drilling with a certain bit geometry, bit weight, and number of rotations to measure depth of penetration.

As general guidance, a DRI of 65 indicates good drillability, and a value of 37 indicates poor drillability. Tests using standard rotary drills indicated that for a DRI of 65, the average penetration rate is 39 cm/min ± 4 cm, and for a DRI of 37 the average penetration rate is 25 cm/min ± 2 cm. The results between those two values are linear. Tests with percussion drills vary by manufacturer. As an example, for a DRI of 65, one manufacturer's drill might penetrate at a rate of 92 cm/min, while a second manufacturer's drill might achieve 192 cm/min. Therefore, based on the DRI there can be no general statement about penetration rates for percussion drills.

Drilling Time

Knowing the depth of hole and drilling rate allows the calculation of the time required for the drill to penetrate the rock (item 3 in Fig. 12.16).

Fixed Time

Fixed drilling time consists of changing steel (adding drill steel, and pullback and uncoupling of steel), blowing or cleaning the hole, moving the drill, and aligning the steel over the next hole (see Table 12.8).

TABLE 12.8 Fixed drilling times.

	Equipment and site condition			
Operation	Percussion drilling, clean bench (min)	Percussion drilling, uneven bench/terrain (min)	Downhole drilling, even bench (min)	Rotary drilling, even bench (min)
Add one steel	0.4	0.4	2.2	2.0
Pull one steel	0.6	0.6	2.5	2.8
Pull last steel	—	—	0.6	1.0
Move	1.4	2.2–2.9	6.0	7.0
Align				2.0

TABLE 12.9 Average weights for drill steel.

Size (in.)	Length (ft)	Weight (lb)
1.50	10	53
1.50	12	64
1.75	10	60
1.75	12	71

Changing Steel If the drilling depth is greater than the length of the drill steel, it will be necessary to add steel during the drilling process and to remove steel when coming out of the hole. For track-mounted percussion drills, the two standard steel lengths are 10 ft and 12 ft. Average weights for these steel lengths are given in Table 12.9. A driller needs about 0.5 min or less to add or remove a length of steel.

The single-pass capability (length of steel) of rotary drills varies considerably, in the range of 20 to 60 ft. The dimensions (diameter and length) of the steel increase with drillhole diameter. Nearly all large rigs have mechanized steel handling, and the time to change a piece of steel is approximately constant for all diameters, but varies with length. One study of the time to add and remove drill rods on rotary drills using 20-ft lengths of steel found it took an average of 1.1 min to add and 1.5 min to remove the rods.

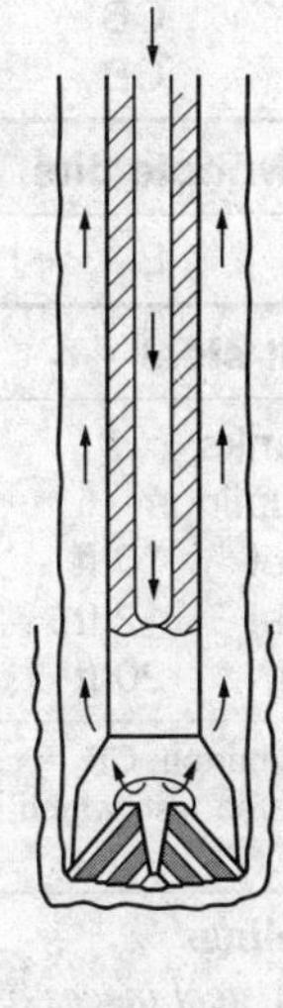

Blow Hole After the grinding of the rock is completed, it is good practice to blow out the hole (item 5 in Fig. 12.16) to ensure all cuttings are removed. However, some drillers prefer to simply drill an extra foot and pull the drill out without blowing the hole clean.

Move the Drill The time required to move (item 6 in Fig. 12.16) between drill hole locations is a function of the distance (blasting pattern) and terrain. Small track-mounted percussion drills can move at only 1 to 3 mph. Track-mounted rotary drills with their high masts can move at a maximum speed of about 2 mph. It should be remembered that hole spacing is often less than 20 ft, and the operator is maneuvering to place the drill over an exact spot, so travel speed is slow. There are some drill rigs equipped with global positioning system (GPS) technology that enables the operator to accurately place the drill over the prescribed hole location. If a high-mast drill must traverse unlevel ground, it may be necessary to lower the mast before moving and then raise it again when the move is accomplished. This will significantly lengthen the movement time.

Align Once over the drilling location the mast or steel must be aligned (item 7 in Fig. 12.16). In the case of a large high-mast drill, the entire machine is leveled using hydraulic jacks. This usually takes about 1 min.

TABLE 12.10a Igneous rock: Average life for drill bits and steel in feet.

Drill bits		Igneous rock				
(in.)	Type	High silica LA < 20 (rhyolite) (ft)	High silica 20 < LA < 50 (granite) (ft)	Medium silica LA < 50 (granite) (ft)	Low silica LA < 20 (basalt) (ft)	Low silica LA > 20 (diabase) (ft)
3	B	250	500	750	750	1,000
3	STD	NR	NR	NR	NR	750
$3\frac{1}{2}$	STD	NR	NR	NR	750	1,500
$3\frac{1}{2}$	HD	200	575	1,000	1,400	2,000
$3\frac{1}{2}$	B	550	1,200	2,500	2,700	3,200
4	B	750	1,500	2,800	3,000	3,500
Rotary bits						
5	ST	NR	NR	NR	NR	NR
$5\frac{7}{8}$	ST	NR	NR	NR	NR	NR
$6\frac{1}{4}$	ST	NR	NR	NR	NR	NR
$6\frac{3}{4}$	ST	NR	NR	NR	NR	800
$6\frac{3}{4}$	CB	NR	NR	1,500	2,000	4,000
$7\frac{7}{8}$	CB	NR	1,700	2,400	3,500	6,000
Downhole bits						
$6\frac{1}{2}$	B	500	1,000	1,800	2,200	3,000
Drill steel						
Shanks		2,500	4,500	5,800	5,850	6,000
Couplings		700	700	800	950	1,100
Steel	10 ft	1,450	1,500	1,600	1,650	2,200
Steel	12 ft	2,200	2,600	3,000	3,500	5,000
5 in.	20 ft	25,000	52,000	60,000	75,000	100,000

B = button, CB = carbide button, HD = heavy duty, ST = steel tooth, STD = standard, NR = not recommended, LA = The LA abrasion test, which is widely used as an indicator of the relative quality or competence of mineral aggregates.

couplings
Short steel pieces of pipe having interior threads (see Fig. 12.4). Couplings are used to hold pieces of drill steel together or to the shank. The percussion energy is transferred through the mated steel, not the coupling; therefore, the coupling must allow the two pieces of drill steel to butt together.

Change Bit

Drill bits, steel, and **couplings** are high-wear items, and the time required to replace or change each affects drilling production. High VHNR values (VHNR ≥ 800) indicate high abrasiveness and therefore increased bit wear, while low values (VHNR ≤ 300) mean prolonged bit life. A time allowance must be made for changing bits, shanks, couplings, and steel (item 8 in Fig. 12.16). Table 12.10 gives the average life of these high-wear items based on drill footage and type on rock. The Table 12.10 data enable calculation of bit change frequency.

TABLE 12.10b Metamorphic rock: Average life for drill bits and steel in feet.

Drill bits (in.)	Type	Metamorphic rock: High silica LA < 35 (quartzite) (ft)	Medium silica low mica (schist) (gneiss) (ft)	Medium silica high mica (schist) (gneiss) (ft)	Medium silica LA < 25 (metalatite) (ft)	Low silica LA > 45 (marble) (ft)
3	B	200	1,200	1,500	800	1,300
3	STD	NR	800	900	400	850
$3\frac{1}{2}$	STD	NR	1,300	1,700	850	1,600
$3\frac{1}{2}$	HD	NR	1,800	2,200	1,200	2,100
$3\frac{1}{2}$	B	450	3,000	3,500	2,000	3,300
4	B	600	3,300	3,800	2,300	3,700
Rotary bits						
5	ST	NR	NR	NR	NR	NR
$5\frac{7}{8}$	ST	NR	NR	NR	NR	1,200
$6\frac{1}{4}$	ST	NR	NR	NR	NR	2,000
$6\frac{3}{4}$	ST	NR	NR	750	NR	4,500
$6\frac{3}{4}$	CB	NR	3,700	4,200	1,200	9,000
$7\frac{7}{8}$	CB	NR	5,500	6,500	2,200	13,000
Downhole bits						
$6\frac{1}{2}$	B	500	2,700	3,200	1,500	4,500
Drill steel						
Shanks		5,000	5,700	6,200	5,550	5,800
Couplings		900	1,000	1,200	750	800
Steel	10 ft	1,700	2,100	2,300	1,500	1,600
Steel	12 ft	3,000	3,300	3,800	2,800	3,000
5 in.	20 ft	50,000	90,000	100,000	85,000	175,000

B = button, CB = carbide button, HD = heavy duty, ST = steel tooth, STD = standard, NR = not recommended, LA = The LA abrasion test, which is widely used as an indicator of the relative quality or competence of mineral aggregates.

Efficiency

Finally, as with all production estimating, the effect of job and management factors must be taken into account (item 11 in Fig. 12.16). Studies of drilling operations in Australia found that the actual portion of drilling time when the bit was actually drilling was from 70 to 75% of the total machine time. Bit activities in some cases required as much as 23% of the total time, while delays for maintenance, breakdown, and survey consumed the remaining.

Experienced drillers working on large projects with good equipment should be able to achieve a 45-min production hour. But as the Australian studies showed, good equipment and the proper bits are critical to realizing

TABLE 12.10c Sedimentary rock: Average life for drill bits and steel in feet.

Drill bits (in.)	Type	High silica fine grain (sandstone) (ft)	Medium silica coarse grain (sandstone) (ft)	Low silica fine grain (dolomite) (ft)	Low silica fine-med. grain (shale) (ft)	Low silica coarse grain (conglomerate) (ft)
3	B	800	1,200	1,300	2,000	1,800
3	STD	NR	850	900	1,500	1,200
$3\frac{1}{2}$	STD	NR	1,500	1,800	3,000	2,500
$3\frac{1}{2}$	HD	850	2,000	2,200	3,500	3,000
$3\frac{1}{2}$	B	2,000	3,100	3,500	4,500	4,000
4	B	2,500	3,500	2,000	5,000	4,800
Rotary bits						
5	ST	NR	1,000	NR	8,000	6,000
$5\frac{7}{8}$	ST	NR	2,500	NR	15,000	13,000
$6\frac{1}{4}$	ST	NR	4,000	4,000	18,000	14,000
$6\frac{3}{4}$	ST	500	6,000	8,000	20,000	15,000
$6\frac{3}{4}$	CB	2,000	8,000	10,000	25,000	20,000
$7\frac{7}{8}$	CB	3,000	10,000	15,000	25,000	20,000
Downhole bits						
$6\frac{1}{2}$	B	2,500	3,500	5,500	7,500	6,000
Drill steel						
Shanks		5,000	5,500	6,000	7,000	6,500
Couplings		1,000	1,200	1,500	2,000	1,750
Steel	10 ft	2,000	2,300	2,500	4,000	3,500
Steel	12 ft	4,500	5,000	6,000	7,500	7,000
5 in.	20 ft	65,000	250,000	200,000	300,000	250,000

B = button, CB = carbide button, HD = heavy duty, ST = steel tooth, STD = standard, NR = not recommended.

good production. If the situation is sporadic drilling with qualified people, a 40-min or lower production hour might be more appropriate. The estimator needs to consider the specific project requirements and the skill of the available labor pool before deciding on the appropriate production efficiency.

EXAMPLE 12.1

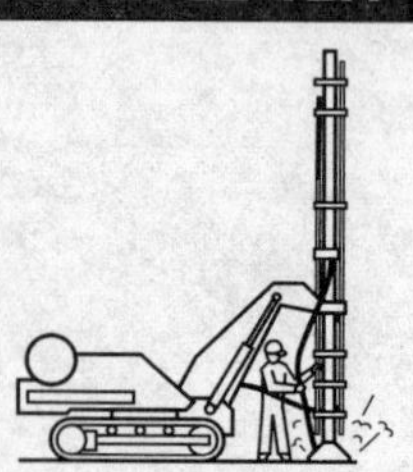

A project utilizing experienced drillers will require drilling and blasting of high-silica, fine-grained sandstone. From field drilling tests, it was determined that a penetration rate of 120 ft/hr can be achieved with a $3\frac{1}{2}$ HD bit on a rotary-percussion drill operating at 100 psi. The drills to be used take 10-ft steel. The blasting pattern will be a 10- × 10-ft grid with 2 ft of subdrilling required. On the average, the specified finish grade is 16 ft below the existing ground surface. Determine the drilling production based on a 45-min hour.

Using the Fig. 12.16 format,

(1)	Depth of hole	(a) 16 ft face	(b) 18 ft drill (16 ft + 2 ft)
(2)	Penetration:	2.00 ft/min	(120 ft ÷ 60 min)
(3)	Drilling time:	9.00 min	(18 ft ÷ 2 ft/min)
(4)	Change steel:	1.00 min	(1 add & 1 remove at 0.5 ea.)
(5)	Blow hole:	0.18 min	(about 0.1 min per 10 ft of hole)
(6)	Move 10 ft:	0.45 min	(10 ft at $\frac{1}{4}$ mph)
(7)	Align steel:	0.50 min	(not a high-mast drill)
(8)	Change bit:	0.09 min	$4\text{ min} \times \left(\dfrac{18\text{ ft per hole}}{850\text{-ft life (Table 12.10c)}}\right)$
(9)	Total time:	11.22 min	
(10)	Operating rate:	1.60 ft/min	(18 ft ÷ 11.22 min)
(11)	Production efficiency:	50 min/hr	
(12)	Hourly production:	72.0 ft/hr	[(45 min/hr) × (1.60 ft/min)]

High-Wear Items

From Table 12.10, the expected life of the high-wear items—the bit, shank, couplings, and steel (see Fig. 12.4)—of the drill can be found. A shank is the short piece of steel that is fixed to the drill chuck and transmits the drill's impact energy to the drill steel. Couplings are used to connect the sections of drill steel together.

EXAMPLE 12.2

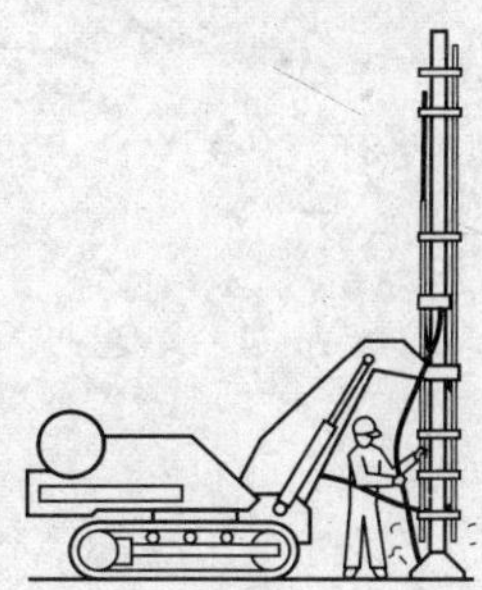

Assuming project conditions and drilling equipment are as described in Example 12.1, what is the expected life, in number of holes that can be completed, for each of the high-wear drilling items?

For an average hole depth of 18 ft, the following number of holes can be completed per each replacement.

	High-wear item	Average life (ft) (Table 12.10c)	Number of 18-ft holes in high-silica, fine-grained sandstone
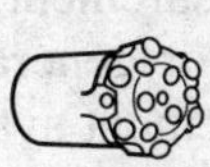	$3\frac{1}{2}$ HD bit	850	850/18 = 47
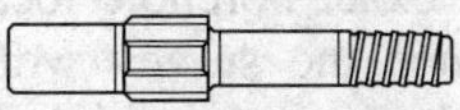	shanks	5,000	5,000/18 = 278
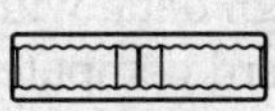	couplings	1,000	1,000/18 = 56
	steel	2,000	2,000/18 = 111

Rock Production

Drilling is only one part of the process of excavating rock, so when analyzing drilling production it is good to consider the cost and output in terms of cubic yards of rock excavated. With the 10 × 10 (ft) pattern and 18 ft of drilling used in Example 12.1, the rock yield is 59.3 cy.

$$\frac{10 \text{ ft} \times 10 \text{ ft} \times 16 \text{ ft}}{27 \text{ cf/cy}} = 59.3 \text{ cy}$$

Though the drilling depth was 18 ft, the excavation depth is only 16 ft. Therefore, each foot of *drill* hole produces 3.3 cy of bank measure (bcy) rock.

$$\frac{59.3 \text{ cy}}{18 \text{ ft}} = 3.3 \text{ cy/ft}$$

If the hourly drilling production is 80.2 ft, then the rock production is 80.2 ft × 3.3 cy/ft equaling 265 cy. This should be matched to the blasting production and to the loading and hauling production. For example, if the loading and hauling capability is 500 cy/hr, it will be necessary to employ two drills.

In calculating cost, it is good practice to make the analysis in terms of both feet of hole drilled and cy of rock produced. Considering only the high-wear items of Examples 12.1 and 12.2, if bits cost \$200 each, shanks \$105, couplings \$50, and a 10-ft steel \$210, what is the cost per cy of rock produced?

Bits	\$200 ÷ 850 ft = \$0.235/ft
Shanks	\$105 ÷ 5,000 ft = \$0.021/ft
Couplings	(2 × \$50) ÷ 1,000 ft = \$0.100/ft
Steel	(2 × \$210) ÷ 2,000 ft = \$0.210/ft
	\$0.566/ft

$$\text{or } \frac{\$0.566 \text{ per ft}}{3.3 \text{ cy per ft}} = \$0.172 \text{ per cy}$$

GPS AND COMPUTER MONITORING SYSTEMS

Current technology for monitoring drilling and blasting operations is progressing from using external and somewhat subjective measurement methods of performance, i.e., time sheets and foreman- or operator-generated drilling footage reports, to onboard data acquisition and management systems. Drills with GPS systems for positioning the drilling bit on the exact borehole location are becoming more prevalent. These systems reduce the survey work required for laying out drilling patterns. The drill pattern design data, with hole attributes of depth and angle, are loaded into the drill's onboard computer system, then based on the GPS positioning information, the drilling position is displayed on an LCD (liquid crystal display) screen, enabling the operator to position the drill without the aid of stakes or markings on the ground. There are also electronic sensor systems that provide production-monitoring capabilities including the recording and displaying of penetration rate, rotary torque and speed, bailing air pressure and volume, pulldown pressure, and hole depth. The bailing air pressure and volume are important to productivity and bit life, as pressure and volume of air are the critical elements for cooling the bit and ensuring that the cuttings are cleared from the face of the bit to prevent regrinding of material. These monitoring systems can provide detailed knowledge about the relative hardness of the strata being drilled. That strata information can be imported into blast design programs to improve explosive loading calculations.

Most of these computer-based monitoring systems are being developed and used in the mining industry. But that technology is finding its way into the construction industry. In the mining industry, there have been reports that this online drill-monitoring capability, which furnishes detailed information on rock quality and fracture zones, can provide a 20 to 25% saving in blasting cost. Accurate rock property data obtained by computerized drill sensors enables a computer-based explosives truck to generate loading designs on the bench. The onboard computer in the explosives truck can formulate and control the loading of the blasthole with the exact quality and type explosive within minutes after the drill has moved to the next blasthole.

DRILLING SOIL

Various types of equipment are used to drill holes in soil and are distinguished from rock-drilling equipment. In the construction and mining industries, holes are drilled into earth for many purposes, including, but not limited to

1. Obtaining soil samples for test purposes
2. Locating and evaluating deposits of aggregate suitable for mining
3. Installing cast-in-place piles or shafts for structural support
4. Enabling the driving of load-bearing piles into hard and tough formations
5. Providing wells for supplies of water or for deep drainage purposes
6. Providing shafts for ventilating mines, tunnels, and other underground facilities
7. Providing horizontal holes through embankments, such as those for the installation of utility conduits

Sizes and Depths of Holes Drilled into Soil

Most holes drilled into soil are produced by rotating bits or heads attached to the lower end of a shaft called a "kelly bar." An external motor (see Fig. 12.17) rotates this bar, which is supported by a truck, a tractor, a skid mount, or a crane.

The sizes of holes drilled into soil may vary from a few inches to more than 12 ft (3.7 m). Drills may be equipped with a device attached to the lower end of the drill shaft, known as an *underreamer,* that will enable a gradual

FIGURE 12.17 Crane-mounted auger-type drill.

increase in the diameter of the hole at the bottom. This enlargement enables a substantial increase in the bearing area under a shaft-type concrete footing.

casing
A steel pipe used as a lining to prevent a hole from collapsing into itself.

When drilling holes through unstable soils, such as mud, sand, or gravel, containing water, it may be necessary to use a **casing**. Sometimes the shaft is drilled entirely through the unstable soil and then a temporary steel casing is installed in the hole to eliminate groundwater and caving. An alternative method is to add sections to the casing as drilling progresses until the hole is completed through the full depth of unstable soil. The remainder of the hole can often then be completed without additional casing. When the hole is filled with concrete, the casing is pulled before the concrete sets.

This type of foundation has been used extensively in areas whose soils are subject to changes in moisture content to considerable depth. By placing the footings below the zone of moisture change, the effects of soil movements due to changes in moisture are minimized.

REMOVAL OF CUTTINGS

Several methods are used to remove the cuttings from the holes in soils. One method of removing the cuttings is to attach a continuous auger to the drill head. The drill head is the actual cutting tool at the bottom of the drill stem. The auger extends from the drill head to above the surface of the ground (see Fig. 12.18). As the drill shaft and the auger rotate, the earth is forced to the top of the hole, where it is removed and wasted. However, the depth of a hole for which this method can be used is limited by the diameter of the hole, the class of soil, and the moisture content of the soil.

Another method of removing the cuttings is to attach the drill head to only a section of the auger. When the auger section is filled with cuttings, it is raised above the surface of the ground and rotated rapidly in reverse to free it of the cuttings.

FIGURE 12.18 Crane-mounted continuous auger-type drill.

A third method of removing the cuttings is to use a combination of a drill head with a cylindrical bucket, whose diameter is the same as the diameter of the hole. As the bucket is rotated, steel cutting blades attached to the bottom of the bucket force the cuttings up and into the bucket. When the bucket is filled, it is raised to the surface of the ground and emptied.

A fourth method of removing the cuttings is to force air and water through the hollow kelly bar and drill shaft to the bottom of the hole and then upward around the drill shaft. The air or water carries the cuttings to the surface of the ground for disposal.

TRENCHLESS TECHNOLOGY

The term trenchless technology is used in reference to the numerous underground construction methods that eliminate or minimize surface disruption. Trenchless methods include directional drilling, horizontal boring, microtunneling, pipe bursting/splitting, pipe jacking, pipe ramming, and vacuum excavation.

Directional Drilling

The technology of directional drilling is not new. The technology was perfected in the oil industry, and that industry has used it with vertical drilling for decades. During the last 20 yr, directional-drilling technology has been adapted to the construction industry for horizontal work such as installing utilities. The technology enables the installation of underground utilities without the necessity of disrupting surface facilities that are already in place and being used. Directional drilling and horizontal boring are similar in job function but differ in procedure and technology. Directional drilling converts the old oil industry technology and uses it for utility work. But there can be problems with alignment accuracy in soils containing rock. Horizontal boring is extremely accurate; however, there is a requirement for boring pits, shoring boxes, and special track installations. Horizontal boring is also a more labor-intensive activity than directional drilling. The cost of directional drilling is typically less than that of horizontal boring. In the case of a 24-in. casing, the price difference can be as much as 25% per linear foot.

Project soil conditions are a critical factor in directional drilling work, with the most important consideration being the presence of rock. Directional-drilling equipment capable of boring through rock is available. However, if rock is expected anywhere along the path of the bore, even for only a few feet in hundreds of feet of total length, the whole bore must be treated as rock drilling. This is because it is not practical to drill part of the distance using standard equipment and to then have to pull that equipment out of the bore and replace it with a special rock-drilling head. Additionally, having to change the drilling head a second time once the rock portion of the bore is completed would further complicate the procedure.

Standard directional-drilling equipment cannot effectively drill through rock. When standard drilling equipment is used and boulders or cobble are encountered, the procedure for advancing the bore is to direct the drilling head to follow a path either above or below the obstruction. Such a procedure results in a bore that does not conform to a uniform grade, a standard requirement for sewer work.

Horizontal directional drilling employs a surface-launched drilling rig (Fig. 12.19) in a two-stage process. In the first stage a small pilot hole is drilled. To drive the pilot bore, a flexible drill steel is used to push a steerable drill head through the ground. The steering control is done manually with steering rods housed within the drill steel, and lasers are used to verify line and grade. A transmitter located behind the drill head enables the head to be tracked by a walkover system on the ground surface. With this system, precise and accurate pilot holes can be achieved when soil conditions are favorable.

FIGURE 12.19 Directional drilling rig.

Typically with directional drilling, the diameter of the casing to be placed is greater than the diameter of the pilot hole; in which case, a second stage is necessary. This second stage is a reaming process. The pilot hole can be reamed simultaneously with pulling the casing (Fig. 12.20). The casing is attached to a puller and swivel, which, in turn, is attached behind the reamer.

FIGURE 12.20 Pipe being drawn directly behind the reamer in a directional drilling second stage.

This arrangement enables the pipe to follow directly behind the reamer as it cuts the soil. A film of pressure grout is pumped into the pilot hole through the reamer. When dry, this grout forms something similar to soil cement due to the mixing with the dirt outside the casing. This grout ensures that there are zero voids outside the casing once it is in place. The size of the reamer can vary considerably, and there are different types of reamers available for different soil conditions, including rock.

Horizontal Boring

A rigid horizontal drilling process where the boring machine sits on guide rails in a boring pit is known as horizontal boring. The pit is excavated to a depth so that the drill, when placed on its tracks, is aimed on the alignment required of the desired hole. As rams push the drill, sections of the sleeve (casing) pipe are pushed into the bore directly behind the drill head. The sleeve pipe should follow no more than 1 in. behind the drill head. After each section of sleeve is pushed into the bore, the pusher rams are retracted and a new sleeve section is joined to those already in the bore. This process is repeated until the bore is complete. Pressure grout is fed through the drill bit to encase the sleeve pipe. The dirt tailings are augured out of the sleeve and back to the pit where they are removed.

As the boring advances, the machine automatically maintains a forward thrust on the casing and the auger. These machines are powered hydraulically, by air, or by electric motors. In the case of either directional drilling or horizontal boring, alignment of the utility pipe in the sleeve is controlled with skates, which serve as spacers that hold the utility pipe at the specified invert. The skates are usually pressure-treated redwood and are located in groups of three, spaced equal distances around the circumference of the utility pipe. They are bound to the pipe with metal straps. Their thickness is dictated by the required final invert of the utility pipe. Spacing of the skates is typically on 15- or 20-ft centers along the utility pipe. Once the utility pipe is braced in place by the skates, sand is blown into the void between the sleeve and the pipe.

With water lines, bends are not a significant problem as the water flows under pressure. As long as a connection can be made to the existing lines, the service will be acceptable. With sewer lines, however, adherence to grade and invert specifications is critical, as sewers usually work by gravity flow. Therefore, understanding the soil properties of a project site and the limitations of the two drilling methods is crucial to successful placement of utility lines.

Microtunneling

In the early 1970s, the Japanese developed microtunneling techniques to replace open sewers in urban areas with underground gravity sewers. The first microtunneling project in the United States occurred in southern Florida in 1984. Microtunneling uses a remotely controlled microtunnel boring machine (MTBM) combined with the pipe-jacking technique to directly install product pipelines underground in a single pass. In the United States, microtunneling has been used to install pipe from 12 in. to 12 ft in diameter. Therefore, the definition for microtunneling does not necessarily refer to tunnel diameter. It is a tunneling process where the workforce does not routinely work in the tunnel.

Pipe Bursting/Splitting

With pipe bursting or splitting technology, at the same time that an existing pipe is purposely broken, by mechanically applying a force from within, a new replacement pipe, of the same or larger diameter, is drawn in behind the bursting tool. To make space for the new pipe the fragments of the old pipe are forced into the surrounding ground. The bursting device may be based on a pneumatic "impact moling" tool, which converts forward thrust into a radial bursting force, or a hydraulic device inserted into the pipe and expanded to exert direct radial force. In the case of pipe splitting instead of bursting, the old pipe is split longitudinally.

Pipe Jacking

Pipe jacking is a nonguided method for directly installing pipes behind a shield machine at the same time as excavation is taking place within the shield. To install a pipeline using this technique, thrust and reception shafts or pits are constructed, usually at manhole positions. Then the pipe is hydraulically jacked into the ground from the thrust (launch) shaft. The excavation of the material ahead of the pipe can be by a small tunnel-boring machine or performed manually. The alignment of the machine is very important, and the longer the jacking shaft, the more accurate the borehole.

A thrust wall is constructed to provide a reaction against which to jack. In poor ground, piling or some special arrangements may have to be employed to increase the reaction capability of the thrust wall. Where there is insufficient depth to construct a normal thrust wall, for example, through embankments, the jacking reaction has to be resisted by means of a structural framework constructed at ground level. This may involve ground anchors or other such methods for transferring horizontal loads.

Pipe Ramming

Pipe ramming is a nonsteerable method of forming a bore by driving a steel casing. Usually the steel casing is open-ended. In a pipe-ramming operation, a ramming tool attached to the rear of a steel pipe drives the pipe into the ground with repeated blows. The method typically requires excavation of two pits. Before ramming, both the pipe and the ramming tool are placed into the launch pit and aligned in the desired direction. Alternatively, the ramming can be launched without a pit, if started at the side of a slope. The soil may be removed from an open-ended casing by augering, jetting, or compressed air. In contrast to pipe jacking, thrust plates or blocks in the insertion pit are not required.

The leading edge of the pipe is almost always open, but when smaller pipes are being installed, it may be closed. The shape of the leading edge provides a small overcut to reduce friction between the pipe and soil and to direct the soil into the pipe interior instead of compacting it outside the pipe. These objectives are usually achieved by attaching a soil-cutting shoe or special bands to the pipe.

Pipe ramming is a viable method for installing steel pipes and casings over distances usually up to 150 ft and having diameters up to 55 in. The method is most useful for shallow installations under railways and roads. The majority of installations are horizontal, although the method can be applied for vertical installations as well.

SAFETY

A 22-year-old rock drill operator with 1 yr of experience was fatally injured while drilling when his clothing became entangled in the rotating drill steel while he was trying to free a dust suction hose. Since 1984, five operators of machine-mounted rock drills have been killed as a result of entanglement in rotating drill steel. Why were these people not instructed in the dangers of their work and how to work safely?

The foundation of an effective safety program rests on a commitment from management. Safety-conscious companies build their safety programs on

- Having all new employees complete a job orientation
- Holding regular safety meetings
- Developing job-specific safety plans
- Investigating all accidents to find the cause—not the blame—to prevent future occurrences

In the case of drilling operations there should be

1. Written safe operating procedures and training programs. The procedures should be regularly reviewed with drill operators. Standard safe operating procedures should emphasize the importance of shutting down the drills when performing tasks near the rotating steel.
2. Written policies for the type of clothing allowed and methods to secure clothing when working around drilling equipment. Never wear loose/bulky clothing around drilling or other equipment with rotating or exposed moving parts.
3. Safe routing of hoses and cables on and around drilling equipment. These must eliminate the necessity of handling hoses or cables in close proximity to rotating/moving equipment.
4. Positioning of drill controls to provide safe performance in all operating conditions. As necessary, the controls should be designed to adapt to the drilling conditions and the operator's size, height, and reach.
5. Provisions for emergency stop/shut-off switches.
6. Requirements to wear proper personal protective equipment (safety glasses, hardhats, boots, and gloves) when operating drilling equipment.

SUMMARY

Drilling may be performed to explore the types of materials to be encountered on a project (exploratory drilling), or it may involve production work such as drilling holes for explosive charges. Other purposes include holes for grouting or rock bolt stabilization work and boring for the placement of utility lines. The rate at which a drill will penetrate rock is a function of the rock, the drilling method, and the size and type of bit. The four critical rock properties that affect penetration rate are (1) hardness, (2) texture, (3) tenacity, and (4) formation.

Most holes drilled in earth use rotating bits or heads attached to the lower end of a shaft called a kelly bar. During the last 20 yr, directional-drilling technology has been adapted to the construction industry for horizontal work such as installing utilities. The technology enables the installation of underground utilities without the necessity of disrupting surface facilities that are already in place and being used. Critical learning objectives include:

- An understanding of the different types of drilling equipment
- An appreciation of the factors that influence the rates of drilling rock
- An ability to prepare a rock-drilling production estimate
- A basic understanding of the new directional-drilling technology that is available

These objectives are the basis for the problems that follow.

PROBLEMS

12.1 A highway cut is being constructed through low-mica schist. Drilling tests indicate that air-track drills with $3\frac{1}{2}$-in. STD bits can penetrate 1.5 ft/min. The drills use 10-ft steel. The blasting pattern will be 8 × 12 ft with 1 ft of subdrilling. The average depth to finish grade is 14 ft. What is the amount drilling time (no fixed time) required to penetrate to the required depth? (10.0 min)

12.2 For the conditions stated in Problem 12.1 and considering that it takes 4 min to change a bit, what portion of total drilling time per hole should be allowed for changing bits? ([4 min × (15 ÷ 1,300′)] Table 12.10b; therefore change bit: 0.05 min)

12.3 A project utilizing experienced drillers will require drilling and blasting of medium-silica granite. No field drilling tests were conducted. It is proposed to use a $6\frac{1}{2}$ B downhole drill at 350 psi. The drills to be used have a single-pass capability of 20 ft and take steel in 20-ft lengths. The blasting pattern will be a 10- × 12-ft grid with 3 ft of subdrilling. On the average, the specified finish grade is 30 ft below the existing ground surface. Determine the drilling production, assuming it takes 30 min to change a $6\frac{1}{2}$-in. downhole hammer. (66.7 ft/hr)

12.4 Deep Hole Drilling, Inc. only uses experienced drillers. Daddy Deep Hole is trying to estimate the drilling production for a new project in medium-silica granite. No field drilling tests were conducted. It is proposed to use a $6\frac{1}{2}$ B downhole drill at 350 psi. The drills to be used take steel in 20-ft lengths. The blasting pattern will be a 14- × 16-ft grid with 2 ft of subdrilling required. On the average, the specified finish grade is 35 ft below the existing ground surface. Assume it takes 30 min to change a $6\frac{1}{2}$-in. downhole hammer and a 50-min hour efficiency. Check Daddy's drilling production number of 66.2 ft/hr.

12.5 A project in medium-silica sandstone is being investigated. Field drilling tests proved that air-track drills with $3\frac{1}{2}$-in. HD bits can achieve a penetration rate of 2.5 ft/min. The drills use 12-ft steel. The blasting pattern will be 6 × 8 ft with 2 ft of subdrilling. The average depth to finish grade is 8 ft. The drilling production must match that of the loading and hauling, which is 210 bcy/hr. Assuming a 50-min production hour, how many drills will be required?

12.6 A highway cut through shale is being constructed. Drilling tests indicate that air-track drills with $3\frac{1}{2}$-in. STD bits can penetrate 3.5 ft/min. The drills use 10-ft steel. The blasting pattern will be 8 × 10 ft with 2 ft of subdrilling. The average depth to finish grade is 15 ft. The project labor scale is \$16 per hour for drillers. The air-track drill and compressor cost \$46 per hour; bits are \$185 each; shanks, \$115; couplings, \$54; and 10-ft steel, \$200. Because of very difficult job conditions, use a 45-min hour to calculate production. What is the hourly drilling production in both feet and bcy? What is the cost of the high-wear items? What is the bcy cost including equipment, labor, and the high-wear items? (109.2 ft/hr; 284 bcy/hr; \$0.232/ft; \$0.307/bcy)

12.7 The president of "Low Bid" Construction Company has just visited one of her jobs. At the project she found most of the rock trucks parked. When she questioned the foreman as to why this new equipment with the capability of hauling 600 cy/hr was not working, the foreman said the drilling could not keep up. The president has told you to solve the problem immediately.

The project is in a dolomite formation. Company drilling experience in this area indicates that compressed-air rotary-percussion drills with $3\frac{1}{2}$-in. button bits can average 1.5 ft of penetration per minute. Drilling efficiency is typically a 50-min hour. These drills use 12-ft drill steel. The best blasting pattern is 10 × 10 ft with 2 ft of subdrilling. The average excavation depth is 8 ft. How many drills will you recommend for use on the job?

12.8 Buffet Inc. has determined that there is an untapped party market at the North Pole. Intensive marketing studies indicate that a "Margaritaville Bar & Grille" would be very successful during the long, dark arctic winters. Joe Cool, now living in Key West, has been employed to manage the construction of this project and has employed you to analyze the foundation-drilling program.

The plan is to construct the establishment 20 ft below the surface of the ice. This will require an extra 2 ft of drilling to ensure that the final floor can be excavated to the required depth after blasting. From field testing with a 90-psi rotary-percussion drill, it has been determined that a direct drilling rate, through the ice, of 175 ft/hr can be achieved with a $3\frac{1}{2}$-HD bit. This drill uses 12-ft drill steel. Studies have indicated that ice acts similar to low silica, LA > 20, diabase. The blasting pattern will be on a 12- × 12-ft grid. The cool climate will affect production.

Because of the high freight cost to the site, $3\frac{1}{2}$- bits cost \$500 each; shanks, \$400; couplings, \$150; and steel, \$600. All prices are FOB (free on board) at the North Pole. What drilling production can be expected and what will be the cost on a per cy basis for the drill's high-wear items that you should tell Mr. Cool to use in the project's construction cost estimate?

12.9 A project in medium-silica granite is being investigated. Field drilling tests proved that air-track drills with $3\frac{1}{2}$-in. B bits can achieve a penetration rate of 1.1 ft/min. The drills use 10-ft steel. The blasting pattern will be 8 × 8 ft with 2 ft of subdrilling. The average depth to finish grade is 16 ft. The drilling production must match that of the loading and hauling operations, which is 300 bcy/hr. Assuming a 40-min production hour drilling, how many drill units will be required?

12.10 The president of "Rock Hog" Construction Company has just visited one of her jobs. At the project, she found most of the rock trucks parked. When she questioned the foreman as to why this new equipment with the capability of hauling 250 cy/hr was not working, the foreman said the drilling could not keep up. The president has told you to solve the problem.

The project is in a high-silica granite formation. Company drilling experience in this area indicates that compressed-air rotary-percussion drills with 4-in. button bits can average 1.3 ft of penetration per minute. Drilling efficiency is typically a 40-min hour. These drills use 12-ft drill steel. The best blasting pattern is 10 × 12 ft with 2 ft of subdrilling. The average excavation depth is 14 ft. How many drills will you recommend for use on the job?

REFERENCES

1. *Advancements in Trenchless Technology*, Samuel T. Ariaratnam, Editor (1999). Proceedings of the Joint Symposium of the GSE and The Northwest Chapter of the NASTT, Edmonton, Alberta, Canada, April.
2. Ariaratnam, Samuel T., and Erez N. Allouche (2000). "Suggested Practices for Installations Using Horizontal Directional Drilling." *Practice Periodical on Structural Design and Construction*, ASCE, Vol. 5, No. 4, pp. 142–149, November.
3. Jimeno, Carlos Lopez, Emilio Lopez Jimeno, Francisco Javier Ayala Carcedo, and Yvonne Visser de Ramiro (Translator) (1995). *Drilling and Blasting of Rocks*, Ashgate Publishing Ltd., Aldershot, Hampshire, GU11 3HR United Kingdom. (English translation, original language Spanish.)

4. *Rock Blasting Terms and Symbols: A Dictionary of Symbols and Terms in Rock Blasting and Related Areas like Drilling, Mining and Rock Mechanics*, Agne Rustan, Editor (1998). Ashgate Publishing Ltd., Aldershot, Hampshire, GU11 3HR United Kingdom.
5. *Standard Construction Guidelines for Microtunneling*, ASCE Standard No. 36-01, American Society of Civil Engineers, Reston, VA.
6. *Trenchless Pipeline Projects—Practical Applications*, Lynn E. Osborn, Editor (1997). Proceedings of the Conference sponsored by the Pipeline Division, ASCE, Reston, VA.
7. Tulloss, Michael D., and Cliff J. Schexnayder (1999). "Horizontal Directional Boring at Lewis Prison Complex," *Practice Periodical on Structural Design and Construction*, ASCE, Vol. 4, No. 3, pp. 119, 120, August.

WEBSITE RESOURCES

1. www.atlascopco.se/rde Atlas Copco is a manufacturer of rock drills and drill rigs.
2. www.halcodrilling.com Halco Group manufactures a comprehensive range of downhole hammers and drill bits.
3. www.pitandquarry.com/pitandquarry The website for *Pit & Quarry* magazine, which is an information source for aggregate producers.
4. www.smc.sandvik.com Sandvik Mining and Construction is a supplier of mechanized underground and surface drilling equipment and tools.
5. www.trenchlessonline.com *Trenchless Technology* magazine online features news based on field reports and technology updates from engineers, consultants, contractors, and manufacturers.

CHAPTER 13

Blasting Rock

When there is a requirement to remove rock, blasting should be considered as it is almost always more economical than mechanical excavation. In the blasting of rock, breakage results from the sustained gas pressure buildup in the borehole caused by the explosion. Every blast must be designed to meet the existing conditions of the rock formation and overburden, and to produce the desired final result. The blast design will affect such considerations as type of equipment and the bucket fill factor of the excavators. The prevention of blasting accidents depends on careful planning and faithful observation of proper blasting practices.

BLASTING

Drilling and blasting is the most frequently used rock excavation technique (Figs. 13.1 and 13.2). It is performed to break rock so that it may be quarried for processing in an aggregate production operation, or to excavate a right-of-way. Blasting is accomplished by discharging an explosive that either has been placed in an unconfined manner or is confined, as in a borehole. Two forms of energy are released when high explosives are detonated: (1) shock and (2) gas. An unconfined charge works by shock energy, whereas a confined charge has a high gas energy output.

The first explosive, gunpowder, was developed in China in the thirteenth century. The first recorded use of an explosive for mining was in 1627 at the Royal Mines of Schemnits at Ober-Biberstollen, Hungary. Holes were drilled in the rock with handheld drills and sledgehammers. Black powder was then loaded into the holes and the charge was ignited, fracturing and displacing the rock. After the days of blasting with black powder were banished by nitroglycerin, there have been steady developments in explosives, detonating and delaying techniques, and in our understanding of the mechanics of rock breakage by explosives.

FIGURE 13.1 Firing a blast on a dam project to excavate rock.

FIGURE 13.2 The result of a blast on a dam project.

When there is a requirement to remove rock, blasting should be considered as it is almost always more economical than mechanical excavation. One reported comparison of mechanical excavation costs versus removal by blasting identified a difference of \$4.05 per cy [6].

There are many types of explosives. A full treatment of each explosive and method is too comprehensive for inclusion in this book. A complete discussion of explosives can be found in handbooks on blasting [2].

GLOSSARY OF BLASTING TERMS

The following glossary defines the important terms that are used in describing blasting operations.

Bench. The horizontal ledge of an excavation face along which holes are drilled for blasting. Benching is the process of excavating using ledges in a stepped pattern.

Bench height. The vertical distance between the base of an excavation and the ledge above where the blastholes are drilled and shot (see Fig. 12.3).

Blasthole. A hole drilled (borehole) into rock to enable the placing of an explosive.

Blasting (shot). The detonation of an explosive to fracture the rock is termed blasting.

Blasting agent. The classification of a particular type of explosive compound from the standpoint of storage and transportation. It is a material or mixture intended for blasting, consisting of an oxidizer and a fuel. It is less sensitive to initiation and cannot be detonated with a No. 8 detonator when unconfined. Therefore, blasting agents are covered by different handling regulations than high explosives.

Burden. The distance from the explosive charge to the nearest free or open rock face is referred to as burden or burden distance (see Fig. 12.3). There can be an apparent burden and a true burden. True burden is in the direction that the displacement of broken rock will move following the firing of the charge.

Deflagration. A rapid chemical reaction in which the output of heat is enough to enable the reaction to proceed and be accelerated without input of heat from another source. The effect of a true deflagration under confinement is an explosion.

Density. An explosive's density is its specific weight, usually expressed as grams per cubic centimeter (g/cc). In the case of some explosives, density and energy are correlated.

Heave. The displacement of rock as a result of the expansion of gases from firing an explosive.

Lead wires. The wires that conduct the electric current from its source to the leg wires of the electric detonator.

Leg wires. The wires that conduct the electric current from lead wires to an electric detonator.

MS delay detonator (millisecond). A detonator that has a built-in delay element. These detonators are commonly available in $\frac{25}{1,000}$-sec increments.

Nitroglycerin. A powerful explosive liquid obtained by treating glycerol with a mixture of nitric and sulfuric acids. Pure nitroglycerin is a colorless, oily, and somewhat toxic liquid. It was first prepared in 1846 by the Italian chemist Ascanio Sobrero.

Prill. In the United States, most ammonium nitrate, both agricultural and blasting grade, is produced by the prill tower method. Ammonium nitrate liquor is released as a spray at the top of a prilling tower. Prills of ammonium nitrate congeal in the upcoming steam and air of the tower. The moisture driven out by the dropping process leaves voids within the prills.

Propagation. The movement of a detonation wave, either in the hole or from hole to hole.

Rounds. A term that includes all the blastholes that are drilled, loaded, and fired at one time.

Sensitiveness. A measure of an explosive's cartridge-to-cartridge propagating ability.

Tamping. The process of compacting the stemming material placed in a blasthole.

TNT. A high explosive whose chemical content is trinitrotoluene, or trinitrotoluol.

COMMERCIAL EXPLOSIVES

Commercial explosives are compounds that detonate on introduction of a suitable **initiation** stimulus. On detonation, the ingredients of an explosive compound react at high speed, liberating gas and heat, thus causing very high-pressure, high-temperature gases. Pressures just behind the detonation front are on the order of 150,000 to 3,980,000 psi (10,340 to 274,340 bars), while temperature can range from 3,000 to 7,000°F (1,600 to 3,900°C). High explosives contain at least one high explosive ingredient. Low explosives contain no ingredients that by themselves can explode. Both high and low explosives can be initiated by a single No. 8 detonator.

Initiation
The act of detonating a high explosive.

Explosives are different in the following ways:

- Strength
- Sensitivity—input energy needed to start the reaction
- Detonation velocity
- Water resistance
- Flammability
- Generation of toxic fumes
- Bulk density

Strength The term strength refers to the energy content of an explosive, which is the measure of the force it can develop and its ability to do work. There are no standard strength measurement methods used by explosives manufacturers.

Strength ratings are misleading and do not accurately compare rock fragmentation effectiveness with explosive type.

Sensitivity The sensitivity of an explosive product is defined by the amount of input energy necessary to cause the product to detonate reliably. Some explosives, such as dynamite, require very little energy to detonate reliably. An electric detonator or blasting cap alone will not reliably initiate bulk-loaded ANFO and some slurries. To detonate these, a primer or booster would have to be used in conjunction with the electric detonator or the cap.

Detonation Velocity Explosive velocity is the speed at which the detonation wave moves through the column of explosive. Commercially available explosives have velocities in the range of 5,000 to 25,000 ft/sec. High-velocity explosives should be used in hard rock, while lower-velocity explosives give better results in softer rock.

Water Resistance The water resistance of an explosive can broadly be defined as its ability to detonate after exposure to water. Water resistance is generally expressed as the number of hours a product can be submerged in static water and still be detonated reliably. A standard water resistance test is used primarily for classification of dynamite explosives. The classifications with regard to water degradation are given in Table 13.1.

Flammability The characteristic of an explosive that describes its ease of initiation from spark, fire, or flame.

Generation of Toxic Fumes The amount of toxic gases produced by an explosive during the detonation process.

Bulk Density The density of an explosive is normally expressed in terms of weight per unit volume. The density determines the weight of explosive that can be loaded into a specific borehole diameter. This is known as the loading density. The specific gravity of an explosive is the ratio of an explosive's weight to the weight of an equal volume of water. For some explosives, density is commonly used as a way to approximate the explosive's strength.

TABLE 13.1 Water resistance classes of dynamite explosives.

Class	Hours submerged and still detonate
1	Indefinitely
2	32 to 71
3	16 to 31
4	8 to 15
5	4 to 7
6	1 to 3
7	Less than 1

There are four main categories of commercial high explosives: (1) dynamite, (2) slurries, (3) ANFO, and (4) two-component explosives. To be a high explosive, the material must be cap sensitive and react at a speed faster than the speed of sound, and the reaction must be accompanied by a shock wave. The first three categories—(1) dynamite, (2) slurries, and (3) ANFO—are the principle explosives used for blasthole charges. Two-component or binary explosives are normally not classified as explosive until mixed. Therefore, they offer advantages in shipping and storage that make them attractive alternatives on small jobs. But their unit price is significantly greater than that of other high explosives.

Dynamite

This nitroglycerin-based product is the most sensitive of all the generic classes of explosives in use today. It is available in many grades and sizes to meet the requirements of a particular job. Straight dynamite (the term straight means that the dynamite contains no ammonium nitrate) is not appropriate for construction applications because it is very sensitive to shock. With straight dynamite, sympathetic detonation can result from adjacent holes that were fired on an earlier **delay**.

delay
The term used to describe noninstantaneous firing of a charge or a group of charges.

The most widely used product in mining, quarries, and construction applications is "high-density extra dynamite," but individual explosive manufacturers have their own trade names for dynamite products. This product is less sensitive to shock than straight dynamite because some of the nitroglycerin has been replaced with ammonium nitrate. The approximate strength of dynamite is specified as a percentage that is an indication of the ratio of the weight of nitroglycerin to the total weight of a cartridge.

Dynamite is used extensively for charging blastholes, especially for smaller size holes. Individual cartridges vary in size from approximately 1 to 8 in. in diameter and 8 to 24 in. in length. A detonator or a Primacord fuse may be used to fire the dynamite. If a detonator is used, one of the cartridges serves as a primer. The detonator is placed within a hole made in this cartridge.

Slurries

This is a generic term for both water gels and emulsions. A slurry explosive is made of ammonium, calcium, or sodium nitrate and a fuel sensitizer along with varying amounts of water.

Water Gels A slurry explosive mixture of oxidizing salts, fuels, and **sensitizers** made water resistant by the cross-linking of gums or waxes is explicitly referred to as a water gel. The primary sensitizing methods are the introduction of air throughout the mixture, the addition of aluminum particles, or the addition of nitrocellulose.

sensitizers
Ingredients used in explosive compounds to promote ease of initiation or propagation of the reactions.

Emulsions A slurry explosive mixture of oxidizing salts and fuels that contains no chemical sensitizers and is made water resistant by an emulsifying agent is explicitly referred to as an emulsion. Emulsions will have a somewhat higher detonation velocity than water gels.

In comparison to ANFO (see next section), slurries have a higher cost per pound and they have less energy. However, in wet conditions they are very competitive with ANFO, because the ANFO is water-sensitive and must be protected in lined holes or a bagged product has to be used. Both of these measures add to the total cost of the ANFO.

An advantage of slurries over dynamite is that the separate ingredients can be hauled to the project in bulk and mixed immediately before loading the blastholes. The mixture can be poured directly into the hole. Some emulsions tend to be wet and will adhere to the blasthole, causing bulk loading problems. Slurries may be packaged in plastic bags for placement in the holes. Because they are denser than water, they will sink to the bottom of holes containing water.

Some slurries may be classified as high explosives, while others, if they cannot be initiated by a No. 8 detonator, are classified as blasting agents. This difference in classification is important for **magazine** storage.

magazine
A building or chamber for the storage of explosives.

ANFO

This explosive is used extensively for construction blasting and represents about 80% of all explosives used in the United States. "ANFO," an ammonium nitrate and fuel oil mixture, is synonymous with dry blasting agents. This explosive is the cheapest source of explosive energy. Because it must be detonated by special primers, it is much safer than dynamite.

Standard ANFO is a mixture of prilled industrial grade ammonium nitrate and 5.7% No. 2 diesel fuel oil. This is the optimum mixture. The detonation efficiency is controlled by the amount of fuel oil. It is less detrimental to have a fuel deficiency, but fuel percentage variances, either surplus or deficiency, affect the blast. With too little fuel, the explosive will not perform properly. With a fuel percentage of 5 there will be a 5.3% energy loss because of excess oxygen and orange nitrous oxide fumes may be produced. Maximum energy output is also reduced when too much fuel is used. The ANFO prills should not be confused with ammonium nitrate fertilizer prills. A blasting prill is porous to better distribute the fuel oil.

Because the ANFO is free flowing, it can be either blown or augured from bulk trucks directly into the blastholes (Fig. 13.3). It is detonated by primers consisting of charges of explosive placed at the bottoms of the holes. Sometimes primers are placed at the bottom and at intermediate depths. Electric detonators or Primacord can be used to detonate the primer.

The detonation velocity and therefore the efficiency of ANFO is dependent on the diameter of the power column. Air-emplaced ANFO in a $1\frac{1}{4}$-in.-diameter column will detonate at velocities of 7,000 to 10,000 ft/sec (fps). If placed in larger-diameter columns the velocity increases—3-in.-diameter column, 12,000 to 13,000 fps; 9-in.-diameter column, 14,000 to 15,000 fps. The placement method also affects the velocity. The velocities for poured ANFO are $1\frac{1}{4}$-in.-diameter column, 6,000 to 7,000 fps; 3-in.-diameter column, 10,000 to 11,000 fps; 9-in.-diameter column, 14,000 to 15,000 fps.

FIGURE 13.3 Loading bulk ANFO into blastholes.

ANFO is not water-resistant. Detonation will be marginal if ANFO is placed in water and shot, even if the interval between loading and shooting is very short. If it is to be used in wet holes, there is a *densified* ANFO cartridge. This product has a density greater than water, so it will sink to the bottom of a wet hole. Standard ANFO has a product density of 0.84 g/cc. The poured density of ANFO is generally between 0.78 and 0.85 g/cc. Cartridges or bulk product sealed in plastic bags will not sink. Another method to exclude the water and to enable the use of bulk ANFO in wet conditions is to preline the holes with plastic tubing that is closed at the bottom. The tubes, whose diameters should be slightly smaller than the holes, are installed in the holes by placing rocks or other weights in their bottoms.

PRIMERS AND BOOSTERS

A prime charge is an explosive ignited by an initiator (detonator or cap), which, in turn, initiates a noncap-sensitive explosive or blasting agent. The resulting detonation is then transmitted to an equally or less sensitive explosive. Often, primers are cartridges of dynamites, highly sensitized slurries, or emulsions together with detonators or a detonating cord. Good priming improves **fragmentation**, increases productivity, and lowers overall cost. The primer should never be tamped.

fragmentation
The degree to which mass rock is broken into pieces by blasting.

Boosters (Fig. 13.4) are highly sensitized explosives or blasting agents, used either in bulk form or in packages and of weights (amounts) greater than those used for primers. Boosters are placed within the explosive column, where additional breaking energy is required.

FIGURE 13.4 Placing an electric detonator in a booster.

To obtain optimum performance of ANFO, the size of the primer must match the borehole diameter. A mismatch in the size of the primer and the borehole diameter will impact the detonation velocity of ANFO.

> The primer, which is the point of initiation of an explosive column, should always be at the point of maximum confinement. This is usually at the bottom of the blasthole.

There are exceptions to where the primer should be placed in the explosive column, and care should be exercised when there are soft seams or fractures in the rock. A bottom primer should be just off the bottom of the hole. When using bulk ANFO, 1 to 2 ft should be placed in the hole before loading the primer. This will ensure good contact between the ANFO column and the primer. Contact between the primer and the explosive column is extremely important.

INITIATING SYSTEMS

Initiator is a term that is used in the explosive industry to describe any device that can be used to start a detonation. An initiation system is a combination of explosive devices and component accessories designed to convey a signal and initiate an explosive charge from a safe distance. The signal function may be electric or nonelectric. Fragmentation, **backbreak**, vibration, and violence of a blast are all controlled by the firing sequence of the individual blastholes. The order and timing of the detonation of individual holes is regulated by the initiation system. When selecting the proper system, one should consider both blast design and safety. Electrical systems are more sensitive to lightning than nonelectric systems, but both are susceptible.

backbreak
Rock broken beyond the limits of the last row of blastholes.

Electric Detonators

The most widely used explosive initiator is the electric detonator (Fig. 13.5). An electric detonator passes an electric current through a wire bridge, similar to an electric lightbulb filament, causing an explosion. The current, approximately 1.5 amps, heats the bridge to incandescence and ignites a heat-sensitive flash compound. The flash compound sets off a primer (initiating charge) that in turn fires a base charge in the detonator (Fig. 13.5). This charge detonates with sufficient violence to fire a charge of explosive.

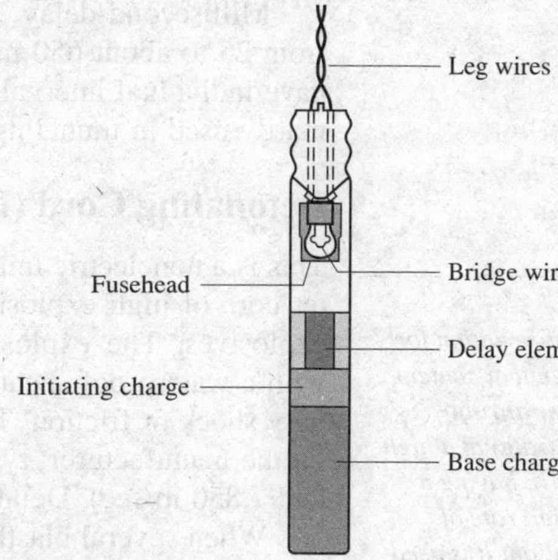

FIGURE 13.5 Electric initiating device.

Electric detonators are supplied with two leg wires in lengths varying from 2 to 100 ft. These wires are connected with the wires from other blastholes to form a closed electric circuit for firing purposes. The leg wires of electric detonators are made of either iron or copper. For ease in wiring a blast, each leg wire on an electric detonator is a different color. There are instantaneous, short-period delay, long-period delay, electronic delay type, and seismic (geophysical exploration) electric detonators. Instantaneous electric detonators are made to fire within a few milliseconds after current is applied.

When two or more electric detonators are connected in the same circuit, they must be products of the same manufacturer. This is essential to prevent a **misfire**, because detonators of different manufacturers do not have the same electrical characteristics. Detonators are extremely sensitive. They must be protected from shock and extreme heat. They are never to be stored or transported with other explosives.

misfire
A charge, or part of a charge, that has failed to fire as planned.

Electric Detonator Safety These devices must be kept away from all sources of electrical energy: batteries, outlets, radios, calculators, pagers, and cellular phones. Electric detonators can be initiated by the batteries, speakers, or even the antennas of such devices.

Nonelectrical Detonators

Nonelectrical detonators (caps) are thin metal cylindrical tubs that are open on one end for the insertion of a **safety fuse**. The tube contains two explosives, one layer upon the other. The bottom layer is called the base charge and is usually an insensitive high explosive. The top layer is the initiating charge and is a sensitive explosive. The ignition powder ensures that the flame is picked up from the safety fuse. At one time, the standard blasting cap contained 2 g of mercury fulminate and was called a No. 8 cap. A No. 6 cap contained 1 g of fulminate. Other caps with less strength had lower numbers. These detonators are sensitive to heat, shock, and crushing. All detonators of this type are instantaneous and, therefore, do not have a delay element.

safety fuse
A fuse containing a low explosive enclosed in a suitable covering. When the fuse is ignited, it will burn at a predetermined speed (35 to 45 sec/ft). It is used to initiate a nonelectric detonator.

Initiation Sequencing

In construction and surface blasting applications, millisecond-delay electric detonators are frequently used. Millisecond-delay blasting can be used both in shooting a single row and in multiple row shots. When each charge breaks its part of rock mass from the burden before the next charge detonates, ground vibration, air blast, and flyrock are minimized, and fragmentation is increased. The delaying procedure effectively reduces the apparent burden for the holes in the succeeding rows.

13.2

Millisecond-delay detonators have individual timing intervals ranging from 25 to about 650 milliseconds (ms). Long-period-delay electric detonators have individual intervals ranging from about $\frac{2}{10}$ sec to over 7 sec. They are primarily used in tunneling and underground mining.

Detonating Cord (Primacord)

PETN
The abbreviation for the chemical content (pentaerythritol tetranitrate) of a high explosive having a very high rate of detonation. It is used in detonating cord.

This is a nonelectric initiation system consisting of a flexible cord having a center core of high explosive, usually **PETN**. It is used to detonate cap-sensitive explosives. The explosive core is encased within textile fibers and covered with a waterproof sheath for protection. Detonating cord is insensitive to ordinary shock or friction. The **detonation rate** of the cord's explosive will depend on the manufacturer; typical rates are between 18,000 and 26,000 ft/sec (5,500 and 7,850 m/sec). Delays can be achieved by attaching in-line delay devices.

detonation rate
The velocity at which a detonation progresses through an explosive.

When several blastholes are fired in a round, the cord is laid on the surface between the holes as a **trunkline**. At each hole, one end of a detonating cord **downline** is attached to the trunkline, whereas the other end of the downline extends into the blasthole. If it is necessary to use a blasting cap and/or a primer to initiate the blast in the hole, the bottom end of the downline may be cut square and securely inserted into the cap.

trunkline
The main line, in the case of a nonelectric system, of detonating cord on the surface extending from the ignition point to the blastholes.

Sequential Blasting Machine

There are programmable blast timing machines for firing electric detonators. These machines enable precise millisecond/microsecond accuracy of blast circuit firing intervals. This provides the blaster the option of many delays within a blast. Since many delays are available, the pounds of explosive fired per delay can be reduced to better control noise and vibration.

downline
Detonating cord lines extending into the blasthole from a trunkline.

ROCK FRAGMENTATION

Rock fragmentation is the most important aspect of production blasting, because of its direct effect on the costs of drilling and blasting, and on the economics of the subsequent operations of loading, hauling, and crushing (see Fig. 13.6). Many variables affect rock fragmentation; rock properties, site geology, in situ fracturing, moisture content, and blasting parameters—the blast design.

There is no complete theoretical solution for the prediction of blast fragmentation size distribution and the precise mechanism of rock fragmentation, as the result of an explosive detonation in a blasthole. However, it is clear that the major mechanisms of rock breakage result from the sustained gas pressure produced in the borehole by the explosion. First, this pressure will cause radial cracking. Such cracking is similar to what happens in the case of frozen water pipes—a longitudinal split occurs parallel to the axis of the pipe. A borehole is analogous to the frozen pipe in that it is a cylindrical pressure vessel. But there is a difference in the rate of loading. A blasthole is pressurized instantaneously. Failure, therefore, instead of being at the one weakest seam, is in *many* seams parallel with the borehole. Burden distance (Figs. 12.3 and 13.7), the direction to the free face, will control the course and extent of the radial crack pattern.

When the radial cracking takes place, the rock mass is transformed into individual rock wedges. If relief is available perpendicular to the axis of the blasthole, the gas pressure pushes against these wedges, putting the opposite sides of the wedges into tension and compression. The exact distribution of such stresses is affected by the location of the charge in the blasthole. In this second breakage mechanism, flexural rupture of the wedge is controlled by the burden distance and bench height. The bench height divided by the burden distance is known as the "stiffness ratio." This is the same mechanism a structural engineer is concerned with when analyzing the length of a column in relation to its thickness.

There is a greater degree of difficulty in breaking rock when the burden distance is equal to the bench height. As bench height increases compared to burden distance, the rock is more easily broken. If the blast is not designed properly and the burden distance is too great, blast energy relief will not be available by this mechanism. When this happens either the blasthole will crater or the stemming will blow out.

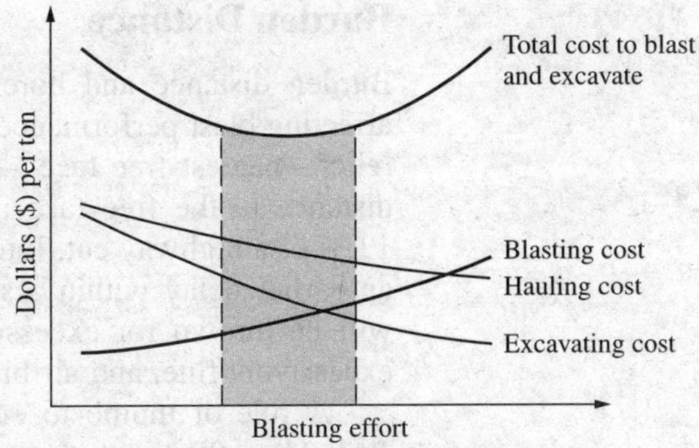

FIGURE 13.6 Optimum blast design.

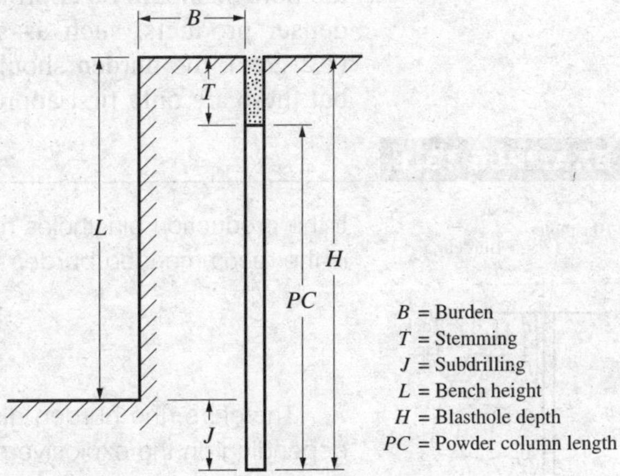

FIGURE 13.7 Blasthole geometry.

BLAST DESIGN

To minimize the total costs of rock excavation one must start with careful planning of the blasts. Every blast must be designed to meet the existing conditions of the rock formation and **overburden**, and to produce the desired final result. Blast design is not an exact science and there is no single solution to the rock removal problem. Rock is not a homogeneous material. There are fracture planes, seams, and changes in bench height to be considered. Wave propagation is faster in hard rock than soft rock. Initial blast designs use idealized assumptions. The engineer develops a blast design realizing that material discontinuities exist in the field. Because of these facts, it must always be understood that the theoretical design is only the starting point for blasting operations on the project.

overburden
The depth of material lying above the rock, which is to be shot.

Empirical formulas provide an estimate of the work that can be accomplished by a given explosive. The application of these formulas that are given in the following sections results in a series of blasting dimensions (burden distance, blasthole diameter, top stemming depth, subdrilling depth, and hole pattern and spacing) suitable for trial shots. Adjustments made from investigating the product of the trial shots should result in the optimum blast dimensions.

Burden Distance

Burden distance and borehole diameter are the two most important factors affecting blast performance (Fig. 13.7). Burden is the shortest distance to stress relief—nearest free face—at the time a blasthole detonates. It is normally the distance to the free face in an excavation, be it a quarry situation (see Fig. 12.3) or a highway cut. Internal free faces can be created by blastholes fired on an earlier delay within a shot. When the burden distance is insufficient, rock will be thrown for excessive distances from the face, fragmentation may be excessively fine, and air blast levels will be high.

A rule of thumb to ensure that the blaster is using the proper burden is $B = 24$ to 30 times the *charge* (explosive) diameter of the blasthole. When blasting with ANFO or other low-density blasting agents (explosive densities near 53 pcf, 0.85 g/cc), and typical rock (density of roughly 170 pcf, 2.7 g/cc), the burden should be approximately 25 times the charge diameter. When using denser products, such as slurries or dynamites with densities near 75 pcf (1.2 g/cc), the burden should be approximately 30 times the charge diameter, but these are only first approximations or guideline values.

EXAMPLE 13.1

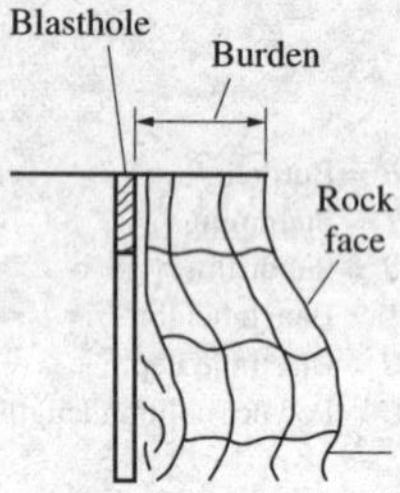

If the production blastholes have a diameter of 0.5 ft (6 in.) by rule of thumb, what is the recommended burden distance?

$$24 \times 0.5 \text{ ft} = 12 \text{ ft and}$$

$$30 \times 0.5 \text{ ft} = 15 \text{ ft}$$

Therefore, the burden distance for the shot should be between 12 and 15 ft, depending on the explosive product used.

An empirical formula for approximating a burden distance to be used for a first trial shot is

$$B = \left(\frac{2\,SG_e}{SG_r} + 1.5\right)D_e \qquad [13.1]$$

where

B = burden in feet

SG_e = specific gravity of the explosive

SG_r = specific gravity of the rock

D_e = diameter of explosive in inches

The explosive diameter will depend on the manufacturer's packaging container thickness, or it will equal the blasthole diameter if a granular or slurry explosive is poured directly into the hole. If the specific explosive product is known, the exact specific gravity information should be used; but in the case of developing a design before settling on a specific product, an allowance should be made.

Rock density is an indicator of rock strength, which, in turn, establishes the amount of energy required to cause breakage. Approximate specific gravities for various rocks are given in Table 13.2.

EXAMPLE 13.2

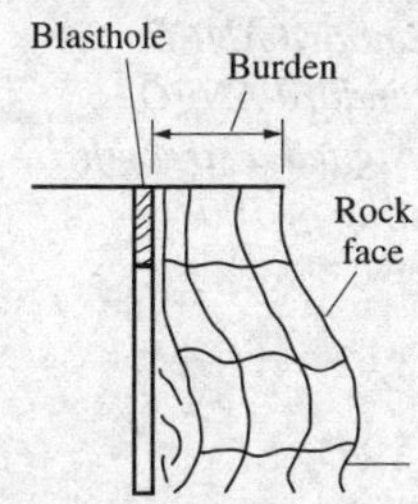

A contractor plans to use bulk ANFO, specific gravity 0.8, to open an excavation in granite rock. The drilling equipment will drill a 3-in. blasthole. What is the recommended burden distance for the first trial shot?

From Table 13.2, specific gravity of granite 2.6 to 2.9, use an average value 2.75:

$$B = \left(\frac{2 \times 0.8}{2.75} + 1.5\right) \times 3 = 6.2 \text{ ft}$$

EXAMPLE 13.3

A contractor plans to use a packaged explosive having a specific gravity of 1.3, to open an excavation in granite rock. The drilling equipment will drill a 3-in. blasthole. The explosive comes packaged in $2\frac{1}{2}$-in.-diameter sticks. What is the recommended burden distance for the first trial shot?

From Table 13.2, specific gravity of granite 2.6 to 2.9, use an average value 2.75:

$$B = \left(\frac{2 \times 1.3}{2.75} + 1.5\right) \times 2.5 = 6.1 \text{ ft}$$

TABLE 13.2 Density by nominal rock classifications.

Rock classification	Specific gravity	Density broken (ton/cy)
Basalt	2.8–3.0	2.36–2.53
Diabase	2.6–3.0	2.19–2.53
Diorite	2.8–3.0	2.36–2.53
Dolomite	2.8–2.9	2.36–2.44
Gneiss	2.6–2.9	2.19–2.44
Granite	2.6–2.9	2.19–2.28
Gypsum	2.3–2.8	1.94–2.36
Hematite	4.5–5.3	3.79–4.47
Limestone	2.4–2.9	1.94–2.28
Marble	2.1–2.9	2.02–2.28
Quartzite	2.0–2.8	2.19–2.36
Sandstone	2.0–2.8	1.85–2.36
Shale	2.4–2.8	2.02–2.36
Slate	2.5–2.8	2.28–2.36
Trap rock	2.6–3.0	2.36–2.53

Explosive density is used in Eq. [13.1] because of the proportional relationship between explosive density and strength. There are, however, some explosive emulsions that exhibit differing strengths at equal densities. In such a case, Eq. [13.1] will not be valid. An equation based on **relative bulk strength** instead of density can be used in such situations.

relative bulk strength *The strength of an explosive compared to standard ANFO. Standard ANFO is assigned a strength rating of 100.*

The relative bulk strength rating of an explosive should be based on test data under specified conditions, but sometimes the rating is based on calculations. Manufacturers will supply specific values for their individual products. The relative energy equation for burden distance is

$$B = 0.67\, D_e \sqrt[3]{\frac{\mathrm{St}_v}{\mathrm{SG}_r}} \quad \textbf{[13.2]}$$

where St_v = relative bulk strength compared to ANFO = 100.

When one or two rows of blastholes are fired in the same shot, the burden distance between rows would be equal. If more than two rows are to be fired in a single shot, either the burden distance of the rear rows must be adjusted or millisecond-delay times must be used to allow the face rock from the front rows to move before the back rows are fired.

Field experiments have shown that 1 to $1\frac{1}{2}$ ms per foot of effective burden is the minimum that can be considered if any relief is to be obtained for firing successive rows. However, for good relief, it is typically found that 2 to $2\frac{1}{2}$ ms per foot of effective burden are required and in some cases where maximum relief is desired 5 to 6 ms per foot is appropriate.

EXAMPLE 13.4

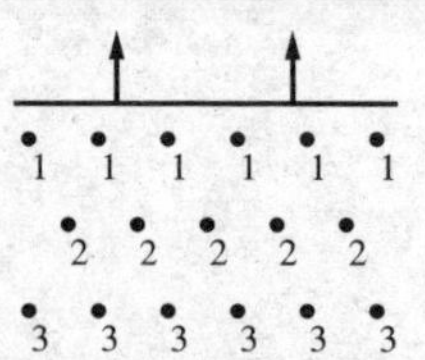

Three successive rows of blastholes will be fired in one shot. The rows are 16 ft apart. It is desired to obtain good relief for firing each row; therefore, how many milliseconds of delay should be allowed between firing the rows?

2 to $2\frac{1}{2}$ ms per foot of effective burden are required.

There should be 32 ms (16 × 2 ms) of delay as a minimum, and 40 ms (16 × 2.5 ms) would be better.

Geological Variations Rock is not the homogeneous material assumed by the empirical formulas; therefore, it is often necessary to employ burden correction factors for specific geological conditions. Table 13.3 provides burden distance correction factors for rock deposition K_d (see Fig. 13.8) and rock structure K_s.

$$B_{\text{corrected}} = B \times K_d \times K_s \quad \textbf{[13.3]}$$

EXAMPLE 13.5

A quarry is in a limestone (SG, 2.6) formation having horizontal bedding with numerous weak joints. From a borehole test-drilling program, it is believed that the limestone is highly laminated with many weakly cemented layers. Because of possible wet conditions, a cartridged slurry (relative bulk strength of 140) will be used as the explosive. The 6.5-in. blastholes will be loaded with 5-in.-diameter cartridges. What is the calculated burden distance?

$$B = 0.67 \times 5 \times \sqrt[3]{\frac{140}{2.6}} = 12.65 \text{ ft}$$

Correction factors from Table 13.3, K_d = 1, horizontal bedding, and K_s = 1.3, numerous weakly cemented layers.

$$B_{\text{corrected}} = 12.65 \times 1 \times 1.3 = 16.4 \text{ ft}$$

TABLE 13.3 Burden distance correction factors.

Rock deposition	K_d
Bedding steeply dipping into cut	1.18
Bedding steeply dipping into face	0.95
Other cases of deposition	1.00
Rock structure	K_s
Heavily cracked, frequent weak joints, weakly cemented layers	1.30
Thin, well-cemented layers with tight joints	1.10
Massive intact rock	0.95

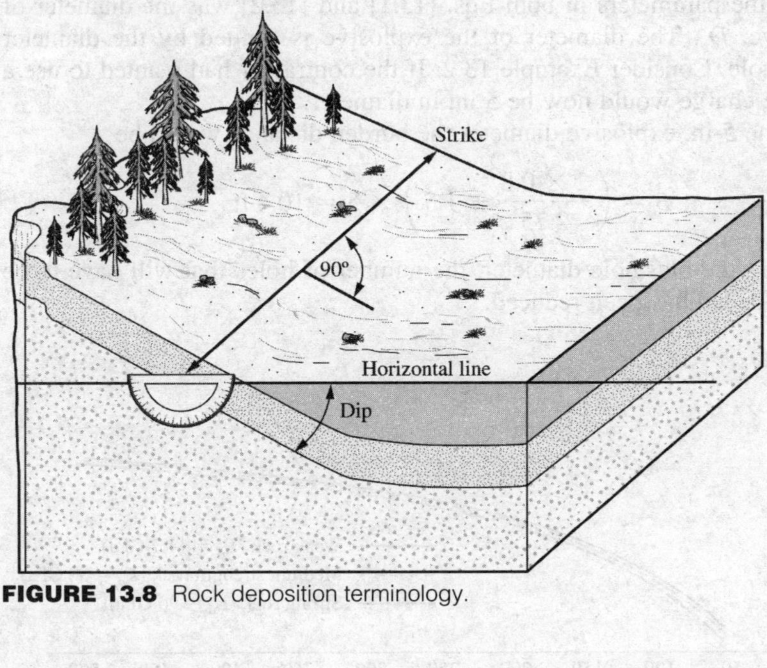

FIGURE 13.8 Rock deposition terminology.

Blasthole Diameter

The diameter of the blasthole will affect rock fragmentation, air blast, **flyrock**, and ground vibration. Larger blastholes have higher explosive loading densities (lb/ft or kg/m). As a result, blast patterns can be expanded while maintaining

flyrock
Rock that is ejected into the air by an explosion.

the same energy factor within the rock mass. This pattern expansion increases production by reducing the total number of holes required.

However, as a general rule, large-diameter blastholes of 6 to 15 in. (15 to 38 cm) have limited applications on most construction projects because of the fragmentation requirements and depth of cut limits. Construction blasthole diameters usually vary from 3.5 to 4.5 in. (90 to 114 mm), and the normal drilling depth is less than 40 ft (12 m). A rule of thumb is to use blasthole diameters smaller than the bench height divided by 60. Therefore, for a bench height of 13 ft, blasthole diameter should be in the order of 2.6 in. $\left(\frac{13 \text{ ft}}{60} \times 12 \text{ in. per ft}\right)$.

nonideal explosives *All commercial explosives are nonideal explosives; they have detonation velocities that change with charge diameter.*

Another consideration in selecting the proper blasthole diameter is the detonation characteristics of the explosive. All **nonideal explosives** have a critical diameter below which they will not reliably detonate. ANFO's performance falls off quickly below 3 in. (76 mm). Figure 13.9 presents the results from an empirical model of the effect charge diameter has on detonation velocity. It is based on field detonation velocity measurements. The blasthole diameter should be large enough to permit the optimum detonation of the explosives.

In some situations, as in a quarry, the blaster can adapt the bench height to optimize the blast, but on a construction project, the existing ground and the specified final project grades set limits on any bench-height modification.

One of the parameters in both Eqs. [13.1] and [13.2] was the diameter of the explosive, D_e. The diameter of the explosive is limited by the diameter of the blasthole. Consider Example 13.2: If the contractor had wanted to use a 5-in. bit, the charge would now be 5 in. in diameter.

Using the 5-in. explosive diameter the burden distance would be

$$B = \left(\frac{2 \times 0.8}{2.75} + 1.5\right) \times 5 = 10.4 \text{ ft}$$

By increasing the blasthole diameter, the number of holes that will have to be drilled and loaded has been reduced.

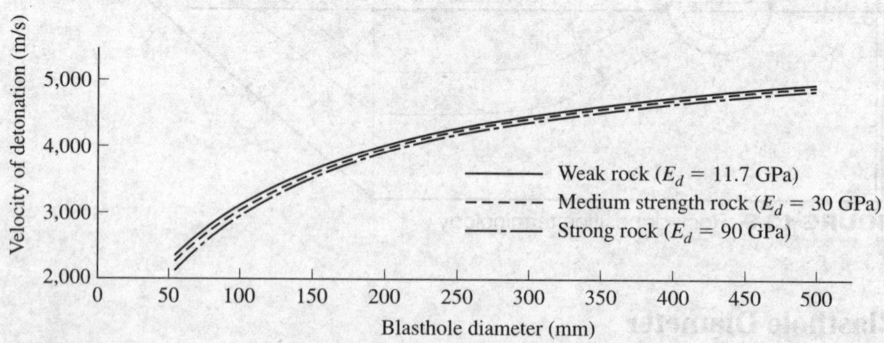

FIGURE 13.9 Model predicted effect of charge (hole) diameter on a blended explosive.

TABLE 13.4 Stiffness ratio's effect on blasting factors.

Stiffness ratio	1	2	3	4 and higher*
Fragmentation	Poor	Fair	Good	Excellent
Air blast	Severe	Fair	Good	Excellent
Flyrock	Severe	Fair	Good	Excellent
Ground vibration	Severe	Fair	Good	Excellent

*Stiffness ratios above 4 yield no increase in benefit.

However, the question of bench height has not been considered in either example. If the bench height was only 13 ft because of the required excavation depth, which size blasthole should be used? The stiffness ratios (SR) for the two blasthole diameters under consideration are

$$\text{3-in. blasthole: } \frac{13 \text{ ft height}}{6.2 \text{ ft burden}} = 2.1 \text{ SR}$$

$$\text{5-in. blasthole: } \frac{13 \text{ ft height}}{10.4 \text{ ft burden}} = 1.3 \text{ SR}$$

Table 13.4 gives the relationship between stiffness ratio and the critical blasting factors. The data in Table 13.4 indicates that for the 5-in. blasthole there will be blasting problems. Even the 3-in. blasthole can only be expected to yield fair results, which indicates that the shot should be redesigned.

It would be good to have an SR value of at least 3. Such a value is necessary for the blast to yield good results. Try a 2-in. blasthole and an explosive diameter of 2 in.

$$B = \left(\frac{2 \times 0.8}{2.75} + 1.5\right) \times 2.0 = 4.2 \text{ ft}$$

$$\text{2-in. blasthole: } \frac{13 \text{ ft height}}{4.2 \text{ ft burden}} = 3.1 \text{ SR}$$

If the rule of thumb blasthole diameter value of 2.6, obtained by dividing the height by 60, had been used, the SR would be 2.4. Therefore, it is clear that rules of thumb are good as magnitude checks or as first approximations, but a full analysis is necessary to develop a blast design.

This means more rows will be required but better fragmentation will result. Drilling cost will be increased but **secondary blasting** and handling cost should be reduced.

secondary blasting *An operation performed, after the primary explosion, to reduce remaining oversize material to a desirable size.*

Three factors affect burden distance: (1) the specific gravity of the rock, (2) the diameter of the explosives, and (3) either the specific gravity of the explosives or the relative bulk strength of the explosive in the case of an emulsion. The underlying geological conditions of a project are a given, and the contractor must work with the given site environment. There are many different commercial explosives having various strengths, but across the complete range of explosive strengths, the effect on calculated burden distance would be very small, only 2 to 3 ft. If the burden distance must be altered to achieve an effective blast that provides good fragmentation and does not cause damage, explosive diameter is the parameter to be adjusted.

Top Stemming Depth

The purpose of top stemming (collar distance) is to confine the explosive energy to the blasthole. To function properly, the material used for stemming must lock into the borehole. It is common practice to use drill cuttings as the stemming material. Very fine cuttings or drill cuttings that are essentially dust will not accomplish the desired purpose. To function properly, the stemming material should have an average diameter 0.05 times the diameter of the hole and it should be angular. It may be necessary in such cases to bring in crushed stone. It is recommended that No. 8 stone be used as stemming for holes of less than 4 in. in diameter and No. 57 stone for holes greater than 4 in. in diameter. Very coarse materials do not make good stemming because they tend to bridge and will be ejected from the hole.

If the stemming depth is too great, there will be poor top breakage from the explosion and backbreak will increase. When the stemming depth is inadequate, the explosion will escape prematurely from the hole (see Fig. 13.10).

Under normal conditions with good stemming material, a stemming distance, T, of 0.7 times the burden distance, B, will be satisfactory but stemming depth can range from 0.7 to 1.3 B. The stemming equation is

$$T = 0.7 \times B \qquad [13.4]$$

FIGURE 13.10 Explosive energy escaping from the blastholes.

Another design approach is to use the ratio of stemming depth to blasthole diameter. An appropriate ratio is from 14:1 to 28:1, depending on the velocities of the explosive and rock, the physical condition of the rock, and the type of stemming used. Greater top stemming depth is required where the velocity of the rock exceeds the detonation velocity of the explosive or where the rock is heavily fractured or low density.

When drilling dust is used for stemming material, the stemming distance will have to be increased by another 30% of the burden distance because the dust will not lock into the hole.

Sometimes it is necessary to deck load a blasthole. Deck loading is the operation of placing inert material (stemming material) in a blasthole to separate several explosive charges in the hole from each other. This is necessary when the blasthole passes through a seam or crack in the rock. In such a situation, if the shot were fired with explosive material completely filling the hole, the force of the blast would blow out through the weak seam. Therefore, when such conditions are encountered, a charge is placed below and above the weak seam, and stemming material is tamped between the charges.

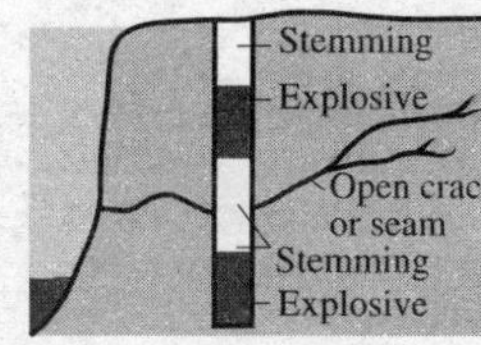

Subdrilling Depth

A shot will normally not break to the very bottom of the blasthole. This can be understood by remembering that the second mechanism of breakage is flexural rupture. To achieve a specified grade by blasting, it is necessary to drill below the desired **floor** elevation (Fig. 13.7). This portion of the blasthole below the desired final grade is termed "subdrilling." The subdrilling distance, J, required can be approximated by the formula

floor
The horizontal, or nearly so, bottom plane of an excavation.

$$J = 0.3 \times B \quad \textbf{[13.5]}$$

Satisfactory subdrilling depths can range from 0.2 to 0.5 B, as results are dependent on the type of rock and its structure.

In many cases, however, the project specifications will limit subdrilling to 10 percent or less of the *bench height*. In blasting for foundations for structures where a final grade is specified, subdrilling of the final cut layer is normally severely restricted by specification. The final layer removed in structural excavations is usually limited to a maximum depth of between 5 to 10 feet (1.5 to 3 m). Typically, subdrilling is not allowed in a final 5-ft (1.5-m) cut layer and is restricted to 2 ft (0.6 m) for the final 10-ft (3-m) cut layer. This is to prevent damage to the rock upon which the foundation will rest. Additionally, the diameter of the blasthole is often limited to 3.5 in. (90 mm).

Hole Patterns and Spacing

The three commonly used blasting patterns are (1) the square, (2) the rectangular, and (3) the staggered (Fig. 13.11). The square drill pattern has equal burden and spacing distances. With the rectangular pattern, the spacing distance between holes is greater than the burden distance. Both of these patterns place the holes of each row directly behind the holes in the preceding row. In the

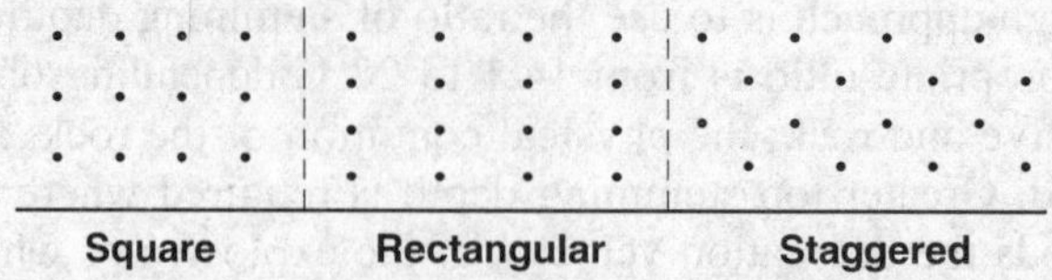

FIGURE 13.11 Common blasting patterns.

staggered pattern, the holes in each row are positioned at the midpoint between the holes in the preceding row. In the staggered pattern, the spacing distance between holes should be greater than the burden distance.

A V-cut firing sequence is used with square and rectangular drilling patterns (Fig. 13.12). The burdens and subsequent rock displacement are at an angle to the original free face when the V-cut firing order is used. The staggered drilling pattern is used for row-by-row firing, where the holes of one row are fired before the holes in the immediately succeeding row.

FIGURE 13.12 V-pattern firing sequence. Numbers indicate firing order.

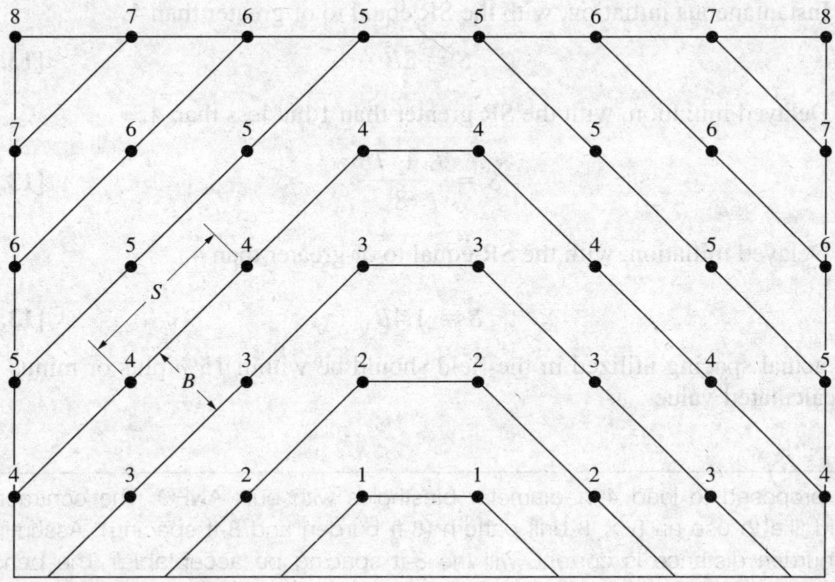

FIGURE 13.13 Spacing distance for a V-pattern firing sequence. Numbers indicate firing order.

Spacing is the distance between adjacent blastholes, measured perpendicular to the burden. Where the rows are blasted one after the other, as when firing a staggered blasthole pattern, the spacing is measured between holes in the same row. Where the blast progresses at an angle to the original free face, as in Fig. 13.12, the spacing is measured at an angle from the original free face (see Fig. 13.13). Spacing distance is calculated as a function of the burden.

> Initiation timing and the stiffness ratio control the proper spacing of blastholes.

When holes are spaced too close and fired instantaneously, there will be venting of the energy, with resulting air blast and flyrock. When the spacing is extended, there is a limit beyond which fragmentation will become harsh. Before beginning a spacing analysis, two questions must be answered concerning the shot: (1) Will the charges be fired instantaneously or will delays be used? (2) Is the stiffness ratio greater than 4? An SR of less than 4 is considered a low bench, and a high bench is an SR value of 4 or greater. This means there are four design cases to be considered:

1. Instantaneous initiation, with the SR greater than 1 but less than 4.

$$S = \frac{L + 2B}{3} \qquad [13.6]$$

where

S = Spacing
L = Bench height

2. Instantaneous initiation, with the SR equal to or greater than 4.

$$S = 2B \qquad \textbf{[13.7]}$$

3. Delayed initiation, with the SR greater than 1 but less than 4.

$$S = \frac{L + 7B}{8} \qquad \textbf{[13.8]}$$

4. Delayed initiation, with the SR equal to or greater than 4.

$$S = 1.4B \qquad \textbf{[13.9]}$$

The actual spacing utilized in the field should be within 15% plus or minus of the calculated value.

EXAMPLE 13.6

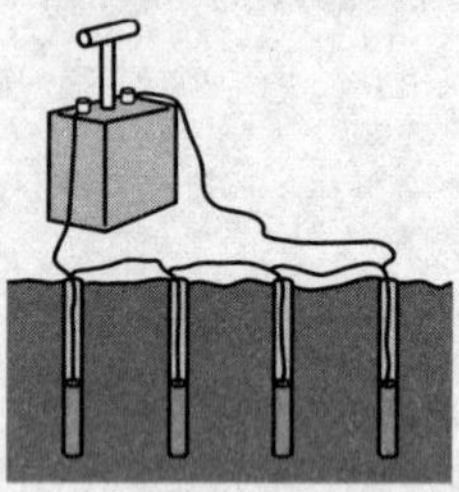

It is proposed to load 4-in.-diameter blastholes with bulk ANFO. The contractor would like to use an 8 × 8 drill pattern (8-ft burden and 8-ft spacing). Assuming the burden distance is correct, will the 8-ft spacing be acceptable? The bench height is 35 ft and each hole is to be fired on a separate delay.

Check stiffness ratio for high or low bench:

$$\frac{L}{B} = \frac{35}{8} = 4.4; \text{ this is } > 4\text{, therefore high bench.}$$

Delay timing; therefore use Eq. [13.9].

$$S = 1.4 \times 8 = 11.2 \text{ ft}$$

Range, 11.2 ± 15%: $9.5 \leq S \leq 12.9$

The proposed spacing of 8 ft does not appear to be sufficient. As a minimum, the pattern should be changed to 8-ft burden × 9.5-ft spacing for the first trial shot in the field.

EXAMPLE 13.7

A project in granite rock will have an average bench height of 20 ft. An explosive having a specific gravity of 1.2 has been proposed. The contractor's equipment can easily drill 3-in.-diameter holes. Assume the packaged diameter of the explosives will be 2.5 in. Delay blasting techniques will be utilized. Develop a blast design for the project.

The specific gravity of granite is between 2.6 and 2.9 (see Table 13.2). Use the average, 2.75.

Using Eq. [13.1], we obtain

$$B = \left(\frac{2 \times 1.2}{2.75} + 1.5\right) \times 2.5 = 5.9 \text{ ft}$$

Use 6 ft for burden distance. Remember that the numbers calculated here will be used in the field. Make it easy for those actually doing the work out in the dust, mud, and those other wonderful conditions. It is probably best to round the calculated numbers to the nearest foot or half foot before giving the design to the driller or blaster.

The stiffness ratio: $\frac{L}{B} = \frac{20}{6} = 3.3$

From Table 13.4, the stiffness ratio is good.

The stemming depth, from Eq. [13.4], is

$$T = 0.7 \times 6 = 4.2 \text{ ft}$$

Use 4 ft for stemming.

The subdrilling depth, from Eq. [13.5], is

$$J = 0.3 \times 6 = 1.8 \text{ ft}$$

Use 2 ft for subdrilling.

The spacing, for an SR greater than 1 but less than 4 and using delay initiation, from Eq. [13.8], is

$$S = \frac{20 + (7 \times 6)}{8} = 7.75 \text{ ft}$$

7.75 ± 15%: The range for S is 6.6 to 8.9 ft.

As a first trial shot, use a 6-ft burden × 8-ft spacing pattern.

Note that we have attempted to design a blast that will require only integer measurements in the field.

POWDER FACTOR

The amount of explosive required to fracture a cubic yard of in situ rock is a measure of the economy of a blast design. Table 13.5 is a loading density chart that enables the engineer to calculate the weight of explosive required for a blasthole. In Example 13.7, the diameter of the explosives was 2.5 in. and the explosive had a density of 1.2. Using Table 13.5, the loading density is found to be 2.55 lb/ft of charge. The powder column length is the total hole length less the stemming, 18 ft in this case [20 ft + 2 ft (subdrilling) − 4 ft (stemming)]. Not considering employment of a primer, the total weight of explosive used per blasthole would be 18 ft × 2.55 lb/ft = 45.9 lb.

The amount of rock fractured by one blasthole is the pattern area times the depth to grade. For the 6- × 8-ft pattern having a 20-ft depth to grade, each hole would have an affected volume of 35.6 cy

$$\left(\frac{6 \text{ ft} \times 8 \text{ ft} \times 20 \text{ ft}}{27 \text{ cf/cy}}\right).$$

The powder factor will be 1.29 lb/cy

$$\left(\frac{45.9 \text{ lb}}{35.6 \text{ cy}}\right).$$

With experience this number provides the engineer a check on the blast design.

TABLE 13.5 Explosive loading density chart in pounds per foot of column for given explosive specific gravity.

Column diam. (in.)	Explosive specific gravity							
	0.80	0.90	1.00	1.10	1.20	1.30	1.40	1.50
1	0.27	0.31	0.34	0.37	0.41	0.44	0.48	0.51
$1\frac{1}{4}$	0.43	0.48	0.53	0.59	0.64	0.69	0.74	0.80
$1\frac{1}{2}$	0.61	0.69	0.77	0.84	0.92	1.00	1.07	1.15
$1\frac{3}{4}$	0.83	0.94	1.04	1.15	1.25	1.36	1.46	1.56
2	1.09	1.23	1.36	1.50	1.63	1.77	1.91	2.04
$2\frac{1}{2}$	1.70	1.92	2.13	2.34	2.55	2.77	2.98	3.19
3	2.45	2.76	3.06	3.37	3.68	3.98	4.29	4.60
$3\frac{1}{2}$	3.34	3.75	4.17	4.59	5.01	5.42	5.84	6.26
4	4.36	4.90	5.45	6.00	6.54	7.08	7.63	8.17
$4\frac{1}{2}$	5.52	6.21	6.89	7.58	8.27	8.96	9.65	10.34
5	6.81	7.66	8.51	9.36	10.22	11.07	11.92	12.77
$5\frac{1}{2}$	8.24	9.27	10.30	11.33	12.36	13.39	14.42	15.45
6	9.81	11.03	12.26	13.48	14.71	15.93	17.16	18.39
$6\frac{1}{2}$	11.51	12.95	14.39	15.82	17.26	18.70	20.14	21.58
7	13.35	15.02	16.68	18.35	20.02	21.69	23.36	25.03
8	17.43	19.61	21.79	23.97	26.15	28.33	30.51	32.69
9	22.06	24.82	27.58	30.34	33.10	35.85	38.61	41.37
10	27.24	30.64	34.05	37.46	40.86	44.26	47.67	51.07

Powder factors for surface blasting can vary from 0.25 to 2.5 pounds per cy (0.1 to 1.1 kg/cm), with 0.5 to 1.0 lb/cy (0.2 to 0.45 kg/cm) being typical values. Higher-energy explosives, such as those containing large amounts of aluminum, will break more rock per unit weight than lower-energy explosives.

In all the design examples up to this point, it has been assumed that only one explosive was used in a blasthole. This is not typically the case; if a hole is loaded with ANFO, it will require a primer to initiate the explosion. When it is expected that the bottom of some holes will be wet, an allowance must be made for a water-resistant explosive, such as a slurry in those locations. In the case of a powder column that is 18 ft, and that will be loaded with ANFO, specific gravity 0.8, a primer will have to be placed at the bottom of the hole. If an 8-in.-long stick of primer, specific gravity 1.3, is used, there will be 216 in. of ANFO and 8 in. of primer.

In this case, assuming all dry holes, the weight of explosives based on a 2.5-in. explosive diameter would be:

ANFO	1.70 lb/ft × (208 in. ÷ 12 in./ft) =	29.46 lb
Dynamite	2.77 lb/ft × (8 in. ÷ 12 in./ft) =	1.85 lb
		31.31 lb

Total weight of explosive per hole is 31.31 lb for 18 ft of powder column.

To check vibration it is necessary to know the total explosive weight. The breakout as to the amount of agent and the amount of primer is also important because the price of the two is considerably different. Primer can cost up to 9 times the unit price of agent. In this case, primer is about 6% by weight of the total amount of explosives.

TRENCH ROCK

When excavating a trench in rock, the diameter or width of the structural unit to be placed in the trench, whether a pipe or conduit, is the principal consideration. When considering necessary trench width, the space required for working and backfill placement requirements must be taken into account. Many specifications will specifically address backfill width. Another important consideration is the size of the excavation equipment.

The geology will have a considerable impact on the blast design. Trenches are at the ground surface; usually, they will be extending through soil overburden and weathered unstable rock into solid rock. This nonuniform condition must be recognized. The blaster must check each individual hole to determine the actual depth of rock. Explosives are placed only in rock, not in the overburden.

If only a narrow trench in an interbedded rock mass is required, a single row of blastholes located on the trench centerline is usually adequate. Equation [13.1] or Eq. [13.2] will provide a first try for hole spacing. The timing of the shot should sequence down the row. The firing of the first hole provides the free face for the progression, which is why the equations are applicable. In the situation where the trench is shallow or there is little overburden, **blasting mats** laid over the alignment of the trench may be necessary to control flyrock. In the case of a wide trench or when solid rock is encountered, a double row of blastholes is common.

blasting mat
A blanket, usually of woven wire rope, used to restrict or contain flyrock.

BREAKAGE CONTROL TECHNIQUES

Controlled blasting employs the use of reduced explosive quantities loaded into holes that are generally smaller in diameter and spaced closer than the main blast. The holes are often placed along the periphery of an excavation or individual shots. Breakage control techniques are used to limit overbreak, reduce fractures within remaining rock walls, and reduce ground vibrations.

Line Drilling

One technique used to control **overbreak** and achieve a definite final excavation surface is line drilling. Line drilling is simply the drilling of a single row of unloaded, closely spaced holes along the perimeter of the excavation. The diameter of these holes ranges from 2 to 3 in. (50 to 75 mm). They are spaced two to four diameters apart. This line of holes provides a weak plane into

overbreak
Rock that is fractured outside the desired excavation limits.

which the blast can break. The maximum practical depth to which line drilling can be done is dependent on the accuracy the drilling can achieve. Seldom can the alignment of the holes be held at depths greater than 30 ft (10 m).

Presplitting

Presplitting is a technique for creating an internal free face, usually at the boundary of an excavation, which will contain stress waves from successively detonated holes (Fig. 13.14). Presplit blastholes can be fired either in advance of other operations or just before any adjacent main blastholes. If the presplit holes are fired just before the main blastholes, there should be a minimum of a 200-ms delay between the presplit blast and the firing of the nearest main blastholes.

Usually, the presplit blastholes are $2\frac{1}{2}$ to 3 in. in diameter and are drilled along the desired surface at spacings varying from 18 to 36 in., depending on the characteristics of the rock. When the rock is highly jointed, the hole spacing must be reduced. There have, however, been some presplit operations using holes as large as $12\frac{1}{4}$ in. in diameter with depths of more than 80 ft. But the maximum depth is limited by the accuracy of the drilling and this usually deteriorates at about 50 ft (15 m). Depths of 20 to 40 ft (6 to 12 m) are more common. The presplit holes are loaded with one or two sticks of explosive at their bottoms, and then have smaller charges, such as $1\frac{1}{4}$- $\times$ 4-in. sticks spaced at 12-in. intervals, to the top of the hole. The sticks may be attached to Primacord with tape, or hollow sticks may be used, which permits the Primacord to

FIGURE 13.14 A presplit rock face.

pass through the sticks with cardboard tube spacers between the charges. It is important that the charges be less than half the presplit hole diameter and they should not touch the walls of the hole.

When the explosives in these holes are detonated ahead of the main blastholes, the webs between the holes will fracture, leaving a surface joint that serves as a barrier to the shock waves from the main blast. This will essentially eliminate breakage beyond the fractured surface.

Presplit Explosive Load and Spacing

The approximate load of explosive per foot of presplit blasthole is given by

$$d_{ec} = \frac{D_h^2}{28} \qquad [13.10]$$

where

d_{ec} = explosive load in pounds per foot
D_h = diameter of blasthole in inches

When this formula is used to arrive at an explosive loading, the spacing between blastholes can be determined by the equation

$$\text{Sp} = 10D_h \qquad [13.11]$$

where Sp = presplit blasthole spacing in inches.

Because of the variations in the characteristics of rocks, the final determination of the spacings of the holes and the quantity of explosive per hole should be determined by tests conducted at the project. Many times field trials will allow the constant in Eq. [13.11] to be increased from 10 to as much as 14.

Presplit blastholes are not extended below grade (floor of the excavation). Instead of subdrilling, a concentrated charge of two to three times d_{ec} should be placed in the bottom of the hole.

Because a presplit blast is only meant to cause fracture, drill cutting can be used as stemming. The purpose is to only momentarily confine the gases and reduce noise. Two to five feet of stemming is normal.

EXAMPLE 13.8

By contract specification, the walls of a highway excavation through rock must be vertically presplit. The contractor will be using drilling equipment capable of drilling a 3-in. hole. What explosive load and hole spacing should be used for the first presplit shot on the project?

Using Eqs. [13.10] and [13.11], one gets

$$d_{ec} = \frac{3^2}{28} = 0.32 \text{ lb/ft}$$

$$\text{Sp} = 10 \times 3 = 30 \text{ in.}$$

Bottom load should be 3 × 0.32 = 0.96 lb.

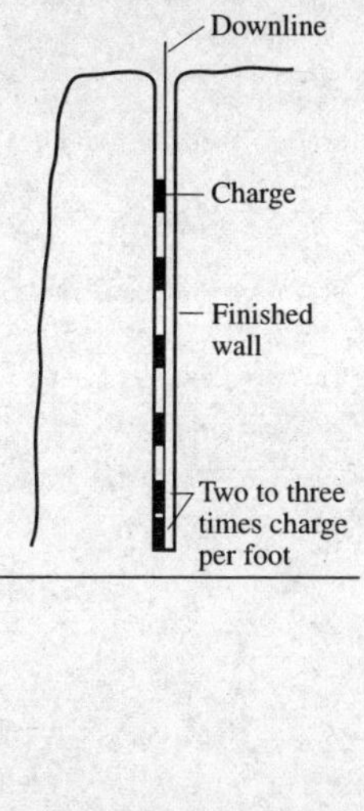

VIBRATION

Rock exhibits the property of elasticity. When explosives are detonated, elastic waves are produced as the rock is deformed and then regains its shape. The two principal factors that will affect how this motion is perceived at any discrete point are the mass of the detonated explosive charge and distance to the charge. Though the vibrations diminish in strength with distance from the source, they can achieve audible and "feelable" ranges in buildings close to the work site. Rarely do these vibrations reach levels that cause damage to structures, but the issue of vibration problems is controversial. The case of old, fragile, or historical buildings is a situation where special care must be exercised in controlling vibrations because there is a danger of significant structural damage. The issue of vibration can cause restraints on blasting operations and lead to additional project cost and time. The determination of "acceptable" vibration levels is in many cases very difficult due to its subjective nature with regard to being a nuisance. Humans and animals are very sensitive to vibration, especially in the low-frequency range (1–100 Hz).

It is the unpredictability and unusual nature of a vibration source, rather than the level itself, that is likely to result in complaints. The effect of intrusion tends to be psychological rather than physiological, and is more of a problem at night, when occupants of buildings expect no unusual disturbance from external sources.

Vibration Strength Levels

When vibration levels from an "unusual source" exceed the human threshold of perception [peak particle velocity (PPV), 0.008–0.012 in./sec], complaints may occur. In an urban situation, serious complaints are probable when PPV exceeds 0.12 in./sec [4], even though these levels are much less than what would result from slamming a door in a modern masonry building. People's tolerance will be improved, provided that the origin of the vibrations is known in advance and no damage results. It is important to provide people with a motivation to accept some temporary disturbance. Appropriate practice is to avoid vibration-causing activities at night.

There are published studies comparing the stresses imposed on structures by typical environmental charges and equivalent particle velocities. A 35% change in outside humidity imposes stress equivalent to almost a 5.0-in./sec particle velocity. A 12°F change in inside temperature imposes stress equivalent to about a 3.3-in./sec particle velocity. A 23-mph wind differential imposes a stress equivalent to about a 2.2-in./sec particle velocity. Typical construction blasting creates particle velocities of less than 0.5 in./sec.

Consequently, it must be remembered that people can perceive very low levels of vibration, and at the same time they are unaware of the silent environmental forces acting on and causing damage to their homes. So even though construction activities cause movements significantly less than those created by common natural occurrences, the impact perceived by humans can cause problems.

Therefore, because blasting operations may cause actual or alleged damages to buildings, structures, and other properties located in the vicinity of the blasting operations, it is desirable to examine, photograph, and document any structures for which charges of damages may be made following blasting operations. To be of value, the survey must be thorough and accurate. Before firing a blast, seismic recording instruments can be placed in the vicinity to monitor the magnitudes of vibrations caused by the shot. The persons responsible for the blasting may conduct the monitoring, or if the company responsible for the blasting carries insurance covering this activity, a representative of the insurance carrier may provide this service.

Vibration Mitigation

Delays in the blast initiation sequence will reduce vibrations because the mass of the individual charges is less than the total that would have been fired without the delays. The U.S. Bureau of Mines has proposed a formula to evaluate vibration and as a way to control blasting operations:

$$D_s = \frac{d}{\sqrt{W}} \qquad [13.12]$$

where

D_s = scaled distance (nondimensional factor)
d = distance from shot to a structure in feet
W = maximum charge weight per delay in pounds

A scaled distance value of 50 or greater indicates that a shot is safe with respect to vibration according to the Bureau of Mines. Some regulatory agencies require a value of 60 or greater.

SAFETY

An accident involving explosives can easily kill or cause serious injury and property damage. The four major causes of blasting-related injuries in surface mining operations between 1978 and 1993 were lack of blast area security (41%), flyrock (28%), **premature** blasts (16%), and misfires (8%). The primary causes of blast area security problems are

premature
A charge that detonates before it is intended to explode.

- Failure by employees and visitors to evacuate the blast area
- Failure to understand the instructions of the blaster or supervisors
- Inadequate guarding of the access roads leading to the blast area

Blast area security accidents are preventable with good training and communications.

Stories of flyrock accidents appear in the news regularly, and the injured or damaged parties are usually innocent parties.

"Blast Catapults Mud, Rocks into Buildings; State Revokes Road Construction Permit" An ammonium nitrate blast sent mud and rock flying for 600 feet damaging four buildings and two cars (*St. Louis Post-Dispatch*, 1997).

"Judge Halts Blasting at Arbor Place Mall" A blasting contractor excavating rock at a construction site sent debris flying and damaged several homes (*The Atlanta Journal and Constitution*, 1998).

"Explosion at Construction Site Damages Homes" A piece of rock fell through the roof of the home of a 72-year-old man when workers set off dynamite to excavate a sewer line. Several other homes were also damaged (Associated Press, 1999, Brentwood, Tenn.).

Improper loading and firing practice contribute to the creation of flyrock. Any irregularity in the geological structure surrounding the borehole can cause an uneven stress field that may result in flyrock. There were 15 reported flyrock incidents in Tennessee during the 1999 to 2003 time frame where the rock traveled more than 500 ft from the blast site. In six of these instances, flyrock travel exceeded 1,000 ft and in one case, flyrock travel approached half a mile.

- A shot involving a partially free face and loaded with a 2.11-lb powder factor threw flyrock 1,500 ft, injuring one person.
- A 5- × 5-ft blast pattern of $4\frac{1}{2}$-in.-diameter holes threw rock about 950 ft, injuring one person.

In shooting charges of explosives, one or more charges may fail to explode; this is referred to as a "misfire." It is necessary to dispose of this explosive before excavating the loosened rock. The most satisfactory method is to shoot these charges if possible.

If electric detonators are used, the leading wires should be disconnected from the source of power prior to investigating the cause of the misfire. If the leg wires to the cap are available, test the cap circuit; and if the circuit is satisfactory, try again to set off the charge.

When it is necessary to remove the stemming to gain access to a charge in a hole, the stemming should be removed with a wooden tool instead of a metal tool. If water or compressed air is available, either one can be used with a rubber hose to wash the stemming out of the hole. A new primer, set on top of or near the original charge, can be used to fire the charge.

The prevention of accidents depends on careful planning and faithful observation of proper blasting practices. Specific Occupational Safety and Health Administration requirements applicable to blasting can be found under Standards: General Provisions 1926.900, Blaster Qualifications 1926.901, Surface Transportation of Explosives 1926.902, Storage of Explosives and Blasting Agents 1926.904, Firing the Blast 1926.909, Inspection after Blasting 1926.910, and Misfires 1926.911.

Explosive manufacturers will provide safety information on their specific products. An excellent source for material on blasting safety practices is the Institute of Makers of Explosives in Washington, D.C.

SUMMARY

In construction, the operation referred to as "blasting" is performed to break rock so that it can be quarried for processing in an aggregate production operation, or to excavate a right-of-way. There is no single solution in designing a blast. Rock is not a homogeneous material. There are fracture planes, seams, and changes in burden to be considered. Blast designs use idealized assumptions. Because of these facts, it must always be understood that the theoretical design is only the starting point for blasting operations in the field. Critical learning objectives include:

- An understanding of the difference between an explosive that exhibits a proportional relationship between explosive density and strength, and one that exhibits differing strengths at equal densities
- An ability to calculate burden distance based on explosive type
- An ability to design an initial blast, including adjusting blasthole size to obtain a satisfactory stiffness ratio
- An understanding of how initiation affects blasthole spacing
- An ability to calculate the powder factor
- An ability to design a presplitting shot
- An ability to use the U.S. Bureau of Mines formula to check blast vibration

These objectives are the basis for the problems that follow.

PROBLEMS

13.1 According to the OSHA standard, what is the no smoking and no open flame zone for an explosives and detonator storage magazine?

13.2 The detonation velocity of ANFO placed in a 3-in.-diameter column or a 9-in.-diameter column is roughly the same. Is this a true statement? What is the expected detonation velocity of poured ANFO in a 3-in.-diameter column?

13.3 To function properly, the stemming material should be angular and should have what average diameter in relation to the hole diameter?

13.4 A contractor will use a packaged emulsion having a specific gravity of 1.2 and relative bulk strength of 135 to open an excavation in granite rock, specific gravity 2.6. The drilling equipment available will drill a 4-in. blasthole. The explosive comes packaged in $3\frac{1}{2}$-in.-diameter sticks. What is the recommended burden distance for the first trial shot? Round the calculated burden distance to the nearest foot. (burden, 9 ft)

13.5 A contractor plans to use bulk ANFO, specific gravity 0.83, to open an excavation in sandstone. The drilling equipment available will drill a $3\frac{1}{2}$-in. blasthole. What is the recommended burden distance for the first trial shot? Round the calculated burden distance to the nearest one-half foot.

13.6 A project in shale will have an average bench height of 28 ft. An emulsion with a relative bulk strength of 118, specific gravity 0.90, will be the explosive used on the project. Dynamite sticks 8 in. long having a specific gravity of 1.3 and a diameter exactly equal to the D_e of the emulsion will be used as a detonator. The contractor's equipment can easily drill $3\frac{1}{2}$-in.-diameter holes. It is assumed that single rows of no more than 10 holes will be detonated instantaneously. The excavation site has structures within 1,500 ft. The local regulatory agency specifies a scaled distance factor of no less than 55. Develop a blast design for the project. Round design dimensions to the *nearest foot*.

13.7 The blasting in Problem 13.6 must be conducted so as to limit overbreakage. Develop a presplitting blast plan.

13.8 A project in limestone will have an average bench height of 20 ft. ANFO, specific gravity 0.90, will be the explosive used on the project. Dynamite, specific gravity 1.4, will be the primer explosive used. The dynamite comes in 8-in.-long $\times$ $1\frac{3}{4}$-in.-diameter sticks. The contractor's equipment can easily drill $3\frac{1}{2}$-in.-diameter holes. It is assumed that single rows of no more than 10 holes will be detonated instantaneously. The excavation site has structures within 900 ft. The holes will be lined. The thickness of the lining material is $\frac{1}{4}$ in. Calculate an appropriate burden distance for the blast design. To ease the fieldwork all design dimensions should be in integer numbers (ft).

13.9 A project in dolomite will have an average bench height of 18 ft. An emulsion relative bulk strength 105, specific gravity 0.8, will be the explosive used on the project. Dynamite sticks 8 in. long having a specific gravity of 1.3 and a diameter exactly equal to the D_e of the emulsion will be used as a detonator. The contractor's equipment can easily drill 3-in.-diameter holes. It is assumed that single rows of no more than 10 holes will be detonated instantaneously. The excavation site has structures within 600 ft. The local regulatory agency specifies a scaled distance factor of no less than 55. Develop a blast design for the project using the maximum permissible integer spacing (burden, 6 ft; stemming, 4 ft; subdrilling, 2 ft; 6 $\times$ 11 pattern; explosives 40.2 lb per hole; only 3 holes can be fired at one time).

13.10 A project in sandstone, having a specific gravity of 2.0, will have an average bench height of 23 ft. An emulsion relative bulk strength 105, specific gravity 0.8, will be the explosive used on the project. Dynamite sticks 1 ft long having a specific gravity of 1.3 and a diameter exactly equal to the D_e of the emulsion will be used as a detonator. The contractor's equipment can easily drill 4-in.-diameter holes. It is assumed that single rows of no more than 30 holes will be detonated instantaneously. The closest structures to the excavation site are one-half mile away. The local regulatory agency specifies a scaled distance factor of no less than 60. Even though an emulsion is being used because of expected wet conditions, the holes will be lined. The thickness of the lining material is $\frac{1}{4}$ in. Develop the blast design spacing, subdrilling, and stemming dimensions for the project. How many holes can be fired instantaneously? The superintendent has told you that good fragmentation will be acceptable.

13.11 A project in limestone will have an average bench height of 18 ft. ANFO, specific gravity 0.84, will be the explosive used on the project. Dynamite, specific gravity 1.3, will be the primer explosive used. The dynamite sticks are 8 in. long by $1\frac{3}{4}$ in. in diameter. The contractor's equipment can easily drill $3\frac{1}{2}$-in.-diameter holes. It is assumed that single rows of no more than 10 holes will be detonated instantaneously. The excavation site has structures within 900 ft. The local regulatory agency specifies a scaled distance factor of no less than 60. The holes will be lined. The thickness of the lining material is $\frac{1}{4}$ in. Develop a blast design for the project.

13.12 The blasting in Problem 13.11 must be conducted so as to limit overbreakage. Develop a presplitting blast plan.

13.13 A materials company is opening a new quarry in a limestone formation. Tests have shown that the specific gravity of this formation is 2.7. The initial mining plan envisions an average bench height of 24 ft based on the loading and hauling equipment capabilities. Bulk ANFO, specific gravity 0.8, and a primer, specific gravity 1.5, will be the explosives used. The contractor's equipment can drill 6-in.-diameter holes. Delayed initiation will be utilized. Develop a blasting plan for a conservative cost estimate (burden, 8 ft; stemming, 6 ft; subdrilling, 2 ft; spacing conservative 9 ft).

13.14 An investigation of a highway project revealed a sandstone formation with steeply dipping bedding into the face. Many weak joints were identified. An analysis of the plan and profile sheets, and the cross sections, showed that the average bench height will be about 15 ft. Bulk ANFO, specific gravity 0.8, and primer, specific gravity 1.3, will be the explosives used on the project. The primer comes in 8-in.-long $\times$ $1\frac{3}{4}$-in.-diameter sticks. The contractor's equipment can easily drill 6-in.-diameter holes. It is assumed that delayed initiation will be utilized.

a. Develop a blast design for the project.

b. If the burden distance is held constant at 5 ft, but the hole spacing is varied in 1-ft increments across the range of *S* developed in part (a), what is the cost per cubic of rock if the ANFO is \$0.166 per lb, and dynamite \$1.272 per lb?

REFERENCES

1. Flinchum, R., and D. Rapp (1993). "Reduction of Air Blast and Flyrock," *Proceedings of the Nineteenth Annual Conference on Explosives and Blasting Technique*. International Society of Explosives Engineers, Cleveland, OH.

2. *ISEE Blasters' Handbook*, 17th ed. (2003). International Society of Explosives Engineers, 30325 Bainbridge Road, Cleveland, OH.
3. Kennedy, David L. (1998). "Multi-valued Normal Shock Velocity versus Curvature Relationships for Highly Non-ideal Explosives," *Proceeding Eleventh International Symposium on Detonation*, Snowmass Village, Colorado, by Los Alamos National Laboratory, NM, pp. 181–192.
4. New, Barry M. (1990). "Ground Vibration Caused by Construction Work," *Tunneling and Underground Space Technology*, Vol. 5, No. 3, Great Britain.
5. Persson, Per-Anders, Roger Holmberg, and Jaimin Lee (1996). *Rock Blasting and Explosives Engineering*, CRC Press, Inc., Boca Raton, FL.
6. Revey, G. F. (1996). "To Blast or Not to Blast?" *Practice Periodical on Structural Design and Construction*, American Society of Civil Engineers, Vol. 1, No. 3, pp. 81–82, August.
7. *Rock Blasting and Overbreak Control* (1991). National Highway Institute, U.S. Department of Transportation, Federal Highway Administration, Pub. No. FHWA-HI-92-001.
8. Siskind, David E. (2000). *Vibrations from Blasting*, International Society of Explosive Engineers, Cleveland, OH.

WEBSITE RESOURCES

1. www.dynonobel.com/dynonobelcom/en/asiapacific/products/blastingguide/ap_products_blasting_guide_Glossary_of_Blasting_Terms.htm The Dyno Nobel is an explosives company headquartered in Oslo, Norway. This is their glossary of blasting terms website.
2. www.ime.org The Institute of Makers of Explosives (IME) is the safety association of the commercial explosives industry in the United States and Canada.
3. www.isee.org The International Society of Explosives Engineers (ISEE) is a professional society dedicated to promoting the safe and controlled use of explosives in mining, quarrying, construction.

CHAPTER 14

Aggregate Production

Many types and sizes of crushing and screening plants are used in the construction industry. The capacity of a crusher will vary with the type of stone, size of feed, size of the finished product, and extent to which the stone is fed uniformly into the crusher. The screening process is based on the simple premise that particle sizes smaller than the screen cloth opening size will pass through the screen and that oversized particles will be retained. After stone is crushed and screened, it is necessary to handle it carefully or the large and small particles may separate.

INTRODUCTION

The amount of processing required to produce suitable aggregate materials for construction purposes depends on the nature of the raw materials available and the desired attributes of the end product. Four functions are required to accomplish the desired results:

1. Particle size reduction—crushing
2. Separation into particle size ranges—sizing/screening
3. Elimination of undesirable materials—washing
4. Handling and movement of the crushed materials—storage and transport

This chapter is devoted primarily to the first three operations—crushing, sizing, and washing. Storage is discussed here, but transport has already been examined in Chapters 9 and 10.

In operating a crushing plant, the drilling pattern, the amount of explosives, the size shovel or loader used to load the stone, and the size of the primary crusher should be matched to ensure that all stone can be economically used. It is desirable for the loading capacity of the shovel or loader in the pit and the

TABLE 14.1 Recommended minimum sizes of primary crushers for use with shovel buckets of the indicated capacities.

Capacity of bucket [cy (cu m)]		Jaw crusher [in. (mm)]*		Gyratory crusher, size of openings [in. (mm)]†	
$\frac{3}{4}$	(0.575)	28 × 36	(712 × 913)	16	(406)
1	(0.765)	28 × 36	(712 × 913)	16	(406)
$1\frac{1}{2}$	(1.145)	36 × 42	(913 × 1,065)	20	(508)
$1\frac{3}{4}$	(1.340)	42 × 48	(1,065 × 1,200)	26	(660)
2	(1.530)	42 × 48	(1,065 × 1,200)	30	(760)
$2\frac{1}{2}$	(1.910)	48 × 60	(1,260 × 1,525)	36	(915)
3	(2.295)	48 × 60	(1,260 × 1,525)	42	(1,066)
$3\frac{1}{2}$	(2.668)	48 × 60	(1,260 × 1,525)	42	(1,066)
4	(3.060)	56 × 72	(1,420 × 1,830)	48	(1,220)
5	(3.820)	66 × 86	(1,675 × 2,182)	60	(1,520)

*The first number is the width of the opening at the top of the crusher, measured perpendicular to the jaw plates. The second two digits are the width of the opening, measured across the jaw plates.

†The recommended sizes are for gyratory crushers equipped with straight concaves.

gyratory crushers *A rock crusher in which a steel center cone rotates eccentrically to crush the material against the outside cylindrical steel wall.*

capacity of the crushing plant to be approximately equal. Table 14.1 gives the recommended minimum sizes of jaws and **gyratory crushers** required to handle the stone being loaded with buckets of the specified capacities.

Many types and sizes of portable crushing (Fig. 14.1) and screening plants are used in the construction industry. When there is a satisfactory deposit of

FIGURE 14.1 Small portable crushing plant.

stone near a project that requires aggregate, it frequently will be more economical to set up a portable plant and produce the crushed stone instead of purchasing it from a commercial source. Larger commercial aggregate plants are, however, the major source of crushed stone for the construction industry in metropolitan areas.

PARTICLE SIZE REDUCTION

GENERAL INFORMATION

Crushers are sometimes classified according to the stage of crushing that they accomplish, such as primary, secondary, and tertiary (see Fig. 14.2). A primary crusher receives the stone directly from the excavation after blasting, and

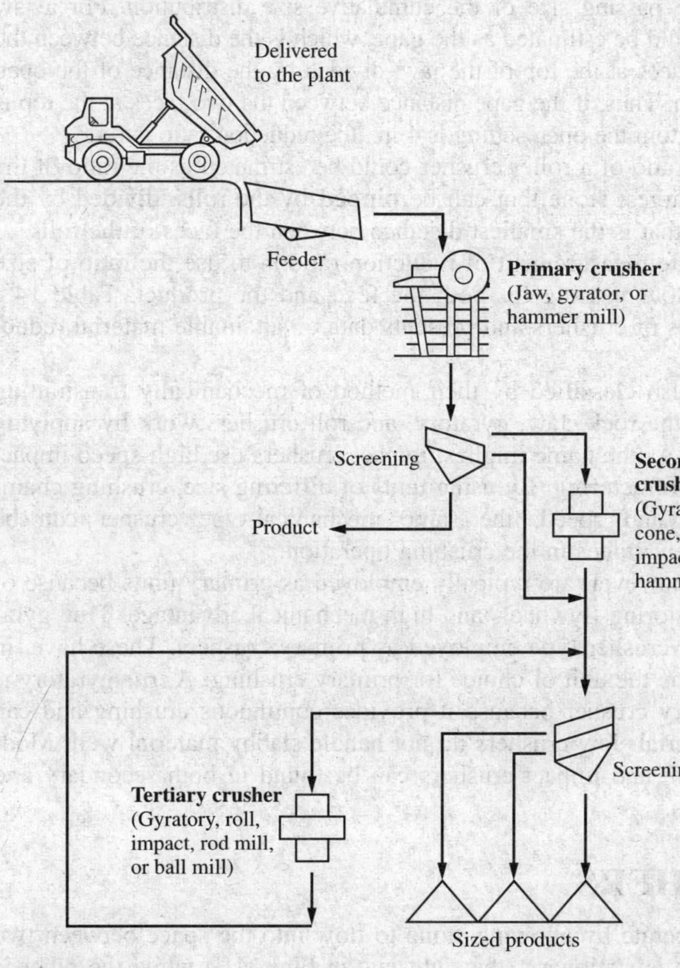

FIGURE 14.2 Aggregate processing steps.

produces the first reduction in stone size. The output of the primary crusher is fed to a secondary crusher, which further reduces the stone size. Some of the stone may pass through four or more crushers before it is reduced to the desired size.

Crushing plants use step reduction because the amount of size reduction accomplished is directly related to the energy applied. When there is a large difference between the size of the feed material entering the crusher and the size of the crushed product, a large amount of energy is required. If there was a concentration of this energy in a single-step process, excessive fines would be generated, and normally there is only a limited market for fines. Fines are a nonrevenue waste material at many plants. Therefore, to minimize the quantity of waste material, the degree of breakage is spread over several stages as a means of closely controlling product size.

As stone passes through a crusher, its reduction in size can be expressed as a reduction ratio: the ratio of crusher feed size to product size. The sizes are usually defined as the 80% passing size of the cumulative size distribution. For a jaw crusher, the ratio could be estimated as the gape, which is the distance between the fixed and moving faces at the top of the jaw, divided by the distance of the open setting at the bottom. Thus, if the gape distance between the two faces at the top is 16 in. and at the bottom the open setting is 4 in., the reduction ratio is four.

nipped
The ability of a set of rollers to draw a particle into the space between them by friction.

The reduction ratio of a roller crusher could be estimated as the ratio of the dimension of the largest stone that can be **nipped** by the rolls, divided by the setting of the rolls that is the smallest distance between the faces of the rolls.

A more accurate measurement of reduction ratio is to use the ratio of size corresponding to 80% passing for both the feed and the product. Table 14.2 lists the major types of crushers and presents data on attainable material reduction ratios.

Crushers are also classified by their method of mechanically transmitting energy to fracture the rock. Jaw, gyratory, and roll crushers work by applying compressive force. As the name implies, impact crushers use high-speed impact force to accomplish fracturing. By using units of differing size, crushing chamber configuration, and speed, the same mechanical-type crusher can be employed at different stages in the crushing operation.

Jaw crushers, however, are typically employed as primary units because of their large energy-storing flywheels and high mechanical advantage. True gyratories are the other crusher type employed as primary crushers. These have, in recent years, become the unit of choice for primary crushing. A *true* gyratory is an excellent primary crusher because it provides continuous crushing and can handle slabby material. Jaw crushers do not handle slabby material well. Models of gyratory, roll, and impact crushers can be found in both secondary and tertiary applications.

JAW CRUSHERS

These machines operate by allowing stone to flow into the space between two jaws, one of which is stationary (the right jaw in Fig. 14.3) while the other is movable. The space between the jaws diminishes as the stone travels downward

TABLE 14.2 The major types of crushers.

Crusher type	Reduction ratio range
Jaw	
Double toggle	
Blake	4:1–9:1
Overhead pivot	4:1–9:1
Single toggle: Overhead eccentric	4:1–9:1
Gyratory	
True	3:1–10:1
Cone	
Standard	4:1–6:1
Attrition	2:1–5:1
Roll	
Compression	
Single roll	Maximum 7:1
Double roll	Maximum 3:1
Impact	
Single rotor	to 15:1
Double rotor	to 15:1
Hammer mill	to 20:1
Specialty crushers	
Rod mill	
Ball mill	

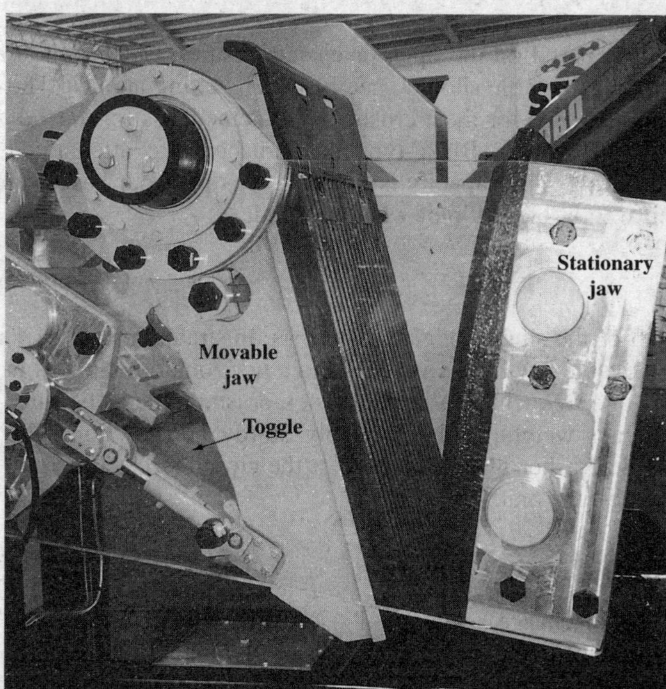

FIGURE 14.3 Cutaway of a single-toggle-type jaw crusher.

under the effect of gravity and the motion of the movable jaw (the left jaw in Fig. 14.3), until the stone ultimately passes through the lower opening. The movable jaw is capable of exerting a pressure sufficiently high to crush the hardest rock. Jaw crushers are usually designed with the toggle as the weakest part. The toggle will break if the crusher encounters an uncrushable object or is subjected to overload. This limits the damage to the crusher.

Double Toggle

A double-toggle jaw crusher, the *Blake* type, has a movable jaw suspended from a shaft mounted on the crusher frame. The rotation of a second shaft, which is eccentric and is located behind the movable jaw, raises and lowers the pitman, actuating two toggles, and these produce the crushing action. As the pitman raises the two toggles, a high pressure is exerted near the bottom of the swing jaw that partially closes the opening at the bottom of the two jaws. This operation is repeated as the eccentric shaft is rotated.

The jaw plates are replaceable. The jaws may be smooth, or, in the event the stone tends to break into slabs, corrugated jaws may be used to reduce the slabbing. The swing jaw may be straight, or it may be curved to reduce the possibility of choking. Double-toggle Blake-type jaw crushers are so large and heavy that they do not lend themselves to portable applications.

Table 14.3 gives representative capacities for various sizes of Blake-type jaw crushers. In selecting a jaw crusher, consideration must be given to the size of the feed stone. The top opening of the jaw should be at least 2 in. wider than the largest stones that will be fed to it.

Capacity tables can be based on the open or closed position of the bottom of the swing jaw; therefore, the table should specify which setting applies. The closed position is commonly used for most crushers and is the basis for the values given in Table 14.3. However, Blake-type jaw crushers are often rated based on the open-size setting. The capacity is given in tons per hour based on a standard material unit weight of 100 lb/cf when crushed.

Single Toggle

When the eccentric shaft of the single-toggle crusher, as illustrated in Fig. 14.4, is rotated, it gives the movable jaw both a vertical and a horizontal motion. This type of crusher is used quite frequently in portable rock-crushing plants because of its compact size, lighter weight, and reasonably sturdy construction. The capacity of a single-toggle crusher is usually rated at the closed size setting and is less than that of a Blake-type unit.

Sizes of Jaw and Roll Crusher Product

While the setting of the discharge opening of a crusher will determine the maximum-size stone produced, the aggregate sizes will range from slightly greater than the crusher setting to fine dust. Experience gained in the crushing

TABLE 14.3 Representative capacities of Blake-type jaw crushers, in tons per hour (metric tons per hour) of stone.*

Size Crusher [in. (mm)]†	Maximum rpm	Maximum hp (kW)	Closed setting of discharge opening [in. (mm)] 1 (25.4)	$1\frac{1}{2}$ (38.1)	2 (50.8)	$2\frac{1}{2}$ (63.5)	3 (76.2)	4 (102)	5 (137)	6 (152)	7 (178)	8 (203)	9 (229)
10 × 6 (254 × 406)	300	15 (11.2)	11 (10)	16 (14)	20 (18)								
10 × 20 (254 × 508)	300	20 (14.9)	14 (13)	20 (18)	25 (23)	34 (31)							
15 × 24 (381 × 610)	275	30 (22.4)		27 (24)	34 (31)	42 (38)	50 (45)						
15 × 30 (381 × 762)	275	40 (29.8)		33 (30)	43 (39)	53 (48)	62 (56)						
18 × 36 (458 × 916)	250	60 (44.8)		46 (42)	61 (55)	77 (69)	93 (84)	125 (113)					
24 × 36 (610 × 916)	250	75 (56.0)			77 (69)	95 (86)	114 (103)	150 (136)					
30 × 42 (762 × 1,068)	200	100 (74.6)				125 (113)	150 (136)	200 (181)	250 (226)	300 (272)			
36 × 42 (916 × 1,068)	175	115 (85.5)				140 (127)	160 (145)	200 (181)	250 (226)	300 (272)			
36 × 48 (916 × 1,220)	160	125 (93.2)				150 (136)	175 (158)	225 (202)	275 (249)	325 (294)	375 (339)		
42 × 48 (1,068 × 1,220)	150	150 (111.9)				165 (149)	190 (172)	250 (226)	300 (272)	350 (318)	400 (364)	450 (408)	
48 × 60 (1,220 × 1,542)	120	180 (134.7)					220 (200)	280 (254)	340 (309)	400 (364)	450 (408)	500 (454)	550 (500)
56 × 72 (1,422 × 1,832)	95	250 (186.3)						315 (286)	380 (345)	450 (408)	515 (468)	580 (527)	640 (580)

*Based on the closed position of the bottom swing jaw and stone weighing 100 lb/cf when crushed.

†The first number indicates the width of the feed opening, whereas the second number indicates the width of the jaw plates.

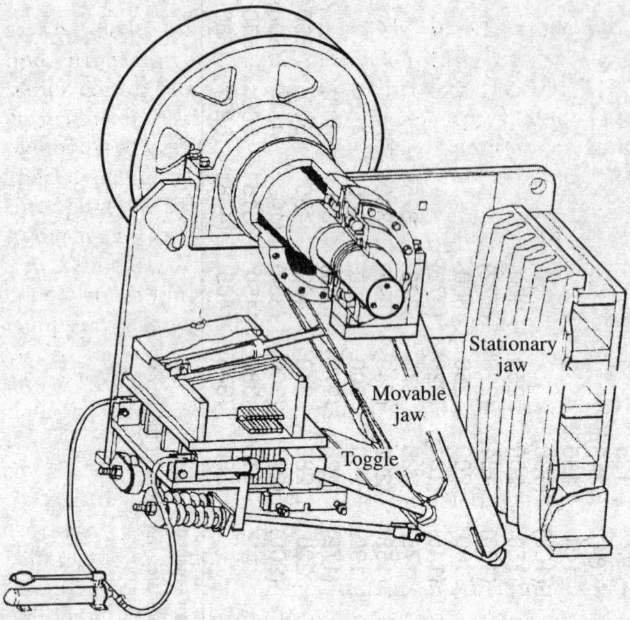

FIGURE 14.4 Single-toggle-type jaw crusher.
Source: Cedarapids, Inc., a Terex Company.

industry indicates that for any given setting for a jaw or roll crusher, approximately 15% of the total amount of stone passing through the crusher will be larger than the setting. If the openings of a screen that receives the output from such a crusher are the same size as the crusher setting, 15% of the output will not pass through the screen. Figure 14.5 presents the percentage of material passing through or retained on screens having the size openings indicated. The chart can be applied to both jaw- and roll-type crushers. To read the chart, select the vertical line corresponding to the crusher setting. Then move down this line to the number that indicates the size of the screen opening. From the size of the screen opening, proceed horizontally to the left to determine the percentage of material passing through the screen or to the right to determine the percentage of material retained on the screen.

EXAMPLE 14.1

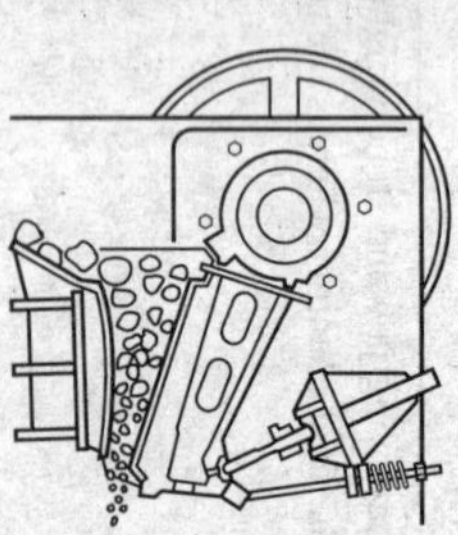

A jaw crusher with a closed setting of 3 in. is fed shot rock at the rate of 50 tons/hr (tph). Determine the amount of stone produced in tph within the following size ranges: in excess of 2 in.; between 2 and 1 in.; between 1 and $\frac{1}{4}$ in.; and less than $\frac{1}{4}$ in.

From Fig. 14.5 the amount retained on a 2-in. screen is 42% of 50, which is 21 tph. The amount in each of the size ranges is determined as

Size range (in.)	Percentage passing screens	Percentage in size range	Total output* of crusher (tph)	Amount produced in size range (tph)
Over 2	100–58	42	50	21.0
2–1	58–33	25	50	12.5
1–$\frac{1}{4}$	33–11	22	50	11.0
$\frac{1}{4}$–0	11–0	11	50	5.5
Total		100%		50.0 tph

* This is a closed system; what is fed into the crusher must equal the product. The example is also based on material weighting 100 lb/cf crushed.

Always read the fine print when using a crusher production chart.

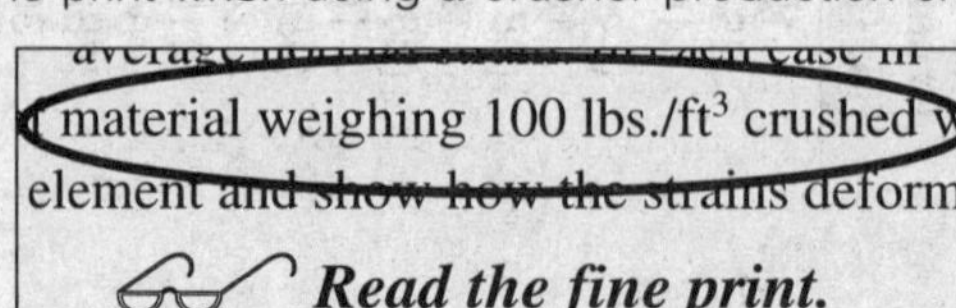

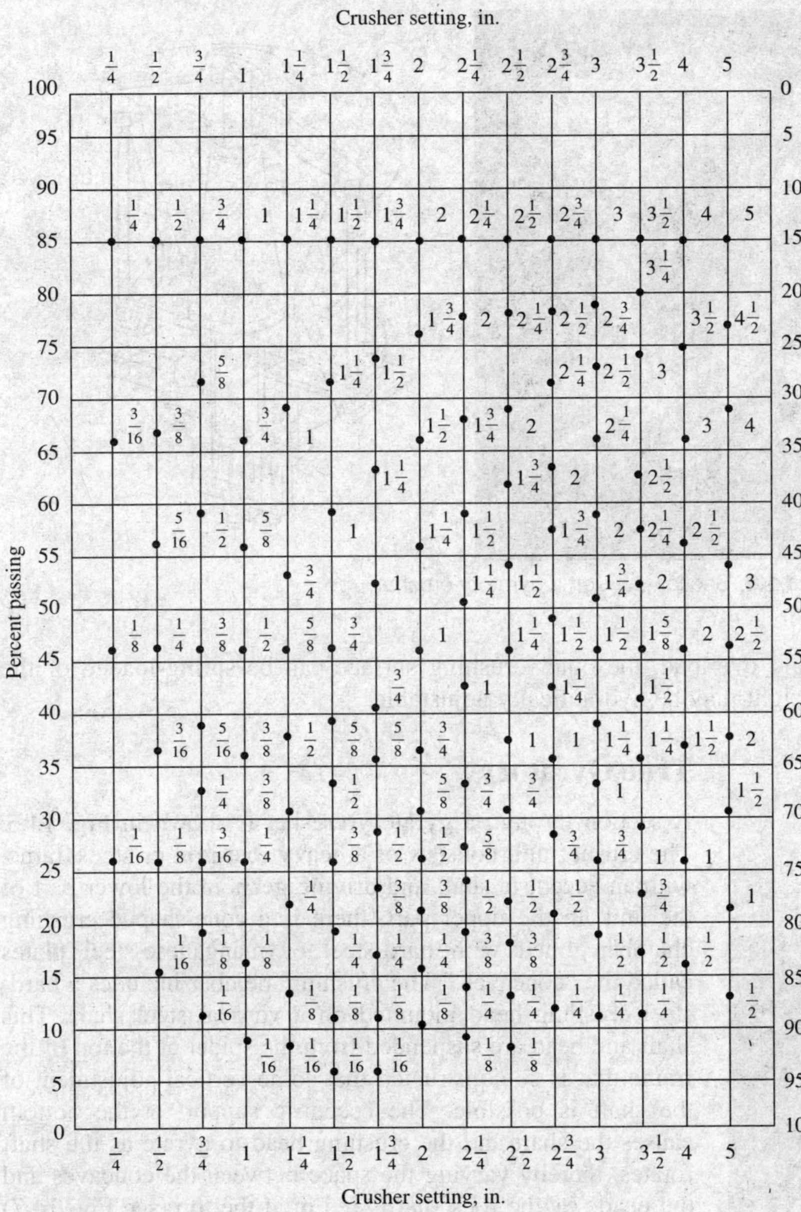

FIGURE 14.5 Analysis of the size of aggregate produced by jaw and roll crushers.
Source: Universal Engineering Company.

GYRATORY CRUSHERS

Gyratories are the most efficient of all primary-type crushers. A gyrating mantle mounted within a deep bowl characterizes these crushers. They provide continuous crushing action and are used for both primary and secondary crushing of hard, tough, abrasive rock. To protect the crusher from uncrushable

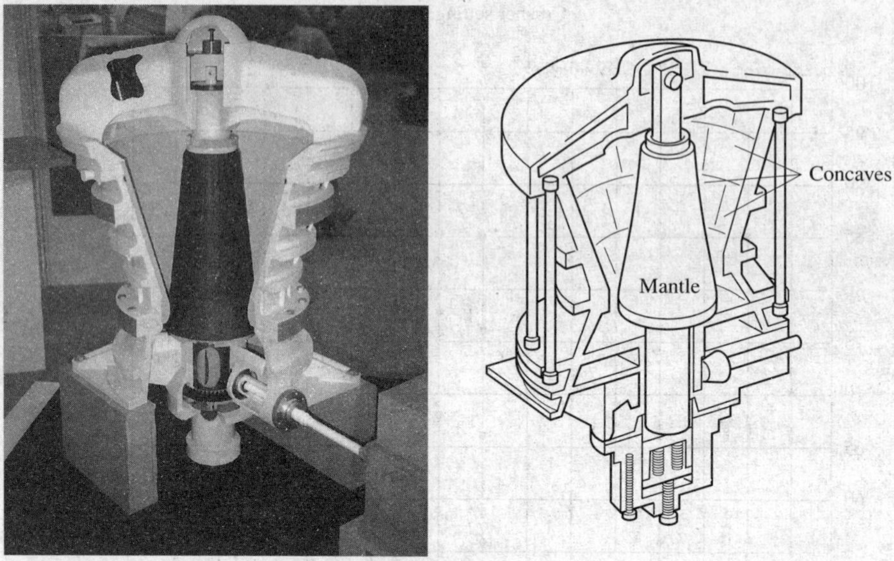

FIGURE 14.6 Section through a gyratory crusher.

objects and overload, the outer crushing surface can be spring-loaded or the mantle height may be hydraulically adjustable.

FIGURE 14.7 Feed chamber over the top of a gyratory crusher with a hoe ram for reducing oversize feed material.

True Gyratory

A section through a gyratory crusher is shown in Fig. 14.6. The crusher unit consists of a heavy cast-iron or steel frame, with an eccentric shaft and driving gears in the lower part of the unit. In the upper part, there is a cone-shaped crushing chamber, lined with hard-steel or manganese-steel plates called the "concaves." The crushing member includes a hard-steel crushing head mounted on a vertical steel shaft. This shaft and head are suspended from the spider at the top of the frame that is so constructed that some vertical adjustment of the shaft is possible. The eccentric support at the bottom causes the shaft and the crushing head to gyrate as the shaft rotates, thereby varying the space between the concaves and the head. As the rock that is fed in at the top (see Fig. 14.7) of the crushing chamber moves downward, it undergoes a reduction in size until it passes through the opening at the bottom of the chamber.

The size of a gyratory crusher is the width of the receiving opening, measured between the concaves and the crusher head. The setting is the width of the bottom opening and may be the open or closed dimension. When a setting is given, it should be specified whether it is the open or closed

FIGURE 14.8 Gyratory crusher in an aggregate plant.

dimension. Normally the capacity of a *true* gyratory crusher is based on an open-size setting. The ratio of reduction for true gyratory crushers usually ranges from 3:1 to 10:1, with an average value around 8:1. A gyratory crusher and the conveyor for delivering the stone are shown in Fig. 14.8.

If a gyratory crusher is used as a primary crusher, the size selected may be dictated by the size of the rock from the blasting operation, or it may be dictated by a desired capacity. When a gyratory crusher is used as a secondary crusher, increasing its speed within reasonable limits may increase the capacity.

Table 14.4 gives representative capacities of gyratory crushers, expressed in tons per hour, based on a continuous feed of stone having a unit weight of 100 lb/cf when crushed. Gyratories with straight concaves are commonly used as primary crushers, whereas those with nonchoking concaves are commonly used as secondary crushers.

Cone Crushers

Cone crushers are used as secondary or tertiary crushers. They are capable of producing large quantities of uniformly fine crushed stone. A cone crusher differs from a true gyratory crusher in the following respects:

1. A shorter cone
2. A smaller receiving opening
3. Rotates at a higher speed, about twice that of a *true* gyratory
4. Produces a more uniformly sized product

Standard models have large feed openings for secondary crushing, and produce stone in the 1- to 4-in. range. The capacity of a standard model is usually rated based on a closed-size setting.

TABLE 14.4 Representative capacities of gyratory crushers, in tons per hour (metric tons per hour) of stone.*

Size of crusher [in. (cm)]	Approximate power required [hp (kw)]	Open-side setting of crusher [in. (mm)]											
		$1\frac{1}{2}$ (38)	$1\frac{3}{4}$ (44)	2 (51)	$2\frac{1}{4}$ (57)	$2\frac{1}{2}$ (63)	3 (76)	$3\frac{1}{2}$ (89)	4 (102)	$4\frac{1}{2}$ (114)	5 (127)	$5\frac{1}{2}$ (140)	6 (152)
							Straight concaves						
8	15–25	30	36	41	47								
(20.0)	(11–19)	(27)	(33)	(37)	(42)								
10	25–40		40	50	60								
(25.4)	(19–30)		(36)	(45)	(54)								
13	50–75				85	100	133						
(33.1)	(37–56)				(77)	(90)	(120)						
16	60–100						160	185	210				
(40.7)	(45–75)						(145)	(167)	(190)				
20	75–125							200	230	255			
(50.8)	(56–93)							(180)	(208)	(231)			
30	125–175								310	350	390		
(76.2)	(93–130)								(281)	(317)	(353)		
42	200–275										500	570	630
(106.7)	(150–205)										(452)	(515)	(569)
							Modified straight concaves						
8	15–25	35	40	45									
(20.0)	(11–19)	(32)	(36)	(41)									
10	25–40		54	60	65								
(25.4)	(19–30)		(49)	(54)	(59)								
13	50–75					95	130						
(33.1)	(37–56)					(86)	(117)						
16	60–100						150	172	195				
(40.7)	(45–75)						(135)	(155)	(176)				
20	75–125							182	200	220			
(50.8)	(56–93)							(165)	(180)	(199)			
30	125–175								340	370	400		
(76.2)	(93–130)								(308)	(335)	(362)		
42	200–275										607	650	690
(106.7)	(150–205)										(550)	(589)	(625)
							Nonchoking concaves						
8	15–25	42	46										
(20.0)	(11–19)	(38)	(42)										
10	25–40	51	57	63	69								
(25.4)	(19–30)	(46)	(52)	(57)	(62)								
13	50–75	79	87	95	103	111							
(33.1)	(37–56)	(71)	(79)	(86)	(93)	(100)							
16	60–100			107	118	128	150						
(40.7)	(45–75)			(96)	(106)	(115)	(135)						
20	75–125				155	169	198	220	258	285	310		
(50.8)	(56–93)				(140)	(152)	(178)	(198)	(233)	(257)	(279)		

*Based on continuous feed and stone weighing 100 lb/cf when crushed.

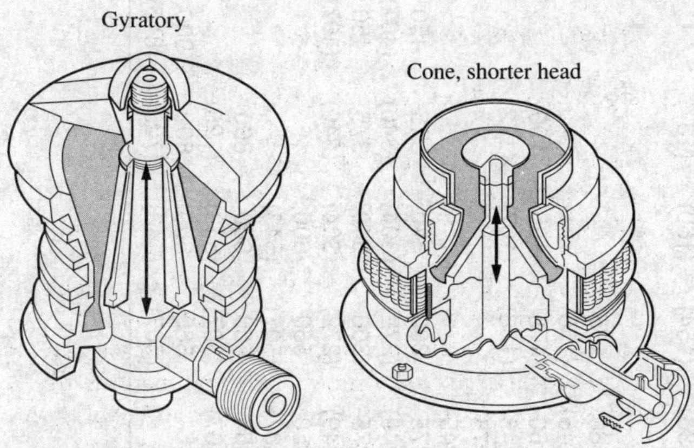

FIGURE 14.9 Sections through a gyratory and a cone crusher.

Attrition models are for producing stone having a maximum size of about $\frac{1}{4}$ in. The capacity of an attrition model cone crusher may not be related to closed-size setting.

Figure 14.9 shows the difference between a gyratory and a standard cone crusher. The conical head on the cone crusher is usually made of manganese steel. It is mounted on a vertical shaft and serves as one of the crushing surfaces. The other surface is the concave that is attached to the upper part of the crusher frame. The bottom of the shaft is set in an eccentric bushing to produce the gyratory effect as the shaft rotates.

The maximum diameter of the crusher head can be used to designate the size of a cone crusher. However, it is the size of the feed opening, the width of the opening at the entrance to the crushing chamber, that limits the size of the rocks that can be fed to the crusher. The magnitude of the eccentric throw and the setting of the discharge opening can be varied within reasonable limits. Because of the high speed of rotation, all particles passing through a cone crusher will be reduced to sizes no larger than the closed-size setting that should be used to designate the size of the discharge opening.

Table 14.5 gives representative capacities for the Symons standard cone crusher, expressed in tons of stone per hour for material having a unit weight 100 lb/cf when crushed.

ROLL CRUSHERS

Roll crushers are used for producing additional reductions in the sizes of stone after the output of the blasting operation has been subjected to one or more stages of prior crushing. A *roll crusher* consists of a heavy cast-iron frame equipped with either one or more hard-steel rolls, each mounted on a separate horizontal shaft.

TABLE 14.5 Representative capacities of Symons standard cone crushers, in tons per hour (metric tons per hour) of stone.*

Size of crusher [ft (m)]	Size of feed opening [in. (mm)]	Minimum discharge settings [in. (mm)]	Discharge setting, [in. (mm)]										
			$\frac{1}{4}$ (6.3)	$\frac{3}{6}$ (9.5)	$\frac{1}{2}$ (12.7)	$\frac{3}{2}$ (15.9)	$\frac{3}{4}$ (19.1)	$\frac{7}{6}$ (22.3)	1 (25.4)	$1\frac{1}{4}$ (31.8)	$1\frac{1}{2}$ (38.0)	2 (50.8)	$2\frac{1}{2}$ (63.5)
2 (0.61)	$2\frac{1}{4}$ (57)	$\frac{1}{4}$ (5.6)	15 (14)	20 (18)	25 (23)	30 (27)	35 (32)						
2 (0.61)	$3\frac{1}{4}$ (82)	$\frac{3}{8}$ (9.5)		20 (18)	25 (23)	30 (27)	35 (32)	40 (36)	45 (41)	50 (45)	60 (54)		
3 (0.91)	$3\frac{7}{8}$ (96)	$\frac{3}{8}$ (9.5)		35 (32)	40 (36)	55 (50)	70 (63)	75 (68)					
3 (0.91)	$5\frac{1}{8}$ (130)	$\frac{1}{2}$ (12.7)			40 (36)	55 (50)	70 (63)	75 (68)	80 (72)	85 (77)	90 (81)	95 (86)	
4 (1.22)	5 (127)	$\frac{3}{8}$ (9.5)		60 (54)	80 (72)	100 (90)	120 (109)	135 (122)	150 (136)				
4 (1.22)	$7\frac{3}{8}$ (187)	$\frac{3}{4}$ (19.0)					120 (109)	135 (122)	150 (136)	170 (154)	177 (160)	185 (167)	
$4\frac{1}{4}$ (1.29)	$4\frac{1}{2}$ (114)	$\frac{1}{2}$ (12.7)			100 (90)	125 (113)	140 (126)	150 (136)					
$4\frac{1}{4}$ (1.29)	$7\frac{3}{8}$ (187)	$\frac{5}{8}$ (15.8)				125 (113)	140 (126)	150 (136)	160 (145)	175 (158)			
$4\frac{1}{4}$ (1.29)	$9\frac{1}{2}$ (241)	$\frac{3}{4}$ (19.0)					140 (126)	150 (136)	160 (145)	175 (158)	185 (167)	190 (172)	
$5\frac{1}{2}$ (1.67)	$7\frac{1}{8}$ (181)	$\frac{5}{8}$ (15.8)				160 (145)	200 (181)	235 (213)	275 (249)				
$5\frac{1}{2}$ (1.67)	$8\frac{5}{8}$ (219)	$\frac{7}{8}$ (22.2)						235 (213)	275 (249)	300 (272)	340 (304)	375 (340)	450 (407)
$5\frac{1}{2}$ (1.67)	$9\frac{7}{8}$ (248)	1 (25.4)							275 (249)	300 (272)	340 (304)	375 (340)	450 (407)
7 (2.30)	10 (254)	$\frac{3}{4}$ (19.0)					330 (300)	390 (353)	450 (407)	560 (507)	600 (543)		
7 (2.30)	$11\frac{1}{2}$ (292)	1 (25.4)							450 (407)	560 (507)	600 (543)	800 (725)	
7 (2.30)	$13\frac{1}{2}$ (343)	$1\frac{1}{4}$ (31.7)								560 (507)	600 (543)	800 (725)	900 (815)

*Based on stone weighing 100 lb/cf when crushed.

Courtesy Nordberg Manufacturing Company.

Single Roll

With a single-roll crusher, the material is forced between a large-diameter roller and an adjustable liner. Because the material is dragged against the liner, these crushers are not economical for crushing highly abrasive materials. But they can handle sticky materials.

Double Roll

Roll crushers with two rollers are so constructed that each roll is driven independently by a flat-belt pull or a V-belt sheave. One of the rolls is mounted on a slide frame to permit an adjustment in the width of the discharge opening between the two rolls. The movable roll is spring-loaded to provide safety against damage to the rolls when noncrushable material passes through the machine.

Feed Size

The maximum size of material that can be fed to a roll crusher is directly proportional to the diameter of the rolls. If the feed contains stones that are too large, the rolls will not grip the material and pull it through the crusher. The angle of nip (grip), B in Fig. 14.10, which is constant for smooth rolls, has been found to be 16°45′.

The maximum-size particles that can be crushed are determined as follows. Referring to Fig. 14.10, these terms are defined:

R = radius of rolls
B = angle of nip (16°45′)
$D = R \cos B = 0.9575R$
A = maximum-size feed
C = roll setting (size of finished product)

Then

$$X = R - D$$
$$= R - 0.9575R = 0.0425R$$
$$A = 2X + C$$
$$= 0.085R + C \qquad [14.1]$$

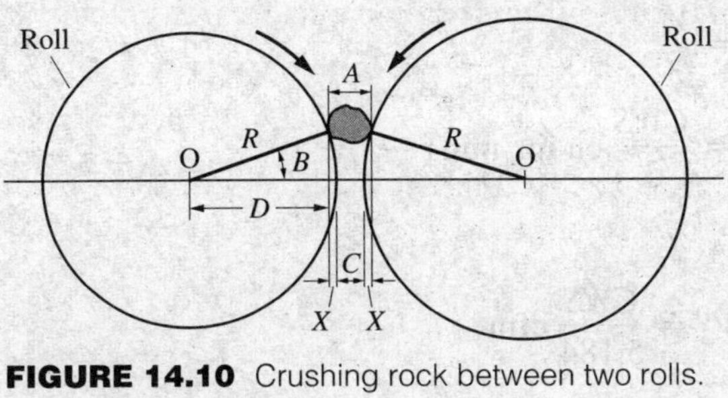

FIGURE 14.10 Crushing rock between two rolls.

EXAMPLE 14.2

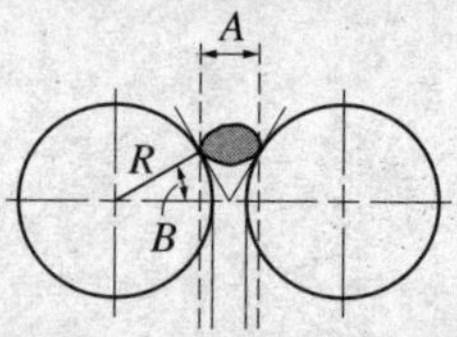

Determine the maximum-size stone that may be fed to a smooth-roll crusher whose rolls are 40 in. in diameter, when the roller setting is 1 in.

$$A = \left(0.085 \times \frac{40 \text{ in.}}{2}\right) + 1 \text{ in.}$$

$$= 2.7 \text{ in.}$$

Capacity

The capacity of a roll crusher will vary with the type of stone, size of feed, size of the finished product, width of rolls, the speed at which the rolls rotate, and the extent to which the stone is fed uniformly into the crusher. Referring to Fig. 14.10, the theoretical volume of a solid ribbon of material passing between the two rolls in 1 min would be the product of the width of the opening times the width of the rolls times the speed of the surface of the rolls. The volume can be expressed in cubic inches per minute or cubic feet per minute (cfm). In actual practice, the ribbon of crushed stone will never be continuous. A more realistic volume should approximate one-fourth to *one-third* of the theoretical volume. An equation that can be use as a guide in estimating the capacity is derived using these terms.

C = distance between rolls in inches
W = width of rolls in inches
S = peripheral speed of rolls in inches per minute
N = speed of rolls in rpm
R = radius of rolls in inches
V_1 = theoretical volume in cubic inches per minute or cfm
V_2 = actual volume, in cubic inches per minute or cfm
Q = probable capacity in tons per hour

Then

$$V_1 = CWS$$

Assume one-third of the theoretical volume, then

$$V_2 = \frac{V_1}{3}$$

$$= \frac{CWS}{3} \text{ cu in./min}$$

Divide by 1,728 cu in. per cf

$$V_2 = \frac{CWS}{5{,}184} \text{ cfm}$$

Assume the crushed stone has a unit weight of 100 lb/cf.

$$Q = \frac{100 \text{ lb/cf} \times (60 \text{ min/hr}) V_2}{2{,}000 \text{ lb/ton}} = 3 \text{ min/cf} \times V_2 \text{ tph}$$

$$= \frac{CWS}{1{,}728} \text{ tph} \qquad [14.2]$$

S can be expressed in terms of the diameter of the roll and the speed in rpm:

$$S = 2\pi RN$$

Substituting this value of S in Eq. [14.2] gives

$$Q = \frac{CW\pi RN}{864} \text{ tph} \qquad [14.3]$$

Table 14.6 gives representative capacities for smooth-roll crushers, expressed in tons of stone per hour for material having a unit weight of 100 lb/cf when crushed. These capacities should be used as a guide only in estimating the probable output of a crusher. The actual capacity may be more or less than the given values.

If a roll crusher is producing a finished aggregate, the reduction ratio should not be greater than 4:1. However, if a roll crusher is used to prepare feed for a fine grinder, the reduction may be as high as 7:1.

TABLE 14.6 Representative capacities of smooth-roll crushers, in tons per hour (metric tons per hour) of stone.*

Size of crusher [in. (mm)]†	Speed (rpm)	Power required [hp (kw)]	Width of opening between rolls [in. (mm)]						
			$\frac{1}{4}$ (6.3)	$\frac{1}{2}$ (12.7)	$\frac{3}{4}$ (19.1)	1 (25.4)	$1\frac{1}{2}$ (38.1)	2 (50.8)	$2\frac{1}{2}$ (63.5)
16 × 16 (414 × 416)	120	15–30 (11–22)	15.0 (13.6)	30.0 (27.2)	40.0 (36.2)	55.0 (49.7)	85.0 (77.0)	115.0 (104.0)	140.0 (127.0)
24 × 16 (610 × 416)	80	20–35 (15–26)	15.0 (13.6)	30.0 (27.2)	40.0 (36.2)	55.0 (49.7)	85.0 (77.0)	115.0 (104.0)	140.0 (127.0)
30 × 18 (763 × 456)	60	50–70 (37–52)	15.0 (13.6)	30.0 (27.2)	45.0 (40.7)	65.0 (59.0)	95.0 (86.0)	125.0 (113.1)	155.0 (140.0)
30 × 22 (763 × 558)	60	60–100 (45–75)	20.0 (18.1)	40.0 (36.2)	55.0 (49.7)	75.0 (67.9)	115.0 (104.0)	155.0 (140.0)	190.0 (172.0)
40 × 20 (1.016 × 508)	50	60–100 (45–75)	20.0 (18.1)	35.0 (31.7)	50.0 (45.2)	70.0 (63.4)	105.0 (95.0)	135.0 (122.0)	175.0 (158.5)
40 × 24 (1.016 × 610)	50	60–100 (45–75)	20.0 (18.1)	40.0 (36.2)	60.0 (54.3)	85.0 (77.0)	125.0 (113.1)	165.0 (149.5)	210.0 (190.0)
54 × 24 (1.374 × 610)	41	125–150 (93–112)	24.0 (21.7)	48.0 (43.5)	71.0 (64.3)	95.0 (86.0)	144.0 (130.0)	192.0 (173.8)	240.0 (217.5)

*Based on stone weighing 100 lb/cf when crushed.

†The first number indicates the diameter of the rolls, and the second indicates the width of the rolls.

Courtesy Iowa Manufacturing Company.

IMPACT CRUSHERS

Impact crushers fracture the feed stone by the application of high-speed impact forces. Advantage is taken of the rebound between the individual stones and against the machine surfaces to fully exploit the initial impact energy. The design of some impact crushers also utilizes shear and compression, in addition to impact action, to fracture the stones. This is accomplished by forcing the stone between the revolving and stationary parts of the crusher. Speed of rotation is important to the effective operations of these crushers as the energy available for impact varies as the square of the rotational speed.

Single Rotor

The single-rotor-type impact crusher (see Figs. 14.11 and 14.12) breaks the stone both by the impact action of the impellers striking the feed material and by the impact that results when the impeller-driven material strikes against the

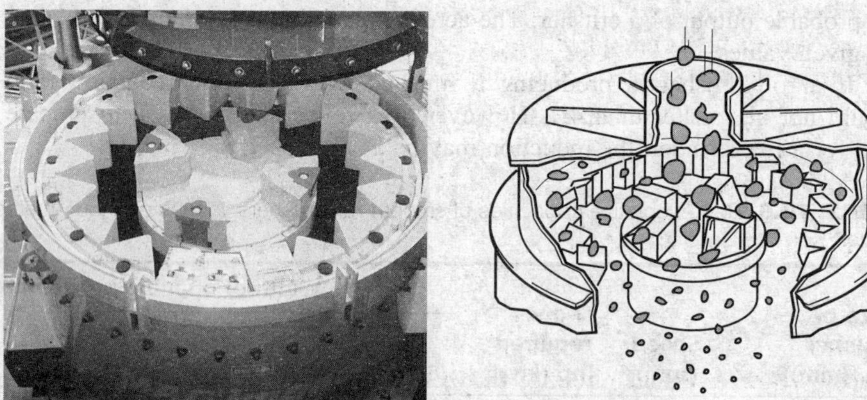

FIGURE 14.11 Interior view of a vertical-shaft impact crusher.

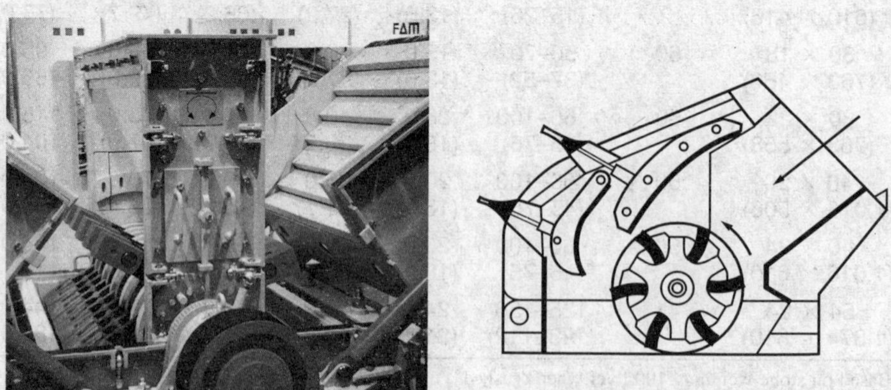

FIGURE 14.12 Interior view of a horizontal-shaft impact crusher.

aprons within the crusher unit. These crushers produce a cubical product but are economical only for low-abrasion feeds. The crusher's production rate is affected by the rotor speed. The speed also affects the reduction ratio. Therefore, any speed adjustment should be done only after consideration is given to both elements, production and final product.

Double Rotor

These units are similar to the single-rotor models and accomplish aggregate-size reduction by the same mechanical mechanisms. They will produce a somewhat higher proportion of fines. With both single- and double-rotor crushers, the impacted material flows freely to the bottom of the units without any further size reduction.

Hammer Mills

FIGURE 14.13 Cutaway of a hammer mill rock crusher showing breaking action.
Source: Cedarapids, Inc., a Terex Company.

The hammer mill, which is the most widely used impact crusher, can be used for primary or secondary crushing. The basic parts of a unit include a housing frame, a horizontal shaft extending through the housing; a number of arms and hammers attached to a spool, which is mounted on the shaft; one or more manganese-steel or other hard-steel breaker plates; and a series of grate bars, whose spacing can be adjusted to regulate the width of openings through which the crushed stone flows. These parts are illustrated in the section through the crusher shown in Fig. 14.13.

As the stone to be crushed is fed into the mill, the hammers, which revolve at a high rpm, strike the particles, breaking them and driving them against the breaker plates, which further reduces their size. Final size reduction is accomplished by grinding the material against the bottom grate bars.

The size of a hammer mill may be designated by the size of the feed opening. The capacity will vary with the size of the unit, the kind of stone crushed, the size of the material fed into the mill, and the speed of shaft rotation. Hammer mills will produce a high proportion of fines and cannot handle wet or sticky feed material.

SPECIAL AGGREGATE PROCESSING UNITS

To produce fine aggregate, such as sand, from stone that has been crushed to suitable sizes by other crushing equipment, rod or ball mills are frequently used. These crushers reduce the material to particle size by tumbling the feed stone with a grinding media such as balls or rods. Motion is imparted to the media through tumbling or rotating the vessel that contains the stone and the media.

FIGURE 14.14 Variable size balls for a ball mill.

Rod Mills

A rod mill is a circular steel shell lined on the inside with a hard-wearing surface. The mill is equipped with a suitable support or trunnion arrangement at each end and a driving gear at one end. It is operated with its axis in a horizontal position. The rod mill is charged with steel rods, whose lengths are slightly less than the length of the mill. Stone is fed through the trunnion at one end of the mill and flows to the discharge at the other end. As the mill rotates slowly, the stone is constantly subjected to the impact of the tumbling rods that produce the desired grinding. A mill can be operated wet or dry, i.e., with or without water added. The size of a rod mill is specified by the diameter and the length of the shell, such as 8 × 12 ft, respectively.

Ball Mill

A ball mill is similar to a rod mill but it uses steel balls, having a distribution of sizes (see Fig. 14.14), instead of rods to supply the impact necessary to grind the stone. Ball mills will produce fine material with smaller grain sizes than those produced by a rod mill.

FEEDERS

Compression-type crushers (jaw crushers) are designed to use particle interaction in the crushing process. An underfed compression crusher produces a larger percentage of oversize material, as the necessary material is not present to fully develop interparticle crushing. In an impact crusher, efficient use of interparticle collisions is not possible with an underfed machine. Gyratory-type crushers do not need feeders.

The capacity of compression- and impact-type crushers will be increased if the stone feed is at a uniform rate. Surge feeding tends to overload a crusher, and then the surge is followed by an insufficient supply of stone. Using a feeder ahead of a crusher eliminates most surge feeding problems that reduce crusher capacity. The installation of a feeder may increase the capacity of a jaw crusher as much as 15%.

There are many types of feeders:

1. Apron
2. Vibrating
3. Plate
4. Belt

Apron feeder. A feeder constructed of overlapping pans that form a continuous belt is referred to as an apron feeder. It will provide a continuous positive discharge of feed material. These feeders have the advantage that they can be obtained in considerable lengths.

Vibrating feeder. There are both simple vibrating feeders activated by a vibrating unit similar to the ones used on horizontal screens (discussed in the next section) and vibrating grizzly feeders. A vibrating grizzly feeder eliminates fines from the material entering the crusher. A vibrating feeder requires less maintenance than an apron feeder.

Plate. By means of rotating eccentrics, a plate can be made to move back and forth on a horizontal plane and can be used to uniformly feed material to a crusher.

Belt feeder. Operating on the same principle as the apron feeder, the belt feeder is used for smaller sizes of material, usually sand or small-diameter aggregates.

SURGE PILES

A stationary crushing plant may include several types and sizes of crushers, each probably followed with a set of screens and a belt conveyor to transport the stone to the next crushing operation or to storage (Fig.14.15). A plant may be designed to provide temporary storage for stone between the successive stages of crushing. This plan has the advantage of eliminating or reducing the surge effect that frequently exists when the crushing, screening, and handling operations are conducted as a liner operation.

The stone in temporary storage ahead of a crusher that is referred to as a "surge pile" can be used to keep at least a portion of a plant in operation at all times. Within reasonable limits, the use of a surge pile ahead of a crusher enables the crusher to be fed uniformly at the most satisfactory rate regardless of variations in the output of other equipment ahead of the crusher. The use of surge piles has enabled some plants to increase their final output production by as much as 20%.

FIGURE 14.15 A multistage crushing plant with surge piles feeding several types of crushers and screening units.

Advantages derived from using surge piles are

1. They enhance uniform feed and ensure high crusher efficiency. Even if there are excavation and hauling disruptions, the crushing plant operations will not be interrupted.
2. Should the primary crusher break down, the rest of the plant can continue to operate.
3. Repairs can be made to either the primary or secondary sections of the plant without complete stoppage of production.

Arguments against the use of surge piles are

1. They take up additional storage area.
2. They require the construction of storage bins or reclaiming tunnels.
3. They increase the amount of stone handling.

The decision to use or not use surge piles should be based on an analysis of the advantages and disadvantages for each plant.

CRUSHING EQUIPMENT SELECTION

In selecting crushing equipment, it is essential that certain information be known prior to making the selection. The information needed includes, but is not necessarily limited to, these items:

1. The kind of stone to be crushed
2. The required capacity of the plant—needed output production

3. The maximum size of the feed stones (information concerning the size ranges of the feed is also helpful)
4. The method of feeding the crushers
5. The specified size ranges of the product

Example 14.3 illustrates a crushing equipment selection process.

EXAMPLE 14.3

Select a primary and a secondary crusher to produce 100 tph of crushed limestone. The maximum-size stones from the quarry will be 16 in. The quarry stone will be hauled by truck, dumped into a surge bin, and fed to the primary crusher by an apron feeder, which will maintain a reasonably uniform rate of feed. The aggregate will be used on a project whose specifications require the following size distributions:

Size screen opening (in.)		
Passing	**Retained on**	**Percent**
$1\frac{1}{2}$		100
$1\frac{1}{2}$	$\frac{3}{4}$	42–48
$\frac{3}{4}$	$\frac{1}{4}$	30–36
$\frac{1}{4}$	0	20–26

Consider a jaw crusher for the primary and a roll crusher for the secondary crushing. The output of the jaw crusher will be screened and the material meeting the specification sizes removed. The remaining material, greater than $1\frac{1}{2}$ in., will be fed to the roll crusher.

Jaw crusher

Assume a setting of 3 in. for the jaw crusher. This will give a ratio of reduction of approximately 5:1, which is satisfactory. The jaw crusher must have a minimum top opening of 18 in. (16-in.-maximum-size feed stone plus 2 in.). Table 14.3 indicates a 24- by 36-in. size crusher has a probable capacity of 114 tph based on stone weighing 100 lb/cf when crushed. Figure 14.5 indicates that the product of the crusher will be distributed by sizes as follows:

Size range (in.)	Percent passing screens	Percent in size range	Total output of crusher (tph)	Amount produced in size range (tph)
Over $1\frac{1}{2}$	100–46	54	100	54.0
$1\frac{1}{2}$–$\frac{3}{4}$	46–26	20	100	20.0
$\frac{3}{4}$–$\frac{1}{4}$	26–11	15	100	15.0
$\frac{1}{4}$–0	11–0	11	100	11.0
Total		100%		100.0 tph

Roll crusher

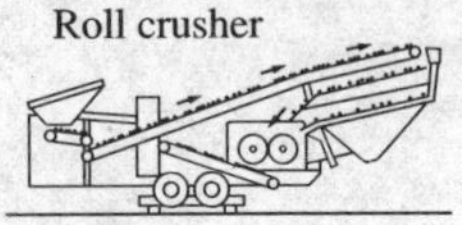

As the roll crusher will receive the output from the jaw crusher, the rolls must be large enough to handle 3-in. stone. Assume a setting of $1\frac{1}{2}$ in. From Eq. [14.1] the minimum radius will be 17.7 in. ($3 = 0.085R + 1.5$). Try a 40- by 20-in. (Table 14.6) crusher with a capacity of approximately 105 tons per hour for a $1\frac{1}{2}$-in. setting.

For any given setting, the crusher will produce about 15% of the crushed stone having at least one dimension larger than the setting. Thus, for a given setting 15% of the stone that passes through the roll crusher will be returned for recrushing. The total amount of stone passing through the crusher, including the returned stone, is determined as follows. Let

$$Q = \text{total amount of stone through the crusher}$$

Then

$$0.15Q = \text{amount of returned stone}$$

$$0.85Q = \text{amount of new stone}$$

$$Q = \frac{\text{amount of new stone}}{0.85}$$

$$= \frac{54 \text{ tph}}{0.85} = 63.5 \text{ tph}$$

The 40- by 20-in. roll crusher will handle this amount of stone easily. The distribution of the output of this crusher by size range will be

Size range (in.)	Percent passing screens	Percent in size range	Total output of crusher (tph)	Amount produced in size range (tph)
$1\frac{1}{2}$–$\frac{3}{4}$	85–46	39	63.5	24.8
$\frac{3}{4}$–$\frac{1}{4}$	46–19	27	63.5	17.1
$\frac{1}{4}$–0	19–0	19	63.5	12.1
Total		85%		54.0 tph

Now combine the output of each crusher by specified sizes.

Size range (in.)	From jaw crusher (tph)	From roll crusher (tph)	Total amount (tph)	Gradation requirement (percent)	Percent in size range
$1\frac{1}{2}$–$\frac{3}{4}$	20.0	24.8	44.8	42–48	44.8
$\frac{3}{4}$–$\frac{1}{4}$	15.0	17.1	32.1	30–36	32.1
$\frac{1}{4}$–0	11.0	12.1	23.1	20–26	23.1
Total	46.0 tph	54.0	100.0		100.0

Plant Layout

The actual layout and erection of the plant are the culminating tasks in the plant design. Plans should initially consider an appropriate equipment configuration within the plant. Special attention should be given to creating a productive, logical flow of material from the point where the trucks enter the plant with the raw material to the point where the trucks leave the plant with the crushed aggregate product. The physical environmental requirements of each piece of equipment, such as foundations, water requirements, and power requirements, should be evaluated to ensure that they are included during erection of the plant.

Drainage is of prime importance in constructing the plant. The plant design should include adequate space around the individual units of the plant. Access must be provided for maintenance personnel to perform repairs and to move cranes and lifting devices for moving the heavy pieces of crushing equipment. There must also be adequate material storage areas.

SEPARATION INTO PARTICLE SIZE RANGES

SCALPING CRUSHED STONE

The term *scalping*, as used in this chapter, refers to a screening operation. Scalping removes, from the main mass of stone to be processed, that stone that is too large for the crusher opening or is small enough to be used without further crushing. Scalping can be performed ahead of a primary crusher, and it represents good crushing practice to scalp all crushed stone following each successive stage of reduction.

Scalping ahead of a primary crusher serves two purposes. It prevents oversize stones from entering the crusher and blocking the opening, and it can be used to remove dirt, mud, or other debris that is not acceptable in the finished product. If the product of the blasting operation contains oversize stones, it is desirable to remove them ahead of the crusher. Scalping should accomplish this removal. The use of a "grizzly," which consists of a number of widely spaced parallel bars, can be used to scalp material ahead of the primary crusher (Fig. 14.16).

The product from the blasting operation can contain an appreciable amount of stone that meets the specified size requirements. In this event, it may be good economy to remove such stone ahead of the primary crusher, thereby reducing the total load on the crusher and increasing the overall capacity of the plant.

It is usually economical to install a scalper after each stage of reduction to remove specification sizes. This stone may be transported to grading screens, where it can be sized and placed in appropriate storage.

FIGURE 14.16 Grizzly scalper ahead of a small crusher.

14.2

SCREENING AGGREGATE

In all but the most basic crushing operation, the crushed rock particles must be separated into two or more particle size ranges. This separation enables the selection of certain material for additional or special processing, or the diversion of certain material so that it bypasses unnecessary processing. The screening process is based upon the simple premise that particle sizes smaller than the screen cloth opening size will pass through the screen and that oversized particles will be retained.

Screen openings can be described by either of two terms: (1) mesh and (2) clear opening. The term "mesh" refers to the number of openings per linear inch. To count the number of openings to an inch, measure from the center of the screen wire to a point 1-in. distance. Clear opening or "space" is a term that refers to the distance between the inside edges of two parallel wires.

Most specifications covering the use of aggregate stipulate that the different sizes shall be combined to produce a blend having a given size distribution. Persons who are responsible for preparing the specifications for the use of aggregate realize that crushing and screening cannot be done with complete precision, and accordingly they allow some tolerance in the size distribution. The extent of tolerance can be indicated by a statement such as: The quantity of aggregate passing a 1-in. screen and retained on a $\frac{1}{4}$-in. screen shall be not less than 30% or more than 40% of the total quantity of aggregate.

Revolving Screens

Revolving screens have several advantages over other types of screens, especially when they are used to wash and screen sand and gravel. The operating action is slow and simple, and the maintenance and repair costs are low. If the aggregate to be washed contains silt and clay, a scrubber can be installed near the entrance of a screen to agitate the material in water. At the same time, streams of water can be sprayed on the aggregate as it moves through the screen.

Vibrating Screens

Vibrating screens consist of one or more layers or "decks" of open mesh wire cloth mounted one above the other in a rectangular metal box (Fig. 14.17). These are the most widely used aggregate production screens. The vibration is obtained by means of an eccentric shaft, a counterweight shaft, or electromagnets attached to the frame or to the screens.

A unit may be horizontal or inclined with slight slope (20° or less) from the receiving to the discharge end. The vibration, 850 to 1,250 strokes per min, causes the aggregate to flow over the surface of the screen. Normally, large amplitude and slower speed are necessary for large screen openings, and the opposite is necessary for small screen openings. In the case of a horizontal screen, the throw of the vibrations must move the material both forward and upward. For that reason, its line of action is 45° relative to the horizontal.

Most of the particles that are smaller than the openings in a screen will drop through the screen, while the oversize particles will flow off the screen at the discharge end. For a multiple-deck unit, the sizes of the openings will be progressively smaller for each lower deck.

A screen will not pass all material whose sizes are equal to or less than the dimensions of the openings in the screen. Some of this material may be retained on and carried over the discharge end of a screen. The efficiency of a screen can be defined as the ratio of the amount of material passing through a screen divided by the total amount that is small enough to pass through, with the ratio expressed as a percent. The highest efficiency is obtained with a single-deck screen, usually amounting to 90 to 95%. As additional decks are installed, the efficiencies of these decks will decrease, being above 85% for the second deck and 75% for the third deck. Wet screenings will increase screening efficiency, but additional equipment is necessary for handling the water.

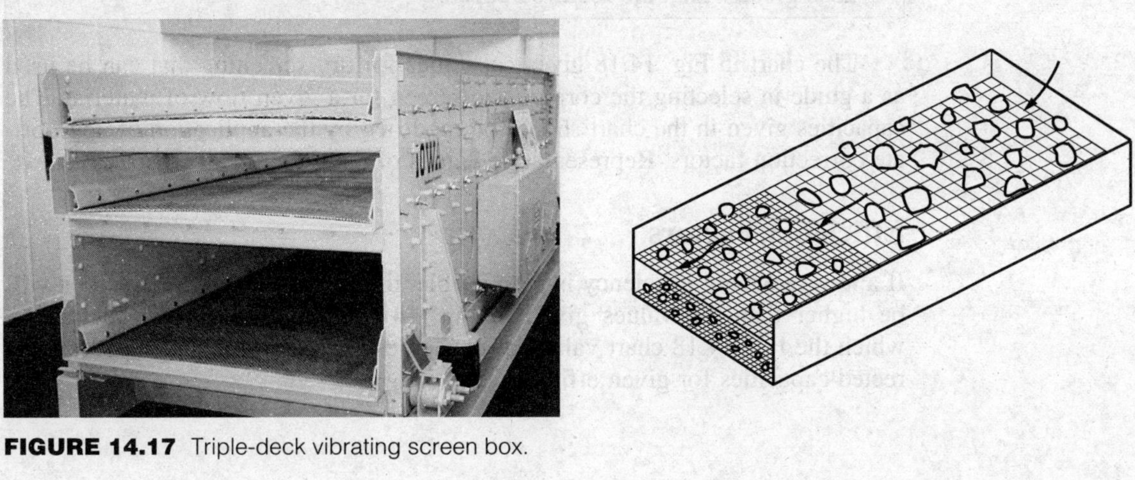

FIGURE 14.17 Triple-deck vibrating screen box.

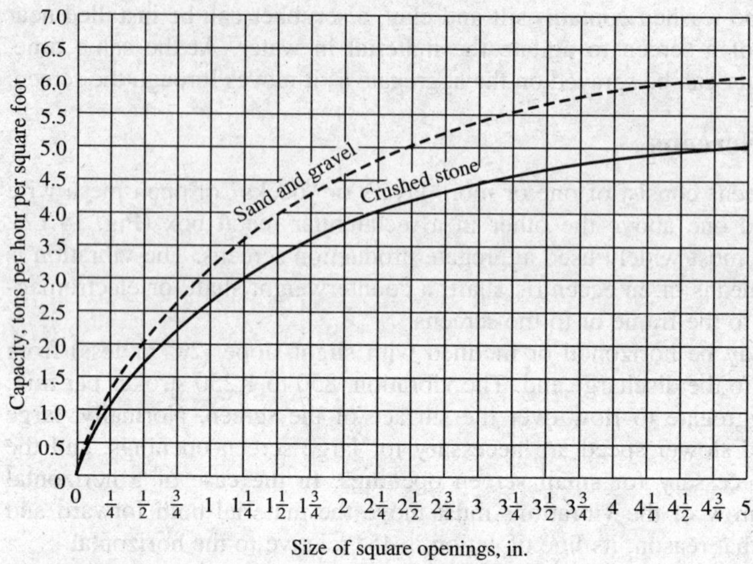

FIGURE 14.18 Screen-capacity chart.

The capacity of a screen is the number of tons of material that 1 sf will separate per hour. The capacity of a screen is *not* the total amount of material that can be fed and passed over its surface, but the rate at which it separates desired material from the feed. The capacity will vary with the size of the openings, kind of material screened, moisture content of the material screened, and other factors.

Because of the factors that affect the capacity of a screen, it will seldom if ever be possible to calculate in advance the exact capacity of a screen. If a given number of tons of material must be passed per hour, it is prudent to select a screen whose total calculated capacity is 10 to 25% greater than the required capacity.

The chart in Fig. 14.18 gives capacities for dry screening that can be used as a guide in selecting the correct size screen for a given flow of material. The capacities given in the chart should be modified by the application of appropriate correction factors. Representative values of these factors are discussed next.

Efficiency Factors

If a low screening efficiency is permissible, the actual capacity of a screen will be higher than the values given in Fig. 14.18. Table 14.7 gives factors by which the Fig. 14.18 chart values of capacities can be multiplied to obtain corrected capacities for given efficiencies.

TABLE 14.7 Efficiency factors for aggregate screening.

Permissible screen efficiency (%)	Efficiency factor
95	1.00
90	1.25
85	1.50
80	1.75
75	2.00

TABLE 14.8 Deck factors for aggregate screening.

For deck number	Deck factor
1	1.00
2	0.90
3	0.75
4	0.60

Deck Factors

This is a factor whose value will vary with the particular deck position for multiple-deck screens. The deck factor values are given in Table 14.8.

Aggregate-Size Factors

The capacities of screens given in Fig. 14.18 are based on screening dry material that contains particle sizes such as would be found in the output of a representative crusher. If the material to be screened contains a surplus of small sizes, the capacity of the screen will be increased, whereas if the material contains a surplus of large sizes, the capacity of the screen will be reduced. Table 14.9 gives representative factors that can be applied to the capacity of a screen to correct for the effect of excess fine or coarse particles.

TABLE 14.9 Aggregate-size factors for screening.

Percent of aggregate less than $\frac{1}{2}$ the size of screen opening	Aggregate size factor
10	0.55
20	0.70
30	0.80
40	1.00
50	1.20
60	1.40
70	1.80
80	2.20
90	3.00

Determining the Screen Area Required

Figure 14.18 provides the theoretical capacity of a screen in tons per hour per square foot based on material weighing 100 lb/cf when crushed. The corrected capacity of a screen is given by the equation

$$Q = ACEDG \quad [14.4]$$

where

Q = capacity of screen, tons per hour
A = area of screen, square feet
C = theoretical capacity of screen in tons per hour per square foot
E = efficiency factor
D = deck factor
G = aggregate-size factor

The minimum area of a screen to provide a given capacity is determined from the equation

$$A = \frac{Q}{CEDG} \quad [14.5]$$

EXAMPLE 14.4

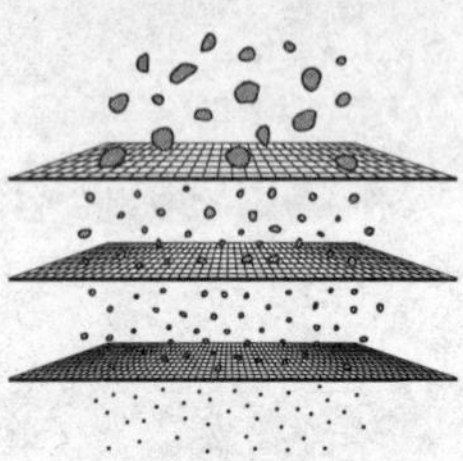

Determine the minimum-size single-deck screen, having $1\frac{1}{2}$-in.-sq. openings, for screening 120 tph of dry crushed stone, weighing 100 lb/cf when crushed. The screen box is 4 ft wide. A screening efficiency of 90% is satisfactory. An analysis of the aggregate indicates that approximately 30% of it will be less than $\frac{3}{4}$ in. in size. The values of the factors to be used in Eq. [14.5] are

Q = 120 tph

C = 3.32 tph/sf (Fig. 14.18)

E = 1.25 (Table 14.7)

D = 1.0 (Table 14.8)

G = 0.8 (Table 14.9)

Substituting these values in Eq. [14.5], we get

$$A = \frac{120 \text{ tph}}{3.3 \text{ tph/sf} \times 1.25 \times 1.0 \times 0.8} = 36.4 \text{ sf}$$

In view of the possibility of variations in the factors used, and to provide a margin of safety, it is recommended to select a screen whose total calculated capacity is 10 to 25% greater than the required capacity.

$$A = 36.4 \text{ sf} \times 1.10 = 40.0 \text{ sf}$$

Therefore, as a minimum, use a 4- by 10-ft (40-sf) screen.

Sand Preparation and Classification Machines

When the specifications for sand and other fine aggregates require the materials to meet specific size gradation requirements, it is frequently necessary to produce the gradations by mechanical equipment. Several types of equipment are available for this purpose. There are mechanical and water flow machines that classify sand into a multiple number of individual sizes. Sand and water are fed to these classifiers at one end of the unit's tank. As the water flows to the outlet end of the tank, the sand particles settle to the bottom of the tank, the coarse ones first and the fine ones last. When the depth of a given size reaches a predetermined level, a sensing paddle will actuate a discharge valve at the bottom of the compartment to enable that material to flow into the splitter box, from which it can be removed and stockpiled.

FIGURE 14.19 Screen classifier for producing specification sand.

Another machine for handling sand is the screw-type classifier. This unit can be used to produce specification sand. A sand screw is erected so that the material must move up the screw to be discharged. In the case of the screw classifier shown in Fig. 14.19, the motor is at the discharge end. Sand and water are fed into the hopper. As the spiral screws rotate, the sand is moved up the tank to the discharge outlet under the motor. Undesirable material is flushed out of the tank with overflowing water.

OTHER AGGREGATE PROCESSING ISSUES

LOG WASHERS

When natural deposits of aggregate, such as sand and gravel, or individual pieces of crushed stone contain deleterious material as a part of the matrix or as deposits on the surface of the aggregate, it will be necessary to remove these materials before using the aggregate. One method of removing the material is to pass the aggregate through a machine called a "log washer." This unit consists of a steel tank with two electric-motor-driven shafts, to which numerous replaceable paddles are attached (Fig. 14.20). When the washer is erected in a plant, the discharge end is raised. The aggregate to be processed is fed into the unit at the lower end, while a constant supply of water flows into the elevated end. As the shafts are rotated in opposite directions, the paddles move the aggregate toward the upper end of the tank, while producing a continuing scrubbing action between the particles. The stream of water will remove the undesirable material and discharge it from the tank at the lower end, whereas the processed aggregate will be discharged at the upper end.

FIGURE 14.20 Log washer for scrubbing coarse aggregate.
Source: Telsmith Division, Barber-Greene Company.

SEGREGATION

After stone is crushed and screened to provide the desired size ranges, it is necessary to handle it carefully or the large and small particles may be separated, thereby destroying the blend in sizes that is essential to meeting graduation requirements. If aggregate is permitted to flow freely off the end of a belt conveyor, especially at some height above the storage pile, the material will segregate by size.

Specifications covering the production of aggregate frequently stipulate that the aggregate transported by a belt conveyor shall not be permitted to fall freely from the discharge end of a belt. The end of the belt should be kept as low as possible and the aggregate should be discharged through a rock ladder, containing baffles, to prevent segregation.

SAFETY

The plant designer has an obligation to provide a workplace that minimizes the possibility of situations that lead to accidents. Ample work room should be provided around all units for personnel movement and the use of tools.

Crushing and screening equipment is designed by the manufacturer considering the safety of all operating personnel. All equipment comes from the manufacturer with protective guards, covers, and shields installed around moving parts. It should not be changed or modified in any manner that eliminates the accident prevention devices. Repeatedly, the Labor Department accident reports detail the results of removing protective covers and guards from rock processing equipment: "He had removed a safety guard and was caught in the machinery." These devices protect operators and others working on or near the machines. Even with safety devices, basic safety practices should be followed at all times.

- Do not remove guards, covers, or shields when the equipment is running.
- Replace all guards, covers, or shields after maintenance.
- Never lubricate the equipment when it is in motion.
- Always establish a *positive lockout* of the involved power source before performing maintenance.
- Block parts as necessary to prevent unexpected motion while performing maintenance or repair.
- Do not attempt to remove jammed product or other blockage when the equipment is running.

A 1998 Department of Labor, Mine Safety and Health Administration accident investigation reports the results of not following this safety procedure. In an accident that was caused by failure to shut off and block the jaw crusher before attempting to dislodge rocks, a crusher operator was fatally injured when he was struck in the face by a sledgehammer head while attempting to free a hang-up in the crusher.

SUMMARY

In operating a crushing plant, the drilling pattern, the amount of explosives, the size of the shovel or loader used to load the stone, and the size of the primary crusher should be coordinated to ensure that all stone can be economically used. Jaw crushers have large energy-storing flywheels and provide a high mechanical advantage. Gyratory and cone crushers have a gyrating mantle mounted within a bowl. A roll crusher consists of a heavy cast-iron frame equipped with either one or more hard-steel rolls, each mounted on a separate horizontal shaft. In impact crushers, the stones are broken by the application of high-speed impact forces. Manufacturers provide capacity charts for their crushers. These are typically based on a standard crushed stone weight of 100 lb/cf.

Following the crushing operation it is almost always necessary to size the product. This sizing or screening enables the plant operator to direct certain selected material to receive additional or special processing, or certain material may be diverted to bypass unnecessary processing. The screening is based on the simple premise that particle sizes smaller than the screen cloth opening size will pass through the screen and that oversized particles will be retained. Critical learning objectives include:

- An understanding and ability to use a manufacturer's crusher capacity chart
- An ability to calculate roll crusher feed size
- An ability to design a crushing plant based on required size distribution specifications
- An ability to calculate required area of screens

These objectives are the basis for the problems that follow.

PROBLEMS

14.1 A jaw crusher, with a closed setting of 3 in., produces 200 tph of crushed stone. Determine the number of tons per hour produced in each of the following size ranges: in excess of $2\frac{1}{2}$ in.; between $2\frac{1}{2}$ and $1\frac{1}{2}$ in.; between $1\frac{1}{2}$ and $\frac{1}{4}$ in.; less than $\frac{1}{4}$ in. (54 tph in excess of $2\frac{1}{2}$ in.; 54 tph between $2\frac{1}{2}$ and $1\frac{1}{2}$ in.; 70 tph between $1\frac{1}{2}$ and $\frac{1}{4}$ in.; 22 tph less than $\frac{1}{4}$ in.)

14.2 A roll crusher, set at 2 in., produces 120 tph of crushed stone. Determine the number of tons per hour produced in each of the following size ranges: in excess of $1\frac{1}{2}$ in.; between $1\frac{1}{2}$ and $\frac{3}{4}$ in.; between $\frac{3}{4}$ in. and $\frac{1}{4}$ in.

14.3 Select a jaw crusher for primary crushing and a roll crusher for secondary crushing to produce 200 tph of limestone rock. The maximum-size stone from the quarry will be 24 in. The stone is to be crushed to the following specifications:

Size screen opening (in.)		
Passing	Retained on	Percent
2		100
2	$1\frac{1}{4}$	30–40
$1\frac{1}{4}$	$\frac{3}{4}$	20–35
$\frac{3}{4}$	$\frac{1}{4}$	10–30
$\frac{1}{4}$	0	0–25

Specify the size and setting for each crusher selected. (48 × 60 jaw with 3-in. setting and 24 × 16 roll with 2-in. setting)

14.4 A jaw crusher and a roll crusher are used in an attempt to crush 140 tph. The maximum-size stone from the quarry will be 18 in. The stone is to be crushed to the following specifications:

Size screen opening (in.)		
Passing	Retained on	Percent
$2\frac{1}{2}$		100
$2\frac{1}{2}$	$1\frac{1}{2}$	30–50
$1\frac{1}{2}$	$\frac{1}{2}$	20–40
$\frac{1}{2}$	0	10–30

Select crushers to produce this aggregate. Can the product of these crushers be processed to provide the desired sizes in the specified percentages? If so, tell how.

14.5 A 30- by 42-in. jaw crusher is set to operate with a $2\frac{1}{2}$-in. opening. The output from the crusher is discharged onto a screen with $1\frac{1}{2}$-in. openings. The efficiency of the screen is 90%. The aggregate that does not pass through the screen goes to a 40- by 20-in. roll crusher, set at $1\frac{1}{4}$ in. Determine the maximum output of the roll crusher in tons per hour for material less than 1 in. There will be no recycle of roll crusher oversize. What amount, in tons per hour, of the roll crusher output is in the range of 1 in. to $\frac{1}{2}$ in., and what amount is less than $\frac{1}{2}$ in.?

14.6 A portable crushing plant is equipped with these units:

One jaw crusher, size 15 × 30 in.

One roll crusher, size 30 × 22 in.

One set of horizontal vibrating screens, two decks, with $1\frac{1}{2}$- and $\frac{3}{4}$-in. openings.

The specifications require that 100% of the aggregate shall pass a $1\frac{1}{2}$-in. screen and 50% shall pass a $\frac{3}{4}$-in. screen. Assume that 10% of the stone from the quarry will be smaller than $1\frac{1}{2}$ in. and that this aggregate will be removed by passing the quarry product over the screen before sending it to the jaw crusher. The crushed aggregate will weigh 110 lb/cf. Determine the maximum output of the plant, expressed in tons per hour. Include the aggregate removed by the screens prior to sending it to the crushers.

14.7 The crushed stone output from a 36- by 42-in. crusher, with a closed opening of $2\frac{1}{2}$ in., is passed over a single horizontal vibrating screen with $1\frac{1}{2}$-in. openings. If the permissible screen efficiency is 90%, use the information in the book to determine the minimum-size screen, expressed in square feet, required to handle the output of the crusher. (45.5 sf minimum)

14.8 The crushed stone output from a 24- by 36-in. jaw crusher, with a closed setting of $2\frac{1}{2}$ in., is passed over the 2- and the 1-in. openings of a vibrating screen. The permissible efficiency is 85%. The stone has a unit weight of 108 lb/cf when crushed. Determine the minimum-size screen, in square feet, required to handle the crusher output. If the screening unit is $3\frac{1}{2}$ feet wide, what will be the nominal screen sizes? Screen lengths should be in integer dimensions of feet.

14.9 The crushed stone output from a 42- by 48-in. jaw crusher, with a closed setting of 3 in., is passed over a vibrating screen having two screens, the first with $2\frac{1}{2}$-in. openings and the second with $1\frac{1}{2}$-in. openings. The permissible efficiency is 85%. The stone has a unit weight of 110 lb/cf when crushed. Determine the minimum-size screen, in square feet, required to handle the crusher output. If the screening unit is 4 ft wide, what will be the nominal screen sizes?

14.10 The crushed stone output from a 36- by 42-in. jaw crusher, with a closed setting of 4 in., is to be screened into the following sizes: $2\frac{1}{2}$ to $1\frac{1}{2}$ in., $1\frac{1}{2}$ to $\frac{3}{4}$ in., and less than $\frac{3}{4}$ in. A three-deck horizontal vibrating screen will be used to separate the three sizes. The stone weighs 115 lb/cf when crushed. If the permissible screen efficiency is 90%, determine the minimum-size screen for each deck, expressed in square feet, required to handle the output of the crusher. If the screening unit is 4 ft wide, what will be the nominal screen sizes?

REFERENCES

1. Barksdale, Richard D. (ed.) (1991). *The Aggregate Handbook*, National Stone Association, Washington, DC.
2. *Cedarapids Pocket Reference Book*, 17th pocket edition (2002). Cedarapids, Inc., 916 16th Street NE, Cedar Rapids, IA.

WEBSITE RESOURCES

1. www.cedarapids.com Cedarapids, Inc., Cedar Rapids, IA, is a manufacturer of aggregate processing equipment.
2. www.icar.utexas.edu International Center for Aggregates Research (ICAR). ICAR is operated jointly by The University of Texas at Austin (UT) and Texas A&M University (TAMU).
3. www.metsominerals.com Metso Minerals Oy, P.O. Box 307, Lokomonkatu 3, FIN-33101 Tampere, Finland, is the global manufacturer of rock size reduction, classification, mineral separation and recovery, and materials handling equipment. Major product brand names include Nordberg, Svedala, Trellex, Lindemann, and Skega.
4. www.mining-technology.com Mining Technology is a British website for the international mining industry. The site includes news on equipment, products, and conference listings, and a comprehensive links page to relevant mining resources.
5. www.nssga.org National Stone, Sand & Gravel Association (NSSGA), Alexandria, VA, is the national trade association representing the crushed stone, sand, and gravel—or aggregates—industries. The association provides support in operating productivity improvements, engineering research, safety and health, environmental concerns, and technical issues.

CHAPTER 15

Asphalt Mix Production and Placement

The ability to easily accommodate a pavement section to staged construction and to recycle old pavements, and the fact that mix designs can be adjusted to utilize local materials are three critical factors favoring the use of asphalt paving materials. An asphalt plant is a high-tech group of machines capable of uniformly blending, heating, and mixing the aggregates and asphalt cement of asphalt concrete, while at the same time meeting strict environmental regulations, particularly in respect to particle emissions. Asphalt pavers consist of a tractor, either track or rubber-tired, and a screed. Mat thickness, which is controlled by the screed, can be maintained by using grade sensors or by tracing an external reference with a shoe or ski.

INTRODUCTION

This chapter deals with the equipment and methods used for the production and placement of asphalt pavements (Fig. 15.1). Although the same or similar equipment may in some instances be used for other purposes, such as rollers for compaction of soils, all machines will be treated in this chapter in relation to asphalt operations.

Asphalt paving materials are produced for the construction of highway, parking lot, and airfield pavements. The ability to easily accommodate a pavement section to staged construction and to recycle old pavements, and the fact that mix designs can be adjusted to utilize local materials are three critical factors favoring the use of asphalt paving materials. When appraising asphalt production and paving equipment, consideration must be given to the types of projects anticipated. Some asphalt mixing plants are operated primarily as producers of multiple mixes for FOB (free on board) plant sales or to serve multiple paving spreads on small jobs. These plants must be able to easily and quickly change their production mix to meet the requirements of multiple customers. Other plants are high-volume producers serving a single paving spread;

FIGURE 15.1 Asphalt paving of a major highway.

this is particularly true of portable plants that are moved from project to project. Equipment is available that is specifically designed to meet the needs of both these situations. A constructor must select the equipment and methods that allow the service flexibility best suited to the specific project types that will be undertaken.

GLOSSARY OF ASPHALT TERMS

The following glossary defines the important terms that are used in describing asphalt equipment and construction processes.

Asphaltic concrete. Asphalt paving material prepared in a hot mix plant and used for construction of highway, parking lot, or airfield pavements having high traffic volumes and axle loads.

Automatic screed controls. A system that overrides the self-leveling action of the asphalt paver screed and enables paving to a predetermined grade and slope, using either a rigid or mobile type reference.

Binder. The asphalt cement material in a paving mix used to bind the aggregate particles together, to prevent the entrance of moisture, and to act as a cushioning agent.

Binder course. A layer of asphalt mix placed between the base and the surface courses of a pavement structure.

Breakdown. The initial compaction of the asphalt mix that takes place directly behind the paver and is intended to achieve maximum density in the shortest time frame.

Friction course. A layer of asphalt mix, usually less than 1 in. thick, placed on a structurally sound pavement to improve skid resistance and smoothness.

Grade or string line. A wire or string erected at a specified grade and alignment that is used as a reference for an automatic control system on a paver, milling machine, or fine-grade machine (trimmer).

Hot bins. Asphalt plant bins used to store dried aggregates prior to proportioning and mixing of the aggregate with the asphalt cement.

Hot elevator. A bucket elevator used to carry hot, dried aggregate from the dryer to the gradation unit of an asphalt plant.

Hot oil heater. A heater for increasing the temperature of heat transfer oil. The heated transfer oil is used as the means to provide temperature control for asphalt plant process operations and for increasing or maintaining the temperature of stored liquid asphalt.

Leveling arms. Two long arms extending forward from each side of the asphalt paver screed and attached to tow points on the tractor. This mechanical connection enables the screed to float on the asphalt mix during placement.

Leveling course. A new layer of asphalt mix placed over a distressed roadway to improve its geometry prior to resurfacing.

Lift (mat). A layer of asphalt mix separately placed and compacted.

Lute. A type of rake used to smooth out minor surface irregularities in the hot asphalt mix behind the paver. It has a straight-edge side and a toothed side

Marshall stability. The maximum load resistance that an asphalt paving material test specimen will develop when tested with the Marshall stabilometer at 140°F (60°C).

Prime coat. An application of liquid asphalt over an untreated base to coat and bond the loose aggregate particles, waterproof the surface, and promote adhesion between the base and the overlying course.

Pugmill. A mechanical device, consisting of paddles attached to rotating shafts, for mixing aggregates and asphalt cement.

RAP (reclaimed asphalt pavement). Asphalt paving material removed from an existing paved surface by cold planing or ripping.

Rubberized asphalt. An asphalt mix containing powdered or shredded rubber that is introduced into the mix to produce a more resilient pavement.

Screed. That part of an asphalt paver that smooths and compacts the asphalt mix.

Slope. The transverse inclination of a roadway or other surface.

Superpave. An asphalt concrete mix-design method for determining the aggregate structure and proportions of aggregate and asphalt cement.

Surface course. The top or riding surface of a pavement structure.

Tack coat. A light application of liquid asphalt usually emulsified with water. Used to help ensure a bond between the surface being paved and the new mat that is being placed.

Thickness controls. Manually operated controls usually located at the outer edge of the asphalt paver screed by which the screed operator can adjust the angle of attack of the screed plate to increase or decrease the mat thickness.

Weigh hopper. A batch plant component usually located under the hot bins in which the aggregates and asphalt cement are weighed prior to discharge into the mixing pugmill.

Wet collector. An asphalt plant dust collection system utilizing water and a high-pressure venturi to capture dust particles from the exhaust gas of a dryer or drum mixer. Wet collected dust cannot be returned to the mix.

Windrow. A linear pile of material placed on a grade or previously placed mat for later pickup or spreading.

Windrow elevator. A mechanical pickup device that travels ahead of an asphalt paver and lifts windrowed asphalt mix into the hopper of a paver. This allows for continuous paver operation. Windrow elevators are usually attached to the paver (see Fig. 15.1).

STRUCTURE OF ASPHALT PAVEMENTS

Asphalt binders offer great flexibility to construct pavements tailored to the requirements of the local situation. Material selection and construction process vary depending on the type of pavement.

Asphalt pavements are constructed as layered systems (see Fig. 15.2). This structure consists of a prepared subgrade, an aggregate subbase, an aggregate base, and an asphalt-bound surface layer. The asphalt surface layer consists of two courses, the surface wearing course and the underlying binder course. The aggregate characteristics and amount of asphalt cement are designed specifically to match the needs of the binder and surfacing courses. Both the asphalt cement and surface course materials are produced at a hot mix plant, where the aggregates and binder are heated and mixed to produce the **hot mix asphalt** concrete.

hot mix asphalt
An asphalt paving mix produced in a plant. The mix aggregates are dried and heated, and combined with hot liquid asphalt cement. The mix is placed at high temperatures.

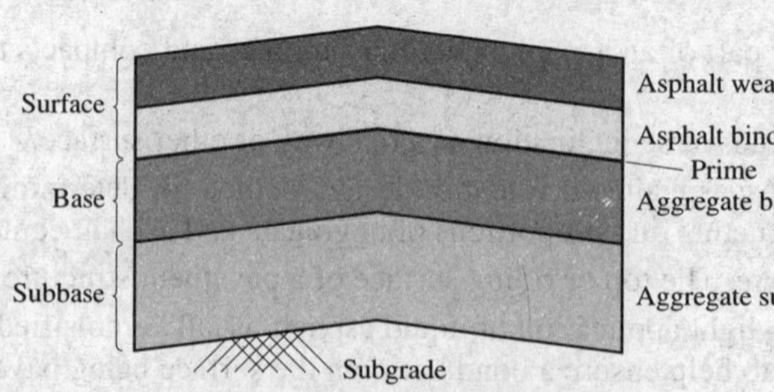

FIGURE 15.2 Typical structure of a flexible pavement.

This pavement structure is still used in many applications; however, to meet the needs of increased traffic demands (higher wheel loadings), there is a tendency to replace the base and subbase layers with asphalt-bound materials. The construction of pavements with full-depth asphalt-bound layers is similar to the construction of pavements with aggregate bases and subbases.

Hot mix asphalt concrete provides a quality pavement for many traffic situations. However, for low-volume road situations, a lower-cost pavement surface alternative can be used. Two popular alternatives are surface treatments and blade-mixed pavements. Both of these surfacing methods provide a wearing layer for the traffic.

A surface treatment is constructed over a base by spraying a layer of liquid asphalt (binder) on the surface, then covering that binder with a layer of aggregates. The surface is then rolled to compact the aggregates into the liquid binder.

A blade mix surface is constructed by placing a windrow of aggregate materials on the prepared base. Liquid asphalt is then metered onto the windrow at the desired application rate. A grader is used to work this material across the base surface until the asphalt is thoroughly mixed with the aggregate. The grader then spreads the blended material, and a roller compacts the surface to the required density.

FLEXIBLE PAVEMENTS

Flexible pavements have a ride surface constructed with a combination of asphalt cement and aggregates. These two basic materials can be proportioned and mixed in many ways to produce a pavement surface suitable to the local conditions. Pavements are designed to meet the following objectives:

1. Support the axle loads imposed by the traffic.
2. Protect the base and subbase from moisture.
3. Provide a stable, smooth, and skid-resistant riding surface.
4. Resist weathering.
5. Provide economy.

The aggregate and asphalt cement that make up the paving material must provide a stable structure capable of supporting the repetitive vertical wheel loads imposed and resisting the kneading mechanism that wheel rotation and movement transmits to the structure. The pavement surface will be subjected to abrasive wear, while the entire section will have to resist structural movement. In addition to providing a structural wearing surface, the asphalt mixture must seal the base, subbase, and subgrade to prevent water intrusion. This is because of the influence of moisture on the strength of underlying soils. Vehicle fuel efficiency, ride quality, and safety are affected by surface texture. Besides moisture effects, the action of heat and cold can be very destructive to pavements. These objectives fix the performance criteria of a pavement section. The purpose of blending aggregates and asphalt binder is to achieve a final product that satisfies all of these objectives.

Aggregates

The load applied to an asphalt pavement is primarily carried by the aggregates in the mix. The aggregate portion of a mix accounts for 90 to 95% of the material by weight. Good aggregates and their proper gradation are critical to the mix's performance. Ideally, an aggregate gradation should be provided that requires the minimum amount of expensive asphalt cement. The asphalt cement fills most of the voids between the aggregate particles as well as the voids in the aggregates.

Generally, the project owner will specify a range of allowable gradations for an asphalt concrete mix. Under the Superpave mix-design system, the allowable aggregate gradations are designated by the nominal maximum aggregate size. The nominal maximum aggregate size is the sieve size that is one size larger than the first sieve that retains more than 10% of the aggregate mass. The maximum aggregate size is one sieve size larger than the nominal maximum aggregate size. The designated aggregate sizes used in Superpave are 37.5, 25, 19.5, 12.5, and 9.5 mm nominal maximum aggregate size. The gradation control points for a 12.5-mm Superpave aggregate designation are shown in Fig. 15.3. For the production of an asphalt concrete mix, a blend of aggregates must be used such that the combined aggregate gradation will fall within the control points, as shown in Fig. 15.4.

Figure 15.3 demonstrates that a full spectrum of aggregate particle sizes is required for an asphalt concrete mix. Individual aggregate particle sizes can range from over 25 mm down to particles that pass through a sieve with openings of 0.0075 mm (No. 200 sieve). If a single stockpile were created with the desired gradation, the larger aggregates would tend to separate from the smaller aggregates. Asphalt concrete made from a segregated stockpile would have pockets of small and fine aggregate and the pavement performance would suffer. To reduce the segregation problem, aggregates are stored in stockpiles of similar sizes. A minimum of three stockpiles is required to store coarse,

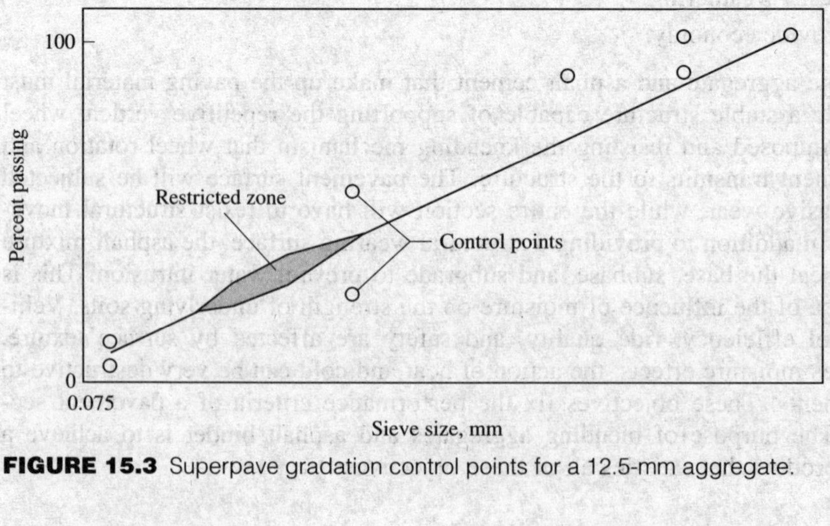

FIGURE 15.3 Superpave gradation control points for a 12.5-mm aggregate.

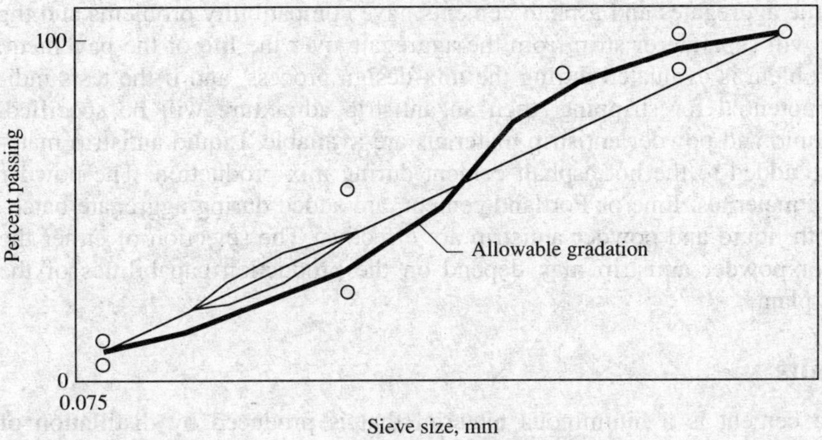

FIGURE 15.4 Example of an allowable 12.5-mm Superpave-designed aggregate blend.

intermediate, and fine aggregates. Frequently a fourth stockpile is used to provide better control of the fine-aggregate characteristics. During the asphalt concrete mix-design process, the needed proportions of each aggregate size are determined. During the mix production, the stockpiles are blended in the predetermined ratios. The method used to combine the aggregates during production depends on the type of production.

Besides gradation, the aggregate properties of cleanliness, resistance to wear and abrasion, texture, **porosity**, and resistance to stripping (debonding between the asphalt and the aggregate) are important.

porosity
The relative volume of voids in a solid or mixture of solids. Porosity is used to indicate the ability of the solid or the mixture of solids to absorb a liquid such as asphalt.

The amount of foreign matter, whether soil or organic material, that is present in the aggregate will reduce the load-carrying ability of a pavement. Visual inspection can often identify an aggregate cleanliness problem, and washing, wet screening, or other methods as discussed in Chapter 14 can be employed to correct the situation.

The effects of aggregate surface texture are manifested in the strength of the pavement structure and in the workability of the mix. Strength is influenced by the ability of the individual aggregate particles to "lock" together under load. This ability to lock is enhanced by angular rough-textured particles. Smooth aggregates, such as "river run" gravels and sands, produce a pavement that exhibits a reduced strength compared to one constructed from aggregates having rough surfaces. When necessary, rounded gravels can be crushed to create more angular surfaces. A mix using aggregates having a rough surface will require slightly more asphalt cement.

The porosity of an aggregate affects the amount of asphalt cement required in a mix. More asphalt must be added to a mix containing porous aggregates to make up for that which is absorbed by the aggregates and is not available to serve as binder. Slag and other manufactured aggregates can be highly porous, which will increase the asphalt cement quantity and cost of a mix. However, because of the ability of these materials to resist wear, their use can be justified based on total lifetime project economics.

Some aggregates and asphalt cements have compatibility problems and the asphalt will separate or strip from the aggregate over the life of the pavement. This problem is evaluated during the mix-design process, and if the tests indicate a potential for stripping, then an antistrip admixture will be specified. Both liquid and powder antistrip materials are available. Liquid antistrip materials are added to the hot asphalt cement during mix production. The powder antistrip materials, lime or Portland cement, are added during aggregate batching. Both liquid and powder antistrip are effective. The selection of either the liquid or powder antistrip may depend on the production capabilities of the asphalt plant.

Asphalts

Asphalt cement is a bituminous material that is produced by distillation of petroleum crude oil. This process may occur naturally, and there are several sources of natural asphalt throughout the world, with Trinidad Lake asphalt being the most famous natural asphalt. However, the vast majority of asphalt is produced by the petroleum industry. Historically, asphalt cement was a waste product that was the "bottom of the barrel" material left over when fuels were extracted from crude oil. As the petroleum industry gained sophistication, the ability to extract higher-value materials from the crude oil increased and altered the quality of the asphalt cements. In addition, different crude oil sources have different chemical compositions, which affects the quality of the asphalt cements. Thus, the quality of an asphalt cement is a function of the original crude oil source and the refining process.

Refined asphalt cement is a viscoelastic material that behaves as a liquid at high temperatures and as an elastic solid at low temperatures. At pavement operating temperatures, asphalt cement has a semisolid consistency. For construction, asphalt cement must be put into a liquid state to mix and coat aggregates. Asphalt cement can either be heated to a liquid condition, approximately 300°F, prior to mixing with the aggregates or converted to a liquid product by dilution or emulsification.

The quality of asphalt cement used for highway construction is controlled through specifications. Specifications have evolved over time. The earliest specification relied on the ability of the asphalt technologist to detect asphalt quality by chewing on a glob of asphalt. Auspiciously, scientific property tests have replaced the old chewing test. These tests evaluate the physical characteristics of the asphalt cement; currently, there are no chemical tests for asphalt cement. There are currently three specification methods for asphalt cement. The most recent, performance grade specifications, are displacing the other methods. However, not all agencies have adopted the performance grade specifications, so construction engineers may be exposed to some of the earlier specifications.

Penetration Grades of Asphalt Cement The earliest codified specification for asphalt cement is the penetration method, AASHTO M20. The primary test of this specification is the penetration test, AASHTO T49. In this test, an asphalt sample, at a temperature of 77°F, is subjected to a needle load, with a 100-g mass and a test time of 5 sec. The distance the needle penetrates into the sample is a measure of the consistency of the asphalt cement. Soft asphalts have high penetration values, while hard asphalts exhibit small penetration values. The asphalt cement penetration grades are 40–50, 60–70, 85–100, 120–150, and 200–300 pen. The tested asphalt cement is placed into one of these grades based on the average of three penetration tests. In addition to the penetration test, there are a variety of other quality tests to which the asphalt cement must adhere to meet the material specifications.

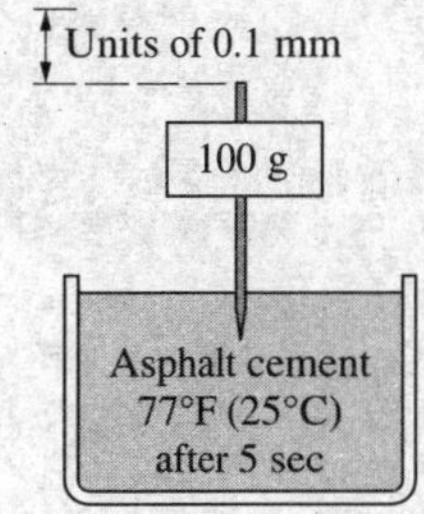

Viscosity Grades of Asphalt Cement The next specification method to gain wide acceptance classified the asphalt cement based on viscosity at 140°F. AASHTO M226 specifies two asphalt cement specifications, AC and AR. The difference between the AC and AR specification is that the AC grade designation is based on testing asphalt cement in an unconditioned state, that is, the asphalt cement is in an "as produced" state. The grade designation for the AR specification is based on testing the asphalt cement after it has been conditioned. The conditioning process simulates the hardening of the asphalt cement during the construction process. The grade designations are AC 2.5, 5, 10, 20, and 40 and AR 500, 1000, 2000, 4000, 8000, and 16000. As in the penetration grading method, there are several other tests that the asphalt cement must comply with for classification into one of these grades.

The viscosity and penetration specifications were effective in controlling the quality of the asphalt cement. However, the tests performed for these specifications could not be used to relate the quality of the asphalt cement to the performance of the pavement. In the late 1980s, the U.S. government sponsored a major research program to enhance pavement construction and performance. One of the products of this research was performance grade specification for asphalt cement.

Performance Grades of Asphalt Cements The performance grade asphalt cement specifications relate engineering measures of asphalt cement properties to specific pavement performance matters. The penetration and viscosity specifications have required test limits at fixed temperatures. The performance grade specifications use fixed test criteria but the test temperatures for evaluating the asphalt cement are selected to reflect the local construction conditions. The grade specifications for the performance grade method are designated based on the temperature application range of the asphalt cement. The grades in the performance grade system are designated as PGhh-ll, where PG identifies the performance grade specification, hh identifies the high temperature application for the asphalt cement in degrees Celsius, and ll identifies the low-temperature designation for the asphalt cement in degrees Celsius. The performance grade designations for asphalt cements are shown in Table 15.1.

Table 15.1 demonstrates that there is a wide range of grades available in the performance grade specifications. Not all these grades will be available in a local area. Refineries limit their production to match the market. Indirectly, the state departments of transportation (DOTs) are the largest asphalt cement

TABLE 15.1 Performance grade asphalt cement classifications.

Lower temperature designation	Upper temperature designation						
	46	52	58	64	70	76	82
−10		X		X	X	X	X
−16		X	X	X	X	X	X
−22		X	X	X	X	X	X
−28		X	X	X	X	X	X
−34	X	X	X	X	X	X	X
−40	X	X	X	X	X		
−46	X	X					

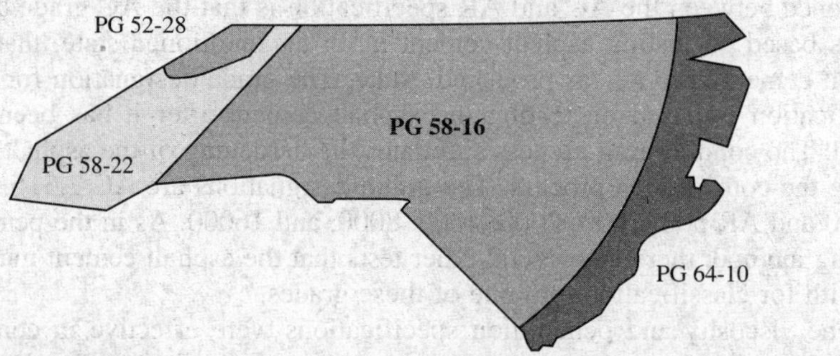

FIGURE 15.5 North Carolina Department of Transportation specified performance grade asphalt cements.

customers. Therefore the refineries seek to meet the DOT specifications. This limits the number of grades of asphalt cements needed. Figure 15.5 shows the four performance grades of asphalt cement specified by the North Carolina Department of Transportation.

Liquid Asphalt Cement

To allow construction without having to heat the asphalt cement to 300°F, two methods have been developed to reduce asphalt cement to a liquid state: cutbacks and emulsions. Asphalt cutbacks are a blend of asphalt cement and a fuel product. Rapid-, medium-, and slow-cure cutbacks can be formulated by diluting the asphalt with gasoline or naphtha, diesel, and fuel oil, respectively. Emulsions are the other option for preparing a liquid asphalt. Asphalt emulsions are produced by using a colloidal mill to break down asphalt cement into very fine "globules" that are introduced into water that has been treated with an emulsifying agent. The emulsifying agent is a soaplike material that enables the asphalt cement globules to remain suspended in the water. Due to cost, safety, and environmental concerns, asphalt emulsions have largely displaced cutbacks when a liquid asphalt is needed.

TABLE 15.2 Types of asphalt emulsions recognized in AASHTO standards.

Particle Charge	Set	Emulsion viscosity* 1	Emulsion viscosity* 2	Residual asphalt cement† h	Residual asphalt cement† s
Anionic	Rapid	RS-1	RS-2 HFRS-2		
	Medium	MS-1 HFMS-1	MS-2 HFMS-2	MS-2h HFMS-2h	HFMS-2s
	Slow	SS-1		SS-1h	
Cationic	Rapid	CRS-1	CRS-2		
	Medium		CMS-2		CMS-2h
	Slow	CSS-1		CSS-1h	

*Refers to the viscosity of the emulsion: 1 has a lower viscosity than 2, as measured by the Saybolt Furol test; HF indicates a high-float emulsion, generally used with dusty aggregates or when placing chip seals on grades.

†The letter h indicates asphalt cement residue of 40 to 90 penetration, s indicates asphalt cement residue with more than 200 penetration. If an h or s is not indicated, penetration is 100 to 200.

Emulsions are manufactured in several grades and types as shown in Table 15.2. In addition to the ASSHTO specified emulsions, many states have specifications for asphalt emulsions suited to their local conditions. State department of transportation specifications control these. Although there are a wide variety of emulsions available, SS-1 and SS-1h are the predominant asphalt emulsions used for prime and tack coats.

The primary applications of emulsion asphalt cements are for tack coats and prime coats during pavement construction, and as the binder for surface treatments, slurry seals, and cold patch materials. For tack and prime coats and surface treatments, the binder is applied by spraying through a distributor truck.

ASPHALT CONCRETE

Asphalt cements are used as the binder in paving mixes. The asphalt cement usually represents less than 10% of the mix by weight. However, it serves the very important functions of bonding the aggregate particles together, preventing the entrance of moisture, and acting as a cushioning medium. All asphalt concrete is a blend of aggregates and asphalt cement. However, by varying the aggregate gradation, different types of asphalt concrete can be produced, such as dense-graded, open-graded, and stone matrix asphalt. The most common type of asphalt concrete is dense-graded mix.

Asphalt concrete mix design refers to the process of selecting the asphalt cement, aggregate structure, and proportions of aggregate and asphalt cement that provide the optimum combination of materials for the pavement loading conditions and environment. Several methods have been developed for mix design, such as **Hveem**, Marshall, and Superpave. The common feature of all asphalt concrete mix-design procedures is that through laboratory evaluation

Hveem
A method of asphalt mix design based on the cohesion and friction of a compacted specimen.

and specifications, the pavement construction requirements are defined. These include the grade of the asphalt cement, the specifications governing the quality of the aggregates, and the quality-control requirements for the in-place asphalt concrete mixture.

Different mix-design procedures will require different quality-control parameters for construction. The most common parameters include

1. Target aggregate gradations
2. Target asphalt content
3. Density of the mix
4. Volume of air voids
5. Volume of voids in the mineral aggregate
6. Volume of voids filled with asphalt
7. Dust to binder ratio

The mix-design process generally consists of estimating a target aggregate gradation. Samples are then prepared at different asphalt contents and tested with the relevant test methods for the mix-design method. Plots are prepared of the mix-design parameters versus asphalt content. The optimum asphalt content is determined, and construction documents are prepared.

ASPHALT PLANTS

GENERAL INFORMATION

Hot-mixed asphalt is produced at a central plant and transported to the paving site in trucks. An *asphalt plant* is a high-tech group of machines capable of uniformly blending, heating, and mixing the aggregates and asphalt cement of asphalt concrete, while at the same time meeting strict environmental regulations, particularly in the area of air emissions. Drum and batch plants are the two most common plant types. Drum mix plants are a newer technology than batch plants and generally are more economical to operate. Drum plants were introduced in the 1970s and dominate the new plant market. About 95% of the new plants are the drum type. However, about 70% of operational plants are batch plants. In 2000 it was reported that 5,000 to 6,000 batch plants are still in operation, but sales of new batch plants are almost at a standstill.

While the mixing process of batch and drum plants is distinctly different, there are many similar elements that vary only in detail between the two types of plants. The similar elements are dust collection, asphalt storage, truck scales, and storage or surge silos. The truck scales are at the plant's product loading location. The empty and loaded weight of the trucks is measured to determine the weight of the load. The truck scales must be calibrated and certified.

BATCH PLANTS

Batch plants which date from the beginning of the asphalt industry, proportion and mix liquid asphalt and aggregates in individual batches (Figs. 15.6 and 15.7). Their primary components in the order of material flow are

- Cold feed system
- Drum dryer
- Hot elevator
- Hot screens
- Hot bins
- Asphalt-handling system
- Pugmill mixer
- Dust collectors
- Surge silo

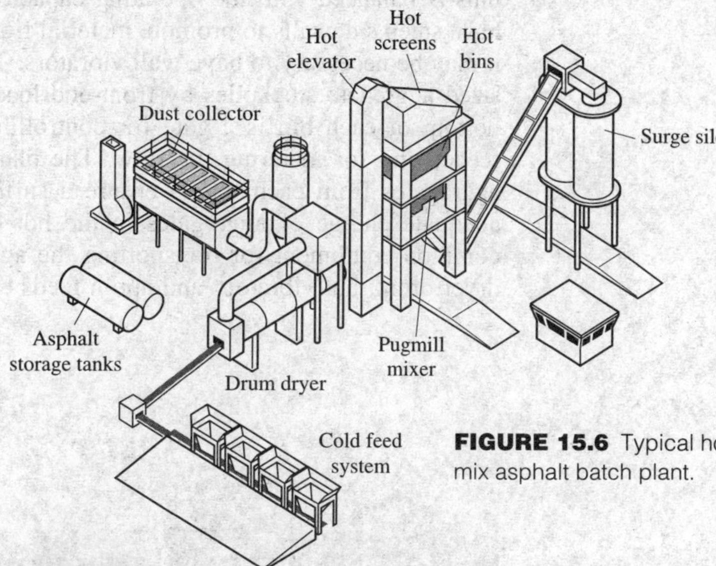

FIGURE 15.6 Typical hot mix asphalt batch plant.

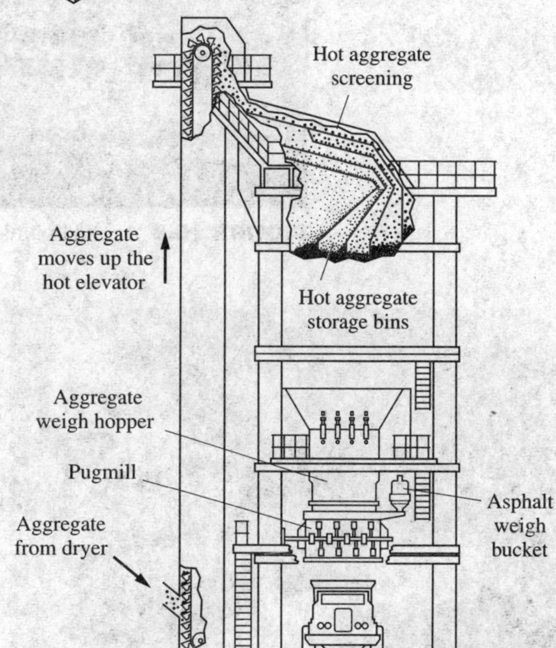

FIGURE 15.7 Batch plant with truck below pugmill mixer.
Source: Barber-Greene, A Division of Caterpillar Paving Products Inc.

Cold Feed System

Cold feed bins provide aggregate surge storage and a uniform flow of properly sized material for mixing. The cold feed system usually consists of three to six open-top bins mounted together as a single unit (see Fig. 15.8). The size of the bins is balanced with the operating capacity of the plant. The individual bins have steep sidewalls to promote material flow. In the case of sticky aggregates, it may be necessary to have wall vibrators. The individual bins can be fed from sized aggregate stockpiles by front-end loader, clamshell, or conveyor. At the bottom of each bin is a gate for controlling material flow (Fig. 15.9) and a feeder unit for metering the flow. The plant operator adjusts the flow of the aggregates from each bin to ensure a sufficient flow of material to keep an adequate charge of aggregates in the hot bins. Belt conveyors are the most common equipment for transporting the aggregates from the cold bins to the dryer drum, but vibratory and apron feeds can be found.

FIGURE 15.8 Asphalt plant with a six-bin cold feed system.

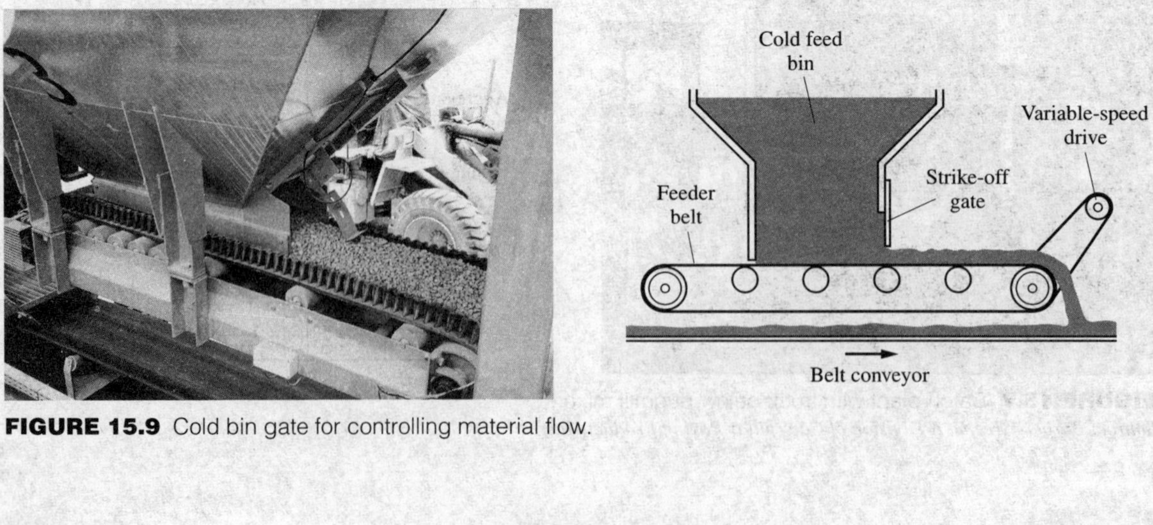

FIGURE 15.9 Cold bin gate for controlling material flow.

Drum Dryer

The purposes of a drum dryer are to heat and dry the aggregates of the mix. Aggregate temperature controls the resulting temperature of the mix. If the aggregate has been heated excessively, the asphalt will harden during mixing. If the aggregates have not been heated adequately, it is difficult to coat them completely with asphalt. Therefore, the aggregate must be heated sufficiently at this step in the process to produce a final mix at the desired temperature.

Aggregates are introduced at the end of the drum dryer opposite the burner and travel through the drum counter to the gas flow (Fig. 15.10). The drum is inclined downward from the aggregate feed end to the burner end. This slope causes the aggregate to move through the drum by gravity. The drum rotates and steel angles, "**flights**," mounted on the inside (Fig. 15.10) lift the aggregate and dump it through the hot gas and burner flame. Finally, the heated aggregate is discharged into the hot (bucket) elevator, the white tower on the left in Fig. 15.7, which carries it to the hot screens at the top of the batch plant tower.

flights
Metal plates of various shapes placed longitudinally inside the shell of a drum dryer or mixer. As the material moves through the drum, the flights first lift the aggregate and then tumble it through the hot gases.

> Every percent increase in aggregate moisture will increase dryer fuel consumption 10%.

Hot Screening

The batch plant vibrating screen unit (in Fig. 15.7, housed in the white upper part of the plant) is usually a $3\frac{1}{2}$-deck arrangement. This enables gradation control of four aggregate sizes into four different hot bins. The screen unit ejects

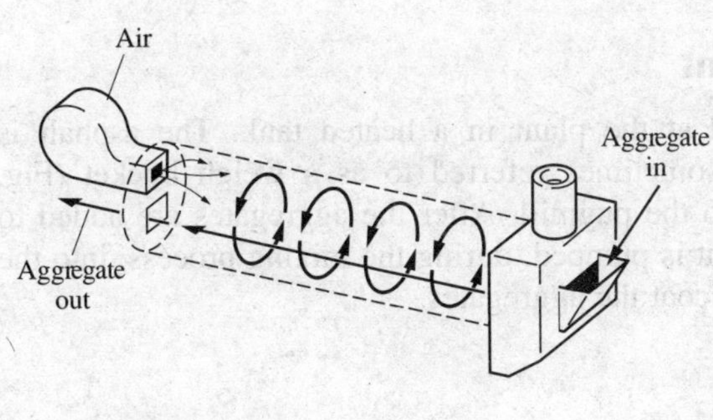

Aggregate flow through a counterflow dryer

Steel angles, "flights," mounted on the inside of a dryer drum

FIGURE 15.10 Typical counterflow dryer.

oversized material out of the production cycle. While the screens provide gradation control, they will not function properly unless the proportioning and flow from the cold feed is correct. If the screens are overloaded, material, which should be passing through a screen and into a hot bin, is carried instead into the bin of the next larger aggregate size. Such a situation destroys the mix formulation and must be avoided.

Hot Bins

The aggregates from the hot screens are stored in the hot bins (in Fig. 15.7, the part of the plant just below the white upper part) until a batch of asphalt concrete is produced. One of the key elements in operating a batch plant is to ensure that the hot bins have sufficient material to feed the pugmill for the production of a batch of asphalt concrete. One of the potential advantages of a batch plant as compared to a drum plant is that the batches are individually blended from the hot bins. This enables the aggregate blend of one batch to be different from the blend of the next batch. However, this is contingent on having the proper-size aggregates available in the hot bins. Frequently, under high-production-rate conditions, the aggregates in the hot bins will not be sized properly for gross changes in the gradations of the mix. Thus, the flexibility of the batch plant is compromised. Large changes in the gradation of the aggregates generally must be accomplished by altering the flow of aggregates from the cold bins so that the proper amount of material is stored in the hot bins.

Weigh Hopper

Aggregate from the hot bins is dropped into a weigh hopper situated below the bins and above the pugmill (Fig. 15.7). To control the weighing of the blended aggregate, the weigh hopper is charged one hot bin at a time. The aggregate weight in the hopper is cumulative, with the mineral filler added last. After charging, the weigh hopper gates are open to discharge the aggregate into the pugmill.

Asphalt-Handling System

The asphalt cement is stored at the plant in a heated tank. The asphalt is pumped to the weigh tank, sometimes referred to as a weigh bucket (Fig. 15.7), ready for discharge into the pugmill. After the aggregates are added to the pugmill, the asphalt cement is pumped, during the mixing process, into the pugmill through spray bars to coat the aggregates.

Pugmill Mixing

Most batch plants use a twin-shaft pugmill for mixing the batch (Fig. 15.11). To achieve uniform mixing a pugmill's live zone should be completely filled with mix. The live zone is from the bottom of the box to the top of the paddle arc.

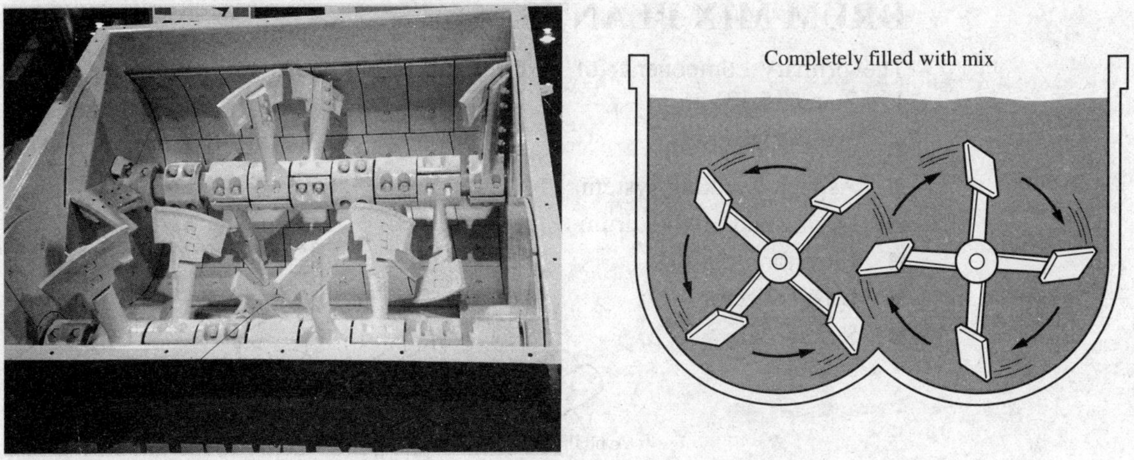

FIGURE 15.11 Twin-shaft pugmill box for mixing asphalt concrete.

The mixing process generally takes about 1 min, 15 sec for loading with the dry materials and 45 sec of mixing time with the asphalt cement. The actual mixing time required is evaluated based on inspection of the resulting coarse aggregate coating. Plant operation specifications require sufficient mixing time to fully coat 90 or 95% of the aggregate, depending on the aggregate size.

The capacity of the plant is a function of the pugmill size and the mixing time. Typical batch quantities range from 1.5 to 5 tons. A batch plant with a 5-ton pugmill can produce 300 tons of mix per hour if a continuous operation can be maintained.

The plant is structured so that the discharge gate of the mixer is sufficiently high to allow truck passage directly below for loading (see Fig. 15.7). Alternatively, a hot elevator can be used to transport the mix to surge silos. These silos enable the plant to operate independent of immediate truck availability. This is particularly advantageous when the plant is serving jobs with different mix designs. Silos also enable the plant operator to premix and store several batches of asphalt concrete to accommodate an uneven distribution of truck arrivals at the plant.

Because of the influence of temperature on the quality of the mix, the purchaser usually specifies the mixing temperature, measured immediately after discharge from the pugmill. The specification range will vary with the type and grade of the asphalt cement. In the case of dense-graded mixes the range across all asphalt cements is from 225 to 350°F. The range for open-graded mixes is from 180 to 250°F. Mixing should be at the lowest temperature that will achieve complete asphalt coating of the aggregates and still allow for satisfactory workability. In some cases, during cold weather or in the case of long haul distances, the asphalt concrete is heated an extra 10°F to allow for temperature loss during the haul.

DRUM MIX PLANTS

The primary components of a drum mix (continuous mix) plant (see Figs. 15.12 and 15.13) are

- Cold feed system
- Asphalt handling system
- Drying and mixing drum
- Elevator
- Dust collector
- Storage silo

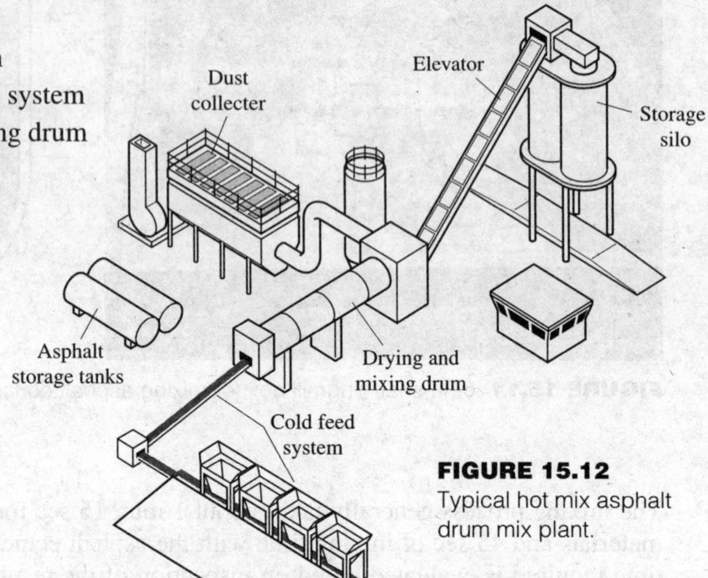

FIGURE 15.12 Typical hot mix asphalt drum mix plant.

Cold Feed System

In a drum plant, all of the drying and mixing is performed within the drum, and the asphalt concrete mix is discharged directly from the drum into storage silos in a continuous manner. There is no opportunity to adjust the aggregate

FIGURE 15.13 Drum mixer of a drum mix asphalt plant.

blend during weighing into a hopper, as can be done with a batch plant. Therefore, the aggregate from each of the cold feed bins must be weighed prior to feeding the material into the drum. Since the aggregates are weighed prior to drying, the moisture content of the aggregates in the cold bins must be monitored and the weights adjusted to ensure the dry mass of the aggregates is correct. Scales mounted on the conveyor belts measure the weights of the aggregates.

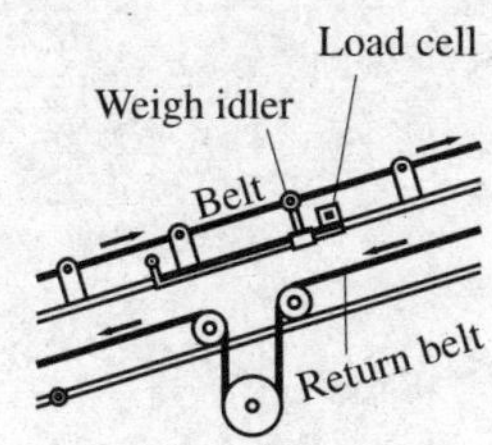

Mixing Drum

The mixing drum (Fig. 15.13) consists of a long tube with flights for tumbling the aggregates and the mix, a burner for heating the aggregates, and a spray bar for applying the asphalt. The basic operation of the drum plant is that the aggregates are metered into one end of the drum. The time the aggregates spend in the drum ranges from 3 to 4 min. During that time the aggregates must be thoroughly dried and heated to the mixing temperature. Near the discharge end of the drum, the asphalt cement is sprayed onto the aggregates. Automatic controls monitor the aggregate quantity and meter in the proper amount of asphalt cement. During drum rotation, the flights in the drum produce a mixing action between the aggregate, RAP (if used), and asphalt cement.

The drum of these plants usually has a slope in the range of $\frac{1}{2}$ to 1 in. per foot of drum length. Rotation speeds are normally 5 to 10 rpm and common diameters are from 3 to 12 ft, with lengths between 15 and 60 ft. The ratio of length to diameter is from 4 to 6. Longer drums are found in recycling applications. The drum slope, length, rotation speed and flights, and the nature of the aggregates control the **dwell time**.

dwell time
The time required for material to pass through a dryer or mixer.

The production rate of the plant is inversely proportional to the moisture content of the aggregates. For example, increasing the moisture content from 3 to 6% in a batch plant with an 8-ft-diameter drum reduces the production rate from 500 to 300 tons/hr.

Originally, drum plants were designed as parallel flow operations, with the aggregates and heated air traveling in the same direction down the drum. Subsequent designs have increased the production rate of drum plants by using counter airflow arrangements in which the heated exhaust exits from the top of the drum where the aggregates are introduced. Counterflow plants can operate at about 12% higher production rates.

Storage Silos

Since drum plants produce a continuous flow of asphalt concrete, the output must be stored in silos (Fig. 15.14) for subsequent dispatch into the trucks. These silos have a bottom dump for directly discharging the asphalt concrete into the trucks. The silos are typically insulated to retain heat. Sophisticated silos can be completely sealed, and even filled with an inert gas to reduce oxidation of the asphalt cement while the asphalt concrete is stored. One of the problems with storage silos is the potential for flow of the asphalt cement from materials at the top of the silo to the bottom. This results in a poor-quality paving mix.

FIGURE 15.14 Storage silos at a drum mix asphalt plant.

DUST COLLECTORS

To avoid contributing to air pollution, asphalt plants are equipped with dust control systems. The two most commonly utilized systems are the water venturi approach and the cloth "**baghouse**" filtration. The wet approach requires the availability of adequate water supplies. This approach introduces water at the point where dust-laden gas moves through the narrow throat of a venturi-shaped chamber. The dust becomes entrapped in the water and is thereby separated from the exhaust gas. A disadvantage of the wet approach is that the collected material cannot be reclaimed for use in the mix.

baghouse
An asphalt plant component that, by the use of special fabric filter bags, captures the fine material particles contained in the dryer or drum exhaust gases.

Dry baghouse filtration mechanically collects the fines from the aggregates and can redirect them back into the mix. The system works by forcing the dust-laden gas through fabric filter bags that hang in a baghouse (Fig. 15.15). By using a reverse pulse of air or by mechanically shaking the bags, the collected dust is removed from the filter. The dust falls into hoppers in the bottom of the baghouse and is moved by augers to a discharge vane feeder. The vane feeder is necessary to keep the baghouse airtight.

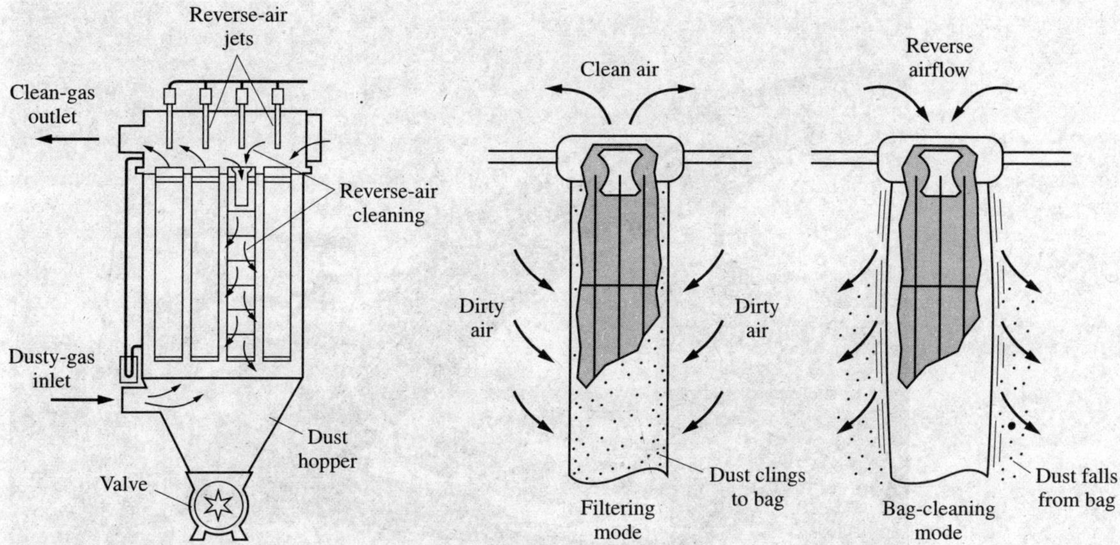

FIGURE 15.15 Schematics of a baghouse dust collector system.

Filter bags are made of fabrics that can withstand temperatures up to 450°F. But care must be exercised when using a baghouse system, as excessive temperatures can melt the bags and/or cause a fire.

ASPHALT STORAGE AND HEATING

When liquid asphalt is combined with the aggregate for mixing, the temperature of the asphalt should be in the range of 300°F. Therefore, both drum mix and batch plants have heating systems to keep the liquid asphalt at the required temperature. If asphalt is delivered at a cooler temperature, the system must be capable of raising the temperature of the delivered asphalt cement. The two methods commonly used for heating liquid asphalt are the direct fire and the hot oil processes.

A direct fire heater consists of a burner that fires into a tube in the asphalt storage tank. With such systems, sufficient asphalt must always be kept in the tank so that the burner tube is always submerged. These systems have a higher thermal efficiency than the hot oil process. The hot oil system is a two-stage heating approach. First, transfer oil is heated, and then the heated oil is circulated through piping within the asphalt tank.

The viscosity of the asphalt cement must be low enough to permit pumping; since different asphalts have different temperature-viscosity relationships, the storage temperature is different for the different grades of asphalts. Typically the storage temperature is in the range of 320°F for soft asphalts to 350°F for hard asphalts. For performance grade asphalts, the producer or supplier should provide the storage temperature that will result in the asphalt cement having the proper viscosity for pumping.

FIGURE 15.16 Milling machine removing an old asphalt pavement.

RECLAIMING AND RECYCLING

Existing asphalt pavements represent a large investment in aggregates and asphalt. By reclaiming these materials, using either cold milling (see Fig. 15.16) or ripping methods, much of that investment can be recaptured. Additionally, cold milling allows restoration of the pavement section without the need to change the grade, thereby eliminating the problems associated with raising curbs and drainage structures as would be necessary if a new mat was placed upon an existing surface.

The reclaimed asphalt paving materials are combined with virgin aggregates, additional asphalt, and/or recycling agents in a hot mix plant to produce new paving mixes. In those cases where the pavement sections were ripped up instead of milled it may be necessary to crush the reclaimed materials in order to reduce the particle size. The new mix design will have to account for both the graduation of the aggregate in the reclaimed asphalt pavement (RAP) and the asphalt content of the RAP.

Recycling in a Batch Plant In a batch plant, RAP can be added to the virgin aggregate in one of four locations

1. Weigh hopper
2. Separate weigh hopper
3. Bucket-elevator
4. Heat transfer chamber

Each of these locations requires superheating the virgin aggregates to provide the heat source for heating and drying the RAP. RAP usually has a moisture content of 3 to 5%, so steam is released as the RAP is heated. The steam

FIGURE 15.17 Double-barrel drum mix asphalt plant.

released from the RAP contains dust that must be filtered. Adding the RAP to the weigh hopper is the simplest method, but the time for heat transfer is limited, and capturing the dust-laden steam requires additional equipment. Using a separate weigh hopper reduces the time to batch slightly and may improve accuracy, but this method has the same disadvantages as adding the RAP directly to the hopper.

Recycling in a Drum Plant Drum plants can recycle reclaimed asphalt pavement by placing a feed collar that introduces the RAP between the portion of the drum that heats the aggregate and the area where the asphalt cement is introduced. Because the RAP contains asphalt cement, it cannot be exposed to direct flame. When using RAP in a mix, the virgin aggregates are superheated to provide sufficient heat to raise the temperature of the RAP to the production temperature. One manufacturer has developed an innovative double-barrel drum design (Fig. 15.17) where the RAP is introduced in an outer drum to avoid exposing the RAP to the flame.

PAVING EQUIPMENT

An asphalt paving operation requires a number of different pieces of equipment. These include

- Sweeper/broom for removing dust from the surface to be paved
- Trucks for transporting the asphalt mix from the plant to the construction site
- Asphalt distributor truck for applying the tack, or seal coats

- Material transfer vehicle (depending on specifications or contractor paving preference)
- Windrow elevator (depending on contractor paving preference)
- Paver
- Rollers

SWEEPER/BROOM

The sweeper/broom is used to remove dust from the surface of existing pavement prior to laying new asphalt. This is done to ensure proper bonding between the new asphalt and the old pavement. When surfacing a prepared **base course** (aggregate or soil cement), the dust layer should be remove either by sweeping with the sweeper or by wetting the base course and recompacting.

base course
That part of the pavement structure placed directly above the subbase.

HAUL TRUCKS

Three basic types of trucks are used to haul the asphalt concrete from the plant to the job site: dump trucks, live bottom trucks, and bottom-dump trucks. Dump trucks and live bottom trucks transfer the mix directly into the paver hopper. Bottom-dump trucks place a windrow of material upon the pavement ahead of the paver (Fig. 15.18) and can only be used with pavers that have an elevator for lifting the mix from the pavement and transferring it to the spreader box of the paver. Since the windrow material loses temperature rapidly, this paving operation is generally limited to the southwestern United States, where the ambient temperatures are high.

Regardless of the type of truck used, the trucks should be insulated and covered to reduce heat loss during transportation. Before the trucks are loaded, the bed is coated with an approved release agent. In the past, a heavy fuel oil like kerosene was used for this. However, this contaminates the asphalt binder and is no longer permitted.

FIGURE 15.18 Bottom-dump truck placing a windrow of material on the pavement ahead of the paver.

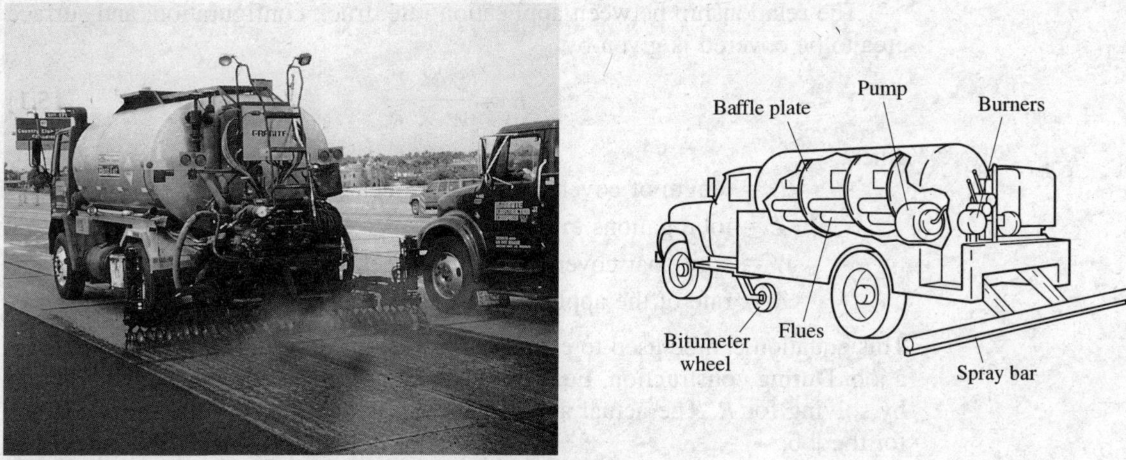

FIGURE 15.19 An asphalt distributor applying a tack coat.

ASPHALT DISTRIBUTORS

When applying an asphalt prime, tack, or seal coat, a specially designed distributor truck is utilized (see Fig. 15.19). An asphalt distributor truck requires constant attention to produce a uniform application. It is critical that the asphalt heater and pump be well maintained. All gauges and measuring devices such as the pump tachometer, measuring stick, thermometers, and bitumeter wheel must be properly calibrated. Spray bars and nozzles should be clean and set at the proper height above the surface receiving the application. The factors that affect uniform application are

- The asphalt spraying temperature
- The liquid pressure across the spray bar length
- The angle of the spray nozzles
- The nozzle height above the surface
- The distributor speed

Asphalt distributors have insulated tanks for maintaining asphalt temperature and are equipped with burners for heating the asphalt to the proper application temperature. Either independently powered or PTO-driven discharge pumps are used to maintain continuous and uniform pressure for the full length of the spray bar. The spray bar nozzles must be set at the proper angle, usually 15 to 30° from the horizontal axis of the spray bar. This is so that the individual spray fans do not interfere or intermix with one another. The height of the nozzle above the surface determines the width of an individual fan. To ensure the proper lap of the fans, the nozzle (spray bar) height must be set and maintained. The relationship between application rate (gallons per square yard) and truck speed is obvious; truck speed must be held constant during the spraying to achieve a uniform application.

The relationship between application rate, truck configuration, and surface area to be covered is given by

$$L = \frac{9 \times T}{W \times R} \quad [15.1]$$

where

L = length of covered surface in feet
T = total gallons to be applied
W = spray bar coverage width in feet
R = rate of the application in gallons per square yard (sy)

This equation can be used to estimate the amount of liquid asphalt required for a job. During construction, Eq. [15.1] can be used to check the application rate by solving for R. The actual application rate is compared to the specifications for the job.

Prior to placement of an asphalt mix on a new base, a prime coat is applied to the base. Normal rates of application for prime vary between 0.20 and 0.60 gal/sy. Prime promotes adhesion between the base and the overlying asphalt mix by coating the absorbent base material, whether it is crushed aggregate, a stabilized material, or an earthen grade. The prime coat should penetrate about $\frac{1}{4}$ in., filling the voids of the base. The prime coat additionally acts as a waterproof barrier preventing moisture that penetrates the wearing surface from reaching the base.

Tack coats are designed to create a bond between existing pavements, be they concrete, brick, or bituminous material, and new asphalt overlays. They are also applied between successive mats during new construction. A tack coat acts as an adhesive to prevent slippage of the two mats. The tack coat is a very thin uniform coating of asphalt, usually 0.05 to 0.15 gal/sy of diluted emulsion. An application that is too heavy will defeat the purpose by causing the layers to creep, as the asphalt serves as a lubricant rather than as a tack.

Seal coats consist of an application of asphalt followed by a light covering of fine aggregate, which is rolled in with pneumatic rollers. Applications rates are normally from 0.10 to 0.20 gal/sy.

ASPHALT PAVERS

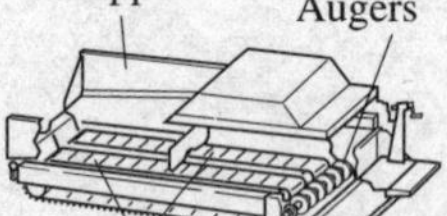

An asphalt paver consists of a tractor, either track (Fig. 15.20) or rubber-tired (Fig. 15.1), and a screed. The tractor power unit has a receiving hopper in the front and a system of slat conveyors to move the mix through a tunnel under the power plant to the rear of the tractor unit. At the rear of the tractor unit, the mix is deposited on the surface to be paved, and augers (Fig. 15.21) are used to spread the asphalt evenly across the front of the trailing screed. Two tow arms, pin connected to the tractor unit, draw the screed behind the tractor. The screed controls the asphalt placement width and depth, and imparts the initial finish and compaction to the hot mix material. There is one manufacturer that offers a paver having two sets of twin-screws, instead of slat conveyors, to move the mix through the tunnel to the rear of the paver (see Fig. 15.22). The use of the screw conveyors is said to reduce mix segregation.

FIGURE 15.20 Track-mounted asphalt paver.

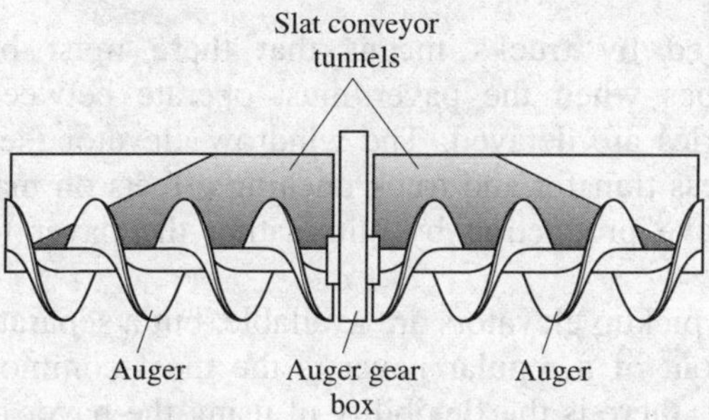

FIGURE 15.21 Drawing of auger arrangement at the back of the conveyor tunnel.

FIGURE 15.22 Asphalt paver having two sets of twin-screws for moving the mix through the tunnel.

Pavers can receive mix directly into their hoppers or can pick up a windrow of material placed in front of the paver. The traditional method for loading the hopper has the truck dump the asphalt mix directly into the paver hopper. Push rollers mounted on the front frame of the paver (see Fig. 15.22), and extending beyond the hopper, push against the rear wheels of the truck. As the paver pushes the truck along, the mix is dumped into the hopper by raising the truck bed or by activating the live bottom of such trailer units.

Loading the hopper with individual truckloads often requires that the paver stop intermittently. This can cause problems with constructing a smooth pavement since frequently the paver must wait between truckloads, and trucks mating up to the paver can jolt it, causing the screed to create a bump in the finished mat.

Windrow Elevators

A paving operation being served by trucks means that there must be paver/truckload transfers and times when the paver must operate between truckloads or even stop if deliveries are delayed. The windrow elevator (see Fig. 15.1) was developed to address transfer and truck queuing effects on mat quality. Elevators can also improve production by eliminating the paver to truck mating.

Pavers with integral windrow pickup elevators are available, but a separate elevator unit attachable to the front of a regular paver is the most common approach. By using an attachment, there is the flexibility of using the paver in both direct load and windrow situations.

The flight system of the elevator continuously lifts the mix into the hopper from the windrow. For efficient operation, the amount of material in one longitudinal foot of windrow cross section must equal the amount required for one longitudinal foot of mat cross section. Minor quantity variations are accommodated by the surge capacity of the paver hopper. Conventional windrow machines have limited width and can only handle fairly narrow rows of material. New elevator designs are being introduced having the capacity to handle wider windrow cross sections. These machines will open windrow operations to smaller, limited-area projects and will enable the use of regular dump trucks for hauling.

Material Transfer Devices

Some contractors use material transfer devices to improve paving quality. A material transfer device can receive multiple truckloads of asphalt mix, remix the material, and deliver it to the paver hopper. Material transfer devices offer several advantages over direct transfer between the haul truck and the paver. By holding several loads of asphalt concrete, the transfer device reduces surge loading of the paver. This enables the paver to continuously operate while the haul trucks transfer the asphalt concrete to the transfer device. The timing of the trucks to the job site is not as critical, since the paver can operate from the supply of material in the transfer device. The material transfer device is self-propelled, so the paver does not need to mate up to and push the haul truck. Finally, by remixing the asphalt concrete, thermal segregation is reduced.

Thermal segregation occurs during the haul because the asphalt concrete at the surface of the load cools while the mix in the center of the load retains heat. The self-leveling screed of the paver is sensitive to the stiffness of the mix, which is largely a function of the temperature of the mix. By remixing the asphalt concrete, a uniform asphalt concrete stiffness is achieved, which allows for smoother pavement construction.

Screed

The "floating" screed is free to pivot about its pin connections. This pin-connected tow-arm arrangement enables the screed to be self-leveling and gives it the ability to compensate for irregularities in the underlying surface. The paver's ability to level out irregularities is controlled by the tractor's wheelbase length and by the length of the screed towing arms. Greater lengths of these two components mean smooth transitions across irregularities and therefore a smooth pavement surface.

Mat thickness, which is controlled by the screed, can be maintained by using grade sensors tracing an external reference with a ski or shoe (Fig. 15.23). Or the screed can be tied to a specified grade by the use of sensors tracing a stringline. When all the forces acting on the screed are constant, it will ride at a constant elevation above the grade or follow the stringline. However, there are factors that can cause the sensor-regulated screed height to vary:

- The screed angle of attack
- The head of asphalt in front of the screed
- The paver speed

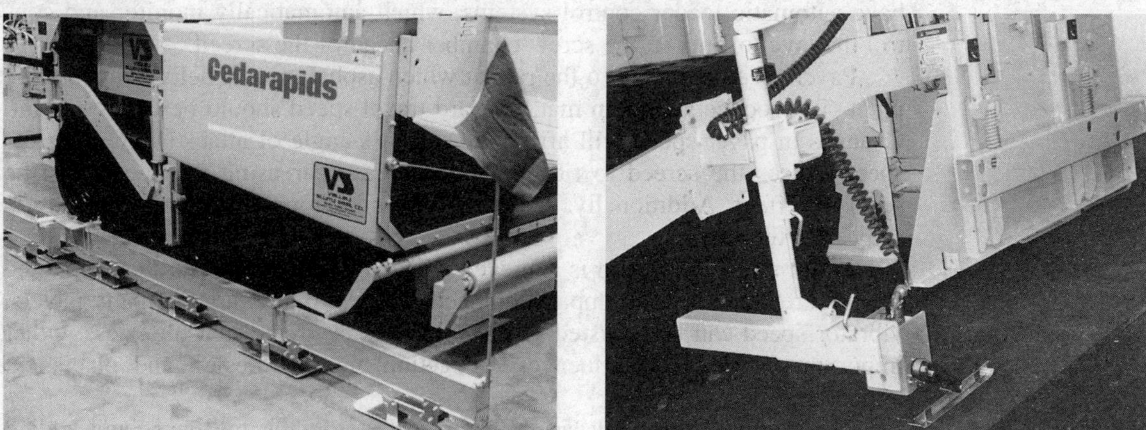

ski shoe

FIGURE 15.23 Screed external-reference grade sensors.

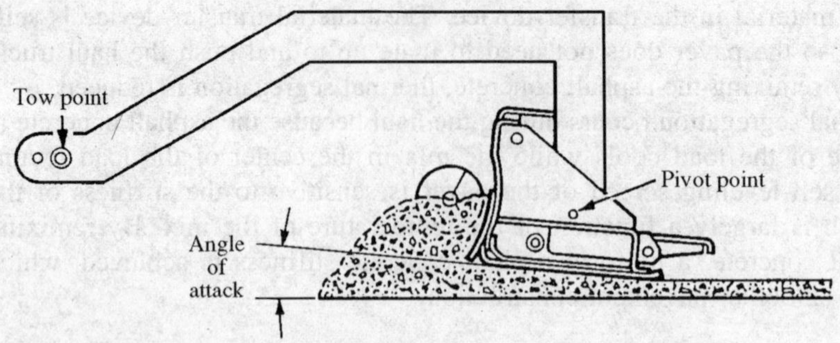

FIGURE 15.24 Screed angle of attack.
Source: Caterpillar Paving Products Inc.

The angle created by the plane of the surface upon which the asphalt is being placed and the plane of the screed bottom is known as the "screed angle of attack" (see Fig. 15.24). This angle is the principal mechanical factor affecting variations in mat thickness. It regulates the amount of material passing under the screed in a given distance. When either the screed or the tow points are vertically displaced, the angle of attack is changed. The screed will immediately begin to move, restoring the original angle, but this correction requires about three tow-arm lengths to be accomplished.

The asphalt material directly in front and across the length of the screed is referred to as the head of material. When this material is not held constant by the hopper drag conveyor and the auger feed, the screed angle of attack will be affected and in turn the mat thickness will change. If the head of material becomes too high, the mat thickness will increase as the screed rides up on the excess material with the paver's forward progress. If the volume decreases, the screed moves down, resulting in reduced mat thickness. Most modern pavers have automatic feeder control systems, which automatically monitor and control the level of mix in the screw chamber ahead of the screed.

Paver speed is linked to the rate at which asphalt mix is delivered from the plant. To produce a smooth mat, forward travel speed should be held constant. Changes in paver speed will affect the screed's angle of attack. Increasing the speed causes the screed to ride down, whereas decreasing the speed has the opposite effect. Additionally, when the paver is stopped, the screed tends to settle into the mat.

Initial mix compaction is achieved by screed vibration. Vibrators mounted on the screed are used to impart compaction force to the mat. On many pavers, vibrator speed can be adjusted to match paver speed and mat thickness. Other important factors that influence compaction are mix design and placement temperature.

The width of a screed can be changed by stopping the paver and adding extensions on one or both sides of the basic screed. Some screeds are hydraulically extendible, enabling the paying width to be varied without stopping the paver. Auger additions may be required when screed width is increased. These are necessary to spread the mix evenly across the full width of the screed and screed extensions. Most screeds, at the end and middle points of their horizontal plane, can be adjusted in the vertical to create either a crown or superelevation.

To prevent material from sticking to the screed at the beginning of a paving operation it is necessary to heat the screed. Built-in diesel or propane burner heaters are commonly used to heat the bottom screed plates. There are also electric-heated screeds having thermostats to monitor screed temperature and to automatically maintain set temperatures. Required heating time will vary with air temperature and the type of mix being placed. About 10 min of heating is normal with burner models, but care must be exercised as overheating can warp the screed.

Paver Production

Continuous paving operations depend on balancing paver production with plant production. The critical choke points in the operation, which must be analyzed and managed, are the *loading of haul units at the plant* and the *haul unit-paver link.*

EXAMPLE 15.1

An asphalt plant can produce 324 tons per hour (tph). A project requires paving individual 12-ft lanes with a 2-in. lift averaging 112 lb/sy-in. What average paver speed will match the plant production? How many 20-ton bottom-dump trucks will be required if the total haul cycle time is 55 min?

$$\frac{324 \text{ tph}}{60 \text{ min/hr}} = 5.4 \text{ ton/min average plant production}$$

$$\frac{2 \text{ in. (thick)} \times 12 \text{ ft (wide)} \times 1 \text{ ft (length)}}{9 \text{ sq ft/sy}} = 2.66 \text{ sy-in./ft paving length}$$

$$\frac{2.66 \text{ sy-in.} \times 112 \text{ lb/sy-in.}}{2{,}000 \text{ lb/ton}} = 0.149 \text{ ton/ft of paving length}$$

$$\frac{5.4 \text{ ton/min}}{0.149 \text{ ton/ft}} = 36.2 \text{ ft/min, average paver speed}$$

$$20 \text{ tons per truck} \times \frac{60 \text{ min/hr}}{55\text{-min cycle}} = 21.8 \text{ tph per truck}$$

$$\frac{324 \text{ tph}}{21.8 \text{ tph per truck}} = 14.9 \text{ trucks}$$

therefore, 15 trucks are required.

Another way to analyze the situation would be to consider time. The paver requires a truck every

$$\frac{\text{20 tons per truck}}{\text{5.4 ton/min, required for paver}} = 3.7 \text{ min}$$

$$\frac{\text{55 min (total truck cycle time)}}{\text{3.7 min (paver requirement)}} = \text{14.9 or 15 trucks are required}$$

This is the minimum number of trucks required. However, haul time is rarely consistent, so extra trucks should be assigned to the project to prevent having to stop the paver.

COMPACTION EQUIPMENT

Because of the relationships between pavement air voids and mechanical stability, durability, and water permeability, asphalt pavements are designed based on the mix being compacted to a specified density. Typically for properly designed pavements the air-voids content should be between 3 and 5%. The factors that affect compaction are mix characteristics, lift thickness, mix temperature, and the operational characteristics of the compaction equipment.

Three basic roller types are used to compact asphalt-paving mixes: smooth drum steel wheel, pneumatic tire (see Fig. 15.25), and smooth drum steel wheel vibratory (see Fig. 15.26) compactors.

The contact surface area of a steel wheel roller is an arc of the cylindrical wheel surface and the roller width. This contact area will decrease as the rolling progresses, and the contact pressure will approach a maximum value.

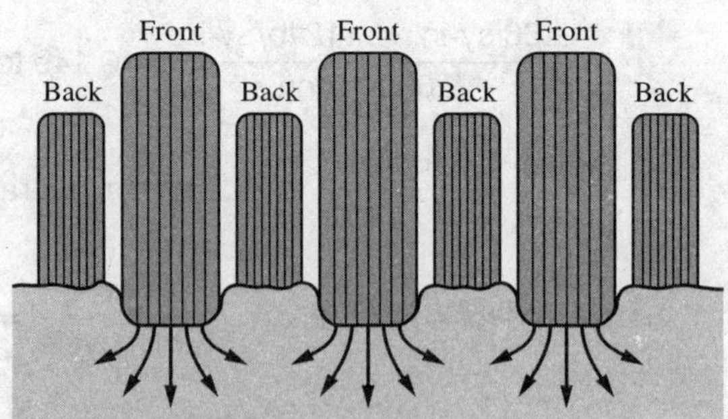

FIGURE 15.25 Pneumatic tire compactor.

FIGURE 15.26 Double-drum steel wheel vibratory compactor.

The contact area of a pneumatic tire is an ellipse, influenced by wheel load, tire inflation pressure, and tire sidewall flexure. With a pneumatic roller, the change in contact area as compaction progresses is less dramatic than in the case of a steel wheel roller.

The compaction capability of pneumatic and vibratory steel wheel rollers can be altered to match the construction conditions. With pneumatic rollers, changing the ballast and tire pressure alters the compaction effort. Adjusting the vibrator frequency and amplitude alters the compaction effort of vibratory steel wheel rollers. These capabilities enhance the operational flexibility of both roller types.

Vibratory Roller Frequency and Amplitude Most vibratory rollers allow the operator to select a vibratory frequency and amplitude. The maximum amount of compaction per pass will be achieved by selecting the highest possible frequency. This is because there are more impacts per foot of travel. But for most mixes the roller should be operated at the lowest possible amplitude. Only in the case of mat thicknesses greater than 3 in. should a higher amplitude be considered. A high-amplitude setting on a thin mat causes the roller to bounce, making effective and uniform compaction difficult. In the case of mats having a thickness of 1 in. or less, vibratory rollers should be operated in the static mode.

Rolling Temperature

Asphalt concrete compaction is achieved by orienting the aggregates into a dense configuration. Hot asphalt cement promotes the reorientation of the aggregates by acting as a lubricant. When the asphalt cools below the softening point, about 180°F, the binder will no longer provide a lubricating action. Below the softening point, the asphalt begins to bind the aggregate into place, making further compaction difficult. The temperature at which the mix was placed is the important factor in determining the time available for compaction. The placing temperature is a function of the production temperature and the heat loss during the haul. To extend the available time, it is good practice to use tarps to cover the mix in the haul trucks. In extreme cases, it may be necessary to use trucks having insulated bodies.

Once the mix passes through the paver, five factors affect the rate of cooling: air temperature, base temperature, mix laydown temperature, layer thickness, and wind velocity. Higher ambient air temperatures allow more time for compaction. Of greater significance is the temperature of the surface upon which the mix is placed. There is a rapid heat transfer between these two surfaces. The higher the temperature of the mix passing through the paver, the more time will be available for compaction. Probably the most significant factor affecting cooling is layer thickness. As layer thickness increases, the time available for achieving density will increase. But thicker layers require more roller passes. High winds can cause the surface of the mat to cool very rapidly.

Rolling should begin at the maximum temperature possible. This maximum temperature is that point at which the mix will support the roller without distorting the asphalt concrete horizontally. Increasing roller drum diameter increases the contact area, which reduces the contact pressure. Hence large-diameter rollers can work on the mat at higher temperatures.

Rolling Steps

Compaction of an asphalt mat is usually viewed in terms of three distinct steps:

- Breakdown rolling
- Intermediate rolling
- Finish rolling

The breakdown step seeks to achieve a required density within a time frame defined by temperature constraints and consistent with paver speed. There is an optimum mix temperature range for achieving proper compaction. If the mix is too hot, the mat will tear and become scarred. If the mix is too cold, the energy requirement for compaction becomes impractical because of the viscous resistance of the asphalt binder. These physical limits define the practical time duration available for the rolling operation.

Sometimes density cannot be achieved by a single roller within the time duration available. In that case, an intermediate rolling step is required to supplement the breakdown step in reaching the required density. Finally, there is a finish-rolling step to remove any surface marks left by the previous rolling or by the paver.

Vibratory steel wheel rollers are usually the roller of choice for breakdown and intermediate rolling because of their adaptability to a range of mixes and differences in mat thickness. With the vibrator turned off, they can also be used for finish rolling. Pneumatic tire rollers can be used for breakdown and intermediate rolling. However, the high temperature of the mix during the breakdown rolling can cause the mix to stick to the rubber tires, so it is more common to use pneumatic rollers for the intermediate rolling. Steel wheel rollers are used for finish rolling. Since vibratory rollers can be used for each of the compaction steps, many contractors use vibratory rollers exclusively for asphalt concrete compaction.

Roller Capacity

The amount of mix that can be produced and delivered to the job sets paver speed. The number and type of rollers on the job must be selected to match the rate of mat placement. Net roller speed, the length of pavement that can be compacted in a unit of time, is influenced by

1. The gross roller speed
2. The number of passes
3. The number of laps
4. The overlap between adjacent laps required to cover the mat width
5. The extension overedge
6. The extra passes for joints
7. The nonproductive travel (overrun for lap change).

Roller capability for a project must match net roller speed with average paver speed. Increasing roller speed reduces compactive effort, so speed alone cannot be used to compensate for production needs. Typical acceptable roller speeds are: 2 to $3\frac{1}{2}$ mph for breakdown, $2\frac{1}{2}$ to 4 mph for intermediate, and 3 to 7 mph for finish rolling.

EXAMPLE 15.2

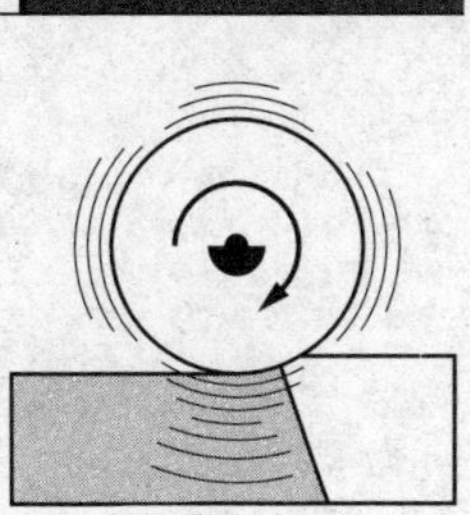

An asphalt plant will produce 260 tph for a project. The mat will be 12 ft wide, 2 in. thick, and will have a density of 110 lb/sy-in. A vibratory roller with a 66-in.-wide drum will be used for compaction. Assume a 50-min hour efficiency factor for the roller. The overlap between adjacent laps and overedges will be a minimum of 6 in. It is estimated that nonproductive travel will add about 15% to total travel. From a test mat, it was found that three passes with the roller are required to achieve density. To account for acceleration and deceleration when changing directions, add 10% to the average roller speed to calculate a running speed. How many rollers should be used on this project?

$$\frac{260 \text{ tph}}{60 \text{ min/hr}} = 4.3 \text{ ton/min average plant production}$$

$$\frac{12 \text{ ft (wide)} \times 2 \text{ in.} \times 110 \text{ lb/sy-in.}}{9 \text{ sf/sy} \times 2{,}000 \text{ lb/ton}} = 0.147 \text{ ton/ft of paving length}$$

Average paving speed,

$$\frac{4.3 \text{ ton/min}}{0.147 \text{ ton/ft}} = 29 \text{ ft/min}$$

Rolling width, 12 ft × 12 in./ft + 2 × 6 in. (each edge) = 156 in.
Effective roller width, 66 in. − 6 in. (overlap) = 60 in.
Number of laps:
First lap 66 in.; 156 − 66 = 90 in. remaining
Additional laps

$$\frac{90 \text{ in.}}{60 \text{ in.}} = 1.5$$

Therefore, three laps will be necessary to cover the 12-ft placement width.

Total number of roller passes, 3 laps × 3 passes per lap = 9 passes

Each pass must cover

29 ft/min, rolling + 29 ft/min, return + maneuver distance

Total roller distance required to match paver speed

$$9 \times 58 \text{ ft/min} \times 1.15 \text{ (nonprod.)} \times 60 \text{ min/hr} = 36{,}018 \text{ ft/hr}$$

Average roller speed

$$\frac{36{,}018 \text{ ft/hr}}{50 \text{ min}} = 720 \text{ ft/min or } 8.2 \text{ mph}$$

Running speed, 8.2 × 1.1 = 9 mph. This speed exceeds the speed limitation for effective compaction, of 2 to $3\frac{1}{2}$ mph for breakdown rolling. Therefore, more than one roller is required. For this example, three rollers would allow operation at 3 mph.

Another solution would be to use a roller having a drum width equal to or greater than 84 in. The use of such a roller would reduce the number of laps to 2, and therefore the number of passes to 6.

Effective roller width, 84 in. − 6 in. (overlap) = 78 inches
Number of laps: First lap 84 in.; 156 − 84 = 72 in. remaining
Additional laps

$$\frac{72 \text{ in.}}{78 \text{ in.}} = 0.9$$

The required running speed would then be 6 mph and the project would require only two such rollers.

SAFETY

Molten asphalt generates fumes that can cause skin diseases and eye and respiratory tract irritation. The National Institute for Occupational Safety and Health (NIOSH) has been evaluating industry-developed technology to control exposures to asphalt fumes during paving operations. NIOSH's involvement was requested by the National Asphalt Pavement Association (NAPA). NIOSH researchers have been assisting manufacturers of asphalt equipment in redesign efforts to reduce emissions. Preliminary results suggest these control systems will capture a significant amount of the asphalt fumes generated during the paving process. NIOSH has published the first of these results in a publication entitled *Engineering Control Guidelines for Hot Mix Asphalt Pavers, Part 1 New Highway-Class Pavers* (DHHS [NIOSH] Publication No. 97-105).

Personnel working around asphalt plants must be trained to be safety conscious [2]. The potential for a serious fire results from the presence of burner fuel for the dryer, liquid asphalt, and the hot oil of the heat transfer system. No open flames or smoking should be permitted in the plant area. All fuel, asphalt, and oil lines should have control valves that can be activated from a safe distance. Both asphalt and hot oil lines should be checked regularly for leaks.

Besides the fire potential, many of the mechanical parts are very hot. These will cause severe burns if personnel should come into contact with them. The mechanical system for moving the aggregates, with its belts, sprockets, and chain drives, presents a hazard to personnel if they should become entangled in the system's moving parts. All of these parts should be covered or protected, but they do require maintenance. Maintenance should only be performed when the plant has been completely shut down. The plant operator must be informed that personnel are working on the plant, and the lockout switch should be tagged.

Even at the paving spread the mix is still very hot and direct contact will result in severe burns. Additionally, trucks backing to the paver have limited rear visibility. Personnel must be made aware of these hazards on a regular basis and necessary safety equipment provided.

SUMMARY

Asphalt mixes offer great flexibility to construct pavements tailored to the requirements of the local situation. Material selection, production, and construction processes vary depending on the type of application. The load applied to an asphalt pavement is primarily carried by the aggregates in the mix. The aggregate portion of a mix accounts for 90 to 95% of the material by weight. Good aggregates and proper gradation of those aggregates are critical to the mix's performance.

Hot-mixed asphalt is produced at a central plant and transported to the paving site in trucks. An asphalt paver consists of a tractor power unit and a trailing screed. Two tow arms, pin connected to the tractor unit, draw the screed behind the tractor. The screed controls the asphalt placement width and depth, and imparts the initial finish and compaction to the material.

Because of the relationships between pavement air voids and mechanical stability, durability, and water permeability, asphalt pavements are designed based on the mix being compacted to a specified density. Three basic roller types are used to compact asphalt-paving mixes: smooth drum steel wheel, pneumatic tire, and smooth drum steel wheel vibratory rollers. Critical learning objectives include:

- An understanding of asphaltic materials
- An understanding of the mixing process and elements of both batch and drum mix asphalt plants
- An understanding of equipment required for asphalt paving operations
- An ability to calculate either coverage or quantity of material for prime, tack, or seal coats
- An ability based on paver speed to calculate required plant production
- An ability based on paver speed to calculate required roller capacity

These objectives are the basis for the problems that follow.

PROBLEMS

15.1 An asphalt plant can produce 300 tph. A project requires paving individual 10-ft lanes with a $1\frac{1}{2}$-in. lift averaging 115 lb/sy-in. What average paver speed will match the plant production? How many 14-ton dump trucks will be required if the total hauling cycle time is 40 min? (52.5 ft/min.; 15 trucks)

15.2 An asphalt plant can produce 240 tph. A project requires paving individual 12-ft lanes with a 1-in. lift averaging 112 lb/sy-in. What average paver speed will match the plant production? How many 10-ton dump trucks will be required if the total hauling cycle time is 30 min?

15.3 An asphalt plant can produce 240 tph. A project requires paving individual $11\frac{1}{2}$-ft lanes with a 2-in. lift averaging 110 lb/sy-in. What average paver speed will match the plant production? How many 10-ton dump trucks will be required if the total hauling cycle time is 55 min?

15.4 An asphalt distributor having an 8-ft-long spray bar will be used to apply a prime coat at a rate of 0.3 gal/sy. The road being paved is 16 ft wide and 2,200 ft in length. How many gallons of prime will be required? (1,174 gal)

15.5 An asphalt distributor having a 12-ft-long spray bar will be used to apply a tack coat at a rate of 0.10 gal/sy. The road being paved is 24 ft wide and 2 miles in length. How many gallons of prime will be required?

15.6 An asphalt plant will produce 180 tph for a highway project. The mat will be 12 ft wide and $1\frac{1}{2}$-in. thick, and will have a density of 110 lb/sy-in. A vibratory roller with a 84-in. wide drum will be used for compaction. Assume a 60-min hour efficiency factor for the roller. The overlap between adjacent laps and overedges will be 6-in. minimum. It is estimated that nonproductive travel will add about 15% to total travel. From a test mat, it was found that three passes with the roller are required to achieve density. To account for acceleration and deceleration when changing directions, add 10% to the average roller speed to calculate a running speed. If it is desired to keep the rolling speed under 3 mph, how many rollers are required for the project?

15.7 A plant will produce 440 tph for a project. The mat will be 12 ft wide, 1 in. thick, and will have a density of 112 lb/sy-in. A vibratory roller with a 60-in.-wide drum will be used for compaction. Assume a 50-min hour efficiency factor for the roller. The overlap between adjacent laps and overedges will be 6-in. minimum. It is estimated that nonproductive travel will add 15% to total travel. From a test mat, it was found that three passes with the roller are required to achieve density. To account for acceleration and deceleration when changing directions, add 10% to the average roller speed to calculate a running speed. What is the required roller speed if only one roller is available?

15.8 An asphalt distributor having a 10-ft-long spray bar will be used to apply a prime coat at a rate of 0.2 gal/sy. The road being paved is 10 ft wide and 5,200 ft in length. How many gallons of prime will be required?

15.9 An asphalt plant will produce 180 tph for a highway project. The mat will be 12 ft wide and $1\frac{1}{2}$-in. thick, and will have a density of 115 lb/sy-in. A vibratory roller with a 66-in.-wide drum will be used for compaction. Assume a 50-min hour efficiency factor for the roller. The overlap between adjacent laps and overedges will be 6-in. minimum. It is estimated that nonproductive travel will add about 15% to total travel. From a test mat, it was found that three passes with the roller are required to achieve density. To account for acceleration and deceleration when changing directions, add 10% to the average roller speed to calculate a running speed. If it is desired to keep the rolling speed under 3 mph, how many rollers are required for the project?

15.10 Cut-back asphalt is a mixture of asphalt cement and:

a. Water
b. Aggregate
c. A volatile

15.11 Asphalt pavements should contain:

a. No air voids
b. Some air voids
c. As many air voids as possible

REFERENCES

1. *Asphalt Construction Handbook*, 6th ed. (1992). Barber-Greene, DeKalb, IL.
2. *Construction of Hot Mix Asphalt Pavements (MS-22)* 2nd ed. (2001). Asphalt Institute, Lexington, Ky.

3. *Engineering Control Guidelines for Hot Mix Asphalt Pavers, Part 1 New Highway-Class Pavers* (DHHS [NIOSH] Publication No. 97-105), National Institute for Occupational Safety and Health, Cincinnati, OH, April. Second printing with minor technical changes.
4. *Hot-Mix Asphalt Paving Handbook* (1991). U.S. Army Corps of Engineers, UN-13(CEMP-ET), July.
5. Roberts, F. L., P. S. Kandhal, E. R. Brown, D. Lee, T. W. Kennedy (1996). *Hot Mix Asphalt Materials, Mixtures, Design and Construction,* 2nd ed., NAPA Research and Educational Foundation, Lanham, MD.
6. Kuennen, Tom (2004). "New Mixes Alter Compaction Technology," *Better Road,* September, pp. 32–46.
7. Scherocman, James A. (1996). "Compacting Hot-Mix Asphalt Pavements, Part I," *Roads & Bridges*, August.
8. Scherocman, James A. (1996) "Compacting Hot-mix Asphalt Pavements, Part II," *Roads & Bridges*, September.

15.2

15.3

WEBSITE RESOURCES

1. www.asphaltinstitute.org The Asphalt Institute is an association of international petroleum asphalt producers, manufacturers, and affiliated businesses. The institute's mission is to promote the use, benefits, and quality performance of petroleum asphalt through engineering, research, and educational activities.
2. www.asphaltalliance.com The Asphalt Pavement Alliance is an industry coalition of the Asphalt Institute, National Asphalt Pavement Association, and the State Asphalt Pavement Association. The alliance is dedicated to enhancing our nation's roads and highways through programs of education, technology development, and technology transfer on topics relating to hot mix asphalt pavements.
3. www.astecinc.com Astec Inc. is a manufacturer of hot mix asphalt plants.
4. www.cmicorp.com CMI Terex Corporation is a major supplier of paving and pavement production equipment.
5. www.asphalt.org The International Society for Asphalt Pavements (ISAP) promotes technical advances in asphalt paving.
6. www.hotmix.org The National Asphalt Pavement Association (NAPA) is a national trade association that represents the hot mix asphalt industry.
7. www.roadtec.com Roadtec is a manufacturer of cold planers, asphalt pavers, and material transfer vehicles.
8. www.roscomfg.com/rosco/index.asp Rosco Manufacturing Company designs and manufactures asphalt distributor trucks, chip spreading equipment, self-propelled brooms, and pneumatic and drum rollers.

CHAPTER 16

Concrete and Concrete Equipment

Concrete consists of Portland cement, water, and aggregates that have been mixed together in the proper proportions and allowed to cure and gain strength. There are two types of concrete-mixing operations: (1) transit mixed and (2) central mixed. Unless the project is in a remote location or is relatively large, the concrete is batched in a central batch plant and transported to the job site in transit-mix trucks. Central-mixed concrete is mixed completely in a stationary mixer and transported to the project in either a truck agitator, a truck mixer operating at agitating speed, or a nonagitating truck.

INTRODUCTION

In building the Bavian canal in 690 B.C., the Assyrians used a mixture of one part lime, two parts sand, and four parts limestone aggregate to create a crude concrete. In 1824 Joseph Aspdin took out a patent in England on "Portland" cement. He named it Portland because its color resembled the limestone on the Isle of Portland in the English Channel. That patent marks the beginning of concrete, as we know it today. This manufactured cement consists of limestone and clay burned at temperatures in excess of 2,700°F. Concrete became widely used in Europe during the late 1800s and was brought to the United States late in that century. Its use continued to spread rapidly as knowledge about it and experience with it grew.

Portland cement concrete is one of the most widely used structural materials in the world for both civil works and building projects (see Fig. 16.1). Its versatility, economy, adaptability, worldwide availability, and especially its low maintenance requirements make it an excellent building material. The term "concrete" is applicable for many products but is generally used with Portland cement concrete. It consists of Portland cement, water, and aggregates that have been mixed together, placed, consolidated, and allowed to cure. The

FIGURE 16.1 Precast concrete deck section for the East span of the Bay Bridge in Oakland, California.

Portland cement and water form a paste, which acts as the glue or binder. When fine aggregate is added (aggregate whose size range lies between the No. 200 mesh sieve and the No. 4 sieve), the resulting mixture is termed mortar. Then when coarse aggregate is included (aggregate sizes larger than the No. 4 sieve but less than 3 in.) concrete is produced. Normal concrete consists of about three-fourths aggregate and one-fourth paste, by volume. The paste usually consists of water-cement ratios between 0.4 and 0.7 by weight. Admixtures are sometimes added for specific purposes, such as to improve workability, to impart color, to retard the initial set of the concrete (e.g., for long hauling distances), to gain rapid hardening and high initial strength (e.g., for post-tensioned elements), to improve flowability (e.g., for self-compacting concrete), and to waterproof the concrete.

The operations involved in the production of concrete will vary with the end use of the concrete, but, in general, the operations include

1. Batching the materials
2. Mixing
3. Transporting
4. Placing
5. Consolidating
6. Finishing
7. Curing

CONCRETE MIXTURES

PROPORTIONING CONCRETE MIXTURES

For successful concrete performance the mixture must be properly proportioned. The American Concrete Institute (ACI) has a number of excellent recommended practices, including one on proportioning concrete mixtures [1]. Detailed treatment of concrete proportioning is beyond the scope of this text, but some practical considerations include:

1. Although it takes water to initiate the hydraulic reaction, as a general rule, the higher the water-cement ratio, the lower the resulting strength and durability of the concrete.
2. The more water that is used (which is not to be confused with the water-cement ratio), the higher will be the slump.
3. The more aggregate that is used, the lower will be the cost of the concrete.
4. The larger the maximum size of coarse aggregate, the less the amount of cement paste that will be needed to coat all the particles and to provide necessary workability.
5. Adequate consolidation produces stronger and more durable concrete.
6. The use of properly entrained air enhances almost all concrete properties with little or no decrease in strength if the mix proportions are adjusted for the air.
7. The surface abrasion resistance of the concrete is almost entirely a function of the properties of the fine aggregate.

FRESH CONCRETE

To the designer, fresh concrete is usually of little importance. To the constructor, fresh concrete is *all-important*, because it is the fresh concrete that must be mixed, transported, placed, consolidated, finished, and cured. To satisfy both the designer and the constructor, the concrete should

1. Be easily mixed and transported.
2. Have minimal variability throughout, both within a given **batch** and between batches.
3. Be of proper workability so as to enable proper consolidation, prevent segregation, completely fill the forms, and provide achievement of a proper finish.

batch
Amount of concrete produced in one mixing operation.

Workability is difficult to define. Like the terms "warm" and "cold," workability depends on the situation. One measure of workability is slump, which is a pseudo-measurable value based on an American Society for Testing and Materials standard test (ASTM C143) [2]. The fresh concrete is placed into a hollow frustrum of a cone, 4 in. in diameter at the top, 8 in. in

TABLE 16.1 Recommended slumps for various types of concrete construction (ACI 211.1) [3].

Types of construction	Slump (in.) Maximum	Slump (in.) Minimum
Reinforced foundation walls and footing	3	1
Plain footings, caissons, and substructure walls	3	1
Beams and reinforced walls	4	1
Building columns	4	1
Pavements and slabs	3	1
Mass concrete	2	1

Measuring slump

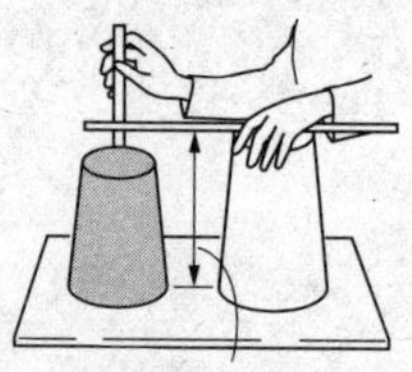

diameter at the bottom, and 12 in. high. After filling and consolidating the concrete in the cone according to a prescribed procedure, the cone is raised from the concrete, allowing the fresh concrete to "slump" down. The amount of slump is measured in inches (or millimeters) from its original height of 12 in., with the stiffest concrete having zero slump and the most fluid concrete having slumps in excess of 8 in. Although the slump only measures one attribute of workability (the ability of fresh concrete to flow), it is the most widely used measure. Table 16.1 gives the recommended slumps for various types of concrete construction [3].

BATCHING CONCRETE MATERIALS

Most concrete batches, although designed on the basis of absolute volumes of the ingredients, are ultimately controlled in the batching process on the basis of weight. Therefore, it is necessary to know the weight-volume relationships of all the ingredients. Then each ingredient must be accurately weighed if the resulting mixture is to have the desired properties. It is the function of the batching equipment to perform this weighing measurement.

Cement

For most large projects, the cement is supplied in bulk quantities from cement transport trucks, each holding 25 tons or more, or from railroad cars. Bulk cement is usually unloaded by air pressure from special truck trailers or rail cars and stored in overhead silos or bins. Cement may be supplied in paper bags, each containing 1 cubic foot (cf) loose measure and weighing 94 lb net. Bagged cement must be stored in a dry place on pallets and should be left in the original bags until used for concrete.

Water

The water that is mixed with the cement to form a paste and to produce hydration must be free from all foreign materials. Organic material and oil may inhibit the bond between the hydrated cement and the aggregate. Many alkalies and acids react chemically with cement and retard normal hydration. The result is a weakened paste, and the contaminating substance is likely to contribute to deterioration or structural failure of the finished concrete. Required water properties are cleanliness and freedom from organic material, alkalies, acids, and oils. In general, water that is acceptable for drinking can be used for concrete.

Aggregates

To produce concrete of high quality, aggregates should be clean, hard, strong, durable, and round or cubical in shape. The aggregates should be resistant to abrasion from weathering and wear. Weak, friable, or laminated particles of aggregate, or aggregate that is too absorptive, are likely to cause deterioration of the concrete.

Moisture Condition For determination of mix proportions, the aggregate should be in a saturated surface dry condition, or an adjustment must be made in the water-cement ratio to compensate for the different amount of water contained in the aggregate.

Mix Proportions

The mix specifications will define specific requirements for the materials that constitute the desired concrete product. Typical requirements include:

1. Maximum size aggregate (e.g., $1\frac{1}{2}$ in.)
2. Minimum cement content (sacks per cubic yard or pounds per cubic yard)
3. Maximum water-cement ratio (by weight or in gallons per sack of cement)

When computing the material quantities for a concrete mix, the following are useful factors and information:

1. The average specific gravity of cement is 3.15.
2. The average specific gravity of coarse or fine aggregate is 2.65.
3. The unit weight of water is 62.4 lb/cf.
4. One cubic foot of water equals 7.48 U.S. gallons.
5. One gallon of water weighs 8.33 lb.
6. Usually the proportion of fine aggregate varies between 25 and 45% of total aggregate volume.

The absolute volume of any ingredient in cf is expressed by

$$\text{Volume (cf)} = \frac{\text{Weight of the ingredient (lb)}}{\text{Specific gravity of the ingredient} \times 62.4\ \text{lb/cf}} \qquad \textbf{[16.1]}$$

Example 16.1 illustrates a method of calculating quantities of cement, aggregate, and water.

EXAMPLE 16.1

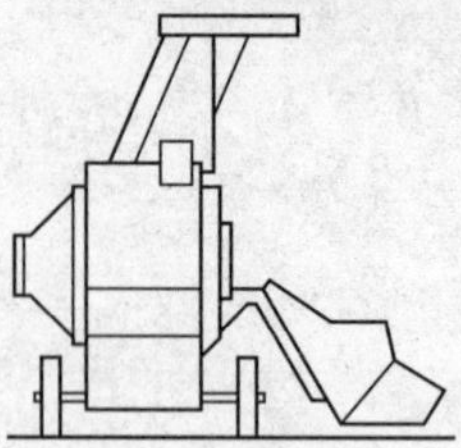

Determine the quantities of materials required per cubic yard to create a concrete mix. The specifications state that the maximum aggregate size is $1\frac{1}{2}$ in., and require a minimum cement content of 6 sacks per cy and a maximum water-cement ratio of 0.65 with 6% air voids:

$$6 \text{ sacks} \times 94 \text{ lb/sack} = 564 \text{ lb cement per cy of concrete}$$

$$\text{Volume cement} = \frac{564 \text{ lb/cy}}{3.15 \times 62.4 \text{ lb/cf}} = 2.87 \text{ cf cement per cy of concrete}$$

Water per cubic yard: 564 lb × 0.65 = 366.6 lb

$$\text{Volume water} = \frac{366.6 \text{ lb/cy}}{1.0 \times 62.4 \text{ lb/cf}} = 5.88 \text{ cf/cy}$$

$$\text{Volume of air: } 27.0 \text{ cf/cy} \times 0.06 = 1.62 \text{ cf/cy}$$

Volume of aggregate:

$$27.0 - 2.87 \text{ (cement)} - 5.88 \text{ (water)} - 1.62 \text{ (air)} = 16.63 \text{ cf/cy}$$

Volume of sand: assume fine aggregate 35%

$$0.35 \times 16.63 \text{ cf/cy} = 5.82 \text{ cf/cy}$$

Weight of fine aggregate:

$$5.82 \text{ cf/cy} \times 2.65 \times 62.4 \text{ lb/cf} = 962.4 \text{ lb/cy}$$

Volume of coarse aggregate:

$$16.63 - 5.82 = 10.81 \text{ cf/cy}$$

Weight of coarse aggregate:

$$10.81 \text{ cf/cy} \times 2.65 \times 62.4 \text{ lb/cf} = 1{,}787.5 \text{ lb/cy}$$

Batching Control

Usually, concrete specifications require the concrete to be batched with aggregate having at least two size ranges (coarse and fine), but up to six ranges can be required. Aggregate from each size range must be accurately measured. The aggregate, water, cement, and admixtures (if used) are introduced into a concrete mixer and mixed for a suitable period of time until all the ingredients are adequately blended together. Most modern concrete plants have performance data on their mixers to show that they can adequately mix an 8-cy batch of concrete in 1 min.

To control batching the Concrete Plant Manufacturers Bureau publishes the Concrete Plant Standards (CPMB 100-96), which outlines plant tolerances [5]. Batching controls are that part of the batching equipment that provides the

TABLE 16.2 Typical batching tolerances (CPMB 100-96).

Ingredient	Individual batchers and cumulative batchers with a tare-compensated control	Cumulative batchers without a tare-compensated control
Cement and other cementitious materials*	±1% of the required weight of materials being weighed or ± 0.3% of scale capacity, whichever is greater	±1% of the required cumulative weight of materials being weighed or ± 0.3% of scale capacity, whichever is greater
Aggregates	±2% of the required weight of material being weighed or ± 0.3% of scale capacity, whichever is greater	±1% of the required cumulative weight of materials being weighed or ± 0.3% of scale capacity, whichever is greater
Water	±1% of the required weight of material being weighed or ± 0.3% of scale capacity, whichever is greater	
Admixtures	±3% of the required weight of material being weighed or ± 0.3% of scale capacity, or ± the minimum dosage rate per 100 lb (45.4 kg) of cement, whichever is greater	±3% of the required cumulative weight of material being weighed or ± 0.3% of scale capacity, or ± the minimum dosage rate per 100 lb (45.4 kg) of cement as it applies to each admixture, whichever is greater

*Other cementitious materials are considered to include fly ash, ground granulated blast furnace slag, and other natural or manufactured pozzolans.

means of controlling the batching device for an individual material. They may be mechanical, hydraulic, pneumatic, electrical, or a combination of these means. Table 16.2 presents the permissible tolerances in accordance with CPMB 100-96. Batch plants are available in three categories: (1) manual, (2) semiautomatic, and (3) fully automatic.

Manual Controls Manual batching occurs when batching devices are activated manually with the accuracy of the batching operation being dependent on the operator's visual observation of a scale or volumetric indicator. The batching devices may be activated by hand, or by pneumatic, hydraulic, or electrical means.

Semiautomatic Controls When activated by one or more starting mechanisms, a semiautomatic batcher control starts the weighing operation of each material and stops automatically when the designated weight has been reached.

Automatic Controls When activated by a single starting signal, an automatic batcher control starts the weighing operation of each material and stops automatically when the designated weight of each material has been reached and interlocked in such a manner that the

1. Charging device cannot be actuated until the scale has returned to zero balance.
2. Charging device cannot be actuated if the discharge device is open.

3. Discharge device cannot be actuated if the charging device is open.
4. Discharge device cannot be actuated until the indicated material is within the applicable tolerances.

MIXING CONCRETE

There are two types of concrete-mixing operations in use: (1) transit mixed and (2) central mixed (see Fig. 16.2). Today, unless the project is in a remote location or is relatively large, the concrete is batched in a central batch plant and transported to the job site in transit-mix trucks, often referred to as *ready-mixed* concrete trucks, or truck mixers (see Fig. 16.3). This type of concrete is controlled by ASTM specification C94 [4], and there is a national organization promoting its use (National Ready Mixed Concrete Association [website 9]). In the European construction culture, concrete for home building and small projects is often produced on the job site using stand-alone mixing machines that range in capacity from small basic units (5 cy/hr) to large advanced assemblies capable of 50 cy/hr.

CONCRETE MIXING TECHNIQUES

When discussing concrete mixing and mixers, *batch* mixing should be distinguished from *continuous* or "flow-type" mixing. Whether on the job site or at

FIGURE 16.2 Central-mixed concrete plant. The mixer drum is seen on the right end of the plant.

FIGURE 16.3 Transit-mix truck (truck mixer) placing concrete for cast-in-place piles.

the ready-mixed plant, concrete is usually mixed by batches. Only for specific applications is mixing continuous, namely, with continuous flow of concrete ingredients charged into one end of the mixer and continuous flow of concrete discharged at the other end.

By their mixing technique, mixers of all types and sizes are either *freefall* mixers or *power* mixers. Freefall mixers (also termed "gravity mixers") mix concrete by lifting the ingredients with the aid of fixed blades inside a rotating drum and then dropping the material by overcoming the friction between the mixture and the blades. Power mixers blend concrete by rapid rotary motion of paddles inside the mixing drum. The size of a mixer is measured by volume. Manufacturers commonly categorize mixers by nominal volume or drum size rather than by the "total volume" or the "dry charge." The nominal volume is the maximum batch capacity, the output of mixer. The total volume of the dry charge is the maximum volume of unmixed ingredients the drum can hold. Maximum batch capacity is commonly $\frac{2}{3}$ to $\frac{3}{4}$ of total drum volume. Thus, for example, a 6-cf mixer, which produces up to 6 cf of fresh concrete, will commonly have a 9-cf total-volume drum. Often a bag-based measure of drum size is also provided by manufacturers (see Table 16.3).

TABLE 16.3 Common sizes of tilting freefall concrete mixers.

Batch capacity (cf)	Batch capacity (bags)
4	$\frac{1}{2}$
6	$\frac{1}{2}$–1
9	1–$1\frac{1}{2}$
12	$1\frac{1}{2}$–2

Mixer output is measured in cf or cy per hour, depending on mixer size, and is determined by the mixer's cycle time. Mixer cycle time is composed of (1) loading time, (2) mixing time, and (3) discharging time. All considered, the estimation of mixer output for self-loading freefall mixers is commonly based on 40 cycles/hr or 1.5 min of total cycle time for mixers of up to 15 cf and 30 cycles/hr, 2 min of total cycle time, for larger mixers. For freefall mixers loaded manually, the number of cycles per hour drops to 15 to 25. For power mixers use 20% more cycles per hour due to shorter mixing time. Actual mixing time is a more complicated issue. Standards usually specify minimum required mixing times, while equipment manufacturers determine effective times according to the mixing intensity that their equipment can achieve. With smaller mixers, those typically used at the job site, mixing time is likely to be longer than for the larger, more advanced machines used in central mixing plants. Typically for central mix plants, the minimum time requirement is 30 seconds per batch.

Freefall Mixers

The drum of a freefall mixer can be filled and emptied by changing its direction of rotation, opening it, or tipping it up. Rotation speed must carefully follow machine-specific instructions and should not be too fast so that the free fall of the mixture is not interrupted by the centrifugal force. Freefall mixing suits concretes that are not too stiff, usually with a 2-in. minimum slump, as commonly used on construction sites. Truck mixers, discussed later, also use freefall mixing. Two common types of freefall mixers are tilting mixers and reversible mixers.

Tilting Mixers These are commonly trailer-mounted or otherwise portable, small to midsize mixers, used either as main concrete mixing equipment on small sites or as ancillary equipment on sites served by a ready-mixed concrete plant. The drum has two axes: one around which the drum rotates and another that serves to change from loading and mixing position (drum opening up) to discharging position (drum opening down). This position change is done manually by a dump wheel (or handle, in the smaller mixers), while drum rotation is electric, gasoline, or diesel powered. Drums are traditionally made of steel, but polyethylene drums for easier cleaning are now offered. Most mixers are of the side-dump type, but end-dump mixers are also manufactured. Material is commonly loaded manually, directly into the drum. There are larger-size self-loading units equipped with a tilting hopper. After being filled manually at ground level, the hopper is tilted up mechanically and dumps the material into the drum. A built-in mechanical drag shovel to facilitate aggregate loading into the hopper is optional on some models. Common sizes of tilting mixers for construction sites are given in Table 16.3; common outputs are in the range of 2 to 10 cy/hr. Larger mixers, in the range of 5 to 15 cy, are used on central-mixed concrete plants, but these are not too common.

Reversible Mixers The drum on a reversible mixer (see Figs. 16.2 and 16.4) has one horizontal axis around which it rotates. There are two openings, one at each end of the drum: one for feeding the ingredients, the other for discharging the mixture. In mixing position, the drum rotates in one direction, while for discharging rotation is reversed. Commonly self-loading, reversible mixers are midsize to large-size units mounted on a two- or four-wheel trailer for transportation between sites. The mixer is equipped with a tilting hopper (see Fig. 16.4), similar to that occasionally found on a tilting mixer, or with a hoist-like hopper that moves up and down on a short inclined set of rails and that dumps the ingredients through a bottom opening into the drum. A built-in mechanical skip to facilitate aggregate loading into the hopper is optional on many models (see Fig. 16.4). Most models of this mixer type are equipped with a water tank and meter, and optionally (the larger models) with a built-in aggregates batcher. The trailer-mounted units, in the range of 8- to 20-cf drum size, are used for on-site service, with common outputs of 6 to 25 cy/hr. The larger, bare units, typically of up to 15-cy drum size, are often used on central mixing plants with much higher outputs.

Power Mixers

Power mixers, also termed paddle mixers, forced mixers, or compulsory mixers, mix concrete by rapid rotary motion of paddles (or mixer heads) moving in centric or eccentric courses inside the mixing drum. To prevent the concrete from sticking to the drum sides and bottom, some of the paddles must constantly

FIGURE 16.4 Reversible concrete mixer with a skip and tilting hopper.

clean the concrete off the sides and bottom and redirect it to the drum's center. The paddles are spring-connected to the drum to prevent them from breaking under the intensive mixing action. The intensive mixing also causes excessive wear to the drum's inner surface, which is not present in freefall mixers. To withstand this wear, the inside of the drum is lined with small, easily replaceable plates that are produced of specialized abrasion-resistant materials (see Fig. 16.5). With their particularly effective mixing, power mixers require a shorter mixing time to obtain homogeneous, high-quality concrete, resulting in greater output. The mixing output of a power mixer is about 50 to 100% higher than that of a same-size freefall mixer. Power mixers can also handle extra-dry mixtures unmanageable by freefall mixers. In Europe, because of these two features, power mixers are generally the preferred type of mixers for the ready-mixed concrete industry and are the dominant mixers in the precast concrete industry. Electrical power mixers are manufactured in a wide variety of types but are commonly categorized as either pan or trough mixers.

Pan Mixers The paddles of a pan mixer (Fig. 16.6) are connected to a vertical shaft inside the pan-shaped drum. In *turbomixers* the vertical shaft is fixed and located in the center of the drum. There are models in which (1) both the drum and paddles rotate in counter directions, (2) the drum is stationary while all the paddles rotate, in the same direction, (3) the drum is stationary but paddles rotate in counter directions. In *planetary mixers* the vertical shaft is rotary and located eccentrically in the stationary drum. In this way, a double motion of the paddles is effected. This motion resembles that of the planets around the

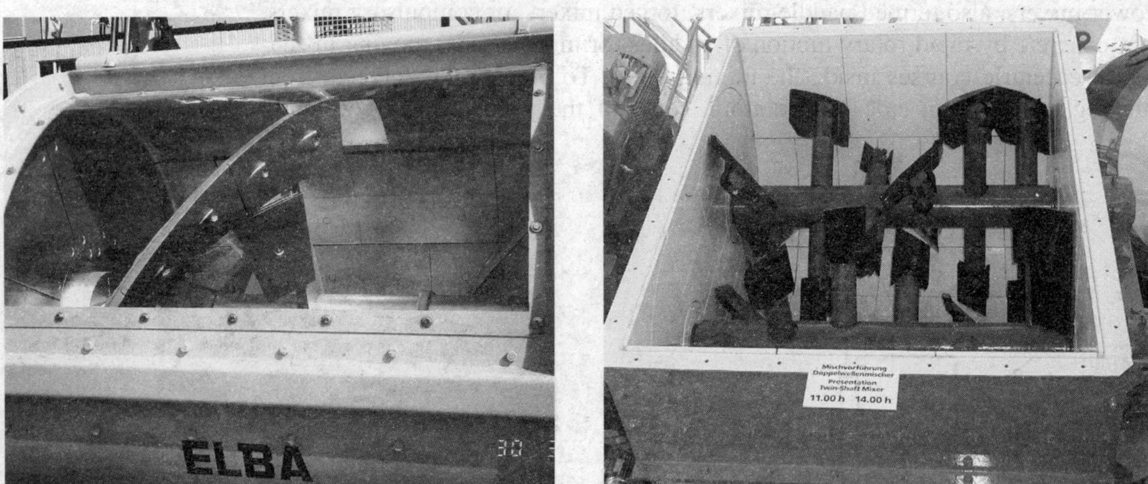

(a) (b)

FIGURE 16.5 Inside view of (a) single-shaft and (b) twin-shaft trough mixers; the sides of the drums are lined with plates to facilitate replacement.

FIGURE 16.6 Inside view of a large pan mixer.

sun (hence the name), and the intensity of the mixing is increased. In *counterflow mixers,* the vertical shaft is fixed and located eccentrically in the rotating drum; the drum and paddles rotate in counter directions. *Evenflow mixers* are identical to counterflow mixers, but for the motion of the drum and paddles, which rotate in the same direction (though in different velocities). There are also various combinations of these mixers; for example, the pan mixer shown in Fig. 16.6 has both centric and planetary paddles rotating in counter directions. Double-shaft pan mixers of various types exist as well. Pan mixers are commonly manufactured in the range of 1- to 5-cy drum size, with outputs of 50 to 300 cy per hr. The size of a large pan mixer may reach a diameter of 17 ft, pan height of 3 ft, and overall height of 6 ft. They are beginning to make an appearance in the United States.

Trough Mixers These are power mixers that have a trough-shaped drum (i.e., resembling a horizontally placed barrel cut by half). The *single-shaft mixer* (Fig. 16.5a) has a horizontal shaft onto which the paddles are connected in a spiral-like arrangement; in some models wave-shaped mixing arms replace the paddles (as seen in Fig. 16.5a). The combination of radial (rotary) and axial (horizontal) movements obtained produces a three-dimensional circulation path that further increases mixing intensity and therefore results in shortened mixing times. Some single-shaft mixer models are used in central mixing plants as climbing mixers; the drum moves up on an inclined track in the course of mixing, and discharge is accomplished by tilting the drum at the top of the track. This plant configuration, possible only with a single-shaft trough mixer, significantly simplifies the conveyance of materials, the aggregates and

TABLE 16.4 Common dimensions of twin-shaft trough concrete mixers.

Drum size (batch capacity) (cy)	Drum length (ft-in.)	Overall mixer length (ft-in.)	Width (ft-in.)	Drum height (ft-in.)	Overall mixer height (ft-in.)
1	4-11	8-0	5-5	3- 3	5- 1
4	7- 5	11-2	8-1	4- 7	6-10
8	9- 8	15-9	10-5	6- 5	9- 1
12	9- 9	16-7	13-3	7-10	10- 6

cement, as drum loading is done at ground level; because climbing is done at the same time as mixing, no increase of overall cycle time results. Typical range of sizes for single-shaft mixers is 2 to 10 cy. There are also *twin-shaft mixers* (Fig. 16.5b), but these are used only as stationary machines, with bottom discharge. The mixing concept is similar to that of the single-shaft mixer, but due to the high degree of turbulence developed in the intersection zone of the two mixing circles, better mixture homogeneity is effectively achieved. Additionally, owing to the design of the twin-shaft mixer and the buildup of the mix between the shafts, wear is considerably reduced as compared to single-shaft or pan-type mixers. As an example of size and output, one manufacturer's rating of twin-shaft mixers is 1- to 12-cy drum size (1.5- to 17.5-cy dry charge) with outputs of 60 to 500 cy/hr. The dimensions of the smallest, typical midsize, and the largest units on this line are given in Table 16.4.

READY-MIXED CONCRETE

Increasingly, concrete is proportioned at a central location and transported to the purchaser in a fresh state, mixed en route. This type of concrete is termed "ready-mixed concrete" or "truck-mixed concrete." It is concrete that is completely mixed in a truck mixer (see Fig. 16.3), with 70 to 100 revolutions at a speed sufficient to mix the concrete completely. Obviously, to be useful, ready-mixed concrete must be available within a reasonable distance from the project. At remote locations and locations requiring large quantities of concrete, concrete plants are generally set up on-site.

The specifications for the batch plant and the transit-mixer transport trucks are covered in detail in ASTM C94 [4]. Of particular importance is the elapsed time from the introduction of water to the placement of the concrete. ASTM C94 allows a maximum of $1\frac{1}{2}$ hours, or before the drum has revolved 300 revolutions, whichever comes first.

Transit-mix trucks, known as truck mixers, are available in several sizes up to about 20 cy (15 cu. m, see Fig. 16.7). The common sizes are in the range of 8 to 12 cy, and the most popular size is 11 cy. Most truck mixers are of the *rear-discharge* type (see Figs. 16.3 and 16.7). There are, however, in the United States, *front-discharge* truck mixers (Fig. 16.8). Owing to their design and configuration, these truck mixers feature lower wheel pressure, and hence

FIGURE 16.7 Large, 15-cu.-m (about 20-cy) truck mixer for transporting ready-mixed concrete.

FIGURE 16.8 Front-discharge truck mixer.

(a) (b)

FIGURE 16.9 Rear-discharge truck mixer with a trailer axle: (a) loaded and (b) empty.

provide a solution for meeting road and bridge load regulations. While more costly, front-discharge mixers have additional advantages: (1) convenient forward-driving approach to discharge location and in-cab driver control of discharge operation and (2) smoother drum rotation (no vibrations) due to elongated shape of drum and even weight distribution, enabling higher road-travel speed and increased rotation velocity and discharge rate. As an alternate solution to meeting road load restrictions, truck mixers (commonly but not solely of the rear-discharge type) may be equipped with an extra, hydraulically operated, foldable trailer axle for additional weight distribution (Fig. 16.9a). This axle is folded up when the mixer is not carrying a load or when there are no wheel-load regulations (Figs. 16.3 and 16.9b).

Truck mixers are not generally employed by construction companies as part of their organizational equipment fleet, but rather are owned by the companies that operate the ready-mixed concrete plants that serve the construction projects. When considering the use of ready-mixed concrete, hauled to the site in large and heavy trucks, attention must be given to on-site travel limitations in terms of both maneuvering space and type of terrain. In planning the placement, a maximum 1 to 1.5 cy/min concrete discharge rate may be considered. In practice, though, discharge time is usually governed by the rate at which concrete is accepted into the pump hopper or directly into the formwork, or by the crane bucket cycle time. Only seldom would the maximum mixer discharge rate be the controlling discharge rate. To increase their independence and operational capacity, truck mixers may be equipped with a boom pump or belt conveyor. This might be a solution, for example, for small jobs requiring single placements of concrete amounts not exceeding the volume of the mixer. The typical reach of truck-mixer-mounted booms is 55 to 90 ft, and of truck-mixer-mounted belts it is 40 to 55 ft. Maximum pumping and conveying rates are 70 to 90 cy/hr.

FIGURE 16.10 Sectional view through the freefall-type drum of a truck mixer.

Truck mixers of all types use freefall mixing, and drums are always of the reversible type (see Fig. 16.10); unlike with reversible mixers, though, the drum has only one opening for both loading and discharge. Truck mixers are capable of thoroughly mixing the concrete with about 100 revolutions of the mixing drum. Mixing speed is generally 8 to 12 rpm. This mixing during transit usually results in stiffening the mixture, and ASTM C94 allows the addition of water at the job site to restore the slump, followed by remixing. This has caused problems and raised questions concerning the uniformity of ready-mixed concrete. ACI 304 [1] recommends that some of the water be withheld until the mixer arrives at the project site (especially in hot weather), then the remaining water be added and an additional 30 revolutions of mixing be required. To offset any stiffening, small amounts of additional water are permitted, *provided the design water-cement ratio is not exceeded*. The uniformity requirements of ready-mixed concrete are given in Table 16.5. To maintain their independence with respect to water addition, both en route and on the job site, truck mixers are equipped with a water tank and water meter. The truck's water tank typically is in the 100- to 200-gal range. The water is also used to wash down the drum and truck chutes after concrete placement, with the aid of a built-in hose.

Ready-mixed concrete may be ordered in several ways, including the following:

1. *Recipe batch*. The purchaser assumes responsibility for proportioning the concrete mixture, including specifying the cement content, the maximum allowable water content, percentage air, and the admixtures required. The

TABLE 16.5 Uniformity requirements for ready-mixed concrete (ASTM C94 [4]).

Test	Requirements, expressed as maximum permissible difference in results of tests of samples taken from two locations in the concrete batch
Weight per cubic foot calculated to an air-free basis	1.0 lb/cf is greater
Air content, volume percent of concrete	1.0%
Slump	
If average slump is 4 in. or less	1.0 in.
If average slump is 4 to 6 in.	1.5 in.
Coarse aggregate content, portion by weight retained on No. 4 sieve	6.0%
Unit weight of air-free mortar based on average for all comparative samples tested	1.6%
Average compressive strength at 7 days for each sample, based on average strength of all comparative test specimens	7.5%

purchaser may also specify the amounts and type of coarse and fine aggregate, and even the origin of the aggregates. Under this approach, the purchaser assumes full responsibility for the resulting strength and durability of the concrete, provided that the stipulated amounts are furnished as specified.

2. *Performance batch*. The purchaser specifies the requirements for the strength and slump or type of concrete-placing equipment to be used (e.g., concrete pump). The manufacturer assumes full responsibility for the proportions of the various ingredients that go into the batch, as well as for the resulting strength and durability of the concrete.
3. *Part performance and part recipe*. The purchaser generally specifies a minimum cement content, the required admixtures, and the strength requirements, enabling the producer to proportion the concrete mixture within the constraints imposed. The manufacturer assumes full responsibility for the resulting strength of the concrete.

Today most purchasers of concrete use the third approach, part performance and part recipe, as it ensures a minimum durability while still allowing the ready-mixed concrete supplier some flexibility to supply the most economical mixture.

CENTRAL-MIXED CONCRETE

This is concrete mixed completely in a stationary mixer and transported to the project in either a truck agitator, a truck mixer operating at agitating speed, or a nonagitating truck. Truck mixers operated as agitators can handle about 20%

more material than when they are used as mixers (for example, a 10-cy mixer would agitate 12 cy of centrally mixed concrete). Plants usually have mixers capable of mixing up 8 cy of concrete in each batch and can produce more than 600 cy of concrete per hour. Some plants have been built with mixers capable of mixing 15 cy of concrete in each batch. The mixer either tilts to discharge the concrete into a truck or a chute is inserted into the mixer to catch and discharge the concrete. To increase efficiency, many large plants have two mixer drums. Plant production capacity, however, is not dependent only on mixing rate but also on the size of the plant-mixer discharge opening, on one hand, and on the size of the intake truck chute or hopper, on the other hand. For example, when concrete is discharged into a dump truck, as may be the case with short hauling distances, output rate may be increased compared to discharge into a truck mixer, provided that the plant mixer is equipped with an optional enlarged discharge door.

In determining the quantities needed and the output for a given plant, one should include any delays in productivity resulting from reduced operating factors such as availability of haul trucks or even the rate at which cement can be delivered to the plant.

Mobile Compact Plants

As an alternative to ready-mixed, central-mixed or transit-mixed, concrete that is transported to the project from the concrete plant, concrete may be produced at the construction site itself. This may typically be the solution for remotely located projects or for projects requiring large amounts of concrete. This may also be the solution for projects located in busy, heavily trafficked urban areas, if unavoidable concrete supply delays are unacceptable. The equipment used for this purpose is usually a smaller-scale self-contained concrete plant, which operates similarly to a full-size permanent central-mix plant, albeit with reduced production capacity. These relocatable compact units are moved between sites either on trucks or, as is common, in a trailer-mounted configuration. They usually need no concrete foundations for site setup, although some preparatory earth works may be needed, depending on plant type, as well as connections to water and power supplies.

Mobile plant units include all the functional components found on permanent plants, although the cement silo is not always an integral part of the unit and may have to be brought separately to the site. The same may be true for the partitions separating the aggregates, depending on aggregate storage method. There are commonly three to five aggregate bins that come in one of the following configurations:

1. *Star silo*. The aggregates are stored on the ground with star-pattern partitions keeping them separated. There is a central shaft around which aggregate discharge openings are arranged and through which the aggregates are conveyed to a weighing station and then on to the mixer. A radial scraper is usually used for piling the aggregates over their discharge openings.

2. *Pocket silo*. The aggregates are stored on the ground inside a vertical pocket-like bin silo. Unlike with star silo, the filling of the bins is commonly done by a loader, and not directly off the supply truck. From the bins, the aggregate moves by gravity for weighing and is then conveyed on to the mixer.
3. *Inline silo*. The aggregates are stored inside an elevated linear silo. Just as with a pocket silo, here too a loader must be used to charge the bins. The aggregates are conveyed for weighing by a belt located underneath the discharge opening line and then on to the mixer.

Modern mobile compact plants are highly automated and can be operated by one worker, or two at the most. They typically use mid- to large-size mixers, of either the freefall or power type, and their concrete production capacity is in the range of 10 to 50 cy/hr. There are higher-capacity integral plants as well, but this is understandably achieved at the expense of compactness and mobility. If the plant is to be employed on the site for a long duration, a less-costly unit may be used that is not necessarily as compact and mobile.

PLACING CONCRETE

Once the concrete arrives at the project site, it must be moved to its final position without segregation and before it has achieved an initial set. This movement can be accomplished in several ways, depending on the horizontal and vertical distance of the movement and other constraints. Methods include buckets, buggies and wheelbarrows, chutes, drop pipes, belt conveyors, and concrete pumps.

BUCKETS

Normally, properly designed bottom-dump buckets (see Fig. 16.11) enable concrete placement at the lowest practical slump. Care should be exercised to prevent the concrete from segregating as a result of discharging from too high above the surface or allowing the fresh concrete to fall past obstructions (such as the forms themselves in the case of columns). Gates should be designed so that they can be opened and closed at any time during the discharge of the concrete.

Buckets come in various sizes, ranging from 0.5- to 4.0-cy, though buckets in the 1.0- to 2.5-cy range are most common. Much larger buckets have been used in dam construction. In building the Hoover Dam in the early 1930s, Frank Crowe, the general superintendent, designed his own 8-cy buckets. Thirty years later at Glenn Canyon Dam, upstream on the Colorado River in Arizona, 13-cy cableway-suspended buckets were used. The weight of a fully loaded bucket, including self-weight, is about 4,800 lb for a 1.0-cy bucket and 9,200 lb for a 2.0-cy bucket. In 2004, the Chinese employed 7.8-cy buckets on the Longtan Dam project. Commonly handled by cranes, buckets should be selected such that they can be lifted safely to the required concrete placing locations by the job-site crane. Quite often, bucket size is determined on the basis of required production rates, as the larger the bucket, the smaller the number of work cycles required to execute a given concrete placement. In these cases, a crane would be selected such that its lifting capacity exceeds the weight of the loaded bucket. If a certain crane-bucket combination satisfies most of the project's production requirements but not all, then for extended-reach locations either a smaller bucket will be used or the bucket in use will be only partly loaded. With any given crane-bucket combination, production rate can be increased by using two buckets instead of one. With bucket loading time thus virtually eliminated, this is particularly beneficial in the case of low- and midrise construction, where bucket loading (and then unloading at the placement location) make up the better part of the overall cycle time.

FIGURE 16.11 Concrete bucket being used for the placement of concrete in formwork for columns.

MANUAL OR MOTOR-PROPELLED BUGGIES

Hand buggy

Power buggy

Hand buggies and wheelbarrows are usually capable of carrying from 4 to 9 cf of concrete, and thus are suitable on many projects, either as the sole or main concrete-placing equipment for small works, or as a complementary means on projects utilizing cranes or pumps as the major concrete-placing equipment. When considering the use of buggies and wheelbarrows, the expense of building ramps, runways, and similar setups should be taken into account, as buggies and wheelbarrows can only operate on smooth and rigid surfaces. Hand buggies are safer than wheelbarrows because they have two wheels rather than one.

Hand buggies and wheelbarrows are recommended for distances less than 200 ft, whereas motor-propelled buggies ("power buggies") can traverse up to 1,000 ft economically. Power buggies basically come in three sizes: 11-cf walk-behind (suitable for tight spaces and restricted areas), 16-cf stand-on (the most common size), and 21-cf stand-on. The smaller-size walk-behind buggy is limited to travel speeds of 3 to 4 mph. The stand-on larger sizes have higher travel speeds, 6 to 7 mph. The power buggy has a single-lever-controlled hydraulic dumping mechanism. The bucket is made of either steel or polyethylene. Steel is preferred only if the buggy is also to be used for hauling hot material, such as hot asphalt; otherwise the polyethylene has an advantage due to easy cleanup of hardened concrete remains.

CHUTES AND DROP PIPES

Chutes are often used to transfer concrete from a higher elevation to a lower elevation. They should have a round bottom, and the slope should be steep enough for the concrete to flow continuously without segregation. Truck mixers are normally equipped with built-in swing (and often extendable) chutes. These chutes are hydraulically operated for direct placing of the mix when the concreting location is within chute reach (as seen in Fig. 16.3). Drop pipes are used to transfer the concrete vertically down. The top 6 to 8 ft of the pipe should have a diameter at least 8 times the maximum aggregate size and may be tapered so that the lower end is approximately 6 times the maximum aggregate size [1]. Drop pipes are used when concrete is placed in a wall or column to avoid segregation caused by allowing the concrete to free-fall through the reinforcement. A *tremie* operation occurs when a drop pipe, which in this case is called a tremie pipe, is used to place concrete under water. The lower end of a tremie pipe is kept continuously immersed in fresh, plastic concrete.

BELT CONVEYORS

Belt conveyors can be classified into three types: (1) portable or self-contained conveyors, (2) feeders or series conveyors, and (3) side-discharge or spreader conveyors. All types provide for the rapid movement of fresh concrete but must have proper belt size and speed to achieve the desired rate of transportation. Particular attention must be given to points where the concrete leaves one conveyor and either continues on another conveyor or is discharged, as segregation can occur at these points. The optimum concrete slump for conveyed concrete is from $2\frac{1}{2}$ to 3 in. [3]. Figure 16.12 shows a belt conveyor mounted on a track carrier being used to place low-slump roller-compacted concrete. The carrier was fed concrete from the plant by a belt conveyor system.

FIGURE 16.12 Low-slump concrete being placed by a belt conveyor mounted on a track carrier.

CONCRETE PUMPS

The pumping of concrete through rigid or flexible lines is not new. Initially developed for the purpose of lining tunnels, German pumping equipment, however, was first introduced in the United States only in the early 1930s, and it was not until the 1970s that concrete pumping gained extensive use in this country. The pump is a fairly simple machine. By applying pressure to a column of fresh concrete in a pipe, the concrete can be moved through the pipe if a lubricating outer layer is provided and if the mixture is properly proportioned for pumping. To work properly, the pump must be fed concrete of uniform workability and consistency. Today concrete pumping is one of the fastest-growing specialty contracting fields in the United States, as perhaps one-fourth of all concrete is now placed by pumping. Pumps are available in a variety of sizes, capable of delivering concrete at sustained rates of 10 to 150 cy/hr and higher. Effective pumping distance varies from 300 to 1,000 ft horizontally, or 100 to 300 ft vertically [7], although occasionally pumps have moved concrete more than 5,000 ft horizontally and 1,000 ft vertically. A standard pump conveyed 6,500 cy of concrete up to a height of 1,745 ft on a project in the Italian Alps in 1994 [15].

Pumps require a steady supply of *pumpable* concrete to be effective. Today there are three types of pumps being manufactured: (1) piston pumps, (2) pneumatic pumps, and (3) squeeze pressure pumps. Most piston pumps currently contain two pistons, with one retracting during the forward stroke of the other to give a more continuous flow of concrete. The pneumatic pumps normally use a reblending discharge box at the discharge end to bleed off the air and to prevent segregation and spraying. In the case of squeeze pressure pumps, hydraulically powered rollers rotate on the flexible hose within the drum and squeeze the concrete out at the top. The vacuum keeps a steady supply of concrete in the tube from the **intake hopper**.

intake hopper
The part of the pump that receives and holds concrete before it is sucked into the pumping cylinder.

There are two main configurations of concrete pumping equipment: (1) pump with a separate pipeline (Fig. 16.13) and (2) a pump and boom combination (Fig. 16.14). The latter is particularly efficient and cost-effective in saving labor and eliminating the need for pipelines to transport the concrete. A third, less common configuration, used mostly in high-rise construction, combines the pump and pipeline with a separate, tower-mounted boom.

FIGURE 16.13 Concrete pump with separate pipeline.

FIGURE 16.14 Truck-mounted pump and boom combination.

Pump with a Pipeline

In this configuration, also termed a *line pump*, the pipeline is a separate system that must be assembled and connected to the pump before pumping operations begin. The pipeline is laid from the location of the pump to the concrete casting area. The pump is located such that the ready-mixed concrete trucks have good access to it. In the case of on-site central-mixed concrete, the pump would be placed with the hopper just under the mixer's discharge opening. On projects that are spread out or for high-rise projects comprising more than one building, several pipelines can be stretched from one pump to various project zones, thus saving the relocation of the pipeline with each change in casting location. A special pipeline gate is used to control which line is being used to any given pumping operation (Fig. 16.15). Another option is to prepare several pipelines and relocate the pump according to the placement location. This latter option requires the use of truck-delivered ready-mixed concrete.

In terms of pump mobility, three types can be distinguished:

1. *Stationary pump*. The pump is mounted on a steel frame and stationed in a fixed location throughout construction. This configuration is suitable for limited-area sites with large and frequent concreting operations.
2. *Trailer pump*. The pump is mounted on a one- or two-axle trailer (see Fig. 16.13) to enable easy on-site relocation as well as movement between sites. It is suitable mainly for spread-out sites. One pump can also serve several adjacent sites, as it can be moved between them according to concreting schedule.

3. *Truck pump*. Not to be confused with the truck-mounted pump and boom combination, this quick setup pump is moved between sites and operated on a standard truck chassis. It is commonly used for one-time concrete operations (such as in small-scale repair or renovation projects, or on a regular construction project prior to setup of the main concrete transportation equipment) when a reach greater than that provided by a pump-truck boom is needed.

FIGURE 16.15 Hydraulic gate splitting the concrete delivery line.

Pipeline *Pipelines* for concrete transportation use pipe diameters of 3 to 8 in. (with 5 and 6 in. the most common sizes). The pipeline is assembled of straight (commonly 10 ft long) and curved steel pipe sections connected to each other by quick couplings. The free end of the line has a flexible rubber pipe 10 to 30 ft long connected to it for better control of the concrete discharge location and for easier handling by the workers who have to direct the spreading of the concrete.

Holding and moving the end of the pipe is hard work, given its weight when filled with concrete. To solve this problem, a special lightweight distribution (or placing) boom can be used. This boom connects to the end of the steel pipe and is supported by a ballasted base. The articulated boom, with a 30- to 50-ft-long operation radius, also enables concreting close to the ballasted base. The entire assembly is crane-lifted for relocation as the concreting progresses.

At the end of each concreting operation, concrete remaining in the line must be cleaned out. This is done by pushing a soft sponge or rubber ball (wash-out ball) through the pipeline with air or water pressure. The line can be cleaned from either end. If cleaned backward (i.e., toward the pump end), no excess concrete will splatter over the just-completed work. With a long line, however, the volume of washed-out concrete may be considerable (e.g., a 500-ft-long, 5-in.-diameter pipeline contains 2.55 cy of concrete). Cleaning from the pump's end should then be considered, with the washed-out concrete used to complete the placement. Alternatively, some provisions for disposing of the waste concrete should be made, such as discharging it into a truck mixer or a bucket for further placement.

Pump and Boom Combination

Also termed a *boom pump*, the pump is mounted on a truck and equipped with a revolving (*slewing*) boom to which a fixed-length delivery line is connected.

The line is made of a steel pipe, commonly of 5 in. diameter. Similar to separate pipelines, here too the free end of the line has a flexible pipe, 9 to 12 ft long (*end hose*), connected to it. In this case, not only does the end hose ease the task of the worker who holds it, but it can also be used to slightly extend the boom's reach.

Hydraulically operated and articulated booms come in lengths of 60 to 200 ft, but the more commonly used booms are in the range of 80 to 140 ft. Most truck-mounted pumps have four or five section booms, but there are also three sections for short booms and six sections for the longest ones. For a given boom length, a greater number of sections provides the boom with greater maneuverability, which may be an advantage in confined areas. Note that "boom length" indicates the boom's maximum *vertical* reach, measured from the ground. Maximum *horizontal* reach, however, measured from the boom's slewing axis, is always shorter, usually by 13 ft for most makes and models of truck-mounted booms. *Effective* horizontal (i.e., net) reach is even shorter, as loss of distance measured from the slewing axis to the front of the truck or to the truck side must be taken into account. Depending on truck make and size, the extended outriggers, by which the truck is stabilized while in operation, may also contribute to the shortening of the effective horizontal reach. It is important that specific boom charts and truck dimensions be consulted when selecting a truck-mounted pump for a given concreting operation. Figure 16.16

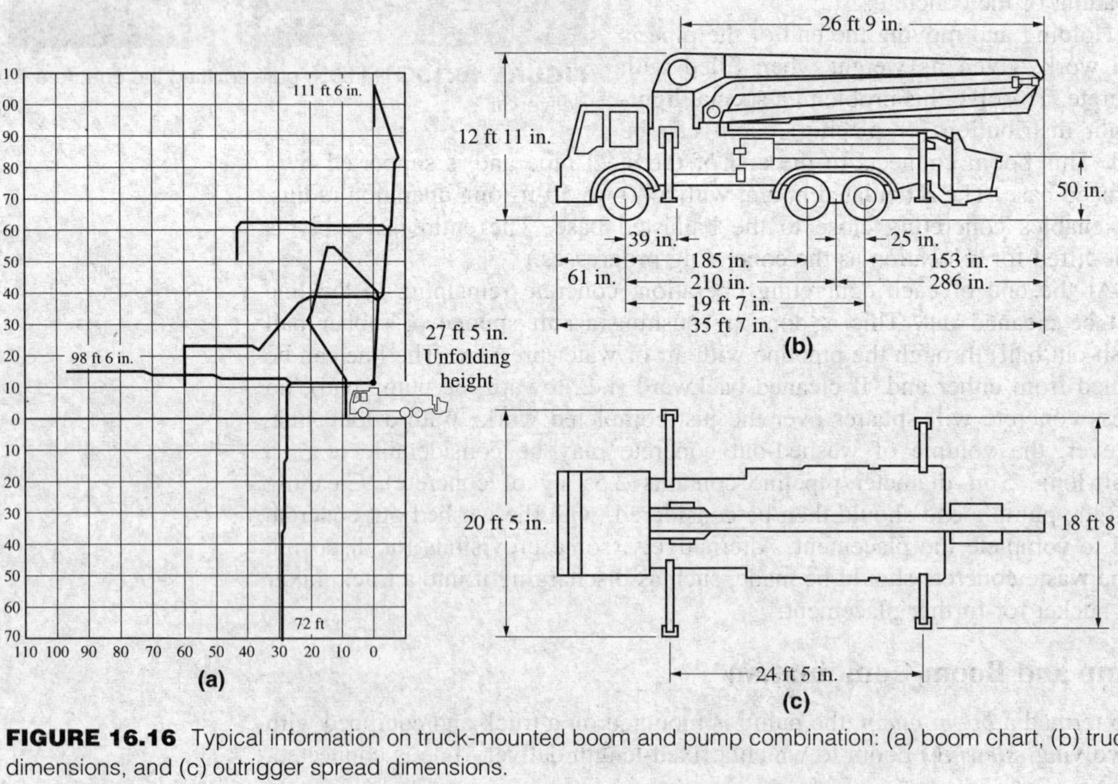

FIGURE 16.16 Typical information on truck-mounted boom and pump combination: (a) boom chart, (b) truck dimensions, and (c) outrigger spread dimensions.

illustrates the kind of information issued by manufacturers of truck-mounted pumps: (a) boom chart; (b) truck dimensions; and (c) outrigger spread dimensions. In this example, the truck is equipped with a four-section, 34-m (111 ft 6 in.) long boom.

Because the reach of boom pumps is limited, these pumps use ready-mixed concrete, with the truck-mounted pump optimally positioned to enable maximum coverage of the concreting area for each concreting operation. If coverage from one location is not possible, the truck is relocated in the course of the operation. Relocation, however, interrupts work continuity and may cause delays, depending on the time it takes to fold the boom, retract the outriggers, move the truck, extend the outriggers, and unfold the boom to continue concrete placing. If relocation was not preplanned, a line of waiting ready-mixed trucks may result as well.

Pump with Pipeline and Tower-Mounted Boom

This configuration combines elements from the two others: the stationary/trailer pump at the (fixed) concrete-loading end, a separate boom at the concrete-placing end, and a pipeline running from the pump to the boom. This is a common solution for high-rise structures that are beyond the reach of a boom pump. The boom—practically the same as the hydraulic articulated placing boom of the boom pump—is mounted on a climbing tower (*mast*). There are two types of masts:

1. A robust steel column runs vertically through openings in the building's floors. The column and the boom mounted on it are either lifted by a crane or self-climb hydraulically (similar to a climbing tower crane, see Chapter 17). Figure 16.17a shows a boom in a folded position mounted on a climbing mast.
2. A lattice mast is similar to the mast of a tower crane. The mast is stationed inside or outside the building and may be either freestanding or tied to the building. The mast can be assembled to its full height right from the beginning or can climb vertically as construction of the building progresses. Figure 16.17b shows a freestanding lattice-mast-mounted boom.

Separate placing booms come in lengths of 80 to 140 ft. Depending on boom length and type of boom-mast connection, some booms come with a short section of ballasted counter-boom (see Fig. 16.17b), while others operate without counterweights. Some booms, termed *detachable booms*, have a quick boom-mast connection that facilitates boom transfer between masts on large projects.

EXAMPLE 16.2

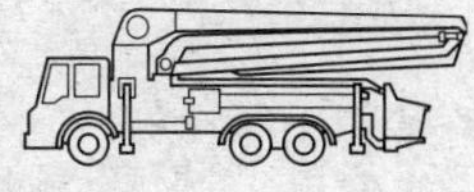

A concrete floor at ground level (slab-on-grade) is to be cast with the boom pump illustrated in Fig. 16.16. A 3-ft-wide strip surrounding the concreting area is needed for the slab's side formwork. What is the maximum effective horizontal reach of the boom, measured from the slab's edge, if the truck is positioned (a) perpendicular to the floor's edge (i.e. boom is operated at the truck's longitudinal axis); (b) parallel to the floor's edge (i.e. boom is slewed 90 degrees)?

(a) (b)

FIGURE 16.17 Mast-mounted concrete placing booms: (a) climbing mast (with folded boom); (b) freestanding lattice mast (with extended boom and counter-boom).

a. Truck positioned perpendicular to the floor's edge.

Maximum horizontal reach (Fig. 16.16a)	98 ft 6 in.	
Distance from slewing axis to truck's front (Fig. 16.16b)	8 ft 10 in.	[(35 ft 7 in.) − (26 ft 9 in.)]
Net horizontal reach (from truck's front)	89 ft 8 in.	
	3 ft 0 in.	(Restricted strip)
Maximum effective horizontal reach (from slab's edge)	86 ft 8 in.	

b. Truck positioned parallel to the floor's edge.

Maximum horizontal reach (Fig. 16.16a)	98 ft 6 in.	
Distance from slewing axis to truck's outriggers (Fig. 16.16c)	10 ft 3 in.	$\left(\frac{20 \text{ ft } 5 \text{ in.}}{2}\right)$
Net horizontal reach (from truck's outriggers)	88 ft 3 in.	
	3 ft 0 in.	(Restricted strip)
Maximum effective horizontal reach (from slab's edge)	85 ft 3 in.	

Compare this result to the pump's "boom length" (that is, the boom's maximum vertical reach) listed on Fig. 16.16a (111 ft 6 in.); that length is 31% greater. This illustrates why nominal boom lengths should not be the sole basis for pump selection.

Pumping Outputs

High placing rates are one of the main advantages of pumping. Pumps now have maximum theoretical outputs of up to 300 cy/hr. These high theoretical outputs, however, are seldom realized. Placing rates are determined in practice by the type and dimensions of the cast element, by the size of the concrete placing crew, and mostly by the rate at which the concrete mixer trucks can deliver concrete to the pump. In a properly organized placing operation, there will always be one truck waiting to position itself at the pump as soon as the previous truck finishes discharging concrete into the pump. If at all possible, there should be room for two trucks to be positioned at the pump hopper—one discharging and the other ready to discharge. This requires adequate access routes for on-site movement of arriving, discharging, and departing trucks. Thus, even under the best of conditions, the placing rate is first and foremost dictated by gross truck discharge cycle times.

Crew size is a limiting factor only when the element being cast or the access to the work area for the workers limits the size of the placing crew (e.g., wall concreting). Otherwise the size of the crew will be determined optimally such that it does not control the placing operation. Another factor that can limit the rate of placing is the formwork, especially for vertical elements. A crucial factor is the lateral hydrostatic pressure of the concrete on wall or column forms [16] (see also Chapter 22); the rate of placing and the imposed hydrostatic pressures from the concrete are commonly much higher with concrete pumping than with crane and bucket placing. This factor must be taken into account when designing the formwork.

The normal maximum theoretical output for the currently available range of boom pumps is 80 to 200 cy/hr, but in the case of the largest pumps the outputs can reach 300 cy/hr. In the case of line pumps, maximum theoretical output ranges from 20 to 140 cy/hr with some of the larger pumps having outputs as high as 260 cy/hr. But even when placing conditions are such that the pump can be fully utilized, it is still advisable to make some allowance for pump downtime (e.g., 10% drop in production rate).

Effective outputs, distinguished from theoretical outputs, are project specific and can vary widely, so it is hard to give average output values, even for a given pump. Under normal conditions, however, the following outputs can be used as initial guidelines for planning, assuming the pump itself is not a limiting factor:

Common building elements of regular dimensions	40 cy/hr
Thick slabs (in excess of 20 in.) and similar elements	60 cy/hr
Mass concreting of large elements (i.e., dams, raft foundations)	80 cy/hr

During construction of a 73-story building, a high-pressure trailer pump having a 138-cy/hr maximum theoretical output was used. On that project the actual outputs achieved were: 70 cy/hr up to the 44th floor, 60 cy/hr up to the 58th floor, and 45 cy/hr up to the top floor.

EXAMPLE 16.3

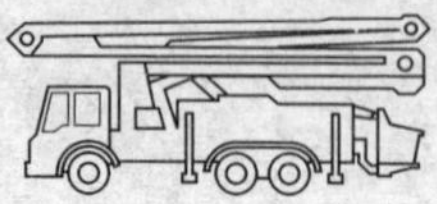

A concrete pump of 100-cy/hr maximum output is used to place 270 cy of concrete. The concrete is supplied by 9 cy transit-mix trucks arriving to the site every 12 min. Barring any interruptions, how many hours will the operation require?

With 90% pump utilization rate (i.e., 10% downtime allowance) the operation would take 270/(100 × 0.9) = 3 hr. Effective output, however, is governed by concrete supply rate:

Number of cycles	5 per hr	(60 min/hr / 12 min)
Output per truck	45 cy/hr	(9 cy × 5)
Placing duration	6 hr	(270 cy / 45 cy/hr)

Basic Pumping Rules

Successful pumping of concrete requires good planning and quality concrete. A common fallacy is to assume that any placeable concrete can be pumped successfully. The basic principle of pumping is that the concrete moves as a cylinder of concrete. To pump concrete successfully, the following rules should be carefully followed:

1. Use a minimum cement factor of 517 lb of cement per cubic yard of concrete ($5\frac{1}{2}$ sacks per cy).
2. Use a combined gradation of coarse and fine aggregate that ensures *no* gaps in sizes that will allow paste to be squeezed through the coarser particles under the pressures induced in the line. This is the most often overlooked aspect of good pumping. In particular, it is important for the fine aggregate to have at least 5% passing the No. 100 sieve and about 3% passing the No. 200 sieve (see gradations given in reference [7]). Line pressures of 300 psi are common, and they can reach as high as 1,000 psi.
3. The use of a minimum 5-in. pipe diameter is desirable.
4. Always lubricate the line with cement paste or mortar before beginning the pumping operating.
5. Ensure a steady, uniform supply of concrete, with a slump of between 2 and 5 in. as it enters the pump.
6. Always presoak the aggregates before mixing them in the concrete to prevent their soaking up mix water under the imposed pressure. This is especially important when aggregates are used that have a high absorption capability (such as structural lightweight aggregate).
7. Avoid the use of reducers in the conduit line. One common problem is the use of a 4- to 5-in. reducer at the discharge end so that workers will have only a 4-in. flexible hose to move around. This creates a constriction and significantly raises the pressure necessary to pump the concrete.

8. Never use aluminum lines. Aluminum particles will be scraped from the inside of the pipe as the concrete moves through and will become part of the concrete. Aluminum and Portland cement react, liberating hydrogen gas that can rupture the concrete with disastrous results.

Employment of Concrete Pumps

The main advantages of concrete pumping are high placing rates, work convenience, and the ability to deliver the concrete while bypassing project site obstructions. A concrete pump, however, is good only for concrete transportation, while most building and heavy civil construction projects require transportation of many other materials and building elements. Such projects, therefore, will require other transportation equipment, such as cranes. Such on-site cranes can then be used for concrete placing as well as other lifting tasks. Thus, concrete pumping should be considered for on-site concrete transportation and placing mainly in the following cases and types of projects:

1. *Projects on which concrete is the primary material to be transported*: Bridges, retaining walls, culverts, and low-rise concrete structures for various uses are examples of projects on which concrete pumps may satisfy all or the majority of material transportation needs, with no or only minor requirements for any further transportation equipment. The pump solution may also suit a specific project phase; for example, the foundations and substructure of a large building. Since concrete is the main transported material at this initial phase, a concrete pump may be the solution even if a crane is to be brought to the site later. The use of the pump in this case will also shorten the duration of the crane's service. Either a truck-mounted boom pump or a line pump should be considered in these cases.
2. *Large concrete structural elements*: Concrete structural elements of large dimensions, such as those encountered during dam and raft foundation construction, or thick slabs and large-size beams may be concreted at a high placing rate using pumping equipment. Using a lower-production-rate means for placing the concrete, such as cranes, may greatly extend the placing duration, while pumping the concrete will reduce the concrete placement time drastically. The simultaneous use of more than one pump for exceptionally large concrete volume placements is common with the number of pumps limited only by site layout, by the production rate of the plants supplying the concrete, and by the ability to secure enough truck mixers. The use of a truck-mounted boom pump should be considered for large volume concrete placements.
3. *Sites with accessibility limitations*:
   a. Difficult on-site conditions, rough terrain, or muddy ground may dictate the use of a line pump.
   b. On-site physical obstructions like bodies of surface water (ponds) cause consideration of either a truck-mounted boom pump or a line pump.
   c. Sites located such that no vehicular access is possible (e.g., in older cities with narrow lanes or stair-based mountainous residential areas); consider either a truck-mounted boom pump or a line pump.

d. Existing buildings (e.g., for repair or renovation works); consider a truck-mounted boom pump.
e. Inside low-roofed spaces (e.g., floor casting in low-clearance warehouses or concrete works in tunnels) consider a truck-mounted boom pump.
f. Sites located in areas with strict limitations concerning interruption to ongoing activity or when no permanent staging area is possible (e.g., bridge construction on an existing highway); consider a truck-mounted boom pump.

4. *Projects on which material transportation equipment limits work productivity*: This is perhaps the most important reason why concrete pumps are used. With the increase in construction operation mechanization and the constant need to speed production—characteristics of modern projects—on-site transportation equipment commonly sets the pace of work progress. Concrete placing by pumps can often increase production and reduce the need for other transportation equipment, therefore consider either a truck-mounted boom or a line pump.

CONSOLIDATING AND FINISHING

CONSOLIDATING CONCRETE

Concrete, a heterogeneous mixture of water and solid particles in a stiff condition, will normally contain a large quantity of voids when placed. The purpose of consolidation is to remove these entrapped air voids and to ensure complete filling of the forms. The importance of proper consolidation cannot be overemphasized, as entrapped air can render the concrete totally unsatisfactory. Entrapped air can be reduced in two ways—use more water or consolidate the concrete. Figure 16.18 shows qualitatively the benefits of consolidation, especially on low-water-content concrete.

Consolidation is normally achieved through the use of mechanical vibrators. Only in case of extremely small/thin elements and particularly wet concrete mixtures is consolidation by hand tools, such as a plastic-head hammer, allowed. Properly vibrated concrete is higher-quality concrete, mainly in terms of strength, but also in terms of reinforcement protection, resistance to aggressive agents, and overall appearance. Vibrators and the vibrating action are characterized and distinguished by the following properties:

1. *Frequency:* The number of vibrations per unit time (commonly minutes).
2. *Amplitude:* The magnitude of the motion in each vibration.
3. *Orientation:* There are vibrators with random motion at all directions, while others have unidirectional motion only.

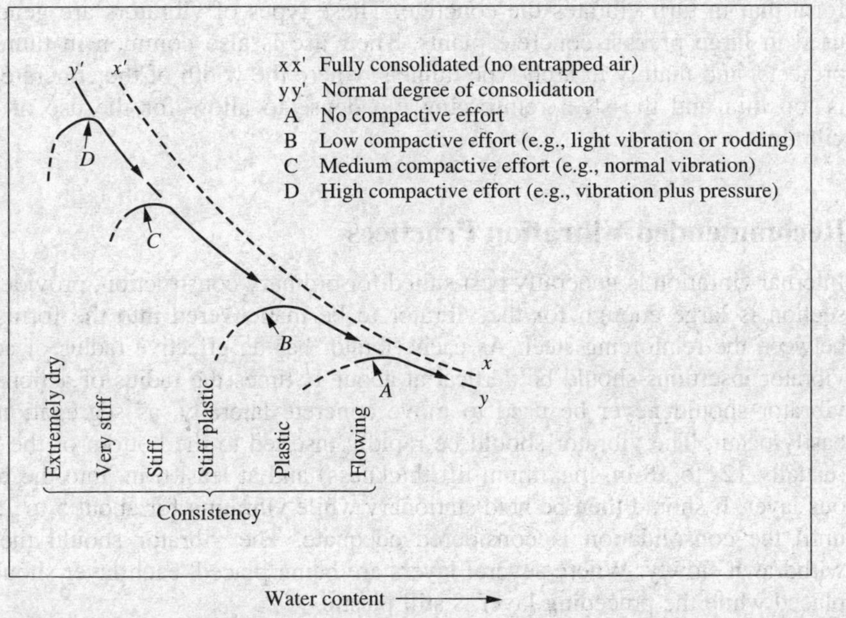

FIGURE 16.18 Effect of compaction effort on concrete quality (from ACI 309R) [8].

In terms of their mode of use, there are three general types of vibrators [8]: (1) internal, (2) surface, and (3) form vibrators. Internal, or spud, vibrators as they are often called, have a vibrating casing or head that is immersed into the concrete and vibrates at a high frequency (often as high as 10,000 to 15,000 vibrations/min) against the concrete. Currently these vibrators are the rotary type and come in sizes from $\frac{3}{4}$ in. to 7 in. (see Fig. 16.19), each with an effective radius of action [8]. They are powered by electric motors or compressed air, as well as by gasoline.

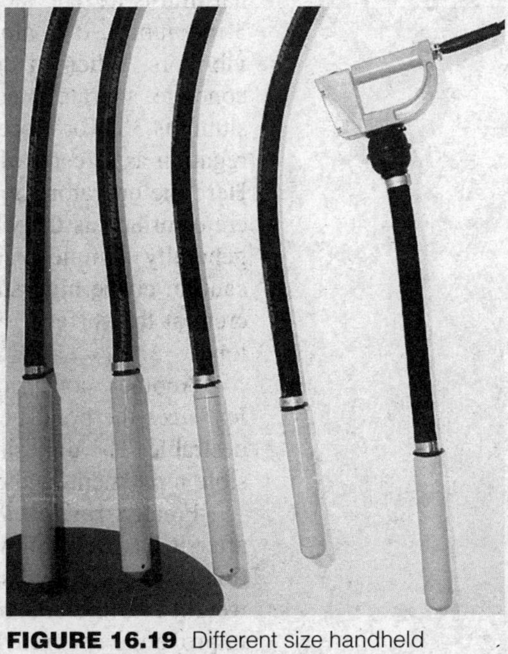

FIGURE 16.19 Different size handheld concrete vibrators.

Surface vibrators exert their effects at the top surface of the concrete and consolidate the concrete from the top down. They are used mainly in slab construction, and there are four general types: (1) the vibrating screed, (2) the pan-type vibrator, (3) the plate or grid vibratory tamper, and (4) the vibratory rolling screed. These surface vibrators operate in the range from 3,000 to 6,000 vibrations/min.

Form vibrators are external vibrators attached to the outside of the form or mold (usually done only when the forms are made of metal). They vibrate the form that in turn vibrates the concrete. These types of vibrators are generally used in large precast concrete plants. Their use is also common in tunneling projects, and mainly in arch-type tunnels, where the width of the concrete arch is too thin and the steel reinforcing too dense to allow for the use of spud vibrators.

Recommended Vibration Practices

Internal vibration is generally best suited for ordinary construction, provided the section is large enough for the vibrator to be maneuvered into the forms and between the reinforcing steel. As each vibrator has an effective radius of action, vibrator insertions should be vertical at about $1\frac{1}{2}$ times the radius of action. The vibrator should never be used to move concrete laterally, as segregation can easily occur. The vibrator should be rapidly inserted to the bottom of the layer (usually 12- to 18-in.-maximum-lift thickness) and at least 6 in. into the previous layer. It should then be held stationary while vibrating for about 5 to 15 sec until the consolidation is considered adequate. The vibrator should then be withdrawn slowly. Where several layers are being placed, each layer should be placed while the preceding layer is still plastic.

Vibration accomplishes two actions. First, it "slumps" the concrete, removing a large portion of air that is entrapped when the concrete is placed. Then continued vibration consolidates the concrete, removing most of the remaining entrapped air. Generally, it will not remove entrained air. The questions concerning overvibration often raised are: When does it occur and how harmful is it? The fact is that on low-slump concrete (concrete with less than a 3-in. slump), it is almost impossible to overvibrate the concrete with internal vibrators. When in doubt as to how much vibration to impart to low-slump concrete, vibrate it some more. The same cannot be said of concrete whose slump is 3 in. or more. This concrete can be overvibrated, which results in segregation as a result of coarse aggregate moving away from the vibrating head. Here the operator should note the presence of air bubbles escaping to the concrete surface as the vibrator is inserted. When these bubbles cease, vibration is generally complete and the vibrator should be withdrawn. Another point of caution concerning surface vibrators is that they too can overvibrate the concrete at the surface, significantly weakening it if they remain in one place too long.

Another concern is the vibration of reinforcing steel. Such vibration *improves* the bond between the reinforcing steel and the concrete, and thus is desirable. The undesirable side effects include damage to the vibrator and possible movement of the steel from its intended position.

Finally, *revibration* is the process whereby the concrete is vibrated again after it has been allowed to remain undisturbed for some time. Such revibration can be accomplished at any time, the running vibrator will sink of its own weight into the concrete and liquefy it momentarily [8]. Such revibration will improve the concrete through increased consolidation.

FINISHING AND CURING CONCRETE

The finishing process provides the desired final concrete surface. Unformed surfaces may require only **screeding** to proper contour and elevation, or a broomed, floated, or troweled finish may be specified.

screeding
The process of shaping the concrete surface to correct elevation by striking off the excess concrete.

It cannot be stated too strongly that *any* work done to a concrete surface after it has been consolidated will weaken the surface. All too often, concrete finishers overlook this fact and manipulate the surface of the concrete, sometimes even adding water, to produce a smooth, attractive surface. On walls and columns, an attractive surface may be desirable and the surface strength may not be too important, but on a floor slab, sidewalk, or pavement, the surface strength is very important. On the latter types of surfaces, only the absolute minimum finishing necessary to impart the desired texture should be permitted, and the use of "jitterbugging" (the forcing of coarse aggregate down into the concrete with a steel grate tool) should not be permitted, as the surface can be weakened significantly.

Furthermore, each step in the finishing operation, from first **floating** to the final floating or **troweling**, should be delayed as long as possible. This duration is limited by the necessity to finish the concrete to the desired grade and surface smoothness while it can still be worked (still in a plastic state).

floating
To create a smooth concrete surface the screeded concrete is worked with a wood or aluminum magnesium float.

troweling
Working the surface after floating with a steel trowel creates a smoother finish.

Floating has three purposes:

1. To embed aggregate particles just beneath the surface;
2. To remove slight imperfections, high spots, and low spots; and
3. To compact the concrete at the surface in preparation for other finishing operations.

In no case should finishing commence if any free bleed water has not been blotted up, nor should neat cement or mixtures of sand and cement be worked into such surfaces to dry them up.

Steel trowel

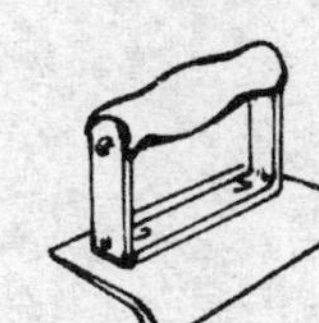

Edger

Troweling

Power trowels (Fig. 16.20) are used to smooth and finish large concrete surfaces such as floor slabs and the upper face of horizontally concreted precast elements. A gasoline engine, diesel engine, or electric motor activates rotating blades, varying in number, size, and rotation speed. Large-size blade systems have an advantage in large open areas due to greater productivity rates, while small-size blade systems are advantageous in tight spaces (e.g., they can finish floor slabs through doorways, against walls, or around obstructions).

Troweling improves the density of the upper concrete layer and seals plastic cracks that developed in the face of the concrete between the time of concrete placing and troweling. The results of concrete quality achieved by troweling depend not only on the equipment and the way it is used, but also on the quality of the concrete and concrete placement. A certain minimum time

(a) Walk-behind single-rotor trowel (b) Ride-on double-rotor trowel

FIGURE 16.20 Power equipment for finishing slabs.

must elapse between concrete placement and troweling, as the concrete must partially "dry up." The operation includes two main steps:

1. Floating is aimed at removing protruding aggregate grains, leveling off swells, filling cavities, and compacting the face of the concrete. At this stage the blades of the power trowel are kept flat.
2. Troweling is aimed at polishing, smoothing, and hardening the face of the concrete. At this stage, the blades of the power trowel are slightly angled.

The second step can be repeated several times, until the desired concrete face is achieved. With each successive operation, the angle of the blades is slightly increased, but the raised edge should not exceed 1 in. above the concrete surface, so as to avoid the formation of slits in the concrete.

Due to its round shape, the power trowel cannot be applied for room corners. Special small-radius disk trowels, vibratory trowels, or hand trowels should then be used.

Power trowels can have a single rotor or double rotors. The rotor is the blade system, often termed the *spider*. Single-rotor trowels are of the walk-behind type (see Fig. 16.20a), while double-rotor trowels are commonly of the ride-on type (see Fig. 16.20b). Both types are fitted with lifting provisions for movement around the job site: a lifting hook for the single-rotor trowel, a lifting system for the much heavier double-rotor trowel. Double-rotor riding trowels weigh much more than single-rotor walk-behind trowels, not only because of the double spiders, but also because of the riding engine and operator deck. They therefore often use special transport dollies for movement around the workplace and trailers for transportation between job sites. Double-rotor trowels may have either overlapping or nonoverlapping spiders.

Common dimensions and capacities for power trowels are

1. Number of blades (per spider): four to six
2. Rotor diameter: 2 to 5 ft

3. Maximum rotation speed: 90 to 180 rpm
4. Engine capacity: 2 to 13 hp single-rotor; 20 to 90 hp double-rotor
5. Height at lifting hook, single-rotor (height-adjustable handle commonly adds 3 to 5 in.): 2 to 3 ft; height, double-rotor: 4 to 5 ft
6. Total length, single rotor (with handle): 5 to 7 ft; length, double-rotor: 6 to 12 ft
7. Width, double-rotor: 3 to 6 ft
8. Weight: 100 to 300 lb single-rotor, 500 to 3,000 lb double-rotor

Curing

Along with placement and consolidation, proper curing of the concrete is extremely important. Curing encompasses all methods whereby the concrete is ensured adequate time, temperature, and supply of water for the cement to continue to hydrate. The minimum time required is normally 3 days, but curing for 28 days or longer is sometimes required. Optimum curing temperatures are between 40 and 80°F. As most concrete is batched with sufficient water for hydration, the only problem is to ensure that the concrete does not dry out. This can be accomplished by ponding with water (for slabs), covering with burlap or polyethylene sheets, or spraying with an approved curing compound.

Curing is one of the least costly operations in the production of quality concrete, and one that is all too frequently overlooked. Concrete, if allowed to dry out during the curing stage, will attempt to shrink. The developing bonds from the cementitious reaction will attempt to restrain the shrinkage from taking place. But the end result is *always* the same; the shrinkage wins out and a crack forms, as the shrinkage stresses are always higher than the tensile strength of the concrete. Proper curing reduces the detrimental effects of cracking and permits the development of the intended concrete strength.

CONCRETE PAVEMENTS

SLIPFORM PAVING

Slipform pavers (see Fig. 16.21) perform the functions of spreading, vibrating, striking off, consolidating, and finishing the concrete pavement to the prescribed cross section and profile with a minimum of handwork. The name "slipform" is derived from the fact that the side forms of the machine slide forward with the paver and leave the slab edges unsupported. Slipform paving is a speedy and economical method for producing smooth and durable concrete pavement.

The basic operating process consists of molding the plastic concrete to the desired cross section and profile under a single relatively large screed. This is accomplished with a full-width screed that is maintained at a predetermined elevation and cross slope with hydraulic jacks that are actuated through an

FIGURE 16.21 Slipform paver spreading, vibrating, striking off, consolidating, and finishing concrete pavement.

automatic control system. The control is referenced to offset grade lines pre-erected parallel to the planned profile for each pavement edge. Concrete is delivered into the front of the paver. It is then internally vibrated as it flows under the machine.

formed
Where the paving machine rides on steel forms that are set to the proper line and grade for the finished slab.

Some pavers operate with the concrete deposited directly on the base (see Fig. 16.22a) and spread it transversely with an auger before striking it off with a moldboard. A considerable amount of continuously reinforced concrete pavement is constructed throughout the country. The general practice, in both **formed** and slipformed paving of continuously reinforced pavements, is to

(a) Concrete dumped directly on the base ahead of the paver

(b) Side conveyor used to place the concrete on the base ahead of the paver

FIGURE 16.22 Placement of concrete ahead of the slipform paver.

place the steel on the base supported at the proper height ahead of the paving operation (see Fig. 16.23). In that case, slipform pavers use an attached belt conveyor (see Fig. 16.22b) to place the concrete in front of the paver as the trucks hauling the concrete cannot maneuver on the base. Most slipform pavers also have automatic dowel inserters that either push or vibrate the dowels into the fresh concrete in the exact position with minimum disturbance to the concrete. All pavers are equipped with a final float-finisher.

Pavement Joints

Typically joints are provided in the concrete pavement at intervals and having dimensions (depth and width) as specified in the project plans. **Contraction joints** or shrinkage joints in the transverse direction across the paving lanes are normally saw cut into the pavement. Manufacturers of saws provide (see Fig. 16.24) machines adaptable to any joint situation. Sawing of contraction joints is done as soon as it is possible to get on the new pavement without damaging the surface.

contraction joints *Joints for the purpose of regulating cracking from the unavoidable and unpredictable contraction of concrete.*

Longitudinal center joints are commonly formed by the use of a continuous polyethylene strip the paver disperses as it moves along. **Expansion joints**, when required, are generally installed by hand methods. Expansion joints are sealed with joint filler to prevent foreign material from entering the joint and rendering it ineffective for its intended purpose.

expansion joints *Separations between adjoining parts of a concrete structure.*

FIGURE 16.23 Reinforcing steel placed on the subbase at the proper height ahead of the paving operation.

FIGURE 16.24 Diamond-bladed saw for cutting pavement joints and freshly cut contraction joint. Note the cut is not made completely through the slab.

Curing Pavement Concrete

Curing is accomplished by spraying a waterproof membrane-curing compound on the pavement. Usually a separate machine follows the paver and applies the compound after all finishing is complete.

Smoothness

The ultimate inspectors for a paving project are the driving public. Drivers are acutely aware of pavement smoothness or roughness. Agencies that purchase paving projects specify roadway smoothness and measure performance with a "profileometer" (see Fig. 16.25). Constructing smooth pavements requires attention to several controlling factors.

FIGURE 16.25 A profileometer being used to measure pavement smoothness.

Concrete Many of the problems occurring in pavements result from the mix. The paver cannot cope with variations in slump and delivery of concrete. The concrete exerts forces on the paver as it spreads and forms the plastic mix; when those forces vary, smoothness is affected. A consistent head of material ahead of the paver is necessary for a quality project (a smooth pavement).

Workability Workability usually means slump, but other components affect workability. Entrained air, while adding to the durability of the concrete, also acts as a lubricating agent within the mix to reduce friction between particles, aiding workability.

Transport Vibration is a vital part of the paving process, but vibration is not meant to be a transport medium to move the plastic concrete. Concrete should be moved by auger, conveyor, or other mechanical means prior to vibration. The plastic mix should be placed uniformly in front of the paver.

Sensors The automatic control system and the stringline that provides intelligence for the operation of the controls cannot be neglected. The stringline must be set accurately, maintained, and checked constantly. Sensor response must be adjusted so that adjustments are not too quick or violent. A dampened response results in a smooth surface. It is recommended that dual-stringline operation of the paving process be used whenever possible. The smoothest pavement results are obtained by use of fixed elevation control on both sides of the paver—dual stringlines.

Attitude The machine attitude is the attack angle of the screed in relation to the concrete. If the angle of attack is forced to change due to "bulldozing" piles of concrete mix, improper sensor adjustment, or varying head of concrete, bumps will result. Such conditions cause uncontrolled lift and drop because of the hydraulic action of the plastic concrete.

Weight and Traction The principle that makes a slipform paver work is the consolidation of concrete in a confined space. The weight of the paver, therefore, is paramount to controlled consolidation of slipformed concrete. Paver weight needs to be uniformly distributed across the width of the concrete surface. Traction is related to weight and power. Paver power must be sufficient to move the loaded machine and provide the necessary energy to the working tools.

ADDITIONAL APPLICATIONS AND CONSIDERATIONS

ROLLER-COMPACTED CONCRETE

Roller-compacted concrete or RCC is a lean-mix concrete of zero-slump consistence. RCC is a design-flexible material that enables the use of local, usually less expensive, aggregates. The material was first used in the United States for dam construction, and then for paving large railroad, truck, and logging yards.

Typically, RCC is delivered to the site in dump trucks or scrapers, though belt conveyor systems have also been used [9]. It is spread with dozers or motor graders, and compacted with vibratory steel-drum rollers. Asphalt pavers have also been used on a few projects to place RCC.

SHOTCRETING

Shotcreting is mortar or concrete conveyed through a hose and pneumatically projected at high velocity onto a surface [10]. The force of the concrete impacting on the surface compacts the mixture. Usually, a relatively dry mixture is used, and thus it is able to support itself without sagging or sloughing, even for vertical and overhead applications. Shotcrete, or gunite, as it is more commonly known, is used most often in special applications involving repair work, thin layers, or fiber-reinforced concrete layers. Figure 16.26 illustrates the use of shotcrete to line a tunnel wall.

There are two methods of producing shotcrete: (1) the dry-mix process and (2) the wet-mix process. In the dry-mix process (see Fig. 16.27) the cement and damp sand are thoroughly mixed and carried to the nozzle, where water is introduced under pressure and completely mixed with the cement and sand before the mixture is jetted onto the surface. This method has been used successfully for more than 50 years.

The wet-mix process involves mixing all the ingredients including water before being delivered through a hose under pressure to the desired surface. Using the wet-mix process, aggregates up to $\frac{3}{4}$ in. in size have been shotcreted.

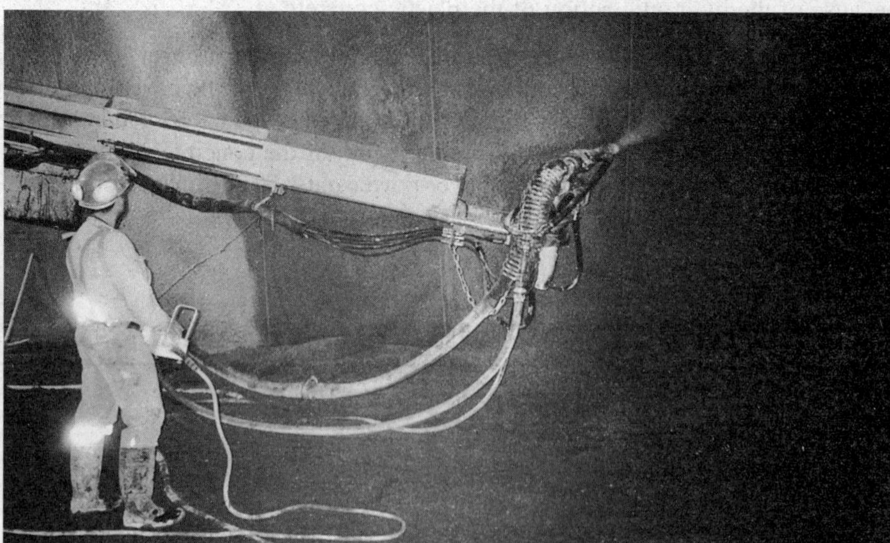

FIGURE 16.26 Using the shotcrete to line a tunnel wall.

FIGURE 16.27 Using a dry-mix process shotcrete mixer.

With either process there is some rebound of the mortar or concrete, resulting in a loss of usually 5 to 10%.

FLY ASH

Fly ash is produced as a by-product from the burning of coal. It is the fine residue that would "fly" out the stack if it were not captured by environmental control devices. Modern coal-burning plants pulverize the coal before burning it so that 100% will pass the No. 200 sieve. In the furnace, the coal is heated to in excess of 2,700°F, melting all the incombustibles. Because of the strong induced air currents in the furnace, the molten residue from the coal becomes spherical in shape and the majority is carried out at the top of the furnace. Captured in massive electrostatic precipitators or bag houses, the fly ash is extremely fine (often finer than Portland cement). Fly ash has been found to be an excellent mineral admixture in Portland cement concrete, improving almost all properties of the concrete. The Environmental Protection Agency has ruled that fly ash must be allowed in all concrete construction involving federal funds [11].

There are basically two types of fly ash: (1) class F from bituminous coal and (2) class C from sub-bituminous and lignitic coal. Their quality is governed by ASTM C618. Portland cement, when it combines with water, releases calcium hydroxide (the white streaks noticeable adjacent to exposed concrete around cracks). This calcium hydroxide does not contribute to strength or durability. By introducing fly ash, the calcium hydroxide and fly ash chemically combine in a process called pozzolanic action. It takes a relatively long time (compared with the cementitious action involved between cement and water), but the resulting concrete is stronger, less permeable, and more durable than before. If time is not a factor in the strength development of the concrete, fly ash may be used to replace part of the Portland cement, resulting in a lower-cost product.

In addition to strength and durability improvements, fly ash also improves the workability of concrete, primarily because of its spherical shape. It has been used to improve the pumpability of concrete mixes, and finishers report that fly ash concrete is easier to finish.

The previous discussion applies to both class F and class C fly ashes. However, *some* class C fly ashes possess significant cementitious properties themselves, probably because of their relatively high calcium contents. Thus, when used as an admixture, extremely high strengths can be obtained, as illustrated by the use of class C fly ash to achieve 7,500 psi concrete in the Texas Commerce Tower in Houston. These fly ashes can replace significant portions of Portland cement with no detrimental effects, as demonstrated with 25% replacement percentages on several concrete pavements in Texas.

Fly ash poses two problems to the concrete designer and constructor. One is that while it has been shown to be a valuable addition to concrete and can result in significant cost saving, it is a by-product from the burning of coal. This means that it can vary in its properties from day to day, and thus a good-quality management program is needed to ensure that only high-quality fly ash is used. Furthermore, because fly ash produced by different plants will vary in quality, it is mandatory that fly ash from each plant be tested prior to its use to ensure that it has the desired quality. Also, adding fly ash means that the concrete batch has five major ingredients rather than four, thus the chances for a mistake being made in the batching are increased. Hence it is desirable to increase the quality control efforts on concrete designs involving fly ash.

PLACING CONCRETE IN COLD WEATHER

When concrete is placed in cold weather, some provision must be made to keep the concrete above freezing during the first few days after it has been placed. Specifications generally require that the concrete be kept at not less than 70°F for 3 days or at not less than 50°F for a duration of 5 days after placement. ACI 306R contains guidelines for concreting in cold weather [12]. Preheating the water is generally the most effective method of providing the necessary temperature for placement.

It should also be kept in mind that since temperature is an important factor in the rate by which fresh concrete sets and hardens (see Chapter 22), concrete placing in cold weather means greater lateral pressures on the vertical forms of walls and columns. This means possibly longer service durations for the formwork until it may be stripped, due to the longer time it takes the concrete to gain its strength.

PLACING CONCRETE IN HOT WEATHER

When the temperature of fresh concrete exceeds 85 to 90°F, the resulting strength and durability of the concrete can be reduced. Therefore, most specifications require the concrete to be placed at a temperature less than 90°F. When concrete is placed in hot weather, the ingredients should be cooled before mixing. ACI 305R contains guidelines for mixing and placing concrete in hot weather [13]. Methods of cooling include using ice instead of water in the mix and cooling the aggregate with liquid nitrogen.

SAFETY

PUMPING CONCRETE

Of all operations involved in the placing of concrete, placing by pumping requires particular caution, given the high pressures inherent to the process. The American Concrete Pumping Association (ACPA) has set safety as its primary concern and the promotion of safe pumping as one of its major objectives. The ACPA has developed a comprehensive safety program and offers its safety materials (guides, manuals, videos, posters, CDs, and other publications) to all interested parties involved in concrete pumping. The following is a list of pumping-related hazards and safety rules:

1. It should be assumed that any part of the pipeline is under pressure until the pump operator has said that the pipeline is not under pressure. No pipe coupling should be released until the pressure has been released by reverse pumping.
2. Compressed-air cleaning of pumping lines is more hazardous than high-pressure water cleaning. It should only be carried out under the supervision of an experienced person, with all instructions and precautions followed exactly.
3. On restricted sites, boom pumps must be positioned such that the boom does not hit nearby structures and other obstructions when slewed, and particularly that people are unlikely to be trapped by the boom. Special care must be taken when operating near power lines (see Fig. 16.14).
4. Rubber end hoses contain highly energized concrete ready to burst out. Combined with its heavy weight and flexibility, control of the hose may be lost and it can whiplash; therefore rubber hose lines should be handled firmly and with care.
5. Outrigger-stabilized boom pumps should be positioned on firm ground and not too close to manholes and excavations. Inclined positioning should be avoided, particularly if the inclination is downward in the direction of the extended boom.

6. Booms should not be used for lifting items such as building materials or toolboxes, or for dragging hoses or pipelines along the ground. Booms should never be used to lift people.
7. Although hopper grilles (grates) are manufactured such that they are strong enough to withstand deflection under the weight of a person, one should never stand or walk on the grille.

SUMMARY

To successfully produce concrete, the mixture must be properly proportioned. The mix specifications will define specific requirements for the materials that constitute the desired concrete product. Ready-mixed concrete or truck-mixed concrete is proportioned at a central location and transported to the purchaser in a fresh state. It is completely mixed in the truck mixer. Central-mixed concrete is completely mixed in a stationary mixer and transported to the project in either a truck agitator, a truck mixer operating at agitating speed, or a nonagitating truck.

Slipform pavers perform the functions of spreading, vibrating, striking off, consolidating, and finishing the concrete pavement to the prescribed cross section and profile with a minimum of final handwork. The basic operating process consists of molding the plastic concrete to the desired cross section and profile under screed. Critical learning objectives include:

- An understanding of the fresh concrete properties that are important to the constructor
- An ability to calculate the quantities required to produce a concrete mix
- An understanding of the methods used to transport fresh concrete
- An understanding of the methods for pumping fresh concrete
- An understanding of the processes used to produce a concrete pavement

These objectives are the basis for the problems that follow.

PROBLEMS

16.1 A concrete batch calls for the following quantities per cubic yard of concrete, based on saturated surface-dry conditions of the aggregate. Determine the required weights of each solid ingredient and the number of gallons of water required for a 6.5-cy batch. Also determine the wet unit weight of the concrete in pounds per cubic foot (cement 3,666 lb; fine aggregate 9,230 lb; coarse aggregate 11,960 lb; water 1,841 lb; wet unit weight 152 pcf).

Cement	6.0 bags
Fine aggregate	1,420 lb
Coarse aggregate	1,840 lb
Water	34 gal

16.2 In Problem 16.1, assume that the fine aggregate contains 7% free moisture, by weight, and the coarse aggregate contains 3% free moisture by weight. Determine the required weights of cement, fine aggregate, coarse aggregate, and volume of added water for an 8-cy batch.

16.3 Determine the quantities of materials required per cubic yard to create a concrete mix. The specifications require a maximum size aggregate of 1 in., a minimum cement content of 5 sacks per cy, and a maximum water-cement ratio of 0.60. Assume 6% air voids (cement 470 lb per cy; water 4.52 cf/cy; fine aggregate 1,068 lb/cy; coarse aggregate 1,986 lb/cy).

16.4 Determine the quantities of materials required per cubic yard to create a concrete mix. The specifications require a maximum size aggregate of $1\frac{1}{2}$ in., a minimum cement content of $5\frac{1}{2}$ sacks per cy, and a maximum water-cement ratio of 0.63. Assume 5% air voids.

16.5 A concrete retaining wall whose total volume will be 735 cy is to be constructed by using job-mixed concrete containing the following quantities per cubic yard, based on surface-dry sand and gravel:

Cement	6 bags
Sand	1,340 lb
Gravel	1,864 lb
Water	33 gal

The sand and gravel will be purchased by the ton weight, including any moisture present at the time they are weighed. The gross weights, including the moisture present at the time of weighing, are

Item	Gross weight (lb/cy)	Percent moisture by gross weight
Sand	2,918	5
Gravel	2,968	3

It is estimated that 8% of the sand and 6% of the gravel will be lost or not recovered in the stockpile at the job. Determine the total number of tons each of sand and gravel required for the project.

16.6 The concrete floors of a 60- by 60-ft building are to be cast with the boom pump shown in Fig. 16.16. The first floor is 9 ft high above ground level, and each additional floor is 9 ft high. The truck can make the pumping only from one side of the building, but there is no restriction whatsoever on the truck's location alongside the building or its distance from it. Lengths of boom sections (taken from boom specifications) are as follows: first section, 26 ft 5 in.; second section, 23 ft 8 in.; third section, 23 ft 9 in.; fourth section, 24 ft 7 in. What is the highest floor level that can be covered fully by the boom?

16.7 A 100-cy/hr maximum output pump is used to place the concrete for a 160-cy slab. The concrete is supplied by 8-cy transit-mix trucks arriving to the site at 10-min intervals. The concrete batch plant is located 5 miles from the site. Average truck speed is 30 mph. Each transit-mix truck takes, on average, 8 min to discharge. The placement crew size is six workers. Average worker output is 4 cy of concrete per hour. (a) How long will the operation take? (b) How can the operation be reorganized more effectively?

REFERENCES

1. ACI Committee 304, "Guide for Measuring, Mixing, Transporting and Placing Concrete," ACI 304R-00, *ACI Manual of Concrete Practice,* Part 2, American Concrete Institute, ACI International, Farmington Hills, MI (published annually).
2. ASTM Committee C9, "Test for Slump of Portland Cement Concrete," *Annual Book of ASTM Standards,* Vol. 04.02 (published annually).
3. ACI Committee 211, "Standard Practice for Selecting Proportions for Normal, Heavyweight, and Mass Concrete," ACI 211.1-91 (reapproved 2002), *ACI Manual of Concrete Practice,* Part 1, American Concrete Institute, ACI International, Farmington Hills, MI (published annually).
4. ASTM Committee C94, "Standard Specification for Ready-Mixed Concrete," *Annual Book of ASTM Standards*, Vol. 04.02 (published annually).
5. *Concrete Plant Standards* (1996). CPMB 100-96, Concrete Plant Manufacturers Bureau, Silver Spring, MD.
6. Ledbetter, Bonnie S., W. B. Ledbetter, and Eugene H. Boeke (1980). "Mixing, Moving, and Mashing Concrete—75 Years of Progress," *Concrete International*, pp. 69–76; November.
7. ACI Committee 304, "Placing Concrete by Pumping Methods," ACI 304.2R-96, *ACI Manual of Concrete Practice,* Part 2, American Concrete Institute, ACI International, Farmington Hills, MI (published annually).
8. ACI Committee 309, "Guide for Consolidation of Concrete," ACI 309R-96, *ACI Manual of Concrete Practice,* Part 2, American Concrete Institute, ACI International, Farmington Hills, MI (published annually).
9. Stewart, Rita F., and Cliff J. Schexnayder (1986). "Construction Techniques for Roller-Compacted Concrete," Roller-Compacted Concrete Pavements and Concrete Construction, Transportation Research Record 1062, Transportation Research Board, National Research Council, Washington, DC, pp. 32–37.
10. ACI Committee 506, "Guide to Shotcreting," ACI 506R-90 (reapproved 1995), *ACI Manual of Concrete Practice,* Part 6, American Concrete Institute, ACI International, Farmington Hills, MI (published annually).
11. EPA (1983). "Guidelines for Federal Procurement of Cement and Concrete Containing Fly Ash," *Federal Register*, Vol. 48, No. 20, pp. 4230–4253, January 28.
12. ACI Committee 306, "Cold Weather Concreting," ACI 306R-88 (reapproved 2002), *ACI Manual of Concrete Practice*, Part 2, American Concrete Institute, ACI International, Farmington Hills, MI (published annually).

13. ACI Committee 305, "Hot Weather Concreting," ACI 305R-99, *ACI Manual of Concrete Practice,* Part 2, American Concrete Institute, ACI International, Farmington Hills, MI (published annually).

14. Crepas, R. A. (1999). *Pumping Concrete: Techniques and Applications*, 3rd ed., Crepas & Associates, Inc., Elmhurst, IL.

15. "World Record" (1994). *International Construction*, pp. 18, 21, September.

16. ACI Committee 347 (2003). "Guide to Formwork for Concrete," ACI 347R-03, American Concrete Institute, ACI International, Farmington Hills, MI.

17. Cooke, T. H. (1990). *Concrete Pumping and Spraying: A Practical Guide*, Thomas Telford, London.

WEBSITE RESOURCES

1. www.aci-int.org American Concrete Institute (ACI), Farmington Hills, MI. The ACI gathers, correlates, and disseminates information for the improvement of the design, construction, manufacture, use, and maintenance of concrete products and structures.
2. www.pavement.com American Concrete Pavement Association (ACPA), Skokie, IL. The ACPA is the national professional organization of the concrete paving industry.
3. www.concretepumpers.com American Concrete Pumping Association (ACPA), Lewis Center, OH. The ACPA promotes concrete pumping as the choice of method of placing concrete in a safe and expedient manner.
4. www.shotcrete.org The American Shotcrete Association (ASA), Farmington Hills, MI. The ASA promotes the training of those in the shotcrete industry in the proper methods, materials, and technique to obtain high-quality shotcrete.
5. www.ascconc.org American Society of Concrete Contractors (ASCC), St. Louis, MO. The ASCC is dedicated to improving concrete construction quality, productivity, and safety. Activities include an extensive *Safety Manual*, concrete and safety hotlines, safety videos, safety bulletins, troubleshooting newsletters, and the *Contractor's Guide to Quality Concrete Construction*.
6. www.crsi.org Concrete Reinforcing Steel Institute (CRSI), Schaumburg, IL. The CRSI is a national trade association representing producers and fabricators of reinforcing steel, epoxy coaters, bar support, and splice manufacturers.
7. www.csda.org Concrete Sawing and Drilling Association (CSDA), St. Petersburg, FL. The CSDA is a trade association of contractors that promotes professional sawing and drilling methods.
8. www.precast.org National Precast Concrete Association (NPCA), Indianapolis, IN. The NPCA pursues quality control programs through plant certification.
9. www.nrmca.org National Ready Mixed Concrete Association (NRMCA), Silver Spring, MD. The NRMCA is an international trade association for producers of ready-mixed concrete, which provides publications on such topics as safety, financial management, maintenance, and driver education.
10. www.portcement.org Portland Cement Association (PCA), Skokie, IL. The PCA represents cement companies in Canada and the United States and serves as the nucleus of the cement industry's work in research, promotion, education, and

public affairs. The PCA's wholly owned subsidiary, Construction Technology Laboratories, Inc., provides a wide range of research, testing, and consulting services.

11. www.pci.org Precast/Prestressed Concrete Institute (PCI), Chicago, IL. The PCI is dedicated to fostering greater understanding of precast/prestressed concrete and to serve as the focal point for the advancement of that industry through technical research and marketing support.
12. www.putzmeister.com Putzmeister America, Sturtevant, WI, is the North American subsidiary of heavy equipment manufacturer, Putzmeister-AG of Germany. The company specializes in concrete and material placing equipment.
13. www.schwing.com Schwing America, St. Paul, MN, manufactures a complete range of truck- and trailer-mounted concrete pumps.

CHAPTER 17

Cranes

Construction cranes are generally classified into two major families: (1) mobile cranes and (2) tower cranes. Because cranes are used to hoist and move loads from one location to another, it is necessary to know the lifting capacity and working range of a crane selected to perform a given service. The rated load for a crane, as published by the manufacturer, is based on ideal conditions. Load charts can be complex documents listing numerous booms, jibs, and other components that can be employed to configure the crane for various tasks. It is critical that the chart being consulted be for the actual crane configuration that will be used.

MAJOR CRANE TYPES

Cranes are a broad class of construction equipment used to hoist and place loads. They are the dominant piece of equipment the world over for construction of both building projects and heavy civil projects. Cranes are the epitome of the growing construction industrialization witnessed in the last decades. Each type of crane is designed and manufactured to work economically in specific site situations; modern-day sites often employ more than one type of crane and more than one crane of the same type.

Construction cranes are generally classified into two major families: (1) mobile cranes and (2) tower cranes. Mobile cranes are the machines of choice in North America, as contractors have traditionally favored them over tower cranes. Tower cranes are usually used in North America only when job-site conditions make mobile crane movement impossible, or for high-rise construction. They are, however, the machines that dominate the construction scene in Europe, whether in the big cities or in rural areas.

The most common mobile crane types are

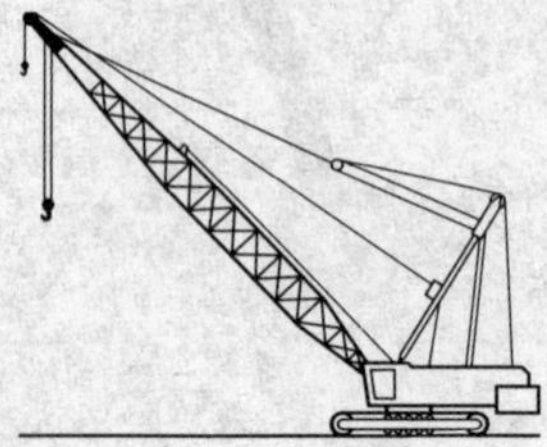

1. Crawler

2. Telescoping-boom truck-mounted

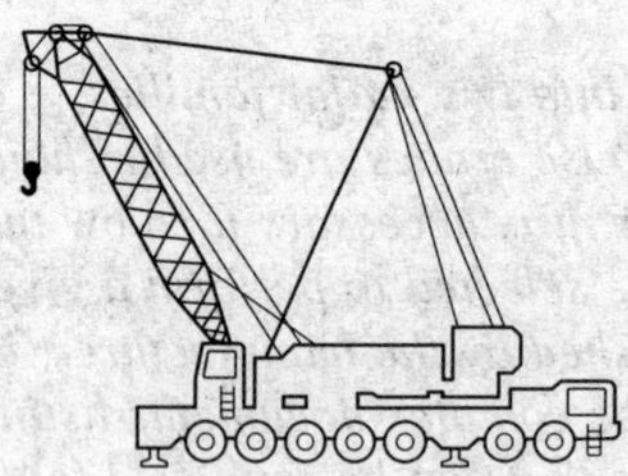

3. Lattice-boom truck-mounted

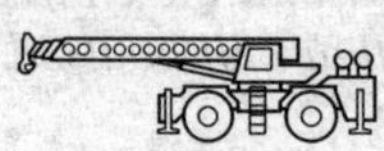

4. Rough-terrain

5. All-terrain

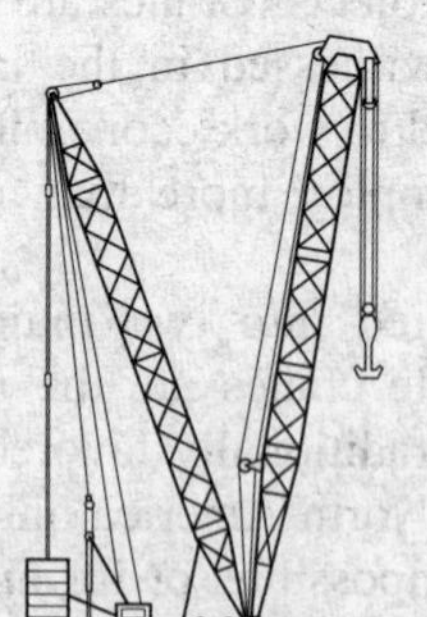

6. Modified cranes for heavy lifting

Some of the mobile-type machines in their basic configuration can have different front-end operating attachments that enable the unit to be used as an excavator or a pile driver, or in other specialized tasks. Such diverse usages are discussed in Chapters 4, 8, 12, 18, and 19.

The most common tower crane types are

1. Top **slewing**
2. Bottom slewing

slewing
Turning or rotating on an axis.

"Hybrid" machines that endeavor to combine features of mobile and tower cranes exist as well, both older types and emerging new models. These, however, constitute merely a small fraction of the huge population of cranes manufactured and used the world over.

The Power Crane and Shovel Association (PCSA) has conducted and supervised both studies and tests that provide considerable information related to the performance, operating conditions, production rates, economic life, and cost of owning and operating cranes. The association has participated in establishing and adopting standards applicable to this equipment. This information has been published in technical bulletins and booklets.

Some of the information from the PCSA is reproduced in this book, with permission of the association. Items of particular interest are cited in the references at the end of the chapter.

MOBILE CRANES

CRAWLER CRANES

The *full revolving superstructure* of this type of unit is mounted on a pair of continuous, parallel crawler tracks. Many manufacturers have different option packages available that enable configuration of the crane to a particular application, standard lift, tower unit, or duty cycle. Units in the low to middle range of lift capacity have good lifting characteristics and are capable of **duty-cycle work** such as handling a concrete bucket. Machines of 100-short-ton* capacity and above are built for lift capability and do not have the heavier components required for duty-cycle work. The **universal machines** incorporate heavier frames, have heavy-duty or multiple clutches and brakes, and have more powerful swing systems. These designs allow for quick changing of drum laggings that vary the torque/speed ratio of cables to the application. Figure 17.1 illustrates a 100-ton crawler crane on a bridge project; Fig. 17.2 shows a large crawler crane (note the inclined lattice mast, not mounted on the smaller crane in Fig. 17.1, which helps decrease compressive forces in the lattice boom).

duty-cycle work
A repetitive lifting assignment of relatively short cycle time.

universal machines
The base machine can be used as a crane or dragline and for pile driving or other such applications.

The crawlers provide the crane with good travel capability around the job site. The crawler tracks provide such a large ground contact area that soil failure under these machines is only a problem when operating on soils having a low bearing capacity. Before hoisting a load, the machine must be leveled and

* A short ton (or U.S. ton) equals 2,000 lb as opposed to a metric or long ton (termed "ton" in the text), which equals 2,240 lb.

FIGURE 17.1 Lattice-boom crawler crane (100 ton) on a bridge project.

ground settlement considered. If soil failure or ground settlement is possible, the machine can be positioned and leveled on mats. When there are good supporting ground conditions, a crawler crane can move with the hoisted load. This ability to carry a hoisted load along with the crane's capacity to travel and work even in poor underfoot conditions is the main advantage a crawler crane has over a wheeled (truck) crane. The distance between crawler tracks affects stability and lift capacity. Some machines have the feature whereby the crawlers can be extended. For many machines, this extension of the crawlers can be accomplished without external assistance.

Relocating a crawler crane between projects requires that it be transported by truck, rail, or barge. As the size of the crane increases, the time and cost to dismantle and load the crane, investigate haul routes, and reassemble the crane will also increase. The durations and costs can become significant for large machines. Relocating the largest machines can require 15 or more truck trailer units. These machines usually have lower initial cost per rated lift capability, compared with other mobile crane types, but movement between jobs is more expensive. Therefore, crawler-type machines should be considered for projects requiring long duration usage at a single site.

Many new models use modular components to make dismantling, transporting, and assembling easier. Quick-disconnect locking devices and pin connectors have replaced multiple-bolt connections.

FIGURE 17.2 Large crawler crane with a rear mast.

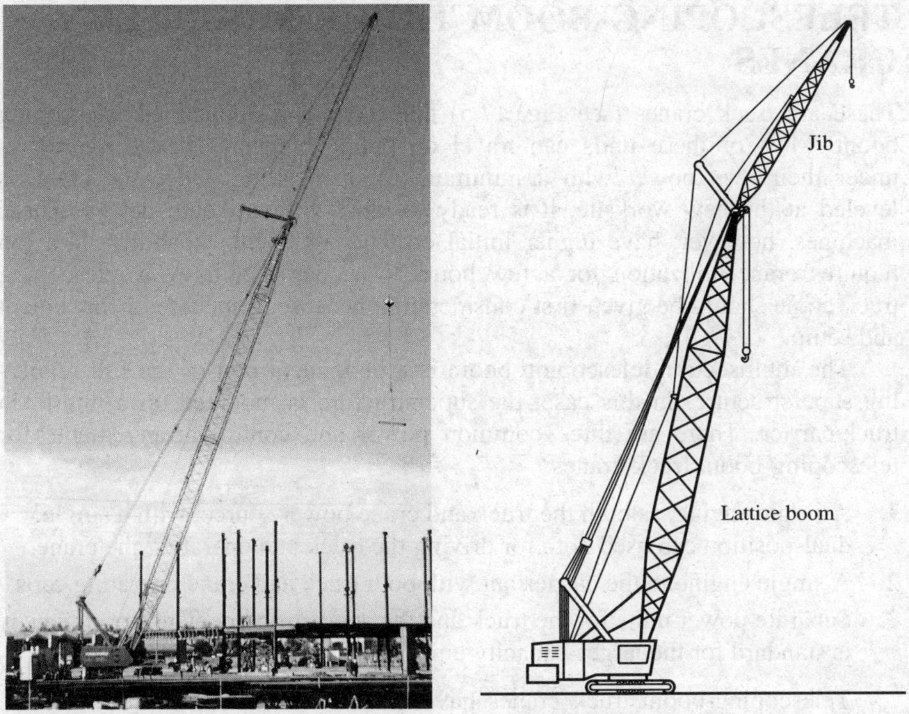

FIGURE 17.3 Lattice-boom crawler crane rigged with a jib extension.

FIGURE 17.4 Rubber-track telescoping-boom crawler crane on an urban project.

Most crawler crane models have a fixed-length lattice boom (Fig. 17.1), which is also the crane type discussed in this section. Many models are optionally rigged with a boom extension known as a "fly jib" or "fixed jib" (Fig. 17.3). A lattice boom is cable-suspended and therefore acts as a compression member, *not* a bending member like a telescoping hydraulic boom. However, new small-size crawler models exist that are equipped with a telescoping boom. Some of these models are equipped with rubber tracks to make them suitable for urban works and movement on pavements (Fig. 17.4).

Common dimensions and capacities for crawler cranes are

1. Maximum boom length: 100 to 400 ft
2. Maximum fly-jib length: 30 to 120 ft
3. Maximum radius (boom only): 80 to 300 ft
4. Minimum radius: 10 to 15 ft
5. Maximum lifting capacity (at minimum radius): 30 to 600 tons
6. Maximum travel speed: 50 to 100 ft/min (0.6 to 1.2 mph)
7. Ground bearing pressure: 7 to 20 psi

TELESCOPING-BOOM TRUCK-MOUNTED CRANES

These are truck cranes (see Fig. 17.5) that have a self-contained telescoping boom. Most of these units can travel on public highways between projects under their own power with a minimum of dismantling. Once the crane is leveled at the new worksite, it is ready to work without setup delays. These machines, however, have higher initial cost per rated lift capability. If a job requires crane utilization for a few hours to a couple of days, a telescoping truck crane should be given first consideration because of its ease of movement and setup.

The multisection telescoping boom is a permanent part of the full revolving superstructure. In this case, the superstructure is mounted on a multiaxle truck/carrier. There are three common power and control arrangements for telescoping-boom truck cranes:

1. A single engine, as both the truck and crane power source, with a single, dual-position cab used both for driving the truck and operating the crane.
2. A single engine in the carrier but with both truck and crane operating cabs.
3. Separate power units for the truck and the superstructure. This arrangement is standard for the larger-capacity units.

outriggers *Movable beams that can be extended laterally from a mobile crane (or from the undercarriage of a bottom-slewing tower crane) to stabilize and help support the unit.*

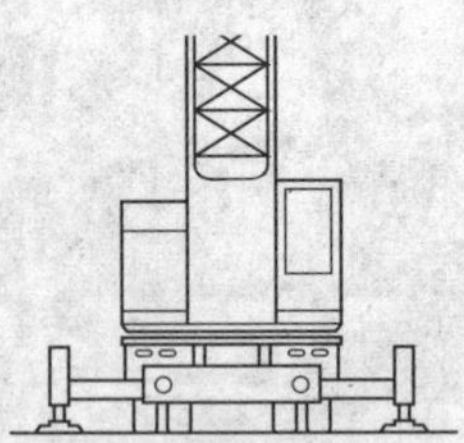

Telescoping-boom truck cranes have extendable **outriggers** for stability. In fact, many units cannot be operated safely with a full reach of boom unless the outriggers are fully extended and the machine is raised so that the tires are clear of the ground. Some models can operate on their tires when there is firm, leveled ground, but their lifting capacity is markedly reduced (by 50% and more, for 3° off level, depending on working radius) compared to working on

FIGURE 17.5 Telescoping truck crane on a building site.

outriggers. In the case of the larger machines, the width of the outriggered vehicle may reach 40 ft, which necessitates careful planning of the operation area. Additionally, these heavy machines transfer, through the outriggers, extremely high loads to the ground. This high ground loading must be considered vis-à-vis the soil-bearing capacity. Large-size timber or steel mats that are used to spread the load over a larger ground area further increase the overall vehicle width. These outrigger space considerations are also a concern when using large lattice-boom truck cranes.

Common dimensions and capacities for telescoping-boom truck cranes are

1. Maximum boom length: 70 to 170 ft
2. Maximum fly-jib length: 30 to 100 ft
3. Maximum radius (boom only): 60 to 140 ft
4. Minimum radius: 10 ft for most models
5. Maximum lifting capacity (at minimum radius): 20 to 120 tons
6. Maximum travel speed: 40 to 70 mph
7. Number of axles: 3 to 4

LATTICE-BOOM TRUCK-MOUNTED CRANES

As with the telescoping-boom truck crane, the lattice-boom truck crane has a fully revolving superstructure mounted on a multiaxle truck/carrier. The advantage of this machine is the *lattice boom*. The lattice-boom structure is lightweight. This reduction in boom weight means additional lift capacity, as the machine predominately handles hoist load and less weight of boom. The lattice boom does take longer to assemble. The lightweight boom will give a less expensive lattice-boom machine the same hoisting capacity as a larger telescoping-boom unit. Figure 17.6 shows a lattice-boom truck crane handling a precast panel on a building project.

The disadvantage of these units is the time and effort required to disassemble them for transport. In the case of the larger units, it may be necessary to remove the entire superstructure. Additionally, a second crane is often required for this task. Some newer models are designed so that the machine can separate itself without the aid of another crane.

Common dimensions and capacities for lattice-boom truck cranes are

1. Maximum boom length: 170 to 470 ft
2. Maximum fly-jib length: 40 to 300 ft
3. Maximum radius (boom only): 130 to 380 ft
4. Minimum radius: 10 to 25 ft

FIGURE 17.6 A large lattice-boom truck crane.

5. Maximum lifting capacity (at minimum radius): 50 to 600 tons
6. Maximum travel speed: 40 to 60 mph
7. Number of axles: 4 to 8

ROUGH-TERRAIN CRANES

These cranes are mounted on two-axle carriers (Fig. 17.7). The operator's cab may be mounted in the upper works, enabling the operator to swing with the load. However, on many models, the cab is located on the carrier. This is a simpler design because controls do not have to be routed across the turntable. In turn, these units have a lower cost. Rough-terrain-type cranes are sometimes referred to as "cherry pickers." This comes from their use during World War II in handling bombs, as the slang name for a bomb was a cherry.

These units are equipped with unusually large wheels and closely spaced axles to improve maneuverability at the job site. They further earn the right to their name by their high ground clearance, as well as the ability of some models to move on slopes of up to 70%. Most units can travel on the highway but have maximum speeds of only about 30 mph. In the case of long moves between projects, they should be transported on low-bed trailers.

Many units now have joystick controls. A *joystick* enables the operator to manipulate four functions simultaneously. The most common models are in the 20- to 50-ton capacity range and typically they are employed as utility cranes. They are primarily lift machines but are capable of light, intermittent duty-cycle work. To utilize their maximum lifting capacity the cranes must be stabilized on outriggers while in operation (as in Fig. 17.7). Since they are commonly used for works requiring frequent on-site relocation, the small to midsize models can also be operated on their wheels, albeit with reduced loading. They can even move slowly (up to 3 mph) with loads, on firm, leveled ground, but only without a fly jib and with the boom in a straight-ahead (i.e., not revolved) position.

FIGURE 17.7 Rough-terrain crane.

Common dimensions and capacities for rough-terrain cranes are

1. Maximum boom length: 80 to 140 ft
2. Maximum fly-jib length: 20 to 90 ft
3. Maximum radius (boom only): 70 to 120 ft
4. Minimum radius: 10 ft for most models
5. Maximum lifting capacity (at minimum radius): 20 to 90 tons
6. Maximum travel speed: 15 to 35 mph
7. Number of axles: 2 for all models

ALL-TERRAIN CRANES

The *all-terrain crane* (Fig. 17.8) is designed with an undercarriage capable of long-distance highway travel. Yet the carrier has all-axle drive and all-wheel steering, crab steering, large tires, and high ground clearance. All-terrain cranes have dual cabs, a lower cab for fast highway travel, and a superstructure cab that has both drive and crane controls. The machine can, therefore, be used for limited pick-and-carry work. Because this crane has both job-site mobility and transit capability, it is an appropriate machine when multiple lifts are required at scattered project sites or at multiple work locations on a single project. Because this machine is a combination of two features, it has a higher cost than an equivalent capacity telescoping truck crane or rough-terrain crane. But an all-terrain machine can be positioned on the project without the necessity of having other construction equipment prepare a smooth travel way as truck cranes would require. Additionally, the all-terrain crane does not need a lowboy to haul it between distant project sites, as would a rough-terrain crane. The concept of the all-terrain crane—relatively new (since the 1980s)—is rapidly gaining popularity both in Europe and North America; consequently manufacturers have gradually been abandoning the manufacture of telescoping-boom truck cranes in favor of all-terrain cranes, as can also be witnessed by the diminishing number of truck crane models available on the market.

Common dimensions and capacities for all-terrain cranes are

1. Maximum boom length: 100 to 270 ft
2. Maximum fly-jib length: 30 to 240 ft
3. Maximum radius (boom only): 70 to 250 ft
4. Maximum radius (with fly jib): 100 to 300 ft (and up to 400 ft for the largest machines)

FIGURE 17.8 All-terrain crane with clam bucket.

5. Minimum radius: 8 to 10 ft
6. Maximum lifting capacity (at minimum radius): 30 to 300 tons (and up to 800 tons for the largest machines)
7. Maximum travel speed: 40 to 55 mph
8. Number of axles: 2 to 6 (and up to 8 or 9 for the largest machines)

MODIFIED CRANES FOR HEAVY LIFTING

These are basically systems that significantly increase the lift capacity of a crawler crane. A crane's capacity is limited by one of two factors: (1) structural strength or (2) tipping moment. If a counterweight is added to prevent tipping when hoisting a heavy load, there is a point when the machine is so overbalanced that without a load it will tip backward. At some point, even with sufficient counterweight, the boom is put into such high compression that it will give way at the butt. Manufacturers, understanding both the need of users to make occasional heavy lifts and the users' reluctance to buy a larger machine for a one-time use, have developed systems that provide the capability while maintaining machine integrity. The three principal systems available are

1. Trailing counterweight
2. Extendable counterweight
3. Ring system

Trailing Counterweight

The base crane does not carry the trailing counterweight. Instead, the additional counterweight is mounted on a wheeled platform behind the crane ("counterweight carrier"), with the platform pin connected to the crane (see Fig. 17.9). The system utilizes a mast positioned behind the *boom*, with the boom suspension lines mounted at the top of the mast. This increases the angle between the boom and the suspension lines, thereby decreasing the compressive forces on the boom. One modern system of this type features a massive hydraulic ram for extension and retraction of the counterweight carrier. The carrier is supported on four articulated self-powered wheel assemblies. Settings and operational limits for capacity and for various load and movement parameters are computer controlled.

Extendable Counterweight

One manufacturer offers a machine with a counterweight system that can be extended away from the rear of the machine to match the leverage with the requirements of the lift.

Ring System

With the ring system, a large circular turntable ring is created outside the base machine (see Fig. 17.10). The heavy counterweight system is supported on this ring. There are auxiliary pin-connected frames at the front and rear of the base

(a) Trailing counterweight (b) Trailing-counterweight-rigged crawling crane

FIGURE 17.9 Crawler crane with trailing counterweight modification for heavy lifting.

(a) Ring system (b) Operating

FIGURE 17.10 Crawler crane ring-system modification for heavy lifting.

machine. These enable both the boom/mast foot and the counterweight to be moved away from the machine. Using rollers or wheels, the auxiliary frames ride the ring. The base crane is really only a power and control source.

New hydraulic self-erecting systems enable a crawler crane to be modified to a ring configuration in only 3 days. This ability makes the system very competitive for jobs lasting more than just a few days.

Because of their size and weight, ring systems essentially render cranes mounted on them to no longer be "mobile." To partly overcome this limitation, one manufacturer offers a feature by which the crane can lift its ring, crawl to a new location, and lower the ring to the ground again.

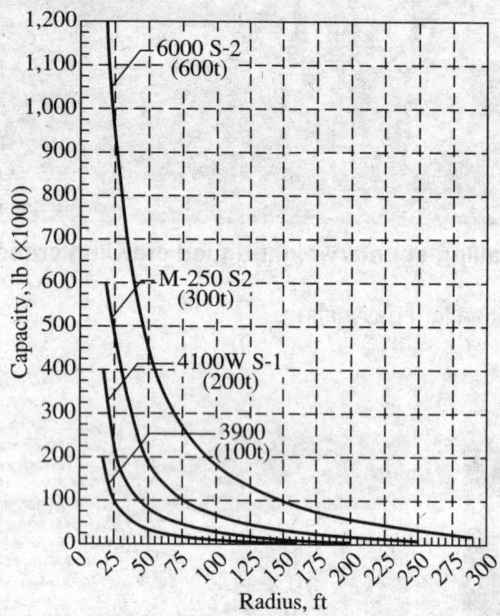

FIGURE 17.11 Safe lifting capacities for four crawler cranes.
Source: Manitowoc Engineering Co.

CRANE BOOMS

Most cranes come equipped with standardized booms designed to optimize their performance over a range of applications. However, cranes may also use optional boom configurations that enable them to adapt to specific lifting conditions. Optional booms and differing boom tops enable a machine to accommodate load clearance, longer reach, or increased lift capacity requirements.

LIFTING CAPACITIES OF CRANES

Because cranes are used to hoist and move loads from one location to another, it is necessary to know the lifting capacity and working range of a crane selected to perform a given service. Figure 17.11 shows typical crane-lifting capacities for four specific crawler cranes of varying sizes. The lifting capacities of units made by different manufacturers will vary from the information in the figure. Individual manufacturers and suppliers will furnish machine-specific information in literature describing their machines.

sheave
Grooved pulley-wheel for changing the direction of a wire rope's pull.

When a crane lifts a load attached to the hoist line that passes over a **sheave** located at the boom point of the machine, there is a tendency to tip the machine over. This introduces what is defined as the *tipping condition*. With the crane on a firm, level supporting surface in calm air, it is considered to be at the point of tipping when a balance is reached between the overturning moment of the load and the stabilizing moment of the machine.

During tests to determine the tipping load for wheel-mounted cranes, the outriggers should be lowered to relieve the wheels of all weight on the supporting surface or ground. The radius of the load is the horizontal distance from the axis of rotation of the crane to the center of the vertical hoist line or tackle with the load applied. The *tipping load* is the load that produces a tipping condition at a specified radius. The load includes the weight of the item being lifted plus the weights of the hooks, hook blocks, slings, and any other items used in hoisting the load, including the weight of the hoist rope located between the boom-point sheave and the item being lifted.

RATED LOADS FOR LATTICE- AND TELESCOPIC-BOOM CRANES

The rated load for a crane, as published by the manufacturer, is based on ideal conditions. Load charts can be complex documents listing numerous booms, jibs, and other components that may be employed to configure the crane for various tasks. It is critical that the chart being consulted be for the actual crane configuration that will be used.

> Interpolation between the published values is *not* permitted; use the next lower value. Rated loads are based on ideal conditions, a level machine, calm air (no wind), and no dynamic effects.

A partial safety factor with respect to tipping is introduced by the PCSA rating standards that state that the rated load of a lifting crane shall not exceed the following percentages of tipping loads at specified radii:

1. Crawler-mounted machines: 75%
2. Rubber-tire-mounted machines: 85%
3. Machines on outriggers: 85%

It should be noted that there are other groups that recommend rating criteria. The Construction Safety Association of Ontario recommends that for rubber-tire-mounted machines a factor of 75% be used.

One manufacturer is producing rubber-tire-mounted cranes having intermediate outrigger positions. For intermediate positions greater than one-half the fully extended length, the manufacturer is using a rating based on 80% of the tipping load. For intermediate positions less than one-half the fully extended length, a rating based on 75% is used. At this time, there is no standard for this type machine.

Load capacity will vary depending on the quadrant position of the boom with respect to the machine's undercarriage. In the case of crawler cranes, the three quadrants that should be considered are

1. Over the side
2. Over the drive end of the tracks
3. Over the idler end of the tracks

Crawler crane quadrants are sometimes defined by the longitudinal centerline of the machine's crawlers instead of by center rotation. The area between the centerlines of the two crawlers is considered over the end, and the area outside the centerlines of the crawlers is considered over the side.

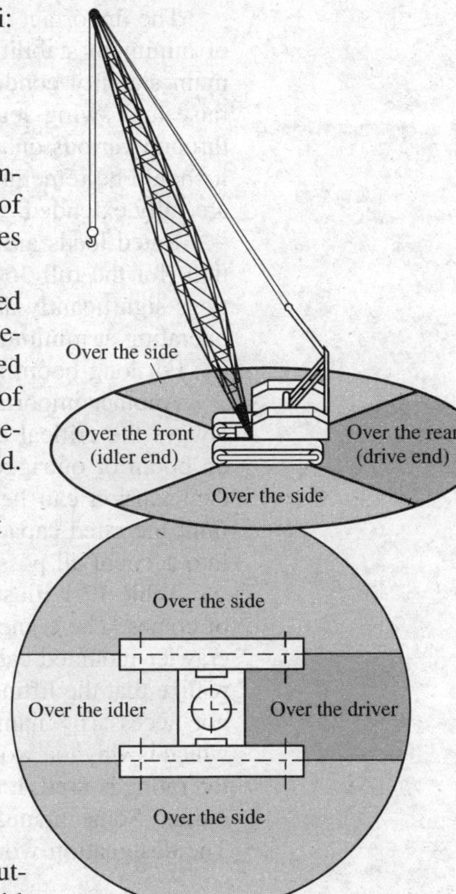

In the case of wheel-mounted cranes, the quadrants of consideration will vary with the configuration of the outrigger locations. If a machine has only four outriggers, two on each side, one located forward and one to the rear, the quadrants are usually defined by imaginary lines running from the superstructure center of rotation through the position of the outrigger support. In such a case, the three quadrants to consider are

1. Over the side
2. Over the rear of the carrier
3. Over the front of the carrier

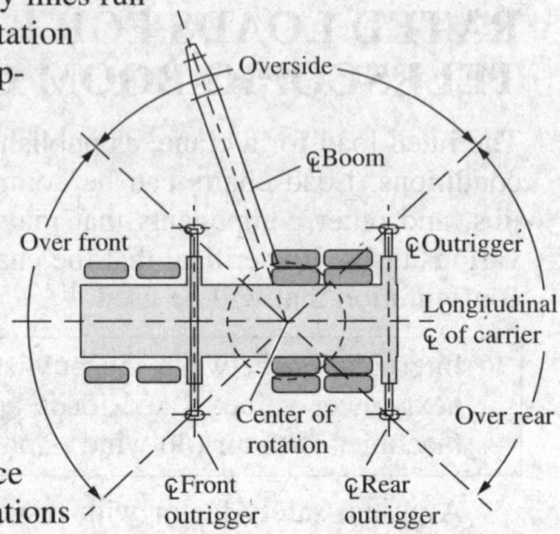

Some wheel-mounted cranes have an outrigger directly in the front, or there can be other machine-specific outrigger configurations. Therefore, the best practice is to consult the manufacturer's specifications concerning how quadrants are defined.

The important point is that the rated load should be based on the direction of minimum stability for the mounting, unless otherwise specified. The minimum stability condition restricts the rated load because the crane must both raise and swing loads. The swinging motion will cause the boom to move through various quadrants, changing the load's effect on the machine. Further, it should be remembered that the rating is based on the fact that the outriggers are fully extended.

Rated loads are based on the assumption that the crane is in a level position (for the full 360° of swing). When a crane is not level, even small variations significantly affect lifting capacity. In the case of a short-boom machine operating at minimum radius, 3° out of level can result in a 30% loss in capacity. For long-boom machines, the loss in capacity can be as great as 50% [9].

Another important consideration with modern cranes is that tipping is not always the critical capacity factor. At short radii, capacity may be dependent on boom or outrigger strength and structural capacity, and at long radii, pendant tension can be the controlling element. Manufacturers' load charts will limit the rated capacity to values below the minimum critical condition taking into account all possible factors.

Table 17.1 illustrates the kind of information issued by the manufacturers of cranes. The crane in this example is described as a 200-ton, nominal rating, crawler-mounted cable-controlled crane with 180 ft of boom. It is important to realize that the lifting capacity by which mobile cranes are identified (although not necessarily named) does not represent a standard classification method, which is why the wording *nominal* is added to the description. In most cases, if a ton rating is used, it refers to capacity with a basic boom and lifting at minimum radius. Some manufacturers use a load moment rating classification system. The designation would be "tm." As an example, a Demag AC 650 (reflecting

TABLE 17.1 Lifting capacities in pounds for a 200-ton, nominal rating, crawler crane with 180 ft of boom.*

Radius (ft)	Capacity (lb)	Radius (ft)	Capacity (lb)	Radius (ft)	Capacity (lb)
32	146,300	80	39,200	130	17,900
36	122,900	85	35,800	135	16,700
40	105,500	90	32,800	140	15,500
45	89,200	95	30,200	145	14,500
50	76,900	100	27,900	150	13,600
55	67,200	105	25,800	155	12,700
60	59,400	110	23,900	160	11,800
65	53,000	115	22,200	165	11,100
70	47,600	120	20,600	170	10,300
75	43,100	125	19,200	175	9,600

*Specified capacities based on 75% of tipping loads.

Source: Manitowoc Engineering Co.

650-ton lifting capacity, in this case referring to a metric ton) is named by a big crane rental company AC 2000 (reflecting 2000 tm max. load moment).

The capacities located in the upper portion of a load chart (Table 17.2), usually defined by either a bold line or by shading, represent structural failure

TABLE 17.2 Lifting capacities in pounds for a 25-ton truck-mounted hydraulic crane.*

Load radius (ft)	Lifting capacity (lb)† Boom length (ft) 31.5	40	48	56	64	72	80
12	50,000	45,000	38,700				
15	41,500	39,000	34,400	30,000			
20	29,500	29,500	27,000	24,800	22,700	21,100	
25	19,600	19,900	20,100	20,100	19,100	17,700	17,100
30		14,500	14,700	14,700	14,800	14,800	14,200
35			11,200	11,300	11,400	11,400	11,400
40			8,800	8,900	9,000	9,000	9,000
45				7,200	7,300	7,300	7,300
50				5,800	5,900	6,000	6,000
55					4,800	4,900	4,900
60					4,000	4,000	4,000
65						3,100	3,300
70							2,700
75							2,200

*Specified crane capacities based on 85% of tipping loads.

†The loads appearing below the solid line are limited by machine stability. The values appearing above the solid line are limited by factors other than machine stability.

conditions. Operators can feel the loss of stability prior to a tipping condition. But in the case of a structural failure, there is no sense of feel to warn the operator; therefore, load charts must be understood and all lifts must be in strict conformance with the ratings.

While the manufacturer will consider crane structural factors when developing a capacity chart for a particular machine, it must always be remembered that operational factors, over which the manufacturer has no control, affect absolute capacity in the field. The manufacturer's ratings can be thought of as valid for a *static* set of conditions. A crane on a project operates in a *dynamic* environment, lifting, swinging, and being subjected to air currents, moisture, and temperature variations. The load chart provided by the manufacturer does not take into account these dynamic conditions. Factors that will greatly affect actual crane capacity on the job are

1. Wind forces on the boom or load
2. Swinging the load
3. Hoisting speed
4. Stopping the hoist

These dynamic factors should be carefully considered when planning a lift.

Rated Loads for Hydraulically Operated Cranes

The rated tipping loads for hydraulic cranes are determined and designated in the same manner as those for cable-controlled cranes. However, in the case of hydraulic cranes, the critical load rating is sometimes dictated by hydraulic pressure limits instead of tipping. Therefore, load charts for hydraulic cranes represent the controlling-condition lifting capacity of the machine, and the governing factor may not necessarily be tipping. The importance of this is that the operator cannot use his physical sense of balance (machine feel) as a gauge for safe lifting capability.

WORKING RANGES OF CRANES

Figure 17.12 shows graphically the height of the boom point above the surface supporting the crane and the load line's distance from the center of rotation based on various boom angles. The figure is for the crane whose lifting capacities are given in Table 17.1.

The maximum length of the boom may be increased to 180 ft. The length of the boom is increased by adding sections at or near the midlength of the boom, usually in 10-ft, 20-ft, or 40-ft increments.

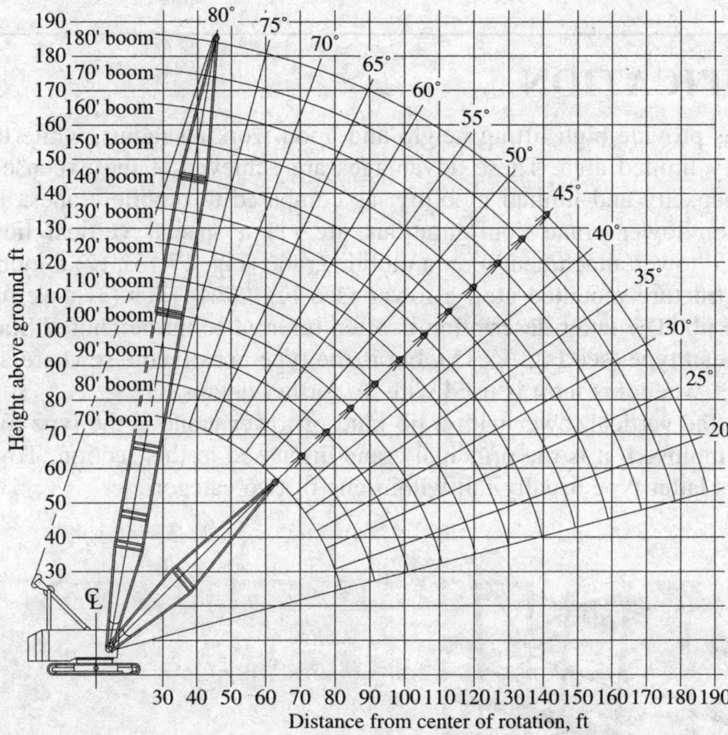

FIGURE 17.12 Working ranges for a 200-ton crawler crane, nominal rating.
Source: Manitowoc Engineering Co.

EXAMPLE 17.1

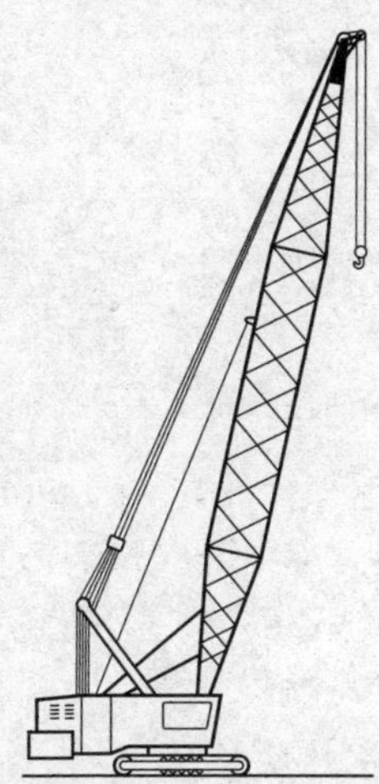

Using the information in Fig. 17.12, determine the minimum boom length that will permit the crane to lift a load 34 ft high to a position 114 ft above the surface on which the crane is operating. The length of the block, hook, and slings that are required to attach the hoist rope to the load is 26 ft. The location of the project will require the crane to pick up the load from a truck at a distance of 70 ft from the center of rotation of the crane. Thus, the operating radius will be 70 ft.

To lift the load to the specified location, the minimum height of the boom point of the crane must be at least 174 ft (114 + 34 + 26) above the ground supporting the crane. An examination of the diagram in Fig. 17.12 reveals that for a radius of 70 ft, the height of the boom point for a 180-ft-long boom is high enough.

If the block, hook, and slings weigh 5,000 lb, determine the maximum net weight of the load that can be hoisted. Using Table 17.1, we find that for a boom length of 180 ft and a radius of 70 ft the maximum total load is 47,600 lb. If the weight of the block, hook, and slings is deducted from the total load, the net weight of the lifted object will be 42,600 lb, which is the maximum safe weight of the lifted object.

TOWER CRANES

CLASSIFICATION

Tower cranes provide high lifting height and good working radius, while taking up a very limited area. These advantages are achieved at the expense of low lifting capacity and limited mobility, as compared to mobile cranes. The three common tower crane configurations are (1) a special vertical boom arrangement ("tower attachment") on a mobile crane (Fig. 17.13), (2) a mobile crane superstructure mounted atop a tower (see Fig. 17.14), or (3) a vertical tower with a jib. The latter description is often referred to in the United States as the European type (see Fig. 17.15), but is the type perceived elsewhere as a "tower crane" when this term is used with no further details.

17.1 Because the vertical tower with a jib kind of tower crane is the type most commonly employed, it is the principal crane discussed in this section. Tower cranes of this latter type usually fall within one of two categories:

FIGURE 17.13 Mobile crane with tower attachment.

FIGURE 17.14 Mobile crane superstructure mounted atop a tower.

FIGURE 17.15 Tower crane: static base, fixed tower, European type.

1. *Top-slewing (fixed tower)* tower cranes (Fig. 17.15) have a fixed tower and a swing circle ("slewing ring" or "crown") mounted at the top, allowing only the jibs (main-jib and counter-jib), tower top, and operator cab to rotate. The tower is assembled from modular sections, and hence the term "sectional tower crane," often used in reference to this type of crane. The crane is stabilized partly at its base (by ballasts or other means of ground anchoring) and partly by ballasts on the counter-jib.
2. *Bottom-slewing (slewing tower)* tower cranes (Fig. 17.16) have the swing circle located under a slewing platform, and both the tower and jib assembly rotate relative to the base chassis. The tower is essentially a telescoping mast, and hence the term "telescopic tower crane" is often applied to this type of crane. The mast is usually a lattice-boom type (always for the midsize and larger cranes), but smaller cranes of this type may have a hollow-section mast similar to the telescopic boom of a mobile crane. The entire ballast is placed on the revolving base platform.

The main differences between these two categories are reflected in setup and dismantling procedures and in lifting height. Bottom-slewing tower cranes are towable between job sites (see Fig. 17.17) and essentially erect themselves using their own motors in a relatively short duration (1 to several hours) using a simple procedure. They are often referred to as "self-erecting" or "fast-erecting" cranes. This is achieved, however, at the expense of service height, as dictated by the telescoping tower (mast) that, because of its revolving base, cannot be braced to a permanent structure. On the other hand, transporting, setting up, and dismantling top-slewing tower cranes requires more time (1 day to 1 week), is more complicated, and is a costlier procedure. Erection of a top-slewing tower crane requires the assistance of other equipment, commonly a

(a) General view (b) Slewing platform and ballast

FIGURE 17.16 Bottom-slewing tower crane.

large-size mobile crane (Fig. 17.18), but the crane can reach greater heights. Consequently, the bottom-slewing models are suitable mainly for shorter-term service on low-rise buildings, while top-slewing cranes commonly serve high-rise buildings on jobs requiring a crane for a long duration.

FIGURE 17.17 Bottom-slewing tower crane in towing position.

(a) Using a telescoping boom truck crane

(b) Using a lattice-boom crawler crane

FIGURE 17.18 Erecting a top-slewing tower crane.

In the United States, tower cranes are usually the machines of choice when

1. Site conditions are restrictive.
2. Lift height and reach are great.
3. There is no need for mobility.
4. Noise limits are imposed.

For these reasons, most tower cranes used in the United States are of the top-slewing category. The nomenclature for a top-slewing tower crane is shown in Fig. 17.19.

In Europe, though, where these machines were extensively introduced after World War II, both top-slewing and bottom-slewing tower cranes are seen on all types of projects: next to low-rise buildings on spacious sites, on road construction jobs, and even on utility projects. Additionally, long rail tracks are assembled to provide crane mobility on the job site. On all these project types, where tower cranes are the choice although not the necessity, economy is accomplished by their low operating cost compensating for possibly higher setup and dismantling costs.

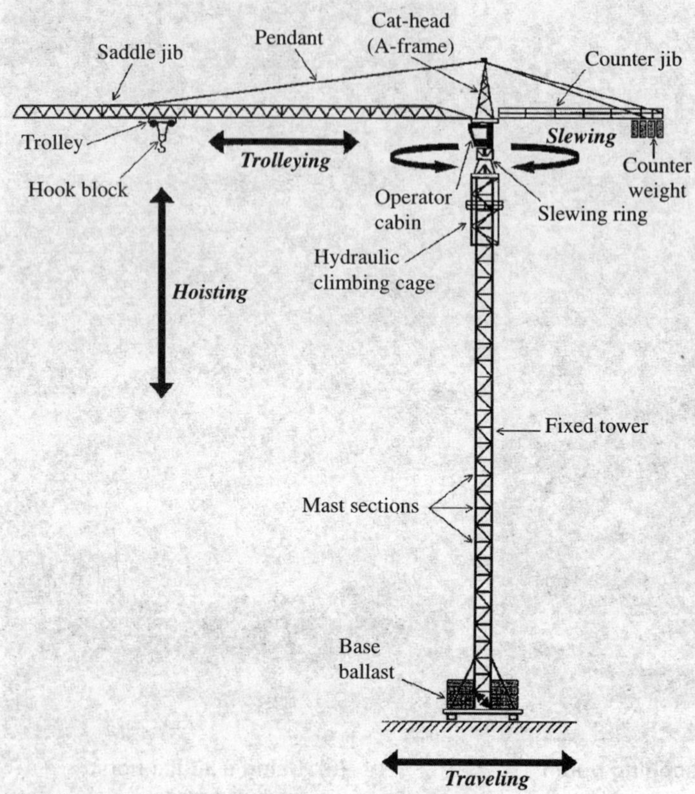

FIGURE 17.19 Nomenclature for a top-slewing tower crane.

OPERATION

Vertical tower cranes can be mounted on a mobile crane substructure, a fixed base, or a traveling base, or can be configured to climb within the structure being constructed.

Mobile Cranes Rigged with Vertical Towers

Crawler- and truck-mounted tower cranes use pinned jibs extending from special booms that are set vertically. A crawler-mounted tower crane can travel over firm, level ground after the tower is erected, but it has only limited ability to handle loads while moving. A truck-mounted tower crane must have its outriggers extended and down before the tower is raised. Therefore, it cannot travel with a load, and the tower must be dismantled before the crane can be relocated. Recent models of truck-mounted tower cranes merge a bottom-slewing tower crane with a mobile truck crane. The resulting crane provides excellent lifting height and outreach, as would a normal tower crane, and yet it is possible to erect and dismantle the crane in less than 15 min.

(a) Placing reinforcing steel for the block **(b) Concrete block completed**

FIGURE 17.20 Engineered foundation block for a fixed-base-type tower crane.

Fixed-Base Tower Cranes

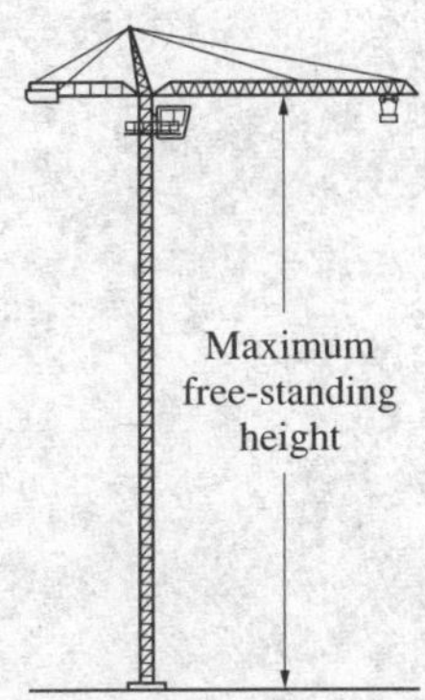

The fixed-base-type crane, commonly of the top-slewing configuration, typically has its tower mounted on an engineered concrete mass foundation, either on fixing angles sunk in the base (see Fig. 17.20) or on its ballasted chassis, which is bolted to the concrete base. On occasion, the tower may be mounted on a ballasted static rail-mounted undercarriage. Usually, at the beginning of the project, a large crawler or mobile crane is used to erect the tower crane to its full height; however, many of these tower cranes have the capability to independently increase their tower height by means of a climbing mechanism. For cranes with this ability, a smaller-size mobile crane could be used as it is not necessary to initially erect the tower to its full height, because additional tower sections (height) can be added later as the work progresses.

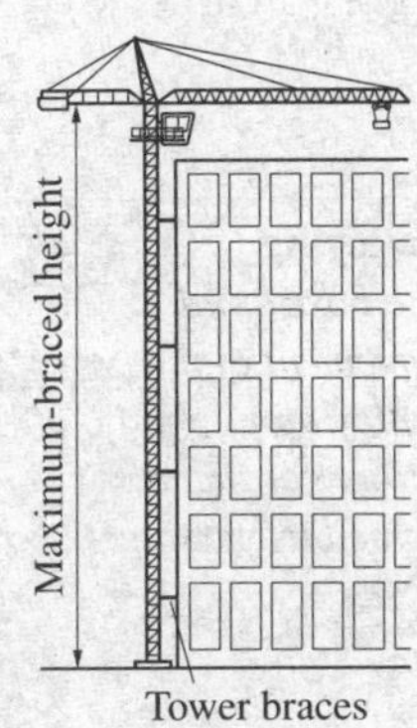

There is a vertical limit known as the *maximum free-standing height* to which fixed-base cranes can safely rise above a base, typically 200 ft for average-size top-slewing cranes, and up to 400 ft for the larger-size cranes. If it is necessary to raise the tower above this limiting height, lateral bracing must be provided. Guy ropes may be used to brace tower cranes, but in the majority of cases the towers are tied to the structure being constructed using engineered steel brackets (anchorage frames) (Fig. 17.21). The cost of bracing a crane to the structure rises sharply with the distance between them, which must be taken into account when planning the exact location of the crane. Even when bracing is provided, there is a *maximum-braced height* tower limit (although 1,000-ft-high top-slewing cranes are not a particular exception). These limits are dictated by the structural capacity of the tower frame and are machine-specific.

(a) To a bridge pier (b) To a building

FIGURE 17.21 Engineered steel anchorage frames tying a top-slewing tower crane to the structure.

Common dimensions and capacities for top-slewing tower cranes are

1. Jib length: 100 to 270 ft
2. Mast section length: 10 to 20 ft
3. Base dimensions: 13 by 13 ft to 27 by 27 ft
4. Tower cross section: 4 by 4 ft to 8 by 8 ft
5. Maximum lifting capacity: 10,000 to 90,000 lb
6. Maximum lifting capacity at end of jib ("jib-nose load"): 2,000 to 13,000 lb
7. Maximum hoisting speed: 150 to 500 ft/min
8. Maximum **trolleying** speed: 100 to 350 ft/min
9. Maximum slewing speed: 0.6 to 1.0 (0.8 for most makes and models) rpm

trolleying *The horizontal movement of the trolley along the jib of a tower crane, also termed traversing.*

A tower that is assembled by the use of other equipment at the beginning of a project cannot have a height greater than its maximum free-standing height, as there is no structure in place to which the lateral bracing can be connected. If, after the building of the structure progresses, it is necessary to raise the crane's tower, the procedure would require removal of the crane's superstructure from the tower frame before a tower section could be added. This would involve other equipment, such as a mobile crane, and would be a very costly proposition.

With the ability to raise themselves by the addition of tower sections, the raising of top-slewing tower cranes to greater vertical heights is fairly easy and economical. These cranes have a special hydraulically operated section termed the "(telescopic) climbing cage" or "climbing frame" for this purpose (see Fig. 17.22).

The erecting operation is a three-step procedure:

1. The crane hoists a new section of tower and moves it next to its tower.
2. The climbing cage hydraulically jacks up the slewing ring and jibs, and the new section of tower is inserted and positioned on the previously erected tower.
3. The hydraulic jacks are released, and the slewing ring and jibs are repositioned and fastened to the extended tower.

FIGURE 17.22 Hydraulic climbing cage of a tower crane.

The same procedure can be exercised in reverse order to lower the crane at the end of the job, as long as the permanent structure now in place does not bar the lowering of the jib. Ensuring the ability to perform this operation may be a consideration in predetermining the exact location of the crane relative to the constructed building.

Climbing cages for any given tower crane make are usually modular and would fit all cranes having the same mast cross section. A company owning a fleet of tower cranes therefore needs only a limited number of cages that move between job sites to raise the cranes as needed.

Fixed-base tower cranes are engineered to withstand high wind speeds. However, winds may affect the stability and operation of the crane through forces imposed on the profile and structure of the crane or on the load being handled. In terms of the crane's profile, large-area boards (e.g., with the contractor's name and logo) placed on the jib increase wind loading. When lifting large-surface-area loads, such as concrete forming panels, the load can act as a wind sail. The cranes are equipped with a special brake, used to hold the jib in position against the wind. This brake operates automatically when the slewing motor is stopped. At the end of the workday, this brake is usually suppressed, so that the jib can **weathervane** in the wind.

weathervane
Slew freely with the wind (for an inoperative crane) so that the area exposed to the wind is minimal.

A fixed-base tower crane can also be of the bottom-slewing type. Unlike top-slewing cranes, though, the tower is not mounted on a concrete block, but on a jacked (outriggered) chassis. A detachable wheeled undercarriage (trailer) serves for relocation, as needed, on the job site and for transportation between sites (see Fig. 17.17). For on-site relocation, the crane can be towed in its full upright operating position or partially folded, and not necessarily fully folded, depending on the terrain (smooth or rough) and its grade. Some models are self-propelled, using the detachable wheeled undercarriage for movement. Under no condition is it allowed to tow the crane while loaded. The mast of the smaller, hollow-section models is often of the folding (not telescoping) type.

Common dimensions and capacities for bottom-slewing tower cranes are

1. Jib length: 70 to 150 ft
2. Number of telescoping/folding mast parts: 2 to 3
3. Base dimensions: 10 by 10 to 15 by 15 ft
4. Total towed transport length: 30 to 50 ft
5. Maximum lifting capacity: 2,000 to 18,000 lb
6. Maximum lifting capacity at end of jib ("jib-nose load"): 1,000 to 5,000 lb
7. Maximum trolleying speed: 60 to 200 ft/min
8. Maximum slewing speed: 0.8 rpm

FIGURE 17.23 Traveling tower crane set on a pair of fixed rails.

Traveling Tower Cranes

The ballasted base of this type of tower crane is set on a pair of fixed rails (see Fig. 17.23). This enables the crane to move along the rails with a load. The advantage is the increased coverage of the work area that can be achieved. There are, however, cases in which the crane is set on rails not for routine traveling but merely to enable its relocation with the progress of the project, as a more economic option compared to crane dismantling and reassembly. Maximum grade for traveling crane rails is model-specific, but is usually not greater than 1%. Traveling tower cranes can be either of the top-slewing or bottom-slewing type, with heights commonly not exceeding 230 ft for the former and 100 ft for the latter. Maximum traveling speed for both types is 65 to 100 ft/min.

When a rail-mounted crane option is considered, the entailed costs of railway procurement and construction (including necessary earthworks over a relatively large area) should be taken into account. The tracks, typically of the width of 13 to 27 ft, may also obstruct the free movement of other equipment and vehicles on the job site. A rail-mounted crane, however, is a likely solution for spread-out projects, in which the construction method does not require extensive crane services in any given zone of the building, or in the case of linear projects such as lock construction.

Climbing Tower Crane

Along with the externally braced tower crane, the climbing tower crane is a common choice of crane for high-rise building construction, and is a lifting mechanism solution for buildings exceeding the maximum-braced height tower crane limit. Structurally supported by the floors of the building that is being constructed, the crane climbs on special climbing collars that are fitted to the building's completed structural floors (Fig. 17.24). The weight of both the crane and the loads lifted is transmitted to the structure of the host building. The crane will have only a relatively short mast because it moves upward with vertical construction progress. This climbing movement, which causes work interruptions, is done incrementally, every few floors, depending on the height of the mast employed. The taller the mast, the less frequent the climbing procedure.

FIGURE 17.24 Internal climbing tower crane; note the scaffolding supporting the floor upon which the crane is positioned.

Vertical movement of the climbing-type tower crane is by a system of hydraulically activated rams and latchings. Normally, the crane is initially mounted on a fixed base, and as the work progresses, it is detached from its chassis and transfered to a climbing frame mounted on the structure. A typical floor section cannot safely support the load imposed by the operating crane; therefore, it is imperative for the structural designer to consider the loads imposed by the crane in the area of this opening. Even when consideration for the crane loads is taken into account in the building design, it will usually be necessary to use shoring for several floors below the tower crane frame (see Fig. 17.24). To avoid the prestrengthening of the host building and the use of temporary shoring, the crane is often placed for climbing inside the elevator shaft commonly serving as the building's structural core. This location also has an advantage in that no openings are left in the floors for completion after the crane has been raised. On the other hand, use of the elevator shaft for this purpose may cause delays in elevator assembly, which is commonly on the critical path of the project's progress schedule. Additionally, it may limit the choice of the forming system for the concrete shaft walls.

At the end of construction, there will be a tower crane at the top of the structure with no means of lowering itself. Removal must be by external methods, such as a mobile crane or by use of a **derrick**. The mobile crane option may be limited by the height of the tower crane and the confined access area next to the building. With a derrick, the tower crane to be dismantled first lifts the derrick to the top of the building. The derrick is placed on the roof such that it can dismantle the tower crane, and then it is dismantled itself with the use of hand tools, and its parts are taken down in the building's elevator.

derrick
Derick was the name of a London hangman of about 1600, and the word refers to the similarity between a gallows and lifting devices of the time. Specifically, it is a lifting apparatus consisting of a mast, held at its head by braces, and used with a hoisting mechanism.

Because of the heights involved and the possible physical interference of the completed structure, the dismantling operation must be carefully planned and given consideration as early as when project equipment is initially selected. If two or more tower cranes were used on the project and there is overlapping hook coverage, it may be possible to use the crane with the higher hook height to dismantle the lower crane. Under circumstances where none of these techniques can be employed, an expensive solution is to use a helicopter.

Jib Configurations

The horizontal jib arrangement (see Figs. 17.15 and 17.19) for a top-slewing crane is commonly known as a "hammer head," though the proper terminology is *saddle jib*. This type of jib is fixed in a horizontal position by pendants. Actually, there are two jibs: (1) the forward main jib with a load block hung from a trolley that moves along the jib to change hook operating radius and (2) an opposite rear counterweight jib—the counter-jib. Some top-slewing crane models come with a retractable saddle jib, which has an advantage on space-restricted sites. The operator's cab is usually directly below the main jib, either at the top of the tower above the slewing ring, or attached to the side of the tower. The operator then has a bird's eye view of the entire site, and loading and unloading zone sight-line obstruction, common with mobile cranes, is minimal. With clear lines of sight, productivity is improved and safety is enhanced. On the other hand, the operator has to daily climb to the cab at the top of the crane. With cranes adjacent to the constructed building, part of the problem is solved by temporary platforms or walkways from one of the upper finished floors to the crane's mast. Another solution, already regulated in a few European countries, is the use of a special crane elevator. Some cranes feature cabs that climb the mast, serving for operator transport (Fig. 17.25). Besides using the cab as an elevator, these cabs enable the operator to position the cab at any height for optimum visibility. The cabs can rotate around the mast in synchronization with the slewing jib.

FIGURE 17.25 Climbing-rotating operator cab for a high tower crane.

When two or more top-slewing cranes are employed on one site, as is often the case, their work envelopes may partly overlap. A certain vertical clear distance (minimum 7 ft) must be maintained between the jib of the upper crane and the lower crane for safety. If the overlapping zone includes the mast of the lower crane, the safety clearance is governed by the height of the top part of the saddle-jib crane (the "cat-head" or the "A-frame"), typically

FIGURE 17.26 Flat-top tower cranes.

13 to 43 ft, and the height of the upper crane must be significantly increased. The height of the cat-head and its contribution to the overall height of the saddle-jib crane may be critical also in a nonoverlapping situation, if maximum crane height is restricted (i.e., near airports, under high-voltage power lines).

A top-slewing crane with a pendant-free, *cantilever jib* ("flat-top" or "top-less" crane, Fig. 17.26) provides a solution for both overlapping and height restriction situations. Flat-top cranes, which started gaining popularity in the early 2000s, offer several other advantages, such as comparable lightweight profile design due to unidirectional action of all forces in the jib regardless of load location and weight, easier and quicker setup, and to erect the required reach of the mobile crane is less.

Bottom-slewing cranes do not have a counter-jib, thus all the counterweight is placed on the rotating tower base (see Fig. 17.16). The operator's cab can be directly below the jib, at the top of the tower, or somewhat lower on the tower. Since the height of bottom-slewing cranes is limited, they do not pose a climbing problem for the operator. Often these bottom-slewing cranes are operated from a control stand at the tower's base or by remote control and not from a top cab. Remote-control operation of top-slewing tower cranes, although not too common, is also possible; when used, it is mainly to overcome the problem of work zones that are not visible to the operator.

If the jib is pinned at its base and supported by cables that are used to control its angle of inclination, it is known as a *luffing jib* (Fig. 17.27). By controlling the jib's inclination the operator varies the crane's hook radius. The term "luffing-jib crane" is commonly used for top-slewing cranes only (often referred to as *luffers*). There are also fixed luffing-jib arrangements for bottom-slewing cranes. These have the jib supported by jib pendants at a fixed angle of inclination (0 to 30°), as determined prior to crane setup, and hook radius is varied by use of a trolley traveling along the angled jib, usually with a reduced

FIGURE 17.27 Luffing-jib tower cranes.

load capacity as compared to horizontal trolleying. The luffing jib in some of these bottom-slewing crane models is also an *articulated jib*, whereby the inner part of the jib (close to the tower) remains horizontal while the rest of the jib can be inclined in the course of operation up to 45° ("obstruction-avoiding jib position"), with the trolley locked at the jib end. A luffing-jib crane has a clear advantage on confined sites (e.g., where existing buildings limit the slewing of a saddle jib) and in the case of working near other cranes (i.e., to avoid overlapping, see Fig. 17.27). With a given crane height it also provides extra height under the hook. Some cities (e.g., Tokyo) prohibit the movement of the crane's jib beyond the limits of the site (termed *oversailing*); under such restrictions, a luffing-jib crane may be the only solution.

TOWER CRANE SELECTION

The use of a tower crane requires considerable planning because the crane is anchored in a fixed location for the duration of the major construction activities. From its fixed position, it must be able to cover all points from which loads are to be lifted and to reach the locations where the loads must be placed. Therefore, when selecting a crane for a particular project, the engineer must ensure that the weight of the loads can be handled at their corresponding required radius. Individual tower cranes are selected for use based on

1. Weight, dimension, and lift radii of the heaviest loads
2. Maximum free-standing height of the crane
3. Maximum braced height of the crane
4. Crane-climbing arrangement
5. Weight of crane supported by the structure

6. Available headroom
7. Area that must be reached
8. Hoist speeds of the crane
9. Length of cable the hoist drum can carry

The vertical movement of material during the construction process creates an *available headroom* clear-distance requirement. This distance is defined as the vertical distance between the maximum achievable crane-hook position and the uppermost work area of the structure. The requirement is set by the dimensions of those loads that must be raised over the uppermost work area during the building process. For practical purposes and safety, hook height above the upper level of construction work should never be less than 20 ft. When selecting a tower crane for a very tall structure, a climbing-type crane may be the only choice capable of meeting the available headroom height requirement.

RATED LOADS FOR TOWER CRANES

Table 17.3 is a capacity chart for a climbing tower crane having a maximum reach of 218 ft. This particular crane can have a stationary free-standing height such that there is 212 ft of clear under-hook height. While hook or lift height

TABLE 17.3 Lifting capacities in pounds for a tower crane.

Jib model	L1	L2	L3	L4	L5	L6	L7	
Maximum hook reach	104′ 0″	123′ 0″	142′ 0″	161′ 0″	180′ 0″	199′ 0″	218′ 0″	Hook reach
	27,600	27,600	27,600	27,600	27,600	27,600	27,600	10′ 3″
	27,600	27,600	27,600	27,600	27,600	27,600	27,600	88′ 2″
	27,600	27,600	27,600	27,600	27,600	27,600	25,800	94′ 6″
	27,600	27,600	27,600	27,600	27,600	25,800	24,200	101′ 0″
	27,600	27,600	27,600	27,600	26,800	24,900	23,400	104′ 0″
		27,600	27,600	27,600	25,200	23,600	22,200	109′ 8″
		27,600	27,600	25,600	23,300	21,800	20,500	117′ 8″
		27,000	27,000	25,100	22,800	21,300	20,100	120′ 0″
Lifting capacities		26,300	26,300	24,300	22,200	20,700	19,500	123′ 0″
in pounds,			24,800	22,800	20,800	19,300	18,300	130′ 0″
two-part line			22,400	20,700	18,700	17,400	16,400	142′ 0″
				19,500	17,600	16,300	15,400	150′ 0″
				18,800	16,800	15,700	14,800	155′ 0″
				17,900	16,200	15,100	14,200	161′ 0″
					15,200	14,200	13,300	170′ 0″
					14,200	13,200	12,400	180′ 0″
						12,300	11,600	190′ 0″
						11,700	10,800	199′ 0″
							10,200	210′ 0″
							9,700	218′ 0″

(Continued on next page)

TABLE 17.3 Lifting capacities in pounds for a tower crane (continued).

Jib model	L1	L2	L3	L4	L5	L6	L7	
Maximum hook reach	100′ 9″	119′ 9″	138′ 9″	157′ 9″	176′ 9″	195′ 9″	214′ 9″	Hook reach
	55,200	55,200	55,200	55,200	55,200	55,200	55,200	13′ 6″
	55,200	55,200	55,200	55,200	55,200	55,200	55,200	48′ 9″
	55,200	55,200	55,200	55,200	55,200	55,200	51,400	51′ 0″
	55,200	55,200	55,200	55,200	55,200	51,500	48,500	53′ 6″
	55,200	55,200	55,200	55,200	51,300	48,300	45,600	56′ 6″
	55,200	55,200	55,200	50,700	47,100	44,600	42,100	60′ 6″
	46,200	46,200	46,200	42,800	39,700	37,400	35,200	70′ 0″
	39,400	39,400	39,400	36,500	34,100	31,900	29,900	80′ 0″
	34,600	34,600	34,600	31,900	29,700	27,700	26,100	90′ 0″
	30,700	30,700	30,700	28,200	26,100	24,100	22,600	100′ 9″
Lifting capacities in pounds four-part line		27,800	27,800	25,600	23,600	21,700	20,300	110′ 0″
		25,400	25,400	23,200	21,300	19,600	18,300	119′ 9″
			23,100	21,100	19,300	17,700	16,400	130′ 0″
			21,300	19,400	17,800	16,300	15,100	138′ 9″
				17,600	16,200	14,700	13,600	150′ 0″
				16,400	15,100	13,800	12,700	157′ 9″
					13,600	12,400	11,400	170′ 0″
					12,900	11,800	10,800	176′ 9″
						11,500	10,600	180′ 0″
						10,700	9,800	190′ 0″
						10,200	9,300	195′ 9″
							9,100	200′ 0″
							8,300	210′ 0″
							8,100	214′ 9″

Counterweights

Jib	L1	L2	L3	L4	L5	L6	L7
105-HP hoist unit AC	37,200 lb	47,600 lb	50,800 lb	37,200 lb	40,800 lb	44,000 lb	54,400 lb
165-HP hoist unit AC	34,000 lb	44,000 lb	47,600 lb	34,000 lb	40,800 lb	40,800 lb	54,800 lb

Source: Morrow Equipment Company, L.L.C.

does not affect capacities directly, there is a relation when hoist speed is considered. Information concerning this relationship between hoist-line speed and load capacity is shown in Table 17.4.

Tower cranes are usually powered by alternating current (AC) electric motors that produce only low-level noises making them an environmentally acceptable solution for working in areas where noise limits are imposed. A crane having a higher motor horsepower can achieve higher operating speeds. When considering the production capability of a crane for duty-cycle work, hoist-line speed and the effect motor size has on speed, as shown in Table 17.4, can be very important selection criteria. This is especially true for high-rise construction, where travel time of the hook, between loading and unloading areas, can be the most significant part of the crane's cycle time, as opposed

TABLE 17.4 Effect of hoist-line speed on lifting capacities of a tower crane.

105 HP-AC, Eddy Current Brake, with four-speed remote controlled gear box, recommended service = 224 amp					
(1) Trolley, two-part line			(2) Trolleys, four-part line		
Gear	Maximum load (lb)	Maximum speed (fpm)	Gear	Maximum load (lb)	Maximum speed (fpm)
1	27,600	100	1	55,200	50
2	15,700	200	2	31,400	100
3	9,300	300	3	18,600	150
4	5,500	500	4	11,000	250

165 HP-AC, Eddy Current Brake, with four-speed remote controlled gear box, recommended service = 250 amp					
(1) Trolley, two-part line			(2) Trolleys, four-part line		
Gear	Maximum load (lb)	Maximum speed (fpm)	Gear	Maximum load (lb)	Maximum speed (fpm)
1	27,600	160	1	55,200	80
2	17,600	250	2	32,200	125
3	10,600	400	3	21,200	200
4	6,200	630	4	12,400	315

Source: Morrow Equipment Company, L.L.C.

to low-rise construction, where travel time is insignificant compared to load rigging and unrigging times. If a project requires operating speeds that are higher than provided by an existing crane, replacing the crane's motors with more powerful ones is an alternative option to bringing in another crane.

Hoist-cable configuration is another factor affecting lifting speed. Tower cranes can usually be rigged with one of two hoist-line configurations, a two-part line or a four-part line. The four-part-line configuration provides a greater lifting capacity than a two-part line within the structural capacity constraints of the tower and jib configuration. The maximum lifting capacity of the crane will be increased by 100% with the four-part-line configuration. However, the increased lifting capacity is acquired with a resulting lost of 50% in vertical hoist speed.

Examination of the Table 17.3 load chart illustrates these points. The first portion of the table is for a crane rigged with a two-part line. Considering an L7 jib model, a crane so rigged could lift 27,600 lb at a radius of 10 ft 3 in., 25,800 lb at a radius of 94 ft 6 in., and 10,200 lb at a radius of 210 ft 0 in. This same crane rigged with a four-part-line arrangement and an L7 jib can lift 55,200 lb at a radius of 13 ft 6 in., 26,100 lb at a radius of 90 ft 0 in., and 8,300 lb at a radius of 210 ft 0 in.

When the operating radius is less than about 90 ft, the crane has a greater lifting capacity with a four-part-line than the two-part-line arrangement. However, when the operating radius exceeds 90 ft, the crane has a slightly greater lifting capacity with the two-part line. This is because of the increased weight of the four-part rigging system (trolley, hook block, and cable) and the fact that structural capacity is the critical factor affecting load-lifting capability. However, when operating at a radius of less than 90 ft, the hoisting system controls the load-lifting capability.

Tower crane load charts are usually structured assuming that the weight of the hook block is part of the crane's dead weight. But the rigging system is taken as part of the lifted load. When calculating loads, the Construction Safety Association of Ontario recommends that a 5% working margin be applied to the computed weight.

EXAMPLE 17.2

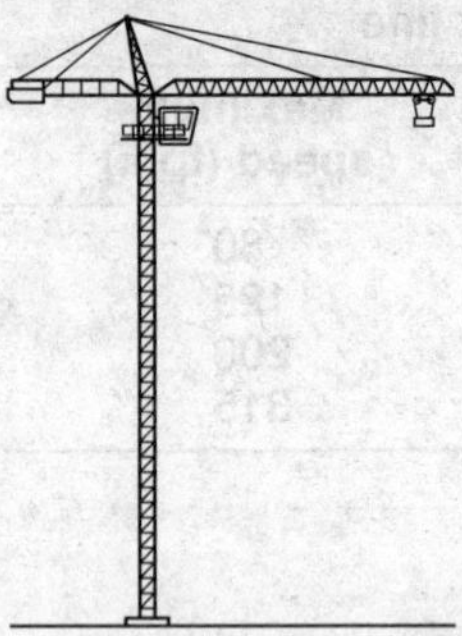

Can the tower crane, whose load chart is shown in Table 17.3, lift a 15,000-lb load at a radius of 142 ft? The crane has an L7 jib and a two-part hoist line. The slings that will be used for the pick weigh 400 lb.

Weight of load	15,000 lb	
Weight of rigging	400 lb	(slings)
	15,400 lb	
	× 1.05	working margin
Required capacity	16,170 lb	

From Table 17.3 the maximum lifting capacity at a 142-ft hook reach is 16,400 lb.

$$16{,}400 \text{ lb} > 16{,}170 \text{ lb}$$

Therefore, the crane can safely make the lift.

EXAMPLE 17.3

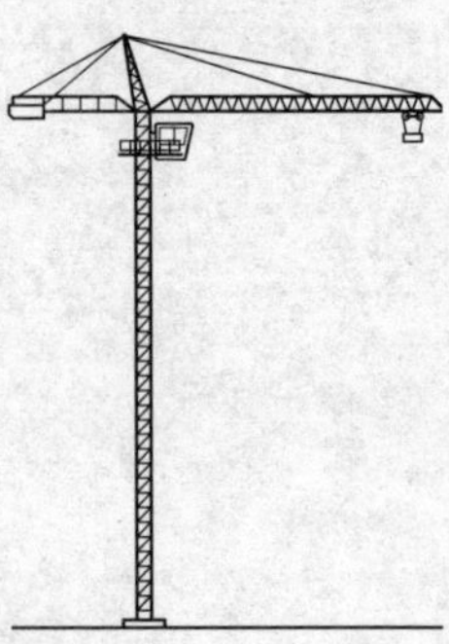

Can the tower crane, whose load chart is shown in Table 17.3, lift a 15,000-lb load at a radius of 138 ft? The crane has an L7 jib and a four-part hoist line. The slings that will be used for the pick weigh 400 lb.

Weight of load	15,000 lb	
Weight of rigging	400 lb	(slings)
	15,400 lb	
	× 1.05	working margin
Required capacity	16,170 lb	

From Table 17.3 the maximum lifting capacity at a 138-ft hook reach is 15,100 lb.

$$15{,}100 \text{ lb} < 16{,}170 \text{ lb}$$

Therefore, the crane cannot make the lift.

RIGGING BASICS

A crane is designed to pick (or lift) a load by means of a hoisting mechanism using ropes. The load must be properly attached to the crane by a rigging system. To properly attach the load, it is necessary to determine the forces that will affect the job, and then to select and arrange the equipment that will move the load safely. The forces involved in rigging will vary with the method of connection and the effects of motion. The rigger must, by proper application of mechanical laws and by resolving load-movement-induced stresses, correctly determine the weight and center of gravity of the load.

Weight

The most important step in any rigging operation is to correctly determine the weight of the load. If this information cannot be obtained from the shipping papers, design plans, catalog data, or other dependable sources, it may be necessary to calculate the weight. It is good practice to verify the load weight as stated in the documents. Weights and properties of structural members can be obtained from

1. Manual of Steel Construction, American Institute of Steel Construction (www.aisc.org)
2. Cold Formed Steel Design Manual, American Iron and Steel Institute (www.steel.org)
3. Aluminum Design Manual, American Aluminum Association (www.aluminum.org)

Center of Gravity

The center of gravity of an object is that location where the object will balance when lifted. When the object is suspended freely from a hook, this point will always be directly below the hook. Thus, a load that is slung above and through its center of gravity will be in equilibrium. It will not tend to slide out of the hitch or become unstable.

One way to determine the center of gravity of an odd-shaped object is to divide the shape into simple masses and determine the resultant balancing load and its location at a point where the weights multiplied by their respective lever arms are in equilibrium. Thus,

$$W_1 \times \ell_1 = W_2 \times \ell_2 \quad \textbf{[17.1]}$$

where W_1 and W_2 are the weights of the larger part and the smaller part, respectively, and ℓ_1 and ℓ_2 are the lever arms of the larger part and the smaller part, respectively (see Fig. 17.28). If we know the sum distance $\ell_1 + \ell_2$, the location of the center of gravity can be computed by Eq. [17.1].

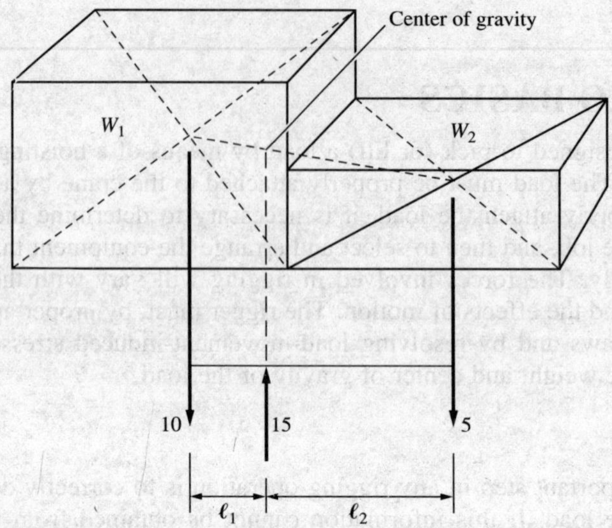

FIGURE 17.28 Calculating center of gravity.

EXAMPLE 17.4

The larger part of an odd-shaped load (see Fig. 17.28) weighs 10 tons. The other part of the shape weighs 5 tons. The larger part has a 4 × 4 × 4-ft square shape. The smaller part is 4 ft wide by 4 ft long by 2 ft high. Determine the center of gravity in the long dimension of the object.

$$10 \times \ell_1 = 5 \times \ell_2; \quad \ell_1 + \ell_2 = 4 \text{ ft}; \quad \Rightarrow \quad \ell_1 = 1.33 \text{ ft}, \; \ell_2 = 2.67 \text{ ft}$$

Therefore, the center of gravity of the object is located 1.33 ft from the center of the larger part toward the smaller part or, if speaking of the total object, the center of gravity in the long dimension is 3.33 ft from the edge of the larger part. The center of gravity in the short dimension is along the centerline of that dimension.

Stresses

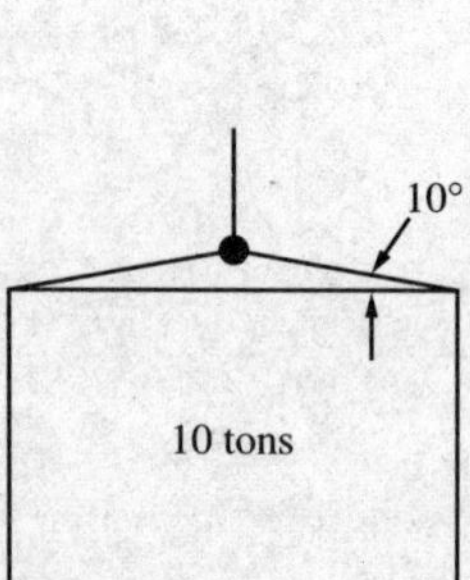

FIGURE 17.29 Stresses induced in a set of slings.

To calculate the stress developed by the load on a rigging arrangement, it must be remembered that all forces must be in equilibrium. If a 10-ton load is supported by a set of slings in such a manner that the individual sling legs make a 10° angle with the load (see Fig. 17.29), the sling is stressed 28.8 tons and there is a 28.4-ton horizontal reaction. Changing the sling angle to 45° will reduce the stress in the sling to 7.1 tons and the horizontal reaction to 5.0 tons.

The general formulas are given by the following equations:

$$N = \frac{T}{2 \sin \alpha} \quad [17.2]$$

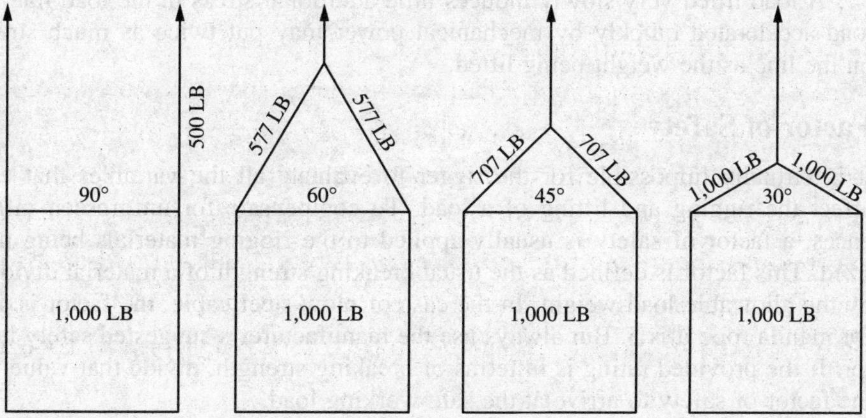

FIGURE 17.30 Small sling angles increase sling stress.

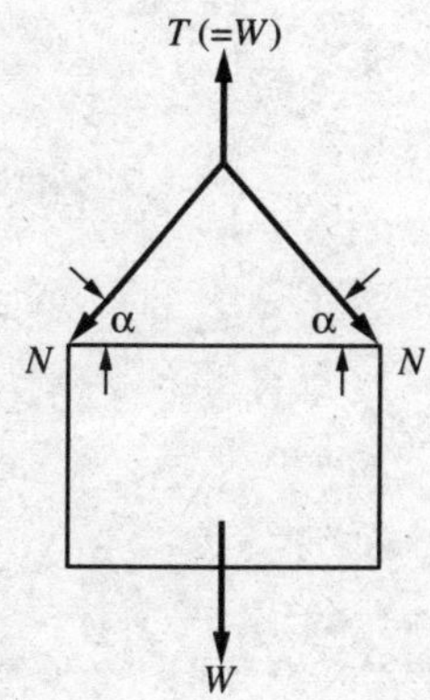

$$N \times \cos \alpha = \frac{T}{2 \tan \alpha} \quad [17.3]$$

where

W = weight of the load
T = tension in the cable
N = stress in the sling
$N \times \cos \alpha$ = horizontal reaction
α = angle between the load and each sling leg

Thus, it is apparent that when the rigging arrangement creates small sling angles, the resulting sling stress produced will be considerably greater than the load (Fig. 17.30). The solution in such situations is to use a supplemental compression member—a spreader bar (see Fig. 17.6). The use of a spreader bar will allow for greater sling angles and reduce the induced sling stresses.

Mechanical Laws

The hook-block and tackle lifting mechanism act as a lever with the fulcrum at the side of the block and the rope fixed at the boom point. These blocks provide a mechanical advantage for lifting the load.

Motion

Lifting operations involve loads in motion. The inertia of an object at rest or in motion, which takes place when there is a decrease or increase in speed, increases the stresses sustained by the rigging. As a hoist line starts to move, the load must accelerate from zero to the normal speed of ascent. This requires an additional force over and above the weight of the load. This force will vary with the rate of change in velocity.

A load lifted very slowly induces little additional stress in the load line. A load accelerated quickly by mechanical power may put twice as much stress on the line as the weight being lifted.

Factor of Safety

It is virtually impossible for the rigger to evaluate all the variables that can affect the rigging and lifting of a load. To compensate for unforeseen influences, a factor of safety is usually applied to the rigging materials being utilized. This factor is defined as the usual breaking strength of a material divided by the allowable load weight. In the case of plow-steel cable, the factor is 10; for manila rope it is 5. But always use the manufacturer's suggested safety factor. If the provided rating is in terms of breaking strength, divide that value by the factor of safety to arrive at the safe working load.

SLINGS

The safety and efficiency of a lift depends on the working attachments that secure the load to the crane hook. Generally called slings (see Fig. 17.6), these working attachments may be wire rope, chain, or nylon-web straps (Fig. 17.31). When using a sling to lift, remember that the capacity of the sling depends on its material and size, the configuration in which it is used, its type of end terminals, and the angle that the legs make with the load.

FIGURE 17.31 Straps used to lift hollow-core precast concrete slab elements.

Wire Rope

The capacity of a wire rope sling is based on the nominal strength of the wire rope. Factors affecting the overall strength of the sling include attachment or splicing efficiency, the wire rope's construction, and the diameter of the hook over which the eye of the sling is rigged. Consider the wire rope sling's ability to bend without distortion and to withstand abrasive wear and abuse. Kinking, for example, causes serious structural damage and loss of strength.

Chain

Only alloy steel chain is suitable for slings used in overhead lifting—predominantly grade 80 or grade 100 alloy chain. Chain slings are ideal for rugged loads that would destroy other types of slings. Chain slings can have one to four legs connected to a master link.

Synthetic Web

Synthetic web slings are good for use on expensive loads, highly finished parts, fragile parts, and delicate equipment. They have less of a tendency than wire rope or chain slings to crush fragile objects. Because they are flexible, they tend to mold themselves to the shape of the load, thus gripping a load securely. Synthetic slings are elastic and stretch more than wire rope or chain, better absorbing heavy shocks and cushioning loads.

Sling Inspection

Frequent sling inspection is essential to safe lifting operations.

Wire Rope Slings If damage from any of the following is visible, consider removing the sling from service:

1. Kinking, crushing, or any other damage resulting in distortion of the rope structure
2. Sever corrosion of the rope or end attachments
3. Severe localized abrasion or scraping

Chain Slings Make a link-by-link inspection for

1. Excessive wear
2. Twisted, bent, or cut links
3. Cracks in the weld area or any portion of the link
4. Stretched links

Synthetic Slings If damage from any of the following is visible, remove the sling from service:

1. Acid or caustic burns
2. Melting or charring of any part of the sling
3. Holes, tears, cuts, or snags

SAFETY

CRANE ACCIDENTS

Crane accident data are limited because typically only deaths and injuries are reported. Property damage incidents are usually not reported, except to insurance carriers. The seriousness of a crane accident, however, is self-evident. Moreover, there are numerous daily cases of "close calls" that could have easily turned into severe or even fatal accidents. In the state of California alone, 158 accidents involving a crane were reported over a period of 3 years, from January 1997 to December 1999.

The enhancement of on-site crane safety requires, first and foremost, that all involved parties (project manager, general superintendent, crane operator, etc.) be well aware of safety hazards—factors that increase the chance of an accident—on the particular job. Potential safety hazards, by which the expected safety level at a given site may be evaluated, preferably before actual construction has started, fall under three categories (factors associated primarily with tower cranes are marked by an asterisk):

1. The "human factor" is reflected mainly in the experience and competence of the operator as well as the signalers, the mode of operator employment (i.e., whether you use your own company operator or hire one through a manpower company) (*), and the attitude of all personnel involved with on-site crane work.
2. "Project factors" are the presence of power lines and the compactness of the site; overlapping of crane work envelopes and oversailing of the crane's jib (*); the length of the workday and working night shifts; work conditions inside the operator cab and the use of optional, advanced operator aids; various visibility interruptions, particularly hidden work zones; and hazardous loads and lifting assignments (see Fig. 17.32).

FIGURE 17.32 Hazardous site: overlapping cranes working near high-voltage power lines.

3. Typical "environmental (i.e., non-project-specific) factors" are winds and severe weather, maintenance standards of the cranes and lifting accessories, and corporate policy toward safety management.

Safety should be a major concern not only when the crane is in operation, but also in other phases of its presence on the project site. This is particularly true for tower cranes during erection and dismantling, climbing, and after-duty hours. During all of these periods the crane is not in its full or "natural" working state. "Natural" working state is when the crane is doing what it is designed and built to do, namely, lift loads. During after-duty hours, when no loads are lifted, the balance of forces is shifted, while no operator is in the cab. A gust of wind, a local structural failure, or a disengagement of brakes that went unnoticed while the cab was unmanned may develop into an accident.

Even more hazardous are erection, climbing, and dismantling operations of tower cranes. A considerable number of tower crane accidents that involved these operations have been reported. These operations differ from the routine state of employment first and foremost because the structure of the crane and its various operation and control systems are not fully configured; the crane is thus in a delicate and constantly changing balance of forces. Additionally, these operations are unique in that they involve other equipment, in close proximity, as well as other personnel. Since these operations are commonly subcontracted, i.e., their executioners are out of the direct control of the construction company and not subject to its internal safety plans and programs, utmost caution should be exercised in the prequalification and selection process of these subcontractors (e.g., thorough examination of experience and accident records). Figure 17.33 shows the aftermath of an accident in which a mobile crane collapsed on the tower crane it was dismantling, a combination of a mechanical failure and human error.

FIGURE 17.33 Accident with a mobile crane during dismantling of a tower crane.

SAFETY PLANS AND PROGRAMS

There should be a corporate-level generalized crane safety program and a project-specific crane safety plan.

Crane Safety Program

The company crane safety plan should address

1. Equipment inspection
2. Hazard analysis—concern for the public, power lines, etc.
3. Crane location
4. Crane movements
5. Definition of lifts—critical, production, and general
6. Determination of responsibility zones and lines of control and reporting
7. Postaccident reporting and investigation procedures.

Crane Safety Plan

On a site- or project-specific basis, the crane safety plan addresses the same topics as the crane safety program. Equipment inspection methods and standards are set forth in detail. Operator aids, such as load indicators and anti-two-block devices, must be inspected daily prior to crane operation. Hazard analysis considers dangers identified in the contract documents and those associated with location of and access to the work—electric lines; ground conditions; and weather, wind, and cold. Additionally, when positioning a crane close to a flight path there must be close coordination with Federal Aviation Administration (FAA) and airport authorities. The FAA requires a permit on construction cranes any time that they will exceed 200 ft in height, and/or when they are placed within 20,000 ft (3.79 mi) of an airport, regardless of height. The FAA requires that FAA FORM 7460-1 "Notice of Proposed Construction or Alteration" be submitted at least 30 days before the following: (1) the date the proposed construction is to begin and (2) the date an application for a construction permit is to be filed.

Crane location discusses crane setup locations with attention to unusual support or interference problems. Crane movement defines procedures for controlling crane movement. For each individual critical lift, a separate written lift plan should be prepared identifying equipment, load properties, personnel, location, and load path. The goal of lift planning is to eliminate as many variables as possible from the lifting equation. Every critical lift should be listed in the crane safety plan. Production lifts do not require a plan for each individual pick but a plan considering appropriate site-specific parameters for each type of production lift should be prepared. General lifts need not be listed.

Advanced Operator Aids

Crane manufacturers provide their machines, at client request, with optional operator aids in excess of the standard aids that are required by safety regulations. Additionally, advanced operator aids that can be used with older machines are offered by independent suppliers. These will upgrade cranes in terms of safety. Examples are digital-display safe load indicators (SLIs), 2D/3D crane operation graphical displays, remote monitoring systems for various load parameters, sensor-based antisway systems, GPS-based weather and wind warning systems, anticollision and zoning systems for computerized control of sites with overlapping cranes, and crane-mounted video cameras enabling the operator to follow the load and view hidden work zones.

ZONES OF RESPONSIBILITY

Because of various project situations, assignment of responsibilities can vary, but there are three general categories of responsibility.

Rigging Personnel

Rigging personnel attach the load to the hook and perform other ground- or structure-based operations. Rigging personnel are responsible for the stability of the load, required tag lines, and load pickup and set-down procedures. That responsibility involves

1. Verifying the actual weight of the load and communicating that information to the crane operator.
2. Attaching (rigging) the load using suitable lifting gear.
3. Signaling or directing the movement of the load by communication with the crane operator.

Crane Operator

The crane operator controls the lift. The operator may abort the lift at any time from initial pickup to final placement. The operator is responsible for all crane movements from the hook upward as well as swing and travel motion. The operator's responsibility involves

1. Confirming from which individual directions will be given. Other than the case of a stop signal, an operator should respond only to the signals from the designated signal person.
2. Confirming that the configuration of the crane is appropriate for the load that is to be lifted and is in conformance with the load chart.
3. Being aware of the site conditions below the ground, at ground level, and above the ground.

4. Confirming the weight of the load.
5. Knowing the location and destination of the load.

Lift Director

The lift director is responsible for the entire lift and must ensure there is full compliance with the Crane Safety Plan and the appropriate lift plan. The lift director is specifically responsible for

1. Ensuring that each of the other parties—riggers, crane operator, and signal persons—understands his or her functions.
2. Ensuring that a signal person is assigned. If multiple signal persons are required, a thorough briefing on the transition between signalers with the crane operator is required.
3. Making a definite and clear assignment of the crane's outrigger configuration and responsibilities to avoid misunderstandings concerning the status of the outrigger operation.
4. Ensuring that everyone knows to whom they should communicate concerns about anything they observe that affects safety.

SUMMARY

Cranes are a broad class of construction equipment used to hoist and place material, building elements and products, and machinery. The most common mobile crane types are (1) crawler, (2) telescoping-boom truck-mounted, (3) lattice-boom truck-mounted, (4) rough-terrain, (5) all-terrain, and (6) modified cranes for heavy lifting. The most common tower crane types are (1) top-slewing and (2) bottom-slewing. Tower cranes provide high lifting height and good working radius, while taking up a very limited area. These advantages are achieved at the expense of lower lifting capacity and limited mobility, as compared to mobile cranes.

The rated load for a crane, as published by the manufacturer, is based on ideal conditions, a level machine, calm air, and no dynamic effects. Load charts can be complex documents listing numerous booms, jibs, and other components that may be employed to configure the crane for various tasks. It is critical that the load chart being consulted be for the actual crane configuration that will be used. *Interpolation between the published load chart values IS NOT permitted*; use the next lower value. Critical learning objectives include:

- An understanding of the basic mobile and tower crane types
- An understanding of load charts and their limitations
- An ability to read mobile and tower crane load charts
- An understanding of responsibilities assigned by lift safety plans and programs

These objectives are the basis for the problems that follow.

PROBLEMS

17.1 Using Fig. 17.11, select the minimum-size crane required to unload pipe weighing 166,000 lb per joint and lower it into a trench when the distance from the centerline of the crane to the trench is 50 ft. (300 ton crane)

17.2 Using Fig. 17.12, select the minimum-length boom required to hoist a load of 80,000 lb from a truck at ground level and place it on a platform 76 ft above the ground. The minimum allowable vertical distance from the bottom of the load to the boom point of the crane is 42 ft. The maximum horizontal distance from the center of rotation of the crane to the hoist line of the crane when lifting the load is 40 ft.

17.3 High Lift Construction Co. has determined that the heaviest load to be lifted on a project they anticipate bidding weighs 14,000 lb. From the proposed tower crane location on the building site, the required reach for this lift will be 150 ft. The crane will be equipped with an L5 jib and a two-part hoist line (see the Table 17.3 load chart). This critical lift is of a piece of limestone facing and will require a 2,000-lb spreader bar attached to a 300-lb set of slings. If assembled in the proposed configuration, can the crane safely make the pick? (17,600 lb capacity > 17,115 lb load. Therefore, the crane can safely make the pick.)

17.4 Low Ball Construction Co. has determined that the heaviest load to be lifted on one of their projects weighs 22,000 lb. From the tower crane location on the building site, the required reach for this lift will be 100 ft. The crane is equipped with an L7 jib and a four-part hoist line (see the Table 17.3 load chart). This critical lift is of a piece of mechanical equipment and will require a 1,000-lb spreader bar attached to a 200-lb set of slings. If assembled in the proposed configuration, can the crane safely make the pick?

17.5 A crane lifts a load with slings, as shown in Fig. 17.29. What is the angle between the load and each sling leg for which the stress in each leg is twice the load?

17.6 A 25-ton truck-mounted hydraulic crane is being used on a construction site. The maximum horizontal distance from the center of rotation of the crane to the hoist line of the crane when lifting a load of 9,000 lb will be 40 ft. Use Table 17.2 crane data to select the minimum-length boom required to safely hoist the load.

17.7 Liftem High Construction Co. has determined that the heaviest lift on a project they are estimating weighs 25,000 lb. The project's tower crane will make this lift (see Table 17.3). From the proposed tower crane location on the building site the required reach for this lift will be 80 ft. The crane will be equipped with an L5 jib and a four-part hoist line. This critical lift is of a piece of HVAC equipment and will require a 1,500-lb spreader bar attached to a 400-lb set of slings. If assembled in the proposed configuration, can the crane safely make the pick?

REFERENCES

1. *Articulating Boom Cranes* (1994). ASME B30.22-1993, an American National Standard, The American Society of Mechanical Engineers, New York.
2. Bates, Glen E., and Robert M. Hontz (1998). *Exxon Crane Guide, Lifting Safety Management System*, Specialized Carriers & Riggers Association, Fairfax, VA.
3. *Below-the-Hook Lifting Devices* (1994). ASME B30.20-1993, an American National Standard, The American Society of Mechanical Engineers, New York.
4. *Crane Handbook* (1990). Construction Safety Association of Ont., Toronto, Canada.
5. "Crane Load Stability Test Code—SAE J765," in *SAE Recommended Practice Handbook* (1980). Society of Automotive Engineers, Inc., Warrrendale, PA.
6. *Crane Safety on Construction Sites* (1998). ASCE Manuals and Reports on Engineering Practice No. 93, American Society of Civil Engineers, Reston, VA.
7. *Cranes Today Handbook 2004* (2004). Directory and Buyers' Guide, 28th ed., Wilmington Publishing, Foots Cray, Sidcup, United Kingdom.
8. *Hammerhead Tower Cranes* (1992). ASME B30.3-1990, an American National Standard, The American Society of Mechanical Engineers, New York.
9. *Mobile Crane Manual* (1993). Construction Safety Association of Ont., Toronto, Canada.
10. *Rigging Manual* (1992). Construction Safety Association of Ont., Toronto, Canada.
11. Shapira, Aviad, and Jay D. Glascock (1996). "Culture of Using Mobile Cranes for Building Construction," *Journal of Construction Engineering and Management*, ASCE, Vol. 122, No. 4, pp. 298–307.
12. Shapira, Aviad, and Clifford J. Schexnayder (1999). "Selection of Mobile Cranes for Building Construction Projects," *Construction Management and Economics* (United Kingdom), Vol. 17, No. 4, pp. 519–527.
13. Shapiro, Howard I., Jay P. Shapiro, and Lawrence K. Shapiro (2000). *Cranes and Derricks*, 3rd ed., McGraw-Hill Book Company, New York.

WEBSITE RESOURCES

1. www.terex-cranes.com American Crane Corporation located in Wilmington, NC is a subsidiary of Terex Lifting.
2. www.connectingcranes.com ConnectingCranes is a portal for comprehensive online services for the lifting industry. It is also the home of *Cranes Today* magazine (Wilmington Publishing).
3. www.csao.org The Construction Safety Association of Ontario, Canada, assists the industry in developing and implementing standards, procedures, and regulations for health and safety.
4. www.groveworldwide.com/default.htm Grove Worldwide, Shady Grove, PA., is part of the Manitowoc Crane Group.
5. www.liebherr.com Liebherr-America, Inc., Newport News, VA. is a Germany-based global manufacturer of mobile and tower cranes.
6. www.liftandtransport.com The website for the magazine *Lifting & Transportation International* (LTI), which covers the lifting and transport of heavy objects.

7. www.linkbelt.com Link-Belt Construction Equipment Company, Lexington, KY. Link-Belt designs and manufactures telescopic and lattice-boom cranes.
8. www.manitowoccranegroup.com/Home/EN/Home.asp The Manitowoc Crane Group, Manitowoc, WI, includes Manitowoc lattice-boom crawler cranes, Potain tower cranes, Grove mobile hydraulic cranes, and National Crane articulating and telescoping cranes that are built, sold, and serviced at multiple locations on five continents.
9. www.potain.com/index.cfm Potain S.A.S., Ecully Cedex, France, is part of the Manitowoc Crane Group. Potain is a global manufacturer of tower cranes based in France.
10. www.cimanet.com Power Crane and Shovel Association (PCSA), Milwaukee, WI. The PCSA is a product-specific group of the Association of Equipment Manufacturers (AEM). The PCSA promotes the standardization and simplification of terminology and classification of cranes for worldwide harmonization.
11. heavyduty.sae.org Society of Automotive Engineers (SAE), Heavy Duty, SAE World Headquarters, Warrendale, PA. The SAE is a resource for technology publications, events, and standards on self-propelled vehicles.
12. www.scranet.org Specialized Carriers and Rigging Association (SCRA), Fairfax, VA. The SCRA provides information about safe transporting, lifting, and erecting oversized and overweight items.

CHAPTER 18

Draglines and Clamshells

A dragline excavator is especially useful when there is need for extended reach in excavating or when material must be excavated from underwater. Clamshell excavators provide the means to excavate vertically to considerable depths. The greatest advantage of a dragline over other machines is its long reach for digging and dumping. The dragline is designed to excavate below the level of the machine. The clamshell bucket is designed to excavate material in a vertical direction. It works like an inverted jaw with a biting motion. The clamshell is capable of working at, above, and below ground level.

INTRODUCTION

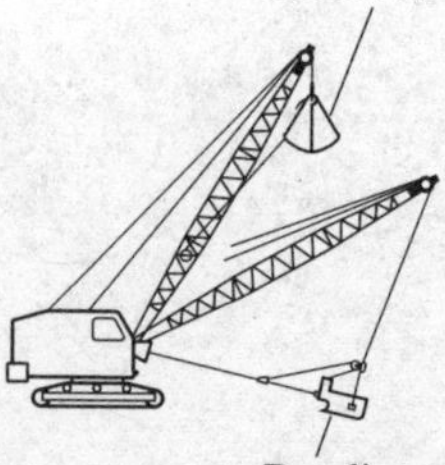

Drag buckets and clamshells are both attachments hung from a lattice-boom crane. The terms dragline and clamshell refer to the particular type of bucket used and to the digging motion of the bucket. A dragline works, as the name implies, by dragging a dragline-type bucket toward the machine. The clamshell bucket is designed to excavate material in a vertical direction. It works like an inverted jaw with a biting motion. With both types of excavators, the buckets are attached to the crane only by cables. Therefore, the operator has no positive control of the bucket, as with hydraulic excavators.

Draglines and clamshells machines belong to a group that is frequently identified as the Power Crane and Shovel Association (PCSA) family [1]. This association has conducted and supervised studies and tests that have provided considerable information related to the performance, operating conditions, production rates, economic life, and cost of owning and operating these machines. The association has participated in establishing and adopting certain standards that are applicable to these machines. The results of the studies, conclusions and actions, and the standards have been published in technical bulletins and booklets. Some of that information published by the Power Crane and Shovel Association is reproduced in this book, with the permission of the association.

DRAGLINES

GENERAL INFORMATION

The dragline is a versatile machine capable of a wide range of operations. It can handle materials that range from soft to medium hard. The greatest advantage of a dragline over other machines is its long reach for digging and dumping. A dragline does not have the positive digging force of a hydraulic shovel or hoe. The bucket breakout force is derived strictly from its own weight. Therefore, it can bounce, tip over, or drift sideward when it encounters hard material. These weaknesses are particularly noticeable with smaller machines and lightweight buckets.

Loading haul units

Draglines are used to excavate material and load it into hauling units, such as trucks or tractor-pulled wagons, or to deposit it in levees, dams, and spoil piles near the pits from which it is excavated. The dragline is designed to excavate below the level of the machine. A dragline usually does not have to go into the excavation or pit to remove material. It works—excavates material—while positioned adjacent to the pit. By the process of casting its bucket into and dragging the bucket out again, it extracts material from the excavation. This is very advantageous when earth is removed from a ditch, canal, or pit containing water.

Building levees

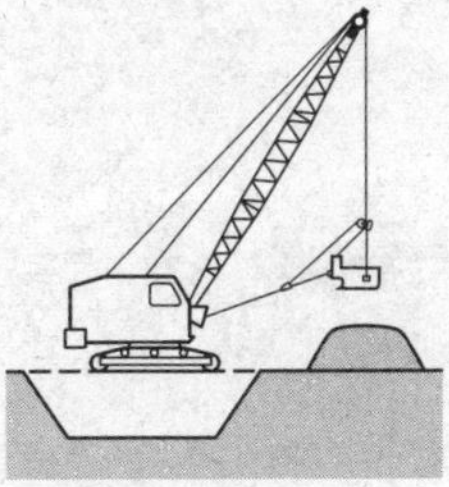

Using a dragline enables the loading of haul units that are positioned at the same level outside of the pit. This is a particular advantage when the material being excavated is wet, as the haul units do not have to go into the excavation and maneuver through the mire. When ground conditions permit, it is better, however, to position the haul units in the pit. Positioning the haul units in the excavation below the dragline will reduce hoist time and increase production.

Frequently it is possible to use a dragline with a long boom to dispose of the earth in one operation if the material can be deposited along the canal or near the pit. This eliminates the need for hauling units, reducing the cost of handling the material.

Crawler-mounted draglines (Fig. 18.1) can operate over soft ground conditions that would not support wheel- or truck-mounted equipment. The travel speed of a crawler machine is very slow, frequently less than 1 mph, and it is necessary to use auxiliary hauling equipment to transport the unit from one job to another. Wheel- and truck-mounted lattice-boom cranes can also be rigged with dragline attachments, but that is not a common practice.

FIGURE 18.1 Crawler-mounted dragline excavating a ditch.

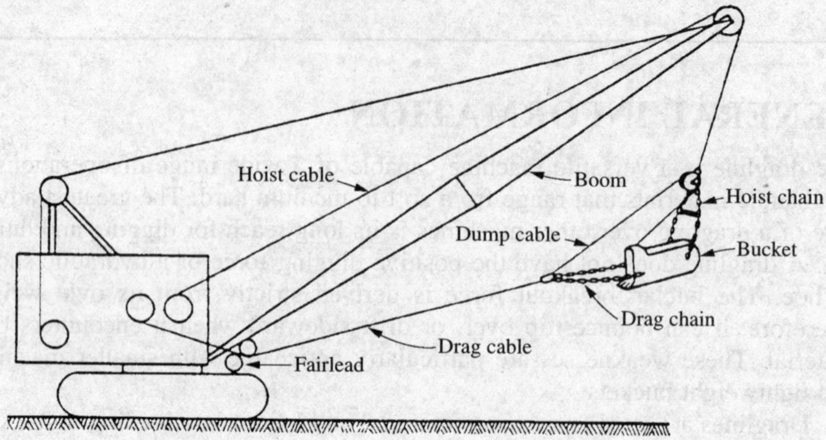

FIGURE 18.2 Basic parts of a dragline.

DESCRIPTION OF A DRAGLINE

Dragline components (Fig. 18.2) consist of a drag bucket and a fairlead assembly. Wire ropes are used for the boom suspension, drag, bucket hoist, and dump lines. The fairlead guides the drag cable onto the drum when the bucket is being drawn in—loaded. The hoist line, which operates over the boom-point sheave, is used to raise and lower the bucket. In the digging operation, the drag cable is used to pull the bucket through the material. When the bucket is raised and moved to the dump point, releasing the tension on the drag cable causes the mouth (open end) of the bucket to drop vertically and gravity then draws the material out of the bucket.

Size of a Dragline

The size of a dragline is indicated by the size of the bucket, expressed in cubic yards. However, most cranes may handle more than one size bucket, depending on the length of the boom utilized and the unit weight of the material excavated. The relationship between bucket size and boom length and angle is presented in Fig. 18.3. The crane boom can be angled relatively low when operating; however, boom angles of less than 35° from the horizontal are seldom advisable because of the possibility of tipping the machine. Because the maximum lifting capacity of a dragline is limited by the force that will tilt the machine over, it is necessary to reduce the size of the bucket when a long boom is used or when the excavated material has a high unit weight. When excavating wet, sticky material and casting onto a spoil bank, the chance of tipping the machine increases because of material sticking in the bucket. In practice, the combined weight of the bucket and its load should produce a tilting force not greater than 75% of the force required to tilt the machine. A longer boom, with a smaller bucket, will be used to increase the digging reach or the dumping radius when it is not desirable to bring in a larger machine.

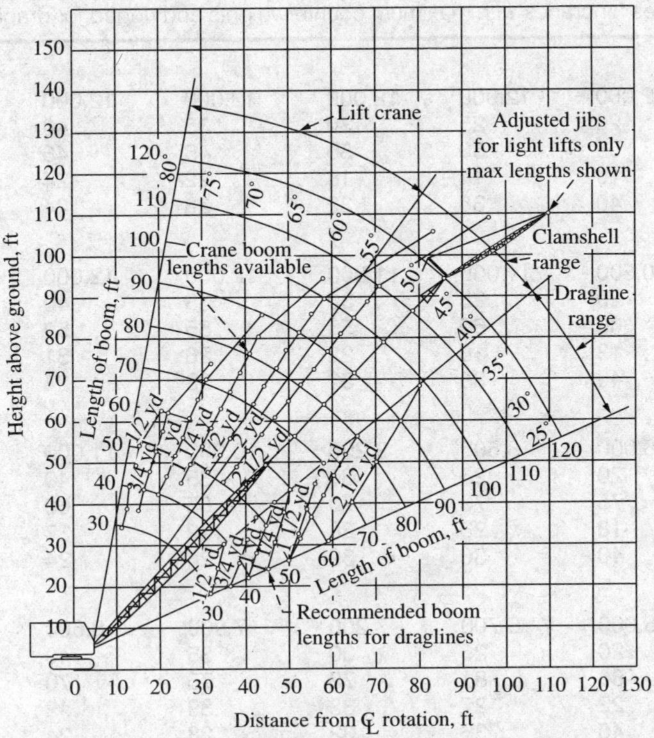

FIGURE 18.3 Working ranges of a lattice-boom machine used with different attachments.

If the material is difficult to excavate, the use of a smaller bucket that will reduce the digging resistance may enable an increase in production.

Typical working ranges for a dragline that will handle buckets varying in size from $1\frac{1}{4}$ to $2\frac{1}{2}$ cy are given in Table 18.1 (see Fig. 18.4 for the dimensions given in the table).

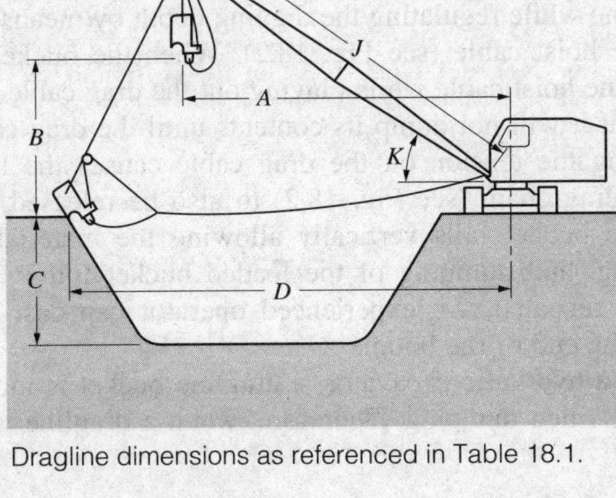

FIGURE 18.4 Dragline dimensions as referenced in Table 18.1.

TABLE 18.1 Typical working ranges for cranes with maximum counterweights and rigged for dragline work.

J, boom length 50 ft						
Capacity (lb)*	12,000	12,000	12,000	12,000	12,000	12,000
K, boom angle (degrees)	20	25	30	35	40	45
A, dumping radius (ft)	55	50	50	45	45	40
B, dumping height (ft)	10	14	18	22	24	27
C, max. digging depth (ft)	40	36	32	28	24	20
J, boom length 60 ft						
Capacity (lb)*	10,500	11,000	11,800	12,000	12,000	12,000
K, boom angle (degrees)	20	25	30	35	40	45
A, dumping radius (ft)	65	60	55	55	52	50
B, dumping height (ft)	13	18	22	26	31	35
C, max. digging depth (ft)	40	36	32	28	24	20
J, boom length 70 ft						
Capacity (lb)*	8,000	8,500	9,200	10,000	11,000	11,800
K, boom angle (degrees)	20	25	30	35	40	45
A, dumping radius (ft)	75	73	70	65	60	55
B, dumping height (ft)	18	23	28	32	37	42
C, max. digging depth (ft)	40	36	32	28	24	20
J, boom length 80 ft						
Capacity (lb)*	6,000	6,700	7,200	7,900	8,600	9,800
K, boom angle (degrees)	20	25	30	35	40	45
A, dumping radius (ft)	86	81	79	75	70	65
B, dumping height (ft)	22	27	33	39	42	47
C, max. digging depth (ft)	40	36	32	28	24	20
D, digging reach	Depends on working conditions and operator's skill with bucket					

*Combined weight of bucket and material must not exceed capacity.

Operation of a Dragline

The excavating cycle begins when the operator swings the empty bucket to the digging position while at the same time releasing the tension on the drag and hoist lines. There are separate drums on the crane unit for the drag and hoist line cables (see Fig. 18.2). Therefore, a good operator can coordinate their motion into a smooth operation. Digging is accomplished by pulling the bucket toward the machine while regulating the digging depth by means of the tension maintained in the hoist cable (see Fig. 18.2). When the bucket is filled, the operator pulls in the hoist cable while playing out the drag cable. The bucket is so constructed that it will not dump its contents until the drag cable tension is released. Releasing the tension on the drag cable causes the tension on the dump cable and drag chain (see Fig. 18.2) to also be released, and the front (open end) of the bucket falls vertically allowing the material to slide out. Hoisting, swinging, and dumping of the loaded bucket follow in that order; then the cycle is repeated. An experienced operator can cast the excavated material beyond the end of the boom.

Compared to a hydraulic excavator, a dragline bucket is more difficult to accurately control when dumping. Therefore, when a dragline is used to load

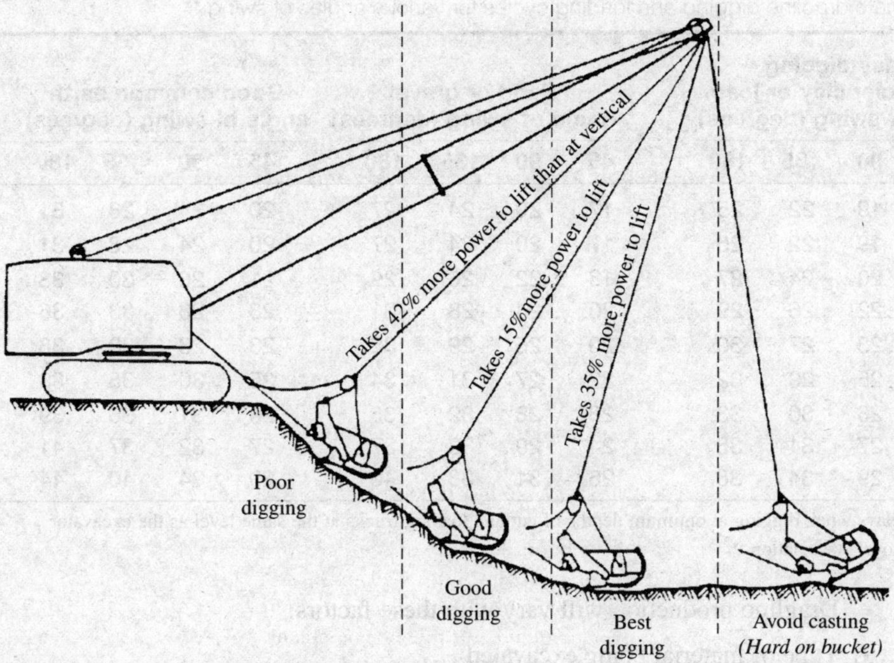

FIGURE 18.5 Dragline digging zones.

haul units, larger units should be used to provide the dragline operator a larger dump target. This will reduce the spillage. A size ratio equal to at least five to six times the capacity of the dragline bucket is recommended.

Figure 18.5 shows the dragline digging zones in relation to excavation production efficiency. Dragline work should be planned to permit most of the digging to be done in the zones that permit the best digging, with the poor digging zone used as little as possible.

DRAGLINE PRODUCTION

The output of a dragline should be expressed in bank measure cubic yards (bcy) per hour. This quantity is best obtained from field measurements. It can be estimated by multiplying the average loose volume per bucket load by the number of cycles (bucket loads) per hour and dividing by 1, plus the swell factor for the material, expressed as a fraction. For example, if a 2-cy bucket, excavating material whose swell is 25%, will handle an average loose volume of 2.4 cy, the bank-measure volume will be 1.92 cy, $\frac{2.4}{1.25}$. If the dragline can make 2 cycles/min, the output will be 3.84 bcy/min (2×1.92) or 230 bcy/hr. This is an ideal 60-min hour peak output that will not be sustainable over the duration of a project.

TABLE 18.2 Approximate dragline digging and loading cycles for various angles of swing.*

Size of dragline bucket	Easy digging light moist clay or loam angle of swing (degrees)				Sand or gravel angle of swing (degrees)				Good common earth angle of swing (degrees)			
(cy)	45	90	135	180	45	90	135	180	45	90	135	180
$\frac{3}{8}$	16	19	22	25	17	20	24	27	20	24	28	31
$\frac{1}{2}$	16	19	22	25	17	20	24	27	20	24	28	31
$\frac{3}{4}$	17	20	24	27	18	22	26	29	21	26	30	33
1	19	22	26	29	20	24	28	31	23	28	33	36
$1\frac{1}{4}$	19	23	27	30	20	25	29	32	23	28	33	36
$1\frac{1}{2}$	21	25	29	32	22	27	31	34	25	30	35	38
$1\frac{3}{4}$	22	26	30	33	23	28	32	35	26	31	36	39
2	23	27	31	35	24	29	33	37	27	32	37	41
$2\frac{1}{2}$	25	29	34	38	26	31	36	40	29	34	40	44

*Time in seconds with no delays when digging at optimum depths of cut and loading trucks at the same level as the excavator.
Source: Power Crane and Shovel Association.

Dragline production will vary with these factors:

1. Type of material being excavated
2. Depth of cut (depth below the base of the dragline's tracks)
3. Angle of swing (the angle created by a set of lines running for the center point of the dragline to the point of excavation and to the point of dump)
4. Size and type of bucket
5. Length of boom
6. Method of disposal, casting or loading haul units
7. Size of the hauling units, when used
8. Skill of the operator
9. Physical condition of the machine
10. Job conditions

Table 18.2 gives approximate dragline digging and loading cycles for various angles of swing.

Optimum Depth of Cut

A dragline will produce its greatest output if the job is planned to permit excavation at the optimum depth of cut. Based on using short-boom draglines, Table 18.3 gives the optimum depth of cut for various sizes of buckets and classes of materials. Ideal outputs of short-boom draglines, expressed in bcy, for various classes of materials, when digging at the optimum depth, with a 90° swing, and no delays are presented in the table. The upper two numbers are the optimum depth of cut in feet and in meters (the value in parentheses); the lower numbers are the ideal output in cubic yards and in cubic meters (the value in parentheses).

TABLE 18.3 Optimum depth of cut and ideal production of short-boom draglines.*

	Size of bucket [cy (cu m)]†								
Class of material	$\frac{3}{8}$ (0.29)†	$\frac{1}{2}$ (0.38)†	$\frac{3}{4}$ (0.57)†	1 (0.76)†	$1\frac{1}{4}$ (0.95)†	$1\frac{1}{2}$ (1.14)†	$1\frac{3}{4}$ (1.33)†	2 (1.53)†	$2\frac{1}{2}$ (1.91)†
Moist loam or light sandy clay	5.0	5.5	6.0	6.6	7.0	7.4	7.7	8.0	8.5
	(1.5)‡	(1.7)‡	(1.8)‡	(2.0)‡	(2.1)‡	(2.2)‡	(2.4)‡	(2.5)‡	(2.6)‡
	70	95	130	160	195	220	245	265	305
	(53)§	(72)§	(99)§	(122)§	(149)§	(168)§	(187)§	(202)§	(233)§
Sand and gravel	5.0	5.5	6.0	6.6	7.0	7.4	7.7	8.0	8.5
	(1.5)	(1.7)	(1.8)	(2.0)	(2.1)	(2.2)	(2.4)	(2.5)	(2.6)
	65	90	125	155	185	210	235	255	295
	(49)	(69)	(95)	(118)	(141)	(160)	(180)	(195)	(225)
Good common earth	6.0	6.7	7.4	8.0	8.5	9.0	9.5	9.9	10.5
	(1.8)	(2.0)	(2.4)	(2.5)	(2.6)	(2.7)	(2.8)	(3.0)	(3.2)
	55	75	105	135	165	190	210	230	265
	(42)	(57)	(81)	(104)	(127)	(147)	(162)	(177)	(204)
Hard, tough clay	7.3	8.0	8.7	9.3	10.0	10.7	11.3	11.8	12.3
	(2.2)	(2.5)	(2.7)	(2.8)	(3.1)	(3.3)	(3.5)	(3.6)	(3.8)
	35	55	90	110	135	160	180	195	230
	(27)	(42)	(69)	(85)	(104)	(123)	(139)	(150)	(177)
Wet, sticky clay	7.3	8.0	8.7	9.3	10.0	10.7	11.3	11.8	12.3
	(2.2)	(2.5)	(2.7)	(2.8)	(3.1)	(3.3)	(3.5)	(3.6)	(3.8)
	20	30	55	75	95	110	130	145	175
	(15)	(23)	(42)	(58)	(73)	(85)	(100)	(112)	(135)

*In cubic yards (cubic meters) bank measure (bcy) per 60-min hour.

†These values are the sizes of the buckets in cubic meters (cu m).

‡These values are the optimum depths of cut in meters (m).

§These values are the optimum ideal outputs in cubic meters (cu m).

Effect of Depth of Cut and Swing Angle on Production

Table 18.3 presents ideal dragline production capability based on digging at optimum depths with a swing angle of 90°. For any other depth or swing angle, the ideal output of the machine must be adjusted by an appropriate depth-swing factor. The effect of the depth of cut and swing angle on dragline production is given in Table 18.4. In Table 18.4, the percentage of optimum depth of cut is obtained by dividing the actual depth of cut by the optimum depth for the given material and bucket. The result of that division must be multiplied by 100 to express the value as a percentage:

$$\text{Percentage of optimum depth of cut} = \frac{\text{Actual depth of cut}}{\text{Optimum depth of cut (Table 18.3)}} \times 100 \qquad \textbf{[18.1]}$$

TABLE 18.4 Effect of the depth of cut and swing angle on dragline production.

Percentage of optimum depth	Angle of swing (degrees)							
	30	45	60	75	90	120	150	180
20	1.06	0.99	0.94	0.90	0.87	0.81	0.75	0.70
40	1.17	1.08	1.02	0.97	0.93	0.85	0.78	0.72
60	1.24	1.13	1.06	1.01	0.97	0.88	0.80	0.74
80	1.29	1.17	1.09	1.04	0.99	0.90	0.82	0.76
100	1.32	1.19	1.11	1.05	1.00	0.91	0.83	0.77
120	1.29	1.17	1.09	1.03	0.98	0.90	0.82	0.76
140	1.25	1.14	1.06	1.00	0.96	0.88	0.81	0.75
160	1.20	1.10	1.02	0.97	0.93	0.85	0.79	0.73
180	1.15	1.05	0.98	0.94	0.90	0.82	0.76	0.71
200	1.10	1.00	0.94	0.90	0.87	0.79	0.73	0.69

CALCULATING DRAGLINE PRODUCTION

Hourly production rates for lattice-boom cranes with dragline attachments are given in Table 18.3. These rates are based on operations at optimum cutting depth, a 90° swing angle, and for specific soil types. The table also assumes maximum efficiency, for example, a 60-min hour. Table 18.4 gives correction factors for different depths of excavation and swing angles. Refer to Table 4.3 for soil conversion factors. Overall efficiency should be based on expected job conditions.

Step 1. Determine an ideal production from Table 18.3, based on proposed bucket size and the type of material.

Step 2. Determine the percent of optimum depth, based on appropriate Table 18.3 data input into Eq. [18.1].

Step 3. Determine the depth of cut/swing angle correction factor from Table 18.4, using the calculated percent of optimum depth of cut and the planned angle of swing. In some cases, it may be necessary to interpolate between Table 18.4 values.

Step 4. Determine an overall efficiency factor based on the expected job conditions. Draglines are seldom productively working for more than 45 min in 1 hr.

Step 5. Determine the estimated production rate. Multiply the ideal production by the depth/swing correction factor and the efficiency factor.

Step 6. Determine soil conversion, if needed (Table 4.3).

Step 7. Determine total hours to complete the work:

$$\text{Total hours} = \frac{\text{Cubic yards moved}}{\text{Production rate/hr}} \qquad \textbf{[18.2]}$$

EXAMPLE 18.1

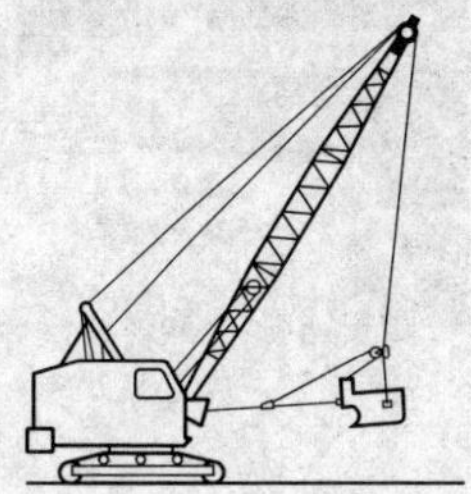

A 2-cy short-boom dragline is to be used to excavate hard, tough clay. The depth of cut will be 15.4 ft, and the swing angle will be 120°. Determine the probable production of the dragline. There are 35,000 bcy of material to be excavated. How long will the project require?

Step 1. Determine the ideal production from Table 18.3, based on a 2-cy bucket size and hard, tough clay material: 195 bcy.

Step 2. Determine the percent of optimum depth of cut, Eq. [18.1]. Optimum depth of cut (Table 18.3): 11.8 ft.

$$\text{Percentage of optimum depth of cut} = \frac{15.4\text{ ft}}{11.8\text{ ft}} \times 100 = 130\%$$

Step 3. Determine the depth of cut/swing angle correction factor from Table 18.4:

Percentage of optimum depth of cut = 130%

Swing angle = 120°

Depth of cut/swing angle correction factor = 0.89

Step 4. Determine an overall efficiency factor based on the expected job conditions. Draglines seldom work at better than a 45-min hour:

$$\text{Efficiency factor} = \frac{45\text{ min}}{60\text{ min}} = 0.75$$

Step 5. Determine production rate. Multiply the ideal production by the depth/swing correction factor and the efficiency factor:

$$\text{Production} = 195 \times 0.89 \times 0.75 = 130\text{ bcy/hr}$$

Step 6. Determine soil conversion, if needed (Table 4.3). Not necessary in this example.

Step 7. Determine total hours, Eq. [18.2]:

$$\text{Total hours} = \frac{35{,}000\text{ bcy}}{130\text{ bcy/hr}} = 269\text{ hr}$$

FACTORS AFFECTING DRAGLINE PRODUCTION

In selecting the size and type bucket to use on a project, one should match the size of the lattice-boom crane and bucket properly to obtain the best action and the greatest operating efficiency. A dragline bucket (see Fig. 18.6) consists of

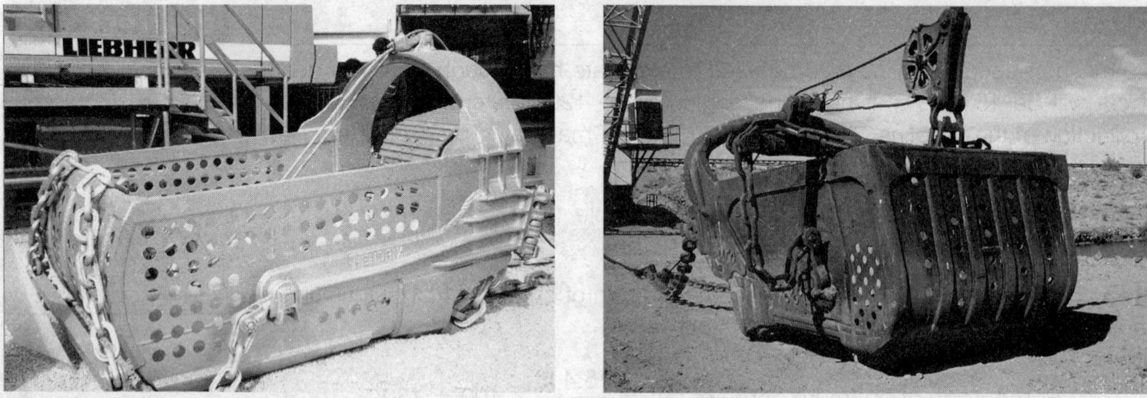

FIGURE 18.6 Dragline buckets.

three parts: the basket, the arch, and the cutting edge. It can be completed with teeth and shrouds (extra metal pieces between the teeth). Buckets are generally available in three types: (1) light duty, (2) medium duty, and (3) heavy duty.

Light-duty buckets, sometimes the term *lightweight* is used, are specifically manufactured for excavating materials that are easily dug, such as sandy loam, sandy clay, or sand. In the appropriate application, they provide greater capacity without sacrificing durability. Medium-duty buckets are used for general excavating service. Sometimes the term *general purpose* is used in reference to medium-duty buckets. Typical applications include excavating clay, soft shale, or loose gravel. Heavy-duty buckets, which have comparatively heavier construction throughout, and are armored for maximum strength and resistance to abrasion, are used for mine stripping, handling blasted rock, and excavating hardpan and highly abrasive materials. Buckets are sometimes perforated to permit excess water to drain from the loads.

Table 18.5 gives representative capacities, weight, and dimensions for different types of dragline buckets.

The normal size of a dragline bucket is based on its struck capacity that is expressed in cubic feet (cf). In selecting the most suitable size bucket for use with a given dragline, it is advantageous to know the loose weight of the material that will be handled. This weight should be expressed in pounds per cubic foot (lb/cf). In the interest of increasing production, it is desirable to use the largest bucket possible. To determine the largest applicable bucket, a careful analysis must be performed to determine that the combined weight of the load and the bucket does not exceed the capacity of the crane. The importance of this analysis is illustrated by referring to the information given in Table 18.1.

EXAMPLE 18.2

Assume that the material to be handled has a loose weight of 90 lb/cf. The use of a 2-cy medium-duty bucket will be considered. If a crane with 80 ft of boom at a 40° angle is to be rigged as a dragline, the maximum safe load is 8,600 lb (Table 18.1).

The approximate weight of the bucket and its load will be

Bucket weight, from Table 18.5	=	4,825 lb
Earth, 60 cf at 90 lb/cf	=	5,400 lb
Combined weight	=	10,225 lb
Maximum safe load	=	8,600 lb

As this weight will exceed the safe load of the dragline, it will be necessary to use a smaller bucket. Try a $1\frac{1}{2}$-cy bucket. The combined weight of the bucket and load will be

Bucket, from Table 18.5	=	3,750 lb
Earth, 47 cu ft at 90 lb/cf	=	4,230 lb
Combined weight	=	7,980 lb
Maximum safe load	=	8,600 lb

If a $1\frac{1}{2}$-cy bucket is used, it may be filled to heaping capacity, without exceeding the safe load limit of the crane.

If a 70-ft boom at a 40° angle, whose maximum safe load is 11,000 lb, will provide sufficient working range for excavating and disposing of the earth, a 2-cy bucket can be used and filled to heaping capacity. The ratio of the output resulting from the use of a 70-ft boom and a 2-cy bucket, compared with a $1\frac{1}{2}$-cy bucket, should be approximately

Output ratio (60 cf/47 cf) × 100 = 127%

Increase in production = 27%

This does not consider the cycle time effect of the different boom lengths.

TABLE 18.5 Representative capacities, weights, and dimensions of dragline buckets.

Size (cy)	Struck Capacity (cf)	Weight of bucket (lb): Light duty	Weight of bucket (lb): Medium duty	Weight of bucket (lb): Heavy duty	Dimensions (in.): Length	Dimensions (in.): Width	Dimensions (in.): Height
$\frac{3}{8}$	11	760	880		35	28	20
$\frac{1}{2}$	17	1,275	1,460	2,100	40	36	23
$\frac{3}{4}$	24	1,640	1,850	2,875	45	41	25
1	32	2,220	2,945	3,700	48	45	27
$1\frac{1}{4}$	39	2,410	3,300	4,260	49	45	31
$1\frac{1}{2}$	47	3,010	3,750	4,525	53	48	32
$1\frac{3}{4}$	53	3,375	4,030	4,800	54	48	36
2	60	3,925	4,825	5,400	54	51	38
$2\frac{1}{4}$	67	4,100	5,350	6,250	56	53	39
$2\frac{1}{2}$	74	4,310	5,675	6,540	61	53	40
$2\frac{3}{4}$	82	4,950	6,225	7,390	63	55	41
3	90	5,560	6,660	7,920	65	55	43

Example 18.2 illustrates the importance of analyzing a job prior to selecting the size crane and bucket to be employed. Random selection of equipment can result in a substantial increase in material handling cost. Efficient handing of materials is a key issue in heavy construction.

EXAMPLE 18.3

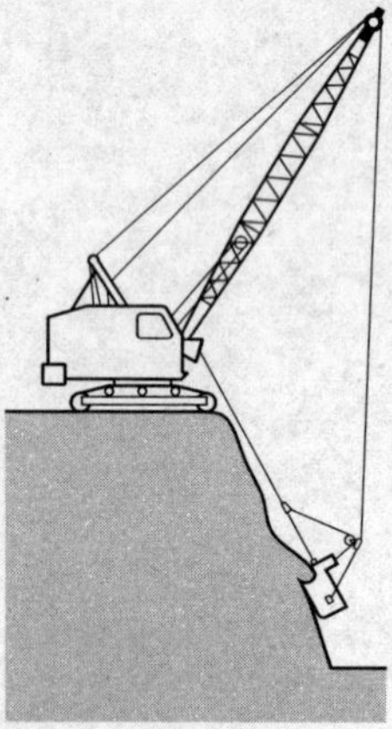

This example illustrates a method of analyzing a project to determine the size of crane rigged as a dragline required for digging a canal. Select a crawler-mounted dragline to excavate 210,000 bcy of common earth having a loose weight of 80 lb/cf. The dimensions of the canal will be

Bottom width	20 ft
Top width	44 ft
Depth	12 ft
Side slopes	1:1

The excavated earth will be cast into a levee along one side of the canal. The levee will have a berm of at least 20 ft between the toe of the levee and the nearest edge of the canal.

The cross-sectional area of the canal will be

$$\frac{(20 + 44)}{2} \times 12 = 384 \text{ sf per running ft}$$

If the earth swells 25% when it is excavated, the cross-sectional area of the levee will be

$$384 \text{ sf} \times 1.25 = 480 \text{ sf per running ft}$$

The levee dimensions will be

Height	12 ft
Base width	64 ft
Crest width	16 ft
Side slope	2:1

The total width from the outside of the levee to the far side of the canal will be

Width of levee =	64 ft
Width of berm =	20 ft
Width of canal =	44 ft
Total =	128 ft

or the machine will have to reach 64 ft, assuming it will turn 180° during the digging and dumping operation.

With a boom angle of 30°, a crane having 70 ft of boom will be required (Table 18.1) to provide the necessary digging and dumping reaches, and to permit adequate dumping height and digging depth.

The project must be completed in 1 yr. Assume that weather conditions, holidays, and other major losses in time will reduce the operating time to 44 weeks of 40 hr each, or a total of 1,760 working hours. The required production per working hour will be 119 bcy. It should be possible to operate with a 150° average swing angle. The efficiency factor should be a 45-min hour.

The required production divided by the efficiency factor is

$$\frac{133}{45/60} = 177 \text{ bcy/hr}$$

Assume the depth-swing factor will be 0.81. The required ideal production is

$$\frac{177}{0.81} = 219 \text{ bcy/hr}$$

Table 18.3 indicates a $1\frac{3}{4}$-cy medium-duty bucket will meet the need. The combined weight of the bucket and load will be

Weight of load, 53 cf at 80 lb per cf	= 4,240 lb
Weight of bucket	= 4,030 lb
Total weight	= 8,270 lb
Maximum safe load, Table 18.1	= 9,200 lb

The equipment selected should be checked to verify whether it will produce the required production:

Ideal output, 210 bcy/hr

Percent of optimum depth, $\frac{12.0}{9.5} \times 100 = 126$

Depth-swing factor, 0.82

Efficiency factor, 45-min hour, or 0.75

Probable output, 210 bcy/hr × 0.82 × 0.83 = 129 bcy/hr

Thus, the equipment should produce the required output of 113 cy, with a slight surplus capacity.

CLAMSHELL EXCAVATORS

GENERAL INFORMATION

The clamshell is a vertically operated bucket capable of working at, above, and below ground level. The clamshell bucket, as the name implies, consists of two scoops hinged together that work like the shell of a clam. Clamshells are used primarily to handle materials such as loose to medium stiff soils, sand, gravel, crushed stone, coal, and shells, and to remove materials from vertical excavations such as cofferdams, pier foundations, sewer manholes, and sheet-lined trenches. In past years, these were hung from a lattice-boom crawler crane (Fig. 18.7); today, there are hydraulic clamshell buckets that are mounted on the stick of hydraulic hoes (see Fig. 9.12d and 18.8). The hydraulic excavator clamshells have limited vertical reach. Therefore, when a deep vertical excavation must be made, a crane with a cable-attached clamshell bucket is still used.

Lattice-Boom Clamshells

The same wire ropes used for the crane (hook) operation can, usually, be used for clamshell operations. However, two additional lines must be added: (1) a secondary hoist line—the closing line—and (2) the tag line. The tag line is a

FIGURE 18.7 Barge-mounted crane using a clamshell bucket.
Source: Kokosing Fru-Con.

FIGURE 18.8 Hydraulic clamshell bucket mounted on the stick of a hydraulic hoe.
Source: Kokosing Fru-Con.

small-diameter cable with a spring tension winder. The tag line prevents the clamshell bucket from twisting while it is being hoisted and lowered. Being spring-loaded, the tag line does not require operator control and does not attach to the crane's operating drums. The winder is usually mounted on the lower part of the boom.

The clamshell bucket is attached to the hoist line of the crane and to a closing line. The closing line is attached to a second cable drum on the crane. The length of the boom determines the height a clamshell can reach. The length of wire rope the crane's cable drums can accommodate limits the depth a clamshell can reach. A clamshell's lifting capacity varies with the crane's length of boom, the operating radius, the size of the clam bucket, and the unit weight of material excavated.

The holding, closing, and tag lines control the bucket. At the start of the digging cycle, the bucket rests on the material to be dug, with its shells open. As the closing line is reeved in, the two shells of the bucket are drawn together, causing them to dig into the material. The weight of the bucket is the only crowding action available for penetrating the material. After the bucket is closed, the holding and closing lines raise the bucket and swing it to the dumping point. The bucket is opened to dump by releasing the tension on the closing line.

CLAMSHELL BUCKETS

The two jaws of a clamshell bucket clamp together by their own weight when the bucket is lifted by the closing line. Clamshell buckets are available in various sizes, and in heavy-duty types for digging, medium-weight-types for general-

(a) (b)

FIGURE 18.9 (a) Wide rehandling clamshell bucket. (b) Heavy-duty clamshell bucket.

purpose work, and lightweight types for rehandling light materials. Manufacturers supply buckets either with removable teeth or without teeth. Teeth are used in digging the harder types of materials but are not required when a bucket is used for rehandling purposes. Figure 18.9 illustrates a rehandling and a heavy-duty digging bucket.

The capacity of a clamshell bucket is usually given in cubic yards. A more accurate capacity is given as water-level, plate-line, or heaped-measure; such volumes are generally expressed in cubic feet. The water-level capacity is the capacity of the bucket if it were hung level and filled with water. The plate-line capacity indicates the capacity of the bucket following a line along the tops of the clams. The heaped capacity is the capacity of the bucket when it is filled to the maximum angle of repose for the given material. In specifying the heaped capacity, the angle of repose usually is assumed to be 45°. The term "deck area" indicates the number of square feet covered by the bucket when it is fully open.

PRODUCTION RATES FOR CLAMSHELLS

Because of the variable factors that affect the operations of a clamshell, it is difficult to give average production rates. The critical variable factors include the difficulty of loading the bucket, the size load obtainable, the height of lift, the angle of swing, the method of disposing of the load, and the experience of the operator. For example, if the material must be discharged into a hopper, the time required to spot the bucket over the hopper and discharge the load will be greater than when the material is discharged onto a large spoil bank.

TABLE 18.6 Representative performance specifications for clamshell rigged crane.

Speeds	
Travel	0.9 mph maximum
Swing	3 rpm maximum
Rated single-line speed	
Lift clamshell	166 fpm
Dragline	157 fpm
Magnet	200 fpm
Third drum (standard travel)	185 fpm
Third drum (independent travel)	127 fpm
Rated line pulls (with standard engine)	
Lift clamshell	29,600 lb SLP
Dragline	31,400 lb SLP
Magnet	24,800 lb SLP
Third drum	25,500 lb SLP

Performance figures are based on machine equipped with standard engine.
SLP stands for Single Line Pull or 1-part line.

Representative performance specifications for a clamshell rigged crane are given in Table 18.6. Example 18.4 illustrates a scenario for estimating the probable output of a clamshell under a particular set of conditions.

EXAMPLE 18.4

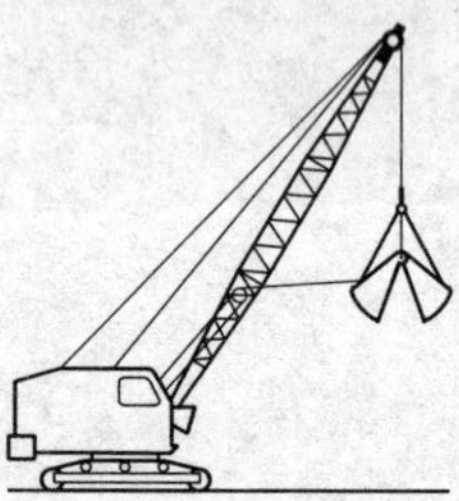

A $1\frac{1}{2}$-cy rehandling-type bucket, whose empty weight is 4,300 lb, will be used to transfer sand from a stockpile into a hopper 25 ft above the ground. The crane's angle of swing will average 90°. The average loose capacity of the bucket is 48 cf. The specifications for the crane unit given in Table 18.6 are applicable to this situation.

Time per cycle (approx.):

Loading bucket		= 6 sec
Lifting and swing load	$\dfrac{25\text{ ft} \times 60\text{ sec/min}}{166\text{ ft/min}}$	= 9 sec*
Dump load		= 6 sec
Swing back to stock pile		= 4 sec
Lost time, accelerating, etc.		= 4 sec
Total cycle time		= 29 sec or 0.48 min

$$\text{Maximum number of cycles per hour } \frac{60\text{ min}}{0.48\text{ min cycle}} = 125$$

$$\text{Maximum volume per hr } (125\text{ cycle} \times 48\text{ cf})/27 = 222\text{ lcy}$$

If the unit operates at a 45-min hour efficiency, the probable production will be

$$222 \times \frac{45}{60} = 167\text{ lcy/hr}$$

If the same equipment is used with a general-purpose bucket to dredge muck and sand from a sheet-piling cofferdam partly filled with water, requiring a total vertical lift of 40 ft, and the muck must be discharged into a barge, the production rate previously determined will not apply. It will be necessary to lift the bucket above the top of the cofferdam prior to starting the swing, which will increase the time cycle. Because of the nature of the material, the load will probably be limited to the water-filled capacity of the bucket, which is 33 cf. The time per cycle should be about

Loading bucket	= 8.0 sec
Lifting $\dfrac{40 \text{ ft} \times 60 \text{ sec/min}}{166 \text{ ft/min}}$	= 14.5 sec
Swinging 90° at 3 rpm $\dfrac{0.25 \text{ rev.} \times 60 \text{ sec/min}}{3 \text{ rev./min}}$	= 5.0 sec
Dump load	= 4.0 sec
Swing back	= 4.0 sec
Lowering bucket $\dfrac{40 \text{ ft} \times 60 \text{ sec/min}}{350 \text{ ft/min}}$	= 7.0 sec
Lost time, accelerating etc.	= 10.0 sec
Total cycle time	= 52.5 sec or 0.875 min

Maximum number of cycles per hr, 60 min ÷ 0.9-min cycle = 67

$$\text{Maximum volume per hr } \frac{67 \text{ cycle} \times 33 \text{ cf}}{27} = 82 \text{ lcy}$$

If the unit operates at a 45-min-hour efficiency, the probable production will be

$$82 \text{ lcy} \times \frac{45}{60} = 62 \text{ lcy/hr}$$

*A skilled operator should lift and swing simultaneously. If this is not possible, as when coming out of a cofferdam, additional time should be allowed for swinging the load.

SAFETY

Keep personnel away from the swing area of dragline and clamshell cranes. These machines must be operated so as not to expose persons in the area to a hazard. The crane operator must not swing the boom and bucket, whether loaded or empty, over the top of trucks and drivers. Title 30 Code of Federal Regulations § 77.409 states

> **Shovels, draglines, and tractors.**
> (a) Shovels, draglines, and tractors shall not be operated in the presence of any person exposed to a hazard from its operation and all such equipment shall be provided with an adequate warning device which shall be sounded by the operator prior to starting operation.

If the machine is working close to a wall or other obstruction that is closer to it than 2 ft, the machine should be fenced off so that people cannot walk between the obstruction and the swinging superstructure of the crane. The crane operator has very limited visibility to the rear of the machine, so limiting access to the pinch point area is important to avoiding crushing accidents.

Another common accident, with cranes equipped as draglines or with clamshell buckets, is where a machine rolls or falls into the excavation it is digging. These accidents often happen when ground conditions deteriorate after a rain. Management must always be cognizant of changing ground conditions.

SUMMARY

The dragline is a versatile machine capable of a wide range of operations. It can handle materials that range from soft to medium hard. The greatest advantage of a dragline over other excavators is its long reach for digging and dumping. However, a dragline does not have the positive digging force of a hydraulic shovel or hoe. The bucket breakout force is derived strictly from its own weight.

The clamshell-equipped crane consists of a clamshell bucket hung from a lattice-boom crane. The clamshell bucket consists of two scoops hinged together. The clamshell is a vertically operated attachment capable of working at, above, and below ground level. The clamshell can dig in loose to medium stiff soils. The length of the boom determines the height a clamshell can reach. Critical learning objectives include:

- An understanding of how a dragline operates and when it should be employed on a project
- An ability to calculate a production estimate for a dragline
- An understanding of how a clamshell crane operates and when it should be employed on a project
- An ability to calculate a production estimate for a clamshell

These objectives are the basis for the problems that follow.

PROBLEMS

18.1 A $2\frac{1}{2}$-cy short-boom dragline is to be used to excavate good common earth. The depth of cut will be 12.6 ft, and the swing angle will be 90°. Determine the probable production of the dragline. There are 60,000 bcy of material to be excavated. How long will the project require?

18.2 Determine the probable production in cubic yard bank measure (bcy) for a 2-cy dragline when excavating and casting good common earth. The average depth of the excavation is 12 ft, and the average angle of swing will be 120°. The efficiency factor will be a 45-min hour. (155 bcy/hr)

18.3 Determine the probable production in cubic yard bank measure (bcy) for a 2-cy dragline when excavating and casting tough clay. The average depth of the excavation is 9 ft, and the average angle of swing will be 140°. The efficiency factor will be a 45-min hour.

18.4 Determine the largest capacity medium-duty dragline bucket that can be used with a dragline equipped with an 80-ft boom when the boom is operating at an angle of 45°. The excavated earth weighs 98 lb/cf loose measure. (Maximum size bucket is $1\frac{3}{4}$ cy.)

18.5 Determine the largest capacity heavy-duty dragline bucket that can be used on a dragline equipped with 70 ft of boom when the boom is operating at an angle of 40°. The excavated earth weighs 92 lb/cf loose measure.

18.6 A $2\frac{1}{2}$-cy dragline with a standard short boom is operated at 60% optimum depth of cut and a 90° angle of swing. It is used to excavate light sandy clay. The average bucket load, expressed in bcy, can be taken to be 0.80 of the struck capacity, cubic feet listed, as listed in Table 18.5. On the assumption that the data from Tables 18.3 and 18.4 are applicable, calculate the peak rate cycle time for the dragline in minutes.

REFERENCES

1. *PCSA-4, Mobile Power Crane and Excavator Standards and Hydraulic Crane Standards*, Power Crane and Shovel Association, A Bureau of Construction Industry Manufacturers Association, 111 East Wisconsin Avenue, Milwaukee, WI.
2. *Crane Load Stability Test Code—SAE J765* (1990). SAE Standards, Society of Automotive Engineers, Inc., Warrendale, PA, October.
3. Stewart, Rita F., and Cliff J. Schexnayder (1985). "Production Estimating for Draglines," *Journal of Construction Engineering and Management*, American Society of Civil Engineers, Vol. 111, No. 1, March, pp. 101–104.
4. Nichols, Herbert L., Jr., and David A. Day (1998). *Moving the Earth: The Workbook of Excavation*, 4th ed., McGraw-Hill, New York.
5. O'Beirne, T., J. Rowlands, and M. Phillips (1997). *Project C3002: Investigation into Dragline Bucket Filling*, Australian Coal Research Limited, August.

WEBSITE RESOURCES

1. www.hendrixmfg.com/_Dragline_Bucket_.php Hendrix Manufacturing Company, Inc. manufactures dragline buckets ranging in size from $\frac{1}{4}$ to 65 cy.
2. www.sae.org/jsp/index.jsp Society of Automotive Engineers (SAE). SAE World Headquarters are in Warrendale, PA.
3. www.aem.org/CBC/ProdSpec/PCSA Power Crane and Shovel Association (PCSA). The PCSA is a bureau of the Association of Equipment Manufacturers. PCSA provides services tailored to meet the needs of the lattice-boom and truck crane industry.

CHAPTER 19

Piles and Pile-Driving Equipment

Piles can be classified on the basis of either their use or the materials from which they are made. On the basis of use, there are two major classifications: (1) sheet and (2) load bearing. Sheet piles are used primarily to retain or support earth. Load-bearing piles, as the name implies, are used primarily to transmit structural loads. In general, the forces that enable a pile to support a load also cause the pile to resist the efforts made to drive it. The function of a pile hammer is to furnish the energy required to drive a pile. Pile-driving hammers are designated by type and size.

INTRODUCTION

This chapter deals with the selection of piles and the equipment used to drive piles (see Fig. 19.1). Timber piles, driven by hand, have been around since man first started building shelters near lakes and rivers. References to cedar timber piles in Babylon can be found in the Bible. Steel piles have been used since 1800 and concrete piles since about 1900. The first "modern" pile-driving equipment was developed by the Swedish inventor Christoffer Polhem in 1740. Load-bearing piles, as the name implies, are used primarily to transmit structural loads, through soil formations with inadequate supporting properties and into or onto a soil stratum capable of supporting the loads. If the load is transmitted to the soil through skin friction between the surface of the pile and the soil, the pile is called a *friction pile*. If the load is transmitted to the soil through the pile's lower tip, the pile is called an *end-bearing pile*. Many piles depend on a combination of friction and end bearing for their supporting strengths.

GLOSSARY OF TERMS

The following glossary defines the terminology used in describing piles, pile-driving equipment, and pile-driving methods.

FIGURE 19.1 Crawler crane driving concrete piles and using hydraulic leads.

Anchor pile. A pile connected to a structure by one or more ties to furnish lateral support or to resist uplift.

Butt of a pile. The larger end of a tapered pile; usually the upper end of a driven pile. Also a general term for the upper portion of a pile.

Cushion. Material inserted between the ram of the pile hammer and the driving cap and for concrete piles also between the driving cap and the pile. This material provides a uniform distribution of impact forces. Typical materials used for cushioning include micarta, steel, aluminum, coiled cable, and wood.

Cutoff. The prescribed elevation at which the top of a driven pile is cut. Also the portion of pile removed from the upper end of the pile after driving.

Downdrag. Negative friction of soil gripping a pile in the case of settling soils. This condition adds load to the installed pile.

Driving cap or *helmet.* A steel cap placed over the pile butt to prevent damage to the pile during driving. It is formed to accept a specific-shaped pile, along with its cushion, if used.

Embedment. The length of pile from the ground surface, or from the cutoff below the ground, to the tip of the pile.

Overdriving. Driving a pile in a manner that damages the pile material. This is often the result of continued hammering after the pile meets refusal.

Penetration. Gross penetration is the downward axial movement of the pile per hammer blow. Net penetration is the gross penetration less the rebound, that is, the net downward movement of the pile per hammer blow.

Pile bent. Two or more piles driven in a group and fastened together by a cap or bracing.

Pile-driving shoe. A metal shoe placed on the pile tip to prevent damage to the pile and to improve driving penetration.

Pile tip. The lower—and in the case of timber piles usually the smaller—end of a pile.

Soldier pile. An H or wide-flange (WF) member driven at intervals of a few feet into which horizontal lagging is placed to support the walls of an excavation.

Tension pile. A pile designed to resist uplift.

PILE TYPES

CLASSIFICATIONS OF PILES

Piles can be classified on the basis of either their use or the materials from which they are made. On the basis of use, there are two major classifications: (1) *sheet* and (2) *load bearing*.

cofferdam
A steel or concrete box erected in the water and then pumped dry to provide a dry work area.

Sheet piles are used primarily to create a rigid barrier for earth and water. Typical uses include cutoff walls under dams, and for **cofferdams**, bulkheads, and trenching. On the basis of the materials from which they are made, sheet pilings can be classified as *steel, prestressed concrete, timber,* or *composite*.

Each type of load-bearing pile has a place in engineering practice, and for some projects more than one type may seem satisfactory. It is the responsibility of the engineer to select the pile type that is best suited for a given project, taking into account all the factors that affect both installation and performance. Considering both the type of the material from which they are made and the method of constructing and driving them, load-bearing piles can be classified as

1. Timber
   a. Treated with a preservative
   b. Untreated

TABLE 19.1 Pile information by type.

Pile type	Typical capacity (ton)	Typical lengths (ft)	Cost range (per linear ft)
Wood	15 to 60	10 to 45	\$8 to \$15
Concrete precast	60 to 120	30 to 60	\$10 to 15
Steel pipe (closed end)	60 to 150	30 to 90	\$16 to \$22
Steel H section	60 to 150+	60 to 200	\$18 to \$30

2. Concrete
   a. Precast-prestressed
   b. Cast-in-place with shells
   c. Augered cast-in-place
3. Steel
   a. H section
   b. Steel pipe
4. Composite
   a. Concrete and steel
   b. Plastic with steel pipe core

Typical capacity, length, and cost information for different types of piles is presented in Table 19.1.

TIMBER PILES

Timber piles are made from the trunks of trees. Such piles are available in most sections of the country and the world. Ordinary lengths are 15 to 45 ft; however, pine piles are available in lengths up to 80 ft and Douglas fir piles are available in lengths in excess of 100 ft from the Pacific Northwest. Typical design loads are 10 to 60 tons.

Treated Timber Piles

If timber piles remain permanently wet (wholly below the water table) they can have a very long service life. The first masonry London Bridge built in 1176 stood on untreated elm piles and lasted 600 yr. They are, however, liable to decay when exposed to a fluctuating water table. In August 1988, an 85-ft section of a bridge over the Pocomoke River in Maryland collapsed after two-pile bents supporting the bridge gave way. Investigation revealed that the untreated timber piles had deteriorated from the combined effects of bacteria, fungi, and aquatic insect larvae infestation, and abrasion by tidal currents. Preservative treatments, such as salt or creosote, are used to reduce the rate of decay and to fight the attacks of marine borers. Standard C3 of the American Wood Preservers' Association (AWPA) requires 20 pounds per cubic foot (pcf) creosote in timbers used in marine waters. These preservatives and the treatment technique should be carefully selected as they may have environmentally detrimental effects. There are new creosote treating processes, AWPA Standard P1/P13, that minimize the amount of residual creosote that occurs on the surface of the treated product. The Environmental Protection Agency has approved creosote-treated timber piling and timbers for marine water applications. Treated piling will last for 50 yr in northern marine waters, and in southern waters they have a useful life of 20 yr. Southern waters are defined as south of Cape Hatteras on the East Coast and below San Francisco on the West Coast.

Untreated Timber Piles

Timber that is exposed to the elements will last only a limited number of years if it is not treated to ward off decay. Although rarely specified, untreated piles can be used as an economical sacrificial member for temporary construction purposes. When untreated piles are used they should be rough-peeled or clean-peeled.

Timber Pile Quality ASTM D 25-99 *Standard Specification for Round Timber Piles* should be used for establishing the acceptance requirements for timber piles delivered to a project. ASTM D 25 provides tables for determining if the timber pile meets the minimum nominal circumference measured 3 ft from the butt and toe of the pile. The straightness of the piles must also be checked. A straight line from the center of the butt to the center of the tip shall lie entirely within the body of the pile. Piles shall also be free from short crooks that deviate by more than 2.5 in. from straightness in any 5-ft length. Sound knots shall be no larger than one-sixth the circumference of the pile located where the knot occurs.

CONCRETE PILES

Concrete piles may be either precast or cast-in-place. Most precast piles are manufactured in established plants and are either prestressed or post tensioned. These piles come in square, cylindrical, or octagonal shapes. Cast-in-place piles are cast on the job site and are classified as cased (driven with a mandrel, driven without a mandrel, or drilled) or uncased (compacted or drilled). Cased-type piles are formed by either driving a hollow steel tube, with a closed end, into the ground or by drilling and then placing the casing, and then filling the casing with concrete. The uncased type is formed by first driving a casing to the required depth. The casing is filled with concrete and then removed, leaving the concrete in contact with the earth.

Precast-Prestressed

Concrete piles manufactured at established plants conform to the Prestressed Concrete Institute *Manual for Quality Control* (PCI MNL-116-85) [5]. Specifications for many projects, such as those used by state highway departments, require the piles to be manufactured at PCI-certified plants. However, for projects with a large quantity of piles, transportation costs from existing manufacturing plants may be significant. Therefore, it may be cost-effective to set up a casting facility on the job or in the general vicinity of the project.

Square and octagonal piles are cast in horizontal forms on casting beds, whereas cylinder piles are cast in cylindrical forms and then centrifugally spun. The square and octagonal piles can be solid or hollow (Fig. 19.2). Solid square piles range in size from 10 to 20 in. Hollow square piles arrange in size from 20 to 36 in. with the core void diameter varying from 11 to 18 in., in agreement with the size of the pile. Octagonal piles, solid or hollow, are from 10 to 24 in. in size. For hollow octagonal piles, the void diameter varies from 11 to 15 in., following the size of the pile.

FIGURE 19.2 Solid and hollow square concrete piles.

After the piles are cast, they are normally steam cured until they have reached sufficient strength to allow them to be removed from the forms. Under controlled curing conditions and utilizing concrete with compressive strengths of 5,000 psi or greater, the piles can be removed from the forms in as little as 24 hr, or as soon as the concrete has developed a compressive strength of 3,500 psi. The piles are then stored and allowed to cure for 21 days or more before reaching driving strengths.

Prestressed concrete piles are reinforced with either $\frac{1}{2}$-in. or $\frac{7}{16}$-in. 270-ksi* high-strength stress-relieved or low-relaxation tendons or strands. The number and type of strand are determined by the design properties of the pile. In addition, the piles are reinforced with spiral reinforcing. The amount of spiral reinforcing is increased at the ends of the pile to resist cracking and spalling during driving. Piling for marine applications may require that the spirals be epoxy coated. Pile dowels may be cast in the tops of the piles for uplift reinforcing, or dowel sleeves may be cast in the top of the pile and the dowels can then be grouted into the pile after driving. Figure 19.3 shows the details for a typical 12-in.-square prestressed concrete pile.

The square and octagonal piles are traditionally cast on beds 200 to 600 ft long in order that multiple piles can be cast and prestressed simultaneously (see Fig. 19.4). The **prestressing** strand will be as long as the casting bed. Bulkheads will be placed in the forms as determined by the desired pile lengths. Utilizing stressing jacks, each strand will be pretensioned to between 20 and 35 kips prior to the concrete placement. Immediately following the concrete placement, the piles are covered with curing blankets and steam is introduced. The steam raises the air temperature at a rate of 30 to 60°F per hour until a maximum temperature of 140 to 160°F is reached. The curing continues and the prestressing forces are not released until the concrete has attained a minimum compressive strength of 3,500 psi.

prestressing
Methods of increasing the load-bearing capacity of concrete by applying increased tension on steel tendons or bars within an element.

Cylinder piles are cast in short sections of up to 16 ft in length with prestressing sleeves or ducts cast into the wall of the pile. Typical outside diameters (O.D.) are 36, 42, 48, 54, and 66 in. with wall thickness varying from 5 in. up to 6 in., depending on the design properties of the pile. Once the concrete is placed in the form, the concrete and the form are centrifugally spun, causing the concrete to consolidate. After the sections have been steam cured and the concrete has obtained sufficient strength, the sections are assembled to the proper pile lengths. The prestressing strand is pulled through the sleeves or ducts and then tensioned with jacks. Finally, the tensioned strands are pressure grouted in the sleeves.

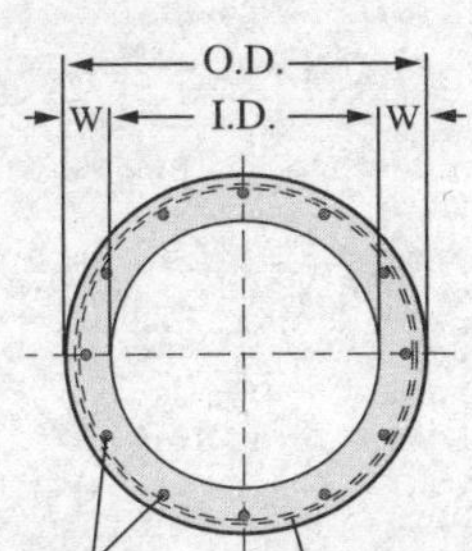

*Note that ksi is kips per square inch, where 1 kip = 1,000 lb.

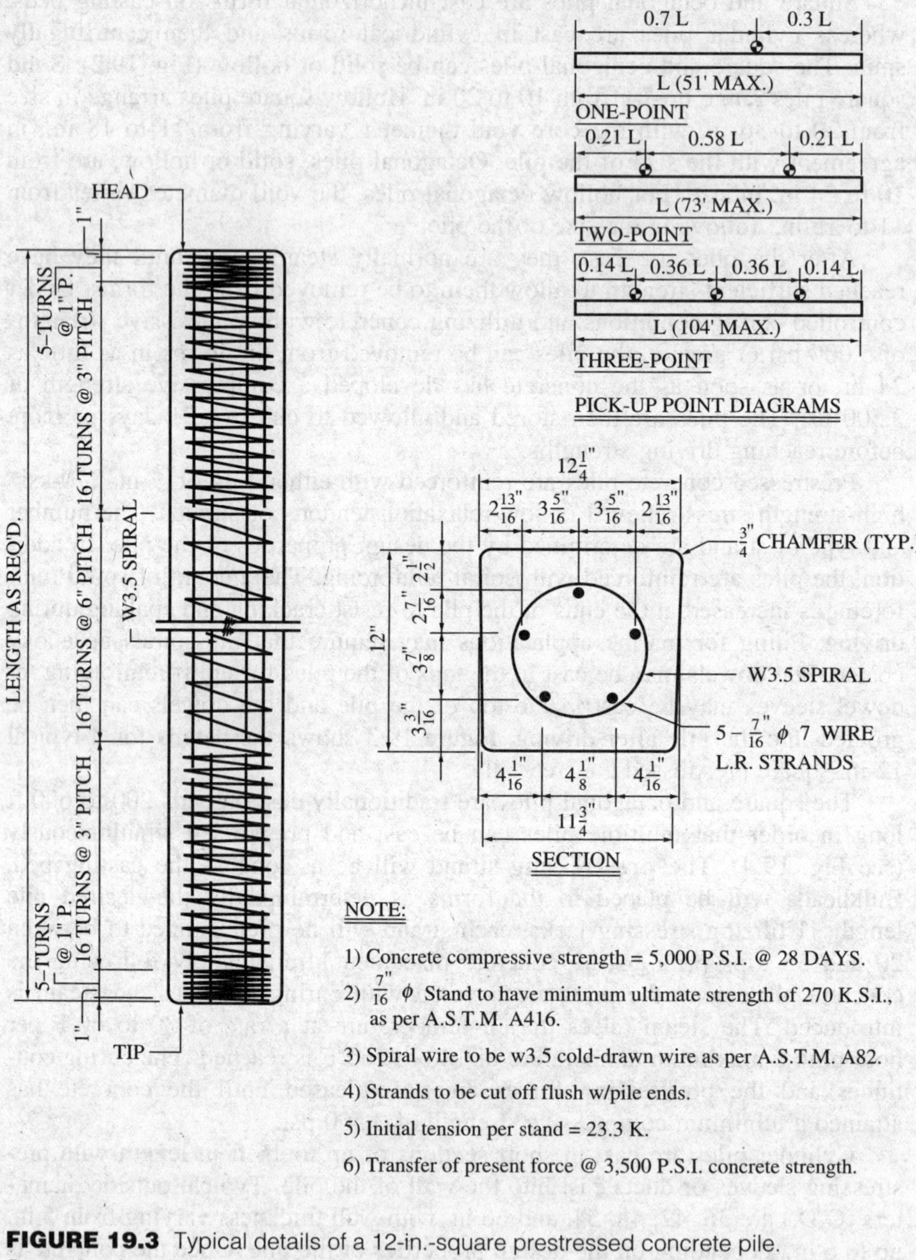

FIGURE 19.3 Typical details of a 12-in.-square prestressed concrete pile.
Source: Bayshore Concrete Products Corporation.

The precast-prestressed piles can be transported by truck to land-based projects or they can be moved by barge in the case of marine projects. Concrete piles must be handled with care to prevent breakage or damage due to

FIGURE 19.4 Casting bed for 12-in. prestressed concrete piles showing bulkheads and strands.
Source: Tidewater Skanska, Inc.

flexural stresses. Long piles should be picked up at several points to reduce the unsupported lengths (see Fig. 19.5). When piles are stored or transported, they must be supported continuously or at the pickup points.

Precast concrete piles can be cast in any desired size and length. To support the trestle bents of the Chesapeake Bay Bridge Tunnel project across the mouth of the Chesapeake Bay in Virginia, 2,500 cylinder piles, 54 in. in diameter, totaling approximately 320,000 linear ft(lf) were used. These cylinder piles were cast at a nearby casting yard constructed specifically for the project. The piles were loaded onto barges that were towed to the bridge site. Driving was performed primarily with a 150-ton barge-mounted Whirley crane.

One of the disadvantages of using precast-prestressed concrete piles, especially for a project where different lengths of piles are required, is the difficulty of reducing or increasing the lengths of piles. It is for this reason that on major piling projects, it is cost-effective to perform a site investigation program and to conduct a test pile program. Such a program aids in predicting the correct lengths for the precast concrete piles.

FIGURE 19.5 Barge-mounted Whirley crane using a three-point pickup to lift 36-in. cylinder pile off a barge and into the leads.
Source: Tidewater Skanska, Inc.

Driving Concrete Piles

Any driven pile must remain structurally intact and not be stressed to its structural limits either during its service life or during driving. In most cases, the highest stress levels occur in a pile during driving. Therefore, pile damage often occurs because of excessive stress levels generated during driving. Control of driving stress is a critical pile driving requirement.

1. Adequate cushioning material between the pile driver's driving cap and the top of the concrete pile is a very economical way of reducing driving stresses in the pile. Usually, a wood cushion is used and may vary in thickness from 4 to 8 in., depending on the length of the pile and the characteristics of the soil. Specifications usually make this a requirement.

   > The heads of concrete piles shall be protected by a pile cushion when the nature of the driving is such as to unduly injure them. When plywood is used, the minimum thickness placed on the pile head prior to driving shall not be less than 4 in (100 mm). A new pile cushion shall be provided if, during driving, the cushion is either compressed to less than one-half the original thickness or begins to burn. The pile cushion dimensions shall be such as to distribute the blow of the hammer throughout the cross section of the pile.

   Pile cushions are not used on steel piles or timber piles. Pile cushions are generally replaced for each pile.
2. Driving stresses are proportional to the ram impact velocity; therefore stresses are reduced by using a hammer with a heavy ram and a low impact velocity or large stroke.
3. Care must be taken when driving piles through soils or soil layers with low resistance. If such soils are anticipated or encountered, it is important to reduce the ram velocity or stroke of the hammer to avoid critical tensile stresses in the pile.
4. In the case of cylinder piles, it is important to prevent the soil plug inside the pile from rising to an elevation above the level of the existing soil on the outside of the pile, thus creating unequal stresses in the pile. This should be monitored, and, in the event the plug does rise inside the pile, it should be excavated down to the level of the existing soil on the outside of the pile.

5. The driving cap should fit loosely around the top of the pile so that the pile can rotate slightly without binding within the driving head. This will prevent the development of torsional stress.
6. The top of the pile must be square or perpendicular to the longitudinal axis of the pile to eliminate eccentricity that causes stress.
7. The ends of the prestressing strand or reinforcing must either be cut off flush with the top of the pile or the driving cap must be designed so that the reinforcing threads through the pile cap in order that the hammer's ram does not directly contact the reinforcing during driving. The driving energy must be delivered to the top of the concrete.

Cast-in-Place Concrete Piles

As the name implies, cast-in-place concrete piles are constructed by placing concrete into a tapered or cylindrical hole previously driven into the ground or into a hole in the ground from which a driven mandrel (steel core) has been withdrawn. In both of these cases, the hole has been created by a driving process that displaced the ground. There are also nondisplacement cast-in-place piles where the soil is removed and the resulting hole filled with concrete.

The cast-in-place displacement-type pile can be of two forms. The first involves driving a temporary steel tube with a closed end into the ground to form a void in the soil, which is then filled with concrete as the tube is withdrawn. The second type is the same except the steel tube is left in place to form a permanent casing. These piles may be tapered or of uniform section. Generally, the tapered piles are cylinder shells with vertical fluting or corrugations. Such tapered piles are commonly referred to as monotube piles. Monotube piles can be either tapered or of a uniform diameter. The monotube pile shell is driven without a mandrel, inspected, and filled with concrete. The desired length of shell is obtained by welding extensions to a standard-length shell.

There are also driven cast-in-situ concrete piles where a casing, closed at the bottom with a plug of dry concrete or gravel or with a shoe, is driven into the ground to the required depth. Concrete is then placed into the casing and a cast-in-situ pile is formed inside the casing. One particular feature of this pile is that as the casing is withdrawn and the concrete compacted, the "green" concrete will key into the borehole and thus provide skin friction between the pile and the soil. However, in the installation of this pile, care must be taken to ensure that the reinforcement cage is positioned correctly to provide adequate cover to all the reinforcement.

The Franki technique (see www.franki.co.gg/) is a patented technique for creating this type of pile. The Franki technique uses a steel drive tube fitted with an expendable steel boot to close the bottom end. Using a pile-driving hammer, the tube is driven to the desired depth. After the tube is driven to the desired depth, the hammer is removed and a charge of dry concrete is dropped into the tube and compacted with a drop hammer to form a compact watertight plug. The tube is then raised slightly and held in that position while repeated blows of the drop hammer expel the plug of concrete from the lower end of the tube, leaving in the tube sufficient concrete to prevent the intrusion of water or soil. Additional concrete is dropped into the tube and forced by the drop hammer to flow out of the bottom of the tube to form an enlarged bulb or pedestal. The final step consists of raising the tube in increments. After each incremental upward movement, additional dry concrete is dropped into the tube and is hammered with sufficient energy to force it to flow out of the tube and fill the exposed hole. This operation cycle is repeated until the hole is filled to the desired elevation. The shaft may be reinforced by inserting a spirally wound steel cage.

There are also nondisplacement cast-in-place concrete piles, one form of which is the augered cast-in-place pile (ACIP). These piles differ from the previously discussed cast-in-place piles since they do not require a shell or pipe. ACIP piles are constructed by rotating a hollow-shaft continuous flight auger into the soil to a predetermined tip elevation or refusal, whichever occurs first. When the desired depth is reached, a high-strength grout is injected under pressure through the hollow shaft. This grout exits through the tip of the auger. A prescribed grout volume is pumped out the end of the auger to build a "grout head" around the auger before it is lifted. The auger is withdrawn in a controlled manner, as grout pumping continues. The head of grout maintains the integrity of the hole, preventing intrusion of soil or water. The soil is removed as the auger is rotated and withdrawn. Reinforcing steel is placed through the fluid grout column. Typical compressive strengths for the grout are 3,000 to 5,000 psi.

FIGURE 19.6 Vibratory hammer driving steel H pile.

STEEL PILES

In constructing foundations that require piles driven to great depths, steel piles probably are more suitable than any other type (see Fig. 19.6). Load-bearing steel piles can be H-sections, hollow box or tubular sections, or pipes. Compared by pile length, steel piles tend to be more costly than concrete piles, but they have high load-carrying capacity for a given weight of pile, which can reduce driving costs. The great strength of steel combined with the small displacement of soil permits a large portion of the energy from a pile hammer to be transmitted to the bottom of a pile. As a result, one can drive steel piles into soils that cannot be penetrated by any other type of pile.

Steel H Section Piles

Compared to concrete piles, steel H piles generally have better driving characteristics and can be installed to great depths. They are used as point-bearing piles with typical lengths of 60 to 200 ft. H piles can be susceptible to deflection on striking boulders, obstructions, or an inclined rock surface. They are widely used because of their ease of handling and driving. A wide range of pile sizes is available, with different grades of steel. Since steel H piles do not cause large soil displacements, they are useful in urban areas or adjacent to structures where **heave** of the surrounding ground may cause problems.

heave
The uplifting of the soil surface between or near driven piles. This is caused by the displacement of soil by the pile volume.

Because the steel H piles can be driven in short lengths, and additional lengths and then welded on top of the previously driven section, they can be utilized more readily in situations where there are height restrictions that limit the length of piles that can be driven in one piece.

Steel Pipe Piles

The steel box girder superstructure of the new Woodrow Wilson Bridge over the Potomac River in Washington, D.C., is supported by 744 steel pipe piles. They range in diameter from 48 to 72 in. and the longest is about 224 ft in length. The foundation piles for the east span of the San Francisco–Oakland Bay Bridge are 8-ft-diameter steel pipe piles (Fig. 19.7).

Pipe piles are most efficient as friction piles as they have substantial surface area that interacts with the surrounding soil to provide significant frictional load resistance. A pipe pile may be driven with the lower end closed with a plate or steel driving point, or the pipe may be driven with lower end open. A closed-end pipe pile is driven in any conventional manner. If it is necessary to increase the length of a pile, two or more sections may be welded together or sections may be connected by using an inside sleeve for each joint. This type of pile is particularly advantageous for use on jobs where headroom for driving is limited and short sections must be added to obtain the desired total length.

An open-ended pipe pile is installed by driving the pipe to the required depth and then removing the soil from the inside. Removal methods include bursts of compressed air, a mixture of water and compressed air, or the use of an earth auger or a small orange-peel bucket. Finally, the hollow pipe is filled with concrete.

FIGURE 19.7 Large steel piles set on a batter utilizing a template and ready for driving.

COMPOSITE PILES

Several types of composite piles are available. These are usually developed and offered to meet the demands of special situations. Two of the most common situations that cause problems, when conventional piles are used, are hard driving conditions and warm marine environments. Hard driving conditions cause problems with applying the energy necessary to drive the pile and at the same time being careful not to destroy the pile. Marine environments subject the pile to marine borer attacks and salt attacks on metal, therefore special piling protective measures are usually specified.

Concrete-Steel Composite Piles

When extremely hard soils or soil layers are encountered, it may be cost-effective to consider the use of a composite concrete and steel pile. The top portion of the pile is a prestressed concrete pile, and the tip would be a steel H pile embedded into the end of the concrete pile. This composite design is suggested for marine applications, where the concrete pile section offers resistance to deterioration and the steel pile tip enables penetration when there are hard driving conditions.

Steel-Concrete Composite Piles

This pile type comprises a steel casing with a hollow spun concrete core, resting on a solid driving shoe. It combines the advantages of a high-quality concrete core with the high tensile strength of the external steel casing. These piles can provide higher durability and improved driveability.

Plastic with Steel Pipe Core Piles

Since the mid-eighties composite plastic piles have been available for special applications [2]. These piles are immune to marine borer attacks, eliminating the need for creosote treatments or special sheathings in marine environments. Their abrasive resistance makes them excellent for fender system use.

Varying the amount of steel in the pile core can alter the piles' physical properties of toughness, resilience, and specific gravity. These piles can be manufactured in a variety of shapes and colors. Plastic pilings 24 in. in diameter and up to 50 ft long have been developed having steel-pipe cores between 12 and 16 in. in diameter. Using a larger pipe section increases stiffness. On a fender system at the Port of Los Angeles, 13-in.-diameter 70-ft-long piles with a 6-in. steel-pipe core were used (Fig. 19.8). A steel-driving tip or shoe is welded to the end of the pile before driving. The tip aids driving and serves to seal the end of the pile. A diagram of a composite plastic and steel pile and a driving shoe is shown in Fig. 19.9.

FIGURE 19.8 Composite plastic and steel core pile fender system at the Port of Los Angeles.
Source: Plastic Pilings Inc.

SHEET PILES

Sheet piles are used primarily to retain or support earth. They are commonly used for bulkheads and cofferdams (see Fig. 19.10) and when excavation depths or soil conditions require temporary or permanent bracing to support the lateral loads imposed by the soil or by the soil and adjacent structures. Supported loads would include any live loads imposed by construction operations. Sheet piles can be made of timber, concrete, or steel. Each of these types can sup-

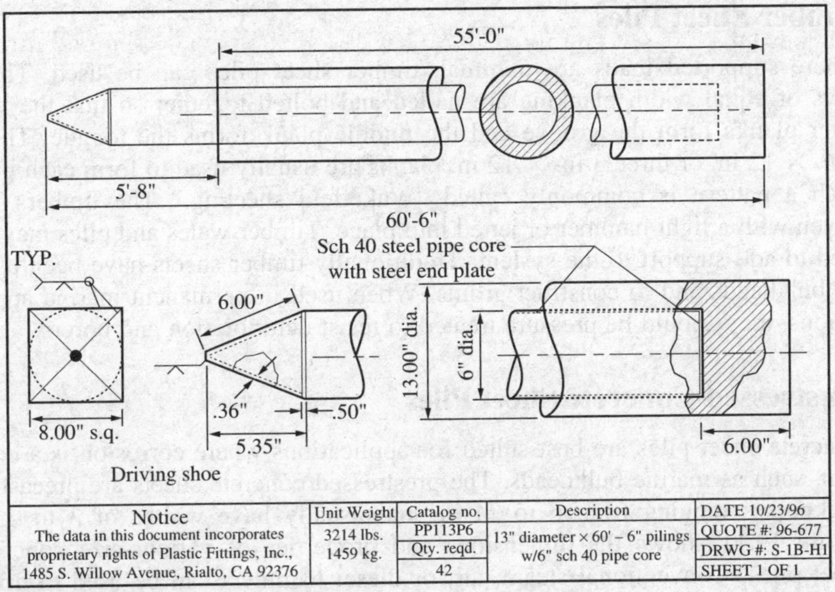

FIGURE 19.9 Diagram of a 13-in.-diameter × 60 ft 6 in. sch 40 pipe core composite plastic and steel pile.
Source: Plastic Pilings Inc.

FIGURE 19.10 Circular sheet piling cells used to create a cofferdam.
Source: Kokosing Fru-Con.

port limited loads without additional bracing or tieback systems. When the depth of support is large or when the loads are great, it is necessary to incorporate a tieback or bracing system with the sheet piles.

Timber Sheet Piles

Where supported loads are minimal, timber sheet piles can be used. Three rows of equal width planking are nailed and bolted together so that the two outer planks form the groove and the middle plank forms the tongue. Three 2 in. × 12 in. or three 3 in. × 12 in. planks are usually used to form each pile. Such a pattern is commonly called "Wakefield sheeting." The timbers are driven with a light hammer or jetted into place. Timber wales and piles may be used to add support to the system. Traditionally timber sheets have been used for bulkheads and to construct groins. When used in permanent marine applications, they should be pressure treated to resist deterioration and borers.

Prestressed Concrete Sheet Piles

Concrete sheet piles are best suited for applications where corrosion is a concern, such as marine bulkheads. The prestressed concrete sheets are precast in thicknesses ranging from 6 to 24 in. and usually have widths of 3 to 4 ft. Figure 19.11 shows the dimensions and properties of various-size concrete sheet piles. Conventional steam, air, or diesel hammers can be used to drive concrete sheet piles. However, jetting is often required to attain the proper tip elevation. Once the piles have been driven, the slots between sections are filled with grout.

Steel Sheet Piles

Steel sheet piling is a rolled section that is interlocked with adjacent sections to form a continuous wall. The hot rolling process includes the formation of a geometrically defined interlock. The interlock produced by the cold rolled process is quite different from the hot rolled interlock and tends to be a looser connection.

In the United States, steel sheet piling is manufactured in both flat and Z section sheets, which have interlocking longitudinal edges. There are more than 10 domestic and foreign manufacturers listing almost 200 sheet piling sections in their catalogs. Most steel sheet piles are supplied in standard ASTM 328 grade steel, but high-strength low-alloy grades ASTM A572 and ASTM A690 are available for use where large loads must be supported or where corrosion is a concern. In marine applications, epoxy-coated piles can be used to resist corrosion.

The flat sections are designed for interlocking strength that makes them suitable for the construction of cellular structures. The Z sections are designed for bending, which makes them more suitable for use in the construction of retaining walls, bulkheads, and cofferdams, and for use as excavation support. Figure 19.12 shows various domestically produced steel sheet piles (PZ, PSA, and PS) and their corresponding properties. The PZ sheets are favored by many contractors for temporary applications because of the "ball and socket" interlock and favorable interlock swing of 10°. The maximum rolling length for PZ sections is 85.0 ft (25.9 m).

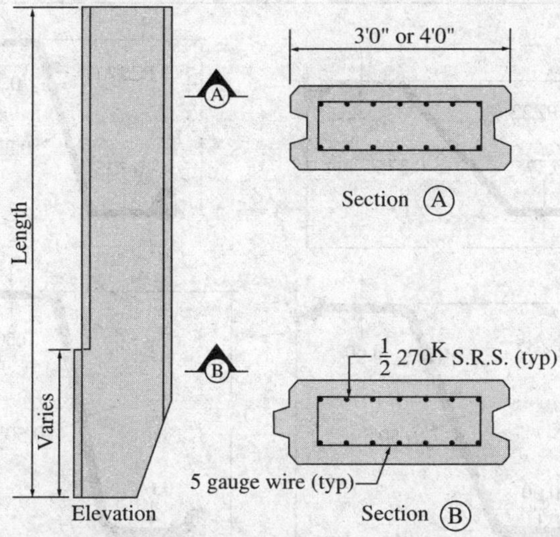

Sheet pile properties
(per L.F. of width)

T Thickness	Area in.^{2}	I, in.^{4}	S, in.^{3}	Max allowable Bending moment (ft kips) $f'c$ = 5,000 psi	(2) $f'c$ = 6,000 PSI	Approx. weight per (1) l.f.
6"	72"	216	72	6.0	7.2	75#
8"	96"	512	128	10.6	12.8	100#
10"	120"	1,000	200	16.6	20.0	125#
12"	144"	1,728	288	24.0	28.8	150#
16"	192"	4,096	512	42.7	51.2	200#
18"	216"	5,832	648	54.0	64.8	225#
20"	240"	8,000	800	66.7	80.0	250#
24"	288"	13,824	1,152	96.0	115.2	300#

I - Moment of inertia.
S - Section modulus.

(1) Weights based on 150 pcf of regular concrete.
(2) Based on 'o' psi allowable stress on tensile face and 0.4f'c allowable stress on compression f

FIGURE 19.11 Prestressed concrete sheet piles and their corresponding properties and weights.
Source: Bayshore Concrete Products Corporation.

Several overseas manufacturers produce steel sheet piles with properties similar to the domestic sheets. Those piles sometimes utilize a different interlocking design than the domestic sheets. Careful attention should be given to this design feature to ensure that sheets will perform in the intended usage and that the interlocking system allows the sheets to be threaded and driven easily.

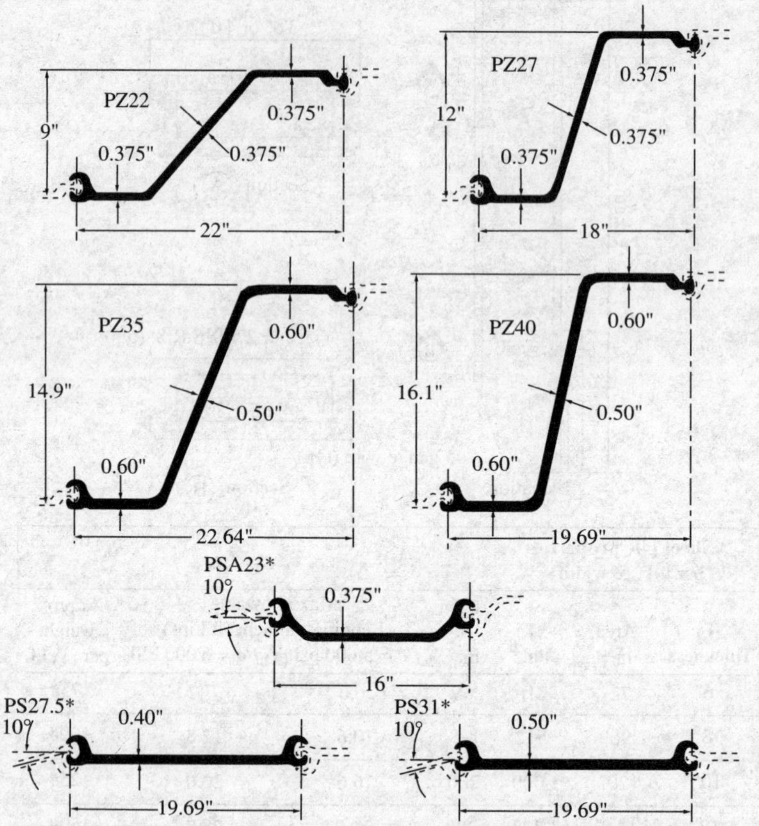

*PSA23, PS27.5, and PS31 when properly interlocked provide a 10 degree swing for lengths up to 70 feet. The dimensions given on this page are nominal.

Properties and Weights

Section designation	Area, sq in.	Nominal width, in.	Weight, lb Per lin. ft of bar	Weight, lb Per sq ft of wall	Moment of inertia, in.4	Section modulus, in.3 Single section	Section modulus, in.3 Per lin. ft of wall	Surface area sq ft per lin. ft of bar Total area	Surface area sq ft per lin. ft of bar Nominal coating area*
PZ22	11.86	22	40.3	22.0	154.7	33.1	18.1	4.94	4.48
PZ27	11.91	18	40.5	27.0	276.3	45.3	30.2	4.94	4.48
PZ35	19.41	22.64	66.0	35.0	681.5	91.4	48.5	5.83	5.37
PZ40	19.30	19.69	65.6	40.0	805.4	99.6	60.7	5.83	5.37
PSA23	8.99	16	30.7	23.0	5.5	3.2	2.4	3.76	3.08
PS27.5	13.27	19.69	45.1	27.5	5.3	3.3	2.0	4.48	3.65
PS31	14.96	19.69	50.9	31.0	5.3	3.3	2.0	4.48	3.65

*Excludes socket interior and ball of interlock.

FIGURE 19.12 Domestically produced steel sheet piles and their corresponding properties and weights.
Source: Bethlehem Steel Corporation.

Sheet piles are driven individually or in pairs, frequently with the use of a vibratory hammer and a guide frame or template. It is usually best to drive in pairs whenever possible. The template or guide (Fig. 19.13) should be at least half as high as the sheets since it is critical to keep the sheets plumb in both planes. It is very difficult to work with damaged sheets, and they should be avoided if possible. Before threading and driving a sheet into the system it is good practice to strike it with a sledgehammer to remove dirt trapped in the interlocks.

Automatic sheet-pile threading devices can be used to thread the new sheet into the previously threaded or driven sheet. These devices allow the sheets to be threaded without the assistance of a person working at the top of the previously placed sheet. If the steel sheet piling is used for temporary support, the individual sheets can be extracted ("pulled") using a vibratory hammer after construction and backfilling have been completed.

FIGURE 19.13 Using a guide frame to drive steel piles with a vibratory hammer.
Source: Kokosing Fru-Con.

When a vibratory hammer is used, the vibrator should be brought to speed before beginning the driving. Similarly, when extracting a pile do not begin pulling until the vibrator has started. In the case of extraction, it is often best to drive the pile a little first, to break the static friction, before beginning the pulling procedure.

Driving Steel Sheet Piling

Before driving any type of piling it is important to carefully study the soil borings in order to anticipate driving conditions, and this is especially important in the case of sheet piling. The presence of boulders or stumps can make driving very difficult. Successful steel sheet pile installation depends on following these commonsense guidelines:

- Work only with undamaged sheets; piles must be straight or driving will be difficult.
- Interlocks must always be free of any dirt, sand, mud, or other debris.
- Always set up a template system. In addition to driving a straight wall, a template or guide system will also aid in keeping sheet pile plumb when excessive driving conditions exist or when an obstruction is encountered.
- Whenever possible, it is recommended to drive sheets in pairs.
- Crane boom length must always be long enough to thread additional sheets—normally at least twice the length of the sheets being driven plus a few feet extra.
- Whenever possible, attempt to drive sheets with the male interlock, ball, or thumb leading. This will aid in eliminating the possibility of the sheet developing a "soil plug" (the interlock filling with soil, sand, or mud).
- Never overdrive. When sheets are bending, bouncing, or vibrating with no penetration, it could be an indication of overdriving, or that the sheet has hit an obstruction. However, it may indicate that a larger vibratory driver/extractor or hydraulic impact hammer is needed.

Many of these guidelines are also applicable to prestressed concrete sheet piles.

DRIVING PILES

THE RESISTANCE OF PILES TO PENETRATION

The forces that enable a pile to support a load also cause the pile to resist the efforts made to drive it. The total resistance of a pile to penetration will equal the sum of the forces produced by skin friction and end bearing. The relative portions of the resistance contributed by either skin friction or end bearing may vary from almost 0 to 100%, depending more on soil type than on the type of pile. A steel H pile driven to refusal in stiff clay should be classified as a skin-friction pile, whereas the same pile driven through a mud deposit to rest on solid rock should be classified as an end-bearing pile.

Numerous tests have been conducted to determine values for skin friction for various types of piles and soils. A representative value for skin friction can be obtained by determining the total force required to cause a small incremental upward movement of the pile using hydraulic jacks with calibrated pressure gauges.

The value of the skin friction is a function of the coefficient of friction between the pile and the soil and the pressure of the soil normal to the surface of the pile. However, for a soil such as some clays, the value of the skin friction may be limited to the shearing strength of the soil immediately adjacent to the pile. Consider a concrete pile driven into a soil that produces a normal pressure of 100 psi on the vertical surface of the pile. This is not an unusually high pressure for certain soils, such as compacted sand. If the coefficient of friction is 0.25, the value of the skin friction will be

$$0.25 \times 100 \text{ psi} \times \frac{144 \text{ sq in.}}{\text{sf}} = 3{,}600 \text{ psf}$$

(pounds per square foot). Table 19.2 gives representative values of skin friction on piles. The data presented in Table 19.2 are intended as a qualitative guide, not as accurate information to be used in any and all cases.

The magnitude of end-bearing pressure can be determined by driving a button-bottom-type pile and leaving the driving casing in place. A second steel pipe, slightly smaller than the driving casing, is lowered onto the button and the force applied through the second pipe to drive the button into the soil is a direct measure of the supporting strength of the soil. This is true because there is no skin friction on the inside pipe.

SITE INVESTIGATION AND TEST PILE PROGRAM

For projects of intermediate to large scale, a thorough site investigation can be very cost-effective. The geotechnical information gathered from borings can be used to determine the soil characteristics and the depths to strata capable of supporting the design loads. The number of blows per foot, from geotechnical

TABLE 19.2 Approximate allowable value of skin friction on piles.*

	Skin friction [psf (kg/sq m)]		
	Approximate depth		
Material	**20 ft (6.1 m)**	**60 ft (18.3 m)**	**100 ft (30.5 m)**
Soft silt and dense muck	50–100 (244–488)	50–120 (244–586)	60–150 (273–738)
Silt (wet but confined)	100–200 (488–976)	125–250 (610–1,220)	150–300 (738–1,476)
Soft clay	200–300 (976–1,464)	250–350 (1,220–1,710)	300–400 (1,476–1,952)
Stiff clay	300–500 (1,464–2,440)	350–550 (1,710–2,685)	400–600 (1,952–2,928)
Clay and sand mixed	300–500 (1,464–2,440)	400–600 (1,952–2,928)	500–700 (2,440–3,416)
Fine sand (wet but confined)	300–400 (1,464–1,952)	350–500 (1,710–2,440)	400–600 (1,952–2,928)
Medium sand and small gravel	500–700 (2,440–3,416)	600–800 (2,928–3,904)	600–800 (2,928–3,904)

*Some allowance is made for the effect of using piles in small groups.

tests such as the standard penetration test, is normally recorded during the soil-sampling operations. That information is valuable in selecting pile lengths, pile types, and sizes, and in estimating capacities. Once a pile type has been selected, or if several types are deemed practical for use on a particular project, a test pile program should be conducted.

Pile lengths to be used for the test pile program are generally somewhat longer than the anticipated lengths, as determined from the boring information. This enables driving of the test piles to greater depths if necessary and allows additional pile length for installing the load test apparatus. Several test piles should be driven and carefully monitored at selected locations within the project area. Dynamic testing equipment and testing techniques can be utilized to gather significant information pertaining to the combined characteristics of the pile, the soil, and the pile-driving equipment. This information will enhance the engineer's ability to predict the required pile lengths for the project and the number of blows per foot required to obtain the desired **bearing capacity.**

bearing capacity *Allowable pile load as limited by the provision that the pressures developed in the materials along the pile and below the pile tip shall not exceed the allowable soil bearing values.*

Depending on the size of the project, one or more of the test piles would be selected for load testing. To apply a load to the selected piles, static test weights, water tanks, or reaction piles and jacks may be employed. Static weights or water tanks permit the loading weight to be incrementally applied directly to the test pile by adding either additional weights or water. In the case of the reaction pile method, steel H piles are used as reaction piles. They are driven in relatively close proximity to the pile to be tested. The magnitude of the test load to be applied and the friction characteristics of the soil determine the number of reaction piles necessary.

FIGURE 19.14 Load test of a 54-in. concrete cylinder pile utilizing reaction piles, beams, and a jack.
Source: Tidewater Skanska, Inc.

A reaction frame or beam is attached to the reaction piles and spans over the top of the test pile in order that the test load can be applied using a hydraulic jack (see Fig. 19.14). The jack is located between the reaction beam and the top of the test pile. As the load is applied by the jack, the reaction beam transfers the load to the test pile, putting it in compression and putting the reaction piles in tension.

Calibrated hydraulic gauges are used to measure the amount of load applied. The test load is applied in increments over a period of time and the pile is continuously monitored for movement. In the case of either type of test, direct load or reaction, the magnitude of the applied test load is normally two to three times the design-bearing capacity of the pile. Any sudden or rapid movement of the pile indicates a failure of the pile.

Once a test pile program has been completed, pile lengths and supporting bearing capacities can be predicted with reasonable accuracy. Where precast concrete piles are determined to be an economical alternative at a particular site, a test pile program can verify with suitable accuracy the required pile lengths. Therefore, when reliable geotechnical and load test data are available, piles can be cast to calculated lengths with little risk of their being either too short or too long.

PILE HAMMERS

Driven piles are usually installed to established criteria: minimum blow count per unit penetration or minimum penetration. Variable subsurface conditions can dictate the use of different pile hammers. The function of a pile hammer is to furnish the energy required to drive a pile. Pile-driving hammers are designated by type and size. The hammer types commonly used include

1. Drop
2. Single-acting steam or compressed air
3. Double-acting steam or compressed air
4. Differential-acting steam or compressed air
5. Diesel
6. Hydraulic—impact and drivers
7. Vibratory drivers

The driving energy for each of the first five pile hammer types listed and for the hydraulic impact hammer is supplied by a falling mass that strikes the top of the pile. The size of a drop hammer is designated by its weight, whereas the size of each of the other hammers is designated by theoretical energy per blow, expressed in foot-pounds (ft-lb).

The following factors should be considered when selecting a method for driving piles.

- Size and the weight of the pile.
- Driving resistance that has to be overcome to achieve the required penetration
- Available space and headroom for equipment
- Noise restrictions

Drop Hammers

A drop hammer is a heavy metal weight that is lifted by a hoist line, then released and allowed to fall onto the top of the pile. Because of the high dynamic forces, a pile cap is positioned between the hammer and the pile head. The pile cap serves to uniformly distribute the blow to the pile head and to serve as a "shock absorber." The cap contains a cushion block that is commonly fabricated from wood.

The hammer may be released by a trip and fall freely, or it may be released by loosening the friction band on the hoisting drum and permitting the weight of the hammer to unwind the rope from the drum. The latter type of release reduces the effective energy of a hammer because of the friction loss in the drum and rope. Leads are used to hold the pile in position and to guide the movement of the hammer so that it will strike the pile axially.

Standard drop hammers are made in sizes that vary from about 500 to 3,000 lb. Hammer weight should be from 0.5 to 2 times the pile weight. The height of drop or fall most frequently used varies from about 5 to 20 ft. The maximum recommended drop height varies with the pile type: 15 ft for timber piles and 8 ft for concrete piles. When a large energy per blow is required to drive a pile, it is better to use a heavy hammer with a small drop than a light hammer with a large drop. The analogy would be trying to drive a large nail with a tack hammer.

Drop hammers are suitable for driving piles on remote projects that require only a few piles and for which the time of completion is not an important factor. A drop hammer normally can deliver four to eight blows per minute.

FIGURE 19.15 Raymond 60× steam hammer driving a 54-in. cylinder pile. *Source: Tidewater Skanska, Inc.*

Single-Acting Hammers

A single-acting steam or air hammer (see Fig. 19.15) has a freely falling weight, called a "ram," that is lifted by steam or compressed air, whose pressure is applied to the underside of a piston that is connected to the ram through a piston rod. When the piston reaches the top of the stroke, the steam or air pressure is released and the ram falls freely to strike the top of a pile. These hammers rely solely on gravity acting on the striking weight through a

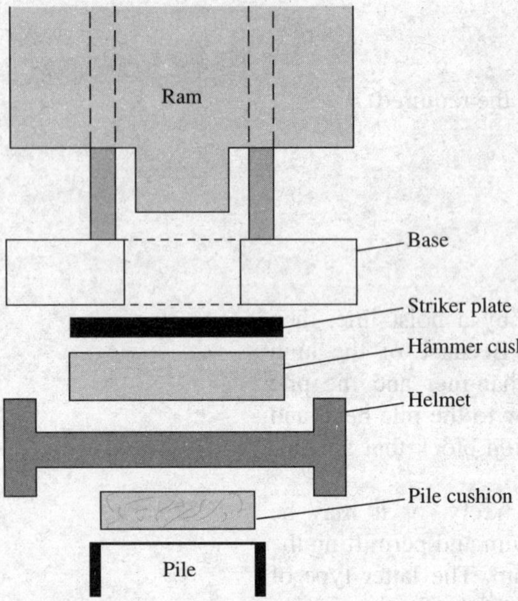

FIGURE 19.16 Arrangement of the hammer cushion, helmet, and pile cushion.

distance to achieve their striking energy. The heavy weight strikes with a low velocity, due to the relatively short fall distance, usually 3 ft; but the fall can vary from 1 to 5 ft for specific hammers. Whereas a drop hammer may strike 4 to 8 blows per minute, a single-acting steam or air hammer will strike 40 to 60 blows per minute when delivering the same energy per blow.

A driving cap is used with a single-acting steam/air hammer. The cap is mated to the case of the hammer. This cap is also referred to as an "anvil" or a "helmet" (see Fig. 19.16). These hammers are usually used in a pile lead, though they can be fitted to operate free hung. The hammers are available in sizes varying from about 7,000 to 1,800,000 ft-lb of energy per blow. Table 19.3 gives data on some of the larger single-acting steam and air hammers.

A useful rule as to a suitable weight for a single-acting hammer is to select a hammer having a weight corresponding approximately to that of the pile. This may not always be possible in the case of heavy reinforced concrete piles, where the hammer is more likely to be half of the weight of the pile. However, it should not weigh less than one-third the weight of the pile. To avoid damage to the pile, the height of drop should be limited to 4 ft. Concrete piles are especially liable to be shattered by a blow from too great a height.

EXAMPLE 19.1

A contractor will use a single-acting air hammer to drive concrete piles. If the pile weighs 21,000 lb, what is the recommended hammer weight? What is the minimum hammer weight that can be used?

Recommended hammer weight = weight of pile = 21,000 lb

Minimum hammer weight = 0.33 weight of pile = 0.33 × 21,000 lb = 6,930 lb

Double-Acting Hammers—Rapid Blow Hammers

With double-acting hammers, the striking ram (piston) is driven by compressed air or steam both when rising and when falling. The air or steam enters a valve box containing a slide valve that sends it alternately to each side of the piston, while the opposite side is connected to the exhaust ports. Thus, with a given weight of ram, it is possible to attain a desired amount of energy per blow with a shorter stroke than with a longer single-acting hammer. Double-acting ham-

TABLE 19.3 Specifications for air or steam pile-driving hammers.

Rated energy	Model	Manufacturer	Type	Style	Blows per min	Wt. of striking parts	Total weight (lb)	Hammer length (ft-in.)	Jaw dimensions	Boiler HP required (ASME)	Steam consump. (lb/hr)	Air consump. (cfm)	Inlet pressure (psi)	Inlet size (in.)
1,800,000	6300	Vulcan	SGL-ACT.	Open	42	300,000	575,000	30'0"	22" x 144"(M)	2,804	43,873	19,485(A)	235	2@6"
1,582,220	MRBS 12500	Menck	SGL-ACT.	Open	36	275,580	540,130	35'9"	CAGE	2,400	52,910	26,500	171	2@6"
867,960	MRBS 8000	Menck	SGL-ACT.	Open	38	176,370	330,690	30'10"	CAGE	1,380	30,860	15,900	171	8"
750,000	5150	Vulcan	SGL-ACT.	Open	46	150,000	275,000	26'3$\frac{1}{2}$"	22" x 120"(M)	1,317	45,426	9,535(A)	175	2@6"
500,000	5100	Vulcan	SGL-ACT.	Open	48	100,000	197,000	27'4"	22" x 120"(M)	1,043	35,977	7,620(A)	150	2@5"
499,070	MRBS 4600	Menck	SGL-ACT.	Open	42	101,410	176,370	27'5"	CAGE	850	19,840	9,900	142	6"
325,480	MRBS 3000	Menck	SGL-ACT.	Open	42	66,135	108,025	25'0"	CAGE	520	12,130	6,000	142	5"
325,000	5650	Conmaco	SGL-ACT.	Open	45	65,000	139,300	23'0"	18$\frac{3}{4}$" x 100"	606	20,907	—	160	3@4"
300,000	3100	Vulcan	SGL-ACT.	Open	60	100,000	195,500	23'3"	18$\frac{3}{4}$" x 88"(M)	900	30,153	6,644(A)	130	3@4"
300,000	560	Vulcan	SGL-ACT.	Open	47	62,500	134,060	23'0"	18$\frac{3}{4}$" x 88"(M)	606	20,897	4,427(A)	150	2@5"
200,000	540	Vulcan	SGL-ACT.	Open	48	40,900	102,980	22'7"	14" x 80"(M)	409	14,126	3,022(A)	130	2@5"
189,850	MRBS 1800	Menck	SGL-ACT.	Open	44	38,580	64,590	22'5"	CAGE	295	7,060	3,700	142	4"
180,000	360	Vulcan	SGL-ACT.	Open	62	60,000	124,830	19'0"	18$\frac{3}{4}$" x 88"(M)	506	17,460	3,736(A)	130	2@4"
180,000	060	Vulcan	SGL-ACT.	Open	62	60,000	128,840	19'0"	18$\frac{3}{4}$" x 88"(M)	506	17,460	3,736(A)	130	2@4"
150,000	5300	Conmaco	SGL-ACT.	Open	46	30,000	62,000	20'9$\frac{1}{2}$"	14" x 80"(m)	234	12,296	2,148(A)	160	4"
150,000	530	Vulcan	SGL-ACT.	Open	42	30,000	57,680	20'5"	10$\frac{1}{2}$" x 54"(M)	234	8,064	1,711	150	3"
120,000	340	Vulcan	SGL-ACT.	Open	60	40,000	98,180	18'7"	14" x 80"(M)	354	12,230	2,628(A)	120	2@3"
120,000	040	Vulcan	SGL-ACT.	Open	60	40,000	87,673	17'11"	14" x 80"(M)	354	12,230	2,628(A)	120	2@3"
93,340	MRBS 850	Menck	SGL-ACT.	Open	45	18,960	27,890	19'8"	CAGE	150	3,530	1,950	142	3"
90,000	030	Vulcan	SGL-ACT.	Open	54	30,000	55,410	16'5"	10$\frac{1}{4}$" x 54"(M)	201	6,944	1,471(A)	150	3"
90,000	300	Conmaco	SGL-ACT.	Open	55	30,000	55,390	16'10"	11$\frac{1}{4}$" x 56"(F)	201	6,944	1,833(A)	150	3"
81,250	8/0	Raymond	SGL-ACT.	Open	40	25,000	34,000	19'4"	10$\frac{1}{4}$" x 25"	172	5,950	—	135	3"
75,000	30X	Raymond	SGL-ACT.	Open	70	30,000	52,000	19'1"	—	246	8,500	—	150	3"
60,000	S-20	MKT	SGL-ACT.	Closed	60	20,000	38,650	15'5"	x 36"	190	—	—	150	3"
60,000	020	Vulcan	SGL-ACT.	Open	59	20,000	43,785	14'8"	10$\frac{1}{4}$" x 54"(M)	161	5,563	1,195(A)	120	3"
60,000	200	Conmaco	SGL-ACT.	Open	60	20,000	44,560	15'0"	11$\frac{1}{4}$" x 56"(F)	161	7,500	1,634(A)	120	3"
56,875	5/0	Raymond	SGL-ACT.	Open	44	17,500	26,450	16'9"	10$\frac{1}{4}$" x 25"	100	4,250	—	150	3"
50,200	200-C	Vulcan	DIFFER.	Open	98	20,000	39,000	13'11"	11$\frac{1}{4}$" x 37"	260	8,970	1,746(A)	142	4"
48,750	016	Vulcan	SGL-ACT.	Open	58	16,250	33,340	13'8"	10$\frac{1}{4}$" x 54"(M)	121	4,182	899(A)	120	3"
48,750	4/0	Raymond	SGL-ACT.	Open	46	15,000	23,800	16'1"	—	85	—	—	120	2$\frac{1}{2}$"
48,750	150-C	Raymond	DIFFER	Open	95–105	15,000	32,500	15'9"	—	—	—	—	120	3"

mers commonly deliver 95 to 300 blows per minute. The lower blow counts are for the larger hammers and the higher counts for the smaller hammers.

In comparison with single-acting hammers of the same overall weight, the ram of the double-acting hammers is much less than that of the single-acting hammer. It is only 10 to 20% of the overall hammer weight, but its effect is increased by the pressure on the upper end of the piston. Ninety percent of the blow energy is derived from the action of the air or steam upon the piston. This type of hammer is not always suitable for driving concrete piles, however. Although the concrete can take the compressive stresses exerted by the hammer, the returning shock wave set up by each blow of the hammer can cause high tensile stresses in the concrete. This can cause the concrete to fail.

The lighter ram and higher striking velocity of the double-acting hammer may be advantageous when driving light- to medium-weight piles into soils having normal frictional resistance. It is claimed that the high frequency of blows will keep a pile moving downward continuously, thus preventing static skin friction from developing between blows. However, when heavy piles are driven, especially into soils having high frictional resistance, the heavier weight and slower velocity of a *single-acting* hammer will transmit a greater portion of the rated energy into driving the piles.

Differential-Acting Hammers

A differential-acting air or steam hammer is a modified single-acting hammer in that the air or steam pressure used to lift the ram is not exhausted at the end of the upward stroke but is valved over the piston to accelerate the ram on the downstroke. The theoretical striking energy is

$$E_r = (W_s + A_{\text{sp}} \times P_i)s \qquad [19.1]$$

Where

E_r = rated striking energy in foot-pounds
W_s = weight of striking parts in pounds
A_{sp} = area of small piston in square inches
P_i = rated operating pressure at the hammer in psi
s = hammer stroke in feet

The number of blows per minute is comparable with that for a double-acting hammer, whereas the weight and the equivalent free fall of the ram are comparable with those of a single-acting hammer. Thus, it is claimed that this type of hammer has the advantages of the single- and double-acting hammers. These hammers require the use of a pile cap with cushioning material and a set of leads.

It is reported that this type hammer will drive a pile in one-half the time required by the same-size single-acting hammer and, in doing so, will use 25 to 35% less air or steam. Table 19.3 includes data for two of these hammers, a Vulcan 200-C and a Raymond 150-C. The values given in the table for rated energy per blow are correct provided the air or steam pressure is sufficient to produce the indicated normal blows per minute.

Diesel Hammers

A diesel pile-driving hammer (see Fig. 19.17) is a self-contained driving unit that does not require an external source of energy such as an air compressor or steam boiler. In this respect, it is simpler and more easily moved from one location to another than a steam hammer. A complete unit consists of a vertical cylinder, a piston or ram, an anvil, fuel- and lubricating-oil tanks, a fuel pump, injectors, and a mechanical lubricator.

After a hammer is placed on top of a pile, the combined piston and ram are lifted to the upper end of the stroke and released to start the unit operating. Figure 19.18 illustrates the operation of a diesel hammer. As the ram nears the end of the downstroke, it activates a fuel pump that injects the fuel into the combustion chamber between the ram and the anvil. The continued downstroke of the ram compresses the air and fuel to ignition heat. The resulting explosion drives the pile downward and the ram upward to repeat its stroke. The energy per blow, which can be controlled by the operator, can be varied over a wide range. Table 19.4 lists the specifications for several makes and models of diesel hammers.

FIGURE 19.17 Diesel hammer driving sheet piling.
Source: Pileco, Inc.

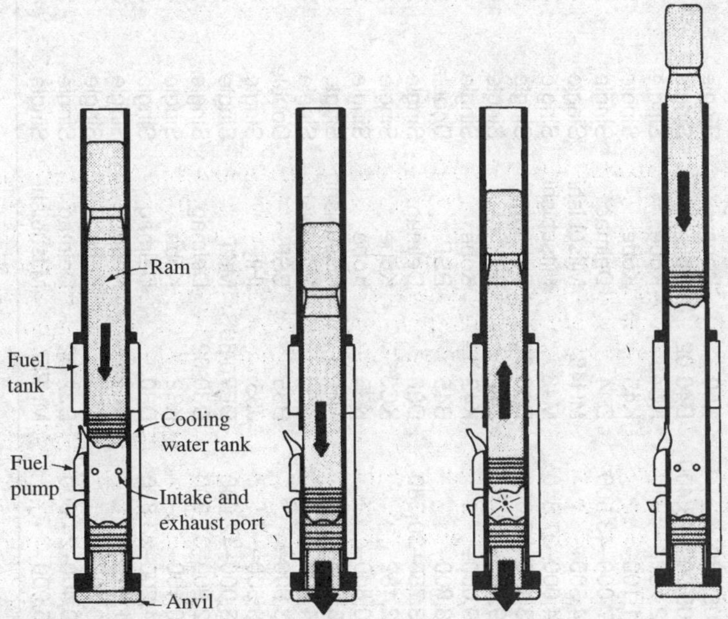

FIGURE 19.18 The operation of a diesel hammer.
Source: L. B. Foster Company.

TABLE 19.4 Specifications for diesel pile-driving hammers.

Energy range (ft-lb)	Model	Manufac-turer	Single/double acting	Blows per min	Piston weight (lb)	Total weight (lb)	Maximum stroke (ft-in.)	Total length (ft-in.)	Width between jaws (in)	Fuel used (gph)
280,000–	K150	Kobe	Single	45–60	33,100	80,500	8'6"	29'8"	CAGE	16–20
161,300–80,600	D62-02	Delmag	Single	36–53	13,670	28,000	12'8"	17'9"	32	5.3
141,000–63,360	MB70	Mitsubishi	Single	38–60	15,840	46,000	8'6"	19'6"	—	7–10
135,200	MH72D	Mitsubishi	Single	38–60	15,900	44,000	8'6"	19'6"	—	7–10
117,175–62,566	D55	Delmag	Single	36–47	11,860	26,300	9'10"	17'9"	32	5.54
105,600–	K60	Kobe	Single	42–60	13,200	37,500	8'0"	24'3"	42	6.5–8.0
105,000–48,400	D46-02	Delmag	Single	37–53	10,100	19,900	10'8"	17'3"	32	3.3
92,752	KC45	Kobe	Single	39–60	9,920	24,700	9'4"	17'11"	—	4.5–5.5
91,100–	K45	Kobe	Single	39–60	9,900	25,600	9'2"	18'6"	36	4.5–5.5
87,000–43,500	D44	Delmag	Single	37–56	9,460	22,440	9'2"	15'10"	32	4.5
84,300	MH45	Mitsubishi	Single	42–60	9,920	24,500	8'6"	17'11"	37	4.0–5.8
84,000–37,840	M43	Mitsubishi	Single	40–60	9,460	22,660	8'10"	16'3"	37	4.0–5.8
83,100–38,000	D36-02	Delmag	Single	37–53	7,900	17,700	10'8"	17'3"	32	3.0
79,500–	J44	IHI	Single	42–70	9,720	21,500	8'2"	14'10"	37	6.86
79,000–	K42	Kobe	Single	40–60	9,260	24,000	8'6"	17'8"	36	4.5–5.5
78,800–	B45	BSP	Double	80–100	10,000	27,500	—	19'3"	36	5.5
73,780–30,380	D36	Delmag	Single	37–53	7,940	17,780	9'3"	14'11"	32	3.7
72,182	KC35	Kobe	Single	39–60	7,720	17,400	9'4"	16'10"	—	3.2–4.3
70,800–	K35	Kobe	Single	39–60	7,700	18,700	9'2"	17'8"	30	3.0–4.0
65,600–	MH35	Mitsubishi	Single	42–60	7,720	18,500	8'6"	17'3"	32	3.4–5.3
64,000–29,040	M33	Mitsubishi	Single	40–60	7,260	16,940	8'0"	13'2"	32	3.4–5.3
63,900–	B35	BSP	Double	80–100	7,700	21,200	—	18'5"	36	4.5
63,500–	J35	IHI	Single	72–70	7,730	16,900	8'3"	14'6"	32	4.76
63,000–42,000	DE70/50B	MKT	Single	40–50	7,000	14,600	10'6"	15'10"	26	3.3
62,900–31,800	D30-02	Delmag	Single	38–52	6,600	13,150	10'7"	17'2"	26	1.7
60,100–	K32	Kobe	Single	40–60	7,050	17,750	8'6"	17'8"	30	2.75–3.5
54,200–23,870	D30	Delmag	Single	39–60	6,600	12,346	8'3"	14'2"	26	2.9
51,518–	KC25	Kobe	Single	39–60	5,510	12,130	9'4"	16'10"	—	2.4–3.2
50,700–	K25	Kobe	Single	39–60	5,510	13,100	9'3"	17'6"	26	2.5–3.0
48,400–24,600	D22-02	Delmag	Single	38–52	4,850	11,400	10'7"	17'2"	26	1.6
46,900	MH25	Mitsubishi	Single	42–60	5,510	13,200	8'6"	16'8"	28	2.4–3.7

Open-end diesel hammers deliver 40 to 55 blows per minute. The closed-end models operate at 75 to 85 blows per minute. In the United States, diesel hammers are almost always used in a set of leads, yet, in other parts of the world, they are commonly used free-hanging. These hammers require a pile cap with "live" cushioning material to protect the pile heads during driving.

A time plot of the force applied to the pile from a diesel hammer is quite different from those for the previously discussed hammers. In the case of a diesel hammer, the force begins to build as soon as the falling ram closes the exhaust ports of the cylinder and compresses the trapped air. With the explosion of the fuel at the bottom of the stroke, there is a force spike that decays as the ram travels upward. The important point is that the loading to the pile spans time and changes magnitude across the time span.

Diesel hammers perform especially well in cohesive or very dense soil layers. The energy per blow increases as the driving resistance of a pile increases. Under normal site conditions, it is usual to select a ratio of ram weight to weight of pile plus cap of 1:2 to 1.5:1. The hammer may not operate well when driving piles into soft ground. Unless a pile offers sufficient driving resistance to activate the ram, the hammer will not operate.

Hydraulic Impact Hammers

Today there is a trend toward the use of hydraulic hammers. They are reported to have an efficiency of 90% or better in delivering energy to the pile. This makes them much more efficient than steam, air, or diesel hammers. A hydraulic hammer operates on the differential pressure of hydraulic fluid instead of compressed air or steam.

Hydraulic Drop Hammer There are hammers where the ram is lifted by hydraulic pressure to a preset height and allowed to free-fall onto the anvil. With these hammers, the height of the drop can be varied to match the pile and ground conditions.

Double-Acting Hydraulic Hammer The ram, which is lifted by hydraulic pressure on a double-acting hydraulic hammer, is also driven hydraulically on its downward stroke. The net energy applied to the pile by the accelerated ram is measured during every blow on a control panel and can be continuously regulated. These hammers can deliver 50 to 60 blows per minute.

Hydraulic Drivers

Press-in hydraulic pile drivers, which can be used for thrusting and extracting steel H piles and steel sheet piles, incorporates a gripping and pushing or pulling technique. This pile driver grips the pile and then pushes the pile down approximately 3 ft. At the end of the downstroke, the pile is released and the gripper slides up the pile 3 ft to begin the process of another push. The equipment can be used in reverse for extracting piles. These drivers develop up to 140 tons of pressing or extracting force, are compact, make minimal noise, and

FIGURE 19.19 A vibratory driver driving steel H pile.

cause very little vibration. They are well suited for driving piles in areas where there is restricted overhead space, since piles can be driven in short lengths and spliced.

Vibratory Pile Drivers

Vibratory pile drivers are especially effective when the piles are driven into water-saturated noncohesive soils. The drivers may experience difficulty in driving piles into dry sand, or similar materials, or into cohesive soils that do not respond to the vibrations. Table 19.5 lists the specifications for several makes and models of vibratory pile drivers.

The drivers are equipped with horizontal shafts, to which eccentric weights are attached. As the shafts rotate in pairs, in opposing directions, at speeds that can be varied in excess of 1,000 rpm, the forces produced by the rotating weights produce vibrations. The vibrations are transmitted to the pile because it is rigidly connected to the driver by clamps. From the pile, the vibrations are transmitted into the adjacent soil. The agitation of the soil materially reduces the skin friction between the soil and the pile. This is especially true when the soil is saturated with water. The combined dead weight of the pile and the driver resting on the pile will drive the pile rapidly. Figure 19.19 shows a vibratory driver driving a steel pile.

Leads are rarely employed with vibratory drivers. The driver is powered either electrically or hydraulically; therefore, a generator or hydraulic power pack is needed as an energy source. Because leads are not required, a smaller crane can usually be employed to handle vibratory driver work.

There are five performance factors that determine effectiveness of a vibratory driver.

1. *Amplitude.* This is the magnitude of the vertical movement of the pile produced by the vibratory unit. It may be expressed in inches or millimeters.
2. *Eccentric moment.* The eccentric moment of a vibratory unit is a basic measure or indication of the size of a driver. It is the product of the weight of the eccentricities multiplied by the distance from the center of rotation of the shafts to the center of gravity of the eccentrics. The heavier the eccentric weights and the farther they are from the center of rotation of the shaft, the greater the eccentric moment of the unit.
3. *Frequency.* This is expressed as the number of vertical movements of the vibrator per minute, which is also the number of revolutions of the rotating shafts per minute. Tests conducted on piles driven by vibratory drivers have indicated that the frictional forces between the piles being driven and the soil into which they are driven are at minimum values when

TABLE 19.5 Specification for hydraulic vibratory pile drivers/extractors.

Dynamic forces (tons)	Model	Manufacturer	Frequency (vpm)	Amplitude (in.)	HP	Max. pull extraction (tons)	Pile clamp force (tons)	Suspended weight (lb)	Shipping weight (lb)	Height (ft-in.)	Depth (ft-in.)	Width	Throat width (in.)
182	V-36	MKT	1600	0.75	550	80	80	18,800	36,300	13–1	1–0	12–0	14
145.4	812	ICE	750–1500	$\frac{1}{2}$–1	330	40	100	14,700	30,200	9–0	2–0	8–0	12
139	S0H1	PTC	1500	1.25	370	44							
100.5	V-20	MKT	1650	0.66	295	40	75	12,500	23,900	5–3	1–2		14
111.4	4000	Foster	1400	0.72	299	40	$\frac{100}{200}$	18,800	32,300	9–10	1–10	9–10	12
78.3	V-16	MKT	1750	0.47	161	50	75	11,700	20,600	5–3	1–2		14
71.0	V-14	MKT	1500	0.32	140	50	75	10,000	29,500	5–3	1–2		14
65.2	416	ICE	800–1600	$\frac{1}{4}$–1	175	40	100	12,200	26,200	8–9	1–10	8–0	12
55	20H6	PTC	1500	0.88	185	22							
48.5	1700	Foster	1400	0.39	147	30	$\frac{80}{100}$	12,900	26,900	7–0	1–10		12
38.8	14H2	PTC	1500	0.85	120	16.5							
36.4	216	ICE	800–1600	$\frac{1}{4}-\frac{3}{4}$	115	20	50	4,500	12,500	6–6	5–0	3–11	12
35.2	1200	Foster	1425	0.34	85	20	60	6,700	11,670	5–0	1–11		12
34.4	7H4	FTC	2000	0.50	115	16.5							
30.0	V-5	MKT	1450	0.50	59	20	31	6,800	10,800	5–4	1–2		14

frequencies are maintained in the range of 700 to 1,200 vibrations per min. In general, the frequencies for piles driven into clay soils should be lower than for piles driven into sandy soils.

4. *Vibrating weight.* The vibrating weight includes the vibrating case and the vibrating head of the vibrator unit, plus the pile being driven.
5. *Nonvibrating weight.* This is the weight of that part of the system that does not vibrate, including the suspension mechanism and the motors. Nonvibrating weights push down on and aid in driving the piles.

A preliminary assessment of the vibratory energy needed to drive a pile section is given by

$$F = 15 \times \frac{(t + 2G)}{100} \qquad \textbf{[19.2]}$$

where

F = centrifugal force in kilonewtons (kN)
t = driving depth in m
G = mass of pile in kg

A penetration rate of 500 mm/min is usually assumed.

SUPPORTING AND POSITIONING PILES DURING DRIVING

batter
A pile driven at an inclination from the vertical to provide more stability.

When driving piles, it is necessary to have a method that will position the pile in the proper location with the required alignment or **batter** and that will support the pile during driving (see Fig. 19.20). The following methods are utilized to accomplish such alignment and support.

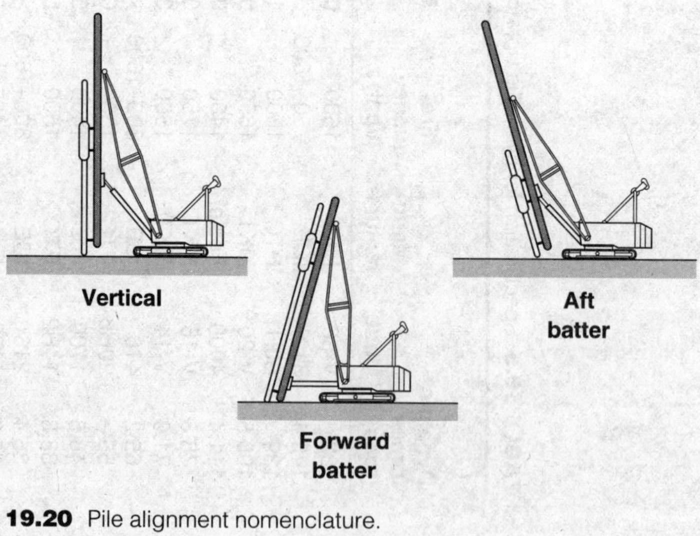

FIGURE 19.20 Pile alignment nomenclature.

FIGURE 19.21 Fixed leads attached to a crawler crane.

Fixed Leads

Fixed leads have a pivot point at the crane's boom top and a brace at their bottom that attaches to the crane (Fig. 19.21). Fixed leads offer good control of the pile position and keep the pile in correct alignment with the hammer so that eccentric impacts that may cause local stress concentrations and pile damage are minimized.

Normally, a set of leads consists of a three-sided steel lattice frame similar in construction to a crane boom with one side open. These are referred to as U-type leads (see Figs. 19.1 and 19.21). Typical technical data for this type of leads can be found at (www.pileco.com/pdf/eng/pileco/u-type.pdf). The open side allows positioning of the pile in the leads and under the hammer. The leads have a set of rails or **guides** for the hammer. Running on the rails, the hammer is lifted above the pile height when a new pile is threaded into the leads. During driving, the hammer descends along the lead rails as the pile moves downward into the ground. When driving batter piles, the leads are positioned at an angle.

guides
The parallel rails of the pile leads that form a pathway for the pile hammer.

Swing Leads

Leads that are not attached at their bottom to the crane or driving platform are known as "swing leads" (Fig. 19.22). The leads and hammer are usually held by separate lines from the crane. Such an arrangement allows the driving rig to position a pile at a location farther away than would be possible with fixed leads. This is not generally the preferred method of driving a pile since it is more difficult to position the pile accurately and to maintain vertical alignment during driving. If for any reason the pile tends to twist or run off the intended alignment, it is difficult to control the pile with swing leads.

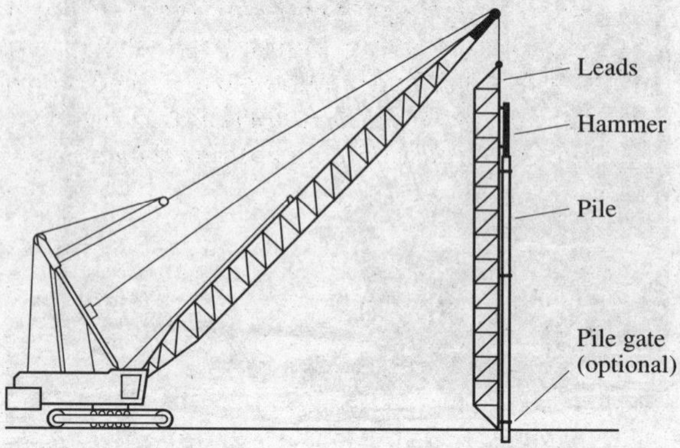

FIGURE 19.22 Swing leads attached to a crawler crane.

Hydraulic Leads

To control pile position, hydraulic leads use a system of hydraulic cylinders connected between the bottom of the leads and the driving rig. This system enables the operator to position the pile very quickly and accurately. Hydraulic leads are extremely useful in driving batter piles, since the system can rapidly and easily adjust the angle of the leads for the required batter. The system is more costly than standard fixed leads.

Templates

Many times a template is used to support and hold the pile in the proper position during driving. Templates are usually constructed from steel pipe or beams and may have several levels of framing to support long piles or piles on a batter (see Fig. 19.7). In marine work, where access to the pile groups is restricted, templates are used regularly. Frequently, a set of leads can be fixed or attached to a template beam, and the combined systems are used to support and guide the pile during driving. Templates or guide frames are commonly used when driving sheet piling. When driving sheet pile cells, circular templates are used to maintain proper alignment (Fig. 19.13).

JETTING PILES

The use of a water jet to assist in driving piles into sand or fine gravel frequently will speed the driving operation. The water, which is discharged through a nozzle at the lower end of a jet pipe, keeps the soil around the pile in agitation, thereby reducing the resistance due to skin friction. Successful jetting requires a plentiful supply of water at a pressure high enough to loosen the soil and to remove it from the hole ahead of the penetration of the pile.

Commonly used jet pipes vary in size from 2 to 4 in. in diameter, with nozzle diameters varying from $\frac{1}{2}$ to $1\frac{1}{2}$ in. There is both low-pressure and high-pressure jetting. Low-pressure jetting has high flow rates and low pressure. Water pressure at the nozzle is between 100 and 300 psi, with the water quantity usually varying from 300 to 500 gallons per minute (gpm). High-pressure jetting may have a flow as low as 50 gpm, but the pressures will reach several thousand psi.

The low-pressure jetting is applicable for any type of pile and is based on reducing friction during driving and avoiding the need for increased driving weight. High-pressure jetting is for driving a pile into hard soils.

Although some piles have been jetted to final penetration, this is not considered good practice primarily because it is impossible to determine the safe supporting capacity of a pile so driven. Most specifications require that piles shall be driven the last few feet without the benefit of jetting. The Foundation Code of the City of New York requires a contractor to obtain special permission prior to jetting piles and specifies that piles shall be driven the last 3 ft with a pile hammer.

SPUDDING AND PREAUGERING

At proposed pile locations where there is evidence of previous construction or when it is suspected the preexisting foundations may interfere with driving operations, it may be prudent to **spud** or preauger to below the depth of the suspected interference. If there is evidence of very hard overlying soil strata, it may be necessary to spud or preauger through the strata prior to beginning the pile-driving operation.

spud
Driving a hard metal point into the ground, then removing and driving the pile.

Preaugering should also be considered where piles are to be driven through an earth fill and into natural soil. This is often the case at bridge abutments where the embankment fill is placed prior to construction of the bridge. The depth of the preaugering should coincide with the depth of the placed fill in order that the piles develop their full bearing capacity in the natural soil.

HAMMER SELECTION

Selecting a suitable hammer or driver for a given project involves consideration of several factors, including (1) the size and type of piles, (2) the number of piles, (3) the character of the soil, (4) the location of the project, (5) the topography of the site, (6) the type of rig (crane) available, and (7) whether driving will be done on land or over water. A pile-driving contractor is usually concerned with selecting a hammer that will drive the piles for a project at the lowest practical cost. As most contractors must limit their ownership to a few representative sizes and types of hammers, a selection should be made from hammers already owned unless conditions are such that it is economical or necessary to purchase or rent an additional size or type driver.

The function of a pile hammer or driver is to furnish the energy required to drive a pile—"energy drives piles." The elementary pile-driving formula is hammer energy equals work of soil resistance; Eq. [19.3] states this relationship and it is illustrated in Fig. 19.23.

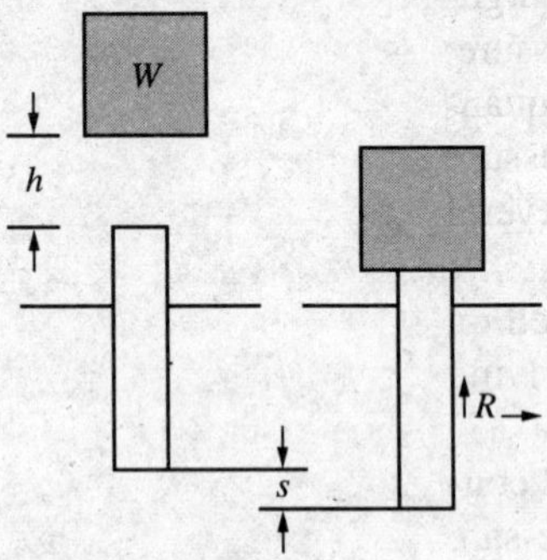

FIGURE 19.23 The fundamental pile-driving formula.

$$Wh = Rs \qquad [19.3]$$

where

W = weight of the falling mass (hammer piston) in pounds
h = height of free fall for mass W in feet
R = soil resistance in pounds
s = penetration of the pile in feet (set of the pile)

The *Engineering News* pile formula used for estimating pile capacities is built on this relationship. The equation for a pile driven with a single-acting hammer is

$$R = \frac{2WH}{S + 0.1} \qquad [19.4]$$

where

R = safe load on a pile in pounds
W = weight of a falling mass in pounds
H = height of free fall for mass W in feet
S = average penetration per blow for last 5 or 10 blows in inches

Note that there is a safety factor of 6 built into Eq. [19.4]. Rearranging Eq. [19.3] and including a requirement that 6 times the load must be supported yields

$$6R = \frac{W(\text{lb}) \times h(\text{ft})}{s(\text{ft})}$$

In the *Engineering News* formula, s is stated in inches; therefore

$$6R = \frac{W(\text{lb}) \times h(\text{ft})}{s(\text{in})} \times \frac{12 \text{ in.}}{\text{ft}} \text{ or}$$

$$R = \frac{2WH}{S}$$

The theoretical energy per blow will equal the product of the weight times the equivalent free fall. The theoretical energy per blow for a differential acting hammer is given by Eq. [19.1]. Since some of this energy is lost in friction as the weight travels downward, the net energy per blow will be less than the theoretical energy. The actual amount of net energy depends on the efficiency of the particular hammer. The efficiencies of the pile hammers vary from 50 to 100%. For a full discussion of driving energy, hammer efficiency, and energy losses see "Energy and Energy Transfer in Pile Driving Systems" (www.vulcanhammer.net/wave/efficnt.htm).

TABLE 19.6 Recommended sizes of hammers for driving various types of piles.*

Length of piles (ft)	Depth of penetration	Weight of various types of piles (lb/lin. ft)						
		Steel sheet†			Timber		Concrete	
		20	30	40	30	60	150	400
Driving through ordinary earth, moist clay, and loose gravel, normal frictional resistance								
25	½	2,000	2,000	3,600	3,600	7,000	7,500	15,000
	Full	3,600	3,600	6,000	3,600	7,000	7,500	15,000
50	½	6,000	6,000	7,000	7,000	7,500	15,000	20,000
	Full	7,000	7,000	7,500	7,500	12,000	15,000	20,000
75	½	—	7,000	7,500	—	15,000	—	30,000
	Full	—	—	12,000	—	15,000	—	30,000
Driving through stiff clay, compacted sand, and gravel, high frictional resistance								
25	½	3,600	3,600	3,600	7,500	7,500	7,500	15,000
	Full	3,600	7,000	7,000	7,500	7,500	12,000	15,000
50	½	7,000	7,500	7,500	12,000	12,000	15,000	25,000
	Full	—	7,500	7,500	—	15,000	—	30,000
75	½	—	7,500	12,000	—	15,000	—	36,000
	Full	—	—	15,000	—	20,000	—	50,000

*Size expressed in foot-pounds of energy per blow.

†The indicated energy is based on driving two steel sheet piles simultaneously. In driving single piles, use approximately two-thirds the indicated energy.

Table 19.6 gives recommended sizes of hammers for different types and sizes of piles and driving resistances. The sizes are indicated by the theoretical foot-pounds of the energy delivered per blow. The theoretical energy per blow given in Tables 19.3, 19.4, and 19.5 is correct provided the hammer is operated at the designated number of strokes per minute.

In general, it is sensible practice to select the largest hammer that can be used without overstressing or damaging a pile. As previously discussed, when a large hammer is used, a greater portion of the energy is effective in driving the pile, and that produces a higher operating efficiency. Therefore, the hammer sizes given in Table 19.6 should be considered as the minimum sizes. In some instances, hammers as much as 50% larger may be used advantageously.

Consideration must also be given to the capabilities of the crane that will be employed to handle the hammer, leads, and pile. The crane must be able to handle this total load at the maximum reach that will be required by the project site conditions. Therefore, while it is important to know the hammer's rated energy and speed (blows per minute) when considering the question of capability to drive the required piles, the total operating weight specification is important when selecting the crane that will support the hammer (see Fig. 19.24).

Working specifications	
Rated energy	42,000 ft-lb (5,807 kg-m)
Minimum energy	16,000 ft-lb (2,212 kg-m)
Stroke at rated energy	10' 3" (312 cm)
Maximum obtainable stroke	10' 5" (318 cm)
Speed (blows per minute)	37-55
Bearing based on EN formula	210 tons (190 tons)
Weights	
Bare hammer	7,610 lbs (3,452 kg)
Ram	4,088 lbs (1,854 kg)
Anvil	545 lbs (247 kg)
Typical operating weight with cap	8,710 lb (3,950 kg)

FIGURE 19.24 Typical pile hammer working specifications.

Dimensions of hammer	
Width (side to side)	20" (508 cm)
Depth	29" (737 cm)
Centerline to front	13 3/4" (349 mm)
Centerline to rear	15 1/4" (387 mm)
Length (hammer only)	16' 1" (490 cm)
Operating length (top of ram to top of pile)	27' 9" (846 cm)

FIGURE 19.25 Typical pile hammer dimensions specification.

The final question to be considered is how much boom will the crane require? This is dependent on the length of the piles and the operating length of the hammer. The hammer specifications will provide this information (see Fig. 19.25). To the sum of these two dimensions should also be added an allowance for positioning the pile under the hammer.

PILE-DRIVING SAFETY

Pile driving is among the most hazardous of all construction operations. Accidents associated with pile driving are usually severe, due to the tremendous forces and weights involved. Here are some rules to help keep you and your workforce safe.

1. All hose connections, including air hammer hoses and steam lines, that lead to the hammer must be securely attached to the hammer using at least $\frac{1}{4}$-in.-diameter chain or cable. This safety prevents the line from whipping in the event the joint at the hammer is broken.
2. Hydraulic hammer hoses can be under high pressure and heat, so you must always protect these lines.
3. Hanging or swinging pile driver leads must have fixed ladders for access. Fixed leads shall be provided with fall protection anchor points, so that the aloft worker may engage fall protection lanyards to the leads.
4. Workers cannot stay on the leads or ladders when piles are being driven.
5. Whenever working under the hammer, a blocking device capable of safely supporting the hammer must be placed in the leads.
6. When the pile driver is being moved, the hammer must be lowered to the bottom of the leads.
7. Tag lines shall be used for controlling unguided piles and flying hammers.
8. All workers must wear hearing protection during pile driving operations.

Specific Occupational Safety and Health Administration requirements applicable to pile driving can be found under Standard 1926.603. Standard 1926.603(c)(2) states, "All employees shall be kept clear when piling is being hoisted into the leads." This was not observed during a driving operation in 2003 and resulted in the death of a 35-year-old construction worker who was standing nearby to position a compressed air hose. The worker was struck by a 2,000-lb steel pile, which fell while it was being raised by the crane.

SUMMARY

Load-bearing piles are used primarily to transmit structural loads through soil formations with inadequate supporting properties into or onto soil stratum capable of supporting the loads. If the load is transmitted to the soil through skin friction between the surface of the pile and the soil, the pile is called a *friction pile*. If the load is transmitted to the soil through the lower tip, the pile is called an *end-bearing pile*.

Piles can be classified on the basis of the materials from which they are made. The common pile types in terms of materials are (1) timber, (2) concrete, (3) steel, and (4) composite. The function of a pile hammer is to furnish the energy required to drive a pile. Pile-driving hammers are designated by type and size. When driving piles, it is necessary to have a method that will position the pile in the proper location with the required alignment or batter, and which will support the pile during driving. Critical learning objectives include:

- An understanding of the different pile types and the advantages of each
- An understanding of the different types of pile-driving hammers and the advantages of each

These objectives are the basis for the problems that follow.

PROBLEMS

19.1 Describe the main function of a pile foundation.

19.2 Piles are made out of different materials. List the major advantages and disadvantages of prestressed concrete piles.

19.3 Use the Web to investigate the advantages and disadvantages of using a cable-operated drop hammer to drive piles.

19.4 Use the Web to investigate the advantages and disadvantages of using a single-acting hammer (air or steam) to drive piles.

19.5 Investigate on the Web the advantages and disadvantages of using a double-acting hammer (air or steam) to drive piles.

19.6 Why are leads used when driving piles?

19.7 A Vulcan model 016 single-acting hammer is used to drive a pile. The stroke of this hammer is 36 in. The average penetration per blow for the last 10 blows is $\frac{1}{4}$ in., what will be the safe rated load using the *Engineering News* equation?

REFERENCES

1. Adams, James, James Dees, and James Graham (1996). "'Clean' Creosote Timber," *The Military Engineer*, Vol. 88, No. 578, pp. 25, 26, June–July.
2. *Construction Productivity Advancement Research (CPAR) Program* (1998). USACERL Technical Report 98/123, U.S. Army Corps of Engineers, Construction Engineering Research Laboratories, September.
3. *Design of Pile Foundations, Technical Engineering and Design Guides as Adapted from the US Army Corps of Engineers, No. 1* (1993). American Society of Civil Engineers, New York.
4. *Guidelines for the Design and Installation of Pile Foundations* (1997). ASCE Standard 20-96, American Society of Civil Engineers, Reston, VA.
5. *Manual for Quality Control,* 4th ed. (1999). PCI MNL-116-99, Prestressed Concrete Institute, 201 N. Wells St., Chicago, IL.

WEBSITE RESOURCES

1. www.usacivil.skanska.com Bayshore Concrete Products, Cape Charles, VA.
2. www.berminghammer.com Berminghammer Corporation Limited, Wellington St. Marine Terminal, Hamilton, Ontario L8L 4Z9, Canada.
3. www.bethsteel.com Bethlehem Steel Corporation, Piling Products, Bethlehem, PA.
4. www.pilebuck.com/paperbooks/dictionary.html A dictionary of pile-driving terminology.
5. www.piledrivers.org/index.php The Pile Driving Contractors Association (PDCA), Glenwood Springs, CO, is an organization of pile-driving contractors.
6. www.usacivil.skanska.com Tidewater Skanska, Inc., based in Virginia Beach, VA, is a subsidiary of Skanska USA Civil. Skanska USA Civil is a business unit of Skanska AB, an international construction company based in Stockholm, Sweden.

CHAPTER 20

Air Compressors and Pumps

Compressed air is used extensively on construction projects for powering rock drills and pile-driving hammers. As air is a gas, it obeys the fundamental laws that apply to gases. The loss in pressure due to friction as air flows through a pipe or a hose is a factor that must be considered in selecting the size of a pipe or hose to transport compressed air to equipment or tools. Most projects require the use of one or more water pumps at various stages during their construction. As with compressed air, when pumping water, friction losses in pipes and hoses must be analyzed before selecting a pump for a given job.

SUPPORT EQUIPMENT

The first air compressors were human lungs; by blowing on cinders man started his fires. Then with the birth of metallurgy man began to melt metals and higher temperatures were needed. A more powerful compressor was required. The first mechanical compressor, the hand-operated bellows, emerged in 1500 B.C. Today compressed air is used extensively on construction projects for powering rock drills, pile-driving hammers (see Fig. 20.1), air motors, hand tools, and pumps. Air compressors are often used to provide the power to operate construction tools and equipment. An air compressor does not drive a pile, it supplies the air that powers the hammer that drives the pile.

Similarly, pumping equipment used to move water is usually not an activity that directly contributes to the building of a project. But the creation of excavations below the water table cannot be efficiently accomplished without effective removal of water. Pumps, going back to the ancient Persian water wheels, which lifted water by an endless system of ropes upon which a series

FIGURE 20.1 Air compressor mounted on the back of a crane for powering a pile hammer.

of clay pots moved, are indispensable to construction work. Those Persian wheels were the models for water lift systems later used in Europe's underground mines.

When dealing with air or water, the principles are the same: there is a quantity of fluid, gas, or liquid, that is being moved and there will be friction losses as the fluid moves from the compressor or pump through pipes and hoses. These friction losses affect the efficiency of the operation and must be considered when engineering the compressors or pumps to meet the project needs.

COMPRESSED AIR

INTRODUCTION

In many instances, the energy supplied by compressed air is the most convenient method of operating equipment and tools. A compressed-air system consists of one or more compressors together with distribution lines to carry the air to the points of use.

When air is compressed, it receives energy from the compressor. This energy is transmitted through a pipe or hose to the operating equipment, where a portion of the energy is converted into mechanical work. The operations of compressing, transmitting, and using air will always result in a loss of energy that will give an overall efficiency of less than 100 percent.

As air is a gas, it obeys the fundamental laws that apply to gases. These laws relate to the pressure, volume, temperature, and transmission of air.

GLOSSARY OF GAS LAW TERMS

The following glossary defines the important terms used in developing and applying the laws that relate to compressed air.

Absolute pressure. The measurement of pressure relative to the pressure in a vacuum—the pressure that would occur relative to absolute zero pressure. It is equal to the sum of the gauge and the atmospheric pressure, corresponding to the barometric reading. The absolute pressure is used in dealing with the gas laws.

Absolute temperature. The temperature of a gas measured above absolute zero. The Kelvin scale is called absolute temperature and the Kelvin is the SI unit for temperature. Kelvin temperatures are written without a degree (°) symbol. In the English system, absolute temperature is in degrees Rankine (R).

Let T_K = Kelvin temperature, T_F = Fahrenheit temperature, and T_R = Celsius temperature. The relationships are then

$$T_K^\circ = \left[\tfrac{5}{9}\left(T_F^\circ - 32^\circ\right)\right] + 273 \qquad \textbf{[20.1]}$$

$$T_R^\circ = T_F^\circ + 456.69^\circ \qquad \textbf{[20.2]}$$

It is common practice to use 460°.

Atmospheric pressure. The pressure exerted at the surface of a body by a column of air in an atmosphere—the pressure in the surrounding air. It varies with temperature and altitude above sea level.

Density of air. The weight of a unit volume of air, usually expressed as pounds per cubic foot (pcf). Density varies with the pressure and temperature of the air. For the standard conditions, weight of air at 59°F and 14.7 psi absolute pressure, the density of air is 0.07658 pcf or 1.2929 kg/m^3.

Fahrenheit temperature. The temperature indicated by a measuring device calibrated according to the Fahrenheit scale. For this thermometer, at a pressure of 14.7 psi, pure water freezes at 32°F and boils at 212°F.

Gauge pressure. The pressure exerted by the air in excess of atmospheric pressure. It is usually expressed in psi or inches of mercury and is measured by a pressure gauge or a mercury manometer.

Pressure. The relationship of a force (F) acting on a unit area (A). It is usually denoted by P and can be measured in atmospheres, inches of mercury, millimeters of mercury, or pascals.

$$P = \frac{F}{A} \qquad \textbf{[20.3]}$$

Standard conditions. Because of the variations in the volume of air with pressure and temperature, it is necessary to express the volume at standard conditions if it is to have a definite meaning. Standard conditions are an absolute pressure of 14.696 psi (14.7 psi is used in practice) and a temperature of 59°F (288 K).

Temperature. Temperature is a measure of the amount of heat contained by a unit quantity of a material. It is that property which governs the transfer of thermal energy, or heat between one system and another.

Vacuum. Its definition is not fixed but it is commonly taken to mean pressures below atmospheric pressure. Therefore, it is a measure of the extent to which pressure is less than atmospheric pressure. For example, a vacuum of 5 psi is equivalent to an absolute pressure of 14.7 − 5 = 9.7 psi.

GAS LAWS

Gases behave differently than solids and liquids. A gas has neither fixed volume nor shape. It assumes the shape of the container in which it is held. There are three properties that interact when working with a gas: pressure, volume, and temperature. Changes in one property will affect the others.

In 1662, Robert Boyle conducted an investigation where he fixed the amount of gas and its temperature during the experiment. He found that when he changed the pressure the volume responded in the opposite direction. This relation is expressed by the equation

$$P_1V_1 = P_2V_2 = K \qquad [20.4]$$

where

P_1 = initial absolute pressure
V_1 = initial volume
P_2 = final absolute pressure
V_2 = final volume
K = a constant

When a gas undergoes a change in volume or pressure with a change in temperature, Boyle's law will not apply.

Nearly a century after Boyle, hot-air balloons were extremely popular in France and scientists were eager to improve the performance of balloons. The prominent French scientist Jacques Charles made detailed measurements on how the volume of a gas was affected by temperature. He found that the volume of a given weight of gas at constant pressure varies in direct proportion to its absolute temperature. Mathematically stated,

$$\frac{V_1}{T_1} = \frac{V_2}{T_2} = C \qquad [20.5]$$

where

V_1 = initial volume
T_1 = initial absolute temperature (A)

V_2 = final volume
T_2 = final absolute temperature
C = a constant

The laws of Boyle and Charles can be combined to give the equation

$$\frac{P_1V_1}{T_1} = \frac{P_2V_2}{T_2} = \text{a constant} \qquad \textbf{[20.6]}$$

Equation [20.6] can be used to express the relations between pressure, volume, and temperature for any given gas, such as air. This is illustrated in Example 20.1.

EXAMPLE 20.1

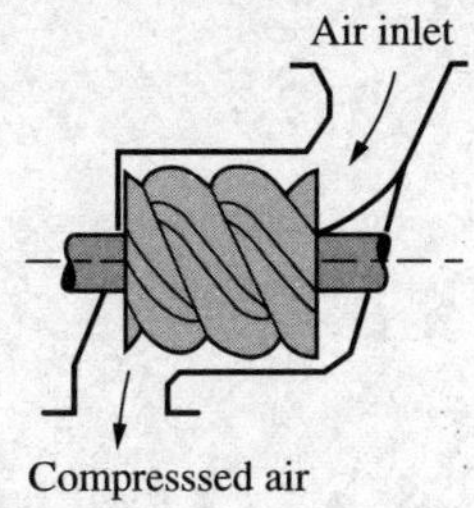

One thousand cubic feet of air, at an initial gauge pressure of 40 psi and temperature of 50°F, is compressed to a volume of 200 cf at a final temperature of 110°F. Determine the final gauge pressure. The atmospheric pressure is 14.46 psi.

$$P_1 = 40 \text{ psi} + 14.46 \text{ psi} = 54.46 \text{ psi}$$

$$V_1 = 1{,}000 \text{ cf}$$

$$T_1 = 460° + 50°\text{F} = 510°\text{A}$$

$$V_2 = 200 \text{ cf}$$

$$T_2 = 460° + 110°\text{F} = 570°\text{A}$$

Rewriting Eq. [20.6] and substituting these values, we get

$$P_2 = \frac{P_1V_1}{T_1} \times \frac{T_2}{V_2} = \frac{54.46 \text{ psi} \times 1{,}000 \text{ cf}}{510°\text{A}} \times \frac{570°\text{A}}{200 \text{ cf}} = 304 \text{ psi}$$

Final gauge: 304.34 psi − 14.46 psi = 289.88 psi or 290 psi

Energy Required to Compress Air

When an air compressor increases the pressure of a given volume of air, it is necessary to furnish energy to the air. Air is drawn into the compressor at pressure P_1 and is discharged at pressure P_2. If a gas undergoes a change in volume without a change in temperature, the process is said to be *isothermal compression or expansion*. When a gas undergoes a change in volume without gaining or losing heat, this is referred to as *adiabatic compression or expansion*. For air compressors used on construction projects, the compression will not be performed under isothermal conditions. Air compressors work at conditions somewhere between pure isothermal (no change in temperature) compression and adiabatic (without gaining or losing heat) compression. The actual compression condition for a given compressor can be determined experimentally.

GLOSSARY OF AIR COMPRESSOR TERMS

The following glossary defines the important terms that are pertinent to air compressors.

Aftercooler. A heat exchanger that cools the air after it is discharged from a compressor.

Centrifugal compressor. A machine in which the compression is effected by a rotating vane or impeller that imparts velocity to the flowing air to give it the desired pressure.

Compression ratio. The ratio of the absolute discharge pressure to the absolute inlet pressure.

Compressor efficiency. The ratio of the theoretical horsepower to the brake horsepower required by a compressor.

Discharge pressure. The absolute pressure of the air at the outlet from a compressor.

Diversity factor. The ratio of the actual quantity of air required for all uses to the sum of the individual quantities required for each use.

Free air. Air as it exists under atmospheric conditions at any given location.

Inlet pressure. The absolute pressure of the air at the inlet to a compressor.

Intercooler. A heat exchanger that is placed between two compression stages to remove the heat of compression from the air.

Load factor. The ratio of the average load during a given period of time to the maximum rated load of a compressor.

Multistage compressor. A compressor that produces the desired final pressure through two or more stages of compression.

Reciprocating compressor. A machine that compresses air by means of a piston reciprocating in a cylinder.

Single-stage compressor. A machine that compresses air from atmospheric pressure to the desired discharge pressure in a single compression operation.

Two-stage compressor. A machine that compresses air in two separate operations. The first operation compresses the air to an intermediate pressure, whereas the second operation further compresses it to the desired final pressure.

AIR COMPRESSORS

The capacity of an air compressor is determined by the amount of free air that it can compress to a specified pressure in 1 minute, under standard conditions (absolute pressure of 14.7 psi at 59°F). The number of pneumatic tools that can be operated from one air compressor depends on the air requirements of the specific tools.

Stationary Compressors

Stationary compressors are generally used for installations where there will be a requirement for compressed air over a long duration of time at fixed locations. One or more compressors may supply the total quantity of air. The installed cost of a single compressor will usually be less than for several compressors having the same capacity. However, several compressors provide flexibility for varying load demands, and in the event of a shutdown for repairs the entire plant does not need to be stopped.

Portable Compressors

Portable compressors are commonly used on construction sites where it is necessary to meet frequently changing job demands, typically at a number of locations on the job site. The compressors may be mounted on rubber tires (see Fig. 20.2) or skids. They are usually powered by gasoline or diesel engines, and most of those used in construction work are of the rotary type. These compressors effect compression by the action of rotating elements.

Rotary Compressors

The working parts of a rotary-screw compressor are two helical rotors. The male rotor has four lobes and rotates 50% faster than the female rotor, which has six flutes with which the male rotor meshes. As the air enters and flows through the compressor, it is compressed in the space between the lobes and the flutes. The inlet and outlet ports are automatically covered and uncovered by the shaped ends of the rotors as they turn. There are also monoscrew compressors that operate in a similar manner. Figure 20.3 shows a rotary-screw air compressor.

FIGURE 20.2 Portable air compressor on a construction project.

FIGURE 20.3 Portable rotary-screw air compressor, 185 cfm.
Source: Sullair Corp.

Rotary-screw compressors are available in a relatively wide range of capacities, with single-stage or multistage compression and with rotors that operate under oil-lubricated conditions or with no oil, the latter to produce oil-free air.

Effect of Altitude on Compressor Capacity

The capacity of an air compressor is rated on the basis of its performance at sea level, where the normal absolute barometric pressure is about 14.7 psi. If a compressor is operated at a higher altitude, such as 5,000 ft above sea level, the absolute barometric pressure will be about 12.2 psi. Thus, at the higher altitude, air density is less and the weight of air in a cubic foot of free volume is less than at sea level. If the air is discharged by the compressor at a given pressure, the compression ratio will be increased, and the capacity of the compressor will be reduced. This may be demonstrated by applying Eq. [20.4].

Assume the 100 cf of free air at sea level is compressed to 100 psi gauge with no change in temperature. Applying Eq. [20.4], we obtain

$$V_2 = \frac{P_1 V_1}{P_2}$$

where

$V_1 = 100$ cf

$P_1 = 14.7$ psi absolute

$P_2 = 114.7$ psi absolute

$$V_2 = \frac{14.7 \text{ psi} \times 100 \text{ cf}}{114.7 \text{ psi}} = 12.82 \text{ cf}$$

At 5,000 ft above sea level,

$V_1 = 100$ cf

$P_1 = 12.2$ psi absolute

$P_2 = 112.2$ psi absolute

$$V_2 = \frac{12.2 \text{ psi} \times 100 \text{ cf}}{112.2 \text{ psi}} = 10.87 \text{ cf}$$

COMPRESSED-AIR DISTRIBUTION SYSTEM

The objective of installing a compressed-air distribution system is to provide a sufficient volume of air to the work locations at pressures adequate for efficient tool operation. Any drop in pressure between the compressor and the point of use is an irretrievable loss. Therefore, the distribution system is an important element in the air supply scheme. The following are general rules for designing a compressed-air distribution system:

- Pipe sizes should be large enough so that the pressure drop between the compressor and the point of use does not exceed 10% of the initial pressure.

- Each header or main line should be provided with outlets as close as possible to the point of use. This permits shorter hose lengths and avoids large pressure drops through the hose.
- Condensate drains should be located at appropriate low points along the header or main lines.

Air Manifolds

Many construction projects require more compressed air per minute than any single job-site compressor will produce. An air manifold is a large-diameter pipe used to transport compressed air from one or more air compressors without a detrimental friction-line loss.

Manifolds can be constructed of any durable pipe. Compressors are connected to the manifold with flexible hoses. A one-way check valve must be installed between the compressor and the manifold. This valve keeps manifold back pressure from possibly forcing air back into an individual compressor's receiver tank. The compressors grouped to supply an air manifold may be of different capacities, but the final discharge pressure of each should be coordinated at a specified pressure. In the case of construction, this is usually 100 psi. Compressors of different types should not be used on the same manifold.

Loss of Air Pressure in Pipe

The loss in pressure due to friction as air flows through a pipe or a hose is a factor that must be considered in selecting the size of a pipe or hose. Failure to use a sufficiently large pipe may cause the air pressure to drop so low that it will not satisfactorily operate the tool to which it is providing power.

Pipe size selection for an air line is a productivity (economics) problem. The efficiency of most equipment operated by compressed air drops off rapidly as the pressure of the air is reduced.

> The manufacturers of pneumatic equipment generally specify the minimum air pressure at which their equipment will operate satisfactorily. However, these values should be considered as minimum and not desirable operating pressures. The actual pressure should be higher than the specified minimum.

Several formulas are used to determine the loss of pressure in a pipe due to friction. Equation [20.7] is a general formula:

$$f = \frac{CL}{r} \times \frac{Q^2}{d^5} \qquad [20.7]$$

where

f = pressure drop in psi
L = length of pipe in feet
Q = cubic feet of free air per second

r = ratio of compression

d = actual inside diameter (I.D.) of pipe in inches

C = experimental coefficient

For ordinary steel pipe, the value of C has been found to equal $0.1025/d^{0.31}$. If this value is substituted in Eq. [20.7], the result is

$$f = \frac{0.1025L}{r} \times \frac{Q^2}{d^{5.31}} \tag{20.8}$$

A chart for determining the loss in pressure in a pipe is given in Fig. 20.4.

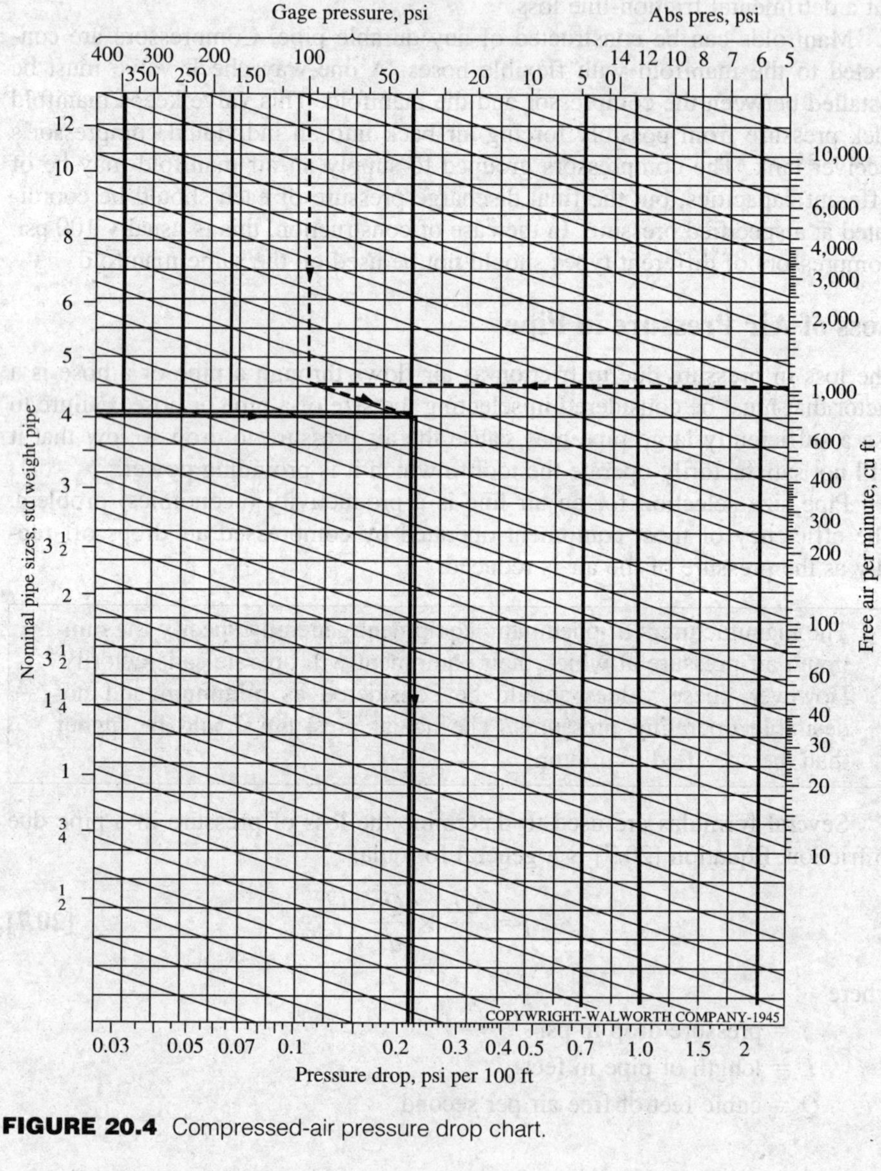

FIGURE 20.4 Compressed-air pressure drop chart.

EXAMPLE 20.2

Determine the pressure loss per 100 ft of pipe resulting from transmitting 1,000 cfm of free air, at 100 psi gauge pressure, through a 4-in. standard-weight steel pipe.

This question can be answered by using the Fig. 20.4 chart.

Enter the chart on the top horizontal scale at 100 psi and proceed vertically downward to a point opposite 1,000 cfm (right vertical scale). From the 1,000 cfm intersect point, proceed parallel to the sloping guide lines to a point opposite the 4-in. pipe (left vertical scale); then proceed vertically downward to the bottom horizontal scale of the chart, where the pressure drop is indicated to be 0.225 psi per 100 ft of pipe.

Table 20.1 gives the loss of air pressure in 1,000 ft of standard-weight pipe due to friction. For longer or shorter lengths of pipe, the friction loss will be in proportion to the length.

The losses given in the table are for an initial gauge pressure of 100 psi. If the initial pressure is other than 100 psi, the corresponding losses may be obtained by multiplying the values in Table 20.1 by a suitable factor. Reference to Eq. [20.8] reveals that for a given rate of flow through a given size pipe the only variable is r, that is, the ratio of compression, based on absolute pressures. For a gauge pressure of 100 psi, $r = 114.7/14.7 = 7.80$, whereas for a gauge pressure of 80 psi, $r = 94.7/14.7 = 6.44$. The ratio of these values of $r = 7.80/6.44 = 1.211$. Thus, the loss for an initial pressure of 80 psi will be 1.211 times the loss for an initial pressure of 100 psi.

To calculate the loss of pressure resulting from the flow of air through fittings, it is common practice to convert a fitting to equivalent lengths of pipe having the same nominal diameter. This equivalent length should be added to the actual length of the pipe in determining pressure loss. Table 20.2 gives the equivalent length of standard-weight pipe for computing pressure losses.

EXAMPLE 20.3

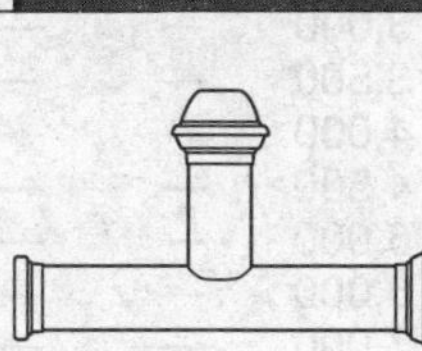

A 6-in. pipe with screwed fittings is used to transmit 2,000 cfm of free air at an initial pressure of 100-psi gauge pressure. The pipeline includes the following items: 4,000 ft of pipe, eight standard on-run tees, four gate valves, and eight long-radius ells. Determine the total loss of pressure in the pipeline.

Size of pipe, 6 in. $V = 2{,}000$ cfm
Length of pipe, 4,000 ft $P_1 = 100$ psi gauge

Using the Table 20.2 data, the equivalent length of pipe will be

Pipe length		= 4,000.0 ft
Gate valves	4 × 3.5 (Table 11.5) =	14.0 ft
Standard on-run tees	8 × 15.2 (Table 11.5) =	121.6 ft
Ells, long-radius ells	8 × 6.1 (Table 11.5) =	48.8 ft
		Total 4,184.4 ft

Table 20.1: 0.99 psi loss per 1,000 ft of pipe. Therefore,

$$\frac{4{,}184.4 \text{ ft}}{1{,}000 \text{ ft}} \times 0.99 \text{ psi} = 4.14 \text{ psi}$$

TABLE 20.1 Loss of pressure in psi in 1,000 ft of standard-weight pipe due to friction for an initial gauge pressure of 100 psi.

Free air per min (cf)	Nominal diameter (in.)												
	$\frac{1}{2}$	$\frac{3}{4}$	1	$1\frac{1}{4}$	$1\frac{1}{2}$	2	$2\frac{1}{2}$	3	$3\frac{1}{2}$	4	$4\frac{1}{2}$	5	6
10	6.50	0.99	0.28										
20	25.90	3.90	1.11	0.25	0.11								
30	68.50	9.01	2.51	0.57	0.26								
40	—	16.00	4.45	1.03	0.46								
50	—	25.10	6.96	1.61	0.71	0.19							
60	—	36.20	10.00	2.32	1.02	0.28							
70	—	49.30	13.70	3.16	1.40	0.37							
80	—	64.50	17.80	4.14	1.83	0.49	0.19						
90	—	82.80	22.60	5.23	2.32	0.62	0.24						
100	—	—	27.90	6.47	2.86	0.77	0.30						
125	—	—	48.60	10.20	4.49	1.19	0.46						
150	—	—	62.80	14.60	6.43	1.72	0.66	0.21					
175	—	—	—	19.80	8.72	2.36	0.91	0.28					
200	—	—	—	25.90	11.40	3.06	1.19	0.37	0.17				
250	—	—	—	40.40	17.90	4.78	1.85	0.58	0.27				
300	—	—	—	58.20	25.80	6.85	2.67	0.84	0.39	0.20			
350	—	—	—	—	35.10	9.36	3.64	1.14	0.53	0.27			
400	—	—	—	—	45.80	12.10	4.75	1.50	0.69	0.35	0.19		
450	—	—	—	—	58.00	15.40	5.98	1.89	0.88	0.46	0.25		
500	—	—	—	—	71.60	19.20	7.42	2.34	1.09	0.55	0.30		
600	—	—	—	—	—	27.60	10.70	3.36	1.56	0.79	0.44		
700	—	—	—	—	—	37.70	14.50	4.55	2.13	1.09	0.59		
800	—	—	—	—	—	49.00	19.00	5.89	2.77	1.42	0.78		
900	—	—	—	—	—	62.30	24.10	7.60	3.51	1.80	0.99		
1,000	—	—	—	—	—	76.90	29.80	9.30	4.35	2.21	1.22		
1,500	—	—	—	—	—	—	67.00	21.00	9.80	4.90	2.73	1.51	0.57
2,000	—	—	—	—	—	—	—	37.40	17.30	8.80	4.90	2.73	0.99
2,500	—	—	—	—	—	—	—	58.40	27.20	13.80	8.30	4.20	1.57
3,000	—	—	—	—	—	—	—	84.10	39.10	20.00	10.90	6.00	2.26
3,500	—	—	—	—	—	—	—	—	58.20	27.20	14.70	8.20	3.04
4,000	—	—	—	—	—	—	—	—	69.40	35.50	19.40	10.70	4.01
4,500	—	—	—	—	—	—	—	—	—	45.00	24.50	13.50	5.10
5,000	—	—	—	—	—	—	—	—	—	55.60	30.20	16.80	6.30
6,000	—	—	—	—	—	—	—	—	—	80.00	43.70	24.10	9.10
7,000	—	—	—	—	—	—	—	—	—	—	59.50	32.80	12.20
8,000	—	—	—	—	—	—	—	—	—	—	77.50	42.90	16.10
9,000	—	—	—	—	—	—	—	—	—	—	—	54.30	20.40
10,000	—	—	—	—	—	—	—	—	—	—	—	—	25.10
11,000	—	—	—	—	—	—	—	—	—	—	—	—	30.40
12,000	—	—	—	—	—	—	—	—	—	—	—	—	36.20
13,000	—	—	—	—	—	—	—	—	—	—	—	—	42.60
14,000	—	—	—	—	—	—	—	—	—	—	—	—	49.20
15,000	—	—	—	—	—	—	—	—	—	—	—	—	56.60

TABLE 20.2 Equivalent length in feet of standard-weight pipe having the same pressure losses as screwed fittings.

Nominal pipe size (in.)	Gate valve	Globe valve	Angle valve	Long-radius ell or on-run of standard tee	Standard ell or on-run of tee	Tee through side outlet
$\frac{1}{2}$	0.4	17.3	8.6	0.6	1.6	3.1
$\frac{3}{4}$	0.5	22.9	11.4	0.8	2.1	4.1
1	0.6	29.1	14.6	1.1	2.6	5.2
$1\frac{1}{4}$	0.8	38.3	19.1	1.4	3.5	6.9
$1\frac{1}{2}$	0.9	44.7	22.4	1.6	4.0	8.0
2	1.2	57.4	28.7	2.1	5.2	10.3
$2\frac{1}{2}$	1.4	68.5	34.3	2.5	6.2	12.3
3	1.8	85.2	42.6	3.1	6.2	15.3
4	2.4	112.0	56.0	4.0	7.7	20.2
5	2.9	140.0	70.0	5.0	10.1	25.2
6	3.5	168.0	84.1	6.1	15.2	30.4
8	4.7	222.0	111.0	8.0	20.0	40.0
10	5.9	278.0	139.0	10.0	25.0	50.0
12	7.0	332.0	166.0	11.0	29.8	59.6

Recommended Pipe Sizes

No book, table, or fixed data can give the correct pipe size for all installations. The correct method of determining the pipe size for a given installation is to make a complete engineering analysis of the project operations.

Table 20.3 gives recommended sizes of pipe for transmitting compressed air for various lengths of run. This information is useful as a guide in selecting pipe sizes.

TABLE 20.3 Recommended pipe sizes for transmitting compressed air at 80 to 125 psi gauge.

Volume of air (cfm)	Length of pipe (ft)				
	50–200	200–500	500–1,000	1,000–2,500	2,500–5,000
	Nominal size pipe (in.)				
30–60	1	1	$1\frac{1}{4}$	$1\frac{1}{2}$	$1\frac{1}{2}$
60–100	1	$1\frac{1}{4}$	$1\frac{1}{4}$	2	2
100–200	$1\frac{1}{4}$	$1\frac{1}{2}$	2	$2\frac{1}{2}$	$2\frac{1}{2}$
200–500	2	$2\frac{1}{2}$	3	$3\frac{1}{2}$	$3\frac{1}{2}$
500–1,000	$2\frac{1}{2}$	3	$3\frac{1}{2}$	4	$4\frac{1}{2}$
1,000–2,000	$2\frac{1}{2}$	4	$4\frac{1}{2}$	5	6
2,000–4,000	$3\frac{1}{2}$	5	6	8	8
4,000–8,000	6	8	8	10	10

Loss of Air Pressure in Hose

Air-line hose is a rubber-covered, pressure-type hose designed for transmitting compressed air. Hose is usually furnished equipped with quick-acting fittings for attaching a tool, a compressor, or another hose. Hose size is based on the amount of air that must be delivered to the tool. When transmitting compressed air at 80 to 125 psi gauge, the Table 20.4 guidance is valid for short hose lengths (25 ft or less). As the required hose length increases, the nominal size must also be increased. The loss of pressure resulting from the flow of air through hose is given in Table 20.5. The Table 20.5 data can be used to design hoses whose length will be greater than 25 ft.

DIVERSITY FACTOR

It is necessary to provide as much compressed air as required. However, to supply the needs of all operating equipment and the many different tools that may be connected to one system (see Fig. 20.5) would require more air capacity than is actually needed. It is probable that all equipment placed on a project will not be in operation during the exact same time periods. An analysis of the job should be made to determine the probable maximum *actual* need, prior to designing the compressed-air system.

As an example, if ten jackhammers are on a job, normally not more than five or six will be consuming air at a given time. The others will be out of use temporarily for changes in bits or drill steel, or moving to new locations. Thus, the *actual* amount of air demanded should be based on five or six drills instead of ten. The same condition will apply to other pneumatic tools.

The *diversity factor* is the ratio of the actual load (cfm) to the maximum calculated load (cfm) that would exist if all tools were operating at the same time. For example, if a jackhammer required 90 cfm of air, ten hammers would require 900 cfm if they were all operated at the same time. However, with only five hammers operating at one time, the demand for air would be 450 cfm. Thus, the diversity factor would be $450 \div 900 = 0.5$.

TABLE 20.4 Recommended hose size, for short hose lengths, when transmitting compressed air at 80 to 125 psi gauge.

Air requirement of tool (cfm)	Hose nominal size (in.)	Typical tools
Up to 15	$\frac{1}{4}$	Small drills and air hammers
Up to 40	$\frac{3}{8}$	Impact wrenches, grinders, and chipping hammers
Up to 80	$\frac{1}{2}$	Heavy chipping and rivet hammers
Up to 100	$\frac{3}{4}$	Rock drills 35 to 55 lb, large concrete vibrators, and sump pumps
100 to 200	1	Rock drills 75 lb and drifters

TABLE 20.5 Loss of pressure, in psi, in 50 ft of hose with ending couplings.

Size of hose (in.)	Gauge pressure at line (psi)	Volume of free air through hose (cfm)													
		20	30	40	50	60	70	80	90	100	110	120	130	140	150
$\frac{1}{2}$	50	1.8	5.0	10.1	18.1										
	60	1.3	4.0	8.4	14.8	23.5									
	70	1.0	3.4	7.0	12.4	20.0	28.4								
	80	0.9	2.8	6.0	10.8	17.4	25.2	34.6							
	90	0.8	2.4	5.4	9.5	14.8	22.0	30.5	41.0						
	100	0.7	2.3	4.8	8.4	13.3	19.3	27.2	36.6						
	110	0.6	2.0	4.3	7.6	12.0	17.6	24.6	33.3	44.5					
$\frac{3}{4}$	50	0.4	0.8	1.5	2.4	3.5	4.4	6.5	8.5	11.4	14.2				
	60	0.3	0.6	1.2	1.9	2.8	3.8	5.2	6.8	8.6	11.2				
	70	0.2	0.5	0.9	1.5	2.3	3.2	4.2	5.5	7.0	8.8	11.0			
	80	0.2	0.5	0.8	1.3	1.9	2.8	3.6	4.7	5.8	7.2	8.8	10.6		
	90	0.2	0.4	0.7	1.1	1.6	2.3	3.1	4.0	5.0	6.2	7.5	9.0		
	100	0.2	0.4	0.6	1.0	1.4	2.0	2.7	3.5	4.4	5.4	6.6	7.9	9.4	11.1
	110	0.1	0.3	0.5	0.9	1.3	1.8	2.4	3.1	3.9	4.9	5.9	7.1	8.4	9.9
1	50	0.1	0.2	0.3	0.5	0.8	1.1	1.5	2.0	2.6	3.5	4.8	7.0		
	60	0.1	0.2	0.3	0.4	0.6	0.8	1.2	1.5	2.0	2.6	3.3	4.2	5.5	7.2
	70	—	0.1	0.2	0.4	0.5	0.7	1.0	1.3	1.6	2.0	2.5	3.1	3.8	4.7
	80	—	0.1	0.2	0.3	0.5	0.7	0.8	1.1	1.4	1.7	2.0	2.4	2.7	3.5
	90	—	0.1	0.2	0.3	0.4	0.6	0.7	0.9	1.2	1.4	1.7	2.0	2.4	2.8
	100	—	0.1	0.2	0.2	0.4	0.5	0.6	0.8	1.0	1.2	1.5	1.8	2.1	2.4
	110	—	0.1	0.2	0.2	0.3	0.4	0.6	0.7	0.9	1.1	1.3	1.5	1.8	2.1
$1\frac{1}{4}$	50	—	—	0.2	0.2	0.2	0.3	0.4	0.5	0.7	1.1				
	60	—	—	—	0.1	0.2	0.3	0.3	0.5	0.6	0.8	1.0	1.2	1.5	
	70	—	—	—	0.1	0.2	0.2	0.3	0.4	0.4	0.5	0.7	0.8	1.0	1.3
	80	—	—	—	—	0.1	0.2	0.2	0.3	0.4	0.5	0.6	0.7	0.8	1.0
	90	—	—	—	—	0.1	0.2	0.2	0.3	0.3	0.4	0.5	0.6	0.7	0.8
	100	—	—	—	—	—	0.1	0.2	0.2	0.3	0.4	0.4	0.5	0.6	0.7
	110	—	—	—	—	—	0.1	0.2	0.2	0.3	0.3	0.4	0.5	0.5	0.6
$1\frac{1}{2}$	50	—	—	—	—	—	0.1	0.2	0.2	0.2	0.3	0.3	0.4	0.5	0.6
	60	—	—	—	—	—	—	0.1	0.2	0.2	0.2	0.3	0.3	0.4	0.5
	70	—	—	—	—	—	—	—	0.1	0.2	0.2	0.2	0.3	0.3	0.4
	80	—	—	—	—	—	—	—	—	0.1	0.2	0.2	0.2	0.3	0.4
	90	—	—	—	—	—	—	—	—	—	0.1	0.2	0.2	0.2	0.2
	100	—	—	—	—	—	—	—	—	—	—	0.1	0.2	0.2	0.2
	110	—	—	—	—	—	—	—	—	—	—	0.1	0.2	0.2	0.2

The approximate quantities of compressed air required by jackhammers and paving breakers are given in Table 20.6. The quantities are based on continuous operation at a pressure of 90 psi gauge.

SAFETY

Extreme care should be exercised when working with compressed air. At close range, it is capable of putting out eyes, bursting eardrums, causing serious skin

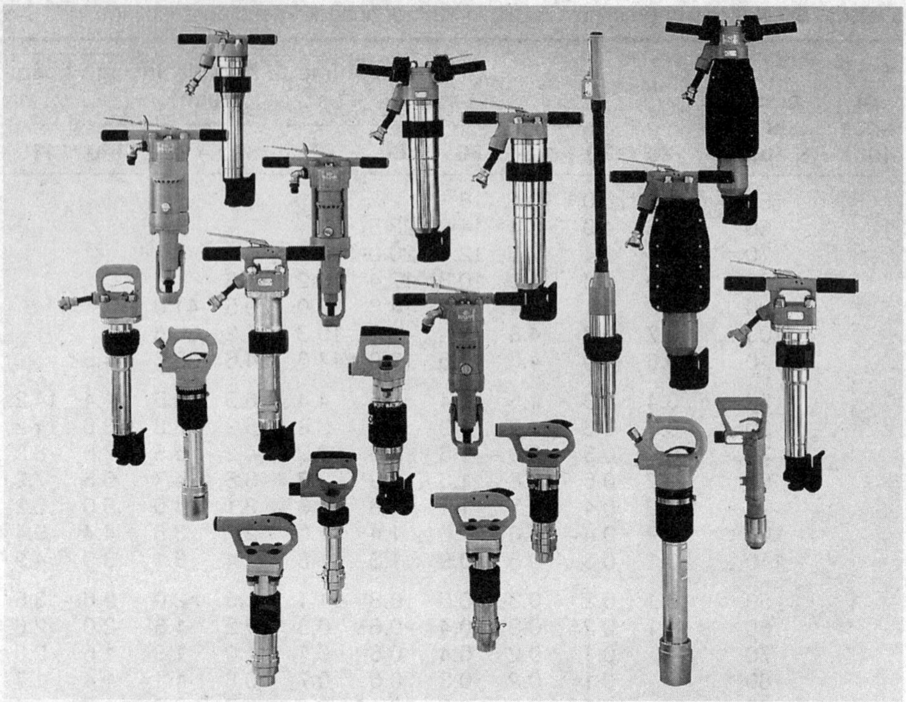

FIGURE 20.5 Air-operated tools that can be connected to an air system.
Source: Sullair Corp.

TABLE 20.6 Quantities of compressed air required by pneumatic equipment and tools.*

Equipment or tools	Capacity or size: Weight (lb)	Capacity or size: Depth of hole (ft)	Air consumption (cfm)
Jackhammers	10	0– 2	15– 25
	15	0– 2	20– 35
	25	2– 8	30– 50
	35	8–12	55– 75
	45	12–16	80–100
	55	16–24	90–110
	75	8–24	150–175
Paving breakers	35	—	30– 35
	60	—	40– 45
	80	—	50– 65

*Air pressure at 90 psi gauge.

blisters, or even killing an individual. Before using an air compressor, it is necessary to check that all the pressure gauges are in good working order.

Pneumatic Tools

Pneumatic power tools can be hazardous when improperly used. Operators must perform a preoperational check of all air hoses, couplings, and connections to determine if leakage or other damage exists. However, it is the responsibility of management to train employees in the proper use of all power tools. Safety requires that employees

- wear appropriate protective clothing and equipment (hearing and eye protection must be worn, gloves and respirators are often appropriate). (OSHA Standard 1926.102, Personal Protective and Life Saving Equipment)
- check that hoses are fastened securely and prevented from becoming accidentally disconnected. A short wire or positive locking device attaching the air hose to the tool will serve as an added safeguard.
- install a clip or retainer to prevent attachments, such as chisels on a chipping hammer, from being unintentionally shot from the barrel.
- turn off the air and disconnect the tool when repairs or adjustments are being made or the tool is not in use.
- inspect the hose to ensure it is in good condition and free from obstructions before connecting a pneumatic tool.
- remove leaking or defective hoses from service. The air hose must be able to withstand the pressure required for the tool.

It is also critical that the surrounding work area be checked for hazards. All underground utilities should be identified prior to the start of excavating with pneumatic tools. In 1986, a laborer was electrocuted when he contacted a 13-kV underground power line while digging with a pneumatic clay spade.

EQUIPMENT FOR PUMPING WATER

INTRODUCTION

Pumps are used extensively on construction projects for operations such as

1. Removing water from pits, tunnels, and other excavations
2. Dewatering cofferdams
3. Furnishing water for jetting and sluicing
4. Lowering the water table for excavations

Construction pumps (see Fig. 20.6) must frequently perform under severe conditions, such as those resulting from variations in the pumping head or from handling water that is muddy or highly corrosive. The required rate of pumping may vary considerably during the duration of a construction project.

The factors that should be considered in selecting pumps for construction applications include

1. Dependability
2. Availability of repair parts
3. Simplicity to permit easy repairs

FIGURE 20.6 Wheel-mounted centrifugal pump.

4. Economical installation and operation
5. Operating power requirements

GLOSSARY OF PUMPING TERMS

The following glossary defines the important terms that are used in describing pumps and pumping operations.

Capacity. The total volume of liquid a pump can move in a given amount of time. Capacity is usually expressed in gallons per minute (gpm) or gallons per hour (gph).

Discharge head. The (total) discharge head is the sum of the static discharge head plus the head losses of the discharge line.

Discharge hose. The hose used to carry the liquid from the discharge side of the pump.

Self-priming. The ability of a pump to separate air from a liquid and create a partial vacuum in the pump. This causes the liquid to flow to the impeller and on through the pump.

Static discharge head. The vertical distance from the centerline of the pump impeller to the point of discharge (see Fig. 20.7).

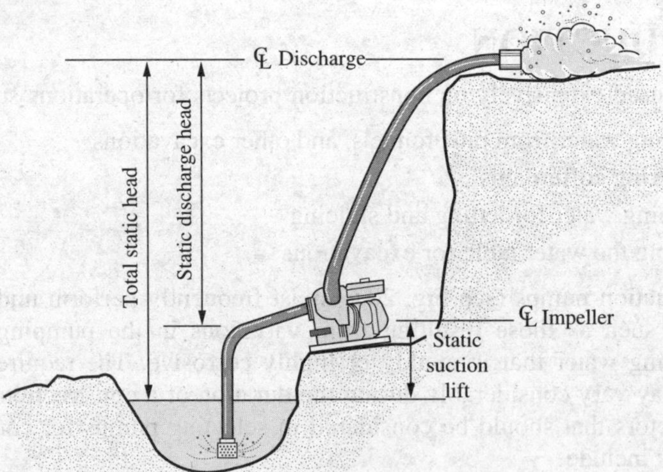

FIGURE 20.7 Dimensional terminology for pumping operations.

Static suction lift. The vertical distance from the centerline of the pump impeller to the surface of the liquid to be pumped (see Fig. 20.7). Suction capability is limited by atmospheric pressure. Therefore, maximum practical suction lift is 25 ft. Decreasing the suction lift will increase the volume that can be pumped.

Strainer. A cover matched to the size of the pump and attached to the end of the suction hose that permits solids of only a certain size to enter the pump body.

Suction head. The (total) suction head is the sum of the static suction lift plus the suction line head losses.

Suction hose. The hose connected to the suction side of the pump. Suction hose is made of heavy rubber or plastic tubing with a reinforced wall to prevent it from collapsing.

Total static head. The suction head plus the discharge head.

CLASSIFICATION OF PUMPS

The pumps commonly employed on construction projects may be classified as (1) displacement—reciprocating and diaphragm—or (2) centrifugal.

Reciprocating Pumps

A reciprocating pump operates as the result of the movement of a piston inside a cylinder. When the piston is moved in one direction, the water ahead of the piston is forced out of the cylinder. At the same time, additional water is drawn into the cylinder behind the piston. Regardless of the direction of movement of the piston, water is forced out of one end and drawn into the other end of the cylinder. This is classified as a double-acting pump. If water is pumped during a piston movement in one direction only, the pump is classified as single acting. If a pump contains more than one cylinder, mounted side by side, it is classified as a duplex for two cylinders, triplex for three cylinders, etc. Thus, a pump may be classified as duplex double-acting or duplex single-acting.

The capacity of a reciprocating pump depends essentially on the speed of the pump cycle and is independent of the head. The maximum head at which a reciprocating pump will deliver water depends on the strength of the component parts of the pump and the power available to operate the pump. The capacity of this type of pump may be varied considerably by varying the speed of the pump.

The advantages of reciprocating pumps are

1. They are able to pump at a uniform rate against varying heads.
2. Increasing the speed can increase their capacity.

The disadvantages of reciprocating pumps are

1. These are heavy and large-size pumps for a given capacity.
2. They impart a pulsating flow of water.

TABLE 20.7 Minimum capacities for diaphragm pumps at 10-ft suction lifts.*

Size	Capacity (gph)
Two-in. single	2,000
Three-in. single	3,000
Four-in. single	6,000
Four-in. double	9,000

*Diaphragm pumps shall be tested with standard contractor's type suction hose 5 ft longer than the suction lift shown.

Source: Courtesy Contractors Pump Bureau.

Diaphragm Pumps

The diaphragm pump is also a displacement type. The central portion of the flexible diaphragm is alternately raised and lowered by the pump rod that is connected to a walking beam. This action draws water into and discharges it from the pump. Because this type of pump will handle clear water or water containing large quantities of mud, sand, sludge, and trash, it is popular as a construction pump. It is suitable for use on jobs where the quantity of water varies considerably, as it will diligently continue pumping air and water mixtures.

The Contractors Pump Bureau specifies diaphragm pumps shall be manufactured in the size and capacity ratings given in Table 20.7.

CENTRIFUGAL PUMPS

A centrifugal pump contains a rotation element, called an impeller (Fig. 20.8), that imparts to water passing through the pump a velocity sufficiently great to cause it to flow from the pump even against considerable pressure. A mass of water may possess energy due to its height above a given datum or due to its velocity. The former is potential energy, while the latter is kinetic energy. One type of energy can be converted into the other under favorable conditions. The kinetic energy imparted to a particle of water as it passes through the impeller is sufficient to cause the particles to rise to some determinable height.

FIGURE 20.8 The impeller of a centrifugal pump.

The principle of the centrifugal pump may be illustrated by considering a drop of water at rest at a height h above a surface. If the drop of water is permitted to fall freely, it will strike the surface with a velocity given by the equation

$$V = \sqrt{2gh} \qquad [20.9]$$

where

V = velocity in feet per second (fps)

g = acceleration of gravity, equal to 32.2 fps at sea level
h = height of fall in feet

If the drop falls 100 ft, the velocity will be 80.2 fps. If the same drop is given an upward velocity of 80.2 fps, it will rise 100 ft. These values assume no loss in energy due to friction through air. It is the function of the centrifugal pump to give the water the necessary velocity as it leaves the impeller. If the speed of the pump is doubled, the velocity of the water will be increased from 80.2 to 160.4 fps, neglecting any increase in friction losses. With this velocity, the water can be pumped to a height given by the equation

$$h = \frac{V^2}{2g} = \frac{(160.4)^2}{64.4} = 400 \text{ ft}$$

This indicates that if a centrifugal pump is pumping water against a total head of 100 ft, the same quantity of water can be pumped against a total head of 400 ft by doubling the speed of the impeller. In actual practice, the maximum possible head for the increased speed will be less than 400 ft. The reduction is caused by increased losses in the pump due to friction. These results illustrate the effect that increasing the speed of an impeller has on the performance of a centrifugal pump.

The energy required to operate a pump is

$$W = \frac{wQh}{e} \qquad \textbf{[20.10]}$$

where

W = energy in ft-lb per min
w = weight of 1 gal of water in pounds
h = total pumping head, in feet, including friction loss in pipe
e = efficiency of pump, expressed decimally

The horsepower required to operate a centrifugal pump is given by

$$P = \frac{W}{33{,}000} = \frac{wQh}{33{,}000e} \qquad \textbf{[20.11]}$$

where

P = power horsepower
33,000 = ft-lb of energy per min for 1 hp

Self-Priming Centrifugal Pumps

On construction projects, pumps frequently must be set above the surface of the water that is to be pumped. Consequently, self-priming centrifugal pumps are well suited for the needs of construction. Such a pump is self-priming to heights of 25 ft when in good mechanical condition.

Effect of Altitude At altitudes above 3,000 ft, there is a definite effect on a pump's performance. As a general rule, a self-priming pump will lose 1 ft of priming ability for every 1,000 ft of elevation. A self-priming pump operated in Flagstaff, Arizona, at an elevation of 7,000 ft will only develop 18 ft of suction lift rather than the normal 25 ft.

Effect of Temperature As the temperature of water increases above 60°F, the maximum suction lift of the pump will decrease. A pump generates heat that is passed to the water. Over a long duration of operation, as the heat increases, a pump located at a height that is very close to the suction maximum can lose prime.

Multistage Centrifugal Pumps

If a centrifugal pump has a single impeller, it is described as a single-stage pump, whereas if there are two or more impellers and the water discharged from one impeller flows into the suction of another, it is described as a multistage pump. Multistage pumps are especially suitable for pumping against a high head or pressure, as each stage imparts an additional pressure to the water. Pumps of this type are used frequently to supply water for jetting, where the pressure may run as high as several hundred pounds per square inch (psi).

Performance of Centrifugal Pumps

Pump manufacturers furnish curves showing the performance of their pumps under different operating conditions. A set of curves for a given pump will show the variations in capacity, efficiency, and horsepower for different pumping heads. These curves are helpful in selecting the pump that is most suitable

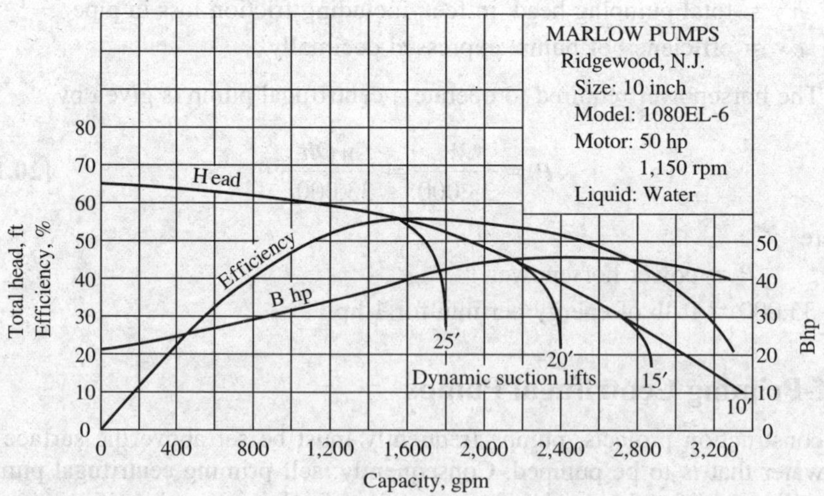

FIGURE 20.9 Performance curves for centrifugal pump.

for a given pumping condition. Figure 20.9 illustrates a set of performance curves for a 10-in. centrifugal pump. For a total head of 60 ft, the capacity will be 1,200 gpm, the efficiency 52%, and the required power 35 brake horsepower (bhp). If the total head is reduced to 50 ft and the dynamic suction lift does not exceed 23 ft, the capacity will be 1,930 gpm, the efficiency 55%, and the required power 44 bhp. This pump will not deliver any water against a total head in excess of 66 ft, which is called the "shutoff head."

Since a construction pump frequently is operated under varying heads, it is desirable to select a pump with relatively flat head capacity and horsepower curves, even though efficiency must be sacrificed to obtain these conditions. A pump with a flat horsepower demand permits the use of an engine or an electric motor that will provide adequate power over a wide pumping range, without a substantial surplus or deficiency, regardless of the head.

The Contractors Pump Bureau publishes pump standards for several types of pumps, including self-priming centrifugal pumps, and these are given in Table 20.8a and 20.8b.

Submersible Pumps

Figure 20.10 shows electric-motor-operated submersible pumps. These are useful in dewatering tunnels, foundation pits, trenches, and similar places. With a submersible pump, there is no suction lift limitation, and of course, no need for a suction hose. Another advantage is that there are no noise problems.

FIGURE 20.10 Electric-motor-operated submersible pump.

TABLE 20.8a Minimum capacities for M rate self-priming centrifugal pumps manufactured in accordance with standards of the Contractors Pump Bureau.

Model 18-M (3-in.)

Total head including friction [ft (m)]		Height of pump above water [ft (m)] / Capacity [gpm (1/min)*]									
		5	(1.5)	10	(3.0)	15	(4.6)	20	(6.1)	25	(7.6)
5	(1.5)	300	(1,136)								
10	(3.0)	295	(1,117)								
20	(6.1)	277	(1,048)	259	(980)						
30	(9.1)	260	(984)	250	(946)	210	(795)	200	(757)		
40	(12.2)	241	(912)	241	(912)	207	(784)	177	(670)	160	(606)
50	(15.2)	225	(852)	225	(852)	202	(765)	172	(651)	140	(530)
60	(18.3)	197	(746)	197	(746)	197	(746)	169	(640)	140	(530)
70	(21.3)	160	(606)	160	(606)	160	(606)	160	(606)	138	(522)
80	(24.4)	125	(473)	125	(473)	125	(473)	125	(473)	125	(473)
90	(27.4)	96	(363)	96	(363)	96	(363)	96	(363)	96	(363)

Model 20-M (3-in.)

Total head including friction [ft (m)]		Height of pump above water [ft (m)] / Capacity [gpm (1/min)*]							
		10	(3.0)	15	(4.6)	20	(6.1)	25	(7.6)
30	(9.1)	333	(1,260)	280	(1,060)	235	(890)	165	(625)
40	(12.2)	315	(1,192)	270	(1,022)	230	(871)	162	(613)
50	(15.2)	290	(1,098)	255	(965)	220	(833)	154	(583)
60	(18.3)	255	(965)	235	(890)	205	(776)	143	(541)
70	(21.3)	212	(802)	209	(791)	184	(696)	130	(492)
80	(24.4)	165	(625)	165	(625)	157	(594)	114	(432)
90	(27.4)	116	(439)	116	(439)	116	(439)	94	(356)
100	(30.5)	60	(227)	60	(227)	60	(227)	60	(227)

Model 40-M (4-in.)

Total head including friction [ft (m)]		Height of pump above water [ft (m)] / Capacity [gpm (1/min)*]							
		10	(3.0)	15	(4.6)	20	(6.1)	25	(7.6)
25	(7.6)	667	(2,525)						
30	(9.1)	660	(2,498)	575	(2,176)	475	(1,798)	355	(1,344)
40	(12.2)	645	(2,441)	565	(2,139)	465	(1,760)	350	(1,325)
50	(15.2)	620	(2,347)	545	(2,063)	455	(1,722)	345	(1,306)
60	(18.3)	585	(2,214)	510	(1,930)	435	(1,647)	335	(1,268)
70	(21.3)	535	(2,025)	475	(1,798)	410	(1,552)	315	(1,192)
80	(24.4)	465	(1,760)	410	(1,551)	365	(1,382)	280	(976)
90	(27.4)	375	(1,419)	325	(1,230)	300	(1,136)	220	(833)
100	(30.5)	250	(946)	215	(815)	195	(738)	145	(549)
110	(33.5)	65	(246)	60	(227)	50	(189)	40	(151)

*Liters per minute.

Source: Courtesy Contractors Pump Bureau.

TABLE 20.8b Minimum capacities for M rate self-priming centrifugal pumps manufactured in accordance with standards of the Contractors Pump Bureau.

Model 90-M (6-in.)

Total head including friction [ft	(m)]	Height of pump above water [ft (m)] 10	(3.0)	15	(4.6)	20	(6.1)	25	(7.6)
		Capacity [gpm (1/min)*]							
25	(7.6)	1,500	(5,678)						
30	(9.1)	1,480	(5,602)	1,280	(4,845)	1,050	(3,974)	790	(2,990)
40	(12.2)	1,430	(5,413)	1,230	(4,656)	1,020	(3,861)	780	(2,952)
50	(15.2)	1,350	(5,110)	1,160	(4,391)	970	(3,672)	735	(2,782)
60	(18.3)	1,225	(4,637)	1,050	(3,974)	900	(3,407)	690	(2,612)
70	(21.3)	1,050	(3,974)	900	(3,407)	775	(2,933)	610	(2,309)
80	(24.4)	800	(3,028)	680	(2,574)	600	(2,271)	490	(1,855)
90	(27.4)	450	(1,703)	400	(1,514)	365	(1,382)	300	(1,136)
100	(30.5)	100	(379)	100	(379)	100	(379)	100	(379)

Model 125-M (8-in.)

Total head including friction [ft	(m)]	Height of pump above water [ft (m)] 10	(3.0)	15	(4.6)	20	(6.1)	25	(7.6)
		Capacity [gpm (1/min)*]							
25	(7.6)	2,100	(7,949)	1,850	(7,002)	1,570	(5,943)		
30	(9.1)	2,060	(7,797)	1,820	(6,889)	1,560	(5,905)	1,200	(4,542)
40	(12.2)	1,960	(7,419)	1,740	(6,586)	1,520	(5,753)	1,170	(4,429)
50	(15.2)	1,800	(6,813)	1,620	(6,132)	1,450	(5,488)	1,140	(4,315)
60	(18.3)	1,640	(6,207)	1,500	(5,678)	1,360	(5,148)	1,090	(4,126)
70	(21.3)	1,460	(5,526)	1,340	(5,072)	1,250	(4,731)	1,015	(3,841)
80	(24.4)	1,250	(4,731)	1,170	(4,429)	1,110	(4,201)	950	(3,596)
90	(27.4)	1,020	(3,861)	980	(3,709)	940	(3,558)	840	(3,179)
100	(30.5)	800	(3,028)	760	(2,877)	710	(2,687)	680	(2,574)
110	(33.5)	570	(2,158)	540	(2,044)	500	(1,893)	470	(1,779)
120	(36.6)	275	(1,041)	245	(927)	240	(908)	240	(908)

*Liters per minute.

Source: Courtesy Contractors Pump Bureau.

For construction applications, a pump made of iron or aluminum is best, as other materials are much more prone to damage when the pump is dropped. The power cord for a submersible pump should have a strain relief protector, as it is rare that someone does not inadvertently lift the pump by the power cord.

There are basically two size categories for submersible pumps; small fractional horsepower size pumps and larger pumps having 1-hp or large power units. The small pumps, typically $\frac{1}{4}$-, $\frac{1}{3}$-, and $\frac{1}{2}$-hp units, are for minor nuisance dewatering applications. The 1-hp and larger pumps are for moving large volumes and/or high head conditions.

LOSS OF HEAD DUE TO FRICTION IN PIPE

Table 20.9 gives the nominal loss of head due to water flowing through new steel pipe. The actual losses may differ from the values given in the table because of variations in the diameter of the pipe and in the condition of the pipe's interior surface.

TABLE 20.9 Friction loss for water, in feet per 100 ft, of clean wrought-iron or steel pipe.*

Flow in U.S. (gpm)	Nominal diameter of pipe (in.)													
	$\frac{1}{2}$	$\frac{3}{4}$	1	$1\frac{1}{4}$	$1\frac{1}{2}$	2	$2\frac{1}{2}$	3	4	5	6	8	10	12
5	26.5	6.8	2.11	0.55										
10	95.8	24.7	7.61	1.98	0.93	0.31	0.11							
15		52.0	16.3	4.22	1.95	0.70	0.23							
20		88.0	27.3	7.21	3.38	1.18	0.40							
25			41.6	10.8	5.07	1.75	0.60	0.25						
30			57.8	15.3	7.15	2.45	0.84	0.35						
40				26.0	12.2	4.29	1.4	0.59						
50				39.0	18.5	6.43	2.2	0.9	0.22					
75					39.0	13.6	4.6	2.0	0.48	0.16				
100					66.3	23.3	7.8	3.2	0.79	0.27	0.09			
125						35.1	11.8	4.9	1.2	0.42	0.18			
150						49.4	16.6	6.8	1.7	0.57	0.21			
175						66.3	22.0	9.1	2.2	0.77	0.31			
200							28.0	11.6	2.9	0.96	0.40			
225							35.3	14.5	3.5	1.2	0.48			
250							43.0	17.7	4.4	1.5	0.60	0.15		
275								21.2	5.2	1.8	0.75	0.18		
300								24.7	6.1	2.0	0.84	0.21		
350								33.8	8.0	2.7	0.91	0.27		
400									10.4	3.5	1.4	0.35		
500									15.6	5.3	2.2	0.53	0.18	0.08
600									22.4	6.2	3.1	0.74	0.25	0.10
700									30.4	9.9	4.1	1.0	0.34	0.14
800											5.2	1.3	0.44	0.18
900											6.6	1.6	0.54	0.22
1,000											7.8	2.0	0.65	0.27
1,100											9.3	2.3	0.78	0.32
1,200											10.8	2.7	0.95	0.37
1,300											12.7	3.1	1.1	0.42
1,400											14.7	3.6	1.2	0.48
1,500											16.8	4.1	1.4	0.55
2,000												7.0	2.4	0.93
3,000													5.1	2.1
4,000														3.5
5,000														5.5

*For old or rough pipes, add 50% to friction values.

Courtesy Contractors Pump Bureau.

The relationship between the head of fresh water in feet and pressure in psi is given by the equation

$$h = 2.31\,p \qquad \textbf{[20.12]}$$

or

$$p = 0.434\,h \qquad \textbf{[20.13]}$$

where

h = depth of water or head in feet
p = pressure at depth h in psi

Table 20.10 gives the equivalent length of straight steel pipe having the same loss in head due to water friction as fittings and valves.

RUBBER HOSE

The flexibility of rubber hose makes its use with pumps very convenient. Such hose may be used on the suction side of a pump if it is constructed with a wire insert to prevent collapse under partial vacuum. Rubber hose is available with end fittings corresponding with those for iron or steel pipe. As a "rule of thumb," total length of hose on a centrifugal pump should be less than 500 ft and less than 50 ft for a diaphragm pump.

Hose size should match the pump size. Using a larger suction hose will increase the pumping capacity, e.g., 4-in. hose on a 3-in. pump, but it can also cause overload of the pump motor. A suction hose smaller in size than the pump will starve the pump and can cause cavitation. This in turn will increase the wear of the impeller and **volute** and lead to early pump failure.

volute
The housing in which the pump's impeller rotates is known as the volute. It has channels cast into the metal to direct the flow of liquid in a given direction.

A discharge hose sized larger than the pump will simply reduce friction loss and can increase volume if long discharge distances are involved. If a small discharge hose is used, friction loss is increased and therefore pumping volume reduced.

Table 20.11 gives the loss in head in feet per 100 ft due to friction caused by water flowing through hose. The values in the table also apply to rubber substitutes.

SELECTING A PUMP

Before a pump for a given job is selected, it is necessary to analyze all information and conditions that will affect its operation and have an understanding of the pumping requirements. Necessary information includes

1. The rate at which the water must be pumped
2. The height of lift from the existing water surface to the point of discharge
3. The pressure head at discharge, if any
4. The variations in water level at suction or discharge
5. The altitude of the project

TABLE 20.10 Length of steel pipe, in feet, equivalent to fittings and valves.

Item	Nominal size (in.)											
	1	$1\frac{1}{4}$	$1\frac{1}{2}$	2	$2\frac{1}{2}$	3	4	5	6	8	10	12
90° elbow	2.8	3.7	4.3	5.5	6.4	8.2	11.0	13.5	16.0	21.0	26.0	32.0
45° elbow	1.3	1.7	2.0	2.6	3.0	3.8	5.0	6.2	7.5	10.0	13.0	15.0
Tee, side outlet	5.6	7.5	9.1	12.0	13.5	17.0	22.0	27.5	33.0	43.5	55.0	66.0
Close return bend	6.3	8.4	10.2	13.0	15.0	18.5	24.0	31.0	37.0	49.0	62.0	73.0
Gate valve	0.6	0.8	0.9	1.2	1.4	1.7	2.5	3.0	3.5	4.5	5.7	6.8
Globe valve	27.0	37.0	43.0	55.0	66.0	82.0	115.0	135.0	165.0	215.0	280.0	335.0
Check valve	10.5	13.2	15.8	21.1	26.4	31.7	42.3	52.8	63.0	81.0	105.0	125.0
Foot valve	24.0	33.0	38.0	46.0	55.0	64.0	75.0	76.0	76.0	76.0	76.0	76.0

Courtesy The Gorman-Rupp Company.

TABLE 20.11 Water friction loss, in feet per 100 ft of smooth bore hose.

Flow in U.S. (gpm)	Actual inside diameter of hose (in.)											
	$\frac{5}{8}$	$\frac{3}{4}$	1	$1\frac{1}{4}$	$1\frac{1}{2}$	2	$2\frac{1}{2}$	3	4	5	6	8
5	21.4	8.9	2.2	0.74	0.3							
10	76.8	31.8	7.8	2.64	1.0	0.2						
15		68.5	16.8	5.7	2.3	0.5						
20			28.7	9.6	3.9	0.9	0.32					
25			43.2	14.7	6.0	1.4	0.51					
30			61.2	20.7	8.5	2.0	0.70	0.3				
35			80.5	27.6	11.2	2.7	0.93	0.4				
40				35.0	14.3	3.5	1.2	0.5				
50				52.7	21.8	5.2	1.8	0.7				
60				73.5	30.2	7.3	2.5	1.0				
70					40.4	9.8	3.3	1.3				
80					52.0	12.6	4.3	1.7				
90					64.2	15.7	5.3	2.1	0.5			
100					77.4	18.9	6.5	2.6	0.6			
125						28.6	9.8	4.0	0.9			
150						40.7	13.8	5.6	1.3			
175						53.4	18.1	7.4	1.8			
200						68.5	23.4	9.6	2.3	0.8	0.32	
250							35.0	14.8	3.5	1.2	0.49	
300							49.0	20.3	4.9	1.7	0.69	
350								27.0	6.6	2.3	0.90	
400									8.4	2.9	1.1	0.28
450									10.5	3.6	1.4	0.35
500									12.7	4.3	1.7	0.43
1,000										15.6	6.4	1.6

Courtesy Contractors Pump Bureau.

6. The height of the pump above the surface of water to be pumped
7. The size of pipe to be used, if already determined
8. The number, sizes, and types of fittings and valves in the pipeline

Example 20.4 illustrates an analysis of pumping conditions and selection of a pump that is compatible with the conditions.

EXAMPLE 20.4

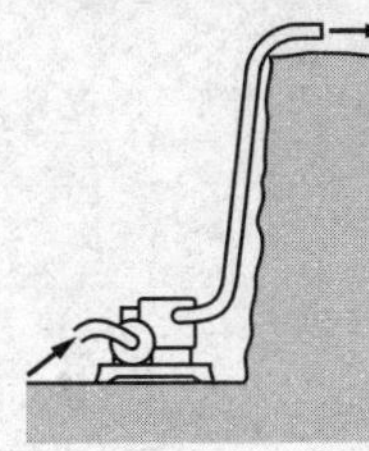

Select a self-priming centrifugal pump, with a capacity of 600 gpm, for the project illustrated in Fig. 20.11. All pipes, fittings, and valves will be 6 in. with threaded connections. Use the information in Table 20.10 to convert the fittings and valves into equivalent lengths of pipe.

Item shown in Fig. 20.11	Equivalent length of pipe, ft
1-ft valve and strainer	76
Three elbows at 16 ft	48
Two gate valves at 3.5 ft	7
One check valve	63
Total	194
Add length of pipe (25 + 24 + 166 + 54 + 10)	279
Total equivalent length of 6-in. pipe	473

From Table 20.9, the friction loss per 100 ft of 6-in. pipe will be 3.10 ft. The total head, including lift plus head lost in friction, will be

Lift, 15 + 54 = 69.0 ft

Head lost in friction, 473 ft at 3.10 ft per 100 ft = 14.7 ft

Total head = 83.7 ft

Table 20.8b indicates that a model 90-M pump will deliver the required quantity of water.

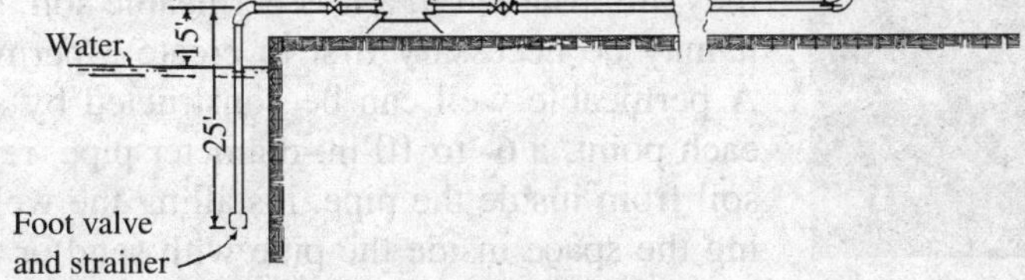

FIGURE 20.11 Pump and pipe installation for Example 20.4.

WELLPOINT SYSTEMS

In excavating below the surface of the ground, constructors often encounter groundwater prior to reaching the required depth of excavation. In the case of an excavation into sand and gravel, the flow of water will be large if some method is not adopted to intercept and remove the water. Dewatering, temporarily lowering the piezometric level of groundwater, is then necessary. After the construction operations are completed, the dewatering actions can be discontinued and the groundwater will return to its normal level. When planning a dewatering activity, it should be understood that groundwater levels change from season to season as a result of many factors [6].

Ditches located within the limits of the excavation can be used to collect and divert the flow of groundwater into sumps from which it can be removed by pumping. However, the presence of collector ditches within the excavation usually creates a nuisance and interferes with the construction operations. A common method for controlling groundwater is the installation of a wellpoint system along or around the excavation to lower the water table below the excavation bottom thus permitting the work to be done under relatively dry conditions.

A *wellpoint* is a perforated tube enclosed in a screen that is installed below the surface of the ground to collect water in order to lower the piezometric level of groundwater. The essential parts of a wellpoint are illustrated in Fig. 20.12. The top of a wellpoint is attached to a vertical riser pipe. The riser extends a short distance above the ground surface, at which point it is connected to a larger pipe called a "header." The header pipe lies on the ground surface and serves as the trunk line to which multiple risers are connected. A valve is installed between each wellpoint and the header to regulate the flow of water. Header pipes are usually 6 to 10 in. in diameter. The header pipe is connectd to the suction of a centrifugal pump. A wellpoint system may include a few or several hundred wellpoints, all connected to one or more headers and pumps.

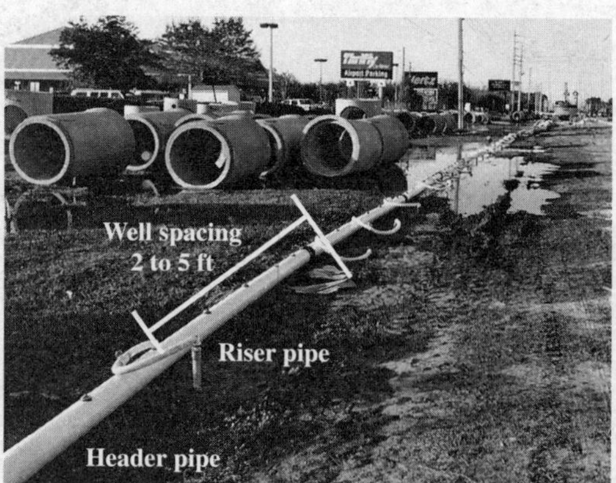

FIGURE 20.12 Parts of a wellpoint system.

The principle by which a wellpoint system operates is illustrated in Fig. 20.13. Figure 20.13(a) shows how a single point will lower the surface of the water table in the soil adjacent to the point. Figure 20.13(b) shows how several points, installed reasonably close together, lower the water table over an extended area.

Wellpoints will operate satisfactorily if they are installed in a permeable soil such as sand or gravel. If they are installed in a less permeable soil, such as silt, it may be necessary first to create a permeable well. A permeable well can be constructed by sinking, for each point, a 6- to 10-in.-diameter pipe, removing the soil from inside the pipe, installing the wellpoint, filling the space inside the pipe with sand or fine gravel, and then withdrawing the pipe. This leaves a volume

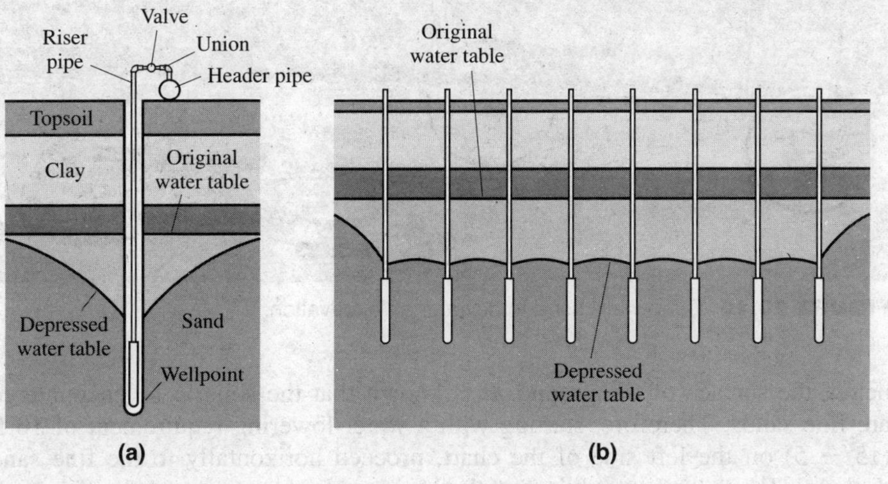

FIGURE 20.13 Water table drawdown resulting from wellpoints.

of sand around each wellpoint to act as a water collector sump and a filter to increase the rate of flow for each point.

Wellpoints can be installed at any spacing, but usually the spacings vary from 2 to 5 ft (see Fig. 20.12) along the header. The maximum height that water can be lifted is about 20 ft. If it is necessary to lower the water table to a greater depth, one or more additional stages of wellpoints should be installed, each stage at a lower depth within the excavation.

Capacity of a Wellpoint System

The capacity of a wellpoint system depends on the number of points installed, the permeability of the soil, and the amount of water present. An engineer who is experienced in this kind of work can perform tests that will provide data to make a reasonably accurate estimate concerning the capacity necessary to lower the water to the desired depth. The flow per wellpoint may vary from 3 to 4 gpm, in the case of fine to medium sands, to as much as 30 or more gpm for course sand. Figure 20.14 presents approximate flow rates to wellpoints in various soil formations.

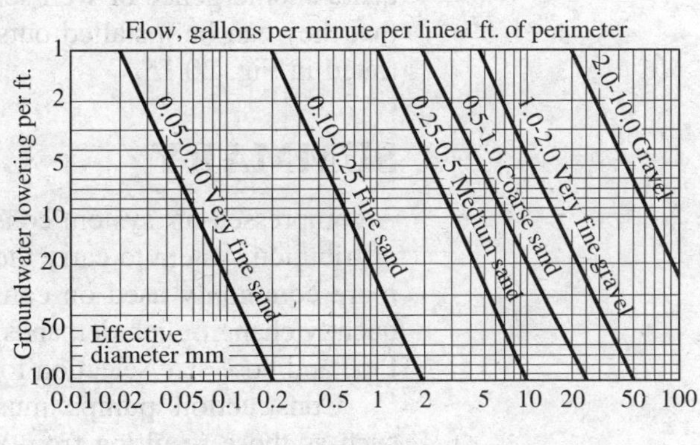

FIGURE 20.14 Approximate flow through various soil formations to a line of wellpoints.
Source: Moretrench America Corporation.

Using the Fig. 20.14 data will aid in the selection of the size of pumps that should be used with a wellpoint system. As an example, consider that it is necessary to dewater a pit that is 15 ft deep and that the water table is 5 ft

FIGURE 20.15 Deep wells used to dewater an excavation.

below the surface of the ground. It is known that the soils to be encountered are fine sands. Therefore, starting with a water-lowering requirement of 10 ft (15 − 5) on the left side of the chart, proceed horizontally to the fine sand diagonal. From the intersection of the horizontal projection and the fine sand diagonal, project a vertical line down to the flow rate numbers on the bottom of the chart. Consequently, a flow of 0.5 gpm per foot of header pipe could be expected for these conditions.

DEEP WELLS

Another method for dewatering an excavation is the use of deep wells. Large-diameter deep wells are suitable for lowering the groundwater table at sites where

- the soil formation becomes more pervious with depth.
- the excavation penetrates or is underlain by sand or coarse granular soils.

In addition, there is a requirement that there be sufficient depth of pervious materials below the level to which the water table is to be lowered for adequate submergence of well screens and pumps. The advantage of deep wells is that they can be installed outside the zone of construction operations, as illustrated in Fig. 20.15.

SUMMARY

A compressed-air system consists of one or more compressors together with a distribution system to carry the air to the points of use. Portable compressors are more commonly used on construction sites where it is necessary to meet frequently changing job demands, typically at a number of locations on the job site. Pipe and hose size selection for an air line requires an analysis of friction losses.

Construction pumps must frequently perform under severe conditions, such as those resulting from variations in the pumping head or from handling water that is muddy, sandy, or trashy. The required rate of pumping may vary considerably during the duration of a construction project. The proper pump solution is to select the equipment that will adequately take care of the pumping needs.

A common method for controlling groundwater is the installation of a wellpoint system along or around the excavation to lower the water table below the excavation bottom, thus permitting the work to be done under relatively dry conditions. Another method for dewatering an excavation is the use of deep wells. Critical learning objectives include:

- An ability to apply the gas laws as appropriate
- An ability to calculate the pressure loss as air flows through a pipe or hose
- An understanding of the diversity factor in calculating actual load on an air system
- An understanding of the types of pumps available
- An understanding of the effect altitude and temperature have on a pump's performance
- An ability to determine pump performance from appropriate charts
- An ability to calculate friction losses
- An ability to select a suitably sized pump based on project conditions and the quantity of water to be moved

These objectives are the basis for the problems that follow.

PROBLEMS

20.1 An air compressor draws in 1,000 cf of air at a gauge pressure of 0 psi and a temperature of 70°F. The air is compressed to a gauge pressure of 100 psi at a temperature of 140°F. The atmospheric pressure is 14.0 psi. Determine the volume of air after it is compressed. (139 cf)

20.2 An air compressor draws in 1,000 cf of free air at a gauge pressure of 0 psi and a temperature of 60°F. The air is compressed to a gauge pressure of 100 psi at a temperature of 130°F. The atmospheric pressure is 12.30 psi. Determine the volume of air after it is compressed.

20.3 Using the Fig. 20.4 chart, determine the pressure loss per 100 ft of pipe resulting from transmitting 200 cfm of free air, at 100-psi gauge pressure, through a 2-in. standard-weight steel pipe. (0.27 psi per 100 ft)

20.4 An air compressor draws in 800 cf of free air at a gauge pressure of 0 psi and a temperature of 85°F. The air is compressed to a gauge pressure of 100 psi at a temperature of 130°F. The atmospheric pressure is 12.20 psi. Determine the volume of air after it is compressed.

20.5 What will be the pressure at the air tool end of an 85-ft-long, 1-in.-diameter hose if the tool requires 120 cfm? The pressure entering the hose is 90 psi. (87.11 psi)

20.6 A 3-in. pipe with screwed fittings is used to transmit 1,000 cfm of free air at an initial pressure of 100-psi gauge pressure. The pipeline includes these items:

900 ft of pipe
Three gate valves

Eight standard on-run tees
Six standard ells

Determine the total loss of pressure in the pipeline.

20.7 If the air from the end of the pipe line of Problem 20.5 is delivered through 50 ft of 1-in. hose to a rock drill that requires 140 cfm of air, determine the pressure at the drill.

20.8 What will be the pressure at the air tool end of a 60-ft-long, 1-in.-diameter hose if the tool requires 120 cfm? The pressure entering the hose is 96 psi.

20.9 A diaphragm pump is best suited for which type of application?

a. Muddy water
b. Slow seepage
c. High volume pumping

20.10 The maximum practical suction lift for a self-priming centrifugal pump is

a. 35 ft
b. 50 ft
c. 25 ft

20.11 A suction hose sized larger than the pump size will

a. Increase capacity
b. Decrease capacity
c. Leak

20.12 A diaphragm pump is classified and operates as which type pump?

a. Rotary
b. Centrifugal
c. Positive displacement

20.13 Pump engine speed affects

a. Pressure
b. Volume
c. Both

20.14 A centrifugal pump is to be used to pump all the water from a small excavation that has been flooded. The dimensions of the volume holding water are 70 ft long, 50 ft wide, and 12 ft deep. The water must be pumped against an average total head of 45 ft. The average height of the pump above the water will be 12 ft. If the excavation must be emptied in 15 hours, determine the minimum model self-priming pump class M to be used based on the ratings of the Contractors Pump Bureau. (349 gpm, Model 40-M, approximately 600 gpm)

20.15 Use Table 20.8 to select a centrifugal pump to handle 300 gpm of water. The water will be pumped from a pond through 460 ft of 6-in. pipe to a point 30 ft above the level of the pond, where it will be discharged into the air. The pump will be set 10 ft above the surface of the water in the pond. What is the designation of the pump selected?

20.16 Select a self-priming centrifugal pump to handle 600 gpm of water for the project illustrated in Fig. 20.11. Increase the height of the vertical pipe from 54 ft to 60 ft. All other conditions will be as shown in the figure. (Model 125-M, approximately 980 gpm)

20.17 Select a self-priming centrifugal pump to handle 300 gpm of water for the project illustrated in Fig. 20.11. Change the size of the pipe, fittings, and valves to 5 in.

20.18 A 3-in. pipe with screwed fittings is used to transmit 600 cfm of free air at an initial pressure of 100-psi gauge pressure. The pipeline includes the following items:

800 ft of pipe
Two gate valves
Six long-radius ells
Four standard on-run tees

Determine the total loss of pressure in the pipeline.

Size of pipe, 3 in. $V = 1{,}000$ cfm
Length of pipe, 800 ft $P_1 = 100$ psi gauge

20.19 What will be the pressure at the air tool end of a 40-ft-long, $\frac{3}{4}$-in.-diameter hose if the tool requires 100 cfm? The pressure entering the hose is 97 psi.

20.20 If the air from the end of a pipeline is delivered through 50 ft of $1\frac{1}{4}$-in. hose to a rock drill that requires 120 cfm of air, determine the pressure at the drill. The pressure at the end of the pipe is 95.35 psi

20.21 Use Table 20.8 to select a self-priming centrifugal pump to handle 300 gpm of water. The water will be pumped from a pond through 360 ft of 4-in. pipe to a point 30 ft above the level of the pond, where it will be discharged into the air. The pump will be set 10 ft above the surface of the water in the pond. What is the designation of the pump selected?

20.22 Select a self-priming centrifugal pump to handle 350 gpm of water for a project as described. The pump will be placed 15 ft above the water source. The water will be drawn from the source using 4-in. steel pipe. There will be a 20-ft vertical section with a foot valve and strainer at its lower end. A 90° elbow will join the vertical pipe to 10 ft of horizontal pipe that connects to the pump. From the pump discharge there is 100 ft of horizontal pipe with a check valve in the line. At the far end of the discharge pipe there is another 90° elbow, which joins the horizontal pipe to 42 ft of vertical pipe. At the tip of this vertical pipe there is still a third 90° elbow and 8 ft of horizontal discharge pipe.

20.23 Select a self-priming centrifugal pump to handle 300 gpm of water for the project illustrated in Fig. 20.11. Change the size of the pipe, fittings, and valves to 5 in.

REFERENCES

1. Powers, J. Patrick (1992). *Construction Dewatering: New Methods and Applications*, 2nd ed., John Wiley & Sons, New York.
2. The Contractors Pump Bureau (CPB). CPB is a bureau of the Association of Equipment Manufacturers, Milwaukee, WI. The Contractors Pump Bureau develops and publishes standards for contractor pumps and auxiliary equipment,
3. *Hand and Power Tools*, OSHA 3080 (2002). U.S. Department of Labor, Occupational Safety and Health Administration, Washington, DC.
4. *Portable Air-Compressor Safety Manual*, Association of Equipment Manufacturers, Milwaukee, WI.
5. *Portable Pump Safety Manual*, Association of Equipment Manufacturers, Milwaukee, WI.
6. Schexnayder, Francis S., and Cliff J. Schexnayder (2001). "Understanding Project Site Conditions," *Practice Periodical on Structural Design and Construction*, ASCE, Vol. 6, No. 1, pp. 66–69, February.
7. *Selection Guidebook for Portable Dewatering Pumps*, Association of Equipment Manufacturers, Milwaukee, WI.

WEBSITE RESOURCES

Compressed Air

1. www.atlascopco.com Atlas Copco, Inc., 70 Demarest Drive, Wayne, NJ. Atlas Copco is a manufacturer of compressors, generators, construction and mining equipment, and pneumatic and electric power tools.
2. www.chicagopneumatic.com Chicago Pneumatic Tool Company, Utica, NY. Chicago Pneumatic manufactures pneumatic tools for the construction and demolition industry.
3. www.portablepower.irco.com Ingersoll-Rand (IR), Mocksville, NC. IR is a manufacturer of a wide range of diesel-driven portable compressors.
4. www.sullair.com Sullair Corporation, 3700 East Michigan Boulevard, Michigan City, IN. Sullair is a manufacturer of rotary-screw compressors.

Pumps

1. www.gormanrupp.com Gorman-Rupp is a manufacturer of self-priming centrifugal, centrifugal, submersible, trash, and air-driven diaphragm pumps and pumping systems for the construction industry.
2. www.griffindewatering.com/frames.htm Griffin Dewatering Corp. performs layout, installation, and operation of construction dewatering systems and Griffin Pump and Equipment, Inc., manufactures a line of construction pumps.
3. www.moretrench.com/services/cd/cdm.html Moretrench American is a dewatering contractor specializing in groundwater control problems.
4. www.thompsonpump.com Thompson Pump & Manufacturing Co. has many diverse lines of pumps including wet prime trash pumps, diaphragm pumps, rotary wellpoint pumps, and high-pressure jet pumps.

CHAPTER 21

Planning for Building Construction

Material procurement, sequencing of operations, site logistics, project schedule, as well as the technical aspects of the project are critical drivers in selecting equipment for building projects. Job-site layout impacts the ability of the general contractor and the subcontractors to operate effectively and efficiently. The project's general contractor is charged by the Occupational Safety & Health Administration Standards with specific duties concerning the maintenance of a job site. Major nuisances can result from construction operations, of which noise, vibration, and illumination are the most significant. Contractors working on projects in urban environments must mitigate the effect of such nuisances on project neighbors.

INTRODUCTION

Various pieces of equipment can be used to support construction processes. To select equipment for vertical—building—projects it is necessary to understand how construction processes for that particular type of work are organized. Material procurement, sequencing of operations, site logistics, project schedule, as well as technical aspects of the project, are critical issues that drive the selection of equipment for building projects. Material procurement typically occurs early in the construction process. Often it takes place even before shop drawings are generated. This is done so that a steel mill has enough time to produce the required steel or a precast concrete manufacturer can manufacture the concrete members to the required strength for handling and incorporation into the structure. Once structural steel members are fabricated or concrete elements cast and cured, they must be transported to the job site. As members arrive at the project site they must either be stored on site or the erection productivity must be matched to the delivery schedule. On an urban project, with a constrained site (Figs. 21.1 and 21.2), the coordination of delivery, storage, and erection activities dictates equipment choices.

FIGURE 21.1 Restricted building site in Washington, D.C., bordered by 14th Street and Pennsylvania Avenue on two sides and existing buildings on the other two sides.

FIGURE 21.2 Stadium project constrained by an existing tennis facility and the athletic field.

At the building site in Fig. 21.1, a concrete batch plant to support construction was installed within the building's basement, as ready-mix trucks could not maneuver to the site during peak traffic hours. Aggregate and cement were delivered to the plant at night when traffic was minimal.

The stadium project shown in Fig. 21.2 had a very constricted triangular-shaped site. The existing athletic field had to be available for events at all times, and university dormitories and a private tennis club abutted the other two sides of the site. The contractor was not permitted to infringe on any of these adjacent properties. Additionally, access to the site was so limited that all of the major concrete elements of the structure had to be cast on the site, as they could not be transported from a precast facility. Therefore, organizing and planning the work was very important. Of particular importance was storage of the concrete columns and beams, and positioning of the necessary cranes. The large crawler crane used to lift the columns had a width, outside edge of track to outside edge of track, of 31 ft (9.5 m). The length of its superstructure, from the rear of the counterweight to the front toe of track, was 39 ft (12 m), so just positioning the crane required considerable space.

SITE LAYOUT

A contractor must consider many factors when laying out a site to support construction operations:

1. Site size compared to building size and configuration.
2. Location of adjacent roads, buildings, and utilities—pedestrian traffic must be kept at a safe distance from the construction site, fencing and barricades may be necessary to block off all or part of a road during construction operations.
3. Soil conditions and excavation requirements—it is important to consider how soil conditions and excavation of a site will change over the duration of a project. The bearing capacity of soils in areas where a crane will operate should be verified. Consideration must be given to the proximity of the crane to the edge of an excavation or foundation wall.
4. Construction sequence and schedule.
5. Location of utilities—the impact of overhead obstructions such as power or communication lines must be considered, precautions must also be taken if cranes must be operated over underground utilities.
6. Equipment requirements—determine size and location of hoisting equipment based on both the physical hoisting requirements and the project schedule.
7. Material quantity, storage, and delivery.
8. Worker parking—parking availability on and around a job site will typically be addressed in the project bid package.
9. Tool and equipment storage.
10. Construction operations facilities and trailers [8].
11. Sanitary facilities—the general contractor usually provides sanitary facilities on a job site.

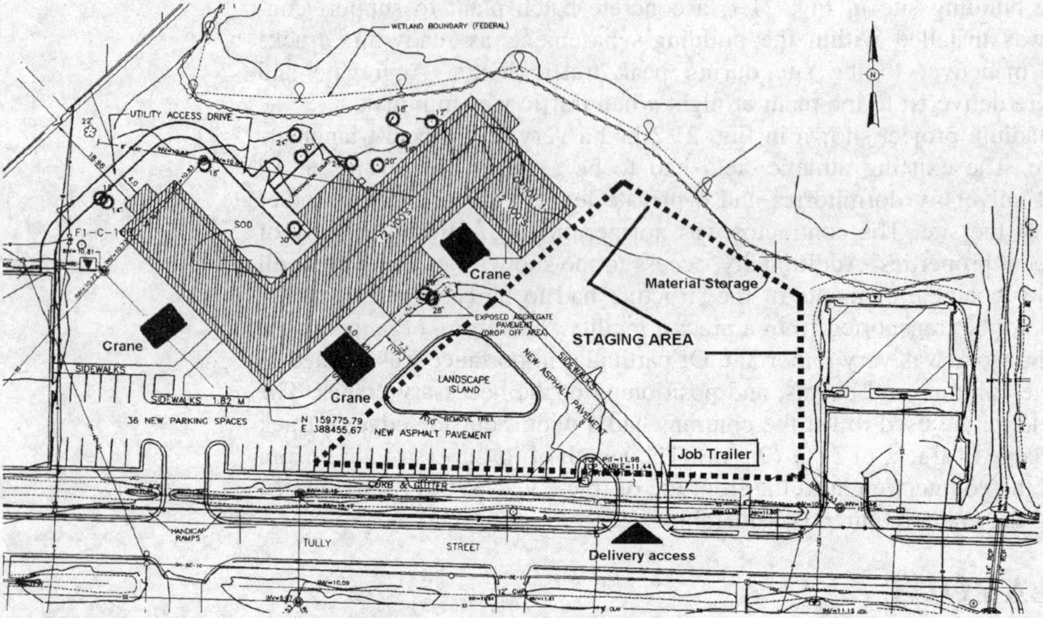

FIGURE 21.3 Site layout plan sketch on a project plan sheet.

The site layout in terms of construction activities, material storage, administrative space, and sanitary facilities can be sketched on a project plan sheet (Fig. 21.3).

Early Coordination

While preparing bids, the general contractor (GC) will define subcontract work packages and seek subcontractor prices. Early interaction between the general contractor and the selected subcontractors is important. The subcontractors will, in turn, seek lower tier subcontractors to perform even more specialized work. As an example, in the case of a steel frame building, a general contractor would normally have a steel subcontractor. The steel subcontractor, typically the fabricator, who is not also an erector, will seek lower tier subcontractors for (1) steel erection, (2) shear studs, (3) metal deck supply and installation, and (4) special fabrications. The general contractor will often require that the major subcontractors, such as the steel subcontractor, provide input for the preliminary project schedule and job-site layout. The job-site layout is key in enabling the subcontractors to operate effectively and efficiently.

Bid Package

The bid package given to each contractor by the project owner will usually include information related to:

- Scope of work
- Job-site conditions and site layout

FIGURE 21.4 Steel stored at the project site for erection.

FIGURE 21.5 Reinforcing steel cages being tied at the project site.

- Space limitations
- Allowable temporary facilities

A prebid meeting may be held to address questions related to the bid package. During this pre-bid stage, the general contractor and the major subcontractors will make a preliminary determination of the equipment requirements necessary for constructing the work. A steel subcontractor would be primarily interested in lifting requirements and the location of an on-site laydown space to accommodate steel deliveries (Fig. 21.4). A concreting subcontractor would also be considering lifting requirements and storage space for forming systems and reinforcing steel (Fig. 21.5).

These subcontractors will need information from the general contractor and have to provide information to the GC about

- Schedule
- Sequence of other trades (subcontractors)

FIGURE 21.6 Telescoping-boom crane erecting steel. Lattice-boom crane erecting steel.

- Crane size and provider
- Placement of crane, job trailers, storage areas, and laydown areas

The job-site layout and equipment used will be controlled by the site constraints. For those subcontractors having lifting requirements, such as the steel and concrete subs, a primary consideration will be crane size, and location or locations. The other two critical pieces of information are (1) laydown area size and (2) laydown area location.

The goal of site layout is to optimize operational processes—steel erection, concrete forming, and placement of the concrete in the forms. The GC and the subs seek to keep the number of crane locations at a minimum. Laydown areas are located as close as possible to the structure. But the decisions are all dependent on project site size and lifting requirements (Fig. 21.6).

Occupational Safety & Health Administration (OSHA) Standard 1926.752 places specific duties on the General Contractor (called the controlling contractor by OSHA).

1926.752(c)

Site layout. The controlling contractor shall ensure that the following is provided and maintained:

1926.752(c)(1)

Adequate access roads into and through the site for the safe delivery and movement of derricks, cranes, trucks, other necessary equipment, and the material to be erected and means and methods for pedestrian and vehicular control. Exception: this requirement does not apply to roads outside of the construction site.

1926.752(c)(2)
A firm, properly graded, drained area, readily accessible to the work with adequate space for the safe storage of materials and the safe operation of the erector's equipment.

As stated in the OSHA Standard it is the responsibility of the GC to provide access points into the construction site. The subcontractors are responsible for communicating any necessary special access requirements to the general contractor. The general contractor usually provides access for workers into the structure of the building; however, movement of subcontractor employees from floor to floor may be the responsibility of the subcontractor (Fig. 21.7).

FIGURE 21.7 Personnel hoist provided for the erection crew.

FIGURE 21.8 Because of site constraints, the cranes had to be located within the stadium footprint in order to place the concrete beams.

The relationship between a building's footprint and the size of the site has a significant impact on the planning and sequencing of construction operations and equipment selection. Space is required for delivery vehicles and for storage of construction materials. Material laydown areas may have to change as construction progresses or construction operations may be dictated by site constraints. The GC will have to much more carefully coordinate all construction processes and the activities of the subcontractors if it is necessary to position the crane within the footprint of the building being constructed, as such a situation often impacts the installation of other building systems (see Fig. 21.8).

The overall project schedule and construction sequencing are dependent on crane location and size (dimensions) of the building being constructed. The Occupational Safety & Health Administration, Standard 1926.753(d) is very specific about worker safety in relation to working under lifted loads.

1926.753(d) Working Under Loads

- Routes for suspended loads shall be preplanned to ensure that no employee is required to work directly below a suspended load except for:
- Employees engaged in the initial connection of the steel; or
- Employees necessary for the hooking or unhooking of the load
- When working under suspended loads, the following criteria shall be met:
- Materials being hoisted shall be rigged to prevent unintentional displacement;

- Hooks with self-closing safety latches or their equivalent shall be used to prevent components from slipping out of the hook; and
- All loads shall be rigged by a qualified rigger.

OSHA Standard 1926.550, Cranes and Derricks further states: All employees shall be kept clear of loads about to be lifted and of suspended loads.

Therefore, other than the employees engaged in the initial connection of the steel or employees necessary for the hooking or unhooking of the load, no one is allowed to be under a hoisted load. This safety requirement will limit work of other trades during erection of steel or placement of concrete. If the size of the building and/or job site allows the erection of steel or concrete placement to progress efficiently without hoisting over other trade workers, it may be possible for other subcontractors to work during the lifting operations.

LIFTING AND SUPPORT EQUIPMENT

The subcontractor responsible for erecting structural members, be they concrete or steel, will typically provide most, if not all, of the erection equipment needed. Typical equipment used for concrete member or structural steel erection includes:

- Cranes
- Manlifts
- Generators
- Welding equipment
- Air compressors

Mobilizing the Crane

Crane mobilization involves transporting the crane to the job site and preparing the crane for use. Mobilizing rough-terrain, all-terrain, and telescoping-boom truck-mounted cranes is a relatively fast process. These cranes can usually be positioned and ready to hoist loads in less than 30 minutes.

Most rough-terrain cranes can travel on surface streets, but have a maximum speed of about 30 mph. For long moves between projects, rough-terrain cranes should be transported on low-bed trailers. All-terrain and truck-mounted cranes are designed to be driven to the job site on surface streets.

Crawler or Tower Cranes Mobilizing a crawler or tower crane is a more involved process. It entails transporting the parts of the crane to the job site and preparing the crane for use. The time and cost to load, haul, assemble, disassemble, and reload a crawler or tower crane on a project need to be carefully considered. Some of the largest crawler cranes need up to 15 trucks to transport them.

Assembling the parts of a crawler or tower crane may take several days. Some crawler cranes can assemble themselves, while others require the use of another crane during their assembly (Fig. 21.9).

FIGURE 21.9 A lattice-boom crawler crane delivered to the project site.

Lift Planning

All crane hoisting must be planned. In the case of general lifts that do not involve **critical lifts** there can be a single set of planning guidance.

critical lift
Any lift involving multiple cranes, or where the weight of the load is close to the capacity of the crane, or there is lift complexity and difficulty, or swing area is restricted.

General lift plans cover activities such as unloading miscellaneous supplies, and the plan would

- Require a designated lift director, who understands the task to be performed, to be in control of the lift. The lift director could be the crane operator, a rigger, or even a carpenter.
- Require that a signal person is assigned and identified to the crane operator.
- Require that the weight of the load be known.
- Document lifting restrictions (weather, temperature, time of day) appropriate to the crane being used and the project site.

Production lifts are repetitive-type lifts that do not fall under the classification of critical lifts. Production lifts are usually treated as a subcategory of general lifts, so their lift plans would modify the general lift plan. The plan for a production lift would

- Provide a physical description of the items to be lifted (size, weight, shape, and center of gravity).
- Describe the appropriate operational factors of the lifts, lifting and swing speeds, and crane travel path.

- Address lift hazards such as rigging and controlling access to the area under the hoisting path.
- Identify lift restrictions that are over and above those of the General Lift Plan.

Critical lifts require a specific lift plan. When the overall risk of a lift is assessed as being significant for whatever reason—load weight, difficulty, complexity, multiple crane lift, restricted area—the lift is classified as critical. A critical lift plan will

- Identify the person in charge of the entire lifting operation.
- Identify the physical properties of the item being lifted, exact weight, dimensions, and center of gravity.
- Identify hoisting equipment by type, rated capacity, boom length, and configuration.
- Identify required rigging including capacity of items and accessories.
- Identify the travel path of the load by the required hoisting, swinging, and travel motions of the crane (see Fig. 21.10).

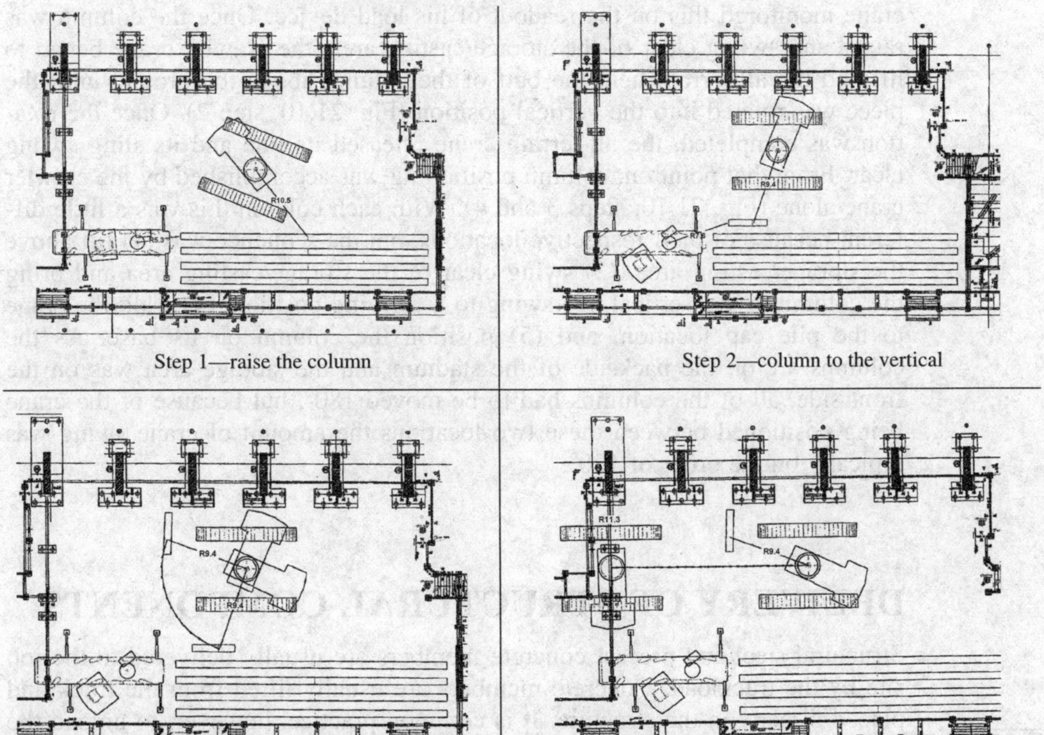

FIGURE 21.10 Lift plan for setting a concrete column using a large crawler crane assisted by an all-terrain crane [7]. The R values in the figures refer to the radius of the lift.

- Include an analysis of weather factor impacts (wind, temperature, visibility) and state weather controlling limits.
- Assign a signaler and provide continuous communications between the signaler and the crane operator.

Raising and positioning the concrete stadium columns that were cast on the project site shown in Fig. 21.2 was an intricate process that required lifting and positioning heavy loads while operating within a very confined project space. Additionally, all of the column hoists were dual picks by a large crawler crane and an all-terrain crane (see Fig. 21.10). All of these lifts would be classified as critical lifts requiring a critical lift plan. Therefore, the projected travel path of load hoisting and swing from lift-off until final placement was plotted for each lift. Figure 21.10 is the travel path plan for lifting the second column (an end column).

At the beginning of the lift, the concrete column is in a horizontal position and is then hoisted off of its shoring towers (Fig. 21.10, step 1). A hole was cast into the column to receive the lifting pin that connected to the slings of the crawler crane. The plan was for the all-terrain crane to always carry a fixed portion of load during this portion of the hoist. The operator of the all-terrain crane monitored this on the readout of his load device. Once the column was raised and swung clear of the storage/casting area, the crawler crane began to lift and the all-terrain held the butt of the column above the ground until the piece was rotated into the vertical position (Fig. 21.10, step 2). Once the rotation was completed, the all-terrain crane released its line and its sling swung clear. From that point on, column positioning was accomplished by the crawler crane alone (Fig. 21.10, steps 3 and 4). With each column this was a little different because of their respective locations, but the sequence was (1) lift above the storage/casting area, (2) swing clear of the storage/casting area and bring the column to the vertical, (3) swing to a carrying position, (4) walk the crane to the pile cap location, and (5) position the column on its base. As the columns are on the backside of the stadium and the storage area was on the front side, all of the columns had to be moved 180°, but because of the crane being positioned between these two locations the amount of crane swing was typically on the order of 120°.

DELIVERY OF STRUCTURAL COMPONENTS

Structural steel and precast concrete members are usually delivered to the job site by the truckload. Concrete members are usually lifted from the truck and placed directly in the structure. It is common practice, however, to unload the steel from the trucks and place it in a laydown area. A crew will then **shakeout** the steel into the order that it will be erected. There will usually be several truckloads of steel in the laydown area at one time to prevent interruption of the erection work. The size and location of the laydown area are important factors to be considered with structural steel erection.

shakeout
Sorting of delivered steel so that it can be erected in the proper sequence.

Delivery Management

For projects where delivery space is limited, extra coordination will be required to accomplish quick unloading. Requirements might include (1) an off-site staging area for delivery trucks to wait until directed to proceed to the site or for trailers to be dropped until they are needed (Fig. 21.11), (2) radio communication between delivery drivers and the job site, and (3) a flagman to direct traffic to or through the site to the delivery point. Such delivery controls are also necessary in the case of large-volume concrete placements when ready-mix trucks must cycle through to the concrete pump or to a placement location.

FIGURE 21.11 A trailer load of precast concrete building facing panels.

Steel Laydown Area

The steel laydown area needs to be flat, firm, and well drained. Cribbing is usually placed under the structural steel members for ease of rigging and to keep the steel clean. Steel erectors generally prefer to have a minimum of two truckloads of steel in the laydown area to ensure that erection can continue without interruption. A typical structural steel project will require a laydown area of between 2,500 and 10,000 sf (50 by 50 ft to 100 by 100 ft). This area allows the steel members to be appropriately spaced and organized for efficient erection. Some projects, where large, built-up trusses are used, will require a much larger laydown area for their fabrication on the job site (Fig. 21.12).

FIGURE 21.12 Assembly of a large roof truss on ground at the job site.

STEEL ERECTION

The American Institute of Steel Construction (AISC) Code of Standard Practice clearly states that "The Erector shall be responsible for the means, methods and safety of erection of the Structural Steel frame." The two primary safety concerns are fall protection for the personnel doing the erecting and stability of the frame during the erection process before all elements of the structure are in place. The Occupational Safety & Health Administration Standard 1926.760(e) also addresses fall protection as follows:

Fall protection provided by the steel erector shall remain in the area where steel erection activity has been completed, to be used by other trades, only if the controlling contractor or its authorized representative

- Has directed the steel erector to leave the fall protection in place; and
- Has inspected and accepted control and responsibility of the fall protection prior to authorizing persons other than steel erectors to work in the area.

The issue of stability is set in motion with the placement of the first structural member, which is usually a column. As the first columns are placed, ironworkers will bolt them to their footings. Prior to erection, the erector must consider column stability in accordance with safety standards and the AISC Code of Standard Practice. Main girders are hoisted after the columns, followed by the beams, which are hoisted and connected to the girders. At least two bolts per connection are used to temporarily fasten each connection. Teamwork and communication are important when erecting steel. The ironworkers typically use hand signals (Fig. 21.13) to communicate with the crane operator.

plumb
Refers to the vertical alignment of the structure.

Temporary bracing is used to provide temporary lateral stability to the steel structure and to **plumb** the frame. Turnbuckles are a common part of the bracing system. A turnbuckle is a device that consists of a link with screw threads at both ends that is turned to bring the ends closer together or farther apart. With the use of several turnbuckles connected to the temporary bracing, the steel frame is shifted until it was vertically aligned—plumb.

FIGURE 21.13 Ironworkers directing the positioning of a steel beam.

After a section of the structural frame is assembled and vertically aligned, ironworkers permanently fasten the structural steel members with additional bolts and welds. Structural steel has to be protected from fire. Steel will not burn, but it will become weak when exposed to intense heat. Building codes regulate the need for fireproofing and its required locations. There are several methods for fireproofing, including spray-on materials. Spray-on material may be Portland cement or a gypsum-based product and can be applied directly to structural steel members (Fig. 21.14).

TILT-UP CONSTRUCTION

Concrete tilt-up wall construction is a method of building concrete interior and exterior walls without the use of vertical formwork and that minimizes the required duration of lifting—crane—equipment time. The process is used to reduce the cost of materials and labor in wall construction. This building process works very well for shell-type buildings. The panels are formed, constructed, erected, and supported very quickly, as compared to other building methods [9].

FIGURE 21.14 Worker applying spray-on fireproofing to steel members.

Panel Layout

With tilt-up panel construction the layout of the panels can be accomplished in many ways. All the panels can be molded on the slab-on-grade if there is sufficient room. If there is not sufficient room, some of the panels can be cast on a false slab that is constructed beside the project. If site conditions greatly restrict the available slab area, panels can be constructed on top of one another.

Each panel has a particular function and place where it belongs in the structure. The goal is to place the panels as close as possible to the spot where they will be erected. This does not mean that they have to be directly over or near their final placement spot, but they are laid out in a pattern that minimizes crane time in erecting the panels. The panels are usually erected in a clockwise rotation starting from a chosen building corner. To ensure a proper placement sequence, the placing of certain panels at the correct place and time may involve picking and lifting some panels over or around others that remain on the slab.

Planning the panel layout is a critical part of this construction technique. The planning resembles a game of chess and the proper strategy ensures that the erection of the panels runs smoothly. The goal is to be able to erect the panels in an efficient manner.

FIGURE 21.15 Forms for tilt-up panels on a slab-on-grade.

Forms

After the building slab is completed, the forms for the wall panels are positioned on the completed slab (Fig. 21.15). This operation includes the panel forming, and the placement of reinforcing steel, cutouts, lift and brace points, ledger plates, and electrical conduit.

A panel placement map is used to draw the panels on the slab. There should be spacing between the panels to allow for work in and around the panels. After the panels are drawn on the slab, the wall forms are laid in place. Steel "L"-shaped forms are usually used. Anchor bolts through the base of the "L" are used to keep the forms in place during the concrete placement.

Cutouts for windows, doors, overhead doors, and guttering are formed inside the wall forms (see Fig. 21.15). Chamfering is used on panel edges so that when the panels are lifted and placed, the concrete is not damaged along its edges.

To keep the wall panels from adhering to the concrete slab a commercial bond breaker is used. Once the bond breaker is applied the crew has to be very careful not to contaminate the surface. All materials brought into the area must be thoroughly cleaned. It is important to progress from the application of the bond breaker to the placement of the concrete as quickly as possible so that nothing can contaminate the coating.

During the form-setting process, inserts are placed in the wall sections for lifting the panel and attaching temporary lateral supports—braces. Each insert has its own particular use. Coil-type inserts are used to attach braces that run between the floor slab and the panel once it is positioned vertically. The number of inserts depends on the size of the panel. There are also lift inserts placed in each panel so that the crane can handle the cast panel. Lift inserts include a plastic half-moon void former, which has two holes in it to attach a metal ring clutch so the panel can be lifted. The lift insert must not be tipped more than 20° in any direction or protrude above the surface of the concrete because the lifting unit must bear on the void that is formed. The void that is formed must match the smooth surface of the plastic former. After the concrete is placed, the former is taken out and the void is checked and cleared of debris to see that it is smooth. Once the inserts are set into the panel forms, the reinforcing steel and other metal items are placed.

Concrete Placement

The next step in producing tilt-up concrete wall panels is concrete placement. Concrete is often placed using a concrete pump truck. By using a pump truck that has a boom, all of the panels on most projects can be reached without having to relocate the pump truck. Using a handheld remote to control the boom and the flow of concrete, the operator can easily maneuver the boom trunk as needed.

After the concrete is placed and finished, it must be allowed to cure to sufficient strength for lifting. The holes for the bracing bolts and ring clutches in the inserts and electrical devices can be cleaned out as soon as the panel can be walked upon.

Erecting Panels

The final phase in the concrete tilt-up panel construction process is the panel erection. Once the panels have cured, the forms are removed and the temporary construction braces attached. These are metal poles having extendable ends. They are used to support the panel temporarily when it is positioned ver-

FIGURE 21.16 Crane erecting tilt-up panels.

tically. They are removed when the panel is permanently fixed in place. These braces are bolted to the panels and then laid on top of the panel until the panel is lifted.

Hoisting the panels requires a mobile crane. Since the crane is a major cost item, it is imperative that the proper lifting sequence be followed so that the crane can move as many panels as possible without having to be repositioned. Repositioning represents a time delay to the lifting crew.

Once the crane approaches a panel, the rigging system is attached. Usually a spreader bar, with evenly spaced pulleys having wire-rope lifting bails, is used for the picks (Fig. 21.16). The lift bail on each end of each wire rope is attached to the ring clutch on the lifting insert embedded in the panel. The panel is then lifted from the casting slab. The top of the panel is raised first, and slowly the panel is moved to the vertical.

Once the panel is leveled and plumbed, the braces are attached to the slab (Fig. 21.17). It is not recommended to hoist panels if the wind is blowing in excess of 20 mph (33 km/h), due to the fact that the panel will act like a large sail and the crane could easily be overturned. Also, with the exception of the erecting crew, no one is allowed within an area $1\frac{1}{2}$ times the panel height dimension, so that if an incident should occur, there is a safe work perimeter. After the braces have been firmly attached to the slab, the crane rigging system is detached from the panel. A laborer on a manlift (Fig. 21.18) can accomplish this task.

Using tilt-up concrete walls is an innovative construction technique. It is a fairly simple process of placing concrete into a horizontal form, letting it cure, and then lifting the panels into place. But the process requires careful planning

FIGURE 21.17 Preparing to attach the tilt-up panel braces to the slab.

FIGURE 21.18 Laborers on a manlift.

and preparation. A great deal of attention to detail is necessary. The crane movement between picks must also be carefully considered.

CONTROL OF CONSTRUCTION NUISANCES

The major nuisances associated with construction are noise, vibration, dust, and illumination for nighttime work. Contractors working on projects in urban environments must mitigate the effect of such nuisances on project neighbors. Noise problems are normally caused by the operation of heavy equipment and specifically by vehicle and machine backup-alarms. Vibration problems are primarily a result of pile driving, blasting operations, or the use of vibratory rollers. Fugitive dust will be generated by construction operations and contract specifications often require that the contractor adhere to a dust control plan. While good illumination is necessary for the work to proceed at night and for safety, proper illumination can be very intrusive to project neighbors.

CONSTRUCTION NOISE

The human ear does not judge sound in absolute terms, but instead senses the intensity of how many times greater one sound is to another. A decibel is the basic unit of sound level; it denotes a ratio of intensity to a reference sound.

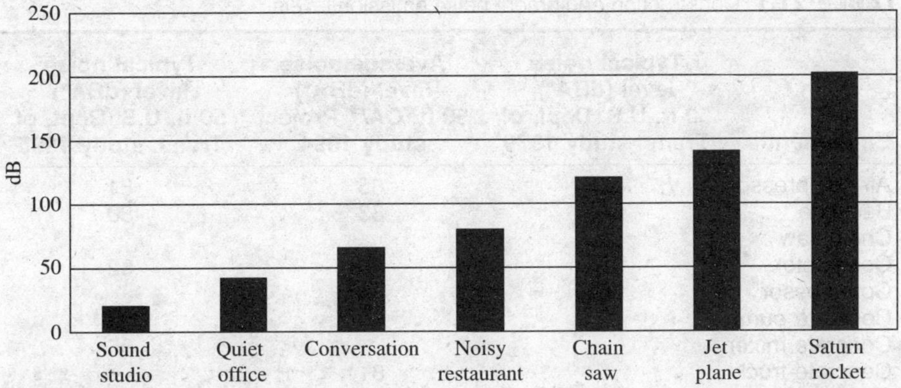

FIGURE 21.19 Representative sound levels.

Most sounds that humans are capable of hearing have a decibel (dB) range of 0 to 140. A whisper is about 30 dB, conversational speech 60 dB, and 130 dB is the threshold of physical pain (Fig. 21.19).

Sound and noise are not the same thing, but sound becomes noise when

- It is too loud.
- It is unexpected.
- It is uncontrollable.
- It occurs unexpectedly.
- It has pure tone components.

Noise is any sound that has the potential to annoy or disturb humans or to cause an adverse psychological or physiological effect on humans. Noise generation on most construction projects is the result of equipment operation (Table 21.1) with diesel engines being the primary generators. Equipment components that generate noise include the engine, cooling fan, air intake, exhaust, transmission, and tires.

Backup Alarms

Standard backup alarms emit a consistently loud noise regardless of background noise levels. At night a standard backup alarm seems excessively noisy against the quieter background sound levels. There are adjustable sound backup alarms available that are either ambient-sensitive or manually adjustable. The ambient-sensitive self-adjusting backup alarms increase or decrease their volume based on background noise levels. These alarms work best on smaller equipment such as backhoes and trucks. The alarm self-adjusts to produce a tone that is readily noticeable over ambient noise levels, but not so loud as to be a constant annoyance to neighbors. Manually adjustable alarms are effective in reducing backup alarm noise nuisance, but their use requires that each alarm be set at the beginning of each day and night shift.

TABLE 21.1 Construction equipment noise emission levels.

Equipment	Typical noise level (dBA*) 50 ft, U.S. Dept. of Trans. study 1979	Average noise level (dBA*) 50 ft, CA/T Project study 1994	Typical noise level (dBA*) 50 ft, U.S. Dept. of Trans. study 1995
Air compressor		85	81
Backhoe	84	83	80
Chain saw			
Compactor	82		82
Compressor	90	85	
Concrete pump			82
Concrete mixer			85
Concrete truck		81	
Concrete vibrator			76
Crane, derrick	86	87	88
Crane, mobile		87	83
Drill rig		88	
Dump truck		84	
Generator	84	78	81
Gradall		86	
Grader	83		85
Hoe ram		85	
Impact wrench			85
Jackhammer		89	88
Paver	80		89
Pump	80		85
Slurry machine		91	
Slurry plant			
Truck	89	85	88
Vacuum excavator			

*dBA, A-weighted decibel unit of sound pressure adjusted to the range of human hearing.

Blue Angel Certification

Due to the strict environmental requirements common in Europe, manufacturers have developed machines for that market that are significantly quieter than similar models sold in the United States. The German government gives a "Blue Angel" certification to machines that meet strict "environmentally friendly" requirements (www.blauer-engel.de/englisch/navigation/body_blauer_engel.htm). Although these machines are available in Europe, it would be extremely difficult to purchase a "Blue Angel" machine in the United States. But these machines definitely demonstrate that the technology is currently available to decrease the noise levels of some construction equipment as much as 15 dBA.

NOISE MITIGATION

Of interest in terms of community noise impact is the overall noise resulting from a construction site. The noise of each individual piece of equipment and sometimes the highest noise source is not always the number one priority.

Noise control is directed toward modification of a perceived sound field. It strives to change the impact to the receiver so that the sounds conform to a desired level. Mitigation of undesired sounds should consider source control, path control, and receptor control.

Source Control

The most effective methods of eliminating construction nuisances are control techniques applied at the source. Source controls, which limit noise, vibration, and dust emissions, are the easiest to oversee on a construction project. Mitigation at the source reduces the problem everywhere, not just along one single path or for one receiver. Construction equipment is a major noise and nuisance generator on nearly all construction projects. The specification of equipment noise emission limits forces the use of modern equipment having better engine insulation and mufflers. Alternative construction methods and equipment can be used to lessen potential construction noise impacts (i.e., cast-in-place piles rather than driven piles, top-down rather than open cut and cover construction, or rubber-tired equipment rather than steel-tracked equipment).

Plan Construction Operations Noise can often be controlled by restricting the movement of equipment into and through the construction site. Long-term noise impacts are generated along haul routes when there are large quantities of materials to be moved. Project planning should consider rerouting truck traffic away from residential streets.

Use Modern Equipment It may be necessary to buy new, quieter equipment when operating in an area where noise is a problem. Sometimes project specification will limit equipment noise emissions, forcing the use of modern equipment having better engine insulation and mufflers.

Operate at Minimum Power Noise emission levels tend to increase with equipment operating power. This is a critical issue with older machines and equipment such as vac-trucks (Fig. 21.20). Require that such equipment operate at the lowest possible power levels.

FIGURE 21.20 Crew using a vac-truck system to locate utilities.

Use Quieter Alternative Equipment Electric- or hydraulic-powered equipment is usually quieter than a diesel-powered machine. Contractors should consider the use of alternative equipment when noise might be an issue. For example, use electric tower cranes instead of diesel-powered tower or mobile cranes.

Path Control

Path control of construction noise, light, vibration, and dust nuisances should be implemented when source controls prove insufficient in adequately minimizing impacts on abutting sensitive receptors. This situation can result due to the close proximity or because of the very nature of the construction work. Thus, having exhausted all possible mitigation methods of controlling a nuisance at the source, the second line of attack is controlling noise, light, vibration, or dust radiation along their transmission paths. When barriers are used, they should provide a substantial reduction in noise levels, should be cost effective, and should be implementable in a practical manner without limiting project access.

Once established, only reflection, diffraction insulation, or dissipation can modify an airborne sound field. In other words, it is necessary to increase the distance from the source or to use some form of solid object to either destroy part of the sound energy by absorption, or to redirect part of the energy by wave deflection. The three techniques for path mitigation are therefore:

- Distance
- Reflection
- Absorption

Move Equipment Farther Away from the Receiver Doubling the distance between the source and the receiver allows a 3 to 6 dBA reduction to be achieved. It is important to recognize that a 6-dB reduction of sound pressure represents a noticeable change in noise level.

Enclose Especially Noisy Activities or Stationary Equipment Enclosures can provide a 10- to 20-dBA sound reduction. Additionally, the visual impact of construction activities has an effect on how construction sounds are perceived. An important noise mitigation issue, therefore, is the audiovisual sensing factor. Enclosures address both the absolute audio and the visual perception issue (Fig. 21.21).

FIGURE 21.21 Slurry plant enclosed for audiovisual sensing and dust control.

Erect Noise Barriers or Curtains Barriers can provide a 5- to 20-dBA sound reduction. These may be very temporary systems mounted on jersey bases for easy relocation or more semipermanent walls designed to last several years on long-duration projects. The design of a noise barrier should involve a structural and wind load analysis. Another temporary noise barrier option is acoustical curtains (Fig. 21.22). Depending on the application, these quilts can reduce sound levels about 10 dBA. Curtains are typically installed in vertical segments. A wide range of these products is available in modular "off-the-shelf" panel sizes. All seams and joints should have a minimum overlap of 2 in. and be tightly sealed.

FIGURE 21.22 Enclosure around a building project.

Receptor Control

Receptor control of a nuisance must be undertaken when all other approaches to mitigation have failed. It should be remembered that the critical receiver might not be a human. Certain precision equipment is sensitive to very low levels of ambient noise and vibration. Additionally, the response of human beings, either singularly or as a group, is a problem due to the individuality of each person. No one individual is likely to exhibit the same reaction to a noise stimulus on two successive days. The receptor problems usually involve individuals located very close to the nuisance-generating activity, in which case it may be easier and more effective to improve the individual's environment instead of controlling all emitted noise, vibration, or dust.

Community Relations Early communication with the general public is vital. Inform the public of any potential construction noise impacts and the measures that will be employed to reduce these impacts. Establish and publicize a responsive complaint mechanism for the duration of the project. The establishment of good rapport with the community can provide high benefits at low cost.

Window Treatment Program In general, window openings are the weak link in a structure's external façade, allowing noise infiltration into a building. A good window treatment can provide an incremental 10-dBA sound reduction in a building.

LIGHTING

Lighting of the work area is important for both quality and safety. Yet, temporary lighting and the resulting glare can create nuisance problems. The central issue is adequate illumination of the work area without simultaneously creating intolerable glare. Therefore, it is good practice to develop a lighting plan for all operations that will be performed during nondaylight hours or in enclosed spaces.

The steps to follow in developing a lighting plan are

1. Assess the work zone to be illuminated.
2. Select the type of light source.
3. Determine recommended lighting levels.
4. Select lighting fixture locations.
5. Determine luminaire wattage.
6. Select luminaire and aiming points.
7. Check for adequacy and glare.

DUST

With urban construction projects being very close to people's living space, dust can be a problem. If it is believed that fugitive dust will be generated by construction operations, the contractor should prepare a dust control plan. The plan should cover

- Earthwork—watering, prewet sites.
- Disturbed surface areas—watering, chemical stabilizers, wind fences, windscreens, berms, or stabilization with vegetation or gravel.
- Open storage stockpiles—watering, chemical stabilizers, wind fences, windscreens, berms, coverings, and enclosures.
- Unpaved roads—watering, chemical stabilizers, stabilization with gravel, restriction of vehicle speed.
- Paved road track out—limit or restrict access; stabilized, gravel, or paved construction entrance pad; wheel wash station; vacuum/wet-broom public roadway.
- Hauling—maintain minimum freeboard, tarps.
- Demolition—watering, prewetting.
- Work limits during high winds—cease work temporarily for certain wind directions.

VIBRATION

Construction activities can cause varying degrees of vibration that spread through the ground. Though the vibrations diminish in strength with distance from the source, they can achieve audible and feelable ranges in structures very close to the work site. Rarely do these vibrations reach levels that cause damage to structures, but the issue of vibration problems is very controversial. The case of old, fragile, or historical buildings is a possible exception where special care must be exercised in controlling vibrations because there is a danger of significant structural damage.

Humans and animals are very sensitive to vibration, especially in the low-frequency range (1–100 Hz). Vibrations from construction work are normally the result of blasting, impact pile driving, demolition, drilling, or the use of vibratory rollers.

Mitigation of Construction Vibrations

The mitigation techniques for reducing vibration impacts are similar to those utilized to reduce noise nuisances. The series of questions that should be addressed concerning vibration effects are

1. Will vibrations be caused?

2. Are sensitive people or structures in the vicinity?

3. Is damage/intrusion possible?

4. Can site-specific trials be conducted to assess possible damage/intrusion?

Answering these questions requires a clear understanding of construction equipment location and construction processes in relation to critical receptors. If the answer to question number three is yes, it will be necessary to modify the construction method.

Project Layout and Access Route heavily loaded trucks away from residential streets. Establish designated haul routes so that the fewest possible homes are affected. Place operating equipment on the construction site as far as possible from vibration-sensitive receptors.

SUMMARY

Project planning is more than just the scheduling of work activities. Attention must be directed to the best location for project lifting equipment in terms of all project activities. However, lift planning based only on locating lifting equipment so that the requirements for making the picks can be satisfied is not the proper approach. Lift planning must consider the impact of equipment positioning in regard to all project activities. Equipment location affects a project globally. The major nuisances associated with construction are noise, vibration, and illumination. Contractors working on projects in urban environments must mitigate the effect of such nuisances on project neighbors. Critical learning objectives include:

- The ability to lay out a project site considering access and construction requirements

- An understanding of the relationship between site-imposed activity restrictions and the project schedule
- An understanding of project nuisance generators and methods to mitigate such nuisances

These objectives are the basis for the problems that follow.

PROBLEMS

21.1 Watering is usually the easiest and most common method of controlling work-site dust. It can sometimes be very difficult to supply sufficient water, therefore, dust suppressants should be considered. Use the Web to identify the environmental considerations relating to the use of chemical dust suppressants.

21.2 Analyze the effectiveness of using a tower crane or a mobile crane for a water treatment plant project. The project team has decided that the best approach for placing concrete on the project would be pumping. The complexity of the concrete and formwork and the accessibility issues to many portions of the project made pumping a better alternative than hook and bucket.

This decision being made, the team moved on to how it would be best to transport the forms and rebar about the cramped job site. Much of the work would require working inside sedimentation facilities and tanks where cranes would clearly be needed. The two company vice presidents discussed the use of crawler cranes for the project. Using this type of crane would require that the work start in one area and proceed to the outside of the project. This would dictate a very sequential project. One of the vice presidents saw this project as one for which a tower crane might be the appropriate crane alternative.

It was found that the crawler cranes were the cheaper alternative considering equipment cost; however, phase construction would be necessary for access reasons. This would mean that only certain areas of the project could be worked at any given time, while with the tower crane it would be possible to build the entire project simultaneously. Research the issues further and present your crane selection with supporting reasons.

21.3 What are the OSHA requirements concerning the routing of concrete buckets being lifted by a crane?

21.4 Reproduce the site plan given here and sketch your proposed construction layout. This is the plan for a small two-story steel frame building that is approximately 145 ft in width and 135 ft in depth. Total height to the top of the steel columns is 25 ft. Indicate a construction entrance, position of the office trailer, steel laydown area, and crane location. What would be the required reach of the crane based on your selected crane position?

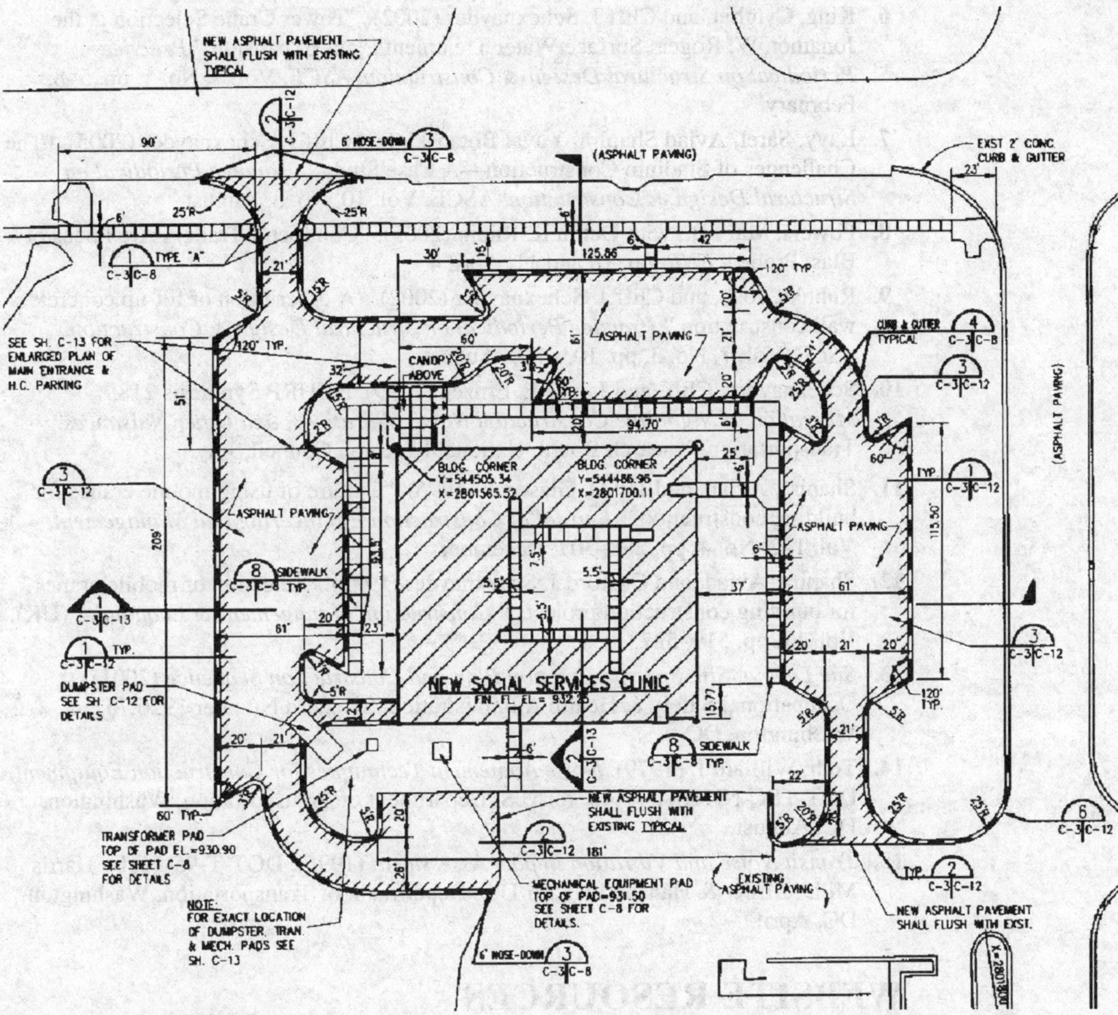

REFERENCES

1. *AISC Code of Standard Practice for Steel Buildings and Bridges* (2000). American Institute of Steel Construction, Inc. Chicago, IL.
2. David, Scott A., and Cliff J. Schexnayder (2000). "Hoisting Y Columns at the Phoenix Airport Parking Garage Expansion," *Practice Periodical on Structural Design & Construction, ASCE,* Vol. 5, No. 4, pp. 138–141, November.
3. *Crane Safety on Construction Sites* (1998). ASCE Manuals and Reports on Engineering Practice No. 93, American Society of Civil Engineers, Reston, VA.
4. *Central Artery/Tunnel Project Noise Control Review* (1994). Harris Miller Miller & Hanson Inc., for Bechtel/Parsons Brinckerhoff, Boston, MA, April.
5. James, David M., and Cliff J. Schexnayder (2003). "Issues in Construction of a Unique Heavily-Reinforced Concrete Structure," *Practice Periodical on Structural Design & Construction, ASCE,* Vol. 8, No. 2, pp. 94–101, May.

6. King, Cynthia, and Cliff J. Schexnayder (2002). "Tower Crane Selection at the Jonathon W. Rogers Surface Water Treatment Plant Expansion," *Practice Periodical on Structural Design & Construction, ASCE,* Vol. 7, No. 1, pp. 5–8, February.
7. Lavy, Sarel, Aviad Shapira, Yuval Botanski, and Cliff J. Schexnayder (2005). "The Challenges of Stadium Construction—A Case Study," *Practice Periodical on Structural Design & Construction,* ASCE, Vol. 10, No. 3, August.
8. Powers, Mary B., with Debra K. Rubin (2005). "Contractor Trailers Are Focus of Blast Probe," *ENR,* pp. 10 and 11, April 4.
9. Ruhnke, Josh, and Cliff J. Schexnayder (2002). "A description of tilt-up concrete wall construction," *Practice Periodical on Structural Design & Construction, ASCE,* Vol. 7, No. 3, pp. 103–110, August.
10. Schexnayder, Cliff, and James E. Ernzen (1999). NCHRP Synthesis 218, *Mitigation of Nighttime Construction Noise, Vibrations, and Other Nuisances*, Transportation Research Board, National Research Council, July.
11. Shapira, Aviad, and Jay D. Glascock (1996). "Culture of using mobile cranes for building construction," *Journal of Construction Engineering and Management*, Vol. 122, No. 4, pp. 298–307, December.
12. Shapira, Aviad, and Clifford J. Schexnayder (1999). "Selection of mobile cranes for building construction projects," *Construction Management & Economics* (UK), Vol. 17, pp. 519–527.
13. *Site Layout, Site-Specific Erection Plan and Construction Sequence* (2001). Occupational Safety & Health Administration, Standard Number 1926.752. Washington, DC.
14. Toth, William J. (1979). *Noise Abatement Techniques for Construction Equipment*, DOT-TSC-NHTSA-79-45, for U.S. Department of Transportation, Washington, DC, August.
15. *Transit Noise and Vibration Impact Assessment* (1995). DOT-T-95-16, by Harris Miller Miller & Hanson Inc., for U.S. Department of Transportation, Washington, DC, April.

WEBSITE RESOURCES

1. www.aisc.org/ContentManagement/ContentDisplay.cfm?ContentID=4211 The *Code of Standard Practice for Steel Buildings and Bridges*, 2000, deals with standard practices relating to the design, fabrication, and erection of structural steel. American Institute of Steel Construction, Inc.
2. www.osha.gov/pls/oshaweb/owastand.display_standard_group?p_toc_level=1&p_part_number=1926 Occupational Safety & Health Administration, Standard 1926, Washington, DC. Provides the complete standard online.

CHAPTER 22

Forming Systems

Formwork accounts for 30 to 70% of the construction cost for concrete framed buildings, hence its importance as a major component of the equipment array used on such construction projects. The selection and use of forming systems are inseparable from the selection and operation of the on-site cranes and other lifting and concrete-placing equipment. Forming systems—for slabs, walls, columns, and other repetitious concrete elements of the structure—are designed and fabricated for many reuses. The constructor's project engineer is extensively involved with various planning aspects of their selection, ordering, erection, stripping, and reuse. Construction companies procure forming systems by either direct purchase or short-term rentals. The labor cost associated with any selected system is an essential part of the economic calculations and comparison of forming alternatives.

CLASSIFICATION

Formwork for concrete can be classified into two main types: conventional formwork and industrialized formwork. Conventional formwork is assembled in situ from standard elements new for each use and is disassembled after each use. Industrialized forms are mostly factory-fabricated products and are used many times as one unit without being disassembled and assembled again for each use. Table 22.1 gives the main characteristics of and shows the differences between these two types of formwork approaches. Although the term formwork relates to both types, it is common practice to single out industrialized formwork by using the term forming systems.

As opposed to conventional formwork of traditional, nonmechanized construction, forming systems—both the old and the emerging new and innovative ones—reflect the industrialization of construction. With the modularization and mechanization of forming systems, they have become an integral component of today's on-site construction equipment. Because the forms are part of a linked

TABLE 22.1 Comparison of conventional and industrialized formwork.

Feature	Conventional formwork	Industrialized formwork
Typology	Temporary structure	Construction equipment
Material	Traditionally wood, but may be steel or aluminum	Mainly steel or aluminum with plywood decking, but also wood
Weight	Not more than 100 lb per element (to be carried by two workers) and commonly in the range of 20 to 60 lb (to be carried by one worker)	In the range of 5 to 20 lb/sf
Handling	Manual	By crane (or other mechanical means)
Number of uses	Single (though the standard elements making up the formwork can be reused many times, depending on the material)	Multiple (up to several hundreds for steel formwork)
Costs	Low production cost but high cost per use	High initial fabrication cost but low cost per use
Site labor input	High	Low
Main use	Nonrepetitive concrete elements	Repetitive concrete elements
Design/ planning	Design is done at the site or at the construction-company level	Design is done at the manufacturer level, while planning (ordering, cost, schedule) is done at the site or at the construction-company level
Structural analysis and testing	Elements are joined together by hinged connections (no transfer of moments); elements can be checked separately by analytical methods	Elements are joined together by fixed connections, welded (can transfer moments); the entire panel/spatial unit is often checked by load testing

production procedure, just like an excavator and a haul truck, forming systems and lifting equipment must be compatible to achieve an efficient building process. The size, weight, and handling requirements of the forming system often dictate the type and capacity of cranes and other on-site lifting equipment. It is in this context that forming systems are addressed in this chapter.

Compared to traditional formwork, industrialized forming systems generally excel in achieving cost savings through considerable savings in erection and dismantling time. Forming costs are reduced thanks to the quick turnaround of the systems and the reduced number of operations required to erect the forms. Additionally, the ability to achieve a smooth, high-quality concrete surface finish with these systems reduces the required finish work, thus providing another cost savings and speeding up the entire building process. With the high production rates achievable with these systems there are significant labor cost savings, and time-related project costs are minimized.

FORMWORK AND THE PROJECT ENGINEER

While commonly not a formwork expert, the project engineer is involved with various aspects of formwork planning and design:

- Engineering design of conventional formwork for simple elements.
- Selecting conventional formwork members based on engineering principles.
- Selecting the desired industrialized forming system to suit the project's construction method and the lifting equipment that will be used. Conversely, adapting the equipment to the forming system selected.
- Studying available sizes and calculating the number and sizes of pieces needed according to the geometry of the concrete element to be formed. This is often done in collaboration with the form system supplier.
- Providing construction data, such as the desired rate of concrete placement, to the formwork designer.
- Coordinating the transport to the site, storage, and multiple lifting for erection and dismantling of the forming system. Site layout is greatly affected by the storage and handling requirements of forming systems, as demonstrated in Fig. 22.1. The schedule of the cranes on site is often dictated by formwork lifting needs.

FIGURE 22.1 Site layout is greatly affected by storage and handling requirements of forming systems, particularly in the case of a restricted site.

- Monitoring the erection of the formwork to ensure compliance with the designer's shop drawings and specifications (conventional formwork) or with the supplier's plans and instructions (forming systems).
- Performing a quality inspection of formwork elements and accessories (e.g., for the detection of damaged timber elements, open joints, and defective welding in the case of steel members).
- Inspecting critical construction details (e.g., sills for better load distribution beneath vertical shoring, limiting extensions of screw jacks in shoring towers).
- Inspecting the formwork during concrete placement and monitoring the concreting process (e.g., to avoid excessive piling of concrete on slab decks, or to ensure alignment of climbing forms).
- Providing guidance as to and monitoring of stripping times of various concrete elements, to comply with the standards and/or results of concrete strength testing. Particular attention to and involvement in the stripping sequence of forms from combined elements (e.g., slab and beams system) and from large-span elements, where engineering knowledge of the structural behavior of the concrete element during and after form stripping is required.

Figure 22.2 illustrates the close interrelationships between project engineering, site management, and the formwork supplier. On this project, the forms supplier maintained a fully equipped three-room office at the project site. Several site layout plans showing the work envelopes of all cranes employed on

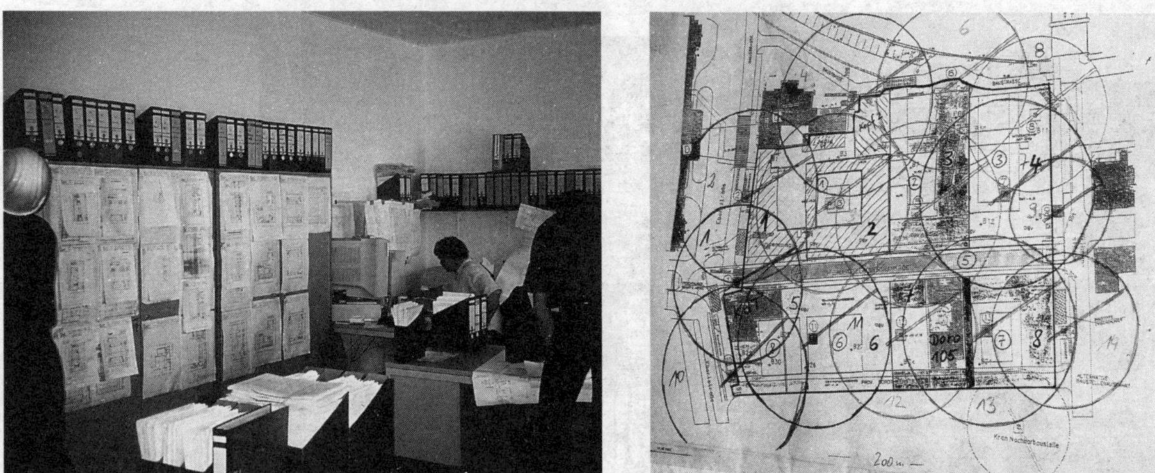

(a) Project site office of the forms system supplier

(b) Form supplier's copy of the project site plan showing crane work envelopes

FIGURE 22.2 Close interrelationship between project engineering/management and the forms system supplier on the new Parliament House project in Berlin, Germany.

the site were displayed on the walls of this office. Close coordination is required to decide which crane is needed or available to unload trucks transporting forms to the site.

FORMWORK DESIGN

Formwork design at the construction site is done almost exclusively for conventional formwork, and therefore is outside the scope of this chapter, which primarily addresses industrialized formwork. There are other books dealing extensively with conventional formwork design [see Refs. 3, 4, 7]. The project engineer should, however, have a basic understanding of formwork design and knowledge of available technical literature that guides formwork designers and users. The fundamental concepts and methods used in conventional formwork design are the same as those applied by manufacturers of prefabricated forming systems. Furthermore, the line dividing conventional formwork and forming systems is not always clear; some modular panels for slab decking, for example, can be regarded as either conventional formwork or industrialized forming systems.

Vertical Loads on Horizontal Forms

Horizontal formwork experiences two types of loadings, dead loads and live loads. Dead loads include the weight of the fresh concrete, reinforcement, and the formwork itself. The combined weight of the fresh concrete and the steel reinforcement is commonly taken as 150 lb/cf, but will be modified if lighter or heavier concrete is used, or in the case of reinforcement made of lighter composite materials (such as FRP—fiber-reinforced polymer). The weight of the forms increases from the uppermost form component downward in the structure. For design purposes, however, the total weight of all the components is used in most cases, and would normally be in the range of 5 to 10 lb/sf of horizontal form. Live loads include the weight of workers, equipment that travels or rests on the formwork (e.g., vibrators, buggies, concrete placing booms), runways, and material stored on the formwork. The formwork should be designed for a live load of not less than 50 lb/sf of horizontal form, but this value should be increased in case of heavier and/or dynamic loading conditions, such as those resulting from motor-propelled (power) buggies.

Lateral Pressure of Concrete on Vertical Forms

When concrete is placed in vertical forms, it produces a horizontal pressure on the surface of the forms that is proportional to the density and depth of the concrete in a liquid or semiliquid state. As the concrete sets, it changes from a liquid to a solid, with a corresponding reduction in the horizontal pressure. The time required for the initial set of the concrete varies with temperature, with a longer time duration required when lower temperatures are experienced, but set time is also affected by the use of admixtures that delay the setting of concrete (retarders). Thus, the maximum lateral pressure produced on the forms varies directly with the rate at which the forms are filled with concrete and with the retarding effect of admixtures, and inversely with the temperature of the concrete.

The American Concrete Institute (ACI), which has devoted considerable time and study to formwork construction practices, recommends the following formulas for determining the maximum pressure produced on the forms by internally vibrated concrete having a slump of 7 in. or less [1].

For wall forms with *R* less than 7 ft/hr and a placement height not exceeding 14 ft:

$$P_m = C_w C_c\left(150 + \frac{9{,}000\,R}{T}\right) \quad \textbf{[22.1]}$$

with a minimum of $600C_w$ lb/sf, but in no case greater than *wh*.

For wall forms with *R* less than 7 ft/hr where placement height exceeds 14 ft, and for all walls with *R* of 7 to 15 ft/hr:

$$P_m = C_w C_c\left(150 + \frac{43{,}400}{T} + \frac{2{,}800\,R}{T}\right) \quad \textbf{[22.2]}$$

with a minimum of $600C_w$ lb/sf, but in no case greater than *wh*.

For columns:

$$P_m = C_w C_c\left(150 + \frac{9{,}000\,R}{T}\right) \quad \textbf{[22.3]}$$

with a minimum of $600C_w$ lb/sf, but in no case greater than *wh*.

where

P_m = maximum lateral pressure in lb/sf
R = rate of filling the forms in ft/hr
T = temperature of concrete during placement, in °F
C_w = unit weight coefficient (in the range of 0.8 to 1.0; reflects variability in the weight of concrete)
C_c = chemistry coefficient (1.0, 1.2, or 1.4; reflects cement type or blend variability)
w = unit weight of concrete in lb/cf
h = depth of fluid concrete from top of placement to point of consideration in form in ft

ACI reference [1] provides guidance for selection of the C_w and C_c coefficients.

In all other cases, namely, where any of the conditions of Eqs. [22.1] to [22.3] are not met (e.g., slump of more than 7 in., or for walls with *R* more than 15 ft/hr), the formwork should be designed for the lateral pressure *p* in lb/sf, as obtained from the full liquid head (hydrostatic pressure) of the concrete:

$$p = wh \quad \textbf{[22.4]}$$

Figure 22.3 is a schematic drawing illustrating the nature of these pressure formulas. Figure 22.3a shows a full liquid head pressure, as would be obtained by Eq. [22.4]. Figure 22.3b shows a situation where the concrete began to harden, thus exerting a pressure at the lower part of the form that is less than the full liquid head pressure (Eqs. [22.1] to [22.3]). The depth of fluid concrete from top of placement is h_1, and is obtained by combining Eq. [22.4] with the formula that is relevant of Eqs. [22.1] to [22.3]. Figure 22.3c shows a situation similar to Fig. 22.3b, but with a lower rate of filling the forms; this results in yet a lower maximum pressure and a smaller height of liquid head pressure, $h_2 < h_1$.

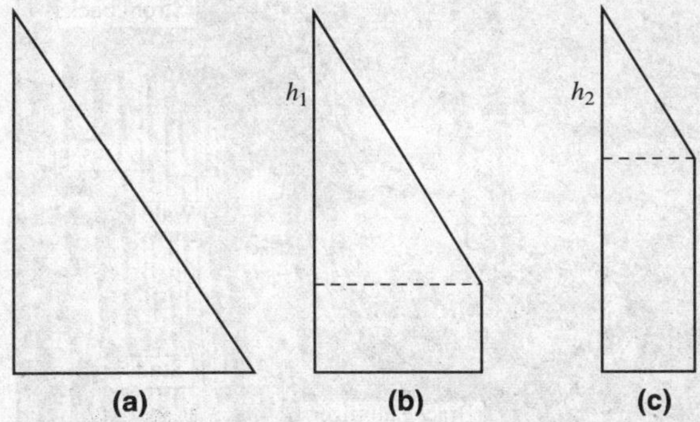

FIGURE 22.3 Lateral pressure of concrete on vertical forms.

Formwork: A Layered System

Forms for most concrete elements, whether conventional or industrialized, are essentially made of *layers* (see Figs. 22.4 and 22.5): (1) sheathing, the first layer, which comes in contact with the concrete; (2) one to three more layers (e.g., joists and stringers in a slab formwork; or studs, **wales**, and strongbacks in wall formwork) that provide the structure that supports the sheathing; and (3) the final layer that either accepts and absorbs the loads (e.g., **ties** in a wall formwork) or transfers the loads to a stable supporting surface (e.g., vertical shores in a slab formwork).

wales
Long horizontal members, usually double, used to hold studs in position (Fig. 22.5).

ties
Tensile elements, commonly rods, used to hold forms together so they will not spread apart when concrete is placed in them.

The essence of formwork design is the determination of the spacing of each layer's elements, for example, in slab formwork: the spacing of the joists that support the sheathing, then the spacing of the stringers that support the

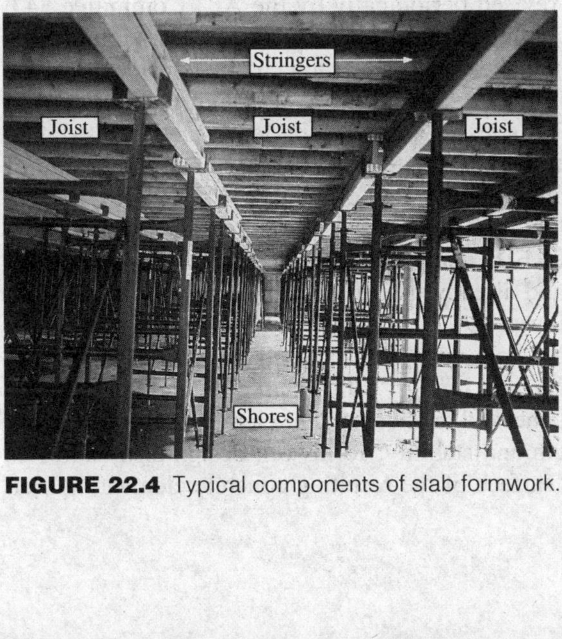

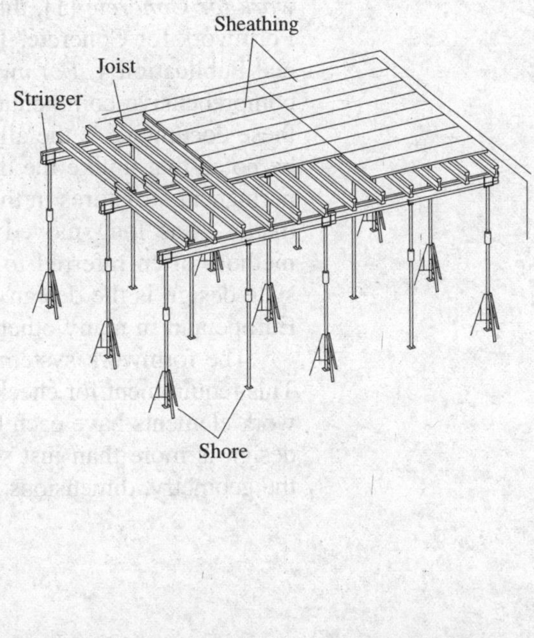

FIGURE 22.4 Typical components of slab formwork.

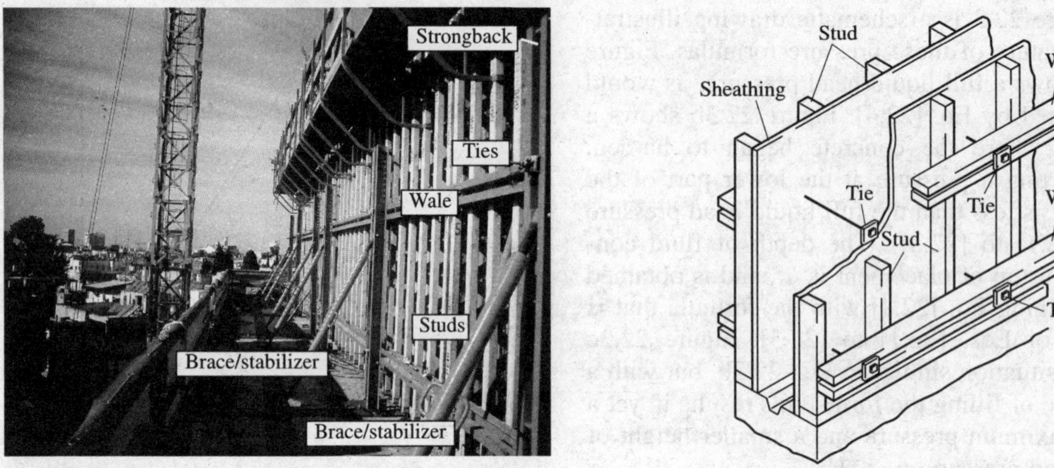

FIGURE 22.5 Typical components of wall formwork.

joists, and finally the spacing of the vertical shores that support the stringers. A top-down calculation approach is commonly used, that is, starting with the sheathing and ending with the final supports. Sometimes the span between elements is referenced rather than spacing of elements, for example, the span of the sheathing element, which is the spacing of the joists; the span of the joists, which is the spacing of the stringers; and so forth. The spans, or spacing, are determined on the basis of statical calculation such that each element in the formwork meets the strength requirements taking into account the loads acting on it, the form material type, and the element's cross-sectional properties. Thus elements are checked, as applicable, for bending, shear, deflection, tension, compression, and local bearing.

In the United States, the document that sets the basis for these statical calculations is the American Concrete Institute standard 347R-03, *Guide to Formwork for Concrete* [1], that is revised periodically by the ACI Committee 347, Formwork for Concrete. The committee also supervises the preparation of Special Publication 4, *Formwork for Concrete* [4], which can be considered as the comprehensive commentary on the ACI standard. The design method used in these documents is the allowable stresses and working loads method. It should be noted that unlike the design method for temporary structures, the design of permanent structures in the United States—and elsewhere in the world, for that matter—has long moved to the limit state design and partial safety factors method, often referred to as LRFD (load and resistance factor design). Limit state design is the design method already used today for formwork in Western Europe and in many other countries [9].

The formwork system as a whole must withstand all loads exerted upon it. This requirement for checking *overall stability* follows after the individual formwork elements have each been checked as independent elements. But formwork design is more than just strength and stability; formwork design also addresses the geometry, dimensions, and surface of the resulting concrete element, as well as the economics of the formwork system. Thus it should be reiterated that the ultimate goal of formwork design is to find the most economical forming alternative among all those meeting strength and quality requirements.

In the case of forming systems, the design of an element must also account for several additional characteristics that are not common to conventional formwork. Forming system design and construction must incorporate provisions for moving the form elements between projects, providing sufficient bracing of large-size panels to withstand lifting and handling stresses, incorporation of quick fasteners and connectors to speed erection and removal, and various mechanized and hydraulic features as well as special provisions for lifting (usually by crane).

FORMWORK ECONOMICS

Formwork accounts for 30 to 70% of the construction cost for concrete structure building frames. With high-clearance construction, for example, concrete slabs and beams at heights of 20 ft or more above the ground or between floors, the share of formwork cost in the overall construction cost for the building frame may be even higher.

Formwork cost is principally comprised of material and labor cost. Industrialized forms may also accrue transportation and lifting-equipment costs, but this depends on the individual contractor's cost accounting practices. Some contractors charge lifting equipment expenses to a general project overhead account and not directly to the forming operation.

Material Cost

The material cost calculation for forms owned by the construction company is different from that for forms procured on a rental basis. If forms are purchased, material cost is generally calculated

$$C_F = \frac{P_F \times \text{USCRF}(n,i)}{N_Y} \qquad \textbf{[22.5]}$$

where

C_F = material cost for one use
P_F = purchase cost
N = overall number of uses before disposal
N_Y = annual number of uses
n = useful life (years), $n = N/N_Y$
i = annual interest rate
USCRF (uniform series capital recovery factor), see Eq. [2.6].

Equation [22.5] assumes that the forms are used throughout their full useful life, and there is no salvage value at the end of that service. If the entire cost of the forms is to be charged to one project or to a series of projects only, then n will be the overall duration of this service (i.e., smaller than the full theoretical useful life of the form), and N will be the total number of uses on these projects. If a salvage value, L_n, is expected after n years, then the cost of material is given by:

$$C_F = \frac{(P_F - L_n) \times \text{USCRF}(n,i) + L_n \times i}{N_Y} \qquad \textbf{[22.6a]}$$

Another approach is (the mathematics are the same, just the arrangement of terms is different):

$$C_F = \frac{P_F \times \text{USCRF}(n,i) - L_n \times \text{USSFF}(n,i)}{N_Y} \qquad \textbf{[22.6b]}$$

where

USSFF (unified series sinking fund factor), see Eq. [2.4].

This is the case, for example, when a form supplier provides the purchasing contractor with the option to sell back a forming system, at the end of a designated project, for a predetermined price. It is an arrangement typically utilized for specialized and costly systems (e.g., automatic climbing systems for wall forming) where no further use is guaranteed to the contractor after the current project for which they were purchased. This is actually more like a lease agreement with a buy-back provision.

In the case where $N \leq N_Y$, namely the useful life is no more than 1 yr (commonly for lumber elements), the time value of money effect on cost is negligible and Eq. [22.5] becomes:

$$C_F = \frac{P_F}{N} \qquad \textbf{[22.7]}$$

Long-lasting steel elements, and to a lesser extent aluminum elements, require periodic routine repair/maintenance (e.g., paintwork, welding, correcting the shape and flatness of metal surfaces that have irregularities), and that expense should be added to the material cost:

$$C_M = \frac{T_M \times \text{USSFF}(f,i)}{N_Y} \qquad \textbf{[22.8]}$$

where

C_M = maintenance expense for one use

T_M = periodic maintenance expense every f years

Equation [22.8] should be used only if a relatively large number of routine repair/maintenance operations are periodically carried out over the form's useful life, and with the understanding that by using Eq. [22.8], we neglect the fact that no such maintenance is carried out at the end of the form's useful life. Otherwise, a more accurate calculation is required. See the more detailed discussion on the cost of capital and on cost factors in Chapter 2.

Note that these routine repair/maintenance costs do not pertain to normal maintenance, such as cleaning and oiling of forms, which does not require skilled labor or costly materials and does not materially extend the useful life of the form. Such costs are considered in the same manner as the operating cost of a machine. Maintenance associated with standard wood elements in conventional formwork is almost always of an ongoing nature, and therefore Eq. [22.8] is not used to account for such cost. Proprietary timber elements in forming systems, on the other hand, are much like steel and aluminum elements, where periodic maintenance may be required.

Forming systems may undergo modification—anywhere from minor alterations to major reconfiguration—to adjust them to their next round of reuse on another project. In this case, modification cost may be calculated as follows:

$$C_R = \frac{R \times \text{PWCAF}(k,i) \times \text{USCRF}(n,i)}{N_Y} \qquad [22.9]$$

where

C_R = average modification cost for one use
R = modification expense after k years
PWCAF (present worth compound amount factor), see Eq. [2.2].

Note that C_R is *average* because only the reuses after the modification are affected by it, whereas Eq. [22.9] *distributes* the cost of modification over the entire span of use of the forms.

If forms are rented, material cost is calculated simply on the basis of rental time. When estimating these costs, the overall duration of the forms' service should be considered. In high-clearance construction, for example, shoring towers may be employed for several weeks just for one use, as they are the first element to be erected and the last to be removed. Monthly rental rates of form elements and forming systems are commonly in the range of 2 to 6% of the purchase cost, depending on the type of element/system, availability, and various business considerations. Oiling and cleaning are the responsibility of the renter, while routine repair/maintenance is commonly the supplier's responsibility. Transportation expense may be added to the rental cost.

EXAMPLE 22.1

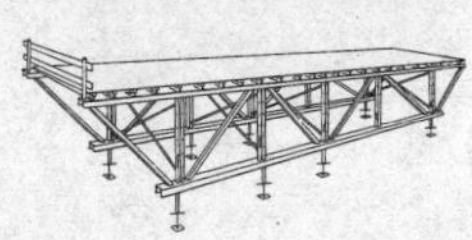

A construction company considers the purchase of a set of flying table forms to be used 110 times on a new project at a rate of 10 times per month. Further reuse of the forms on another project is uncertain, and thus the entire cost of the forms is to be charged to the current project. The forms cost \$23/sf. Another option is to rent the forms for \$0.90/sf per month. Should the company buy or rent the forms?

If purchased, material cost per use is given by Eq. [22.7]:

$$C_{F1} = \frac{\$23/\text{sf}}{110} = \$0.21/\text{sf}$$

If rented, the forms are to be used 110/10 = 11 months, and material cost is

$$C_{F2} = \frac{\$0.90/\text{sf per month} \times 11 \text{ months}}{110} = \$0.09/\text{sf}$$

Therefore, the company should favor rental over purchase in this case.

Note that unlike purchase cost per use, which decreases with the number of reuses, rental cost per use is not affected by the overall number of reuses but rather by the duration of each use (or by the number of uses per month).

EXAMPLE 22.2

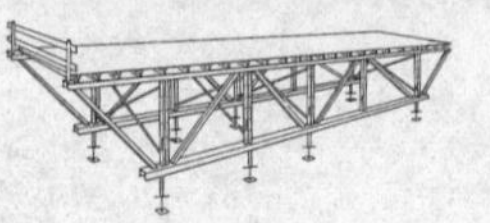

For the conditions described in Example 22.1, what is the number of uses for which purchase and rental costs become even?

If capital cost was neglected, then the breakeven number of uses could be obtained by using Eq. [22.7]:

$$N = \frac{P_F}{C_{F2}} = \frac{\$23}{\$0.09} = 256$$

Therefore, if the number of uses is expected to be greater than 256, the company should favor purchase over rental.

However, with this number of uses, the length of service far exceeds 1 yr (based on 10 uses per month there would be 120 uses per year), and it no longer meets the condition $N \leq N_Y$ for which Eq. [22.7] can be used instead of Eq. [22.5]. Thus, capital cost should be considered in this case, and for an annual interest rate of 6% the number is obtained by using Eq. [22.5]:

$$C_{F2} = \frac{P_F \times \text{USCRF}(N/N_Y, i)}{N_Y}$$

P_F = \$23/sf from Example 22.1

N_y = 120 uses

$$\text{USCRF}\,(N/N_y, i) = \frac{i(1+i)^n}{(1+i)^n - 1} \text{ or } \frac{0.06(1.06)^{N/120}}{(1.06)^{N/120} - 1}$$

then

$$\$0.09 = \frac{\$23 \times \text{USCRF}(N/N_Y, i)}{120}$$

$$\frac{\$0.09 \times 120}{\$23} = \frac{0.06(1.06)^{N/120}}{(1.06)^{N/120} - 1}$$

$$0.4696 = \frac{0.06(1.06)^{N/120}}{(1.06)^{N/120} - 1}$$

$$0.4696 \times [(1.06)^{N/120} - 1] = 0.06(1.06)^{N/120}$$

$$0.4696(1.06)^{N/120} - 0.4696 = 0.06(1.06)^{N/120}$$

$$(1.06)^{N/120} = 1.1465$$

$$N = 281 \text{ or } n = 2.34 \text{ years}$$

Indeed, the difference between these two results—256 and 281—expresses the cost of capital in this case.

The overall number of uses possible for the form, N, is the one factor for which information is commonly missing or inaccurate, that is, uncertainty is high. There should be a distinction between elements whose useful life is up to 1 yr (commonly lumber and low-grade plywood) and elements and forms whose useful life is longer than 1 yr. It should be noted that quite often, the overall number of uses in practice is governed more by the length of each use than by the maximum theoretical number of uses. For example, if a given steel form is used once a week, then it may easily reach its maximum number of uses, commonly taken as 300. However, if the cycle time of a given steel element is 2 months per each use (e.g., a shoring tower in high-clearance construction), then it is unlikely that it will be used its maximum theoretical number of 180 uses, as this means the element must remain in service for 30 years. Average values of overall number of uses, N, are given in Table 22.2; average values of yearly number of uses, N_Y, under common work conditions, are given in Table 22.3.

TABLE 22.2 Overall number of uses of forms and form elements.

Type of form or form element	Overall number of uses
Plywood as sheathing in conventional formwork	10–20*
Plywood as sheathing in modular and industrialized forms (coated plywood; unprotected edges)	20–40†
Plywood as sheathing in modular and industrialized forms (coated plywood; protected edges)	40–80†
Lumber beams in conventional formwork	15–20‡
Steel beams in conventional formwork	180
Proprietary timber beams in conventional formwork	50
Proprietary timber beams in industrialized forms	120
Steel elements in industrialized forms (table forms and wall forms)	300
Steel forms (wall forms and tunnel forms)	300
Shoring towers in high-clearance construction	60–90§

*Lower values for uncoated plywood; higher values for coated plywood.

†Depending on quality of plywood.

‡Depending on cross-section dimensions and on role of beams as stringers or joists.

§These low values result from the low yearly number of uses of shoring towers in high-clearance construction due to the much longer employment time of these shoring towers as compared to vertical shores in regular-height slabs.

TABLE 22.3 Typical number of form uses per year.

Type of form	Number of uses per year
Industrialized wall forms	75
Tunnel and half-tunnel forms	75
Table forms	50
Elements in conventional slab formwork	12
Shoring towers in high-clearance construction	4–6*

*Depending on height of supported concrete element (slab, beam).

One of the factors affecting the time duration of each form use is the time required until the forms can be safely removed. This is particularly true with slab forms. Stripping times depend on the structural scheme of the concrete element, on its span, and on the development of concrete strength as monitored by testing. See the ACI formwork guide [1] for specific values.

When computing the cost of forms, the results usually pertain to the net area of the forms in contact with the concrete, which is commonly referred to as the *contact surface area*. To obtain realistic costs, these results must be increased to reflect the ratio between the actual or gross area of the forms and their net area. Table 22.4 gives common multiplication coefficients.

Labor Cost

Labor cost is easy to calculate once labor productivity is determined. In many cases, however, labor productivity data for the operations of erecting and stripping formwork are hard to obtain. Typical labor productivity data are given in Table 22.5. These, however, are merely a general indication of average time requirements.

> Because of the inherent difficulty in generalizing productivity rates in construction, due mainly to the varied work environment and worker skills, care should always be exercised when applying productivity data from one case to another.

Among the factors affecting labor productivity in formwork erection and stripping are labor skill and experience, crew size and organization, timely support by lifting equipment, availability of spare parts, weather, and how the

TABLE 22.4 Ratio of gross to net area of forms.

To obtain gross area of forms	Multiply net area by
Industrialized wall forms	1.20
Table forms for slabs	1.10
Tunnel and half-tunnel forms	1.15

TABLE 22.5 Labor productivity for erection and stripping of forms.

Forms	Labor productivity (labor hr/sq m*)	Labor productivity (labor hr/sf*)
Walls		
Conventional formwork	0.8–1.1	0.07–0.10
Modular forms, small panels	0.6–0.9	0.06–0.08
Industrialized forms, large panels	0.2–0.5	0.02–0.05
Slabs		
Conventional formwork	0.6–0.8	0.06–0.07
Modular forms, small panels	0.3–0.4	0.03–0.04
Table forms	0.2–0.4	0.02–0.04
Walls and slabs†		
Tunnel and half-tunnel forms	0.2–0.3	0.02–0.03

*Net area (contact surface area)

†Labor productivity per overall form area

workers are compensated (hourly employment or by piece work). One major advantage forming systems have over conventional formwork in this regard is the industrialization of the forming process, that is, erection and stripping are done similarly to the production process in a factory. Thus, the repetitive process is stabilized after experience and competence have been gained, and uncertainty in the prediction of labor productivity is therefore reduced.

Labor productivity for many forming operations can be found in various publications (e.g., R. S. Means [2]), some of which also detail common crew sizes and other pertinent data. Construction companies typically estimate based on their own internal productivity databases, which they have developed from their own experience and accumulated field data. Form manufacturers can also provide labor productivity data, which may be valuable, particularly in the case of unique forming systems. These data, however, reliable as they may be, are commonly based on work forming large areas, regular shapes of the formed concrete element, and an optimal work environment. To use such data, allowances should be made to reflect the reality of a particular project.

The following equation is the simple formula by which labor cost is calculated:

$$C_W = W \times S_W \qquad [22.10]$$

where

C_W = labor cost in \$/sf

W = work input in labor hr/sf

S_W = worker wages in \$/labor hr

Note that although labor productivity and labor cost are commonly measured per unit of contact area, as also given in Eq. [22.10], they may be measured differently (e.g., per element—or complete tower—in high multitier shoring towers).

EXAMPLE 22.3

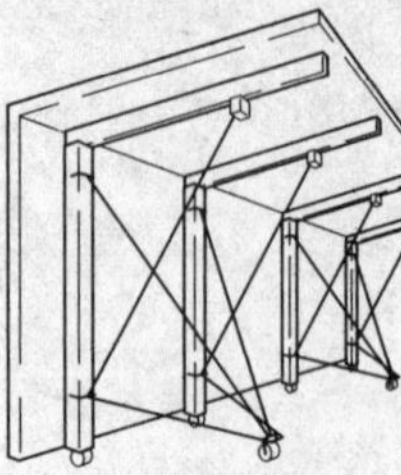

A construction company considers the purchase of a set of half-tunnel forms for \$20/sf. The forms are to be used 200 times for the forming of 1,000 sf of walls and 1,000 sf of slabs per use on a series of residential buildings over a period of 4 yr, and then they will be sold. Salvage value is expected to be 10% of original purchase price. No maintenance costs are expected. Labor productivity is estimated at 0.025 labor hr/sf. Hourly wages are \$22. Consider 5% annual interest rate. What is the average formwork cost (material and labor) per use for this project?

Actual area of forms: (1,000 sf slab + 2* × 1,000 sf wall) × 1.15 (Table 22.4)
= 3,450 sf

Purchase price:	3,450 sf at \$20/sf = \$69,000
Average number of uses per year:	200/4 = 50
Salvage value:	\$69,000 × 10% = \$6,900

Material cost per use, Eq. [22.6a]:

$$C_F = \frac{\$69{,}900 \times 0.9 \times \text{USCRF}\ (4, 5\%) + 6{,}900 \times 0.05}{50}$$

$$= \frac{\$62{,}100 \times 0.28201 + \$345}{50} = \$357.16$$

Labor cost per sf, Eq. [22.10]:	0.025 labor hr/sf at \$22/hr = \$0.55/sf
Labor cost per use:	3,450 sf × \$0.55/sf = \$1,897.50
Formwork cost per use:	\$357.16 + \$1,897.50 = \$2,254.66

*Note that for every 1 sf of wall, 2 sf of forms—both sides of the wall—are required.

VERTICAL SYSTEMS

Vertical forming systems must address three matters stemming from the height dimension of the formed concrete elements: (1) lateral pressure of the fresh concrete on the forms, (2) overall stability of the entire forming system, and (3) accessibility of workers to the top of the formwork. There is an apparent difficulty in dealing with such issues in industrialized systems that tend to be modular and uniform. This is particularly true with respect to concrete pressure, which changes with the height of *liquid* concrete in the forms. Vertical formwork is designed based on a rate of placement and the resulting lateral pressure curve—the pressure of liquid concrete and timing of initial set. Conventional formwork can more readily be adapted to this pressure curve by increasing the spacing between the horizontal beams or the rows of ties that absorb the lesser lateral pressure toward the top of the form. Industrialized forming systems for walls and columns, however, solve this problem by giving higher priority to uniformity and modularity over the potential saving in better correspondence to the pressure differences.

The two other problems—stability and accessibility—are addressed by industrialized solutions. Industrialized forming makes use of higher-strength materials, a high mechanization level, and capacities of lifting equipment.

These features are demonstrated by advanced bracing systems, various patented retracting and form-climbing mechanisms, and inclusion of work platforms that are an integral part of the form system.

Wall Forms

Forming systems for walls can be grouped into four main families:

1. Hand-set forms
2. **Ganged forms**
3. Large-panel forms
4. Large custom-made forms

Ganged forms
Prefabricated panels joined to make a much larger unit.

Hand-Set Forms These are small-panel modular forms that can be used without crane lifting assistance. The panel is composed of sheathing connected permanently to a protecting frame and crossbeams. The panels connect to each other, through holes and slots in the frames, by screws, wedges, or various patented provisions. There are also provisions for the connection of horizontal alignment beams, mounting of work platforms, and insertion of wall ties. With manual work in mind, panel weight is limited to 100 lb and its maximum size is about 20 sf. Common dimensions are 1 to 3 ft wide and 3 to 10 ft high. There are all-steel panels, all-aluminum panels, or plywood-sheathing panels whose frame and crossbeams are made of steel or aluminum. Typical maximum permissible fresh concrete pressure is 800 to 1,200 psf (40 to 60 kN/sq m).

Ganged Forms Often termed *modular forms*, these are essentially hand-set forms or larger panels that are ganged together, often with the aid of additional beams (wales or wales and strongbacks), to form one rigid large-size panel that is used as a full, crane-lifted industrialized form (Fig. 22.6). Screws, clamps, or other means connect the panels such that they can be disassembled after having served their purpose on a particular project, and then reassembled in a different configuration for another project. Common ganged forms use plywood sheathing; high-quality, coated, and edge-protected plywood boards ensure a great number of reuses before they have to be replaced in their frames. Beam systems are made of all-aluminum crossbeams and wales, all-steel crossbeams and wales, or aluminum crossbeams and steel wales. Strongbacks, though not often used, are usually made of steel. The typical weight is in the range of 6 to 14 lb/sf. With proprietary components and hardware, ganged form systems provide solutions for extensions, outside and inside corners, bulkheads, pilasters, and work platforms (see Fig. 22.6), as required, so that the form functions as a complete industrialized

FIGURE 22.6 Ganged forms used for wall construction.

product. The size of the panel is limited only by the size of the formed concrete element; the height, however, is limited by the maximum allowable lateral pressure exerted by the concrete on the form, as specified by the form manufacturer. For any given form height, calculations must be made so that concrete placement rate is adjusted to the strength of the form. Typical maximum permissible fresh concrete pressure is 1,600 psf (80 kN/sq m).

Large-Panel Forms These are all-steel forms (Fig. 22.7) prefabricated as large units whose parts are all welded together. The crane-lifted form frequently comes complete with integral work platforms, guardrails, ladders, and proprietary screw ties. Because everything is integral to the form, it can be regarded as the ultimate industrialized wall form. On the other hand, these forms do not offer the same level of flexibility as ganged forms, and therefore are economic primarily for buildings with a great number of identical repetitive concrete elements. Large-panel forms come in various sizes; a typical large element is one floor (≈10 ft) high and up to 20 ft long. With a typical weight of 16 to 20 lb/sf, a 200-sf panel may weigh up to 4,000 lb. These panels are built to withstand any reasonable liquid concrete pressure developed, even those resulting from high placement rates.

Large Custom-Made Forms These are made-to-measure large-size forms composed of standard elements to fit the specific needs of a given project (Fig. 22.8). They are built on-site. The sheathing is made of plywood whose edges, unlike with hand-set or ganged forms, are not protected by frames (and are

FIGURE 22.7 All-steel large-panel wall forms in residential construction.

FIGURE 22.8 Large custom-made wall forms with timber studs and steel wales and strongbacks.

therefore replaced more frequently). The beam system is composed of either two layers (studs and wales, or wales and strongbacks) or three layers (studs, wales, and strongbacks). Typically these large custom systems have aluminum joists and wales, or wales and strongbacks, or timber studs with steel wales and strongbacks (Fig. 22.8). Because these are large forms, they utilize the same special accessories and hardware (e.g., for corners, intersecting walls, and bulkheads) as found with large-panel forms that are factory-manufactured. As with ganged forms, here too the size of the form is limited only by the size of the formed concrete element. However, because these are commonly heavier forms, in terms of weight per sf, as compared to ganged forms, the lifting capacity of the crane should be checked. For example, a typical element of 600 sf may weigh as much as 10,000 lb. There is practically no maximum concrete pressure limit, as the forms are built with element spacing and tie strength to suit the desired rate of concrete placement.

All of these wall form types share the problem of how to erect the outside forms of external walls in buildings that are more than one floor in height. The common solution is an external scaffold and work platform assembly connected to the wall one floor below. On this external scaffold and assembly, the outside wall form is positioned and braced for overall stability and alignment (see Fig. 22.8). The scaffold is bolted to inserts in the previous placed concrete or connected by special mountings that utilize the holes left in the concrete by the wall ties. It is advisable to use these scaffolds such that their lower part,

FIGURE 22.9 Wall form system including main work platform, upper concreting platform, and finishing trailing platform.

which exerts pressure on the wall, is laid against the floor one level below. In the case of most modern wall systems, and particularly for high-rise construction systems, these scaffolds are often an integral part of the outside wall form, and the entire assembly is hoisted as one unit. The work platform is also used to retract the form from the completed concrete wall (stripping), cleaning the forms, and preparing them for reuse. Most manufacturers offer a complete system that includes, in addition to the main work platform, two additional, narrower platforms: an upper platform for the concrete placement operation, and a lower (*trailing*) platform for various poststripping and finishing work. Figure 22.9 shows a complete system of wall forms with a three-level work platform.

Specialized Wall Forms

Based usually on their standard wall-forming systems, manufacturers offer various special forming systems for cases in which standard systems do not provide adequate solutions.

Core Walls The forming of walls for elevator shafts and stairwells poses special difficulties in that the inner form is essentially a closed box that must somehow shrink in size to allow for removal. Several patented solutions are offered by manufacturers, which essentially cause the form to shrink once the crane begins to raise it. Another concern is the selection of the core forms in relation to the location of a climbing tower crane or a climbing mast of a boom pump inside the core. Not all types of core forms can be fitted into such a confined space, which is further restricted by the presence of a mast. Thus, if a certain type of core forms is favored (e.g., one that will allow for rapid progress in the construction of the core, as is often the case in high-rise buildings), the crane or boom mast will have to be located elsewhere. Conversely, if the crane or the boom mast or both is placed inside the core, it may affect the type of core forms selected for the project.

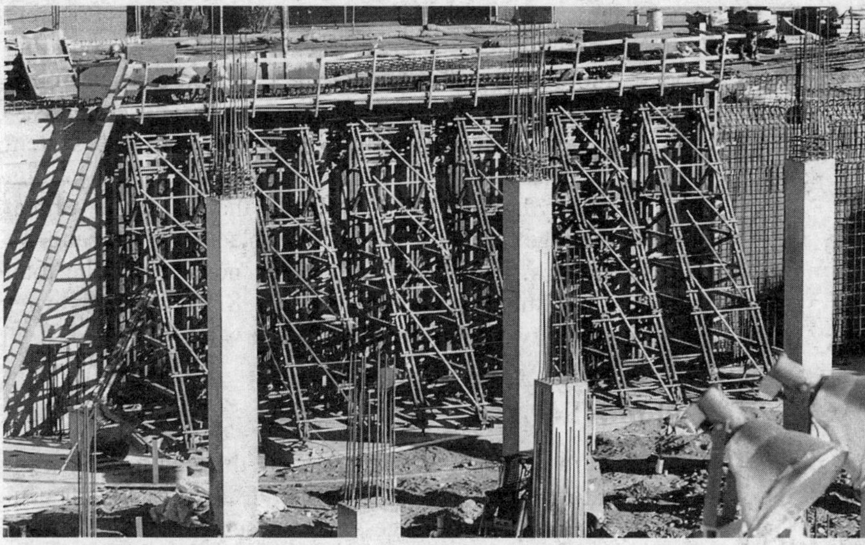

FIGURE 22.10 Single-sided wall forming system.

Single-Sided Walls Single-sided walls are walls concreted against excavation slopes, pier walls, or any other existing permanent surface, such that only one side of the concreted wall needs to be formed. In the case of forms on one side only, there is no way to use regular wall ties. Common solutions are therefore based on the use of diagonal braces that are anchored to the ground (Fig. 22.10). With the increase of urban construction that seeks to maximize land utilization by the use of underground spaces, single-sided walls have become more widespread, resulting in a growing variety of solutions offered by manufacturers.

Self-Climbing Forms Self-climbing forms (Fig. 22.11) are wall-forming systems that raise themselves hydraulically from one floor to the next without the use of a crane. These systems offer four advantages: (1) a saving in crane time, (2) the ability to function in high winds, (3) an exceptionally fast progress pace, and (4) increased safety. There are several self-climbing methods, but none should be confused with **slipforming**, as the self-climbing form operates by retracting from the hardened concrete wall of the current level before it climbs to the next level. After its initial setup in the structure, with the aid of the crane, the modern self-climbing system operates as a self-contained unit that includes weather-protected work platforms, material storage and rebar laydown areas, various worker utilities and facilities, and—as required—concrete placing booms, all making this system virtually a *vertical plant*.

22.1

slipforming
A forming system that continuously moves with the concrete placement.

Because of their advanced technology and sophisticated features, modern self-climbing forms are used mainly for high-rise buildings (and other tall structures such as chimneys and bridge pylons) where the height of the structure and the resulting number of reuses compensates for the high initial cost of

FIGURE 22.11 Self-climbing forms in high-rise construction of the Park Tower in Chicago.

the system. One typical use is for constructing concrete building cores when these are raised ahead of the building floors. Another typical use is for the complete vertical frame of the building—cores and peripheral walls and columns—where the entire floor is cast together. Such is the system used on the 70-floor, 950-ft high building shown in Fig. 22.11. Four hydraulic cylinders shifted the 240 sy of vertical formwork, together with a concrete placing boom and all work platforms, in only 15 min to the next level. The total weight of the unit being moved was 88,000 lb. The tower progressed by two complete floors each week.

Column Forms

Compared to walls and slabs, it is more difficult to industrialize forms for columns. Whereas concrete slabs and walls usually come in relatively large-area flat surfaces, and are thus geometrically given to standardization and industrialization, concrete columns come in a wider variety of geometries and sizes. Yet the impact of moving from conventional to industrialized forming for columns, in terms of cost per cf of concrete, is greater than for slabs and walls, given the higher ratio of forming contact area to concrete volume in columns. Table 22.6 gives example values for typical elements. Another important factor driving industrialization of column forms is the high level of repetitiveness of columns, particularly in high-rise buildings.

These conflicting factors—a great variety of shapes and dimensions, on the one hand, and high repetitiveness with relatively large forming area, on the other hand—result generally in widespread use of industrialized forms for columns. Yet the use of prefabricated made-to-measure forms is more extensive in columns than in walls as compared to the use of standard forms.

Standard Column Forms These are commonly derivatives of wall-forming systems, which provide solutions to unique column-forming problems. Advanced standard column form systems offer provisions for various size options, versatile solutions for clamps, couplers and external ties, telescoping panel stabilizers (used on two perpendicular form sides), ladders and integrated safety cages, and platforms at the top of the forms, which in many cases can be at a great height. Most standard systems offered by manufacturers are adjustable forms for square and rectangular columns. There are basically two form configurations:

1. The form is made of four panels that can be combined in an overlapping pattern to achieve the required size. Typical size adjustment is in the range of 8 to 50 in. No clamps are used and the panels are externally bolted. These are commonly all-steel forms, or steel or aluminum frames onto

TABLE 22.6 Ratio of forming contact area to volume for typical concrete elements.

Element and dimensions	Contact-area-to-volume ratio (sf/cf)
Slab	
8 in. deep	1.50
12 in. deep	1.00
16 in. deep	0.75
20 in. deep	0.60
Wall	
8 in. thick	3.00
12 in. thick	2.00
16 in. thick	1.50
20 in. thick	1.20
Column	
8 × 8 in.	6.00
12 × 12 in.	4.00
16 × 16 in.	3.00
20 × 20 in.	2.40

which replaceable plywood sheets are connected. Panels are manufactured in a small variety of lengths (e.g., 2, 7, and 10 ft, or 1, 2, 4, 8, and 12 ft) that, combined together, allow height variations in increments of 1 or 2 ft, commonly up to a maximum of 25 ft. Fast concrete placement in the forms is made possible by designing and fabricating the forms to withstand high concrete pressures, typically up to maximum of 1,600 to 2,400 psf (80 to 120 kN/sq m).

2. The form is made of sheathing-and-stud panels with adjustable steel or aluminum clamping wales; different column sizes require different panels. The sheathing is made of plywood and the studs are steel, aluminum, or timber. A typical system with proprietary timber girders as studs can sustain concrete pressures of up to 2,000 psf (100 kN/sq m) with square or rectangular cross-sections up to 4 × 4 ft and no height limit.

In both configurations, the panels are commonly set into two L-shaped halves that are handled each as one unit for form closing, stripping, and lifting. Some systems offer hinged connection of the two L-shaped halves along one corner to speed form placement and stripping operations. Other advanced systems have panel design and stripping mechanisms that allow lifting of the complete four-panel form as one unit. To save crane time, these systems can be hand-mounted on wheels for horizontal movement and relocation. The weight of a complete four-panel unit of such systems may reach 2,000 lb for a 10-ft high form. External ties and various other locking devices, which eliminate the use of internal ties through the dense reinforcement common in concrete columns, is an important feature of these systems.

Standard forms for circular columns are made of steel. The form is made of two halves with quick latching provisions. Common diameters for columns are in the range of 10 to 30 in., but some manufacturers offer form diameters up to 96 in. Common form heights are in the range of 1 to 10 ft, with a small variety of different heights to achieve the desirable height in 10- or 12-in. increments. The typical weight of a 10-ft-high half-form segment is 280 to 330 lb for 12 in. diameter and 420 to 500 lb for 24 in. diameter. Typical maximum permissible fresh concrete pressure is as high as 3,000 psf (150 kN/sq m).

Custom-Made Forms When a project involves construction of a series of columns, a typical forming solution is to factory fabricate made-to-measure steel forms. The larger the repetitive number of columns, the more worthwhile the investment in sophisticated forms, with provisions for quick and easy locking, or for adjustment to slight changes in dimensions, as often is the case in high-rise buildings. The robustness of the form to withstand numerous cycles of erection, stripping, and relocation is another consideration. Figure 22.12 exemplifies such forms. These are made-to-measure steel forms for a large number of circular columns with round capitals. They were fabricated to fit the project's specific column design. The savings in labor, the rapid work operations, and the high-quality concrete face achieved were sufficient reasons to invest in expensive forms.

Another case is that of a small series of columns, where certain requirements necessitate the use of factory-prefabricated forms. Figure 22.13a shows a 20-ft-high steel form for bridge pier construction that was used only 12 times. The owner specified, in the project's documents, a requirement for all-welded, steel-face forms, in order to secure a smooth, high-quality concrete surface finish, as well as round edges, as specified by the architect. Meeting

(a) Forms and finished columns **(b) Capital forms lifted by crane**

FIGURE 22.12 Steel custom-made forms for large series of circular columns.

(a) Factory-fabricated steel forms **(b) Site-fabricated timber forms**

FIGURE 22.13 Custom-made forms for small series of columns.

these requirements would have been much more difficult to accomplish using conventional timber formwork. This form is made of four detachable parts for work convenience, and its overall weight is 3 tons.

An additional case, similar to custom-made wall forms, is on-site prefabricated column forms, made usually of standard timber elements, as shown in Fig. 22.13b. This form is built such that adjustment to varying cross-section dimensions is fairly simple. Short external diagonal ties were used at the corners so that no ties had to be inserted through the reinforcing steel cages of the column.

HORIZONTAL SYSTEMS

Common forming systems for slabs typically come in two configurations: hand-set forms and table (*flying*) forms. A third configuration, column-mounted forms (also termed "drawer" forms), may provide a good solution for high-clearance slabs, as they eliminate the need for high falsework. Drawers, however, are suitable only for a limited number of building designs, due to their dependence on an orthogonal grid of closely spaced columns, and therefore their use is not as widespread.

Hand-Set Slab Forms

These are small-panel modular forms. With respect to the definition of industrialized forms, as compared to conventional forms (see Table 22.1), hand-set slab forms can be viewed as semi-industrialized forms: the sheathing and the joists are prefabricated as one unit, while the stringers and the vertical shores are separate elements. Some systems use no stringers at all, in which case, their panels are supported directly by the shores. The modular panels are lightweight and hand-carried, but they can be stacked for crane lifting and transportation from one location (e.g., building floor) to another. Hand-set slab forms have traditionally used welded all-steel panels; these have been mostly replaced by plywood sheathing connected to a protecting steel frame with crossbeams functioning as joists. In recent years, aluminum has gradually been replacing the steel for the frame and crossbeams, resulting in much lighter panels, which in turn has reduced erection and dismantling time.

Unlike wall forms, which can be stripped shortly after concrete placement (12 hr according to [1]), minimum stripping times for slab forms are much longer. This has created an incentive for finding a way to increase panel utilization by shortening the cycle time for each use. The common solution is in the form of a special drophead (Fig. 22.14) for the vertical prop shores (Fig. 22.15). Several patented drophead configurations exist, all providing the ability to strip the modular panels and the separate stringers without disturbing the prop shores. The slab is then supported by the props until such shoring is no longer necessary. Other solutions that enable early stripping of the panels are also offered by various manufacturers.

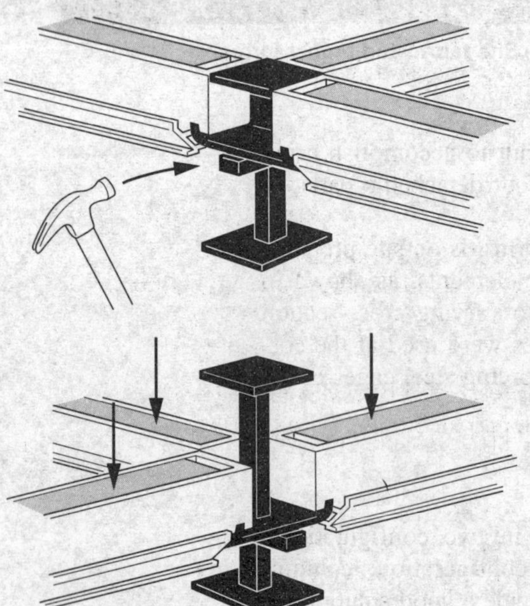

FIGURE 22.14 Drophead.

Figure 22.15 shows a typical plywood-and-aluminum hand-set slab form system with dropheads. The drophead of the shown system is released with a hammer blow, causing the formwork to drop 2 to 3 in. for the removal of the panels and the aluminum stringer beams. These can then be immediately moved to the next forming cycle. To utilize the system in this manner, the constructor uses two sets of props with one set of panels and beams. The panel size for this system is 60 × 30 in., but smaller sizes, used mainly for infill areas, are available as well. Aluminum stringer beams are 90 in. long, creating three panel-width bays between the props. No element is heavier than 34 lb for convenient manual handling. Provisions for infill areas, wall offset, forming around columns, and other irregularities are offered.

FIGURE 22.15 Plywood-and-aluminum hand-set slab forms with drophead system.

Table Forms

The table is a slab form that basically comprises sheathing, joists, stringers, and vertical shoring all in one piece that is craned as a complete unit from one casting location to the next. Table sheathing—or the face of the form—is commonly made of plywood. The table is typically structured such that its length and width can be modified for any particular project by the use of overlapping or telescoping joists (for width) and stringers (for length). Once the structural part has been set to accommodate the desired dimensions, the made-to-measure plywood cover is fastened onto it. Metal sheathing is sometimes used, but this is not as widespread. It is not as flexible for size alteration as plywood, but it has the advantage of enabling the effective use of blowers to heat the concrete and thereby speeding the curing and enabling early stripping and reuse of the forms.

There are two basic table systems:

1. Joist-and-stringer systems (commonly two single or double stringers) supported by single post shores or by shoring towers. The joists and the stringers are either made of the same material, commonly all-steel elements or all-timber elements, or the joists are timber while the stringers are made of steel. Vertical shoring, whether single posts or towers, is made either of steel or aluminum.

FIGURE 22.16 Aluminum table form (truss system) supported on truss jacks.

2. Truss systems in which vertical shoring is provided by trusses—two per table. These also serve as stringers (i.e., there are no separate stringers). The trusses and joists are either all aluminum (more common) or all steel. Adjustable legs or screw jacks under the trusses are used to achieve the full table height (Fig. 22.16).

Both system types use jacking mechanisms (of post shores, shoring towers, or legs supporting the trusses) to level the table to its required height and for stripping the table. Jacks are either mechanical or hydraulic. The standard height of table forms in full service position is in the range of 9 to 12 ft. Truss heights are 4 to 6 ft.

Table sizes vary greatly, from as little as 100 sf to as much as 500 sf. Common dimension ranges of joist-and-stringer systems are 6 to 10 ft width and 13 to 17 ft length; those of truss systems are 7 to 12 ft width and 5 to 50 ft length. Whereas standard tables are always rectangular, tables can also be custom manufactured to accommodate nonorthogonal shapes.

Table weight is determined mainly by the form's size but also by the type of vertical shoring and the material of which the elements are made. Heavy tables, made mostly of steel, weigh as much as 13 lb/sf, while light tables, made either of aluminum joists and trusses or timber joists and stringers and aluminum shores, weigh as little as 8 lb/sf. The larger-size tables are always all-aluminum light truss systems. Thus, an average-size table typically weighs 1,000 to 2,000 lb, while the largest tables may reach as much as 4,000 lb.

When planning the location of the crane on the site, the weight of the table and, often even more importantly, its dimensions must be considered. While table positioning is done by the crane in a straight downward movement, the removal of the form (after it has been lowered and released from the concrete) is a trickier operation. There are three removal methods:

1. A specially designed lifting device (a C frame) is used to lift the table (Fig. 22.17). No rolling out or dismantling platform is needed when a C frame is used. With this method the protrusion of the form from the building façade in the course of removal is minimal, and so is the clearance required for a crane located adjacent to the building (Fig. 22.18). The possible disadvantage of this method is the weight of the C frame, which together with its chain assembly may reach 2,500 lb. The C frame typically has a maximum capacity of 5,000 lb; very long tables, however, cannot be removed by this method.
2. The table is moved out to a dismantling platform cantilevering from the floor below, and then hooked to the crane and lifted upward. With this method, the protrusion of the form from the building façade and the resulting crane clearance required are the maximum (practically the entire length of the table). Horizontal movement of the table to the platform is done by wheels, which are attached to the shoring legs or to the truss jacks, by separate roller systems, or by special trolleys. With truss systems, the trolleys may also be used to jack down the table prior to wheeling it out. Four workers are usually required to push out a mid- to large-size table, unless motorized trolleys are used.
3. The table is moved out, but without the use of platforms. Rigging to the crane is done in two steps. First the table is rigged while it projects out from the building about one-third of its length, at which time it is still balanced by the length remaining inside the building. Then, while the crane carries part of the weight, the table is pushed farther out, and the second rigging takes place when about two-thirds of the table cantilevers out of the building. This method is particularly suitable for long truss systems.

There are cases in which tables are used more than once on the same level (e.g., large-area halls). In such cases, the tables are commonly moved by a

FIGURE 22.17 Table form lifted by a crane using a C frame.

FIGURE 22.18 Table form being removed by a crane located near the façade of the building.

special motorized trolley system that is also equipped with devices to lower the table before movement and then lift it again in its next casting location. Some trolley systems for this horizontal movement can be operated by only one worker.

The use of tables is particularly convenient when the building frame is supported on bearing walls. Table erection between two such walls requires only the covering of the gap (minimum space) between the table form and the wall on each side (necessary to allow stripping). This forming of the gap space is accomplished using wood or other light material fillers. If, however, the building frame is based on columns, special provisions must be made to form the strips created by the column lines. This is commonly done by extendable table joists, but there are other solutions, even conventional forming. Irrespective of the building's structural system, conventional formwork may also be needed for small areas of the ceiling that cannot be table formed, such as in stairwells where it is not possible to remove the tables after concrete placement. Other solutions for such areas are the use of thin precast concrete panels that function as stay-in-place forms (forms that remain as part of the finished building) or the use of full-depth precast concrete elements.

The casting of building floors will sometimes include short monolithic vertical elements at the perimeters of the slab. The presence of such vertical lips above the floor can block the removal of the tables. One solution is to use truss systems that are removable by a C-frame device, as long as the combined truss and deck height (i.e., without the truss jacks) is smaller than the opening left at the façade. Another solution is to use tables of the joist-and-stringer system-type that have collapsible legs. A similar case is that of a spandrel beam that, in addition to posing a geometric obstacle, has to be formed and cast with the rest of the ceiling. The table form is then supplemented by a beam form to accommodate the change in the cross section of the ceiling. The beam form is supported either by steel profiles hanging from the table's stringers or by steel beams connected to the table's trusses and supported, as needed, by an additional row of post shores. In both cases, the cantilevered beam form system, which also includes a working deck protruding farther out, acts to overturn the table before concrete is placed. Therefore, the table has to be anchored to the floor (Fig. 22.19).

FIGURE 22.19 Tables with extension spandrel beam forming and working deck chained to the floor to withstand overturning.

While tables are usually manufactured and used based on common room heights, they can also provide deck-forming solutions for elevated slabs at greater heights. On the Hong Kong Convention Center project, 36-ft-high tables were used [5]. Supported on aluminum shoring towers and steel beams, these massive tables—45 ft long and 25 ft wide—each weighed 21,000 lb. To allow quick and easy movement to the next casting location on the same level, an innovative pneumatic system was used to reposition the tables on hover-pads, and only three workers were needed to move the tables.

COMBINED VERTICAL AND HORIZONTAL SYSTEMS

Wall-and-slab forming systems come in two configurations: half-tunnel forms and full-tunnel forms (not to be confused with forming systems for tunnels). Both are commonly referred to as tunnel forms, or simply tunnels.

Tunnel Form Systems

Tunnel forms, used to form walls and slabs that are concreted in one operation, are suitable for repetitive use in an orthogonal-wall layout, such as may be found in high-rise housing projects, and particularly for cellular-type construction such as hotels, office buildings, and hospitals (Fig. 22.20). The concrete walls then become part of the structural frame of the building, with little or no use of light-material partitions.

Tunnel forms are made entirely from steel, including the face of the form. The concrete produced has a smooth, high-quality finish. If half tunnels are used (Fig. 22.21), they are connected together by special latches to form one full-tunnel unit. To cast a wall, two tunnel forms are needed, one on each side of the wall. To cast an external wall, a tunnel form is used on the inside and a large-panel wall form is used on the exterior side. A vertical form panel may also be connected to the tunnel at its rear to allow simultaneous casting of back walls perpendicular to the main orientation of the bearing walls.

Tunnel forms are generally manufactured in fixed modular sizes. These modular sizes are combined to achieve the desirable room length, often with the aid of short extension panels. Variable heights are available in a small number of standard increments, but the form can be adjusted by replacing the lower section of the wall form with a higher segment. To accommodate room widths that are greater than the width of the tunnel form, filler panels can be used, supported by special extendible beams, or a table form is placed between two half tunnels. Custom-size tunnels for specific projects can also be purchased from manufacturers; this is economical for the contractor as long as there are sufficient reuses.

FIGURE 22.20 Cellular-type building construction.

The combined horizontal and vertical area of regular tunnels is 100 to 350 sf, and their weight is 14 to 16 lb/sf. Thus, a full large-size tunnel (e.g., 9 ft

FIGURE 22.21 Two tunnel form halves are connected to form a full-unit form.

high, 12 ft wide, 20 ft long) may weigh as much as 3 tons, which must be considered when selecting the crane for the project and determining its location in relation to the building. Similar to the case with table forms, the tower crane used to handle the formwork cannot be placed too close to the façade of the building; otherwise its mast will block the removal of the forms directly in front of its position. On the other hand, the crane's work envelope must cover not only the building's footprint but also at least half a tunnel form's length or more if forms are to be pulled out from the façade opposite the side of the crane (or any other façade but the one closest to the crane).

Each form use involves casting the walls and ceiling for the current level. The completed ceiling is also the floor for the next level. Along the lower edge of the wall forms there are jacks used to raise the entire form before concrete placement; this is done so that the form can be lowered for stripping after the cast. Additionally the jacks provide the capability to level the form. Because of this need to raise the form above the floor surface so that it can later be removed, an approximately 4-in.-high section of wall for the subsequent building floor will be cast with the wall and ceiling operation of the level below.

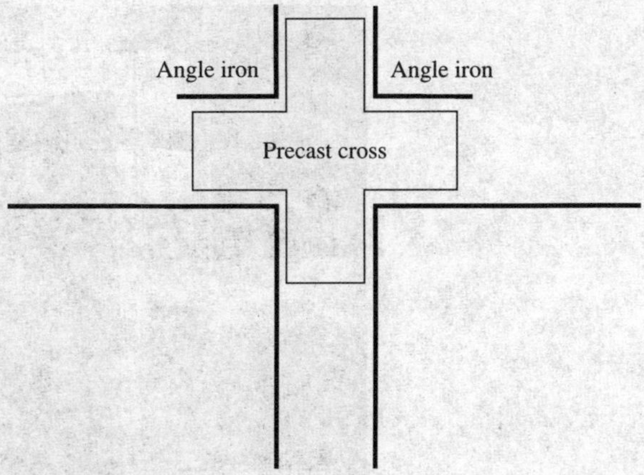

FIGURE 22.22 Precast concrete cross-shaped spacer.

This casting of a 4-in.-high stem wall is accomplished by the use of small precast concrete crucible (cross-shaped) blocks placed in the space between two adjacent tunnel forms (Fig. 22.22). These crucible blocks, which are about 2 in. in thickness so that the concrete can easily be placed in the wall form, serve as the forms' top spacers, holding the forms at the proper wall width. On top of the crucible blocks, two angle irons are placed so as, to create a form for the stem wall concrete. The angle irons also serve as leveling guides for the ceiling's concrete placement. In a similar manner, precast T-shaped elements are used for external walls.

To remove half-tunnel forms, the latches connecting the two form halves are first opened. One of the forms is lowered and at the same time slightly detached from the wall. Then it is pulled out of the building. The process is repeated for the second form. The removal of full tunnels poses a more difficult problem, as the forms are locked in the hardened concrete box, and must somehow contract so that they can be pulled out. This need to shrink the form for removal is essentially what drives the design of the full tunnel and what results in the limited variations in tunnel form configurations offered by manufacturers.

There are basically two tunnel designs and contracting configurations:

1. The complete three-sided (side-top-side) tunnel form is one continuous unit (Fig. 22.23). Diagonals that run from the lower part of the side form to the deck form have hydraulic pistons that can retract, thereby shortening the length of the diagonals. When this happens, the entire tunnel contracts (see Fig. 22.23): the upper deck form curves down into the middle while detaching from the concrete slab, and the side forms bend in at the top while also detaching from the concrete walls. A mechanical arrangement that operates similarly is available from some manufacturers.
2. The deck form has two sections fixed to the wall forms on both sides of the tunnel and a narrow detachable section that can move up and down with the aid of short articulated diagonals, or knee struts (Fig. 22.24). When the tunnel is lifted by the crane, the struts lock in their "up" position and the

FIGURE 22.23 Hydraulic contraction of a full-tunnel form for stripping.

FIGURE 22.24 Inside view of tunnel form with hydraulic knee struts and heaters hanging by metal rods from the top of the form.

detachable section is housed in the deck to create a uniform plane. The struts are then secured for service by means of wedges and safety pins. For stripping, the wedges and safety pins are removed; the struts are unlocked by hydraulically bending their knees and the detachable section is lowered slightly, then the side forms are moved inward and released from the concrete.

FIGURE 22.25 Full-tunnel forms used in hotel construction.

Tunnels—half and full—are removed from the building frame in very much the same way as table forms (see the description in the Table Forms section of this chapter). For removal that is aided by pushing the tunnel out, the forms are equipped with special wheels or rollers. Dismantling platforms can be seen in Fig. 22.25.

When considering the use of half-tunnel forms versus full-tunnel forms, several operational factors should be taken into account. Table 22.7 summarizes the important operational factors that should be considered.

To speed the construction process and enhance utilization of the forms, the concrete is commonly heated after placement to hasten the curing process and to thereby permit early stripping and reuse of the forms. Blowers placed inside each tunnel is a common heating method (note the cylindrical heaters hanging from the top of the form in Fig. 22.24). The blowers heat the steel forms, and by convection through the steel the concrete is heated as well. To reduce heat loss, tarpaulin curtains are hung at the open end of the tunnel (see Fig. 22.25). The work schedule for using tunnel forms is very inflexible as compared to that when using other forming systems. With large-panel wall forms or with table forms, there is no particular importance to the sequence of form erection,

TABLE 22.7 Comparison of half tunnel and full tunnel form systems.

Factor	Half tunnels	Full tunnels
Cost of forms (per sf)	Lower	Higher
Required lifting capacity of cranes	Lower	Higher
Number of lifting cycles and resulting crane time	Higher	Lower
Overall erection and dismantling work input	Higher	Lower
Need for reshoring of lower floor	Yes	No

and if one unit is damaged, work can continue using the other units. Tunnel forms, on the other hand, necessitate a fixed erection sequence, since there is no way to insert a form in between two other erected forms. There is thus a need to erect each form in its turn, and if one form is damaged, the entire erecting process will be halted.

To maximize utilization, a 24-hr work cycle should be maintained. A set of tunnels is the number of forms (tunnels and gable wall forms) required for one concreting operation. The number of sets employed in a given project is the result of the desired work pace and the minimal curing time required before the forms can be stripped. For example, if a high-rise residential building with four apartments per floor is to be constructed at the rate of one floor per week and the tunnels can be stripped and reused daily (i.e., overnight curing), then one full-apartment set is needed. If, on the other hand, the tunnels used on Monday can be stripped only on Wednesday (i.e., a 48-hr use cycle), then two full-apartment sets are needed in this case. In fact, with a 5-workday week, 1 day is left for various completion and general works and for contingencies. It should be borne in mind that to concrete the first apartment in this cycle, tunnels are also needed for the other side of the separation walls between the apartments (i.e., one full-tunnel-form set and at least the part of a second, which is necessary to complete the other side of the wall). Hence a one-apartment set is, in fact, bigger (in terms of form area) than the contact area (walls and slabs) of one apartment. It also means that the amount of work on the first apartment (i.e., first day in the weekly cycle) is the greatest and that of the fourth apartment is the smallest. The tunnels used to concrete separating walls are left in place when work moves to the next apartment.

SHORING TOWERS

Formwork for cast-in-place concrete in high-clearance construction is commonly based on multitier shoring towers, also termed *load towers* or *support towers*, which are essentially *frame-based* systems, to distinguish them from *tube and coupler* systems. A few other solutions exist (e.g., the *drawer* forms mentioned earlier), but they are not often used. Thus, shoring towers of various heights are an inseparable part of construction processes on commercial, residential, industrial, public, and civil engineering projects. The demand for their use has grown even higher since erection and dismantling convenience renders such towers a shoring solution for low-clearance construction as well, where single props have traditionally been used. Furthermore, towers often serve as temporary supports in precast concrete construction, and even as access scaffolds.

Since shoring towers are made up of hand-carried elements and are assembled anew for each use, they may be regarded as conventional forms (see Table 22.1). Their industrialized nature, however, is distinct: (1) they are modular; (2) they are made of steel or aluminum and can be reused a large number of times; (3) when erected to great heights (and often for lower heights as well), they are usually preassembled in short modules on the ground (Fig. 22.1) and then craned to their final location very much the same as any other industrialized forming system; and (4) they often make up the vertical shoring of industrialized table forms. The importance of shoring towers also stems from the fact that their impact on the cost of forming for high-clearance situations is significant.

Shoring towers come in a wide variety of configurations, assembly methods, and load-bearing capacities. They are made of painted steel, galvanized steel, or aluminum. There has been a clear course of development in recent years in favor of aluminum towers, with increased leg-bearing capacity. Shoring towers can be classified into four categories:

1. *Standard* towers of a typical 8,000-lb/leg carrying capacity are used mainly for light construction and maintenance work.
2. *Heavy-duty* towers have a safe working load in the range of 10,000 to 20,000 lb/leg; this is the class of towers mainly used in building construction.
3. *Extra-heavy-duty* towers have carrying capacities of up to 50,000 lb/leg and are mostly used for heavy civil projects.
4. *Ultra-high-load* towers have working loads of up to 100,000 lb/leg (200 tons per 4-leg tower).

Towers with a 20,000- to 30,000-lb/leg working load, namely of the third category, are occasionally used on building construction projects. It should be borne in mind, though, that with the typical building construction loads, on the one hand, and the size of the elements commonly used as stringers and joists (which limit tower spacing), on the other hand, a 25,000-lb/leg tower is likely to be extremely underutilized on most building construction cases.

As spatial structures, proprietary tower systems differ from each other in their (1) basic-frame configuration; (2) tier configuration (i.e., the fashion in which the basic frames, often with additional members, form the tower tier); and (3) whole-tower configuration (i.e., the fashion in which the tiers connect to each other). Thus, in addition to classification by carrying capacity, towers may be labeled according to their configuration. Table 22.8 presents four families of shoring towers [13]. Example towers from each family are shown in Fig. 22.26. Table 22.9 presents typical technical shoring tower data for the four tower families.

The main advantageous feature of shoring towers is their ability to provide vertical support to, practically, any desired height. One source [6] cites 200 ft as the height limit of shoring towers. Figure 22.27, however, shows multitier aluminum towers providing vertical shoring for a 240-ft-high slab in a high-rise residential building. The floor area of this building increases at the 25th floor, and the increased area is maintained until the 30th floor. A temporary structure of such height must be carefully designed; it also must be braced extensively—internally and to the permanent structure—to withstand buckling and to maintain overall stability. The decision to use such tall towers was adopted after a careful investigation of various alternatives, conducted on a similar project with a 200-ft-high slab [12], found the towers-based solution to be the most economical under the given conditions.

TABLE 22.8 General classification of shoring tower types [13].

Tower family	Tier configuration	Typical basic-frame configuration	Notes
A	Each tier is made up of two parallel frames connected by two pairs of crossbraces	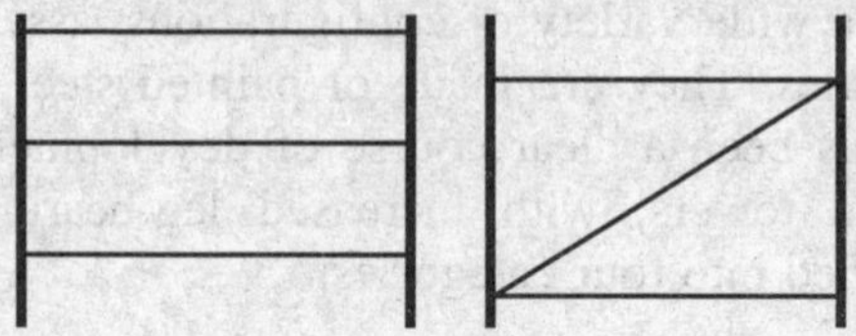	Separate towers may be interconnected by crossbraces to produce a larger tower assembly or a continuous tower row
B	Typical tower section is made up of four telescopic props, connected by sets of four ledger frames (of the same width, or two of each width)	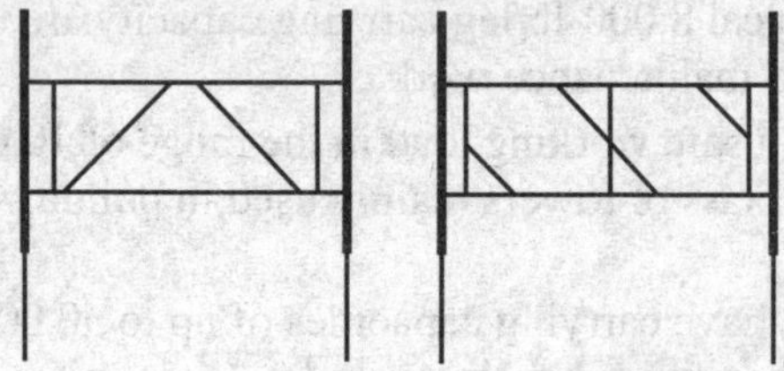	Props can also be used separately as single post shores; separate towers can be interconnected by ledger frames to produce a larger tower assembly or a continuous tower row
C	Each tier is made up of two parallel frames; tiers are turned 90° in relation to each other	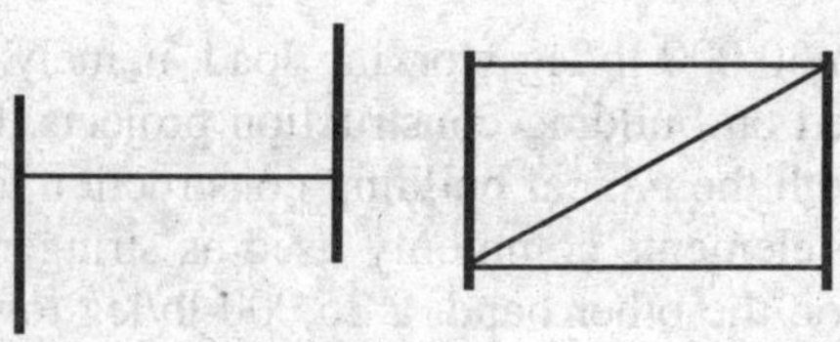	Towers are always square
D	Each tier is made up of four frames (of the same width, or two of each width) connected to each other	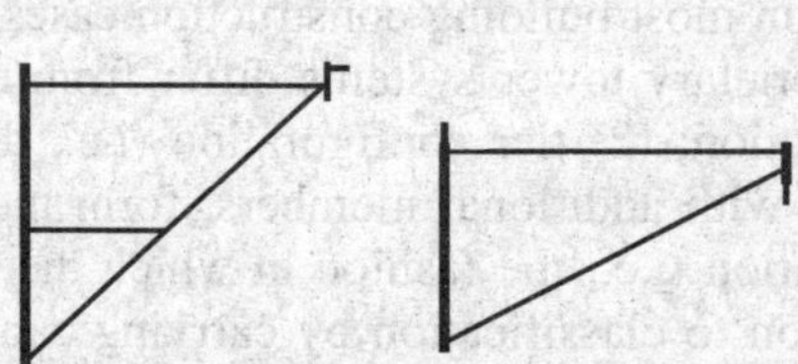	Assembly of triangular (in addition to square or rectangular) towers is possible in some models; separate towers can be interconnected to produce a larger tower assembly in some models

The main problem experienced when attempting to estimate the cost of formwork for situations where multitier shoring towers are used lies in the difficulty of predicting the labor hours necessary for erection and dismantling of the towers. The higher the total shoring height involved, the smaller is the amount of published or internal company work productivity data available. The cost of erecting and dismantling extremely high towers may be so great that the cost of the concrete and the concrete placing is an almost inconsequential component of the total concrete construction cost. Work studies of erection and dismantling of shoring towers [10, 11, 12] have found that (1) while labor productivity decreases with shoring height, the relationship is not linearly pro-

(a) Family A

(b) Family B

(c) Family C

(d) Family D

FIGURE 22.26 Multitier shoring towers.

portional, and (2) when discussing labor productivity, the number of tiers comprising the tower is more indicative of productivity than the total vertical reach. These findings help us in estimating labor productivity based on case histories. For example, if the labor hours required for a five-tier-high tower are given, it can predict that the labor hours required for a similar tower with one

TABLE 22.9 Comparative data of shoring tower families [13].

Technical data	Family A	Family B	Family C	Family D
The vertical dimension: data pertaining to tower height				
Range of tier heights (B: prop lengths) in. (cm)	35–83 (90–210)	59–246 (150–625)	20–71 (50–180)	20–59 (50–150)
Weight of basic frame (B: prop and ledger frame) lb (kg)	6–14 (14–31)	3–15 (7–34)	3–16 (7–36)	3–8 (7–17)
Gang lifting of whole tower sections	Yes	Yes	Yes	No
Standard component types per tier (B: per ledger-frame tier)	2	2	2	1
Overall standard components per tier (B: per ledger-frame tier)	4	8	2–8	4
The horizontal dimension: data pertaining to tower layout				
Range of horizontal measurements in. (cm)	24–181 (60–460)	22–150 (55–380)	39–59 (100–150)	39–79 (100–200)
Working load per leg ton (kN)	5.0–9.0 (45–80)	5.0–6.7 (45–60)	5.6–6.7 (50–60)	5.6–7.9 (50–70)
Combined slab and beam support	Yes	No	No	No
Assembly as ganged table forms	Yes	Yes	No	No

FIGURE 22.27 Form support using 240-ft-high aluminum shoring towers with carrying capacity of 18,000 lb/leg.

additional tier will be slightly more than 20% higher. To further help with estimation of production rates, prediction models based on work studies can be developed for various kinds of towers [11]. The formulas given by Eqs. [22.11] and [22.12], which are based on such work studies, enable us to predict erection and dismantling labor hours for one model of A-type *steel* towers, and Eqs. [22.13] and [22.14] enable us to predict erection and dismantling labor hours for one model of A-type *aluminum* towers (all manual work). These are not formulas that can be applied to all kinds of towers, but are valid only for the models described.

A-type steel towers, erection:

$$W = 0.085T^2 + 0.335T + 0.17 \qquad [22.11]$$

A-type steel towers, dismantling:

$$W = 0.030T^2 + 0.120T + 0.04 \qquad [22.12]$$

A-type aluminum towers, erection:

$$W = 0.0035T^3 + 0.024T^2 + 0.36T + 0.30 \qquad [22.13]$$

A-type aluminum towers, dismantling:

$$W = 0.0028T^3 + 0.016T^2 + 0.15T + 0.01 \qquad [22.14]$$

where

W = labor hours (labor hours per tower)

T = number of tiers.

Note that, while the number of tiers comprising the tower affects labor productivity considerably more than tower height itself, tower height can by no means be ignored. The above-listed equations were developed based on the common practice in which frames of the largest height-measure available are used, within the constraints imposed by the exact total tower height required: 5 ft 3 in. average tier height for Eqs. [22.11] and [22.12] (maximum frame height available in this tower model: 6 ft); 6-ft-average tier height for Eqs. [22.13] and [22.14] (maximum frame height available in this tower model: 7 ft).

EXAMPLE 22.4

What is the overall predicted work input for erection and dismantling of a tower array consisting of 20 A-type steel shoring towers rising to the height of 52 ft 6 in.? Equations [22.11] and [22.12] are applicable for the proposed towers.

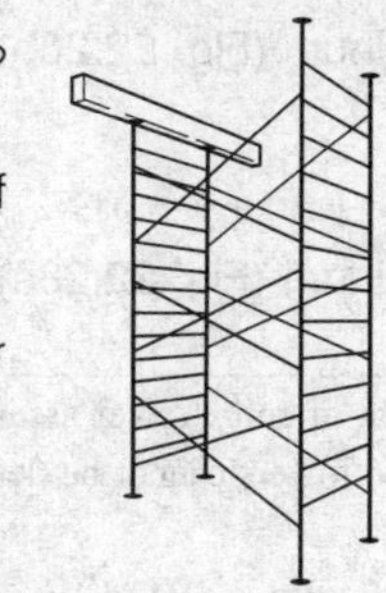

With an average tier height of 5 ft 3 in., a 52 ft 6 in. tower is composed of 10 tiers.

Erecting labor Eq. [22.11]: $W = 0.085 \times 10^2 + 0.335 \times 10 + 0.17 = 12.02$ hr

Dismantling labor Eq. [22.12]: $W = 0.030 \times 10^2 + 0.120 \times 10 + 0.04 = 4.24$ hr

Total labor: $W = 16.26$ hr per one tower

Therefore, 20 × 16.26 = 325.2 labor hr for the 20 tower array.

Table 22.10 presents labor productivity for various types and heights of shoring towers. The limited nature of productivity data has already been stressed; therefore the Table 22.10 data must be used with caution. The following construction parameters should be considered when adopting labor hour results, obtained in one case, to another situation (several parameters pertain specifically to shoring towers, while others are general and apply to any formwork):

- Erection and dismantling method (with/without crane assistance, horizontal preassembly on the ground or vertical in situ assembly). Crane assistance and preassembly would incur lower inputs.
- Distance of staging area from erection area. The greater the transfer distance of tower parts, the greater the labor hours required.
- Restricted work site. The more constricted the assembly zone, the greater the required labor hours.
- Shape and size of overall tower array (small number of towers supporting irregular concrete element with geometric constraints versus great number of towers in a regular and repetitive pattern).
- Complexity of bracing system, including number of required connections of towers to each other and to the permanent structure.
- Experience of workers and their familiarity with the particular tower system used.

TABLE 22.10 Example labor productivity of shoring towers.

Tower type	Tower height* (ft)	Number of tiers	Tower cross section (ft)	Construction parameters	Labor productivity (labor hours per tower)		
					Erection	Dismantling	Total
A Steel	20 33 52	3 6 10	6 x 4 6 x 4 6 x 4	Very large sample Manual work In situ assembly	1.94 5.24 12.02	0.67 1.84 4.24	2.61 7.08 16.26
A Aluminum (Fig. 22.26a)	26 43 57	4 7 9	8 x 6 8 x 6 8 x 6	Very large sample Manual work In situ assembly	2.35 5.19 8.03	1.05 2.81 4.70	3.40 8.00 12.73
D Steel (Fig. 22.26d)	57	14	6 x 4	24 towers Manual work In situ assembly	16.17	6.47	22.64
B Aluminum (Fig. 22.26b)	87	(15)†	9 x 8	7 towers Crane assisted Horizontal preassembly	41.28	24.77	66.05
A Aluminum (Fig. 22.26a)	200	8	8 x 6	7 towers Crane assisted Horizontal preassembly	85.54	83.07	168.61

*Height of entire tower assembly, measured from bottom of base plates to bottom of stringers lying on tower heads.

†B-type towers cannot be defined in terms of tiers; this is the number of sets of four ledger frames along the tower.

- Employment mode of workers. Lower labor productivity can be expected from workers contracted on an hourly-wage base, greater productivity from workers contracted on piecework basis.
- Safety regulations. When working at height, the harnessing of workers for safety can slow the tower erection rate.
- Length of workday. Working overtime and/or under bad lighting conditions reduces productivity.
- Weather conditions affect production (extreme high/low temperatures, rain).

Labor productivity controls the economy of shoring tower solutions involving multitier towers, therefore measures taken by the constructor at all design, selection, planning, and execution stages that result in reduced labor requirements will improve economy. The following recommendations should be considered:

- Selection of tower type: to reduce labor requirements, select towers with high average tier size (height), as determined by (1) size of largest frame and (2) variety of tier sizes available.
- Selection of tier size (for a given tower type): select the largest frames possible, to minimize number of tiers for a given tower height; when two (same type) towers rising to the same height are made up of a different number of tiers, erection of the tower with the smaller number of tiers will incur lower labor requirements.

- When using various tier sizes, place the larger (i.e., heavier) tiers first (at the bottom of the tower) and the smaller tiers last (at the top of the tower).
- Select stringers and joists such that tower spacing will be maximized; towers are often utilized only to a small percentage of their load-carrying capacity.
- Select towers to correspond to the expected loads. Towers with capacities exceeding the expected loads provide no advantage and tend to be heavier (resulting in higher labor requirements) and/or costlier.
- There is an advantage in selecting a tower type with which the work crew is familiar; tower types often differ greatly, and experience acquired with one type does not guarantee efficient work with another.
- Organize for optimal crew size, with clear task allocations for each worker. Too small a crew makes it hard to maintain individual tasks, and the shifting of workers from one task to another increases labor requirements; too large a crew means wasted resources. Optimal crew size depends to some extent on tower type and erection method, as well as on transfer distance of tower parts; it is usually governed, however, by tower height. As a rule of thumb, low towers (up to three tiers) require two workers, midrise towers (up to six tiers) require three workers, and towers higher than six tiers will be efficiently erected by four workers.
- Tower parts should be placed as close as possible to the erection zone; this must be considered before parts are supplied and unloaded at the site.
- Towers allowing horizontal preassembly on the ground may have an advantage when a crane is available to assist with the erection of large assemblies. At the same time, it should be borne in mind that with some tower types, horizontal preassembly can incur higher labor requirements compared to other types erected in situ.
- Bracing (connecting towers to each other and/or to the permanent structure) the towers must be based on engineering analysis rather than intuition; inadequate bracing may lead to failure, and therefore intuition often results in excess bracing, which increases labor hours needlessly.
- When shoring a linear concrete element (e.g., beam, as opposed to slab), towers that can be assembled as triangles may have an advantage, as labor requirements for three-leg towers are roughly 80% of that required for four-leg towers of the same type.

SAFETY

Formwork is, in general, a major safety concern, due mainly to its nature as temporary work. As such, formwork safety is addressed extensively by regulatory bodies, various safety agencies, manufacturers, contractors, construction practitioners, and researchers. Industrialized forming systems hold the potential for yet greater risks, as they add another aspect due to issues stemming from their size and weight, as well as their requirement for handling by cranes and other lifting equipment. One example is that of using the crane to aid with the stripping of large-panel wall forms by nonvertical pulling (instead of using the crane for strictly vertical lifting once the wall form has been completely stripped). This dangerous practice often results in the sudden detachment of the form panel from the concrete wall and its striking of nearby workers. Another example is that of the dangerous practice of using the crane's hook to pull table forms out of the building instead of moving the table horizontally by other means until it protrudes from the building façade such that the crane can lift it vertically, or using a C-frame device.

At the same time, industrialized forming systems hold the potential for curbing risks associated with formwork fabrication. This is because the essence of industrialization is that it moves the fabrication process off of the construction site, while shifting on-site work primarily to assembly. A factory environment is much more controllable than that of a construction site. Also, the quality of factory-fabricated forms is usually higher than for those built on-site, and therefore the forms are not as susceptible to failures.

The awareness of form manufacturers to safety issues is much higher than that of the on-site workforce, resulting in a constant search for improvement of their products' quality and safety. Hence, wall- and column-forming systems come furnished with integral work platforms, guardrails, ladders, and protective climbing cages. We also see advanced rigging provisions, built-in lift eyes located at exactly the points in the form where they are needed, and forming systems that make use of hydraulics to raise themselves without the use of a crane, which can be affected by winds.

In the United States, the main bodies that address formwork-related safety and publish regulations, guides, and recommendations are the U.S. Department of Labor's Occupational Safety & Health Administration (OSHA) [website 3], the American Concrete Institute (ACI) [website 1], and the Scaffolding, Shoring and Forming Institute, Inc. (SSFI) [website 2]. OSHA formwork-related regulations are listed under "Requirements for Cast-in-Place Concrete" (Subpart Q) of the "Safety and Health Regulations for Construction" (Part 1926). ACI refers to formwork safety in its *Guide to Formwork for Concrete, ACI 347R-03* [1] (3.1—Safety precautions). The SSFI publishes its recommendations in codes of safe practices sorted by type of formwork/shoring. In addition, the American National Standards Institute and the American Society of Safety Engineers together publish ANSI/ASSE A10.9-2004, "Concrete and Masonry Work Safety Requirements," which has a section (7) on formwork. These publications relate mostly to formwork safety in general, and their instructions and recommendations should be followed strictly, with no regard to the type of formwork—conventional or industrialized—that is practiced. There are additionally a small number of publications—mainly by SSFI—that focus on specific form types of the kind treated in this chapter (e.g., flying deck forms, shoring towers), and these must also be followed strictly when applicable.

The mechanical handling of formwork on site, a characteristic of industrialized forming systems, is a critical operation. All parties involved in crane handling of the form should be aware of form's weight and the proper handling method. An example is the stripping of wall forms when the form may have to be lifted over the partially completed works with possible protruding rebar for subsequent levels of work.

> It should be borne in mind by all parties involved in formwork design, manufacturing, and construction that in addition to the very real moral and legal responsibility to maintain safe conditions for laborers and the public, safe construction is, in the final analysis, more economic than any short-term cost savings from cutting corners on safety provisions [1].

SUMMARY

Forming systems used in constructing the many repetitious concrete elements of a structure are designed and fabricated for many reuses. The constructor's project engineer, while not usually responsible for designing the major formwork elements, is extensively involved with various planning aspects of their selection, ordering, erection, stripping, and reuse. Industrialized forms are typically factory-fabricated products that are used numerous times as one unit without being disassembled and assembled again.

The modularization and mechanization of forming systems has made these units integral components of on-site construction equipment. Compared to traditional formwork, industrialized forming systems generally excel in achieving cost savings through considerable savings in erection and dismantling time. Formwork cost is principally comprised of material and labor cost. Labor cost is controlled by labor productivity, something that is often very hard to determine. Typical labor productivity data are given in Table 22.5 but these should be used with caution as they only provide an indication of average time requirements.

Shoring towers of various heights are definitely an inseparable part of concrete construction and have been found very useful for other construction processes on commercial, residential, industrial, public, and civil engineering projects. They are made up of hand-carried elements and are assembled anew for each use. Shoring towers come in a wide variety of configurations, assembly methods, and load-bearing capacities. They are made of painted steel, galvanized steel, or aluminum, but there has been a clear course of development in favor of aluminum towers, with increased leg-bearing capacity.

Due to its nature as a temporary work, formwork presents several distinct safety concerns. Because a formwork failure can result in loss of life to workers both above and below the forms, the safety of forming systems is addressed extensively by regulatory bodies and various safety agencies. Industrialized forming systems hold the potential for yet greater risks, as they add another aspect to the safety issue stemming from their size and weight, as well as their close involvement with cranes and other lifting equipment. Critical learning objectives include:

- An understanding of the project engineer's role in the utilization of formwork systems
- An understanding of the magnitude of the pressure fresh concrete exerts on formwork
- An ability to calculate the cost of forming systems
- An understanding of different types of forming systems available to successfully complete a project
- An understanding of the safety issues inherent with the employment of forming systems

These objectives are the basis for the problems that follow.

PROBLEMS

22.1 A construction company is bidding on a project comprising five high-rise buildings to be erected one after the other. The company considers the use of advanced, hydraulically operated tunnel forming systems to be purchased and used 60 times on each building over a 1-yr period of time. The tunnel forms cost $28/sf. No salvage value is expected at the end of the 5-yr project. As the two last buildings to be constructed slightly differ in their design, the forms will have to be modified at the cost of $5/sf. Periodical maintenance is expected every 120 uses at a cost of 5% of the purchase cost. Consider an annual interest rate of 4%. What is the expected average material cost per sf for each use?

22.2 A construction company is investigating two forming options for a new hotel project. Option A is the use of large-panel forms for the walls and table forms for the slabs. Option B is tunnel forms for walls and slabs. In both cases, the equipment is to be rented, for the total area of 2,000 sf of walls and 2,000 sf of slabs. Overall, 160,000 sf of walls and 160,000 sf of slabs will be formed. Costs, labor productivity, and work durations are as follows:

Parameter	Option A	Option B
Construction duration (= rental period), months	12	10
Rental rate, $/sf of form per month:		
Wall forms	$0.37/sf	
Table forms	$0.46/sf	
Tunnel forms		$1.44/sf
Labor productivity		
Wall forms	0.04 hr/sf	
Table forms	0.03 hr/sf	
Tunnel forms		0.03 hr/sf

Due to the higher weight of the tunnels, this option requires greater lifting capacity, resulting in an additional $35,000 of crane cost. There would be no difference between the two options in the quality of the concrete. Hourly wages are $21. The overhead of the project amounts to $90,000 per month. What is the most economical option?

22.3 A crew of four workers is scheduled to erect 40 A-type aluminum shoring towers. Equations [22.13] and [22.14] are applicable for these towers. The towers are 48 ft high. Length of the workday is 8 hr.

a. What duration (in work days) is erection estimated to take?

b. What duration (in work days) is dismantling estimated to take?

c. How would the answers for (a) and (b) change if the high towers were twice the described height?

REFERENCES

1. *ACI 347R-03, Guide to Formwork for Concrete* (2003). American Concrete Institute, Farmington Hills, MI.
2. *Building Construction Cost Data* (published annually). R. S. Means Co., Kingston, MA.
3. Hanna, A. S. (1999). *Concrete Formwork Systems,* Marcel Dekker, New York.
4. Hurd, M. K. (1995). *Formwork for Concrete, SP-4,* 6th ed., American Concrete Institute, Farmington Hills, MI.
5. "Innovations in Formwork" (1997). *International Construction*, 36(6), pp. 49, 50, 55, 56, 58, June.
6. Johnston, R. S. (1996). "Design guidelines for formwork shoring towers," *Concrete Construction,* 41(10), 743–747.
7. Peurifoy, R. L., and G. D. Oberlender (1996). *Formwork for Concrete Structures*, 3rd ed., McGraw-Hill, New York.
8. Shapira, A. (1995). "Rational design of shoring-tower-based formwork," *Journal of Construction Engineering and Management,* ASCE, 121(3), pp. 255–260.
9. Shapira, A. (1999). "Contemporary trends in formwork standards—a case study," *Journal of Construction Engineering and Management,* ASCE, 125(2), pp. 69–75.
10. Shapira, A. (2004). "Work inputs and related economic aspects of multitier shoring towers," *Journal of Construction Engineering and Management,* ASCE, 130(1), pp. 134–142.
11. Shapira, A., and D. Goldfinger (2000). "Work-input model for assembly and disassembly of high shoring towers," *Construction Management and Economics,* 18(4), pp. 467–477.
12. Shapira, A., Y. Shahar, and Y. Raz (2001). "Design and construction of high multi-tier shoring towers: case study," *Journal of Construction Engineering and Management,* ASCE, 127(2), pp. 108–115.
13. Shapira, A., and Y. Raz (2005). "Comparative analysis of shoring towers for high-clearance construction," *Journal of Construction Engineering and Management,* ASCE, 131(3), pp. 293–301.

WEBSITE RESOURCES

1. www.aci-int.org American Concrete Institute (ACI), Farmington Hills, MI. The American Concrete Institute disseminates information for the improvement of the design, construction, manufacture, use, and maintenance of concrete products and structures, including formwork for concrete.
2. www.ssfi.org Scaffolding, Shoring & Forming Institute, Inc. (SSFI), Cleveland, OH. SSFI is an association of companies that produce scaffolding, shoring, and forming products in North America. It develops engineering criteria and standard testing procedures for scaffolding, shoring, and forming and disseminates current information relative to their proper and safe use.
3. www.osha.gov Occupational Safety and Health Administration, U.S. Department of Labor. OSHA Assistance for the Construction Industry is listed in www.osha.gov/doc/index.html and includes "Standards," a link to the 29 CFR1926, Safety and Health Standards for Construction.
4. www.aluma.com Aluma Systems, Aluma Enterprises, Inc., Toronto, Ontario, Canada. Aluma Systems is a provider of aluminum-based concrete forming and shoring solutions, industrial scaffolding services, and construction expertise.
5. www.doka.com Conesco Doka, Ltd., Little Ferry, NJ, is a part of the Doka Group. Doka is a European manufacturer and a worldwide supplier of concrete formwork products.
6. www.efco-usa.com Efco Corporation, Des Moines, IA. Efco is an American manufacturer of systems for concrete construction.
7. www.outinord-americas.com Outinord Universal, Inc., Miami, FL. Outinord is a European manufacturer and a worldwide supplier of all-steel concrete forming systems.
8. www.patentconstruction.com Patent Construction Systems, Paramus, NJ, a member of the Harsco Corporation. Patent is an American supplier of scaffolding, concrete forming, and shoring products.
9. www.peri-usa.com Peri Formwork Systems, Inc., Hanover, MD. Peri is a European manufacturer and supplier of formwork, shoring, and scaffolding systems.
10. www.safway.com Safway Services, Inc., Waukesha, WI, a company of ThyssenKrupp Services AG. Safway is an American manufacturer of scaffolds and shoring systems.
11. www.symons.com Symons Corporation, Des Plaines, IL, a Dayton Superior Company. Symons is an American manufacturer of concrete forms.
12. www.wacoscaf.com Waco Scaffolding & Equipment, Cleveland, OH. Waco is an American manufacturer of scaffolding, forming, and shoring products.

APPENDIX A

Alphabetical List of Units with Their SI Names and Conversion Factors

To convert from	to	Symbol	Multiply by
Acre (U.S. survey)	square meter	m^2	4.047×10^3
Acre-foot	cubic meter	m^3	1.233×10^3
Atmosphere (standard)	pascal	Pa	1.013×10^5
Board foot	cubic meter	m^3	$2.359 \div 10^3$
Degree Fahrenheit	Celsius degree	°C	$t_{°C} = (t_{°F} - 32)/1.8$
Degree Fahrenheit	Absolute	°A	$°A = (t_{°F} + 459.67)$
Foot	meter	m	$3.048 \div 10$
Foot, square	square meter	m^2	$9.290 \div 10^2$
Foot, cubic	cubic meter	m^3	$2.831 \div 10^2$
Feet, cubic, per minute	cubic meters per second	m^3/s	$4.917 \div 10^4$
Feet per second	meters per second	m/s	$3.048 \div 10$
Foot-pound force	joule	J	1.355×1
Foot-pounds per minute	watt	W	$2.259 \div 10^2$
Foot-pounds per second	watt	W	1.355×1
Gallon (U.S. Liquid)	cubic meter	m^3	$3.785 \div 10^3$
Gallons per minute	cubic meters per second	m^3/s	$6.309 \div 10^5$
Horsepower (550 ft. lb/sec)	watt	W	7.457×10^2
Horsepower	kilowatt	kW	$7.457 \div 10$
Inch	meter	m	$2.540 \div 10^2$
Inch, square	square meter	m^2	$6.452 \div 10^4$
Inch, cubic	cubic meter	m^3	$1.639 \div 10^5$
Inch	millimeter	mm	2.540×10
Mile	meter	m	1.609×10^3
Mile	kilometer	km	1.609×1
Miles per hour	kilometers per hour	km/h	1.609×1
Miles per minute	meters per second	m/s	2.682×10
Pound	kilogram	kg	$4.534 \div 10$
Pounds per cubic yard	kilograms per cubic meter	kg/m^3	$5.933 \div 10$
Pounds per cubic foot	kilograms per cubic meter	kg/m^3	1.602×10

To convert from	to	Symbol	Multiply by
Pounds per gallon (U.S.)	kilograms per cubic meter	kg/m^3	1.198×10^2
Pounds per square foot	kilograms per square meter	kg/m^2	4.882×1
Pounds per square inch (psi)	pascal	Pa	6.895×10^3
Ton (2,000 lb)	kilogram	kg	9.072×10^2
Ton (2,240 lb)	kilogram	kg	1.016×10^3
Ton (metric)	kilogram	kg	1.000×10^3
Tons (2,000 lb) per hour	kilograms per second	kg/s	$2.520 \div 10$
Yard, cubic	cubic meter	m^3	$7.646 \div 10$
Yards, cubic, per hour	cubic meter per hour	m^3/h	$7.646 \div 10$

Note: All SI symbols are expressed in lowercase letters except those that are used to designate a person, which are capitalized.

Sources: *Standard for Metric Practice,* ASTM E 380-76, IEEE 268-1976, American Society for Testing and Materials, 1916 Race Street, Philadelphia, PA 19103.

National Standard of Canada Metric Practice Guide, CAN-3-001-02-73/CSA Z 234.1-1973, Canadian Standards Association, 178 Rexdale Boulevard, Rexdale, Ontario, Canada M94 IRS.

APPENDIX B

Selected English-to-SI Conversion Factors

In general, the units appearing in this list do not appear in the list of SI units, but they are used frequently, and it is probable that they will continue to be used by the construction industry. The units meter and liter may be spelled metre and litre. Both spellings are acceptable.

Multiply USC (English) unit	by	To obtain metric unit
Acre	0.4047	Hectare
Cubic foot	0.0283	Cubic meter
Foot-pound	0.1383	Kilogram-meter
Gallon (U.S.)	0.833	Imperial gallon
Gallon (U.S.)	3.785	Liters
Horsepower	1.014	Metric horsepower
Cubic inch	0.016	Liter
Square inch	6.452	Square centimeter
Miles per hour	1.610	Kilometers per hour
Ounce	28.350	Grams
Pounds per square inch	0.0689	Bars
Pounds per square inch	0.0703	Kilograms per square centimeter

APPENDIX C

Selected U.S. Customary (English) Unit Equivalents

Unit	Equivalent
1 acre	43,560 square feet
1 atmosphere	14.7 lb per square inch
1 Btu	788 foot-pounds
1 Btu	0.000393 horsepower-hour
1 foot	12 inches
1 cubic foot	7.48 gallons liquid
1 square foot	144 square inches
1 gallon	231 cubic inches
1 gallon	4 quarts liquid
1 horsepower	550 foot-pounds per second
1 mile	5,280 feet
1 mile	1,760 yards
1 square mile	640 acres
1 pound	16 ounces avoirdupois
1 quart	32 fluid ounces
1 long ton	2,240 pounds
1 short ton	2,000 pounds

APPENDIX D

Selected Metric Unit Equivalents

Unit	Equivalent
1 centimeter	10 millimeters
1 square centimeter	100 square millimeters
1 hectare	10,000 square meters
1 kilogram	1,000 grams
1 liter	1,000 cubic centimeters
1 meter	100 centimeters
1 kilometer	1,000 meters
1 cubic meter	1,000 liters
1 square meter	10,000 square centimeters
1 square kilometer	100 hectares
1 kilogram per square meter	0.97 atmosphere
1 metric ton	1,000 kilograms

INDEX

Abrasion drills, 340, 346
Absolute pressure, 641
Absolute temperature, 641
Access roads, 680–81. *See also* Haul routes
Accidents. *See also* Safety
  annual costs, 10
  blasting, 401–2
  crane, 62–63, 572–73
  crusher, 438–39
  dozer, 195–96
  dragline and clamshell, 597–98
  excavator, 255–56
  falls, 688
  grader, 330
  loader, 284
  pile driving, 636–37
  planning to avoid, 61–63
  scraper, 249
  trenching, 288–89
  truck, 317
Acoustical curtains, 697
Additional reference for Compacted cubic yard, 71–72
Adiabatic compression, 643
Adjustable backup alarms, 693
Adjusted basis, 30, 31
Advanced operator aids for cranes, 575
Aftercoolers, 644
Aggregate production
  basic considerations, 407–9
  crushing (*see* Crushing)
  safety issues, 438–39
  screening and separation, 431–37
  washing and handling, 437–38
Aggregates
  in flexible pavements, 448–50
  importance of assessing, 90
  requirements for concrete, 487, 488, 489, 513
  types, 91
Aggregate-size factors, 435
Air, compressed. *See* Compressed air
Air compressors, 644–46
Air hammers, 621–22, 623
Air manifolds, 647
Airports, using cranes near, 574
Air-track drills, 343–44
All-terrain cranes, 534, 541–42
Altitude effects
  on compressor capacity, 646
  on engine power, 155, 156–58
  on water pumps, 660
Aluminum contamination of concrete, 513
Aluminum Design Manual, 567
American Aluminum Association, 567
American Association of State Highway and Transportation Officials, 96, 97
American Concrete Institute, 710, 746
American Institute of Steel Construction, 567
American Iron and Steel Institute, 567
American National Standards Institute, 746
American Society for Testing and Materials (ASTM), 92
American Society of Agricultural Engineers, 200
American Society of Safety Engineers, 746
Ammonium nitrate, 375, 378
Amplitude, 121, 628
Anchorage frames, 555, 556
Anchor piles, 601
ANFO, 378–79, 396
Angle blades, 180, 181
Angle of swing (shovels), 262–64
Angling of dozer blades, 179
Antistrip materials, 450
Apatite, 349
Application rate of water, 108
Apron feeders, 427
Aprons of scrapers, 228–29
Articulated jibs, 562
Articulated rear-dump trucks, 298–99
Asphalt. *See also* Paving equipment
  flexible pavements, 447–53
  overview, 443–44
  pavement structures, 446–47
  production plants, 454–65
  terminology, 444–46
Asphalt cements, 450–53
Asphalt concrete, 444, 453–54
Asphalt distributors, 467–68
Atmospheric pressure, 641
Attachments, future development of, 7
Attitude of pavers, 523
Attrition model cone crushers, 419
A-type shoring tower labor costs, 742–43
Augered cast-in-place piles, 610
Augers
  in asphalt pavers, 468, 469
  attached to drills, 361
  on trimmers, 333–34
Auger scrapers, 226–27
Automatic batching controls, 489–90
Automatic screed controls, 444
Automatic sheet-pile threading devices, 617
Automation, 7, 18
Available headroom for tower cranes, 563
Available power, 148–54
Average annual investment method, 41–42
Average-end-area method, 69–71
Average haul distance, 78, 79, 80–84
Average haul grades, 78–80
Backbreak, 380
Backfilling
  defined, 116
  with dozers, 183
  utility trench problems, 128
Backhoe-loaders, 289–90
Backhoes. *See* Hoes
Backtrack loading, 243, 244
Backup alarms, 693
Baghouse filtration systems, 462–63
Balance lines, 77–78, 79
Ball mills, 425–26
Balloon compaction tests, 103
Bank cubic yards
  defined, 98
  scraper capacities in, 227
  shovel bucket capacities in, 261
Bank volume, 72
Barometric pressure, 157, 646
Barriers to reduce noise, 697
Basalt, 354
Base charges, 380, 381
Base machines, future applications, 7
Base materials, 122
Base stabilization, 133
Basis, 30
Batching of concrete, 485, 486–90
Batch plants, asphalt, 455–59, 464–65
Batter piles, 630
Bearing capacity, 619
Bechtel, Warren A., 4
Belt conveyors, 504–5
Belt feeders, 427
Benches, 374
Bench height
  defined, 374
  for drilling, 339
  as factor in fragmentation, 383
  modifying, 388
Benching of trenches, 288
Best Manufacturing Company, 4
Bidding
  cost calculations for, 39–47
  package contents, 678–79
  pricing earthwork, 84–86
  procedures, 12
Binary explosives, 377
Binder courses, 444, 446
Binders, 116, 444
Bits. *See* Drill bits
Blade-mixed pavements, 447
Blades (dozer)
  for land clearing, 197–98
  major types, 178–82, 185
  volumetric load, 185–87
Blade-to-blade dozing, 184

Blake crushers, 412, 413
Blast-hole drills, 340
Blastholes. *See also* Drilling
  defined, 374
  diameter calculations, 387–90
  patterns and spacing, 347–48, 391–95
  presplitting, 398–99
  stemming, 390–91
Blasting. *See also* Drilling
  blast design factors, 383–85
  breakage control, 397–99
  defined, 374
  initiating and delay devices, 380–82
  major types of explosives, 375–79
  overview, 372–74
  patterns, 347–48
  powder factors, 395–97
  primers and boosters, 379–80
  rock fragmentation, 382–83
  safety issues, 401–2
  terminology, 339, 374–75
  trenches, 397
  vibration from, 400–401
Blasting agents, 374
Blasting caps, 381
Blasting mats, 397
Blind spots, 249
Blowers in tunnel forms, 736, 737
Blowing out drill holes, 353
Blue Angel certification, 694
Bolted connections, 5–6
Bond breakers, 690
Bonding capacity, 52
Book value, 31, 43
Booms. *See also* Cranes; Pile driving
  concrete pumps with, 507–9
  options for, 537, 544, 548
  selecting for pile driving, 636
  sizes for draglines, 582, 583, 584
  telescoping, 534, 537, 545–48
Boosters, 379, 380
Borrow pits, 92
Bottom-dump buckets, 502, 503
Bottom-dump trucks
  basic features, 300–301
  dump times, 308
  hauling asphalt with, 466
Bottom-slewing tower cranes
  basic features, 551–53, 557–58
  common dimensions and capacities, 558
  as common type, 535
  counterweights, 552, 561
Boulder Dam, 5–6
Boulders, 92
Bowls (scraper), 228–29
Boyle's law, 642
Braces
  for shoring towers, 745
  for single-sided wall forms, 723
  in tilt-up concrete walls, 690–91, 692
  for tower cranes, 555
Brake horsepower, 151
Breakage control techniques, 397–99
Breakdown, 444
Breakdown rolling, 476–77
Breaking characteristics of rock, 350
Brittleness values, 352
Brush disposal, 199
Bucket cylinders, 266, 267
Buckets
  attached to drills, 362
  clamshell, 593, 594–95
  concrete, 502–3
  dragline, 582–83, 589–90
  front shovel size ratings, 257–58
  of gradalls, 332
  hoe size ratings, 267–68
  loader types, 275–77
  matching crushers to, 408
  matching to hauling unit capacities, 259–60, 303, 304–5, 310
  special-purpose, 269, 276
Bucket teeth. *See* Teeth
Bucyrus Foundry and Manufacturing Company, 3, 4
Buggies (concrete), 504
Building construction planning. *See* Construction planning
Bulk density of explosives, 376
Bulkers, 135, 136
Bulldozers. *See* Dozers
Burden, 338, 374
Burden distance
  in blast design calculations, 384–87, 388, 389–90
  correction factors, 386, 387
  defined, 338, 339
  effects on fragmentation, 382–83
Burning debris, 199
Butt of piles, 601
Button bits, 341–42
Buy-rent-lease decisions, 49–52
Calcite, 349
Calyxite, 346
Cantilever jibs, 561
Capacity of pumps, 656
Carbide-insert bits, 341, 345
Carmichael & Fairbanks, 4
Cash assets, 51
Cash flow diagrams, 22, 23
Casings
  for cast-in-place piles, 609–10
  for drilling unstable soils, 360
  pulling with directional drilling and boring, 364–65
Cast-in-place piles, 609–10
Cast iron, steel replacing, 5
Caterpillar, Inc.
  altitude effect guidelines, 157
  automation of equipment, 7
  Computer-Aided Earthmoving System, 7, 8
  dozer production curves, 189–90, 191
  ripper performance and production charts, 206, 212
  scraper performance charts, 230, 231
  scraper specifications, 232
Cave-ins, preventing, 288–89
Cement, 134–38, 486, 489. *See also* Concrete
Cementatious reactions, 133
Center of gravity, crane loads, 567, 568
Central-mixed concrete, 500–502
Centrifugal compressors, 644
Centrifugal pumps, 656, 658–63
C-frames
  for cranes, 731
  for dozer blades, 178, 179
Chain elevators, 225, 226
Chain loading, 243, 244
Chain of mistakes, 62–63
Chain slings, 571
Chamfered edges, 690
Charles' law, 642
Chatter, 326
Cherry pickers, 540
Chesapeake Bay Bridge Tunnel, 607
Chipping debris, 199
Churn drills, 340
Chutes (concrete), 504
Clamshell doors, 301
Clamshells
  attachments and control, 593–94
  bucket features, 593, 594–95
  as hoe attachment option, 269
  production estimates, 595–97
  safety issues, 597–98
Classification
  of heavy equipment, 13–14
  of piles, 602–3
  rock hardness, 348–49, 350
  of soils and rocks, 92–93, 96, 97
  trucks, 296–97
  water pumps, 657–58
Clays
  cementing reactions of lime with, 133
  compactors for, 118
  defined, 92
  hauling in trucks, 302
  responses to dozing, 185, 186
  skin friction on piles, 619
  weight and swell, 99
Clearing blades, 197–98
Clear opening, 432
Climbing cages, 557
Climbing mixers, 495–96
Climbing tower cranes, 559–60
Climbing towers, 509, 510
Closed-end diesel hammers, 627
Closed-end pipe piles, 611
Closing lines, 594
Coarse aggregate, 91
Coarse-grained soil classification, 93
Coefficient of traction, 153, 155, 156, 174
Cofferdams, 602, 613
Cohesion, 91
Cohesionless materials, dozing, 185, 186
Cohesive materials, 91
Cold feed systems, 456, 460–61
Cold-Formed Steel Design Manual, 567
Cold milling, 464
Cold weather, concrete placement in, 526
Column forms, 724–27
Combined cost per hour values, 48
Combined vertical-horizontal formwork, 733–38
Commercial explosives, 375–79. *See also* Explosives
Committee on European Construction Equipment (CECE) standards, 257–58
Common excavation, 94
Compacted cubic yards, 98
Compacted volume, 72
Compaction curves, 101, 102, 106
Compaction of pavements, 472, 474–78
Compaction of soils
  adjusting, 106–11
  in cement stabilization operations, 137
  goals, 115–16

Compaction of soils—*Cont.*
importance, 101, 102
testing, 102–5, 106
Compaction wheels, 126
Compactors
in cement stabilization operations, 137
common types, 117, 119–28
dynamic, 129–31
methods used by, 116–18
production estimates, 128–29
terminology, 116
Company rental systems, 18
Competition in construction industry, 8–9
Composite piles, 611–12, 613
Compound interest, 19–20
Compressed air
compressor types, 644–46
distribution systems, 646–52
diversity factor, 652–53
early development, 639
overview, 640–41
safety issues, 653–55
terminology, 641–42, 644
Compression ratio, 644
Compressors, 644–46
Compton effect, 104
Computer-Aided Earthmoving System, 7, 8
Computer systems
crane operator aids, 575
for drilling, 358–59
in earthmoving equipment, 7, 8
for record keeping, 18
Concaves, 416
Concrete. *See also* Formwork
asphalt, 444, 453–54
batching, 486–90
central-mixed, 500–502
consolidating, 514–16
contaminants in, 486–87, 513
development, 483–84
finishing and curing, 517–19, 737
mixing techniques, 490–96
pavements, 519–23
placement methods, 502–14, 690
pressure on forms, 707–9, 718
proportions and workability, 485–86
ready-mixed, 490, 496–500
roller-compacted, 523–24
safety issues, 527
shotcreting, 524, 525
for tower crane bases, 555
weather conditions and, 526–27
"Concrete and Masonry Work Safety Requirements," 746
Concrete columns, lift planning for, 685, 686
Concrete piles
cast in place, 609–10
driving, 608–9
types, 604–7, 614, 615
Concrete Plant Standards, 488–89
Concrete pumps, 505–14
Concrete-steel composite piles, 612
Cone crushers, 417–19, 420
Conglomerate, 356
Consolidated average hauls, 83–84
Consolidating concrete, 514–16
Constant speed clearing, 199–200
Constricted sites, 676, 677
Construction and Health Regulations, 11
Construction braces, 690–91
Construction equipment. *See* Equipment
Construction industry, economic and regulatory environment, 8–12
Construction planning. *See also* Earthwork construction planning
basic considerations, 12–14, 60–63, 675–77
lifting and support equipment, 683–86
nuisance control, 692–99
site layout, 677–83
steel erection, 688
structural component delivery, 686–87
tilt-up panel construction, 689–92
Consumption rates for fuel, 35
Contact surface area, 716
Contaminants in concrete, 486–87, 513
Contracting environment, 11–12
Contraction joints, 521
Contraction of full-tunnel forms, 735–36
Contractors Pump Bureau, 658, 661, 662
Contract turnover, 12
Contract volume, 12
Control points, Superpave aggregate size, 448, 449
Conventional formwork, 703–5
Conveyors, 334, 504–5
Core drills, 340
Core wall forms, 722
Correction factors. *See also* Efficiency factors
burden distance, 386, 387
dozer production, 190, 192, 193–94
Corrugations, grading, 326
Corundum, 349
Corvée laborers, 1
Cost estimating, 12. *See also* Production estimates
Cost of capital, 25–26
Counterflow mixers, 495
Counterflow plants, 461
Counter-jibs, 560
Counterweights
bottom-slewing tower cranes, 552, 561
on crawler cranes, 542, 543
Couplings (drill), 354
Crane elevators, 560
Cranes. *See also* Formwork; Pile driving
for concrete buckets, 503
crawler, 534, 535–37
dynamic compaction with, 129
estimated repair and maintenance costs, 45
handling formwork with, 731, 734, 746
lattice-boom truck-mounted, 534, 539–40
lifting capacities, 544–48, 563–66
modified, 534, 542–44
overview of major types, 533–35
planning use of, 679–80, 683–86
rigging, 567–71
rough-terrain and all-terrain, 534, 540–42
safety issues, 62–63, 572–76, 682–83
selecting for pile driving, 635
telescoping-boom truck-mounted, 534, 538–39
tower crane operation, 554–62
tower crane selection, 562–63
tower types, 550–53
working ranges, 548–49
Crawler cranes
estimated repair and maintenance costs, 45
features and types, 535–37
illustrated, 534
lifting quadrants, 545
mobilizing, 683
modified, 542–44
with towers, 550, 554
working ranges, 549
Crawler dozers. *See also* Dozers
basic features, 172–75
land clearing with, 196–99
scraper loading with, 224, 243–47
transmissions, 175–76
Crawler loaders. *See* Track loaders
Crawler-mounted draglines, 581
Crawler-mounted excavators, 45
Crawler tractors
coefficient of traction, 155, 156
drawbar pull, 152, 154, 159
rolling resistance, 144
Crawler-tractor "towed" two-axle scrapers, 223
Creosote, 603–4
Crew size, 511, 745
Critical lifts, 684, 685–86
Crosshaul, 75, 80
Cross-ripping, 213
Cross section views, 65–67
Crucible blocks, 735
Crude oil, asphalt from, 450
Crushing
basic considerations, 407–9
equipment selection, 408, 428–31
feeders, 426–27
gyratory crusher features, 415–19
impact crusher features, 424–25
jaw crusher features, 410–14, 415
rod and ball mills, 425–26
roll crusher features, 419–23
screening and separation after, 432–37
stages of, 409–10
surge piles from, 427–28
Cummins diesel engines, 4
Curing
in cement stabilization, 137–38
of concrete, 519, 522, 737
in lime stabilization, 134
Curtains, noise, 697
Curved-shank rippers, 209
Cushion blades, 181, 182
Cushions for pile driving, 601, 608
Custom-made column forms, 726–27
Custom-made wall forms, 720–22
Cut areas. *See also* Earthwork construction planning
calculating, 72–75
cross section views, 66, 69
Cutbacks, asphalt, 452
Cut height, effects on shovel production, 261–62
Cutoff, 601
Cutter mandrels, 333
Cuttings from drilling, 340, 361–62
Cycle times
concrete mixers, 492
dozers, 187
hoes, 271, 272, 273
loaders, 279
scrapers, 243–45
trucks, 296, 303, 308, 315
Cylindrical stone rollers, 115, 116

Dam construction, concrete buckets for, 502–3
Data collection devices, 19
Datum lines, 77
Da Vinci, Leonardo, 2
Davis-Bacon Act, 9–10
Dead loads on formwork, 707
Debris gathering and disposal, 199, 202, 203
Decibels, 692–93
Deck factors, 435
Deck loading, 391
Deep wells, 670
DeereTrax, 19
Deflagration, 374
Delayed firing, 377, 394
de Lesseps, Ferdinand, 1–2
Delivery of structural components, 686–87
Densified ANFO, 379
Density
  of air, 641
  of explosives, 374, 386, 395, 396
  of rock, 385
  as soil compaction measure, 115–16
Depreciation
  defined, 28, 39–40
  tax savings, 30–34, 42–43
  time value method, 40–41
Depth of cut
  dragline, 586–87, 588
  hoes, 271
Derricks, 559
Design-bid-build method of contracting, 12
Design-build contracts, 12
Design of formwork, 707–11
Detachable booms, 509
Detonating cord, 382
Detonation rate, 382
Detonation velocity, 376, 378
Detonators, 380–81
Dewatering systems, 668–70
Diabase, 354
Diagonals, in full-tunnel forms, 735–36
Diamond, hardness of, 349
Diamond bits, 340, 346
Diamond drills, 346, 347
Diaphragm pumps, 658
Diesel engines
  advantages, 148
  fuel consumption rates, 36
  importance to equipment development, 6
  introduction in construction equipment, 4
  noise, 695
Diesel hammers, 625–27
Differential-acting steam/air hammers, 624
Differing site condition clauses, 64
Digitizing tablets, 67
Dipping planes, 209, 350
Direct-drive machines, 152, 175
Direct fire heaters, 463
Directional drilling, 362–65
Discharge head, 656
Discharge hoses, 656, 665
Discharge pressure, 644
Discharge rates (truck mixers), 498
Discounted present worth analysis, 26–28
Disk harrows, 110
Dismantling tower cranes, 559–60
Displacement pumps, 657–58
Dispose-replace question, 19
Distance, increasing to reduce noise, 696
Distribution systems
  asphalt, 467–68
  compressed air, 646–52
Ditches, cutting, 321–22, 325
Diversity factor, 644, 652–53
Dolomite, 356
Double-acting hammers, 622–24, 627
Double-acting pumps, 657
Double-roll crushers, 421
Double-rotor impact crushers, 425
Double-rotor trowels, 518
Double-toggle jaw crushers, 412, 413
Downdrag, 602
Downgrade loading, 248
Downhole bits, 340, 354, 355, 356
Downhole drilling, 345
Downlines, 382
Down payments, 51
Down stationing, 78
Dozers
  blades, 178–82
  economical haul distance, 77
  land-clearing jobs, 196–203
  major tasks, 182–84
  overview of types, 171
  performance characteristics, 172–77
  production estimates, 185–95
  rock ripping, 204–13
  safety issues, 195–96
  sample replacement analysis, 47–48
  scraper loading with, 224, 243–47
  spreading embankment material, 85
Draglines
  basic features, 580, 582–83
  major functions, 581
  operation, 584–85
  production estimates, 585–93
  safety issues, 597–98
Drainage of crushing plants, 431
Drawbar pull
  defined, 152
  determining, 154
  for direct-drive dozers, 175
  performance charts, 158
  usable, 176–77
Drawer forms, 727–28
Dredges, 2
Drifters, 343
Drill bits
  average life by rock types, 354, 355, 356
  basic types, 341–42
  changing time, 353–54
  defined, 338
Drilling. *See also* Blasting
  bit types, 341–42
  for breakage control, 397–98
  cuttings removal, 361–62
  methods, 346–50
  overview, 337–38
  production estimates, 350–58
  rock drill types, 342–46
  safety, 367
  soils, 359–62
  terminology, 338–40
  for trenchless applications, 362–67
Drilling rate index, 352
Drilling rates, 341
Drills, 340, 342–46
Drill steel
  average life by rock types, 354, 355, 356
  changing time, 352–53
  defined, 341
  impact of extensions on production, 347
Driving caps
  defined, 602
  fitting of, 609
  with single-acting hammers, 622
Drop hammers, 621, 627
Dropheads, 728, 729
Drop pipes, 504
Drum dryers, 457
Drum mix plants, 454, 460–61, 462, 465
Drying actions (soil), 109–10
Dry-mix process (shotcreting), 524, 525
Dry unit weight of soil, 95
Dual cabs on all-terrain cranes, 541
Dumping, for maximum scraper production, 248
Dump times
  in scraper production estimates, 240–42
  trucks, 303, 308, 314
Dump trucks, 298–301, 466. *See also* Trucks
Duplex pumps, 657
Dust
  from cement stabilization, 136
  collectors in asphalt plants, 462–63
  control plans, 698
Duty-cycle work (cranes), 535
Dwell time, 461
Dynamic compactors, 129–31
Dynamic factors on cranes, 548
Dynamite, 377

Earthmoving blades, 197
Earthwork construction planning. *See also* Construction planning; Geotechnical materials
  basic considerations, 60–64
  estimating material quantities, 13, 67–75
  graphical presentations, 64–67
  mass diagrams, 75–84
  material categorization, 94
  pricing approaches, 84–86
  volume change calculations, 96–101
Earthwork volume sheets, 72–75
Eastside project, 63
Eccentric moment of vibratory pile drivers, 628
Economical haul distance, 77–78, 240, 241
Economical load time, 240, 241
Eells, Dan P., 4
Effective grade
  determining speed from, 160–62
  total resistance as, 146, 147
Efficiency factors
  dozer production, 194
  drilling, 355–56
  front shovels, 263, 264
  graders, 328
  scraper production, 245
  screening, 434, 435
  truck production, 309
Efficiency of compressors, 644
Egress from trenches, 288
Eisenhower, Dwight D., 4, 5, 6
Ejectors, 228–29
Elasticity of rock, 400
Electric detonators, 380–81
Electric motors for tower cranes, 564–65

Elevating crane cabs, 560
Elevating scrapers, 225–26, 229
Elevators, personnel, 681
Elevator shafts, 559, 722
Embedments, 602
Emissions from asphalt production, 479
Empty vehicle weight, 161, 234
Emulsions, 377, 452–53
Enclosures to reduce noise, 696
End areas, volume calculations based on, 67–71
End-bearing piles, 600, 618
End-of-lease options, 51
Engineering News pile formula, 634
Environmental factors in crane safety, 573
Equations
  air pressure drop, 647, 648
  asphalt distribution, 468
  average annual investment method, 41
  average end area, 69
  average haul grade, 79
  burden distance, 384, 386, 388
  center of gravity, 567
  compactor production, 128–29
  concrete pressures, 708
  dozer production, 187, 189
  dynamic compaction effects, 130
  excavator production, 255
  explosive load, 399
  formwork labor costs, 742–43
  formwork material costs, 711–13
  grade resistance, 145–46
  grader production, 328, 329
  grader spreading and mixing, 326
  hoe production, 271
  hole spacing, 393–94
  horsepower and torque, 150
  labor costs, 717
  land-clearing production, 199–200
  net present worth, 27
  oil consumption, 37
  percentage optimum depth of cut (dragline), 587
  pile driving, 624, 630, 634
  power, 150
  presplit blasthole spacing, 399
  pusher cycle times, 243
  pusher-scraper production, 245, 246
  rimpull, 153
  rock layer characteristics, 208
  rock ripping production, 211
  roll crusher capacity, 422–23
  rolling resistance, 144
  Rome cutting formula, 201
  scaled distance, 401
  scraper production cycle, 232
  scraper travel time, 238
  screen capacity, 436
  shrinkage, 98
  single payments, 20
  sling stress, 568–69
  soil weight-volume relationships, 95–96
  straight-line depreciation, 31–32
  subdrilling depth, 391
  swell, 98–99
  tax depreciation shield, 33
  taxes, insurance, and storage costs, 43
  temperature and pressure, 641, 642, 643
  tire speed and loading, 311
  top stemming, 390
  total resistance, 141, 146
  trapezoidal computations, 68
  truck cycle times, 308
  truck haul time, 306
  truck load time, 306
  truck production, 309
  truck return time, 308
  uniform payments, 21–25
  usable force, 155
  volume of concrete ingredients, 487
  volumetric load (dozer blades), 186, 187
  water content adjustments, 100, 107–8
  water pressure and head, 665
  water pump, 658–59
  work, 148
Equipment classification, 13–14. *See also* Classification
Equipment costs. *See also* Production estimates
  buy-rent-lease question, 49–52
  evaluating alternatives, 25–27
  major factors affecting, 17–18
  operating costs, 34–39
  ownership costs, 28–34
  payment methods and, 20–25
  record keeping for, 18–19
  replacement decisions, 47–49
Equipment evolution, 1–7
Equipment-intensive projects, 9
Equipment power. *See* Machine power
Equipment records, 18–19
Equipment spreads
  elements of, 8–9
  production estimates, 12–13, 84–86
Equipment use plans, 12–14. *See also* Construction planning
Equivalence, 20
Erie Canal, 2
Estimating. *See* Production estimates
European-type cranes, 550–51
Evenflow mixers, 495
Excavation categorization, 94
Excavators. *See also* Clamshells; Draglines
  altitude effects, 157
  backhoe-loaders, 289–90
  basic features, 253–55
  blasting versus, 373
  compaction wheels on, 126
  early development, 3–4
  estimated repair and maintenance costs, 45
  front shovel features, 253–54, 257–58
  front shovel production variables, 260–64
  front shovel selection, 258–60
  hoe features, 254, 264–68
  hoe production estimates, 271–73
  hoe selection, 268–71
  Holland loaders, 290, 291
  loader buckets and attachments, 275–77
  loader production estimates, 279–83
  loader types and features, 274
  matching to truck capacities, 259–60, 303, 304–5, 309–10
  modern hydraulic, 2
  multiple attachments for, 7, 268–69
  trenching machines, 284–89
  vac units, 290–91
Expansion joints, 521
Experience, as guide to blade load estimates, 186
Explosives, 375–79. *See also* Blasting
Extendable counterweight cranes, 542
Extra-heavy-duty shoring towers, 739
Extralong undercarriages, 173

Face of rock, 347
Factory-fabricated column forms, 726–27
Fahrenheit temperatures, 641
Fairlead assemblies, 582
Falls, preventing, 688
Families, shoring tower, 739, 740, 741, 742
Fast-erecting cranes, 551
Fatalities
  2002 work zone total, 61
  crushers, 438–39
  drilling, 367
  excavator, 255–56
  graders, 330
  loaders, 284
  pile driving, 637
  trenching, 288
  trucks, 317
Federal Aviation Agency (FAA) crane regulations, 574
Feeders, 426–27
Feed size (roll crusher), 421–22
Fender systems, 612
Field measurements
  compaction, 103
  rock ripping production, 211
  volumetric load (dozer blades), 186–87
Fill areas, 66, 69, 72–75. *See also* Earthwork construction planning
Fill factors
  defined, 255
  front shovel buckets, 258
  hoe buckets, 268
  loader buckets, 276–77
Fine aggregate, 91
Fine grading
  in cement stabilization operations, 137
  parking lots, 321
  production estimates, 329
Fine-grained soils, 93
Fines, 410
Finishing concrete, 517–19
Finishing equipment. *See also* Graders
  gradalls, 330–32
  graders, 320–30
  trimmers, 332–34
Finish rolling, 476–77
Finish work, 173
Fire hazards, 203, 479
Fireproofing steel frames, 688, 689
Fittings, loss of air pressure in, 649, 651
Fixed-base tower cranes, 555–58. *See also* Cranes; Tower cranes
Fixed drilling times, 352–53
Fixed jibs, 537
Fixed leads, 631
Flammability of explosives, 376
Flat-top tower cranes, 561
Flexible pavements, 447–53
Flights, 229, 457
Floating, 517, 518
Floor elevation, 391
Fluorite, 349
Fly ash
  in concrete, 525–26
  as stabilization agent, 131–32, 135
Fly jibs, 537

Flyrock, 387, 401–2
Flywheel horsepower, 151
Fontana, Giovanni, 2
Forklift attachments, 276
Formations (rock), 350
Formed paving, 520
Formulas. *See* Equations
Form vibrators, 516
Formwork. *See also* Concrete
 basic types, 703–5
 combined vertical-horizontal systems, 733–38
 as concrete placement factor, 511
 costs, 711–18
 design, 707–11
 horizontal systems, 727–33
 project engineers' involvement with, 705–7
 safety issues, 745–47
 shoring towers, 738–45
 for tilt-up concrete walls, 689–90
 vertical systems, 718–27
Formwork for Concrete, 710
Foundations for tower cranes, 555
Four-part-line configurations, 565–66
Fragmentation, 379, 382–83
Franki technique, 609–10
Free air, 644
Freefall mixers, 491, 492–93, 499
Freezing, preventing during cement
  stabilization, 138
Frequencies of vibratory equipment, 121, 628–30
Friction courses, 444
Friction losses
 in pumps, 659
 in water pipes, 664–65, 666
Friction piles, 600
Front-discharge truck mixers, 496–98
Front shovels
 basic features, 253–54, 257–58, 259
 height and position effects on production,
  261–64
 production cycles, 260–61
 selecting, 258–60
Fuel
 calculating costs, 35–36, 44
 use in explosives, 378
Full-tunnel forms, 733–38
Functional identification, 13–14

Ganged wall forms, 719–20
Gas laws, 641–43
Gasoline engines, 4, 36
Gauge pressure, 641
Gears, 5
General contractors, site layout duties, 677–83.
 See also Construction planning
General lift plans, 684
General-purpose buckets, 275
GeoGauge, 105
Geophones, 207, 208
Geospatial engineering, 7
Geotechnical materials. *See also* Earthwork
  construction planning; Rock; Soils
 blasting correction factors for, 386–87
 categorization, 94
 dozer production correction factors for, 193
 impact on machine power requirements,
  140–41
 major types, 91–93
 responses to dozing, 185, 186
 volume changes during construction, 96–101
 weight-volume relationships, 94–96
Glen Canyon Dam, 502
Global positioning systems (GPS)
 current and future applications, 7
 for placing drill holes, 353, 358–59
Gneiss, 349, 355
Government actions, economic risks from,
  9–10
Gradalls, 330–32
Grade assistance, 145, 235–36
Grade correction factors, 194
Grade lines, 445
Grade resistance, 141, 145–46
Graders
 altitude effects, 157
 basic features, 320–22
 in cement stabilization operations, 137
 major functions, 324–27
 mixing soil with, 111
 production estimates, 328–29
 safety issues, 329–30
 scarifying soil with, 110
Grades
 average haul, 78–79, 80
 resistance and assistance formulas, 141,
  145–46
 scraper production and, 235–36
 truck performance on, 299
Grade sensors, 471
Grain-size analysis, 101
Grain-size curve, 91
Grain structure of rock, 350
Granite, 354
Granular materials, 107
Graphical presentations, 64–67
Grapples, 269
Gravel. *See also* Aggregate production;
  Crushing
 classification, 93
 compactors for, 118
 defined, 92
 hauling in trucks, 302
 weight and swell, 99
Gravimetric capacity, 141, 301, 302–3
Grizzlies, 431, 432
Gross horsepower, 151
Gross penetration of piles, 602
Gross vehicle weight, 161, 234–35
Groundwater diversion, 668–70
Grout, 365, 610
Guides for pile hammers, 631
Guide to Formwork for Concrete, 710, 746
Gunpowder, 372
Gypsum, 349
Gyratory crushers
 features and types, 415–19
 matching to shovel buckets, 408
 as primary crushers, 410
 reduction ratios, 411, 417

Half-tunnel forms, 733, 734, 735, 737
Hammer heads, 560
Hammer mills, 425
Hammers (pile driver)
 diesel, 625–27
 differential-acting, 624
 double-acting, 622–24, 627
 drop, 621, 627
 hydraulic, 627
 selecting, 633–36
 single-acting, 621–22, 623
 vibratory, 616, 617
Hand buggies, 504
Hand-set slab forms, 728, 729
Hand-set wall forms, 719
Hand signals, 688
Hard asphalts, 451
Hardness, drilling rate and, 348–49, 350
Hardwoods, clearing, 201
Harnischfeger, Henry, 4
Harrows, 110
Haul distances
 calculating, 80–84
 in loader production estimates, 280
 in mass diagrams, 77–78, 79
 in scraper production estimates, 240, 241
Haul routes
 coefficient of traction, 155–56
 grade changes, 147–48
 impact on scraper production, 235–37, 238
 impact on truck production, 315–16
 layout, 148
 maintaining, 326
 rolling resistance, 143, 235, 236
 truck selection and, 299
Haul times, 306–8, 314
Haul trucks. *See also* Trucks
 matching to excavator capacities, 259–60,
  303, 304–5, 309–10
 placing hoes above, 273
Haul-unit exchange, 261
Header pipes, 668
Head loss, 664–65, 666
Head of material, 472
Headroom requirements, for tower cranes, 563
Heaped capacity
 clamshell buckets, 595
 front shovel buckets, 258
 hoe buckets, 267, 268
 loader buckets, 276
 scrapers, 227
 trucks, 301–2
Heat exchangers, 644
Heating systems
 asphalt storage, 463
 screeds, 473
 tunnel forms, 736, 737
Heave, 374, 611
Heavy-duty clamshell buckets, 595
Heavy-duty dragline buckets, 590, 591
Heavy-duty rock bodies, 298
Heavy-duty shoring towers, 739
Height of cut, effects on shovel production,
  261–62
Helmets, 602, 622
Hertz, 105
High explosives
 in detonating cord, 382
 low explosives versus, 375
 major types, 377–79
High-output diesel engines, 6
High-pressure jetting, 633
High-strength steels, 6
High-wear items
 in drills, 340, 354, 355, 356, 357
 as operating cost, 38–39, 46–47
 of rippers, 209

Hoes
  basic features, 254, 264–66, 267
  bucket size ratings, 267–68
  production estimates, 271–73
  selecting, 268–71
Hoist-cable configurations (tower cranes), 565–66
Hole patterns and spacing, 347–48, 391–95
Holland loaders, 290, 291
Holt, Benjamin, 5
Holt Manufacturing Company, 4–5
Hong Kong Convention Center, 732–33
Hoover Dam, 5–6, 502
Hoppers
  asphalt plants, 458, 464–65
  concrete mixers, 492–93
  concrete pumps, 505
  pavers, 470
Horizontal boring, 362, 365
Horizontal drilling, 343
Horizontal formwork loads, 707
Horizontal formwork systems, 727–33
Horizontal preassembly for shoring towers, 745
Horizontal-shaft impact crushers, 424
Horsepower, 150, 151, 659
Hoses
  compressed air safety, 655
  concrete pumping, 527
  drilling safety precautions, 367
  loss of air pressure in, 652, 653
  pile driving safety precautions, 637
  proper connections, 11
  sizing, 652
  use with water pumps, 656, 657, 665, 666
Hot bins, 445, 458
Hot elevators, 445
Hot mix asphalt
  batch plants, 455–59
  defined, 446
  drum mix plants, 460–61, 462, 465
Hot oil heaters, 445, 463
Hot weather, concrete placement in, 527
Hourly fuel costs, 35–36
H section piles, 610–11
Human hearing thresholds, 692–93
Human labor, in early construction industry, 1–2
Hveem, 453
Hybrid cranes, 535
Hydraulic cranes, 548
Hydraulic drills, 343–44
Hydraulic drivers, 627–28
Hydraulic excavators. *See* Excavators
Hydraulic impact hammers, 627
Hydraulic leads, 632
Hydraulic ram attachments, 269
Hydrostatic powertrains, 175, 176
Hydrostatic trenchers, 285
Hydrovac units, 290–91

Idler wheels, 173
Idle time, with pusher-scraper combinations, 245–46
Igneous rocks
  defined, 92
  impact on drill bits and steel, 354
  rippability, 205, 206
Illumination plans, 697–98
Impact compactors, 125–26
Impact crushers, 411, 424–25
Impellers of centrifugal pumps, 658
Impervious deposits, 129
Industrialized formwork, 703–5, 746. *See also* Formwork
Inflation pressures, 124, 125
Initiation (explosive)
  defined, 375
  hole spacing and, 393–94
  systems for, 380–82
Injury rates, 10. *See also* Accidents; Safety
Inlet pressure, 644
Inline silos, 502
In situ strata, 129
Inspection of slings, 571
Instantaneous initiation, 393–94
Insurance, 34, 43
Intake hoppers, 505
Intercoolers, 644
Interest
  calculating payments with, 20–25
  cost-of-capital rate, 25–26
  defined, 19
Interlocking steel piles, 614–17
Intermediate rolling, 476–77
Internal combustion engines, 4, 148–54
Internal rental systems, 18
Internal vibrators, 515, 516
International Harvester dozer production guidelines, 189
Interstate highways, 6
Investment options, 25–27
Isothermal compression, 643

Jackhammers, 342, 654
Jacks, supporting formwork with, 730, 735
Jackson Lake Dam, 129
Jaw crushers
  features and types, 410–14
  matching to shovel buckets, 408
  reduction ratios, 410, 411
  sizes of output, 415
Jetting piles, 632–33
Jibs
  extensions to lattice booms, 537
  on tower cranes, 550–51, 560–62
Joints in pavement, 521
Joist-and-stringer table form systems, 729, 730
Joists in formwork, 709–10
Joysticks, 540

Kelly bars, 359
Kelvin scale, 641
Kerosene, 466
K/G blades, 197, 201
Knee struts, 735–36

Labor costs. *See also* Production estimates
  formwork, 716–18, 740–45
  regulation of, 9–10
Labor laws, 9–10
Ladder-type trenching machines, 286–87
Lampson Thumper, 130
Land clearing, 196–203
Large custom-made wall forms, 720–22
Large-panel wall forms, 720
Large trucks, 304
Laser blade controls, 183
Lattice booms
  with clamshells, 593–94
  on crawler cranes, 534, 537
  load ratings, 545–48
  on truck-mounted cranes, 534, 539–40
Lattice masts, 509, 510
Laydown areas, steel components, 687
Layers of formwork, 709–11
Leads, for pile driving, 631–32, 637
Lead wires, 374
Leasing, 27, 51–52
Leg wires, 374, 381
LeTourneau, R. G., 5
Leveling arms, 445
Leveling courses, 445
Lift directors, 576
Lifting capacities
  cranes, 544–45, 563–66
  hoes, 270–71
Lifting equipment. *See* Cranes
Lift inserts, in tilt-up concrete walls, 690
Lift planning, 574, 684–86
Lifts
  controlling thickness with dozers, 183
  defined, 103, 445
  effect of thickness on scraper production, 248
  for rock fills, 118
Light-duty dragline buckets, 590, 591
Lighting plans, 697–98
Lime-fly ash stabilization, 131–32, 135
Lime modification, 134
Lime stabilization, 133–34
Limits of soils, 96, 97
Limit state design, 710
Line drilling, 397–98
Line pumps, 506
Linked-system production rate, 84
Lips on floors, 732
Liquid asphalt cements, 452–53
Liquid asphalt curing, 137
Liquid limit of soils, 96
Live bottom trucks, 466
Live loads on formwork, 707
Live zones of pugmills, 458
Load and resistance factor design, 710
Load-bearing piles, 602–3
Loaded vehicle weight, 161
Loaders
  altitude effects, 157
  basic features, 274
  buckets and attachments, 275–77
  operating specifications, 277, 278
  production estimates, 279–83
Load factors, 644
Load-growth curves (scrapers), 233, 234, 239–40
Loading density, 376, 395, 396
Load moment rating classification systems, 546–47
Load ratings (cranes), 545–48, 563–66
Loads, working under, 682–83
Load times
  in scraper production cycle, 232
  in scraper production estimates, 238–40, 241
  in truck production estimates, 306, 313–14
Load towers, 738
Load weights, 234–35, 302, 306
Log washers, 437, 438
Longtan Dam, 503
Long-term leases, 51

Loose cubic yards, 98
Los Angeles Aqueduct, 4–5
Loss of head, 664–65, 666
Loss of pressure, compressed air systems, 647–49, 650, 651, 652, 653
Low explosives, 375
Low-ground-pressure undercarriages, 173
Low-pressure jetting, 633
Lubricant costs, 36–37, 44–45
Luffing jibs, 561–62
Lutes, 445

Machine power
  available, 148–54
  basic considerations, 140–41
  performance charts, 158–64
  required power factors, 141–48
  usable, 155–58
Machine time records, 19
Magazine storage, 378
Maintenance, 52. *See also* Repairs
Maintenance costs of formwork, 712–13
Major repairs. *See* Repairs
Mandrels, 333
Manifolds (compressed air), 647
Manlifts, 692
Manual batching (concrete), 489
Manual labor, early history, 1–2
Manually operated compactors, 122, 123, 127–28
Manual of Steel Construction, 567
Manufacturer's blade ratings (dozers), 185–86
Manufacturer's production estimates (dozers), 189
Marble, 355
Market value cost of capital, 28
Marshall stability, 445
Mass diagrams, 75–84
Mass excavations, 63
Mass ordinate, 74
Masts, 509, 510. *See also* Tower cranes
Material costs in formwork, 711–16
Materials. *See* Geotechnical materials; Rock; Soils
Material transfer devices, 470–71
Material type correction factors, 193
Material weight correction factors, 193
Maximum braced height (cranes), 555
Maximum dry density, 101, 105
Maximum free-standing height (cranes), 555
Maximum haul distance, 77–78
Medium-duty dragline buckets, 590, 591
Membrane curing, 134
Mesh, 432
Metalatite, 355
Metamorphic rocks
  defined, 92
  impact on drill bits and steel, 355
  rippability, 205, 206
Microtunneling, 365–66
Middlesex Canal, 2
Miller Park Stadium accident, 62–63
Millisecond-delay detonators, 374, 382
Minimum attractive rate of return, 25
Minimum stability condition of cranes, 546
Mining
  automated rock-hauling systems, 7
  computerized drilling systems, 359
  first use of explosives, 372
Miscellaneous ownership costs, 34
Misfires, 381, 402
Mixing
  in cement stabilization operations, 136, 137
  concrete proportions, 487–88
  to dry soil, 111
  techniques for concrete, 490–96
Mixing drums (asphalt), 461
Mobile compact concrete plants, 501–2
Mobile cranes
  common types illustrated, 534
  crawler types, 535–37
  lattice-boom truck-mounted, 539–40
  lifting capacities, 544–48
  modified, 542–44
  popularity, 533
  rough-terrain and all-terrain, 540–42
  telescoping-boom truck-mounted, 538–39
  with towers, 550, 554
  working ranges, 548–49
Modification, stabilization versus, 134
Modification costs, formwork, 713
Modified cranes, 534, 542–44
Modified Proctor test, 101, 102, 103
Modular wall forms, 719–20
Moh hardness classifications, 348, 349
Moisture content. *See* Water content of soils
Moldboards, 322–23, 324–27
Monighan Machine Company, 4
Monotube piles, 609
Motor graders. *See* Graders
Motors for tower cranes, 564–65
MS delay detonators, 374, 382
Mulgrew-Boyce Company, 4
Mulholland, William, 4–5
Multilane trimmers, 333
Multipurpose buckets, 275
Multipurpose tools, hoes as, 7, 268–69
Multishank rippers, 210–11
Multistage centrifugal pumps, 660
Multistage compressors, 644

National Institute for Occupational Safety and Health (NIOSH), 479
Natural asphalt, 450
Nebraska tests, 154
Net penetration of piles, 602
Net present worth, 27
Net vehicle weight, 161
Net volume (earthwork), 71–72
Nipping, 410, 421
Nitroglycerin, 375
Noise
  controlling, 694–97
  as nuisance, 692–94
  tower crane limits, 564
Nominal maximum aggregate size, 448
Nonelectric detonators, 381
Nonideal explosives, 388
Nonvertical pulling by cranes, 746
Nonvibrating weight, 630
North Carolina DOT asphalt performance grades, 452
Nuclear compaction tests, 104–5
Nuisance control, 692–99
Number of passes, with graders, 327, 328
Number of uses, formwork, 715–16
Nylon cord tires, 6
Occupational Safety and Health Act, 9–10
Occupational Safety and Health Administration (OSHA)
  blasting safety standards, 402
  creation of, 10–11
  formwork regulations, 746
  pile driving safety standards, 637
  rollover protection mandate, 10
  standards for general contractors, 680–81
  standards for working under loads, 682–83
  trench safety regulations, 288–89
Octagonal piles, 605
Off-balance-sheet financing, 52
Off-highway trucks
  body configuration, 298
  bottom-dumping, 300
  major uses, 296, 297
Oil changes, 44
Open-end diesel hammers, 627
Open-end pipe piles, 611
Operating costs (equipment), 34–39, 43–47
Operating efficiency. *See* Efficiency
Operating specifications (loaders), 277, 278
Operating techniques for rock ripping, 213
Operational identification, 13–14
Operators
  dozer production correction factors for, 193
  fatigue factors, 152
  future phaseout, 7
  location on tower cranes, 560
  wages, 9–10, 35
  zones of responsibility for cranes, 575–76
Optimization of equipment spreads, 9
Optimum depth of cut (dragline), 586, 587
Optimum water content, 91, 101, 106
Order-of-magnitude drilling rates, 351
Organic matter, 92
OSHA requirements. *See* Occupational Safety and Health Administration (OSHA)
OSHA Safety and Health Standards, 11
Otis, William S., 3–4
Outriggers on cranes, 538–39, 546
Overall number of uses, formwork, 715–16
Overbreak, 397
Overburden, 383
Overdriving, 602, 617
Overhauls, 34, 37–38, 43
Overinflated tires, 311, 327
Overlapping tower cranes, 560–61
Overloading equipment, 141, 302–3
Oversailing, 562
Oversize material
  aggregate screening and separation, 431–37
  as consideration in blasting, 347
  impact on excavator production, 261
  scalping, 431
Overvibration of concrete, 516
Ownership and operating (O&O) costs, 17–18
Ownership costs (equipment)
  in bids, 39–43
  impact of heavy ripping projects, 212, 213
  major elements, 28–34

Pad drum vibratory compactors, 117, 118, 122
Painted Rock Dam, 124
Panama Canal, 3
Panel layout, 689
Pan mixers, 494–95

Parallelogram ripper linkages, 210
Parallel ripper linkages, 210
Park Tower, 724
Particle size reduction. *See* Crushing
Path control for noise, 695–97
Patterns of drilling, 347–48
Pavements
  asphalt concrete, 453–54
  asphalt production plants, 454–65
  basic structure, 446–47
  concrete, 519–23
  cross section views, 67
  defined, 91
  flexible, 447–53
  overview, 443–44
  terminology, 444–46
Paving breakers, 654
Paving equipment
  asphalt pavers, 468–74
  compactors, 474–78
  concrete, 519–23
  distributors, 467–68
  overview, 465–66
  sweepers and trucks, 466
  terminology, 444–46
Payload capacity, 141
Payments, calculating, 20–25
Penalties, for OSHA violations, 11
Penetration grades of asphalt cement, 450–51
Penetration of piles, 602, 618
Penetration rates (drilling), 348–50, 351, 352
Percussion drills, 340, 342–44
Performance batches, 500
Performance charts
  general uses, 158–62
  retarder, 162–64
  scrapers, 229–31
  trucks, 307
  water pumps, 660–61
Performance grades of asphalt cement, 451–52
Personnel hoists, 681
PETN, 382
Pile bents, 602
Pile caps, 621
Pile driving. *See also* Cranes
  composite piles, 611–12
  concrete piles, 604–10
  hammers, 620–30, 633–36
  jetting piles, 632–33
  penetration resistance, 618
  pile classification, 602–3
  pile support and positioning, 630–32
  safety issues, 636–37
  sheet piles, 612–17
  site studies and testing, 618–20
  spudding and preaugering, 633
  steel piles, 610–11
  terminology, 601–2
  timber piles, 603–4
Pile-driving shoes, 602
Pile tips, 602
Piling-up production rates, 202, 203
Pilot holes, 363
Pipe bursting/splitting, 366
Pipe jacking, 366
Pipelines (concrete), 507, 509
Pipe piles, 611
Pipe ramming, 366–67
Pipes (compressed air)
  loss of pressure in, 647–49, 650, 651
  sizing, 651
Pipes (water), loss of head in, 664–65, 666
Pitch
  dozer blades, 179
  moldboards, 323, 324
Pitted surfaces, grading, 326
Planetary mixers, 494–95
Planimeters, 67–68
Planing, 324–25
Planning. *See* Construction planning; Earthwork construction planning
Plan views, 65
Plasticity, 91
Plasticity index, 96
Plastic limit of soils, 96
Plastic piles, 612, 613
Plate feeders, 427
Plate-line capacity, 595
Plumbing steel frames, 688
Plywood sheathing, 715, 719
Pneumatic-tired rollers, 117, 118, 122–25
Pneumatic tools, 654–55. *See also* Compressed air
Pocket silos, 502
Pocomoke River bridge, 603
Polhem, Christoffer, 600
Ponding, 109
Porosity of aggregate, 449
Porosity of soil, 95
Portable compressors, 645
Portable crushers, 408–9
Portland cement, 134–38, 486, 489. *See also* Concrete
Positive lockout of power sources, 438
Powder columns, 339
Powder factors, 395–97
Power, 150
Power buggies, 504
Power cords, 663
Power Crane and Shovel Association (PCSA)
  bucket size ratings, 257–58
  crane studies, 535
  dragline and clamshell studies, 580
  optimum shovel cut heights, 262
Power lines, crane safety and, 572
Power mixers, 491, 493–96
Power-shift transmissions, 6, 175, 176
Power trowels, 517–19
Pozzolanic reactions, 133
Preassembly for shoring towers, 745
Preaugering, 633
Prebid meetings, 679
Prefabricated column forms, 726–27
Premature blasts, 401
Present worth compound amount factor (PWCAF), 20, 24, 27, 713
Presplitting, 398–99
Pressure (defined), 641
Pressure grout, 365
Pressure loss
  compressed air systems, 647–49, 650, 651, 652, 653
  water pipes, 664–65, 666
Prestressed concrete piles, 604–9, 614, 615
Previous experience, as guide to blade load estimates, 186
Prewetting, 137, 248
Prills, 375
Primacord, 382, 398–99
Primary crushers, 409, 410–17
Prime coats, 445, 468
Primers (explosive), 378, 379–80, 397
Proctor tests, 101, 102
Production costs, relation to O&O costs, 18
Production cycles. *See also* Cycle times
  front shovels, 260–61
  scrapers, 232–33
Production estimates
  clamshells, 595–97
  compactors, 128–29
  dozers, 185–95
  dragline, 585–93
  drilling, 350–58
  equipment spreads, 12–13, 84–86
  excavators, 255
  factors causing variances, 12
  front shovels, 260–64
  hoes, 271–73
  land clearing jobs, 199–203
  loaders, 279–83
  pavers, 473–74
  rock ripping, 211–13
  scrapers, 233–47
  trenching machines, 287
  trucks, 305–10
Production lifts, planning, 684–85
Productivity. *See also* Production estimates
  estimating, 13
  formwork labor, 716–17, 740–45
  truck size and, 303–5
Product Link, 19
Profileometers, 522
Profile views, 65, 66, 76
Project average hauls, 83–84
Project engineers, involvement with formwork, 705–7
Project factors in crane safety, 572
Propagation, 375
Property taxes, 34
Proportioning concrete, 485
Pugmills
  concrete, 493–96
  defined, 445
  mixing, 458–59
Pulverization, in lime stabilization, 134
Pumps (concrete)
  with boom, pipeline, and tower, 509, 510
  boom type, 507–9
  development, 505
  with pipelines, 506–7
  rates of output, 511–12
  rules for, 512–13
  safety issues, 527
  when to use, 513–14
Pumps (water)
  basic functions, 655–56
  centrifugal, 658–63
  displacement, 657–58
  early development, 639–40
  friction losses for, 664–65, 666
  selecting, 665–67
  terminology, 656–57
Purchase expense, 28–29
Pusher-loaded scrapers, 223–25, 243–47
Push-pull scrapers, 224, 225

Quadrant positions for cranes, 545–46
Quantity surveys, 63
Quantity take–off, 13
Quartz, 349
Quartzite, 355
Quick estimating, rock ripping production, 211

Radial cracking in blastholes, 382–83
Radial ripper linkages, 210
Railroads, 3–4
Rails, tower cranes on, 558
Rakes, 198
Rammer attachments, 269
Rammer compactors, 127, 128
Rams. *See* Hammers (pile driver)
Rated lifting capacity
  cranes, 545–48, 563–66
  hoes, 270–71
Reaction piles, 619–20
Ready-mixed concrete, 490, 496–500
Reaming, 364–65
Rear-discharge truck mixers, 496, 498
Rear-dump trucks
  dump times, 308
  fleet performance example, 312–17
  performance charts, 307
  types, 298–99
Rear-vision cameras, 330
Receptor control of noise, 697
Recipe batches (concrete), 499–500
Reciprocating compressors, 644
Reciprocating pumps, 657
Reclaimed asphalt pavement, 445, 464–65
Rectangular hole patterns, 391, 392
Reducers, 512
Reduction ratios of crushers, 410, 411, 417, 423
Refraction seismography, 205, 207–9
Regulation (government), 9–10. *See also* Occupational Safety and Health Administration (OSHA)
Rehandling clamshell buckets, 595
Reinforcing steel, 605
Relative bulk strength, 386
Release agents, 466
Remote-controlled cranes, 561
Remote-controlled rollers, 122, 123
Rent
  of formwork, 713–14
  internal systems, 18
  as machine procurement option, 49–50
  use of money, 19–25
Repairs
  adding to machine value, 43
  calculating annualized costs, 24–25
  as operating cost, 37–38, 45
  as ownership cost, 34
  on rented equipment, 50
Replacement analysis, 18, 47–49
Replacement of high-wear items, 38–39, 46–47
Required power, 141–48
"Requirements for Cast-in-Place Concrete," 746
Resale data, 29
Resistance to penetration (pile driving), 618, 619
Retarder performance charts, 162–63, 231
Return times, 308, 314
Reversible mixers, 493
Revibration, 516
Revolving screens, 432–33
Rhyolite, 354
Rigging of cranes, 567–71, 575
Rigid-frame rear-dump trucks, 298, 299
Rimpull
  calculating, 153–54
  defined, 152
  performance charts, 158–62, 230
Ring system cranes, 542–44
Ripping
  attachments for, 209–11, 322
  development of capabilities, 204–5
  to enhance scraper production, 247–48
  evaluating rock for, 205–9
  high-wear item costs, 47
  production estimates, 211–13
Riser pipes, 668
Risk factors
  financial, 9–12
  impact on cost of capital, 25
  safety planning for, 61–63
Risks. *See* Accidents; Safety
Road building, as impetus for equipment development, 4, 6
Rock
  basic types, 91–92, 94
  density and specific gravity, 385
  hardness, 348–49, 350
  impact of type on drill life, 354, 355, 356
  importance of assessing, 90
  removing before soil stabilization, 136
  terminology, 91
  weight and swell, 99
Rock buckets, 276
Rock crushing. *See* Crushing
Rock drills, 342–46
Rock fragmentation. *See* Fragmentation
Rock hauling, impact on trucks, 298
Rock lifts, 118
Rock ripping. *See* Ripping
Rod mills, 425–26
Roll crushers
  features and types, 419–23
  reduction ratios, 410, 411, 423
Roller-compacted concrete, 523–24
Rollers
  ancient, 115, 116
  asphalt compaction, 474–77
  major types, 117, 118, 119–25
  production estimates, 128–29
Rolling resistance
  defined, 141
  elements of, 142–44
  scraper production and, 235, 236
  truck tires, 310–11, 316
Rollover protective structures, 10
Roman road building, 115
Rome K/G clearing blades, 197, 201
Rome Plow Company land-clearing production estimates, 200–203
Rotary bit life, by rock type, 354, 355, 356. *See also* Drill bits; Drilling
Rotary compressors, 645–46
Rotary drills, 344, 345
Rotary-percussion drilling, 345
Rotary stabilizers, 132
Rough-terrain cranes, 534, 540, 683
Rounds of explosives, 375
Rubber hose, 665, 666
Rubberized asphalt, 445
Saddle jibs, 560
SAE engine power ratings, 156–57
SAE standards
  bucket size ratings, 257–58
  dozer blade ratings, 185
  heaped capacity of scrapers, 227
  loader bucket capacity, 276
Safe load indicators, 575
Safety
  asphalt production, 479
  blasting, 381, 401–2
  clamshell and dragline, 597–98
  compressed air, 653–55
  concrete, 527
  crane, 570, 572–76
  crushing, 438–39
  dozer, 195–96
  drilling, 367
  as economic risk, 9–10
  excavator, 255–56
  fall protection, 688
  formwork, 745–47
  gradall, 332
  grader, 329–30
  land clearing, 203
  loader, 284
  pile driving, 636–37
  planning for, 61–63
  scraper, 249
  trenching, 288–89
  truck, 317
"Safety and Health Regulations for Construction," 746
Safety features, evolution of, 7
Safety fuses, 381
Safety regulations, development of, 10–11. *See also* Occupational Safety and Health Administration (OSHA)
Salvage value, 24, 29, 30
Sand
  classification, 93
  compactors for, 118
  defined, 92
  hauling in trucks, 302
  responses to dozing, 185, 186
  skin friction on piles, 619
  weight and swell, 99
Sandcone compaction tests, 103
Sand preparation and classification machines, 437
Sandstone, 356
San Francisco-Oakland Bay Bridge, 611
Scaffolding, 721–22
Scaffolding, Shoring and Forming Institute, Inc., 746
Scaled distance, 401
Scalping, 431
Scarification, 134, 135, 136
Scarifiers, 110, 322, 323–24
Scrapers
  altitude effects, 157
  automated, 7, 8
  basic types, 223–27
  economical haul distance, 77
  operating parts, 228–29
  operational considerations, 247–49
  performance charts, 160, 163, 229–31
  production cycles, 232–33

Scrapers—*Cont.*
production estimates, 233–47
strengths of, 222–23
volumetric capacities, 227–28
Screed angle of attack, 472
Screeds, 444, 445, 471–73
Screening aggregate, 432–37, 457–58
Screw conveyers, 468, 469
Screw-type classifiers, 437
Seal coats, 468
Seat belts, 195–96, 249
Secondary blasting, 389
Secondary crushers, 409, 410, 417–25
Sedimentary rocks
defined, 92
impact on drill bits and steel, 356
rippability, 205, 206
Segregation, thermal, 471
Segregation of aggregate, 438, 448
Seismic-velocity method, 211–13
Seismographs, 205, 207–9
Self-adjusting backup alarms, 693
Self-climbing wall forms, 723–24
Self-erecting tower cranes, 551
Self-loading scrapers, 223, 225–27
Self-priming pumps, 656, 659–60
Self-propelled pneumatic rollers, 123
Self-propelled vibratory-plate compactors, 127
Semiautomatic batching controls, 489
Semi-U (SU) blades, 181, 182, 185
Sensitivity of explosives, 375, 376
Sensitizers, 377
Sequencing of initiation, 381–82
Sequential blasting machines, 382
Service life, 28
Sewer lines, laying, 365
Shakeout, 686
Shale, 356
Shanks
of percussion drills, 340, 357
of rippers, 209–11
Shapiro, Howard I., 62
Sheathing
in formwork, 709–10, 715
of trenches, 288
Sheaves, 544
Sheepsfoot rollers, 117, 118, 119
Sheet piles, 602, 612–17
Shielding of trenches, 288
Shores in formwork, 709–10
Shoring of trenches, 288
Shoring towers, 738–45
Short-boom draglines, 586, 587
Shotcreting, 524, 525
Shot drills, 340, 346, 347
Shot rock
defined, 222
loading with front shovels, 257
oversize material in, 261
Shrinkage factor, 98
Shrinkage of concrete, 519
Shrinkage of soils, 91
Shutoff head, 661
Shuttle loading, 243, 244
Sideboards, 229–31, 302–3
Side-by-side dozing, 184
Side casting, 179
Side draft, 325
Side-dump buckets, 276
Side spillage, reducing, 184
Siever J-values, 352
Sight lines of graders, 330
Silos
in asphalt plants, 459, 461, 462
in concrete plants, 501–2
Silt
classifying, 93, 97
compaction effects, 106
compactors for, 118
defined, 92
skin friction on piles, 619
Single-acting hammers, 621–22, 623
Single-acting pumps, 657
Single payment compound amount factor
(SPCAF), 20
Single payments, calculating, 20–21
Single-powered axle scrapers, 224
Single-roll crushers, 421
Single-rotor impact crushers, 424–25
Single-shaft trough mixers, 494, 495–96
Single-sided wall forms, 723
Single-stage compressors, 644
Single-toggle crushers, 411, 412, 414
Site investigation for pile driving, 618–20
Site layout, 677–83, 706
Site restrictions, 676, 677
Size ratings (shovels and hoes), 257–58, 267–68
Skates, 365
Skin friction
penetration resistance and, 618, 619
reducing, 628, 632–33
Slat conveyors, 468
Slewing, 535
Slings, 568–69, 570–71
Slipforming, 723
Slipform paving, 519–21
Slopes. *See also* Grades
defined, 445
total resistance, 147
working with graders, 321–22
Sloping of trenches, 288
Slot dozing, 184
Slump, 485–86
Slurries, explosive, 377–78
Slurry plant enclosures, 696
Small trucks, 303
Smooth-drum vibratory compactors, 117, 118, 121
Smoothness of concrete pavements, 522–23
Smooth-roll crusher capacities, 423
Snell's law, 205
Soft asphalts, 451
Soil failure beneath cranes, 535–36
Soils. *See also* Compaction of soils
adjusting water content, 100, 106–11
basic types, 91–94
compaction methods, 116, 117
compaction tests, 102–5, 106
drilling, 359–62
importance of assessing, 90, 101
limits, 96, 97
terminology, 91
volume changes during construction, 96–101
weight-volume relationships, 94–96
Soil stabilization
agents, 109, 133–38
overview, 131–32
Soldier piles, 602
Source control of noise, 695
Spacing of drill hole rows, 340, 393
Spandrel beams, 732
Specialized wall forms, 722–24
Special-purpose buckets, 269, 276
Specialty crushers, 411
Specific gravity of explosives, 376, 396
Specific gravity of soil, 95
Speed
as advantage of wheel dozers, 174
controlling factors for trucks, 308
determining from effective grade, 160–62
drawbar pull versus, 158, 159
of graders, 327, 328
of pavers, 472
as pusher-loaded scraper advantage, 223
rimpull versus, 158–60
tower crane operation, 564–65
Spillage, reducing, 182, 184
Spiral reinforcing, 605
Spotters for fill operations, 249
Spray bars, 467
Spray-on fireproofing, 688, 689
Spreader bars, 569, 691
Spreading
with dozers, 183
with graders, 326
of soil stabilizers, 135, 136
Spreads. *See* Equipment spreads
Spudding and preaugering, 633
Spud vibrators, 515
Square hole patterns, 391, 392
Square piles, 605
Stabilization agents, 109, 133–38
Stabilizers, rotary, 132
Staggered hole patterns, 392
Stairwells, forms for, 722
Standard column forms, 724–26
Standard conditions, 642
Standard model cone crushers, 417, 419
Standard Proctor test, 101, 102, 103
Standard rate approach, 18
Standard shoring towers, 739
Standby trucks, 316
Star silos, 501
Static discharge heads, 656
Static steel-wheel rollers, 117
Static suction lift, 657
Stationary compressors, 645
Stationary pumps, 506
Stations, 64
Station yards, 75, 82
Steam hammers, 621–22, 623
Steam-powered equipment, 3–4
Steel-concrete composite piles, 612
Steel elements in formwork, 715
Steel erection, planning for, 688
Steel laydown areas, 687
Steel piles, 610–11, 614–17
Steels, high-strength, 6
Steel wheel rollers, 474
Stemming, 339, 390–91
Stem wall casting, 735
Step reduction, 409–10
Stick cylinders, 266, 267
Stiffness ratio, 383, 389, 393
Stingers, 197

Storage costs, 34, 43
Straight blades, 180, 181, 185
Straight dynamite, 377
Straight-line depreciation, 31–32
Straight-shank rippers, 209
Strainers, 657
Strain relief protectors, 663
Strength of explosives, 375–76
Stresses
  from crane loads, 568–70
  pile driving, 608–9
  on structures, 400
Striker bars, 340
Stringlines, 445, 523
Stripping
  of asphalt from aggregate, 450
  with dozers, 182
  volume calculations, 71, 72, 74
Stripping of forms
  for hand-set slabs, 728
  labor costs, 716–17
  safety issues, 746
  timing, 706, 716
  for tunnel forms, 735–37
Strongbacks, 719, 721
Struck capacity
  front shovel buckets, 258
  hoe buckets, 268
  scrapers, 227
  trucks, 301, 302
Structural component delivery, planning for, 686–87
Structural excavations, 63
Stump disposal, 199
Stump removal, 202
Subbase, 116
Subcontractors, 678, 679–80
Subdrilling
  defined, 340
  determining depth, 391
  illustrated, 339
  purpose of, 350–51
Subgrade, 116
Subgrade stabilization, 133
Submersible pumps, 661–63
Suction head, 657
Suction hoses, 657, 665
Suez Canal, 1–2
Superchargers, 157–58
Superpave mixes, 445, 448
Support towers, 738
Surety companies, 52
Surface courses, 445, 446
Surface treatments, 447
Surface vibrators, 514–16
Surge feeding, 427
Surge piles, 427–28
Surge silos, 459
Surveyors, impact of GPS systems on, 7
Suspended loads, working under, 682–83
Sweeper/brooms, 466
Swell factor
  calculating, 98–99
  rule-of-thumb, 100–101
  in scraper volumes, 227
Swing angles
  effects on dragline production, 586, 588
  effects on shovel production, 262–64
Swing leads, 631
Symons cone crushers, 419, 420
Sympathetic detonation, 377
Synthetic web slings, 571

Table forms, 729–33, 746
Tack coats, 445, 468
Tag lines, 593–94
Talc, 349
Tamping explosives, 375
Tamping rollers, 117, 118, 119–20
Tandem-powered scrapers, 224, 225
Tax code depreciation schedules, 32–33
Taxes
  depreciation savings, 30–34, 42–43
  on fuel, 35
  leased equipment deductions, 51
  as ownership cost, 34, 43
Teeth
  on clamshell buckets, 595
  on dragline buckets, 590
  as high-wear items, 39
  on loader buckets, 275
  of scarifiers, 324
Telescoping booms
  crawler mounted, 537
  load ratings, 545–48
  truck mounted, 534, 538–39
Temperature
  absolute, 641
  for asphalt compaction, 475–76
  defined, 642
  effects on water pumps, 660
Templates for pile driving, 617, 632
Tenacity of rock, 350
Tension piles, 602
Tertiary crushers, 409, 410, 417–25
Test piles, 619–20
Texture of rock, 350
Theoretical fuel costs, 36
Thermal segregation in asphalt, 471
Thickness controls, 446
Threading devices (sheet piles), 617
Throttle load factors, 36, 44
Thrust walls, 366
Tier size selection, shoring towers, 744
Ties, in formwork, 709
Tilting mixers, 492–93
Tilt of dozer blades, 178, 179
Tilt-up panel construction, 689–92
Timber piles, 603–4, 614
Timbers in formwork, 715
Time factors in fuel costs, 36, 44
Time of delivery prices, 44
Time value method, 40–41
Time value of money, 19–25, 46
Tipping condition, 544
Tipping loads, 544–45
Tips of rippers, 209
Tires
  ballast for, 173
  coefficient of traction, 155, 156
  of graders, 327
  introduction of nylon cord in, 6
  as operating cost, 38, 45–46
  overloading, 302–3
  of pneumatic rollers, 122–25
  rolling resistance, 143, 144
  sample cost calculations, 23–24, 45–46
  selection for trucks, 310–12
TNT, 375
Toggles, 412–14
Ton-miles-per-hour limits, 311–12
Topaz, 349
Top-slewing tower cranes
  basic features, 551–53, 554, 557
  with cantilever jibs, 561
  common dimensions and capacities, 556
  as common type, 535
Topsoil stripping volume calculations, 71, 72, 74
Top stemming depth, 390–91
Torque, 150, 151
Torque converters
  basic features, 152
  on dozers, 175
  introduction of, 6
Total assistance, 236–37
Total cycle times. *See* Cycle times
Total head, 657
Total resistance
  calculating, 141, 146–47
  determining speed from, 160–62
  scraper production and, 236–37
Total soil volume, 95
Towed impact compactors, 125–26
Tower cranes
  load ratings, 563–66
  mobilizing, 683
  operation, 554–62
  popularity, 533
  selecting, 562–63
  types, 550–53
Towers, concrete booms on, 509, 510
Towers (shoring), 738–45
Toxic fumes of explosives, 376
Track loaders
  basic features, 274
  cycle times, 279
  operating specifications, 278
  production estimates, 282–83
Track-mounted drills, 343–44, 353
Tracks, of crawler dozers, 172–73
Traction
  coefficient, 153, 155, 156, 174
  impact on usable drawbar pull, 176–77
  improving, 173
  as pusher-loaded scraper disadvantage, 224
Tractors, 157, 300–301
Tractor scrapers, 160, 163. *See also* Scrapers
Trailer pumps, 506
Trailing counterweight cranes, 542, 543
Trailing platforms, 722
Trains, collisions with, 330
Transition points, on mass diagrams, 77
Transit mixers, 490, 491, 496–99
Transmissions
  basic functions, 151–52
  of dozers, 175, 194
Transportation, impact on equipment development, 2
Transportation Research Board (TRB), 263, 264
Trapezoidal computations, 68–69
Travel distance, impact on machine selection, 148
Traveling tower cranes, 558
Travel speeds, 238, 281–82. *See also* Speed
Travel time, 238, 279–80
Treated timber piles, 603–4
Tree count method, 200–203

Trees
  clearing, 197–98, 200–203
  disposal, 199
Tremie operations, 504
Trench boxes, 270, 288, 289
Trenching machines, 284–89
Trenchless technology, 362–67
Trench rock, 397
Trimmers, 332–34
Triplex pumps, 657
Trolley systems, 732
Trough mixers, 494, 495–96
Troweling, 517–19
Truck pumps, 507
Trucks
  altitude effects, 157
  asphalt, 466
  capacities, 301–3
  concrete, 490, 491, 496–99
  dumping types, 298–301
  economical haul distance, 77
  fleet performance calculations, 312–17
  overview of types and advantages, 296–97
  production estimates, 305–10
  safety issues, 317
  size-productivity relation, 303–5
  tires, 310–12
True gyratory crushers, 416–17
Trunklines, 382
Truss table form systems, 730
Tungsten carbide, 341
Tunnel form systems, 733–38
Turbocharging, 6, 157–58
Turbomixers, 494
Turnbuckles, 688
Turning scrapers, 241–42
Turnover of contracts, 12
Turns with graders, 327
Twin-shaft pugmills, 459
Twin-shaft trough mixers, 495–96
Two-component explosives, 377
Two-linked earthwork systems, 84, 85
Two-part-line configurations, 565–66
Two-stage compressors, 644

Ultra-high-load shoring towers, 739
Unadjusted basis, 30, 31
Unclassified excavation, 94
Undercarriage configurations (dozers), 173
Undergrowth density, impact on clearing time, 201–2
Underinflated tires, 310, 311
Underreamers, 359–60
Unified soil classification system, 93
Uniformity of concrete batches, 499, 500
Uniform payments, calculating, 21–25
Uniform series, compound amount factor (USCAF), 22
Uniform series capital recovery factor (USCRF), 22, 24, 711
Uniform series present worth factor (USPWF), 22, 27
Uniform series sinking fund factor (USSFF), 22, 24, 712
Unit costs
  dozer production, 188–89, 195
  pusher-scraper production, 246–47
Unit weight of materials, 95, 302
Universal (U) blades, 181, 182, 185
Universal machines, 535
Untreated timber piles, 604
Up stationing, 78
Upward drilling, 343
Usable drawbar pull, 176–77
Usable power of machines, 155–58
U.S. Army Corps of Engineers, Computer-Aided Earthmoving System, 7, 8
Utility trench backfills, 128
U-type leads, 631

Vac excavators, 290–92
Vac-trucks, 695
Vacuum, 642
Variance in bids, 12
Velocity equation, 658–59
Vertical formwork systems, 718–27
Vertical lips on floors, 732
Vertical-shaft impact crushers, 424
Vibrating feeders, 427
Vibrating screens, 433–34
Vibration
  from blasting, 400–401
  planning to control, 698–99
Vibrators for consolidating concrete, 514–16
Vibratory compactors
  major types, 117, 118, 120–22, 123
  use on pavements, 475
Vibratory pile drivers, 616, 617, 628–30
Vickers test, 349
Vines, impact on clearing time, 201–2
Viscosity grades of asphalt cement, 451
Visibility correction factors, 194
Void ratio, 95
Volume of contracts, 12
Volume sheets, 72–75, 81
Volumes of material
  calculating for earthwork projects, 67–75
  changes during construction, 96–101
  relationship to soil weight, 94–96
Volumetric load
  of dozer blades, 185–87
  methods of expressing, 96–98, 141
  of scrapers, 227–28
Volutes, 665
V-pattern blasting, 392, 393

Wages, 9–10, 35
Wakefield sheeting, 614
Wales, 709–10, 721
Walk-behind compactors, 122, 123, 127–28
Wall-and-slab forming systems, 733–38
Wall forms, 689–90, 719–24
Wastegates, 334
Waste materials in aggregate production, 410
Water
  adding to ready-mix, 499
  effects on concrete strength, 485
  requirements for concrete, 486–87, 489
Water content of soils
  adjusting, 100, 106–11
  calculating, 95
  on haul roads, 326
  importance, 101, 102
  soil limits, 96, 97
  testing, 102–5, 106
Water distributors, 109
Water gels, 377
Water jets, 632–33
Water-level capacity, 595
Water lines, laying, 365
Water pumps. *See* Pumps (water)
Water resistance of explosives, 376
Water table, 129
Water trucks, 85, 109
Water venturi filtration, 462
Watt, James, 149–50
Weather
  concrete placement and, 526–27
  as consideration for cement stabilization, 138
  impact on soil moisture, 110–11
Weathervaning of tower cranes, 557
Web slings, 571
Webs of clearing blades, 197
Weigh hoppers, 446
Weight capacity, 302, 306
Weight hoppers, 458
Weight of loads, as crane rigging factor, 567
Weight of solids formula, 95
Welded equipment, 5–6
Wellpoint systems, 668–70
Wet collectors, 446
Wet-mix process (shotcreting), 524
Wheelbarrows, 504
Wheel dozers
  basic features, 172, 174–75
  transmissions, 176, 177
Wheel loaders
  basic features, 274, 275
  cycle times, 279
  fuel consumption rates, 35
  multiple attachments for, 7
  operating specifications, 278
  production estimates, 279–82
Wheel-mounted cranes
  estimated repair and maintenance costs, 45
  lattice-boom, 539–40
  lifting quadrants, 546
  major types illustrated, 534
  rough-terrain and all-terrain, 540–42
  telescoping-boom, 538–39
Wheel-mounted excavators, 45
Wheel-mounted hoes, 265
Wheel tractor scrapers, 160, 163
Wheel-type trenching machines, 285–86
White curing compound, 137
Williams-Steiger Act, 10
Wind loads, 557
Windows, noise control through, 697
Windrow elevators, 446, 470
Windrows
  of debris, 199, 202, 203
  to mix soils, 111
  moving with graders, 324–26
  of paving materials, 446
Wire rope slings, 571
Woodrow Wilson Bridge, 611
Work, calculating, 148–49
Workability of concrete, 485–86, 523
Working capital assets, 51
Working ranges of cranes, 548–49
Working ranges of draglines, 583

Yankee Geologists, 3

Zones of application, 248
Zones of responsibility, 575–76
Z sections, 614